Chiral Separations by Capillary Electrophoresis

CHROMATOGRAPHIC SCIENCE SERIES

A Series of Textbooks and Reference Books

Editor: JACK CAZES

1. Dynamics of Chromatography: Principles and Theory, *J. Calvin Giddings*
2. Gas Chromatographic Analysis of Drugs and Pesticides, *Benjamin J. Gudzinowicz*
3. Principles of Adsorption Chromatography: The Separation of Nonionic Organic Compounds, *Lloyd R. Snyder*
4. Multicomponent Chromatography: Theory of Interference, *Friedrich Helfferich and Gerhard Klein*
5. Quantitative Analysis by Gas Chromatography, Josef Novák
6. High-Speed Liquid Chromatography, *Peter M. Rajcsanyi and Elisabeth Rajcsanyi*
7. Fundamentals of Integrated GC-MS (in three parts), *Benjamin J. Gudzinowicz, Michael J. Gudzinowicz, and Horace F. Martin*
8. Liquid Chromatography of Polymers and Related Materials, *Jack Cazes*
9. GLC and HPLC Determination of Therapeutic Agents (in three parts), Part 1 *edited by Kiyoshi Tsuji and Walter Morozowich*, Parts 2 and 3 *edited by Kiyoshi Tsuji*
10. Biological/Biomedical Applications of Liquid Chromatography, *edited by Gerald L. Hawk*
11. Chromatography in Petroleum Analysis, *edited by Klaus H. Altgelt and T. H. Gouw*
12. Biological/Biomedical Applications of Liquid Chromatography II, *edited by Gerald L. Hawk*
13. Liquid Chromatography of Polymers and Related Materials II, *edited by Jack Cazes and Xavier Delamare*
14. Introduction to Analytical Gas Chromatography: History, Principles, and Practice, *John A. Perry*
15. Applications of Glass Capillary Gas Chromatography, *edited by Walter G. Jennings*
16. Steroid Analysis by HPLC: Recent Applications, *edited by Marie P. Kautsky*
17. Thin-Layer Chromatography: Techniques and Applications, *Bernard Fried and Joseph Sherma*
18. Biological/Biomedical Applications of Liquid Chromatography III, *edited by Gerald L. Hawk*
19. Liquid Chromatography of Polymers and Related Materials III, *edited by Jack Cazes*
20. Biological/Biomedical Applications of Liquid Chromatography, *edited by Gerald L. Hawk*
21. Chromatographic Separation and Extraction with Foamed Plastics and Rubbers, *G. J. Moody and J. D. R. Thomas*

22. Analytical Pyrolysis: A Comprehensive Guide, *William J. Irwin*
23. Liquid Chromatography Detectors, *edited by Thomas M. Vickrey*
24. High-Performance Liquid Chromatography in Forensic Chemistry, *edited by Ira S. Lurie and John D. Wittwer, Jr.*
25. Steric Exclusion Liquid Chromatography of Polymers, *edited by Josef Janca*
26. HPLC Analysis of Biological Compounds: A Laboratory Guide, *William S. Hancock and James T. Sparrow*
27. Affinity Chromatography: Template Chromatography of Nucleic Acids and Proteins, *Herbert Schott*
28. HPLC in Nucleic Acid Research: Methods and Applications, *edited by Phyllis R. Brown*
29. Pyrolysis and GC in Polymer Analysis, *edited by S. A. Liebman and E. J. Levy*
30. Modern Chromatographic Analysis of the Vitamins, *edited by André P. De Leenheer, Willy E. Lambert, and Marcel G. M. De Ruyter*
31. Ion-Pair Chromatography, *edited by Milton T. W. Hearn*
32. Therapeutic Drug Monitoring and Toxicology by Liquid Chromatography, *edited by Steven H. Y. Wong*
33. Affinity Chromatography: Practical and Theoretical Aspects, *Peter Mohr and Klaus Pommerening*
34. Reaction Detection in Liquid Chromatography, *edited by Ira S. Krull*
35. Thin-Layer Chromatography: Techniques and Applications, Second Edition, Revised and Expanded, *Bernard Fried and Joseph Sherma*
36. Quantitative Thin-Layer Chromatography and Its Industrial Applications, *edited by Laszlo R. Treiber*
37. Ion Chromatography, *edited by James G. Tarter*
38. Chromatographic Theory and Basic Principles, *edited by Jan Åke Jönsson*
39. Field-Flow Fractionation: Analysis of Macromolecules and Particles, *Josef Janca*
40. Chromatographic Chiral Separations, *edited by Morris Zief and Laura J. Crane*
41. Quantitative Analysis by Gas Chromatography, Second Edition, Revised and Expanded, *Josef Novák*
42. Flow Perturbation Gas Chromatography, *N. A. Katsanos*
43. Ion-Exchange Chromatography of Proteins, *Shuichi Yamamoto, Kazuhiro Naka-nishi, and Ryuichi Matsuno*
44. Countercurrent Chromatography: Theory and Practice, *edited by N. Bhushan Man-dava and Yoichiro Ito*
45. Microbore Column Chromatography: A Unified Approach to Chromatography, *edited by Frank J. Yang*
46. Preparative-Scale Chromatography, *edited by Eli Grushka*
47. Packings and Stationary Phases in Chromatographic Techniques, *edited by Klaus K. Unger*
48. Detection-Oriented Derivatization Techniques in Liquid Chromatography, *edited by Henk Lingeman and Willy J. M. Underberg*
49. Chromatographic Analysis of Pharmaceuticals, *edited by John A. Adamovics*
50. Multidimensional Chromatography: Techniques and Applications, *edited by Hernan Cortes*
51. HPLC of Biological Macromolecules: Methods and Applications, *edited by Karen M. Gooding and Fred E. Regnier*

52. Modern Thin-Layer Chromatography, *edited by Nelu Grinberg*
53. Chromatographic Analysis of Alkaloids, *Milan Popl, Jan Fähnrich, and Vlastimil Tatar*
54. HPLC in Clinical Chemistry, *I. N. Papadoyannis*
55. Handbook of Thin-Layer Chromatography, *edited by Joseph Sherma and Bernard Fried*
56. Gas–Liquid–Solid Chromatography, *V. G. Berezkin*
57. Complexation Chromatography, *edited by D. Cagniant*
58. Liquid Chromatography–Mass Spectrometry, *W. M. A. Niessen and Jan van der Greef*
59. Trace Analysis with Microcolumn Liquid Chromatography, *Milos Krejcl*
60. Modern Chromatographic Analysis of Vitamins: Second Edition, *edited by André P. De Leenheer, Willy E. Lambert, and Hans J. Nelis*
61. Preparative and Production Scale Chromatography, *edited by G. Ganetsos and P. E. Barker*
62. Diode Array Detection in HPLC, *edited by Ludwig Huber and Stephan A. George*
63. Handbook of Affinity Chromatography, *edited by Toni Kline*
64. Capillary Electrophoresis Technology, *edited by Norberto A. Guzman*
65. Lipid Chromatographic Analysis, *edited by Takayuki Shibamoto*
66. Thin-Layer Chromatography: Techniques and Applications: Third Edition, Revised and Expanded, *Bernard Fried and Joseph Sherma*
67. Liquid Chromatography for the Analyst, *Raymond P. W. Scott*
68. Centrifugal Partition Chromatography, *edited by Alain P. Foucault*
69. Handbook of Size Exclusion Chromatography, *edited by Chi-San Wu*
70. Techniques and Practice of Chromatography, *Raymond P. W. Scott*
71. Handbook of Thin-Layer Chromatography: Second Edition, Revised and Expanded, *edited by Joseph Sherma and Bernard Fried*
72. Liquid Chromatography of Oligomers, *Constantin V. Uglea*
73. Chromatographic Detectors: Design, Function, and Operation, *Raymond P. W. Scott*
74. Chromatographic Analysis of Pharmaceuticals: Second Edition, Revised and Expanded, *edited by John A. Adamovics*
75. Supercritical Fluid Chromatography with Packed Columns: Techniques and Applications, *edited by Klaus Anton and Claire Berger*
76. Introduction to Analytical Gas Chromatography: Second Edition, Revised and Expanded, *Raymond P. W. Scott*
77. Chromatographic Analysis of Environmental and Food Toxicants, *edited by Takayuki Shibamoto*
78. Handbook of HPLC, *edited by Elena Katz, Roy Eksteen, Peter Schoenmakers, and Neil Miller*
79. Liquid Chromatography–Mass Spectrometry: Second Edition, Revised and Expanded, *Wilfried Niessen*
80. Capillary Electrophoresis of Proteins, *Tim Wehr, Roberto Rodríguez-Díaz, and Mingde Zhu*
81. Thin-Layer Chromatography: Fourth Edition, Revised and Expanded, *Bernard Fried and Joseph Sherma*
82. Countercurrent Chromatography, *edited by Jean-Michel Menet and Didier Thiébaut*
83. Micellar Liquid Chromatography, *Alain Berthod and Celia García-Alvarez-Coque*

84. Modern Chromatographic Analysis of Vitamins: Third Edition, Revised and Expanded, edited by *André P. De Leenheer, Willy E. Lambert, and Jan F. Van Bocxlaer*
85. Quantitative Chromatographic Analysis, *Thomas E. Beesley, Benjamin Buglio, and Raymond P. W. Scott*
86. Current Practice of Gas Chromatography–Mass Spectrometry, *edited by W. M. A. Niessen*
87. HPLC of Biological Macromolecules: Second Edition, Revised and Expanded, *edited by Karen M. Gooding and Fred E. Regnier*
88. Scale-Up and Optimization in Preparative Chromatography: Principles and Bio-pharmaceutical Applications, *edited by Anurag S. Rathore and Ajoy Velayudhan*
89. Handbook of Thin-Layer Chromatography: Third Edition, Revised and Expanded, *edited by Joseph Sherma and Bernard Fried*
90. Chiral Separations by Liquid Chromatography and Related Technologies, *Hassan Y. Aboul-Enein and Imran Ali*
91. Handbook of Size Exclusion Chromatography and Related Techniques: Second Edition, *edited by Chi-San Wu*
92. Handbook of Affinity Chromatography: Second Edition, *edited by David S. Hage*
93. Chromatographic Analysis of the Environment: Third Edition, *edited by Leo M. L. Nollet*
94. Microfluidic Lab-on-a-Chip for Chemical and Biological Analysis and Discovery, *Paul C.H. Li*
95. Preparative Layer Chromatography, *edited by Teresa Kowalska and Joseph Sherma*
96. Instrumental Methods in Metal Ion Speciation, *Imran Ali and Hassan Y. Aboul-Enein*
97. Liquid Chromatography–Mass Spectrometry: Third Edition, *Wilfried M. A. Niessen*
98. Thin Layer Chromatography in Chiral Separations and Analysis, *edited by Teresa Kowalska and Joseph Sherma*
99. Thin Layer Chromatography in Phytochemistry, *edited by Monika Waksmundzka-Hajnos, Joseph Sherma, and Teresa Kowalska*
100. Chiral Separations by Capillary Electrophoresis, *edited by Ann Van Eeckhaut and Yvette Michotte*

Chiral Separations by Capillary Electrophoresis

Ann Van Eeckhaut
Vrije Universiteit Brussel
Brussels, Belgium

Yvette Michotte
Vrije Universiteit Brussel
Brussels, Belgium

CRC Press is an imprint of the
Taylor & Francis Group, an **Informa** business

CRC Press
Taylor & Francis Group
6000 Broken Sound Parkway NW, Suite 300
Boca Raton, FL 33487-2742

First issued in paperback 2019

ISBN-13: 978-1-4200-6933-4 (hbk)
ISBN-13: 978-0-367-38489-0 (pbk)

Library of Congress Cataloging-in-Publication Data

Chiral separations by capillary electrophoresis / editors, Ann Van Eeckhaut, Yvette Michotte.
p. cm. -- (Chromatographic science series ; 100)
Includes bibliographical references and index.
ISBN 978-1-4200-6933-4 (hardcover : alk. paper)
1. Chiral drugs. 2. Chirality. 3. Enantiomers. 4. Electrophoresis. I. Eeckhaut, Ann Van. II. Michotte, Yvette. III. Title. IV. Series.

RS429.C484 2009
541'.372--dc22 2009031932

Visit the Taylor & Francis Web site at
http://www.taylorandfrancis.com

and the CRC Press Web site at
http://www.crcpress.com

Contents

Preface xi
Editors xiii
Contributors xv

Chapter 1 Pharmacological Importance of Chiral Separations 1
Ann Van Eeckhaut and Yvette Michotte

Chapter 2 Principles of Chiral Separations by Free-Solution Electrophoresis 23
Radim Vespalec

Chapter 3 Cyclodextrin-Mediated Chiral Separations 47
Gerald Gübitz and Martin G. Schmid

Chapter 4 Factors Influencing Cyclodextrin-Mediated Chiral Separations 87
Anne-Catherine Servais, Jacques Crommen, and Marianne Fillet

Chapter 5 Macrocyclic Antibiotics as Chiral Selectors 109
Salvatore Fanali, Zeineb Aturki, Giovanni D'Orazio, and Anna Rocco

Chapter 6 Chiral Separations Using Proteins and Peptides as Chiral Selectors 139
Jun Haginaka

Chapter 7 Chiral Ligand Exchange Capillary Electrophoresis 163
Vincenzo Cucinotta and Alessandro Giuffrida

Chapter 8 Chiral Separations by Micellar Electrokinetic Chromatography 195
Joykrishna Dey and Arjun Ghosh

Chapter 9 Chiral Microemulsion Electrokinetic Chromatography 235
Kimberly A. Kahle and Joe P. Foley

Chapter 10 Chiral Separations in Nonaqueous Media 271
Ylva Hedeland and Curt Pettersson

Chapter 11 Quantitative Analysis in Pharmaceutical Analysis 313
Ulrike Holzgrabe, Claudia Borst, Christine Büttner, and Yaser Bitar

Chapter 12 Analysis of Chiral Drugs in Body Fluids... 341
Pierina Sueli Bonato, Valquiria Aparecida Polisel Jabor, and Anderson Rodrigo Moraes de Oliveira

Chapter 13 Chiral CE–MS... 363
Serge Rudaz, Jean-Luc Veuthey, and Julie Schappler

Chapter 14 Enantioseparations by Capillary Electrochromatography 393
Michael Lämmerhofer

Chapter 15 Recent Applications in Capillary Electrochromatography 459
Debby Mangelings and Yvan Vander Heyden

Chapter 16 Chiral Separations with Microchip Technology 501
Markéta Vlčková, Franka Kalman, and Maria A. Schwarz

Index.. 515

Preface

In 1848, Louis Pasteur, a French chemist renowned for his work on the vaccine for rabies, discovered molecular chirality. He found that the sodium ammonium salt of racemic tartaric acid crystallized as a mixture of nonsuperimposable mirror-image crystals. Pasteur called these molecules dissymmetric. The terms "chiral" and "chirality" were first introduced by Lord Kelvin in 1894: "I call any geometrical figure or group of points chiral and say that it has chirality, if its image in a plane mirror ideally realized, cannot be brought to coincide with itself." The enantiomeric terminology to refer to nonsuperimpossible mirror images was already introduced in 1856 by Carl Friedrich Naumann, a German crystallographer. However, this terminology was only accepted and widely used in the context of stereochemistry in the 1890s.

Louis Pasteur was also the first to have observed enantioselectivity in a biochemical process. In 1857, he discovered the enantioselective fermentation of tartaric acid by microorganisms. At the end of the nineteenth century and the beginning of the twentieth, other studies confirmed the pharmacological differences between enantiomers. Despite these early insights, most of the studies have been performed only since the 1980s when technology for chiral analysis advanced. These new technologies allowed, for example, the large-scale production of pure enantiomers and the enantioselective analysis of biological samples.

Capillary electrophoresis has proven to be an effective tool for chiral separation. The aim of this book is to provide a general overview of the principles of chiral separation by capillary electrophoresis and of the different chiral selectors available. In addition, pharmaceutical and biomedical applications and newer techniques, such as capillary electrophoresis coupled to mass spectrometry and microchip technology, are also included. In two chapters, the possibilities of capillary electrochromatography are explained, including the different chiral columns available and some recent pharmaceutical and biomedical applications.

We gratefully acknowledge the contributors without whom this book would not have existed and the staff at CRC Press for all their help.

Yvette Michotte
Ann Van Eeckhaut

Editors

Yvette Michotte is the head of the Department of Pharmaceutical Chemistry and Drug Analysis of the School of Pharmacy at the Vrije Universiteit Brussel, Belgium. She is also heading the research group of experimental neuropharmacology.

Her main research topics are the development and validation of analytical methods for the quantification of compounds for label claim control of drug substances and drug products, and for the quantification of drugs, metabolites, and endogenous compounds in biological matrices. One of her principal topics is the analysis of drugs, neurotransmitters, and neuropeptides in microdialysates from *in vivo* neuropharmacological experiments. The analytical techniques used are micro- and nano-liquid chromatography, coupled to amperometric and mass spectrometric detection (LC–MS/MS), and capillary electrophoresis.

Ann Van Eeckhaut graduated in pharmaceutical chemistry at the Vrije Universiteit Brussel, Belgium, in 1997. She received her PhD in pharmaceutical sciences in 2004 in the Department of Pharmaceutical Chemistry and Drug Analysis of the same university under the supervision of Prof. Yvette Michotte. Her PhD thesis was entitled "Chiral separations by cyclodextrin-mediated capillary electrophoresis." Her current interests as a postdoctoral researcher are oriented to the development and validation of methods for the quantification of small molecules and peptides in biological samples, especially toward the analysis of neuropeptides in microdialysates using miniaturized liquid chromatography coupled to tandem mass spectrometry.

Contributors

Zeineb Aturki
Institute of Chemical Methodologies
Consiglio Nazionale delle Ricerche
Monterotondo Scalo, Italy

Yaser Bitar
Faculty of Pharmacy and Food Chemistry
Department of Pharmaceutical Chemistry and Quality Control
University of Aleppo
Aleppo, Syria

Pierina Sueli Bonato
Faculty of Pharmaceutical Sciences of Ribeirão Preto
University of São Paulo
Ribeirão Preto, Brazil

Claudia Borst
Institute of Pharmacy and Food Chemistry
University of Würzburg
Würzburg, Germany

Christine Büttner
Institute of Pharmacy and Food Chemistry
University of Würzburg
Würzburg, Germany

Jacques Crommen
Department of Analytical Pharmaceutical Chemistry
Institute of Pharmacy
University of Liège
Liège, Belgium

Vincenzo Cucinotta
Department of Chemical Sciences
University of Catania
Catania, Italy

Joykrishna Dey
Department of Chemistry
Indian Institute of Technology
Kharagpur, India

Giovanni D'Orazio
Institute of Chemical Methodologies
Consiglio Nazionale delle Ricerche
Monterotondo Scalo, Italy

Anderson Rodrigo Moraes de Oliveira
Faculty of Philosophy, Sciences and Letters of Ribeirão Preto
University of São Paulo
Ribeirão Preto, Brazil

Salvatore Fanali
Institute of Chemical Methodologies
Consiglio Nazionale delle Ricerche
Monterotondo Scalo, Italy

Marianne Fillet
Department of Analytical Pharmaceutical Chemistry
Institute of Pharmacy
University of Liège
Liège, Belgium

Joe P. Foley
Department of Chemistry
Drexel University
Philadelphia, Pennsylvania

Arjun Ghosh
Department of Chemistry
Indian Institute of Technology
Kharagpur, India

Alessandro Giuffrida
Department of Chemical Sciences
University of Catania
Catania, Italy

Gerald Gübitz
Department of Pharmaceutical Chemistry
Institute of Pharmaceutical Sciences
Karl-Franzens-University
Graz, Austria

Jun Haginaka
School of Pharmacy and Pharmaceutical Sciences
Mukogawa Women's University
Nishinomiya, Hyogo, Japan

Ylva Hedeland
Department of Medicinal Chemistry
Division of Analytical Pharmaceutical Chemistry
Biomedical Centre
Uppsala University
Uppsala, Sweden

Ulrike Holzgrabe
Institute of Pharmacy and Food Chemistry
University of Würzburg
Würzburg, Germany

Valquiria Aparecida Polisel Jabor
Faculty of Pharmaceutical Sciences of Ribeirão Preto
University of São Paulo
Ribeirão Preto, Brazil

Kimberly A. Kahle
Merck & Co. Inc.
West Point, Pennsylvania

Franka Kalman
Solvias AG
Basel, Switzerland

Michael Lämmerhofer
Department of Analytical Chemistry and Food Chemistry
University of Vienna
Vienna, Austria

Debby Mangelings
Department of Analytical Chemistry and Pharmaceutical Technology
Vrije Universiteit Brussel
Brussels, Belgium

Yvette Michotte
Department of Pharmaceutical Chemistry, Drug Analysis and Drug Information
Vrije Universiteit Brussel
Brussels, Belgium

Curt Pettersson
Department of Medicinal Chemistry
Division of Analytical Pharmaceutical Chemistry
Biomedical Centre
Uppsala University
Uppsala, Sweden

Anna Rocco
Institute of Chemical Methodologies
Consiglio Nazionale delle Ricerche
Monterotondo Scalo, Italy

Serge Rudaz
School of Pharmaceutical Sciences
University of Geneva
University of Lausanne
Geneva, Switzerland

Julie Schappler
School of Pharmaceutical Sciences
University of Geneva
University of Lausanne
Geneva, Switzerland

Martin G. Schmid
Department of Pharmaceutical Chemistry
Institute of Pharmaceutical Sciences
Karl-Franzens-University
Graz, Austria

Maria A. Schwarz
Department of Chemistry
University of Basel
Basel, Switzerland

and

Solvias AG
Basel, Switzerland

Anne-Catherine Servais
Department of Analytical Pharmaceutical Chemistry
Institute of Pharmacy
University of Liège
Liège, Belgium

Yvan Vander Heyden
Department of Analytical Chemistry and Pharmaceutical Technology
Vrije Universiteit Brussel
Brussels, Belgium

Ann Van Eeckhaut
Department of Pharmaceutical Chemistry, Drug Analysis and Drug Information
Vrije Universiteit Brussel
Brussels, Belgium

Radim Vespalec
Brno, Czech Republic

Jean-Luc Veuthey
School of Pharmaceutical Sciences
University of Geneva
University of Lausanne
Geneva, Switzerland

Markéta Vlčková
Department of Chemistry
University of Basel
Basel, Switzerland

1 Pharmacological Importance of Chiral Separations

Ann Van Eeckhaut and Yvette Michotte

CONTENTS

1.1 General Terminology of Isomerism 2
1.2 Chirality 3
1.3 Stereoselectivity in Pharmacodynamics 5
1.4 Stereoselectivity in Pharmacokinetics 7
1.4.1 Absorption 8
1.4.2 Distribution 8
1.4.3 Metabolization 9
1.4.3.1 Substrate Stereoselectivity 9
1.4.3.2 Product Stereoselectivity 10
1.4.3.3 Substrate–Product Stereoselectivity 11
1.4.4 Renal Excretion 12
1.5 Interspecies Differences 12
1.6 Patient-Specific Factors 12
1.6.1 Age 12
1.6.2 Gender 13
1.6.3 Genetic Factors 13
1.6.4 Disease State 14
1.7 Drug-Related Factors 14
1.7.1 Input Rate 14
1.7.2 Interactions 15
1.8 Single Enantiomer versus Racemate—"Chiral Switches" 15
1.9 The Need for Analytical Tools 17
1.10 Introduction to Capillary Electrophoresis 17
References 18

1.1 GENERAL TERMINOLOGY OF ISOMERISM [1,2]

Isomers are compounds that have the same stoichiometric molecular formula. However, they can differ in either connectivity or spatial orientation. There are two major classes of isomers, namely, constitutional isomers and stereoisomers.

Constitutional isomers have a different binding pattern of the atoms in the three-dimensional space. They often vary substantially in their physical and chemical properties. For example, 1-propanol and 2-propanol bear the same molecular formula $C_4H_{10}O$, but their structural formula differs (Figure 1.1). Another example is cortisone, prednisolone, and aldosterone, all of which have $C_{21}H_{28}O_5$ as molecular formula [3].

Stereoisomers have identical constitution, but differ in the spatial arrangement of certain atoms or groups. Stereoisomers can be subdivided into two distinct groups: optical isomers or enantiomers and diastereoisomers. Enantiomers are nonsuperimposable mirror images of each other (Figure 1.1). A pair of enantiomers will be identical in atomic constitution and bonding. However, they exhibit a different configuration of nonequivalent groups around a stereogenic center (in most cases a

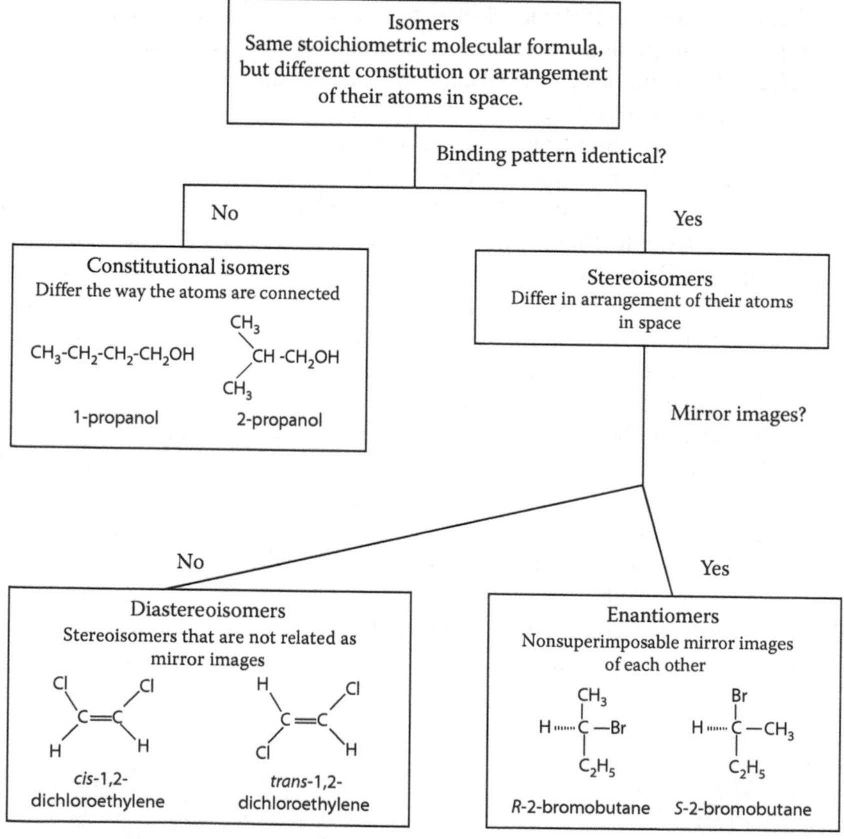

FIGURE 1.1 Overview of isomerism.

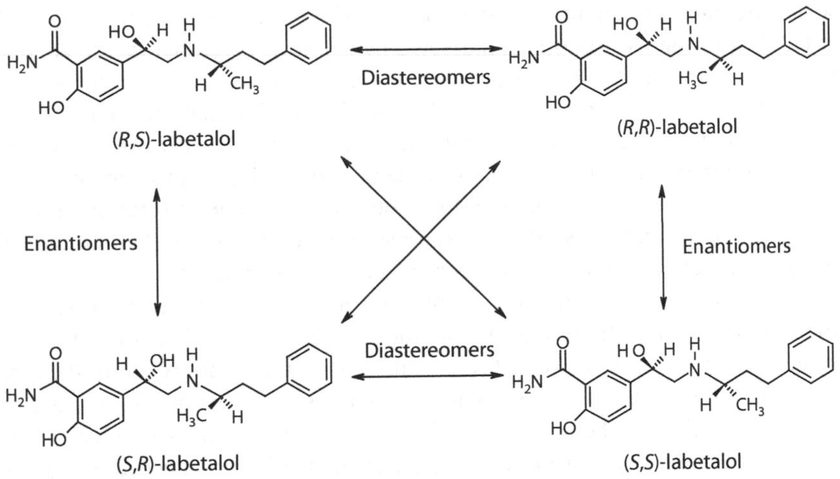

FIGURE 1.2 The relationship between the diastereoisomers of labetalol.

carbon atom). They possess identical physicochemical properties in achiral media. However, each enantiomer rotates the plane of polarized light in opposite direction, the degree of rotation being identical (same number of degrees, but in opposite direction). A second type of stereoisomers, referred to as diastereoisomers, are molecules that are not mirror images of each other. Diastereoisomers, unlike enantiomers, do exhibit different physicochemical properties. One example of diastereoisomerism is the *cis–trans* isomers (Figure 1.1). Another example are molecules that contain at least two chiral centers. In molecules with n (more than 1) chiral centers, the maximal possible number of stereoisomers is 2^n. Such a molecule possesses n pairs of enantiomers and the remaining stereoisomers are diastereoisomers. An example is given in Figure 1.2. The adrenoreceptor blocking agent labetalol has two chiral centers and therefore, it possesses four stereoisomers of which two pairs of enantiomers. Another well-known example is ephedrine and pseudoephedrine.

1.2 CHIRALITY

The term chirality originates from the Greek word "χειρ," meaning hand, and it describes drugs which have the property of molecular asymmetry and therefore exist as a pair of nonsuperimposable mirror images. In 1848, Louis Pasteur initiated the study of the physical and chemical properties of asymmetric molecules. He discovered that tartaric acid exists as two crystalline salts, which rotate the plane of polarized light in opposite directions [4].

The chiral center is usually, but not obligatory, a carbon atom. A chiral center is then formed when four different groups are linked to the same carbon atom. Less commonly, other atoms including phosphor, sulfur, and nitrogen may also form a chiral center. A racemate is an equimolar mixture of a pair of enantiomers. It does not show optical activity. Both enantiomers, on the other hand, have the ability to rotate the plane of polarized light. Both with the same number of degrees, although in opposite direction. The enantiomer that rotates the plane of polarized light clockwise is called dextrorotatory and this is indicated by a d- or (+)-sign before the chemical name. The enantiomer that rotates the plane counterclockwise is termed levorotatory and is designated by the prefix l- or (–). This method of designation is based upon a physical property of the molecule and it does not provide any information concerning the absolute configuration or three-dimensional arrangement of atoms around the chiral center [5,6]. Since both the direction and magnitude of rotation may vary with experimental conditions, like for example the solvent, one should look carefully at using the direction of rotation as a stereochemical descriptor [5].

Currently, the International Union of Pure and Applied Chemistry (IUPAC) recommends the Cahn–Ingold–Prelog method of designation. This model uses the absolute structural configuration of the enantiomers. It does not relate to the direction of rotation of polarized light. The enantiomers are classified *R* (rectus) or *S* (sinister), according to the sequence-rule procedure [7,8]. In this sequence rule, the atoms attached to a chiral center are assigned priorities according to some rules. The most important are the atomic number and mass. In Figure 1.3, a carbon atom is depicted to which four different ligands (a, b, c, d where a>b>c>d) are attached. If the left molecule is viewed remote from the group with the lowest priority d, the path from a to b to c follows a clockwise direction. This molecule is assigned as the *R* configuration. In the molecule, on the right-hand side, the path from a to b to c traces a counterclockwise direction. This molecule is assigned as the *S*-enantiomer.

A third way of designation, the D/L convention, is commonly used by biochemists to describe the stereochemical configuration of sugars and amino acids. This designation should not be confused with the d/l nomenclature. Sugars are designated as D or L corresponding to the configuration of the chiral carbon farthest from the carbonyl group, using D- and L-glyceraldehyde as reference molecules. In the case of α-amino acids, the molecule is viewed along the C–H axis between the hydrogen and the asymmetric carbon atom. If the clockwise order of the other three groups is –COOH, –*R*, $-NH_2$, then the amino acid belongs to the D-series; otherwise it belongs to the L-series.

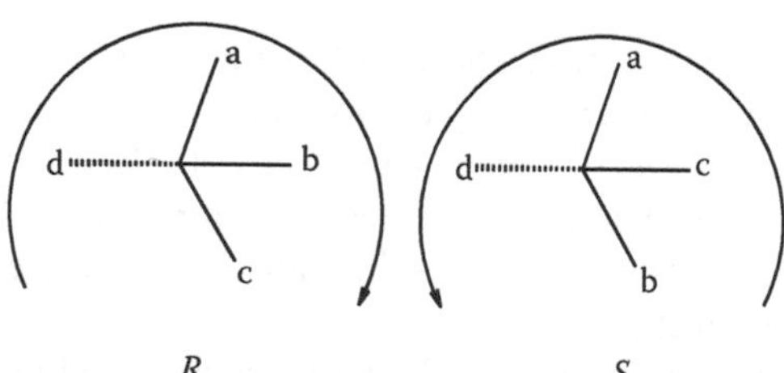

FIGURE 1.3 The *R*/*S* convention. A carbon atom is depicted to which four different ligands a, b, c, d are attached, where a>b>c>d. When the left molecule is viewed remote from the group with the lowest priority d, the path from a to b to c follows a clockwise direction. This molecule is assigned the *R* configuration. In the molecule on the right, the path from a over b to c traces a counterclockwise direction. This molecule is assigned the *S*-enantiomer.

Enantiomers have the same physicochemical properties (e.g., melting point, solubility, etc.) in achiral media. Any passive, nonstereoselective pharmacological process will be identical for both enantiomers. Even so, they should be considered as different chemical entities. Since the molecules they interact within biological systems are chiral, differences in potency, pharmacological activity, and pharmacokinetic profile can occur between enantiomers. At the molecular level, biological systems are homochiral. The essential amino acids are all L-enantiomers, whereas carbohydrates have the D-configuration. Therefore, when a chiral, exogenous compound is introduced into the body, physiological processes will show a high degree of chiral distinction. Each enantiomer will interact differently with chiral targets as receptors, enzymes, and ion channels. Consequently, this could lead to different effects observed for the enantiomers. If the chiral center of the molecule is involved in the interaction with these targets, large differences between the enantiomers can be observed. As a result, both enantiomers of a chiral compound can differ in their pharmacological and toxicological properties. The more active enantiomer is called eutomer and the less active, the distomer. However, these terms can be misleading as seen in the example of propoxyphene. Dextropropoxyphene is the eutomer for its analgesic properties, but it is the distomer when the antitussive properties are considered [9]. The ratio of the activity of the eutomer and distomer, the eudismic ratio, is a measure of the stereoselectivity [10].

Despite the early understanding that enantiomers often differ in potency, the stereospecific aspects of chiral drugs were completely ignored till the 1980s [11–13]. A major reason was that the available synthesis methods yielded racemates and that full-scale production of pure enantiomers was not judged to be feasible, as well for technical as for economical reasons [11]. In most cases, the pharmacological and clinical evaluation was therefore carried out with racemates and using achiral analysis methods. In 1984, Ariëns described this practice as "sophisticated nonsense in pharmacokinetics and clinical pharmacology" [10]. Since then there was an emerging growth in chemical technologies, allowing the production of pure enantiomers. For their development of catalytic asymmetric synthesis Knowles, Noyori, and Sharpless, were awarded the Nobel Prize in Chemistry in 2001. Also large advances were made in the development of stereoselective analysis methods, which opened opportunities to study the importance of stereospecificity in pharmacology [11,12].

The goal of this chapter is to give an overview of all potential differences in pharmacodynamic properties and pharmacokinetic effects between enantiomers of a chiral drug. To illustrate this, several examples will be given. However, we do not aim at given a complete overview of all the existing literature.

1.3 STEREOSELECTIVITY IN PHARMACODYNAMICS

Pharmacodynamics is the study of the biochemical and physiological effects of drugs and their mechanisms of action. Four possible reasons can be given for differential effects of enantiomers *in vivo*.

Firstly, most of the pharmacological activity can reside in one enantiomer. The other enantiomer, which can be regarded as an impurity, can be inactive or even be responsible for toxic effects. Thus, the racemate can contain up to 50% of impurity or ballast. Examples are escitalopram and levodopa. Escitalopram is the active *S*-enantiomer of citalopram, a selective serotonin reuptake inhibitor. It is approximately 100 times more potent as reuptake inhibitor than its *R*-enantiomer [14–17]. While the *R*-enantiomer is practically devoid of uptake inhibition potency, it does appear to interact with histamine receptors. Therefore, the single enantiomer drug escitalopram will be less sedating than the racemate [16]. In addition, it has been shown that *R*-citalopram counteracts the binding of escitalopram to the allosteric site of the serotonin transporter [18]. Serious side effects limited the use of racemic dopa for the treatment of Parkinson's disease. These adverse effects included nausea, vomiting, anorexia, involuntary movements, and granulocytopenia. The use of levodopa instead of the racemate was found to halve the required dose, to reduce toxicity, and to improve the motor function. The occurrence of granulocytopenia was related to the dextrorotary enantiomer [3].

Secondly, the two enantiomers can have almost identical qualitative and quantitative pharmacological activity. The enantiomers of promethazine, for example, have nearly identical antihistaminic properties and toxicity [5,19,20]. Similarly, both enantiomers of flecainide and propafenone have equivalent antiarrhythmic potency [5,20,21].

Thirdly, the enantiomers can possess similar pharmacological properties; however, they can differ in potency. Most chiral drugs belong to this category. Examples of this class of chiral drugs are warfarin, verapamil, levocetirizine, and (*R*,*R*)-methylphenidate. *S*-warfarin has a five times larger anticoagulant potency than its antipode *R*-warfarin [15,22,23]. *S*-verapamil is approximately 20 times more potent than its *R*-enantiomer with regard to negative dromotropic effects on atrioventricular node conduction [24]. Levocetirizine has an approximately 30-fold higher affinity for human H_1-receptor than dextrocetirizine [25]. (*R*,*R*)-methylphenidate, used for the treatment of attention-deficit hyperactivity disorder, is approximately 10 times more potent than (*S*,*S*)-methylphenidate in the inhibition of dopamine and noradrenaline from striatal and hypothalamic synaptosomes respectively [26].

Fourthly, enantiomers can have qualitatively different pharmacological activities. For example, dextropropoxyphene is used as an analgesic while levopropoxyphene has antitussive properties. Another example is the mixed adrenoreceptor blocker labetalol. The (*S*,*R*)-isomer has α-blocking effects, while the β-blocking activities are due to the (*R*,*R*)-isomer [19,27].

According to these pharmacodynamic differences seen between enantiomers, in several cases the use of a single enantiomer instead of the racemate could be suggested. However, it is not always clear-cut as seen in the example of ibuprofen. Although *R*-ibuprofen is not active, it is the precursor of the active *S*-enantiomer via a unidirectional metabolic chiral inversion pathway (Figure 1.4). *R*-ibuprofen is metabolized to an intermediary *R*-CoA thioester which is then epimerized to an *S*-CoA thioester and in turn converted to *S*-ibuprofen. This results in a nonsignificant difference in activity *in vivo* between the racemate and the individual enantiomers [3,20,28–31]. Another example is the well-known thalidomide. Racemic thalidomide was introduced as a relatively safe sedative drug in the late 1950s, but soon thereafter

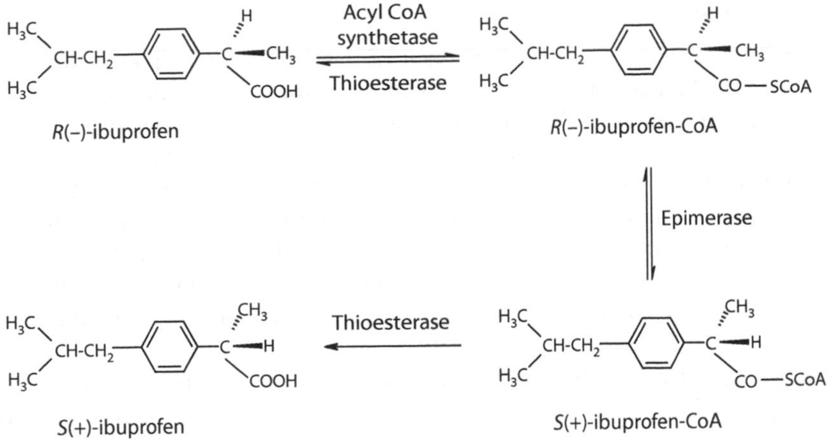

FIGURE 1.4 Stereospecific inversion of *R*- to *S*-ibuprofen. (Adapted from Ali, I. et al., *Chirality*, 19, 453, 2007. With permission; From Tan, SC. et al., *Enantiomer*, 4, 195, 1999. With permission.)

it was withdrawn because it caused serious fetal malformations [11,32]. Studies have shown that the sedative effect of thalidomide is related to the *R*-enantiomer and that probably the *S*-enantiomer is responsible for the teratogenic effect [32]. These findings have led to the wrong statement in literature that the thalidomide tragedy could have been avoided if only the *R*-enantiomer would have been used. Since rapid racemization occurs *in vivo*, their difference in toxicity cannot be exploited [2]. Indeed, the enantiomers of thalidomide undergo spontaneous hydrolysis and fast chiral interconversion at physiological pH [20,32]. Since 1998, the use of thalidomide is again approved by the Food and Drug Administration. However, now it is used for the treatment of the debilitating and disfiguring lesions associated with erythema nodosum leprosum (ENL), a complication of Hansen's disease, commonly known as leprosy. Recently, thalidomide was also approved for the treatment of multiple myeloma. Known or possible pregnancy is of course an absolute contraindication to the use of thalidomide [32].

1.4 STEREOSELECTIVITY IN PHARMACOKINETICS

Besides pharmacodynamic considerations, the pharmacokinetic properties must also be taken into consideration. Pharmacokinetics refers to how the body affects the drug and how the body handles absorption, distribution, metabolization, and excretion of drugs in relation to their pharmacological, therapeutic, or toxic response [19]. The pharmacokinetic importance of drug chirality depends on the process under consideration. Stereochemistry does not influence passive processes such as diffusion across membranes. However, when an enzyme or transporter system is involved, chiral discrimination can be observed [33,34]. Enantiospecificity in pharmacokinetics is generally quite low and the contribution to the eudismic ratio is usually quite small. However, the clinical implications with respect to dosages and routes of administration may be very important [6].

1.4.1 Absorption

Absorption is the process of the movement of the drug from the place of administration into the bloodstream. This process requires the transport of drug molecules through cell membranes. Three possible mechanisms are described: passive diffusion through the membrane, facilitated diffusion via channels or carriers, and active transport via carriers and pump systems. The absorption of most drugs is a passive process; however, drugs can also be absorbed via active processes like receptor-mediated transport. Active absorption may favor one enantiomer or the two enantiomers may differ in absorption characteristics. L-Dopa, for example, is actively absorbed from the gastrointestinal tract. D-Dopa, on the other hand, enters the bloodstream via passive diffusion [20].

The absorption rate is also an important parameter. An example is ibuprofen that is already described earlier. The chiral inversion of inactive *R*-ibuprofen to the active *S*-enantiomer in the gastrointestinal tract depends on the absorption rate: the longer the drug stays in the gastrointestinal tract (i.e., low absorption rate), the greater will be the extent of inversion [35].

1.4.2 Distribution

Once the drug is available in the bloodstream, it will distribute throughout the different compartments of the body. Two physicochemical properties are essentially responsible for the distribution of drugs: plasma- and tissue protein binding and the partition coefficient. Partitioning into various sites is a physical property and is therefore not considered as enantioselective. However, binding to proteins, which consists of chiral building blocks, can show stereoselectivity.

Stereoselective binding of chiral drugs to serum albumin and α_1-acid glycoprotein has already been shown. For example, the affinity for the binding of the essential amino acid *S*-tryptophan to the benzodiazepine and indole site of human serum albumin is about 100 times greater than that of *R*-tryptophan. An interesting example is the antihypertensive drug propranolol. *S*-propranolol binds to α_1-acid glycoprotein to a slightly greater extent than *R*-propranolol [19,27,35]. However, the stereoselective binding to human serum albumin is opposite of that observed for α_1-acid glycoprotein [20]. The overall stereoselectivity in the binding of propranolol to human serum resembles that seen for α_1-acid glycoprotein, meaning that the free fraction of the *R*-enantiomer is higher than that of the pharmacological active *S*-enantiomer [27].

Stereoselective distribution may also occur as a result of drug–lipid interactions. Hanada et al. [36] have observed stereoselective binding of verapamil and disopyramide to phosphatidylserine, a tissue-binding site for basic drugs, with *R*-verapamil and *R*-disopyramide being preferentially bound [20,21,36].

Stereoselective interactions with tissue uptake transporter systems and storage mechanisms have also been reported [20]. van Bree et al. [37] have studied the transport of baclofen and its enantiomers across the blood–brain barrier in rats. They showed that the blood–brain barrier clearance of *R*-baclofen was approximately four times higher than that of the *S*-enantiomer. On the basis of their obtained results, they hypothesized that *S*-baclofen is preferentially transported into the central nervous system. Barraud de Lagerie et al. [38] demonstrated that in mice mefloquine enantiomers undergo efflux in a stereoselective manner and that this stereoselectivity was due to the efflux proteins P-glycoprotein and/or breast cancer resistance protein. Alves et al. [39] studied the stereoselective disposition of the antiepileptic drug licarbazepine in mice. Although both enantiomers had comparable systemic exposure, the extent of brain penetration for *S*-licarbazepine was found to be two times higher than for the *R*-enantiomer. Further investigations into the exact mechanism still have to be performed.

1.4.3 Metabolization

Much of the stereoselectivity observed in pharmacokinetics is caused by stereoselective metabolic pathways. Both phases I and II metabolic reactions are capable of discriminating between enantiomers. Three different groups can be distinguished and are discussed in the following.

1.4.3.1 Substrate Stereoselectivity

Substrate stereoselectivity is the preferential enzymatic metabolization of one enantiomer. The oxidation of warfarin in human, for example, is stereoselective for the *S*-enantiomer of the drug yielding (*S*)-7-hydroxywarfarin [23,33].

Both enantiomers of omeprazole are prodrugs, which are converted within the parietal cell to the active proton pump (H^+, K^+-ATPase) inhibitor, an achiral sulfenamide. Therefore, once within the parietal cell, *R*- and *S*-omeprazole are equipotent inhibitors of H^+, K^+-ATPase. However, in vitro experiments have shown that there is a significant stereoselectivity in the metabolism of the enantiomers of omeprazole. The three main metabolites are omeprazole sulfone, 5-hydroxy-omeprazole, and 5-*O*-desmethyl omeprazole. The hydroxy and 5-*O*-desmethyl metabolites of both enantiomers are mainly formed by CYP2C19, whereas sulfoxidation is catalyzed by CYP3A4. *S*-omeprazole (also called esomeprazole) is metabolized to a relatively greater extent via CYP3A4 compared to the *R*-enantiomer, which is almost completely metabolized by CYP2C19. In addition, the intrinsic clearance of esomeprazole is approximately three times lower than that of *R*-omeprazole, mainly due to its lower intrinsic clearance through CYP2C19 to the 5-hydroxy metabolite (Figure 1.5). Consequently, increased plasma concentrations of esomeprazole and an increased inhibitory effect on acid secretion are observed. As a result, at equal doses, esomeprazole achieves 70%–90% higher steady-state serum concentrations than the racemic omeprazole [40–44].

Substrate stereoselectivity also includes enantiomer inversion and first-pass metabolism. Examples of enantiomer inversion have already been described in Section 1.3. Enantiomer inversion has been extensively described for several 2-arylpropionic acid derivatives, the so-called profens. In Section 1.3, the chiral inversion of *R*-ibuprofen to the active *S*-ibuprofen is, for example, shown. A well-known example of stereoselectivity in first-pass metabolization is verapamil. This L-type calcium channel blocker is less potent after oral administration than following single intravenous administration. It has been shown that after oral

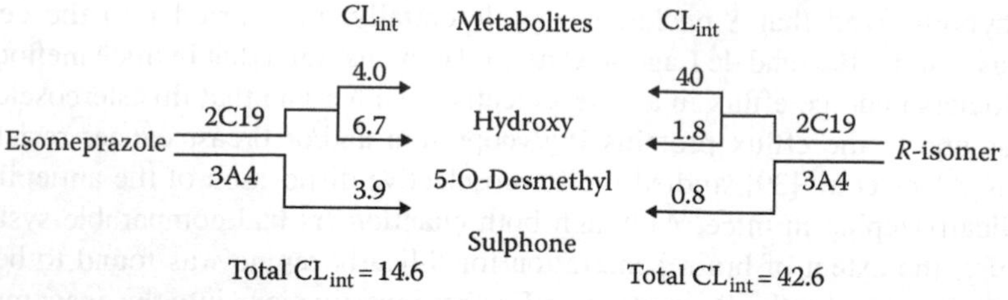

FIGURE 1.5 Metabolism scheme illustrating the intrinsic clearance values (CL_{int}) for the different metabolic pathways of esomeprazole and *R*-omeprazole from in vitro experiments on human liver microsomes [40,43]. (Reproduced from Andersson, T. et al., *Clin. Pharmacokinet.*, 40, 411, 2001. With permission.)

administration of racemic verapamil, the proportion of the pharmacologically more active *S*-enantiomer of the total plasma verapamil concentration is substantially lower than that of the *R*-enantiomer [5,24,45]. This can be explained by stereoselective differences in the first-pass metabolism of both enantiomers following single oral administration of racemic verapamil [46].

1.4.3.2 Product Stereoselectivity

In this case, an achiral drug is transformed into chiral metabolites. In Figure 1.6, it is shown how some reductive or oxidative metabolization reactions can form new chiral centers.

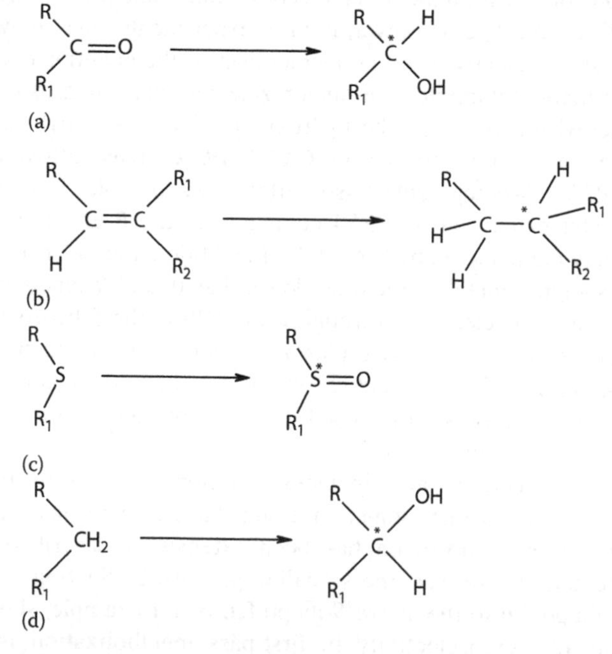

FIGURE 1.6 Introduction of new chiral centers via reductive and oxidative metabolization pathways. (a) Reduction of carbonyl to alcohol, (b) hydrogenation of alkenes to alkanes, (c) sulfoxidation, and (d) hydroxylation. Asterisks indicates chiral center.

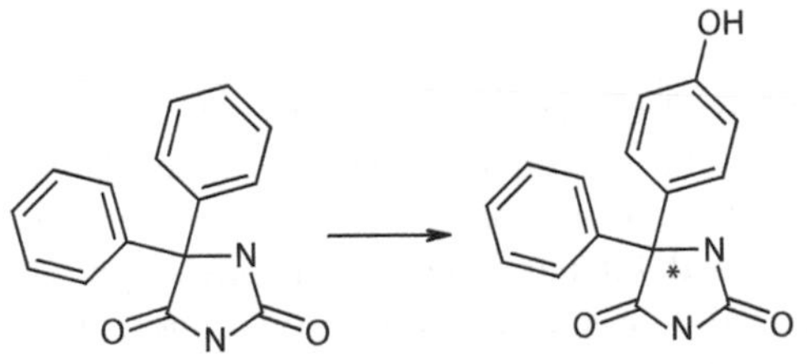

FIGURE 1.7 *p*-Hydroxylation of phenytoin. Asterisks indicates chiral center.

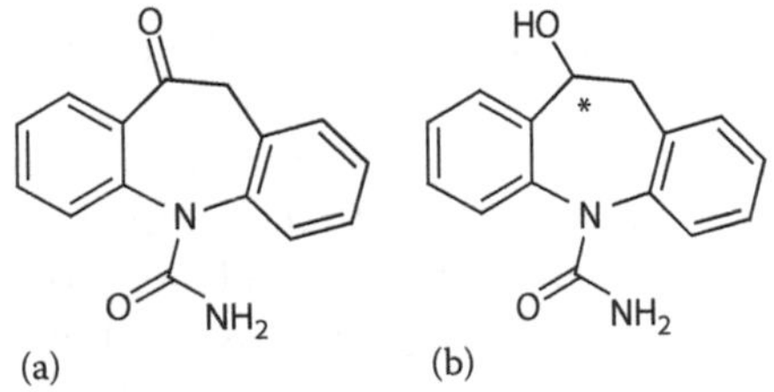

FIGURE 1.8 Structure of (a) oxcarbazepine and (b) licarbazepine. Asterisks indicate the chiral center.

An example of product stereoselectivity is the aromatic *p*-hydroxylation of one of the phenyl rings of phenytoin, creating a chiral center at C-5 of the hydantoin ring (Figure 1.7) [31,34]. A second example is the antiepileptic drug oxcarbazepine. In humans, this achiral prodrug is rapidly reduced in the liver to the pharmacologically active and chiral metabolite 10-hydroxycarbazepine or licarbazepine (Figure 1.8). This metabolic reduction is stereoselective, resulting in a 4:1 ratio between the active *S*-enantiomer and *R*-licarbazepine [39,47].

1.4.3.3 Substrate–Product Stereoselectivity

Substrate–product stereoselectivity means the stereoselective metabolization of one of the enantiomers whereby another chiral center is introduced and diastereoisomers are thereby formed. Warfarin is metabolized via two principal routes: aromatic hydroxylation of the coumarin ring (see substrate stereoselectivity) and ketone reduction in the side chain that produces a new chiral center (Figure 1.9). The latter pathway shows marked substrate stereoselectivity for the *R*-enantiomer, as well as product stereoselectivity, namely, the formation of alcohols of the *S*-configuration [23,33,48].

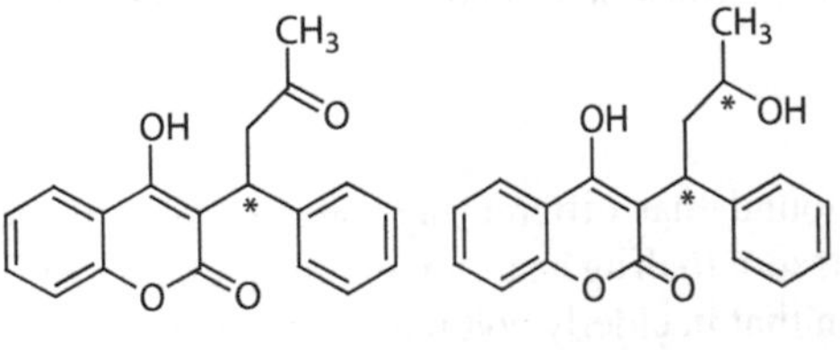

FIGURE 1.9 Structure of warfarin and warfarin alcohol. Asterisks indicate the chiral center.

1.4.4 Renal Excretion

Renal excretion consists of glomerular filtration, with or without additional tubular secretion and/or tubular reabsorption. Since glomerular filtration is a passive process, stereoselectivity may only be observed with tubular secretion that occurs via active processes or with tubular reabsorption. Although tubular reabsorption is usually passive, it may also be active. However, only few examples of active tubular reabsorption exist [25]. In contrast, tubular secretion is an active process consisting of uptake from blood across the basolateral membrane of renal epithelial cells followed by efflux into urine across the apical membrane [25]. Apparent stereoselectivity in renal clearance may also be observed as a consequence of preferential protein binding, since only the unbound fraction of the drug is available for glomerular filtration [20,25].

Cetirizine is mainly eliminated unchanged in urine through both glomerular filtration and tubular secretion. The higher renal clearance of dextrocetirizine is partly due to the higher unbound fraction available for glomerular filtration compared to levocetirizine. However, the major difference in renal clearance between both enantiomers is determined by stereoselectivity in tubular secretion. The exact mechanism leading to a higher tubular secretion of the distomer is not known. It may be due to the higher free fraction available for secretion and/or to a higher affinity for renal drug transporters [25].

1.5 INTERSPECIES DIFFERENCES

An important aspect to remember is that some drugs can display significant interspecies differences. An interesting example is the proton pump inhibitor omeprazole. Initial *in vivo* experiments to assess the plasma concentrations following oral administration and the effects on stimulated acid secretion of both omeprazole enantiomers were carried out in rats as well as in dogs. No differences between the enantiomers could be detected in the dog. However, in the rat the *R*-enantiomer was significantly more active than the *S*-enantiomer or the racemate. In contrast, the *S*-enantiomer exhibited the highest bioavailability and oral potency in inhibiting gastric acid secretion in humans [41].

1.6 PATIENT-SPECIFIC FACTORS

Various factors related to the drug and the patient may affect the stereospecificity seen in the pharmacodynamics and kinetics. Factors related to the drug include the dosing rate, interaction between both enantiomers, and interaction between the enantiomers and other drugs. These factors will be discussed in the next section. Patient-related factors include age, gender, disease state, and the patient metabolic profile [27,49].

1.6.1 Age

Some studies have found that different age groups may be associated with age-specific levels of exposure to drug enantiomers [49]. One example is here given. Tan et al. [50] have shown that in elderly volunteers the free fraction of *S*-ibuprofen, and not of the *R*-enantiomer, was increased when compared to young volunteers. They also observed a decrease in the clearance of *S*-ibuprofen, particularly via oxidative pathways, in these elderly volunteers. This higher availability of the active *S*-enantiomer may contribute to the incidence of adverse reactions observed in the elderly population.

1.6.2 Gender

Not so much information can be found concerning the gender differences on the level of stereoselective dispositions [49]. Although the antihypertensive effect of labetalol was the same for men and women, the total plasma concentrations were 80% higher in women. This discrepancy could be explained by stereoselective differences in the pharmacokinetics of the labetolol isomers in both groups. Indeed, while the concentration of the (*S*,*R*)-isomer which has α-blocking properties and the two relatively inactive (*S*,*S*)- and (*R*,*S*)-isomers were between 60% and 80% higher in women, the concentrations of the main β-blocking (*R*,*R*)-isomer were the same in both groups. This explains why similar antihypertensive effects were observed in men and women [27,51].

1.6.3 Genetic Factors

Genetic factors are known to be significant in drug disposition, particularly for those agents whose metabolism is associated with the debrisoquine oxidation genetic polymorphism [5]. Approximately 10% of the Caucasian population is deficient in the expression of the cytochrome P450 isozyme CYP2D6. They are called poor metabolizers. CYP2D6 activity ranges from complete deficiency in these poor metabolizers, over extensive metabolism (homozygous and heterozygous) to ultrarapid metabolism [52]. The administration of a racemate, which is substrate for CYP2D6, to one of these subtypes may lead to different drug concentrations and enantiomeric compositions depending on the phenotype. Therefore, the therapeutic response may be influenced by the genetic make-up of the patient. Scordo et al. [52] have studied the influence of CYP2D6 genetic polymorphism on the steady-state plasma concentrations of the enantiomers of fluoxetine and norfluoxetine. They observed a significantly higher *S*-norfluoxetine/*S*-fluoxetine ratio in homozygous compared to heterozygous CYP2D6 extensive metabolizers. In addition, very low concentrations of *S*-norfluoxetine and very low *S*-norfluoxetine/*S*-fluoxetine and *S*/*R*-norfluoxetine ratios were seen in the CYP2D6 poor metabolizers included in their study. These findings led to the suggestion that CYP2D6 plays a role in the stereoselective conversion of *S*-fluoxetine to *S*-norfluoxetine [52].

Genetic polymorphism also exists for CYP2C19 expression, with approximately 3%–5% of Caucasian, 4%–7% of the black Africans, and 15%–20% of Asian populations being poor metabolizers with no CYP2C19 function [12,51]. This isoenzyme is also called mephenytoin hydroxylase. An example is the proton-pump inhibitor omeprazole, which showed a significant intervariability in its pharmacokinetics and also regarding its effect on gastric acid secretion. This difference in response was especially pronounced between slow and rapid metabolizers, since omeprazole is mainly metabolized by CYP2C19. After repeated administration of esomeprazole or racemic omeprazole to extensive metabolizers, the area under the curve (AUC) of esomeprazole was found to be twofold higher than that of omeprazole. In poor metabolizers the plasma concentrations of esomeprazole were lower than those of omeprazole. Thus, the use of the single enantiomer drug esomeprazole results in less overall variability in the pharmacokinetics. This demonstrates the lower influence of CYP2C19 on the metabolism of esomeprazole as compared with omeprazole (see also Section 1.4.3.1) [12,40,42–44].

1.6.4 Disease State

Disease states may cause changes in enantiomer ratios of pharmacokinetic parameters [49]. For example, Jamali et al. [53] have observed that surgery for wisdom tooth removal resulted in substantial decrease in the serum concentration of both ibuprofen enantiomers and in delayed absorption of the enantiomers. In addition, dental extraction resulted in a reversal of the serum ibuprofen enantiomer concentrations, resulting in lower concentrations of the active *S*-enantiomer. The authors suggested that this may be as a result of a change in the metabolic chiral inversion of *R*-ibuprofen to the pharmacological more important *S*-enantiomer. Thus, dental patients may experience a delay of onset of the pain relief and even treatment failure may occur in patients taking a normal dose (200–600 mg) of ibuprofen after surgery.

Differences in both pharmacodynamic effects and pharmacokinetic profile of drugs undergoing significant and enantioselective first-pass metabolism can be observed in patients with liver cirrhosis [5]. Neugebauer et al. [54] have shown that in patients with liver cirrhosis, the hepatic first-pass extraction of the antihypertensive drug carvediol is markedly reduced, eliminating the difference in bioavailability between the two enantiomers which is observed in healthy subjects.

Similarly, in renal disease patients, Imamura et al. [55] have demonstrated that the stereoselective serum protein binding of the beta-adrenergic blocking agent alprenolol, seen in healthy volunteers, was significantly altered. Alprenolol mainly binds to α_1-acid glycoprotein, whose levels are increased in case of renal disease. Consequently, the binding percentage of alprenolol in serum was higher in renal disease patients compared to healthy volunteers. Since the pharmacologically active *S*-enantiomer has a higher binding constant for this protein than *R*-alprenolol, the free concentration of *S*-alprenolol decreased more markedly than that of the *R*-enantiomer.

1.7 DRUG-RELATED FACTORS

1.7.1 Input Rate

An alteration in the oral input rate may affect the plasma concentrations of both enantiomers of a racemic drug. A significant effect on first-pass metabolism and subsequently on the clinical response can be observed, especially for drugs with a high hepatic extraction rate, like verapamil and metoprolol [27]. For example, significant different *R*/*S* ratios of verapamil in plasma were observed when racemic verapamil was administered either in an immediate release or sustained release formulation [56,57]. Mistry et al. [58] have studied the influence of the input rate on the stereoselective pharmacokinetics of metoprolol. They have shown that the *S*/*R* ratio for plasma metoprolol concentrations show significant differences in the absorption phase (1–4 h) versus the terminal elimination phase (8–16 h) when fast input of the drug is obtained. However, slow input displayed no significant difference in *S*/*R* ratio between the absorption and elimination phase.

1.7.2 Interactions

The two enantiomers of a racemic drug may interact with each other at different pharmacokinetic or pharmacodynamic levels. The *R*-enantiomer of propafenone, a class Ic antiarrhythmic drug, for example, reduces the metabolism of the *S*-enantiomer, leading to a significantly reduced oral clearance of *S*-propafenone in the presence of the *R*-enantiomer [21]. When the enantiomers of disopyramide, a class Ia antiarrhythmic drug, were administered separately to humans, no differences in clearance, renal clearance, or volume of distribution were observed. However, when racemic disopyramide was given, an important pharmacokinetic interaction was observed, resulting in lower plasma and renal clearance, a longer half-life, and a smaller volume of distribution for the *S*-enantiomer [6,45].

In addition to enantiomer–enantiomer interactions, a racemic drug may interact stereoselectively with other drugs. For instance, the anticoagulant warfarin exhibits enantiomer–enantiomer interactions and in addition its elimination is altered in a stereospecific manner by cimetidine [19,21]. *R*-warfarin inhibits the hydroxylation of *S*-warfarin, the pharmacologically more active enantiomer [59]. In addition, cimetidine decreases the rate of metabolism of *R*-warfarin and not that of the *S*-enantiomer [60,61]. This results in an accumulation of *R*-warfarin which in turn further suppresses the metabolism of *S*-warfarin. As a result, the active *S*-enantiomer is accumulated and an increased anticoagulant activity is observed [19].

1.8 SINGLE ENANTIOMER VERSUS RACEMATE—"CHIRAL SWITCHES"

Besides de novo synthesis of an enantiomerically pure drug, like the cholesterol lowering drug rosuvastatine and the antiepilepticum tiagabine, there is also the possibility of switching from the existing racemic drug to the single enantiomer of that drug. Chiral switches are defined as drugs that have already been claimed, approved and/or marketed as racemates or a mixtures of diastereoisomers, but have since been redeveloped as single enantiomers [2,62,63].

Since the early 1980s, the importance of stereochemistry and the use of single enantiomers instead of the racemate have been debated [4,64]. The European Medicines Evaluation Agency (EMEA) states in its guideline on investigation of chiral active substances that if a new racemate appears promising, both enantiomers should also be studied as early as possible to assess the relevance of stereoisomerism for effects and fate *in vivo* [65].

Four reasons for considering the enantiomers instead of the racemate are proposed [64]. Firstly, the use of a single enantiomer may allow the reduction of the dose while either maintaining or improving outcome. Secondly, the assessment of the dose–response relationship of a single enantiomer may be simpler than for a racemate. Thirdly, pharmacodynamic and pharmacokinetic variability between patients may be reduced and fourthly, any toxicity from the inactive enantiomer may be minimized [4,45,63,66]. Another reason, commercially, is that when patents on racemic drugs expire, pharmaceutical companies may have the opportunity to extend patent coverage through development of the chiral switch enantiomers with desired bioactivity.

Although the aforementioned reasons appear to present compelling arguments in favor of single enantiomers, there have been commercial failures as well as successes [66]. In the case of bupivacaine, for example, improved tolerability was gained using levobupivacaine. In 1979, seven cases of sudden cardiovascular collapse with difficult resuscitation or death were reported after intravenous injection of racemic bupivacaine. Levobupivacaine is significantly less cardiotoxic than the racemate and has the same anesthetic profile. Therefore, the racemate was replaced by the single enantiomer [66–68]. On the other hand, not all chiral switches were as promising as levobupivacaine. The anorectic agent fenfluramine was one of the first to undergo the chiral switch process with the marketing of the dextrorotary *S*-enantiomer, dexfenfluramine. The levorotatory *R*-enantiomer was responsible for most of the side effects seen with the racemate. Both fenfluramine and dexfenfluramine were, however, withdrawn from the market because of a perceived risk of pulmonary hypertension [2,14,66,68].

Besides switching from a racemate to an active enantiomer, the drug can also be replaced by the active, chiral metabolite and, if possible, by the active enantiomer of this metabolite. An example is the switch of hydroxyzine to its racemic active metabolite cetirizine. Cetirizine is formed from oxidation of the primary alcohol of hydroxyzine (Figure 1.10) [69,70]. Levocetirizine, the active enantiomer of cetirizine, is also on the market. A second example is eslicarbazepine, the *S*-enantiomer of licarbazepine. Eslicarbazepine is a new drug currently undergoing clinical development for the treatment of epilepsy [39]. Licarbazepine is the active metabolite of oxcarbazepine (Figure 1.8). More information concerning these compounds is already given in Section 1.4.3.2.

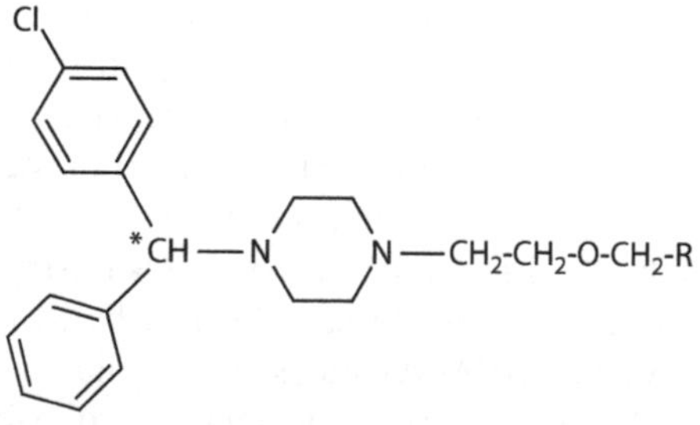

FIGURE 1.10 Structures of hydroxyzine (R=CH_2OH) and cetirizine (R=COOH). Asterisks indicate the chiral center.

1.9 THE NEED FOR ANALYTICAL TOOLS

The knowledge of chirality of drugs exists already for a long time. However, until the development of analytical tools that were able to discriminate between enantiomers, not much research could be done. Enantioselective analysis methods were required not only for pharmacodynamic and pharmacokinetic studies, but also for toxicological studies, in development and during quality control of drug substances and drug products. Thus, there was a real need for stereospecific detection, identification, and quantification of individual enantiomers of drugs and their metabolites in various biological media [19]. Even if a drug is given as a pure enantiomer, methods that can discriminate between enantiomers will be required because, for example, racemization can occur [19].

On analytical scale, a large variety of chromatographic methods, such as liquid chromatography (LC), gas chromatography (GC), and thin layer chromatography (TLC) were first developed [71,72]. At present, LC still dominates chromatographic enantiomeric analysis in industry [73]. However, over the lasts two to three decades capillary electrophoresis (CE) has proven to be a powerful alternative to chromatography. CE offers a tremendous flexibility for enantiomeric separations, because a wide variety of chiral additives are available. Compared to other analytical techniques like LC, CE offers several advantages including simplicity, short analysis times, high efficiencies, different separation mechanisms, small volumes, and low running costs. Nowadays capillary electrochromatography (CEC) offers some alternative mechanisms for solving individual separation problems. CEC combines features of both CE and LC: the efficiency of CE with the selectivity of stationary phase. Therefore, it represents a very useful extension to CE [71,74–76]. In Chapters 14 and 15, this technique is further explained.

1.10 INTRODUCTION TO CAPILLARY ELECTROPHORESIS

Electrophoresis as a separation technique was first described by Tiselius in 1937, for which he was awarded a Nobel Prize in 1948. Tiselius' basic concept of using a tube for electrophoretic separation received little notice until Hjerten described the first CE apparatus in 1967 [77,78]. Even then, CE stayed relatively unknown until Jorgenson and Lukacs advanced the technique by using 75 μm internal diameter capillaries [78–81].

The European Pharmacopoeia defines CE as a physical method of analysis based on migration, inside a capillary tube, of charged analytes dissolved in an electrolyte solution, under the influence of a direct-current electric field. The separation of compounds is based upon their charge to mass ratio. The migration of ionic species through a capillary depends on two forces: the electrophoretic mobility or the mobility of the ion and the electroosmotic flow or the mobility of the solvent. The advantages of CE, compared to LC, GC, and TLC, are its simplicity and its applicability for the separation of a wide range of compounds using the same instrument and, in most cases, the same capillary while changing only the composition of the background electrolyte. In addition, CE possesses high resolving power due to its plug flow and minimal diffusion [82]. Although, CE has also a few disadvantages.

Because of the low injection volume, the concentration sensitivity is low and stacking procedures are therefore sometimes needed. Compared to LC and GC, precision is worse due to the different injection mechanism. However, instrument manufacturers and scientists have worked on improving the system performance. In recent years, many fully validated CE methods have been described and CE is becoming a well-established technique not only in academia but also in industry [83].

REFERENCES

1. Caldwell, J, Wainer, IW. 2001. Stereochemistry: Definitions and a note on nomenclature. *Hum Psychopharmacol* 16: S105–S107.
2. Agranat, I, Caner, H, Caldwell, A. 2002. Putting chirality to work: The strategy of chiral switches. *Nature Rev Drug Discov* 1: 753–768.
3. Cordato, DJ, Mather, LE, Herkes, GE. 2003. Stereochemistry in clinical medicine: A neurological perspective. *J Clin Neurosci* 10: 649–654.
4. Caldwell, J. 2001. Do single enantiomers have something special to offer? *Hum Psychopharmacol* 16: S67–S71.
5. Hutt, AJ, Tan, SC. 1996. Drug chirality and its clinical significance. *Drugs* 52: 1–12.
6. Agrawal, YK, Bhatt, HG, Raval, HG, Oza, PM, Gogoi, PJ. 2007. Chirality—A new era of therapeutics. *Mini Rev Med Chem* 7: 451–460.
7. Cahn, RS, Ingold, C, Prelog, V. 1966. Specification of molecular chirality. *Angew Chem Int Ed* 5: 385–415.
8. Prelog, V, Helmchen, G. 1982. Basic principles of the CIP-system and proposals for a revision. *Angew Chem Int Ed (English)* 21: 567–583.
9. Ariëns, EJ. 1986. Stereochemistry: A source of problems in medicinal chemistry. *Med Res Rev* 6: 451–466.
10. Ariëns, EJ. 1984. Stereochemistry, a basis for sophisticated nonsense in pharmacokinetics and clinical pharmacology. *Eur J Clin Pharmacol* 26: 663–668.
11. Waldeck, B. 2003. Three-dimensional pharmacology, a subject ranging from ignorance to overstatements. *Pharmacol Toxicol* 93: 203–210.
12. Andersson, T. 2004. Single-isomer drugs—True therapeutic advances. *Clin Pharmacokinet* 43: 279–285.
13. Ariëns, EJ. 1991. Racemic therapeutics-ethical and regulatory aspects. *Eur J Clin Pharmacol* 41: 89–93.
14. Baumann, P, Zullino, DF, Eap, CB. 2002. Enantiomers' potential in psychopharmacology—A critical analysis with special emphasis on the antidepressant escitalopram. *Eur Neuropsychopharmacol* 12: 433–444.
15. Rentsch, KM. 2002. The importance of stereoselective determination of drugs in the clinical laboratory. *J Biochem Biophys Methods* 54: 1–9.
16. Rivas-Vazquez, RA, Perel, JM. 2003. The role of stereochemistry and chiral pharmacology in psychotropic drug development. *Prof Psychol-Res Pract* 34: 210–213.
17. Hyttel, J, Bøgesø, KP, Perregaard, J, Sánchez, C. 1992. The pharmacological effect of citalopram residues in the (S)-(+)-enantiomer. *J Neural Transm Gen Sect* 88: 157–160.
18. Yevtushenko, VY, Belous, AI, Yevtushenko, YG, Gusinin, SE, Buzik, OJ, Agibalova, TV. 2007. Efficacy and tolerability of escitalopram versus citalopram in major depressive disorder: A 6-week, multicenter, prospective, randomized, double-blind, active-controlled study in adult outpatients. *Clin Ther* 29: 2319–2332.
19. Islam, MR, Mahdi, JG, Bowen, ID. 1997. Pharmacological importance of stereochemical resolution of enantiomeric drugs. *Drug Safety* 17: 149–165.
20. Hutt, AJ. 2007. Chirality and pharmacokinetics: An area of neglected dimensionality? *Drug Metabol Drug Interact* 22: 79–112.

21. Mehvar, R, Brocks, DR, Vakily, M. 2002. Impact of stereoselectivity on the pharmacokinetics and pharmacodynamics of antiarrhythmic drugs. *Clin Pharmacokinet* 41: 533–558.
22. Takahashi, H, Echizen, H. 2001. Pharmacogenetics of warfarin elimination and its clinical implications. *Clin Pharmacokinet* 40: 587–603.
23. Lewis, RJ, Trager, WF, Chan, KK, Breckenridge, A, Orme, M, Roland, M, Schary, W. 1974. Warfarin. Stereochemical aspects of its metabolism and the interaction with phenylbutazone. *J Clin Invest* 53: 1607–1617.
24. Busse, D, Templin, S, Mikus, G, Schwab, M, Hofmann, U, Eichelbaum, M, Kivisto, KT. 2006. Cardiovascular effects of (R)- and (S)-verapamil and racemic verapamil in humans: A placebo-controlled study. *Eur J Clin Pharmacol* 62: 613–619.
25. Strolin Benedetti, M, Whomsley, R, Mathy, FX, Jacques, P, Espie, P, Canning, M. 2008. Stereoselective renal tubular secretion of levocetirizine and dextrocetirizine, the two enantiomers of the H1-antihistamine cetirizine. *Fundam Clin Pharmacol* 22: 19–23.
26. Heal, DJ, Pierce, DM. 2006. Methylphenidate and its isomers—Their role in the treatment of attention-deficit hyperactivity disorder using a transdermal delivery system. *CNS Drugs* 20: 713–738.
27. Mehvar, R, Brocks, DR. 2001. Stereospecific pharmacokinetics and pharmacodynamics of beta-adrenergic blockers in humans. *J Pharm Pharm Sci* 4: 185–200.
28. Ali, I, Gupta, VK, Aboul-Enein, HY, Singh, P, Sharma, B. 2007. Role of racemization in optically active drugs development. *Chirality* 19: 453–463.
29. Hutt, AJ, Caldwell, J. 1983. The metabolic chiral inversion of 2-arylpropionic acids—A novel route with pharmacological consequences. *J Pharm Pharmacol* 35: 693–704.
30. Walle, T, Walle, UK. 1986. Chirality 4. Pharmacokinetic parameters obtained with racemates. *Trends Pharmacol Sci* 7: 155–158.
31. Williams, DA. 2002. Drug metabolism. In *Foye's Principles of Medicinal Chemistry*, eds. D, Williams and T, Lemke. Philadelphia, PA: Lippincott Williams & Wilkins, pp. 174–233.
32. Eriksson, T, Björkman, S, Höglund, P. 2001. Clinical pharmacology of thalidomide. *Eur J Clin Pharmacol* 57: 365–376.
33. Caldwell, J. 1995. Stereochemical determinants of the nature and consequences of drug metabolism. *J Chromatogr A* 694: 39–48.
34. Caldwell, J. 1996. Importance of stereospecific bioanalytical monitoring in drug development. *J Chromatogr A* 719: 3–13.
35. Jamali, F, Mehvar, R, Pasutto, FM. 1989. Enantioselective aspects of drug action and disposition—Therapeutic pitfalls. *J Pharm Sci* 78: 695–715.
36. Hanada, K, Akimoto, S, Mitsui, K, Mihara, K, Ogata, H. 1998. Enantioselective tissue distribution of the basic drugs disopyramide, flecainide, and verapamil in rats: Role of plasma protein and tissue phosphatidylserine binding. *Pharm Res* 15: 1250–1256.
37. van Bree, JB, Heijligers-Feijen, CD, de Boer, AG, Danhof, M, Breimer, DD. 1991. Stereoselective transport of baclofen across the blood-brain barrier in rats as determined by the unit impulse response methodology. *Pharm Res* 8: 259–262.
38. Barraud de Lagerie, S, Comets, E, Gautrand, C, Fernandez, C, Auchere, D, Singlas, E, Mentre, F, Gimenez, F. 2004. Cerebral uptake of mefloquine enantiomers with and without the P-gp inhibitor elacridar (GF1210918) in mice. *Br J Pharmacol* 141: 1214–1222.
39. Alves, G, Figueiredo, I, Falcão, A, Castel-Branco, M, Caramona, M, Soares-DA-Silva, P. 2008. Stereoselective disposition of S- and R-licarbazepine in mice. *Chirality* 20: 796–804.
40. Andersson, T, Hassan-Alin, M, Hasselgren, G, Röhss, K, Weidolf, L. 2001. Pharmacokinetic studies with esomeprazole, the (S)-isomer of omeprazole. *Clin Pharmacokinet* 40: 411–426.

41. Olbe, L, Carlsson, E, Lindberg, P. 2003. A proton-pump inhibitor expedition: The case histories of omeprazole and esomeprazole. *Nat Rev Drug Discov* 2: 132–139.
42. Andersson, T, Rohss, K, Bredberg, E, Hassan-Alin, M. 2001. Pharmacokinetics and pharmacodynamics of esomeprazole, the S-isomer of omeprazole. *Aliment Pharmacol Ther* 15: 1563–1569.
43. Abelo, A, Andersson, TB, Antonsson, M, Naudot, AK, Skanberg, I, Weidolf, L. 2000. Stereoselective metabolism of omeprazole by human cytochrome P450 enzymes. *Drug Metab Dispos* 28: 966–972.
44. Lindberg, P, Keeling, D, Fryklund, J, Andersson, T, Lundborg, P, Carlsson, E. 2003. Review article: Esomeprazole-enhanced bio-availability, specificity for the proton pump and inhibition of acid secretion. *Aliment Pharmacol Ther* 17: 481–488.
45. Triggle, DJ. 1997. Stereoselectivity of drug action. *Drug Discov Today* 2: 138–147.
46. Vogelgesang, B, Echizen, H, Schmidt, E, Eichelbaum, M. 1984. Stereoselective 1st pass metabolism of highly cleared drugs—Studies of the bioavailability of L-verapamil and D-verapamil examined with a stable isotope technique. *Br J Clin Pharmacol* 18: 733–740.
47. Volosov, A, Xiaodong, S, Perucca, E, Yagen, B, Sintov, A, Bialer, M. 1999. Enantioselective pharmacokinetics of 10-hydroxycarbazepine after oral administration of oxcarbazepine to healthy Chinese subjects. *Clin Pharmacol Ther* 66: 547–553.
48. Hermans, JJR, Thijssen, HHW. 1992. Stereoselective acetonyl side chain reduction of warfarin and analogs. Partial characterization of two cytosolic carbonyl reductases. *Drug Metab Dispos* 20: 268–274.
49. Brocks, DR. 2006. Drug disposition in three dimensions: An update on stereoselectivity in pharmacokinetics. *Biopharm Drug Dispos* 27: 387–406.
50. Tan, SC, Patel, BK, Jackson, SH, Swift, CG, Hutt, AJ. 1999. Ibuprofen stereochemistry: Double-the-trouble? *Enantiomer* 4: 195–203.
51. Johnson, JA, Akers, WS, Herring, VL, Wolfe, MS, Sullivan, JM. 2000. Gender differences in labetalol kinetics: Importance of determining stereoisomer kinetics for racemic drugs. *Pharmacotherapy* 20: 622–628.
52. Scordo, MG, Spina, E, Dahl, ML, Gatti, G, Perucca, E. 2005. Influence of CYP2C9, 2C19, and 2D6 genetic polymorphisms on the steady-state plasma concentrations of the enantiomers of fluoxetine and norfluoxetine. *Basic Clin Pharmacol Toxicol* 97: 296–301.
53. Jamali, F, Kunz-Dober, CM. 1999. Pain-mediated altered absorption and metabolism of ibuprofen: An explanation for decreased serum enantiomer concentration after dental surgery. *Br J Clin Pharmacol* 47: 391–396.
54. Neugebauer, G, Gabor, M, Reiff, K. 1992. Disposition of carvedilol enantiomers in patients with liver cirrhosis: Evidence for disappearance of stereoselective first-pass extraction. *J Cardiovasc Pharmacol* 19: S142–S146.
55. Imamura, H, Komori, T, Ismail, A, Suenaga, A, Otagiri, M. 2002. Stereoselective protein binding of alprenolol in the renal diseased state. *Chirality* 14: 599–603.
56. Bhatti, MM, Lewanczuk, RZ, Pasutto, FM, Foster, RT. 1995. Pharmacokinetics of verapamil and norverapamil enantiomers after administration of immediate and controlled-release formulations to humans: Evidence suggesting input-rate determined stereoselectivity. *J Clin Pharmacol* 35: 1076–1082.
57. Karim, A, Piergies, A. 1995. Verapamil stereoisomerism: Enantiomeric ratios in plasma dependent on peak concentrations, oral input rate, or both. *Clin Pharmacol Ther* 58: 174–184.
58. Mistry, B, Leslie, JL, Eddington, ND. 2002. Influence of input rate on the stereospecific and nonstereospecific first pass metabolism and pharmacokinetics of metoprolol extended release formulations. *Chirality* 14: 297–304.

59. Kunze, KL, Eddy, AC, Gibaldi, M, Trager, WF. 1991. Metabolic enantiomeric interactions: The inhibition of human (S)-warfarin-7-hydroxylase by (R)-warfarin. *Chirality* 3: 24–29.
60. Choonara, IA, Cholerton, S, Haynes, BP, Breckenridge, AM, Park, BK. 1986. Stereoselective interaction between the R enantiomer of warfarin and cimetidine. *Br J Clin Pharmacol* 21: 271–277.
61. Niopas, I, Toon, S, Rowland, M. 1991. Further insight into the stereoselective interaction between warfarin and cimetidine in man. *Br J Clin Pharmacol* 32: 508–511.
62. Caner, H, Groner, E, Levy, L, Agranat, I. 2004. Trends in the development of chiral drugs. *Drug Discov Today* 9: 105–110.
63. Agranat, I, Caner, H. 1999. Intellectual property and chirality of drugs. *Drug Discov Today* 4: 313–321.
64. Hindmarch, I. 2001. The enantiomer debate: Current status and future directions. *Hum Psychopharmacol* 16: S101–S104.
65. Directive 75/318/EEC. 1993. Investigation of Chiral Active Substances.
66. Leonard, BE. 2001. An introduction to enantiomers in psychopharmacology. *Hum Psychopharmacol* 16: S79–S84.
67. McClellan, KJ, Spencer, CM. 1998. Levobupivacaine. *Drugs* 56: 355–362.
68. Tucker, GT. 2000. Chiral switches. *Lancet* 355: 1085–1087.
69. Presa, IJ. 1999. H1 antihistamines: A review. *Alergol Immunol Clin* 14: 300–312.
70. Nelson, WL. 2002. Antihistamines and related antiallergic and antiulcer agents. In *Foye's Principles of Medicinal Chemistry*, eds. D, Williams and T, Lemke. Philadelphia, PA: Lippincott Williams & Wilkins, pp. 794–818.
71. Gübitz, G, Schmid, MG. 2001. Chiral separation by chromatographic and electromigration techniques. A review. *Biopharm Drug Dispos* 22: 291–336.
72. Ward, TJ. 2002. Chiral separations. *Anal Chem* 74: 2863–2872.
73. Chankvetadze, B. 2001. Enantioseparation of chiral drugs and current status of electromigration techniques in this field. *J Sep Sci* 24: 691–705.
74. Blaschke, G, Chankvetadze, B. 2000. Enantiomer separation of drugs by capillary electromigration techniques. *J Chromatogr A* 875: 3–25.
75. Amini, A. 2001. Recent developments in chiral capillary electrophoresis and applications of this technique to pharmaceutical and biomedical analysis. *Electrophoresis* 22: 3107–3130.
76. Gübitz, G, Schmid, MG. 2000. Recent progress in chiral separation principles in capillary electrophoresis. *Electrophoresis* 21: 4112–4135.
77. Hjertén, S. 1967. Free zone electrophoresis. *Chromatogr Rev* 9: 122–219.
78. Lele, M, Lele, SM, Petersen, JR, Mohammad, AA. 2001. Capillary electrophoresis: General overview and applications in the clinical laboratory. In *Clinical and Forensic Applications of Capillary Electrophoresis*, eds. JR, Petersen and AA, Mohammad. Totowa, NJ: Humana Press, pp. 3–20.
79. Jorgenson, JW, Lukacs, KD. 1981. Free-zone electrophoresis in glass-capillaries. *Clin Chem* 27: 1551–1553.
80. Jorgenson, JW, Lukacs, KD. 1981. High-resolution separations based on electrophoresis and electroosmosis. *J Chromatogr* 218: 209–216.
81. Jorgenson, JW, Lukacs, KD. 1981. Zone electrophoresis in open-tubular glass-capillaries. *Anal Chem* 53: 1298–1302.
82. Issaq, HJ. 2002. Thirty-five years of capillary electrophoresis: Advances and perspectives. *J Liq Chrom Rel Technol* 25: 1153–1170.
83. Altria, KD, Elder, D. 2004. Overview of the status and applications of capillary electrophoresis to the analysis of small molecules. *J Chromatogr A* 1023: 1–14.

2 Principles of Chiral Separations by Free-Solution Electrophoresis

Radim Vespalec

CONTENTS

2.1 Analytical Classification of Chiral Compounds 23
2.2 Indirect Separations 25
2.3 Direct Separations 26
2.3.1 Models of Direct Chiral Discrimination 26
2.3.2 Electrophoretic Theory 28
2.3.3 Application of the Theory 33
2.3.4 Stability Constants 34
2.3.4.1 Raw Data and the Use of Constants 35
2.3.4.2 Methods of Calculation 38
2.4 Chiral Selectors 40
2.5 Closing Comment 42
List of Symbols 43
References 45

2.1 ANALYTICAL CLASSIFICATION OF CHIRAL COMPOUNDS

Separation selectivity is a prerequisite to a successful chemical analysis. Therefore, the attainment of reasonable separation selectivity is its necessary starting step. The difference between the migration speeds of analytes toward their solvent is the very cause for the separation selectivity in electrophoretic methods. Difference in charge-to-volume ratio of analytes is the principle feature that causes different migration speeds of the analytes in their solvent. The shape and solvation layer of the charged species are additional characteristics affecting the speed [1,2]. The layer of solvent molecules firmly adheres to the dissolved species and migrates with it. In this way, the layer increases the volume of the migrating dissolved species and masks their shape difference. Hence, crystallographic radii of ions and molecules underestimate their migration volumes. Water is a standard electrophoretic solvent. Different hydrophilicity of either analyzed species or their large building blocks gives therefore the chance to affect the mobility of the dissolved species by means of the composition of the background electrolyte (BGE) solvent [2,3]. The macroscopic liquid flow is a secondary process that frequently occurs in electrophoretic separation capillaries. This flow does not affect the difference between the migration speeds of analytes toward their solvent, disregarding the origin of the flow; however, it must be taken into account in the calculation of separation selectivity due to its effect on the overall migration velocity of the zones of analytes [1–3].

In chiral separations, the analytes are sterically different forms of chiral species. Principal difficulty of their separation consists in the almost general incapability of common achiral separation systems to evoke different migration speeds of these forms, with the exception of some pairs of diastereoisomers. The main reason for the incapability is the identical charge-to-volume ratio of sterically different forms of chiral analytes at any solution pH except for some pairs of diastereoisomers that are weak electrolytes. Till now, no separation was reported for sterically different forms of chiral species with one chiral center in any achiral separation system disregarding the center is discrete (located on one atom) or structural (spread over some group of atoms featured by mutually rigid steric arrangement). Additional difficulty results from the fact that each chiral center in a compound doubles the number of sterically different forms of the compound. Carbon is the most common discrete chirality center in organic species; however, chirality on other atoms like N, P, and S exists, too. Enantiomers are sterically different forms of compounds with one discrete chirality center; atropoisomers are sterically different forms of structurally chiral compounds. Chiral selectors effective in chiral separations of enantiomers are effective in chiral separations of atropoisomers, too; the same holds for the analytical methods utilizing chiral selectors. Therefore, chiral compounds with one center of chirality are discussed together and enantiomers serve as their representatives in this chapter.

Diastereoisomers differ in shape and in acid–base properties of their ionizable groups [4]. The shape difference of diastereoisomers is less favorable for their separation than the difference in their acid–base properties because the shape difference is hidden by the cover-up effect of the solvation layer. The discussed differences are the highest when the chiral centers of the diastereoisomeric compound are bonded together and markedly decrease with increasing distance of the centers. Unfortunately, it is seldom possible to reach a mutual bonding of chiral atoms in intentionally synthesized diastereoisomers, e.g., in indirect chiral separations. If the chiral centers are separated by three or more atoms, the shape and the charge differences decrease to such an extent that the difference in the migration speeds of such diastereoisomers ceases to be measurable. This holds for both chromatography [4] and electrophoresis [5]. Special additives, e.g., linear polymers, are sometimes of help [6] in addition to chiral selectors that are effective in chiral separations of enantiomers. Analytical behavior of chiral compounds with three and more centers depends on their mutual distances. Without a chiral selector, separation of some of their sterically different forms may be possible only if two of the centers are sufficiently close to each other. In such a case, the chiral compound behaves like one pair of diastereoisomers in achiral separation systems. The higher is the number of chiral centers in a molecule, the easier it is to find a chiral selector that interacts with the molecule fragments bonded to one of the chirality centers [5]. However, full discrimination of sterically different forms of a molecule with three centers is so difficult that a mixture of chiral selectors (a chiral selector array) needs to be used as a rule.

From the analytical point of view, a chemical entity is considered stable if it does not change measurably during an analysis. In the theories and models described in this chapter, species both chemically and chirally (free of spontaneous racemization) stable are considered. There are two analytically applicable types of interaction between a chiral selector and a pair of enantiomers. The first type yields kinetically stable interaction products that can be isolated as chemical entities separable in achiral separation systems. Chiral separations of this type are described in Section 2.2. Section 2.3 deals with the chiral separations based on a reversible interaction between a chiral selector and enantiomers. The principal aim of the theory and practice of such separations is to find the conditions under which the enantiomers migrate at different speeds. We have neither technical means nor a method that would enable us to measure the migration speed of one ion toward its solvent. The only possibility is to work with a large assembly of ions that embody a macroscopically observable property, e.g., the absorption of a light beam passing through their solution. A narrow plug of the ions is introduced in the inlet of a separation capillary and we measure the time at which the plug maximum reaches a detection point, e.g., a detection window of the capillary. The conversion of the migration time into mobility is based on the assumption that each contribution to the plug velocity is constant throughout the traveled distance. The terms ionic or effective mobility are used for the calculated value if the bulk solution does not move towards the capillary wall. The apparent mobility is obtained if the bulk solution moves at a constant velocity against the capillary wall. If the plug velocity toward the capillary wall is not constant throughout the traveled distance for any reason, the calculated value represents a mean value characteristic only for the respective experiment.

2.2 INDIRECT SEPARATIONS

A chemical reaction is an interaction between two species, yielding stable and mutually separable products. Chiral separations based on a chemical reaction between a chiral selector and chiral analytes are relevant only to chiral species with one center of chirality. According to this conception, the chemical form of enantiomers permanently transforms prior to separation. This is why the chiral separations utilizing the derivatization of the enantiomer pairs with a chiral selector are called indirect separations. In an indirect separation, a pair of enantiomers, (*R*)A and (*S*)A, converts into a pair of diastereoisomers, (*R*)AC and (*S*)AC. If the chemical reaction of an enantiomeric pair with a chiral selector should give two diastereoisomeric products, (*R*)AC and (*S*)AC, the selector must be a chirally pure enantiomer. If not, four reaction products result from the reaction between a chiral selector that contains two sterically different forms (*R*)C and (*S*)C, and a chiral analyte (*R*, *S*)A:

$$(R,S)\mathrm{A}+(R,S)\mathrm{C}\rightarrow(R)\mathrm{A}(R)\mathrm{C}+(S)\mathrm{A}(R)\mathrm{C}+(R)\mathrm{A}(S)\mathrm{C}+(S)\mathrm{A}(S)\mathrm{C} \tag{2.1}$$

In achiral separation systems, the symmetrical reaction products (*R*)A(*R*)C and (*S*)A(*S*)C have identical analytical properties. The same holds for the asymmetrical products (*R*)A(*S*)C and (*S*)A(*R*)C [4]. Thus, two zones, one consisting of the symmetrical and the other of the asymmetrical products, are obtained in achiral separation systems if a chirally impure chiral selector is utilized in the derivatization reaction. Consequently, both qualitative and quantitative correctness of the analysis is lost. With a chirally impure chiral selector (*R*,*S*)C, a correct analysis can be obtained only if the BGE is enriched with a chiral selector or with a mixture of chiral selectors effective in direct chiral separations. However, in this case the indirect separation loses its main advantage and the derivatization reaction becomes a senseless and complicating step. Obviously, general difficulties and disadvantages of achiral derivatization reactions occur in the derivatization reactions with chiral selectors, too. An irremovable drawback of the indirect chiral separations is their inherent incapability to fully discriminate sterically different forms of compounds that contain two and more chiral centers.

The indirect separations were very valuable for chiral separations of enantiomeric pairs in the past century when the set of accessible chiral selectors was very limited and the electrophoretic chiral separations did not exist. At that time, the only alternatives to achiral separations of derivatized enantiomers were their direct chiral separations requiring very expensive chiral stationary phases and time-consuming procedures. Therefore, the family of the derivatization agents for the indirect chiral separations in chromatography is much more abundant than the family of derivatization agents recommended in the electrophoretic literature, see, e.g., Ref. [5]. The development of the electrophoretic direct chiral separations started in the last decade of the past century. The development, which is summarized in one monograph [2] and in numerous reviews mostly cited in Refs. [2,5], led to a full dominance of the direct chiral separations. As a consequence, the practical attractiveness of the indirect chiral separations quickly decreased. Nowadays, their use is only justified if the diastereoisomers provided by the derivatization step offer an additional profit, e.g., an easier detection of the separated zones or increased separation selectivity.

2.3 DIRECT SEPARATIONS

2.3.1 Models of Direct Chiral Discrimination

Direct discrimination of enantiomers and other sterically different forms of chiral species is conditioned by a special constituent of the separation system that interacts differently with the sterically different forms of chiral species. In separation science, for such a constituent we use the term chiral selector. There are two models representing the mechanism of the direct chiral discrimination. A common feature of these models is a sterically rigid arrangement of the whole chiral selector or at least of the chiral selector part that induces the discrimination. Due to this rigidity, the strength of the interaction of a chiral selector with sterically different forms of a chiral species can be different.

Dalgliesh introduced the concept of a three-point interaction between the binding site present in a chiral selector and a chiral analyte (Figure 2.1) [7]. The revised version of the concept [8] is applicable to chiral selectors of any type, size, charge, aggregation state, and chemical form, in which the selectors interact with chiral analytes. The interactions 1–1′, 2–2′, and 3–3′ participating in chiral discrimination may be of any kind (inclusion in a cavity, coulombic interaction, hydrogen bonding, π–π

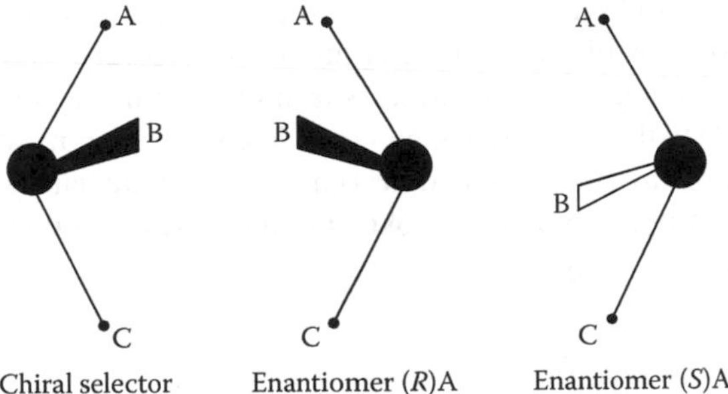

FIGURE 2.1 The diagram of the three-point interaction between a chiral selector and two enantiomers, according to Dalgliesh. For details, see text.

interaction, hydrophobic interaction, steric hindrance, etc.), any type (attractive or repulsive), and any strength. The one and only restriction valid for the participating interactions is trivial: the attractive interaction or the sum of the attractive interactions must be stronger than the repulsive interaction or the sum of the repulsive interactions. The interaction points may be identical to building blocks, functional groups, or individual atoms of chiral selectors. The same holds for chiral species. Several a priori conclusions or expectations evidently follow from Dalgliesh's concept. All of them have been verified in practice. The applicability of a chiral selector increases with a variety and a number of its potential binding sites and interaction points, hence with the selector complexity and size. The more promising and widely applicable chiral selectors are therefore relatively complicated species featuring either a rigid structure or many potential interaction points in mutually rigid arrangement. Similarly, the chiral separation of the analytes that embody a complicated structure and a high number of various interaction points is easier than the chiral separation of simple species. The simpler is the chiral species, the more difficult it is to find an effective chiral selector as well as to reach reasonable separation selectivity with it. For example, the chiral separation of 1,2-diaminocyclohexane, which is an important raw compound for chiral syntheses, has not been reported yet. This is why the chiral separations of simple species generally require high separation efficiency. From this point of view, capillary electrophoresis is a preferable and highly recommendable tool for difficult chiral separations and for chiral exploratory studies.

The second concept of the direct chiral discrimination is adopted from biochemistry. It combines the concept of the shape fit of one enantiomer for a chiral selector cavity with the Dalgliesh's concept of interaction points [7] (Figure 2.2). Intuitively,

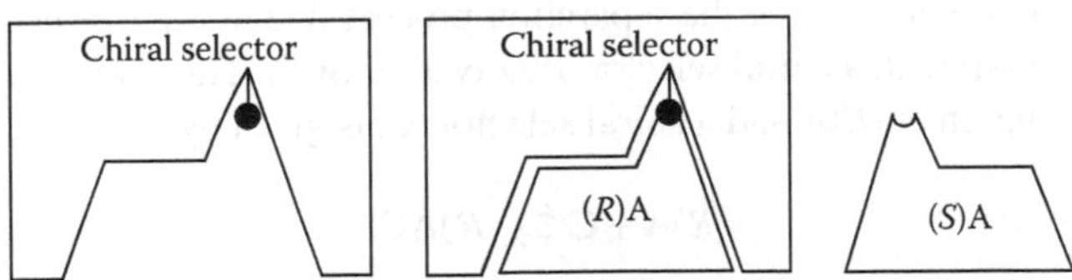

FIGURE 2.2 The diagram of the chiral discrimination based on the shape fit between a chiral selector cavity and the enantiomers, (R)A and (S)A. For details, see text.

this model represents chiral separations featured by high separation selectivity or even by separation specificity. Such separations are typical of proteins and other biopolymers functioning as chiral selectors (affinity chiral selectors). The identical model is applicable to the imprinted polymers. Mere shape identity of one enantiomer and a chiral selector cavity as assumed in some literature are hardly realistic. Such a model misses the driving force that evokes the interaction between the enantiomer and the chiral selector cavity.

2.3.2 Electrophoretic Theory

Direct chiral separations utilize the difference in strong and reversible complexation of the enantiomers with a ligand. In electrophoresis, this approach was first applied by Stepanova and Stepanov for the separation of calcium from strontium using citric acid as the interacting ligand [9]. Twenty-three years later, the approach was implemented by Wren and Rowe in the interpretation of the chiral separation of the propranolol enantiomers with β-cyclodextrin as the chiral selector [10]. In subsequent articles, the authors showed the effect of organic solvents on the propranolol separation [11] and the validity of the theory to the separations of other chiral drugs with methyl-β-cyclodextrin [12]. The essence of the theory proposed by Wren and Rowe is expressed in a mathematical form by Equations 2.2, 2.5 through 2.8, 2.10, and 2.11. In the theory, the quantities characterizing the interacting species are absent. This makes the theory generally applicable disregarding the chemical nature, size, charge, and other chemical aspects of both the chiral species and the chiral selectors used for their analyses. The only implicit requirement is trivial and results from the nature of the electrophoretic transport. At least one of the interacting species must be at least partly charged. The studies on the structure and stoichiometry of the complexes arising during the chiral separations have confirmed that the 1:1 interaction stoichiometry describes a vast majority of chiral separations [2]. This confirmation is strongly supported by a rich family of published calculations of stability constants in which the 1:1 complexation is commonly postulated. For the list of the constants published approximately till the half of 1999, see, e.g., Ref. [5]. Different stoichiometries occur exceptionally and should be expected only in the chiral separations of large and complex chiral species [2]. In the articles reporting more complex stoichiometry, e.g., 1:2 [13], the presented theory describes the elementary steps of the overall separation process [14,15].

In the direct chiral separation, the analytes enter the separation in the chemical form they embody in the sample supplied for an analysis. If the sample and the BGE differ in pH, their ionization state can be changed. The chemical form of the analytes changes only transiently during the separation process due to their reversible interaction (complexation) with a chiral selector dissolved in BGE. The reversible interaction between an enantiomer (*R*)A and a chiral selector, C, is given by:

$$(R)\mathrm{A} + \mathrm{C} \leftrightarrows (R)\mathrm{AC} \tag{2.2}$$

In the BGE containing a chiral selector, C, the free and bonded (complexed) chemical forms of the enantiomer (*R*)A, (*R*)A, and (*R*)AC, respectively, exist simultaneously and migrate with mobilities μ_{RA} and μ_{RAC}, respectively. The mobilities μ_{RA} and μ_{RAC} always differ provided that at least one of the forms (*R*)A or (*R*)AC is charged. The mobility of the charged (*R*)A would be equal to that of (*R*)AC only in the case that the charge-to-mass ratios of the species (*R*)A and C forming the complex (*R*)AC would be identical. However, this is impossible because in that case the coulombic repulsion between the species (*R*)A and C bearing the electrophoretic charges of the same polarity would prevent the complexation. Thus, the free form of the enantiomer (*R*)A tends to be separated from the complexed form, (*R*)AC. Then, the mixed zone of (*R*)A, which consists of the forms (*R*)A and (*R*)AC, migrates with the Tiselius' effective mobility, $\mu_{A,eff}$, which is defined for *i* coexisting forms of A as [1]

$$\mu_{A,eff} = \sum \mu_{A,i}\, x_{A,i} \tag{2.3}$$

The $\mu_{A,i}$ is the ionic mobility of the coexisting form *i*, $x_{A,i}$ is the molar fraction of *i*. Evidently,

$$\sum x_{A,i} = 1 \tag{2.4}$$

Considering the equilibrium (Equation 2.2), the effective mobility of (*R*)A in the BGE containing the chiral selector C, $\mu_{RA,eff}$ is:

$$\mu_{RA,eff} = \mu_{RA} x_{RA} + \mu_{RAC} x_{RAC} \tag{2.5}$$

The faster are the complex forming and the complex decomposition processes, the lower is the broadening of the migrating zone by the equilibrium (Equation 2.2). The broadening disappears, if both processes are much faster than the separation process [16].

The stability constant is a thermodynamic characteristic of the equilibrium (Equation 2.2). The stoichiometric stability constant, K_{RA}, is defined by the equilibrium concentrations of (*R*)A, C, and (*R*)AC, [(*R*)*A*], [*C*], and [(*R*)*AC*], respectively

$$K_{RA} = \frac{[(R)AC]}{[(R)A][C]} \tag{2.6}$$

Equation 2.5 acquires a favorable form for a practical application, if Equations 2.4 and 2.6 are substituted in it. For (*R*)A, the resulting form of the Equation 2.5 is

$$\mu_{RA,eff} = \frac{\mu_{RA} + \mu_{RAC} K_{RA}[C]}{1 + K_{RA}[C]} \tag{2.7a}$$

Analogously

$$\mu_{SA,eff} = \frac{\mu_{SA} + \mu_{SAC} K_{SA}[C]}{1 + K_{SA}[C]} \tag{2.7b}$$

is obtained for the effective mobility of the enantiomer (*S*)A. Yet, the separation of the enantiomers (*R*)A and (*S*)A has not been reported in an achiral separation system (in BGE without a chiral selector). It demonstrates that the shape difference of the solvated enantiomers does not induce measurable difference in their mobilities. Therefore, the mobilities of the enantiomers (*R*)A and (*S*)A, μ_{RA} and μ_{SA}, respectively, can be replaced by the common symbol μ_A. The chiral selectors routinely used in the direct chiral separations are never enantiomers. Thus, diastereoisomers, which are only sterically different forms of chiral compounds separable in achiral separation system, never originate in direct chiral separations. Chiral selectors are usually larger than the enantiomers (*R*)A and (*S*)A and their complexation contributes to the masking of the shape difference between the enantiomers (*R*)A and (*S*)A. The inseparability of the enantiomers in absence of a chiral selector therefore demonstrates that the complexes (*R*)AC and (*S*)AC would not be separable in achiral separation systems, too. It is therefore fully substantiated to replace the mobilities of the complexes (*R*)AC and (*S*)AC, μ_{RAC} and μ_{SAC}, respectively, with the common symbol μ_{AC}. Using both simplifications, the equations for the dependence of the effective mobilities of the enantiomers (*R*)A and (*S*)A on the equilibrium concentration of the chiral selector, $[C]$, are

$$\mu_{RA,eff} = \frac{\mu_A + \mu_{AC} K_{RA}[C]}{1 + K_{RA}[C]} \tag{2.8a}$$

$$\mu_{SA,eff} = \frac{\mu_A + \mu_{AC} K_{SA}[C]}{1 + K_{SA}[C]} \tag{2.8b}$$

It follows that the only reason for the different effective mobilities of the separated enantiomers, $\mu_{RA,eff}$ and $\mu_{SA,eff}$, at given equilibrium concentration of the chiral selector, $[C]$, is the difference between the stability constants K_{RA} and K_{SA}, $K_{RA} \neq K_{SA}$. Dependences of $\mu_{RA,eff}$ and $\mu_{SA,eff}$ on the equilibrium chiral selector concentration, $[C]$, given by Equations 2.8a and b for a hypothetical separation, are shown in Figure 2.3. The Equations 2.8a and b demonstrate that the equilibrium concentration of the chiral selector, $[C]$, is a crucial parameter if we aim to control the effective mobilities of the enantiomers, $\mu_{RA,eff}$ and $\mu_{SA,eff}$.

Meaningful separation selectivity, *S*, of two analytes A and B, is positive. Therefore, the separation selectivity in electrophoresis is always defined as

$$S = \frac{|v_{A,app} - v_{B,app}|}{0.5(v_{A,app} + v_{B,app})} \tag{2.9a}$$

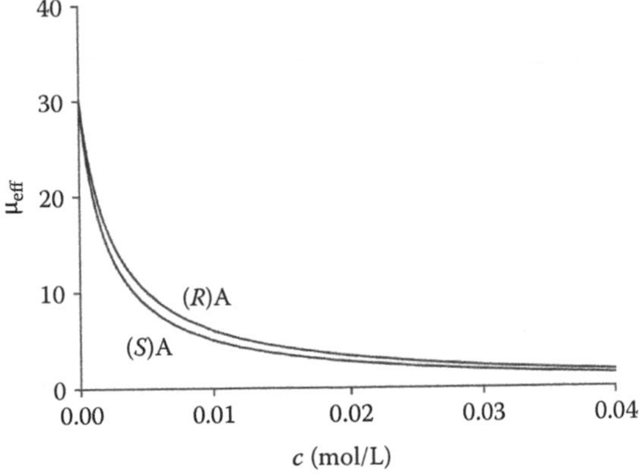

FIGURE 2.3 The dependences of $\mu_{RA,eff}$ and $\mu_{SA,eff}$ on the equilibrium chiral selector concentration, [*C*], calculated from Equations 2.8a and b using the arbitrary chosen raw data $\mu_A = 30$ mobility units, $\mu_{AC} = 0$ mobility units, $K_{RA} = 400$ L/mol and $K_{SA} = 500$ L/mol.

see, e.g., Refs. [1–3]. In the most general case, the apparent velocities of the analytes A and B, $v_{A,app}$ and $v_{B,app}$, respectively, are the sums of the velocities of the electroosmotic transport of the analytes toward the BGE solvent and the electroosmotic and the hydrodynamical flow velocities of BGE toward the capillary wall. Velocity is a vector, thus, its size and direction have to be considered for each velocity constituent. The electrophoretic velocities of the mixed zones of the enantiomers (*R*)A and (*S*)A, $v_{RA,eff}$ and $v_{SA,eff}$, respectively, are the products of their effective mobilities, $\mu_{RA,eff}$ and $\mu_{SA,eff}$, and the intensity of the electric field, *E*, applied on the capillary, $\mu_{RA,eff}\,E$ and $\mu_{SA,eff}\,E$, respectively. The electroosmotic flow velocity equals the product of the electroosmotic coefficient, μ_{eo}, and the intensity of the external electric field, *E*, $\mu_{eo}\,E$. The velocity of the hydrodynamic flow, v_{hd}, is independent of the applied electric field. The discussion about the influence of the hydrodynamic and the electroosmotic flow on the separation selectivity can be found elsewhere, see, e.g., Ref. [1]. The hydrodynamic flow should always be avoided in electrophoretic separations in free solution due to a deterioration of separation efficiency. In absence of the hydrodynamic flow, the effective mobilities of the separated enantiomers, $\mu_{RA,eff}$ and $\mu_{SA,eff}$, together with the electroosmotic coefficient, μ_{eo}, are decisive for the separation selectivity, see, e.g., Refs. [1–3].

$$S = \frac{\left|\mu_{RA,eff} - \mu_{SA,eff}\right|}{0.5(\mu_{RA,eff} + \mu_{SA,eff}) + \mu_{eo}} \tag{2.9b}$$

In the absence of electroosmosis, $\mu_{eo} = 0$, only the effective mobilities of the enantiomers (*R*)A and (*S*)A determine the separation selectivity, *S*. Therefore, the absolute value of the difference between the effective mobilities of the enantiomers (*R*)A and (*S*)A, $|\Delta\mu_{eff}|$ is a key quantity for *S* disregarding the presence or absence

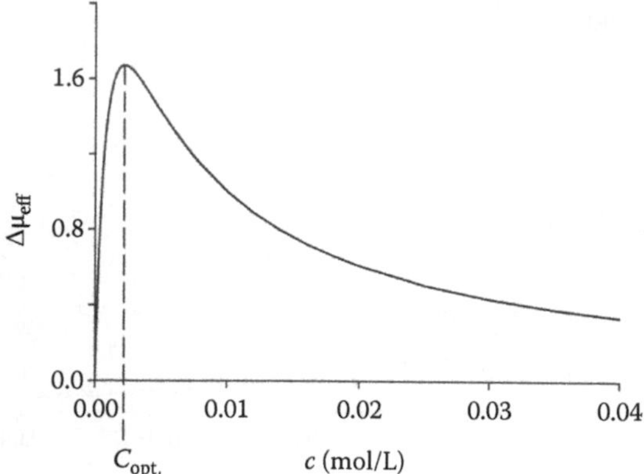

FIGURE 2.4 The dependence $\Delta\mu_{eff} = \mu_{RA,eff} - \mu_{SA,eff} = f([C])$ calculated from Equation 2.10 using the arbitrary raw data given in Figure 2.3. $[C_{opt}] = 2.24 \times 10^{-3}$ mol/L for the chosen stability constants $K_{RA} = 400$ L/mol and $K_{SA} = 500$ L/mol.

of any BGE flow. If we subtract Equation 2.8b from Equation 2.8a, we obtain $\Delta\mu_{eff} = (\mu_{RA,eff} - \mu_{SA,eff})$:

$$|\Delta\mu_{eff}| = \frac{|\mu_{RA} - \mu_{SA}||K_{RA} - K_{SA}|[C]}{1 + (K_{RA} + K_{SA})[C] + K_{RA}K_{SA}[C]^2} \tag{2.10}$$

A typical shape of the dependence $|\Delta\mu_{eff}| = f([C])$ is presented in Figure 2.4.

The complexation (Equation 2.2) is an exothermal process similar to other spontaneous processes. Therefore, a lower temperature shifts the equilibrium to the right, a higher temperature acts against the complexation. The higher is the stability constant at a given temperature, the higher is its temperature induced shift. Thus, a lowered temperature increases the stability constants and, simultaneously, the separation selectivity due to the increased difference between the effective mobilities of the enantiomers and vice versa. If the temperature shifts evoke dissimilar changes in the separation selectivity, at least one reversible side interaction meaningfully affects the investigated chiral separation. At a given temperature, the highest mobility difference is reached at the optimum chiral selector concentration, $[C_{opt}]$. The value of $[C_{opt}]$ is correlated with the stoichiometric stability constants, K_{RA} and K_{SA}

$$[C_{opt}] = (K_{RA}K_{SA})^{-0.5} \tag{2.11}$$

It is evident from Figure 2.4 that it is unreasonable to apply a higher equilibrium chiral selector concentration than the optimal one. Higher concentrations decrease the separation selectivity and increase the separation time. It benefits the separation speed to decrease the chiral selector concentration to the lowest value that yields sufficient separation selectivity.

2.3.3 Application of the Theory

The theory considers the simplest possible arrangement of the chiral discrimination process. It supposes that the chiral discrimination occurs on one chiral center in infinite homogeneous solution that is free of concentration gradients and side interactions of either analytes or chiral selectors with other BGE constituents. Only one interaction is considered between the chiral selector and each of the separated enantiomers. It follows from this assumption that each of the interacting species must enter the interaction in one chemical form. However, such an arrangement of the separation process is impossible in practice. Therefore, deviations of the practical arrangements of separations from the ideal model have to be taken into account. The ideal model either has to be modified accordingly or the deviations have to lead to adequate corrections of primary experimental data.

A real separation system frequently differs from the model by the fact that before the interaction (Equation 2.2) either the chiral selector or the analytes or both occur in two chemical forms. In this case, it is necessary to consider that these chemical forms interact with each other in any possible combination. In this case, more than one interaction inevitably contributes to the final chiral discrimination [2]. Such a situation is common if the BGE pH is one of the analytical optimization tools and either the chiral selector or the separated chiral compound is a weak electrolyte. Another frequent deviation is caused by side interaction of the chiral selector with an achiral constituent of the separation system, e.g., an inclusion of organic constituents of the BGE buffers or the BGE solvent into the cavity of cyclodextrins or a preferential solvation of a macrocyclic antibiotic with an organic constituent of the BGE solvent. Another side interaction that frequently occurs is the ion pairing of the analytes with the ionic BGE constituents bearing opposite charge. Such BGE constituents are sometimes added to tune electroosmosis, the BGE pH and/or the BGE ionic strength, or to enable analyte detection. Obviously, it is not easy to estimate preliminarily all the effects that may call the validity of the simple basic theory in question. This is why the routine for the optimization of chiral separations should include the investigation of the effects of pH, the buffer composition, the BGE solvent composition as well as any other intentionally added BGE constituents. Special care is necessary if the numerical data from such an optimization are to be used as raw data for the calculation of the stability constants. This holds even if seemingly reasonable and analytically usable data are obtained from the equations valid for the basic theory. The constants obtained in this way are conditional practically with no exceptions and, therefore, they are valid only for the used experimental conditions.

Special attention should be paid to the inner surface of the separation capillary, especially if it is uncoated. Adsorption on the capillary wall can distort the mobilities of both forms of analytes or considerably increase the concentration of the chiral selector that interacts with the analytes. The latter effect can be observed for any selector with positively charged atoms and groups, e.g., macrocyclic antibiotics, and is almost standard in the separations with affinity selectors in uncoated fused silica capillaries. The implicit requirement of homogeneous solution excludes relevance of the model for the separation systems containing micelles, microemulsion particles, or any other heterogeneous phases that either contain or bear a chiral selector. The description of the chiral discrimination process in any two-phase system requires taking into account any equilibrium that may compete with the equilibrium between the chiral selector and the separated enantiomers. In the two-phase system, there are competing interactions that can be easily overlooked: the achiral interaction between either the chiral selector or the enantiomers and any achiral constituent of the second phase, e.g., a detergent forming micelles, a water insoluble organic solvent forming micromicelles or silica gel matrix. If the chiral selector occurs in some two-phase separation system, adequate corrections have to be done to draw credible conclusions or to obtain quantitative data about the chiral discrimination power of the chiral selector for a given pair of enantiomers. Without these corrections, the conclusions and the calculated data hold only for the used experimental system. The comparison of these data with the data obtained in free solution provides misleading conclusions.

If we do not reach a chiral discrimination of a given pair of enantiomers with a chiral selector, we cannot simply conclude that there is no interaction between the chosen chiral selector and the enantiomers. The only evidence that these species do not interact is the independence of the mobilities of the enantiomers on the chiral selector concentration.

2.3.4 Stability Constants

There are three physicochemical characteristics of any reversible chemical process: the stability constant and the rate constants determining the speeds of the product formation, k_1, and the product decomposition, k_2. The theory of electrophoretic chiral separations requires very high rate constants that are comparable with those of, e.g., acid–base equilibria. Very high rate constants guarantee that at any moment the instantaneous system composition is negligibly different from the equilibrium composition. Then, considering the equilibrium (Equation 2.2), the stability constant is the only physicochemical characteristic that is of practical importance. The thermodynamic stability constant, K_{th}, is the basic characteristic of any elementary equilibrium process. The easiest way to obtain this constant is to calculate the ratio of the rate constants for the product formation and decomposition, k_1 and k_2, respectively, considering the law of the active mass action:

$$K_{th} = \frac{k_1}{k_2} \tag{2.12}$$

However, the determination of these rate constants has never been reported yet for the processes leading to electrophoretic or chromatographic chiral separations. For both enantiomers, the stoichiometric stability constant for the complexation between a chiral selector and a chiral analyte defined by Equation 2.6 is a much more easily accessible form of the stability constant. The recalculation of the stoichiometric stability constant, K_{RA}, to the thermodynamic stability constant, K_{th}, is a routine procedure and is described in detail elsewhere, e.g., Ref. [17].

2.3.4.1 Raw Data and the Use of Constants

The validity of the theory of chiral separations in free solution given in Section 2.3.2 is conditioned by three implicit suppositions that are not always mentioned in connection with the theory. In monographs (see e.g., Ref. [1]), the principal description of the electrophoretic transport is ascribed to the migration speed of any charged species determined by the equilibrium between two forces of the opposite directions. The accelerating force, F_a, equals the product of the external electric field applied on the separation capillary (E, given in volts per meter) and the total electrophoretic charge of the migrating species. The electrophoretic charge is the product of the number of the elemental charges of the migrating species, z, and the magnitude of the elemental charge, e. Therefore,

$$F_a = zeE \tag{2.13}$$

A viscous medium decelerates the migrating species by the drag force, F_d, which acts in the direction opposite to that of F_a. The drag force, F_d, is directly proportional to both the medium viscosity, η, and the migrating species velocity, v, given in meters per second (m/s)

$$F_d = k\eta v \tag{2.14}$$

k is the proportionality constant. The electrophoretic theory of chiral discrimination considers no other force or process that would affect the migration speed of the charged species. Evidently, such a migration exists only in infinitely diluted solutions that are never used in electrophoretic separations. The second implicit supposition underlying the theory ascribes the changes of the mobilities of enantiomers, e.g. ($\mu_{RA} - \mu_{RA,eff}$), only to their complexation with the dissolved chiral selector, C. The third implicit supposition assumes that no other solution constituent competes with the considered analyte for the chiral selector. These suppositions are substantial for the model simplicity and allow creating a clear qualitative picture of the whole chiral discrimination process.

Reproducibility is the only requirement placed on the experimental data in analytically optimized separations. There is no need to find out to what extent the presented ideal model of the chiral discrimination process agrees with the reality in the separation system that provides analytically satisfactory and reproducible experimental data. However, the model should agree with the processes and equilibria in a real system, if the experimental data from the system are to be used as raw data for the calculation of stability constants [14,15]. In our case, the calculated stability constants can be considered reliable only if (1) all the processes and equilibria that may affect the measured mobilities are included in the model and if (2) the changes of the effective mobilities are caused only by the change in the concentration of the chiral selector. To check the raw data, we should start with the evaluation of the model's applicability, which is described in the previous section. Without such evaluation, the obtained constants have to be considered conditional; this means valid only for the used experimental conditions. For any other conditions, the calculated constants represent only approximations of the true constants. In addition, the raw data entering the calculation have to be obtained at constant temperature, ionic strength, and viscosity of the BGEs. Further, the measured data are influenced by the changes of the BGE composition due to the variations of the chiral selector concentration. This influence has to be either negligible or subjected to a proper correction. If not, systematic errors are introduced in the raw data. A macroscopic viscosity determined by viscosimetry is relevant only if the BGE consists exclusively of low-molecular-weight constituents [11,18–20]. If the BGE contains some polymeric constituents like proteins, polymerized cyclodextrins, or synthetic linear polymers, a microscopic viscosity of a solution (i.e., the viscosity of the solution in the closest vicinity of a migrating ion) is decisive for the migration speed of small ions. How unimportant the macroscopic viscosity is for the migration speed of the small ions is demonstrated, e.g., by the electromigration of very small DNA fragments in capillaries filled with a cross-linked polyacrylamide gel [1,21] or by analysis with a chiral selector incorporated in the three-dimensional gel polymerized inside the separation capillary [22]. Such capillary fillings are of an infinite macroscopic viscosity.

Equations 2.8a and b and its algebraic modifications are used in any calculation of stoichiometric stability constants K_{RA} and K_{SA}. Except $\mu_{A,eff}$, the mobility of the enantiomers (*R*)A and (*S*)A in absence of the chiral selector C, μ_A, is the only quantity in these equations that is always experimentally accessible. The ionic and effective mobilities of enantiomers need to be measured with the highest possible precision. If the error of the input mobility data reaches 1%, the resulting error of the calculated stability constant can be up to 10% [23]. The mobility of the complexed enantiomer, μ_{AC}, has the physicochemical meaning of the enantiomer mobility in the solution containing an infinitely high concentration of the chiral selector, C. Direct and simple determination of μ_{AC} is therefore impossible with the exception of the uncharged complexes with $\mu_{AC} = 0$. A correct value of μ_{AC} can be calculated from the effective mobilities of the enantiomers (*R*)A and (*S*)A, $\mu_{RA,eff,opt}$ and $\mu_{SA,eff,opt}$, respectively, measured at the optimal concentration of the chiral selector, $[C_{opt}]$, and from the mobility of A, μ_A [23]

$$\mu_{AC} = (\mu_{RA,eff,opt} + \mu_{SA,eff,opt}) - \mu_A \tag{2.15}$$

A good estimate of μ_{AC} is easily obtained, if the mobility of a complex that contains a macromolecular chiral selector is identified with the mobility of the selector. If Equation 2.15 cannot be used, the precise estimation of μ_{AC} is the most difficult and the least precise for the mobility of the charged complexes consisting of a low-molecular-weight analyte and a medium-molecular-weight selector, e.g., vancomycin or cyclodextrin.

The equilibrium concentration of the chiral selector, $[C]$, is a quantity, the exact value of which is never accessible. The only way how to solve this problem is to substitute the equilibrium concentration of the chiral selector, $[C]$, for the total analytical concentration of the chiral selector in BGE, c [24]. The existing equations for the conversion of migration times into mobilities have been derived with the presumption that the migration speed of the respective charged analyte is constant throughout the separation capillary. In direct chiral separations, this requirement is met for migrating analytes if

$$[C] \approx c = \text{Const.} \tag{2.16}$$

holds at any point of the capillary. In a capillary with electroosmosis, the electroosmosis velocity must be constant during the experiment too. The fraction of the bonded selector equals the equilibrium concentration of the formed complex. Thus,

$$c = [C] + [AC] \tag{2.17}$$

Equation 2.16 holds only in the case that, in any point of the capillary, the fraction of the chiral selector complexed with the chiral analyte is so small, that $[AC]$ is negligible compared with $[C]$ and, thus, with the analytical concentration of C, c. In order to meet the requirement

$$[AC] \ll [C], \tag{2.18}$$

low stability of the forming complex AC, K_{AC}, and low equilibrium concentration of the chiral analyte, $[A]$, are desirable. At chosen experimental conditions, the complex stability cannot be affected. Thus, in order to minimize $[AC]$, the only possibility is to reduce $[A]$. This is accomplished by a decrease in the concentration of A introduced into the separation capillary inlet. The limit for such a concentration cut down is given by the minimum detectable concentration of A introduced into the capillary inlet. If light absorption photometry is applied as the detection technique, light absorption capabilities of the analytes and those of all the BGE constituents including the chiral selector have to be considered.

Let us accept that the requirement $[AC] \leq 0.1\ [C]$ fulfils the inequality in Equation 2.18 starting from the capillary inlet and, thus, the approximation in Equation 2.16 is valid. Then, according to Equation 2.6

$$[A] = 0.1 K_{RA}^{-1} \tag{2.19}$$

is the highest concentration of one enantiomer introduced into the capillary inlet, which guarantees the constant migration speed of the enantiomer throughout the capillary. Evidently it holds provided that the stacking of A is absent at the capillary inlet. The total concentration of the chiral selector, c, can be substituted for $[C]$ in the experimentally obtained dependence, if the loaded $[A]$ is not higher than $0.1\ K_{RA}^{-1}$. If the 1:1 racemic mixture of enantiomers is injected, twice higher concentration may be introduced into the capillary. Let us consider, for the sake of an illustration, the chiral separation of the enantiomers (R)A and (S)A with the mean value of their stoichiometric stability constants of 1×10^3 L/mol. The optimal equilibrium concentration of the chiral selector, C, dissolved in BGE is 1 mmol/L. The highest applicable concentration of the 1:1 racemic mixture is 0.2 mmol/L at this chiral selector concentration. If the stability constant is 1×10^4 L/mol, the value of $[A]$ drops to 0.02 mmol/L. However, for many organic species at the photometric detection in the radially illuminated capillary, the injected concentration of 0.02 mmol/L is close to or even below the detection limit. It shows that for the highest stability constants measured by capillary zone electrophoresis according to the existing theories, the sensitivity of the detectors can become a decisive aspect. If we use a racemic mixture instead of pure enantiomers to measure the raw data, the availability of the raw data could be limited due to possible zone broadening in the range of low chiral selector concentrations. The zone broadening decreases separation efficiency and, thus, resolution. Therefore, in the measurements of stability constants above approximately 1×10^3 L/mol, the eligibility of the detectors that markedly increase the width of the electrophoretic zones may be limited even if they offer high detection sensitivity.

2.3.4.2 Methods of Calculation

Equation 2.11 rewritten as

$$K_{\text{mean}} = [C_{\text{opt}}]^{-1} \tag{2.20}$$

gives the geometric mean of the stability constants, $K_{\text{mean}} = (K_{\text{RA}} K_{\text{SA}})^{0.5}$, for a pair of enantiomers provided that the experimental data meet or at least approach the requirements discussed in the previous section. Equations 2.8a and b are the starting point for the determination of the individual stability constants, K_{RA} and K_{SA}. A value of the stability constant as well as a value of the separation selectivity is meaningful only if it is positive. If we rearrange and generalize Equation 2.8a and b accordingly, the equation

$$K_{\text{A}} = \frac{1}{[C]} \left| \frac{\mu_{\text{A}} - \mu_{\text{A,eff}}}{\mu_{\text{A,eff}} - \mu_{\text{AC}}} \right| \tag{2.21}$$

is obtained. This equation is convenient for both numerical calculations and for computer fitting procedures. For the K_{A} calculation, three pairs of $\mu_{\text{A,eff}} = f([C])$ data would be sufficient from the mathematical point of view, if the raw data were ideal, i.e., free of errors. However, in practice we have only "ideal-like" data at best. These data are free of systematic errors and their random errors are very small. Three pairs of such data are sufficient for fast K_{A} estimation that informs on the concentration range to be chosen for the optimization of the chiral selector concentration. The estimate of μ_{A} is obtained from the first experiment without a chiral selector ($[C] = 0$); obviously, Equation 2.21 is not defined for this experiment. At $[C] \neq 0$, one experiment is sufficient for the K_{A} estimation, if a reasonable estimate of μ_{AC} exists and two experiments are necessary, if no μ_{AC} estimation is available. Optimally, the concentrations of the chiral selector chosen for these experiments should differ by one order of magnitude and the optimal chiral selector concentration, $[C_{\text{opt}}]$, lies between them. Estimates of the constants for both enantiomers are obtained if a racemic mixture is introduced into the capillary and at least partially separated zones of the enantiomers are reached in the presence of the chiral selector.

To obtain credible stability constants from "good" data, we usually need at least six pairs of $\mu_{A,eff} = f([C])$ data located around the $[C_{opt}]$ in a sufficiently wide range of concentrations. The required number of measurements depends on the random errors of the raw data. Computer fitting procedures are recommendable for the calculation of the stability constants, if the "good" raw data come from the experiments in which one chemical form of either separated enantiomers or the used chiral selector is not guaranteed. This is often the case, if pH is the optimized parameter. In this case, all possible elementary equilibria must be included in the model. As a consequence, the calculation equation is much more complicated than Equation 2.21. This more demanding calculation is mandatory, if pH-dependent "stability constants" are calculated from Equation 2.21 using raw data measured in two BGEs that differ only in pH. This pH-dependence originates from the different affinity of the chiral selector to particular enantiomers, if at least one of the interacting species is differently charged at different pH values. The examples of the calculations relevant to such so-called duoselective separations are given in the monograph [2]. An analogous concept is applicable to the calculations of stability constants from the chiral separations that are accomplished with two or more chiral selectors (with an array of selectors), e.g., an uncharged and a charged cyclodextrin.

The so-called bad $\mu_{A,eff} = f([C])$ data suffer from both systematic and random errors. Nevertheless, the computer fitting procedures are so versatile that they can treat the "bad" data, too. However, the fitting follows purely mathematical criteria. Therefore, the goal of such a fitting is a mathematically optimized fit regardless of both the physical meaning of the calculated "adjustable constants" K_A, μ_A, and μ_{AC} and their accurate (real) magnitudes. It is evident from Equation 2.21 that for given $\mu_{A,eff}$, the variations of μ_A and μ_{AC} affect the calculated K_A. Thus, the calculated values of any "adjustable constant" K_A, μ_A, or μ_{AC} should be subjected to an independent credibility check. The easiest check is to compare the calculated values of μ_A for both enantiomers with the experimental μ_A. Further, it is possible to compare the optimal chiral selector concentration obtained using Equation 2.11 and the calculated stability constants with the optimal chiral selector concentration read from the experimental dependence $\Delta\mu_{eff} = f([C])$. The calculated value of μ_{AC} can be compared with the value estimated in any other way, e.g., for the second enantiomer, and the physical plausibility of the calculated "adjustable constants" K_A, μ_A, and μ_{AC} can be assessed. For example, if we check up the μ_{AC} value obtained by the fitting of the data from the complexation of a cationic enantiomer with an uncharged chiral selector, the $\mu_{AC} = 0$ or the negative sign of the μ_{AC} meaningfully different from zero safely implies "bad" data. Further, bad data are recognized in the case of the enantiomers of the compound A, if the μ_{AC} values that are reasonably different from zero carry opposite signs.

The 1:1 reversible complexation leading to one product is very frequent in nature and it was investigated by many experimental techniques. In the reported studies, many terms were used for the equilibrium constants characterizing various 1:1 complexations; even richer is the family of the graphical and linearization methods that had been developed for the calculation of the constants, see, e.g., Ref. [15], before computers became a standard scientific outfit. From all the published methods, four linearization methods were selected for the calculation of the stability constants from the mobility data measured by capillary electrophoresis [25,26]. However, only the *X*-reciprocal method that employs the equation

$$\frac{(\mu_{A,eff}-\mu_A)}{[C]}=-K_A(\mu_{A,eff}-\mu_A)+K_A(\mu_{AC}-\mu_A) \tag{2.22}$$

provides the stability constant directly. In this method, the $(\mu_{A,eff}-\mu_A)/[C]$ values are plotted against $(\mu_{A,eff}-\mu_A)$ and the stability constant, K_A, results as the–slope of the obtained linear dependence. According to the recently proposed linearization method that rearranges Equation 2.21 in the form

$$\mu_{A,eff}=\frac{1}{K_A}\left|\frac{\mu_A-\mu_{A,eff}}{[C]}\right|+\mu_{AC} \tag{2.23}$$

the stability constant, K_A, and the mobility of the complex, μ_{AC}, are obtained simultaneously [27]. If $\mu_{A,eff}$ is plotted against $|\mu_A-\mu_{A,eff}|/[C]$, K_A results as the $(\text{slope})^{-1}$ of the obtained linear relationship and μ_{AC} as the intercept. The linearization can be easily done in Excel. If we substitute the demanding statistical procedure reported in [27] for the easy Excel calculation, the effect on the calculated K_A and μ_{AC} values is minimal [28]. This finding is in accordance with a previously reported similar observation [25].

Mathematically, it is possible to calculate logarithm, and, consequently, pK value, only from dimensionless quantities. The standard method how to eliminate a quantity dimension is to replace the quantity having the dimension, e.g., $[A]$, [mol/L], by a dimensionless relative quantity of the same magnitude, e.g., $[A]/c_A^0$. This is reached if the arbitrary reference quantity, e.g., c_A^0, has the dimension of the quantity to be normalized. Evidently, the reference quantity needs to be of the unit magnitude, e.g., $c_A^0=1\,\text{mol/L}$. Applying this approach, we obtain the normalized stoichiometric stability constant, K_A^0, that is dimensionless and of the same magnitude as K_A. K_A^0 can be recalculated to the thermodynamic stability constant, K_A^{th}, if each of the quantities participating in the definition of K_A^0 is recalculated to the standard conditions by means of a respective dimensionless coefficient, γ, e.g., γ_A. Previously, these coefficients were called activity coefficients. Applying these steps, we can write [17]

$$K_A^{th}=K_A^0\frac{\gamma_{AC}}{\gamma_A\gamma_C}=K_A\frac{c_A^0c_C^0}{c_{AC}^0}\frac{\gamma_{AC}}{\gamma_A\gamma_C} \tag{2.24}$$

2.4 CHIRAL SELECTORS

The compounds intended for a routine application as chiral selectors must meet various demands in addition to those discussed above. It is a must that the chiral discrimination capability of the compound working as a chiral selector is not dependent on time at least for one working day. This feature is conditioned by both chemical and chiral stability of the chiral selector. Both the stabilities are required in solid state as well as in BGE. The chiral stability can be easily deteriorated by spontaneous racemization on the chiral atom. The energetic barrier of this process is usually below 100 kJ/mol [29] while the strength of the chemical bond is scarcely below this limit, see, e.g., Ref. [30]. The spontaneous racemization was the reason why the compounds with chiral atoms synthesized roughly 25 years ago as chiral selectors fully lost their chiral discrimination capability within one year. A potential chiral selector must be sufficiently soluble in the BGE. Equation 2.11 shows that a decrease in the stability constant, K_A, by one order of magnitude rises up the required solubility of the respective chiral selector by one order of magnitude. The absence of side-interactions of the chiral selector with other BGE constituents is a requirement that, strictly speaking, is never met in practice. Wide applicability, on one hand, and a high selectivity or even specificity for particular chiral compounds, on the other hand, is required simultaneously for one chiral selector. The ideal chiral selector should not adsorb on the inner capillary wall and its interaction with the analytes should neither deteriorate the separation efficiency nor excessively decrease the separation speed. Lastly, the selector needs to be compatible with the accessible detectors, reasonably priced and nontoxic. Obviously, no chiral selector fully meets these demands.

Cyclodextrins are the most frequently used selectors in electrophoresis. All types of practically important chiral compounds have been separated with native cyclodextrins or their derivatives disregarding the charge of the analyzed chiral species. The complexation of cyclodextrins with analytes does not deteriorate separation efficiency; instead a slight increase in the separation efficiency is sometimes observed. The native cyclodextrins and their many derivatives offer a substantial practical advantage of the optical transparency in the short-wavelength UV-light. As a drawback, some cyclodextrin derivatives are very expensive. Nevertheless, any other type of chiral selector is either complementary or of second choice in practice. The stability constants published for the separations with cyclodextrins till the half of 1999 were mostly of the order of magnitude 10^1–10^2 for the weakly interacting enantiomer [5]. Constants exceeding 1000 were seldom reported and a constant exceeding 10,000 was reported in one separation with β-cyclodextrin. It follows from Equation 2.9b that the separation selectivity in the absence of electroosmosis is close to the mobility difference normalized to the mobility of the faster migrating enantiomer. In the chiral separations with cyclodextrins published till 1999, the highest reached mobility differences normalized in this way were mostly below 0.25 and never exceeded 1.

The linear oligosaccharides constitute a second class of chiral selectors that are optically transparent. In the separation systems with the uncharged linear oligosaccharides, the separation selectivity is generally close to that in the systems with cyclodextrins. The reason for this is the chemical relationship between the building blocks of cyclodextrins and linear oligosaccharides. The uncharged polysaccharides are water insoluble in contrast to the charged linear polysaccharides. Optical transparency of the latter is frequently limited due to the carboxyl groups. Among the chiral selectors soluble in water, there is not one that is UV-transparent below approximately 230–250 nm. Macrocyclic antibiotics are compounds of a medium molecular size that ranges approximately from 700 to 2000 Daltons. The compounds were put on the market as drugs. A bit later, their tendency to bind to carboxylic groups of amino acids inspired the use of these compounds and their derivatives as chiral selectors. Rifamycin is a compound of one macrocyclic ring, the glycopeptidic antibiotics of the molecular weight above 1400 Daltons consist of up to four rings. The BGEs with the macrocyclic antibiotics embody high separation efficiency and a higher selectivity for the discrimination of anions than the BGEs with cyclodextrins. Crown ethers are synthetic macrocyclic oligoethers usually substituted with four carboxylic groups or with some other substituents. The inclusion in their ring is always one of the supposed interactions. With these selectors, the selectivity of the reported separations is close to that of cyclodextrins. Considering other tested selectors types, two of them are worth mentioning here. The ligand-exchange selectors are based on the complexes of divalent copper, zinc, nickel, or cadmium with amino acids or hydroxycarboxylic acids. These selectors were highly popular in the early years of the separation science. Recently, their practical importance decreased due to several competing selector types. Bile salts are used almost exclusively in the micellar separation systems. All the listed selectors are of a low or medium molecular weight. With the exception of cyclodextrins, the stoichiometric stability constants for the complexation of these selectors with chiral analytes occur seldom in the literature.

The affinity chiral selectors may be considered as macromolecular relatives of the macrocyclic antibiotics. Chemically, these biopolymers are usually proteins and glycoproteins. In contrast to the other selectors, the steric structure (conformation) of the affinity chiral selectors strongly depends on the pH and chemical composition of the BGE, see, e.g., Ref. [31]. The conformation markedly affects the capability of the affinity selectors for the chiral discrimination of particular chiral species, see, e.g., Refs. [32,33]. One affinity chiral selector may therefore mimic the enantioselective capabilities of an array of various structurally rigid chiral selectors provided that its conformation is manipulated properly. The relative molecular mass of the affinity selectors is usually of the order of 10^4 Daltons, the stability constants for the complexation with low-molecular-weight analytes range from 10^4 to 10^6 L/mol. Such high stabilities indicate a potential for highly specific separations; unfortunately, they usually indicate decreased separation efficiency, too. The widths and shapes of the detected zones of low-molecular-weight analytes separated with the affinity selectors rather resemble liquid chromatographic than electrophoretic separations.

2.5 CLOSING COMMENT

Several separation techniques dealt with the analytical properties of chiral selectors and with their capability to discriminate sterically different forms of chiral species belonging to different classes of compounds. Basic pieces of knowledge extracted from these studies can be applied in all the separation techniques, if specific features of particular techniques are taken into account [32,34–36]. The field of chiral separations is not completely covered yet. Despite this, chiral separations extensively applied in practice for many years proved to be one of the indispensable means contributing to the present boom in the investigation of life that is chiral from its very nature. It is well-founded to expect that the life science research will ask for the solution of novel and more demanding tasks that either result from or are connected with chirality. This is the main reason why to deepen the understanding of the interactions between chiral selectors and chiral analytes and why to enlarge the set of accessible chiral selectors. The necessity of the deeper understanding of chiral separations is emphasized by the fact that analogy and personal experience are more effective in the a priori estimation of the promising selectors and, sometimes, even in the proposing chiral separations than the existing theories. Life science research will require the investigation of compounds and compound types that are practically out of attention of the present chiral separation science. An example may be the family of more than 50,000 fully synthetic cluster compounds of boron. The compounds are based on the three-center two-electron chemical bond, which does not exist in the compounds occurring in nature, see, e.g., Ref. [37]. The accumulation of these bonds substantially contributes to unusual chemical properties of the cluster compounds of boron and to their capability to provoke remarkable biochemical effects. This capability is demonstrated by the recently discovered potential of some cluster compounds of boron to treat the HIV forms resistant to organic compounds [38]. The majority of the boron compounds are structurally chiral due to the rigidity of their three-dimensional clusters. These electron deficient compounds represent a great challenge for the development of chiral selectors free of spontaneous racemization and featured by chiral discrimination capabilities different from those of the existing chiral selectors.

LIST OF SYMBOLS

$\mu_{A,eff}$ effective mobility of the species A

$\mu_{A,i}$ ionic mobility of the *i*th form of the species A

$x_{A,i}$ molar fraction of the *i*th form of the species A

$\mu_{RA,eff}$ effective mobility of the enantiomer (*R*)A

μ_{RA} ionic mobility of the enantiomer (*R*)A

x_{RA} molar fraction of the enantiomer (*R*)A

μ_{RAC} ionic mobility of the complex (*R*)AC

x_{RAC} molar fraction of the complex (*R*)AC

K_{RA} stoichiometric stability constant for the reversible complexation of the enantiomer (*R*)A with the chiral selector C

$[(R)AC]$ equilibrium concentration of the complex (*R*)AC

$[(R)A]$ equilibrium concentration of the enantiomer (*R*)A

$[C]$ equilibrium concentration of the chiral selector C

μ_{SAC} ionic mobility of the complex (*S*)AC

K_{SA} stoichiometric stability constant for the reversible complexation of the enantiomer (*S*)A with the chiral selector C

$\mu_{SA,eff}$ effective mobility of the enantiomer (*S*)A

$\Delta\mu_{eff}$ difference of the effective mobilities of the enantiomers (*R*)A and (*S*)A

S separation selectivity

$v_{A,app}$ apparent migration velocity of the analyte A, identical with its velocity toward the separation capillary wall

$v_{B,app}$	apparent migration velocity of the analyte B, identical with its velocity toward the separation capillary wall
μ_{eo}	electroosmotic coefficient
v_{eo}	velocity of the electroosmotic flow
v_{hd}	velocity of the hydrodynamic flow
$v_{RA,eff}$	electrophoretic velocity of the mixed zone of the enantiomer (*R*)A
$v_{SA,eff}$	electrophoretic velocity of the mixed zone of the enantiomer (*S*)A
μ_A	ionic mobility of the species A
μ_{SA}	ionic mobility of the enantiomer (*S*)A
μ_{AC}	ionic mobility of the complex AC
$[C_{opt}]$	optimal concentration of the chiral selector C
E	intensity of the electric field given V/m
V	voltage applied on the separation capillary
m	distance given in meters
K_{th}	thermodynamic stability constant for the reversible complexation
F_a	force accelerating a dissolved ion, induced by the electric field applied on the solution
z	number of the elemental charges of the dissolved ion
e	elemental charge
F_d	force decelerating an ion migrating in the solution
k	proportionality constant in Equation 2.14
η	viscosity of the solution
v	velocity of a migrating ion toward its solvent
$\mu_{RA,eff,opt}$	effective mobility of the enantiomer (*R*)A at the optimum concentration of the chiral selector
$\mu_{SA,eff,opt}$	effective mobility of the enantiomer (*S*)A at the optimum concentration of the chiral selector
c	analytical concentration of the chiral selector in BGE
$[AC]$	equilibrium concentration of the complex AC
$[A]$	equilibrium concentration of the species A
K_{mean}	geometrical mean value of the stoichiometric stability constants K_{RA} and K_{SA}
K_A^{th}	thermodynamic stability constant for the equilibrium complexation of the species A with the chiral selector, C
K_A^0	normalized stoichiometric stability constant for the equilibrium complexation of the species A with the chiral selector, C
K_A	stoichiometric stability constant for the equilibrium complexation of the species A with the chiral selector, C
γ_{AC}	factor for the recalculation of the equilibrium concentration of the complex AC from the experimental conditions to standard conditions (activity coefficient γ_{AC})
γ_A	factor for the recalculation of the equilibrium concentration of the species A from the experimental conditions to standard conditions (activity coefficient γ_A)
γ_C	factor for the recalculation of the equilibrium concentration of the chiral selector C from the experimental conditions to standard conditions (activity coefficient γ_C)

REFERENCES

1. Foret, F, Křivánková, L, Boček, P. 1993. *Capillary Zone Electrophoresis*. Weinheim: VCH Verlagsgesellschaft.
2. Chankvetadze, B. 1997. *Capillary Electrophoresis in Chiral Analysis*. Baffins Lane: John Wiley & Sons.
3. Boček, P, Vespalec, R, Giese, WR. 2000. Selectivity in CE. *Anal Chem* 72: 586A–595A.
4. Allenmark, SG. 1988. *Chromatographic Enantioseparations: Methods and Applications*. New York: Ellis Horwood.
5. Vespalec, R, Boček, P. 2000. Chiral separations in capillary electrophoresis. *Chem Rev* 100: 3715–3753.
6. Schützner, W, Caponecchi, G, Fanali, S, Rizzi, A, Kenndler, E. 1994. Improved separation of diastereomeric derivatives of enantiomers by a physical network of linear polyvinylpyrrolidone applied as pseudophase in capillary zone electrophoresis. *Electrophoresis* 15: 769–773.
7. Dalgliesh, CE. 1952. The optical resolution of aromatic amino-acids on paper chromatograms. *J Chem Soc* 47: 3940–3942.
8. Pirkle, WH, Pochabsky, TC. 1989. Consideration of chiral recognition relevant to liquid chromatography of enantiomers. *Chem Rev* 89: 347–362.
9. Stepanova, ND, Stepanov, AV. 1969. The effect of temperature on the efficiency of the electromigration separation of calcium and strontium in solutions of citric acid. *Zh Prikl Khim* (Russian) 42: 1670–1673.
10. Wren, SAC, Rowe, RC. 1992. Theoretical aspects of chiral separation in capillary electrophoresis I. Initial evaluation of model. *J Chromatogr* 603: 235–241.
11. Wren, SAC, Rowe, RC. 1992. Theoretical aspects of chiral separation in capillary electrophoresis II. The role of organic solvent. *J Chromatogr* 609: 363–367.
12. Wren, SAC, Rowe, RC. 1993. Theoretical aspects of chiral separation in capillary electrophoresis III. Application to β-blockers. *J Chromatogr* 635: 113–118.
13. Bowser, MT, Chen, CDY. 1998. Higher order equilibria and their effect on analyte migration behavior in capillary electrophoresis. *Anal Chem* 70: 3261–3270.
14. Rossotti, FJC, Rossotti, H. 1961. *Determination of Stability Constants*. New York: McGraw-Hill.
15. Connors, KA. 1987. *Binding Constants—The Measurement of Molecular Complex Stability*. New York: John Wiley & Sons.
16. Giddings, JC. 1991. *Unified Separation Science*. New York: John Wiley & Sons.
17. Vespalec, R, Boček, P. 2000. Calculation of stability constants for the chiral selector-enantiomer interactions from electrophoretic mobilities. *J Chromatogr A* 875: 431–445.
18. Penn, SG, Goodall, DM, Loran, JS. 1993. Differential binding of tioconazole enantiomers to hydroxypropyl-β-cyclodextrin studied by capillary electrophoresis. *J Chromatogr* 636: 149–152.
19. Penn, SG, Bergström, ET, Goodall, DM, Loran, JS.1994. Capillary electrophoresis with chiral selectors: Optimization of separation and determination of thermodynamic parameters for binding of tioconazole enantiomers to cyclodextrins. *Anal Chem* 66: 2866–2873.
20. Fanali, S, Boček, P. 1996. A practical procedure for the determination of association constants of the analyte-chiral selector equilibria by capillary zone electrophoresis. *Electrophoresis* 17: 1921–1924.
21. Li, SFY. 1992. *Capillary Electrophoresis Principle, Practice and Applications*. Amsterdam: Elsevier Science Publishers B.V.
22. Guttman, A, Paulus, A, Cohen, AS, Grinberg, N, Karger, BL. 1988. Use of complexing agents for selective separation in high-performance capillary electrophoresis: Chiral resolution via cyclodextrins incorporated within polyacrylamide gel columns. *J Chromatogr* 448: 41–53.

23. Vespalec, R, Boček, P. 1998. Methods for determination of electrophoretic mobility and stability of complexes originating in solutions during the chiral discrimination process. *Electrophoresis* 19: 276–281.
24. Penn, SG, Bergström, ET, Knights, I, Liu, G, Ruddick, A, Goodall, DM. 1995. Capillary electrophoresis as a method for determining binding constants: Application to the binding of cyclodextrins and nitrophenols. *J Phys Chem* 99: 3875–3880.
25. Rundlett, KL, Armstrong, DW. 1996. Examination of the origin, variation, and proper use of expressions of association constants by capillary electrophoresis. *J Chromatogr A* 721: 173–186.
26. Rundlett, KL, Armstrong, DW. 1997. Methods for the estimation of binding constants by capillary electrophoresis. *Electrophoresis* 18: 2194–2202.
27. Barták, P, Bednář, P, Kubáček, L, Stránsky, Z. 2000. Advanced statistical evaluation of complex formation constant from electrophoretic data. *Anal Chim Acta* 407: 327–336.
28. Barták, P. Personal communication.
29. Krupčík, J, Oswald, P, Májek, P, Sandra P, Armstrong, DW. 2003. Determination of the interconnection energy barrier of enantiomers by separation methods. *J Chromatogr A* 1000: 779–800.
30. Kerr, JA. 1996–1997. Strengths of Chemical Bonds. In *CRC Handbook of Chemistry and Physics*, ed. DR. Linde, 77th edition, pp. 9-51–9-69. Boca Raton: CRC Press.
31. Foster, JF. 1977. Some aspects of the structures and conformal properties of serum albumin. In *Albumin Structure, Function and Uses*, eds. V. Rosenoer, M. Oratz, M. Rothshild, pp. 51–84. New York: Pergamon Press.
32. Vespalec, R, Šustáček, V, Boček, P. 1993. Prospects of dissolved albumin as a chiral selector in capillary zone electrophoresis. *J Chromatogr* 638: 255–261.
33. Šimek, Z, Vespalec, R. 1994. Interpretation of enantioselective activity of albumin used as the chiral selector in liquid chromatography and electrophoresis. *J Chromatogr A* 685: 7–14.
34. Penn, SG, Liu, G, Bergström, ET, Goodall, DM, Loran, JS. 1994. Systematic approach to treatment of enantiomeric separations in capillary electrophoresis and liquid chromatography. I. Initial evaluation using propranolol and dansylated amino acids. *J Chromatogr A* 680: 147–155.
35. Piperaki, S, Penn, SG, Goodall, DM. 1995. Systematic approach to treatment of enantiomeric separations in capillary electrophoresis and liquid chromatography. II A study of the enantiomeric separation of fluoxetine and norfluoxetine. *J Chromatogr A* 700: 59–67.
36. Ferguson, PD, Goodall, DM, Loran, JS. 1996. Systematic approach to treatment of enantiomeric separations in capillary electrophoresis and liquid chromatography. III. A binding constant-retention factor relationship and effects of acetonitrile on the chiral separations of tioconazole. *J Chromatogr A* 745: 25–35.
37. Williams, RE. 1992. The polyborane, carborane, carbocation continuum: Architectural patterns. *Chem Rev* 92: 177–207.
38. Cígler, P, Kozísek, M, Rezácová, P, Brynda, J, Otwinowski, Z, Pokorná, J, Plesek, J, Grüner, B, Dolecková-Maresová, L, Mása, M, Sedlácek, J, Bodem, J, Kräusslich, HG, Král, V, Konvalinka, J. 2005. From nonpeptide toward noncarbon protease inhibitors: Metallacarboranes as specific and potent inhibitors of HIV protease. *Proc Natl Acad Sci USA* 102: 15394–15399.

3 Cyclodextrin-Mediated Chiral Separations

Gerald Gübitz and Martin G. Schmid

CONTENTS

3.1 Introduction ... 47
3.2 Use of Native CDs and Neutral CD Derivatives ... 52
3.3 Use of Charged Cyclodextrin Derivatives ... 53
3.3.1 Negatively Charged CD Derivatives ... 54
3.3.2 Positively Charged CD Derivatives ... 54
3.3.3 Amphoteric CD Derivatives ... 57
3.4 Polymerized CDs ... 58
3.5 Dual Selector Systems ... 58
3.5.1 Dual CD Systems ... 58
3.5.2 CDs and Carbohydrates ... 58
3.5.3 CDs and Surfactants ... 60
3.5.4 CDs and Ion-Pairing Reagents ... 60
3.5.5 CDs and Crown Ethers ... 60
3.5.6 CDs and Macrocyclic Antibiotics ... 61
3.5.7 Inclusion Complexation and Ligand-Exchange ... 61
3.6 Capillary Electrochromatography ... 62
3.6.1 Open-Tubular CEC ... 62
3.6.2 Packed CEC (P-CEC) ... 63
3.6.3 Monolithic Phases ... 63
3.6.3.1 Silica-Based Monoliths ... 63
3.6.3.2 Polymer-Based Monoliths ... 65
3.7 Microfluid Devices ... 66
3.8 Nonaqueous Medium ... 67
3.9 Miscellaneous ... 68
References ... 71

3.1 INTRODUCTION

Cyclodextrins (CDs) represent the most frequently used chiral selectors in electromigration techniques. CDs are oligosaccharides composed by different glucose units connected to each other through α-(1,4)-glucosidic bonds. They can be obtained by enzymatic treatment of starch. α-CD consists of six, β-CD of seven, and γ-CD of

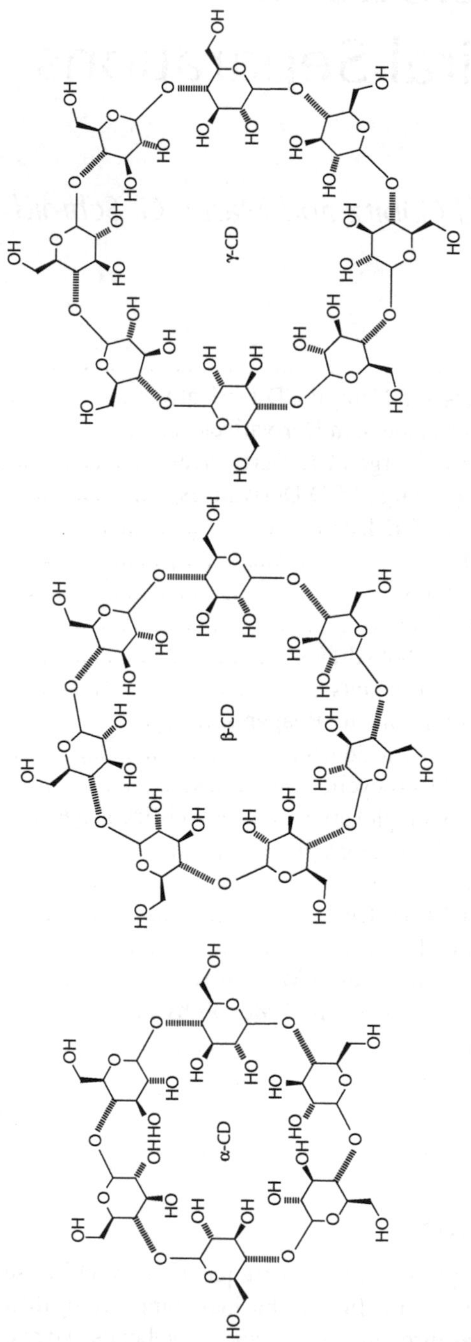

FIGURE 3.1 Formulas of α-, β-, γ-CD.

eight glucopyranose units (Figure 3.1). CDs form a truncated cone with a hydrophobic cavity and hydrophilic outer surface. The three native CDs have the same depth but different widths. The chiral recognition mechanism is based on inclusion of a bulky hydrophobic moiety of an analyte into the cavity. A further prerequisite for chiral recognition is the interaction of polar groups of the analyte close to the chiral center with the hydroxy groups at positions 2 and 3 at the mouth of the CD (Figure 3.2). A recent study combining capillary electrophoresis (CE) and nuclear magnetic resonance (NMR) investigations gives more insight into the chiral recognition mechanisms of CDs [1]. Further articles dealing with theoretical aspects of the chiral recognition mechanism have been published by Dodziuk et al. [2], Bikadi et al. [3], and Zhang et al. [4]. Fanali and Bocek [5] developed a procedure for the determination of association constants of analyte-CD equilibria. The calculation is based on the measurement of the effective mobility of an analyte at different concentrations of chiral selector (zero and two different nonzero concentrations):

$$\mu'_A = \frac{1}{1+K[S]}(\mu_0 + K[S]\mu'_{AS}) \tag{3.1}$$

$$\mu'_{AS} = \frac{\left[\mu'_1 C_1(\mu_0 - \mu'_1) - \mu'_2 C_2(\mu_0 - \mu'_1)\right]}{\left[\mu_0(C_1 - C_2) + \mu'_1 C_2 - \mu'_2 C_1\right]} \tag{3.2}$$

$$K = \frac{\mu_0 - \mu'_1}{\mu'_1 - \mu'_{AS}}\frac{1}{C_1} = \frac{\mu_0 - \mu'_2}{\mu'_2 - \mu'_{AS}}\frac{1}{C_2} \tag{3.3}$$

μ_0 is the actual mobility of analyte A at zero concentration of CD
μ_{AS} is the actual mobility of the complex CD-analyte
$[S]$ is the concentration of the chiral selector; $S = 0$ and C_1 and C_2
μ' are the corrected mobilities

Tökes et al. [6] applied computational molecular modeling approaches for the prediction of the suitability of CDs for a certain compound class.

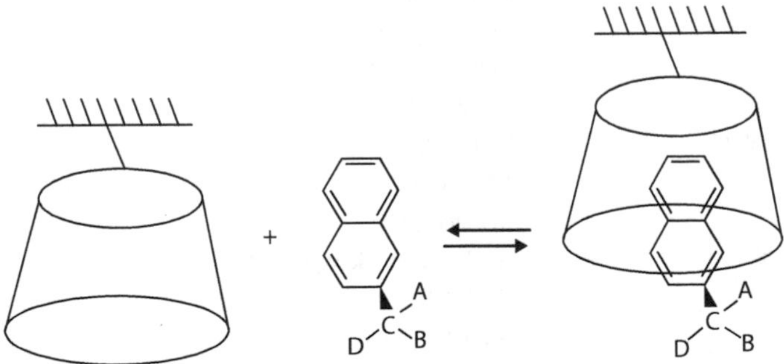

FIGURE 3.2 Inclusion mechanism of an analyte into a CD.

The native CDs have been widely used as chiral selectors for a broad spectrum of compounds. The hydroxy groups in position 2, 3, and 6 are available for derivatization. Thereby the depth of the cavity can be enlarged. Furthermore, the solubility can be increased.

This chapter covers the literature on chiral separations using CDs up to February 2008. Pharmaceutical and biomedical applications are summarized starting from January 2005 to February 2008. For earlier applications, the reader is referred to comprehensive review articles [7–15]. Table 3.1 gives an overview of the recent applications, which are not mentioned in the text below.

TABLE 3.1
Enantiomer Separation of Drugs by CE, MEKC, and CEC

Analyte	Chiral Selector	References
Adrenaline and analogs	β-CD derivatives	[4]
Adrenaline	DM-β-CD	[194]
Adrenaline	β-CD	[4]
Amisulpride	*S*-β-CD	[195]
Amphetamine analogs	DM-β-CD	[196]
Anticholinergic drugs	Heptakis(2,3-di-*O*-methyl-6-*O*-sulfato)-CD	[197]
Atenolol	Highly sulfated CDs	[100,198]
Azoles	Different β-CDs	[199]
Baclofen	Highly sulfated CDs	[200]
Baclofen	Highly sulfated CDs	[201]
Bevantolol	CM-β-CD	[202]
Bupivacaine	HP-β-CD/DM-β-CD	[203]
Butorphanol	Sulfated γ-CD	[204]
β-Blockers	β-CD	[205]
β-Blockers	Different CDs	[206]
Carvedilol	β-CD	[207]
Catecholamines	(HP-β-CD)/heptakis(DM-β-CD)	[208]
Catecholamines	Sulfated CDs	[209]
Cefadroxil	β-CD (monolithic CEC)	[210]
Cetirizine	Diacetylsulfato-β-CD	[211]
Cetirizine	Sulfated β-CD	[212]
Cetirizine	Sulfated β-CD	[213]
Chlorpheniramine	Highly sulfated CDs	[198]
Cinchona alkaloids	DM-β-CD	[214]
Citalopram	CM-γ-CD	[215]
Cycloamine	Sulfated γ-CD	[204]
D-Amino acids	β-CD/sodium deoxycholate	[216]
Dimetinden	CE-β-CD	[217]
Donezepil	S-β-CD	[218]
Doxazosin and intermediate	β-CD	[219]

TABLE 3.1 (continued)
Enantiomer Separation of Drugs by CE, MEKC, and CEC

Analyte	Chiral Selector	References
D-Phenylalanine	D-β-CD	[220]
D-Serine	HP-β-CD	[221]
Ephedrine	DM-β-CD	[222]
Etodolac	HP-β-CD	[223]
Etomitate	S-β-CD, β-CD	[224]
Fluoroquinolones	HP-β-CD	[225]
Frovatripan	SB-β-CD	[226]
Gemfibrozil	TM-β-CD	[227]
Hydroxychloroquine and met.	S-β-CD/HP-β-CD	[228,229]
Ibuprofen	TM-β-CD	[230]
Ibuprofen and met.	TM-β-CD	[231]
Imazaquine	HP-β-CD	[232]
Itraconazole	TM-β-CD	[233]
Ketamin and metab.	Sulfated β-CD	[234]
Lactic acid	HP-β-CD	[235]
Lisuride	γ-CD	[204]
Lorazepam	HP-β-CD	[236]
Metalaxyl	Succ-β-CD	[237]
Metaproterenol	β-CD	[207]
Methadone and metabolites	Highly sulfated γ-CD	[238]
Methoxamine	β-CD	[207]
Methylamphetamine	heptakis(2,6-diacetyl-6-sulfato)-β-CD	[239]
Metyrosine	β-CD	[240]
Moxifloxacin	Highly sulfated γ-CD	[241]
NSAIDs	Propylamino-β-CD	[242]
Omeprazole	M-β-CD	[243]
Omeprazole and metabolites	HDMS-β-CD	[244]
Pheniramine	CE-β-CD	[245]
Propranolol	β-CD	[246]
Propranolol	Highly sulfated CDs	[198]
Rivastigmine	β-CD	[247]
Salbutamol	HDAS-β-CD	[248]
Salbutamol	HP-β-CD/DM-β-CD	[203]
Sitafloxacin	γ-CD/D-Phe/Cu(II)	[249]
Sympathomimetic drugs	β-CD	[205]
Tamsulosin	Sulfated β-CD	[250]
Terbutalin	β-CD	[207]
Timolol	HDMS-β-CD	[251]
Tropa alkaloids	CD-modified microemulsion	[252]

Note: Applications, which are mentioned in the text are not included.

3.2 USE OF NATIVE CDS AND NEUTRAL CD DERIVATIVES

The first application of CDs for chiral separation by electromigration techniques was reported by Smolkova's group [16]. The authors used the technique of isotachophoresis (ITP) employing dimethylated β-CD (DM-β-CD) or trimethylated β-CD (TM-β-CD) as chiral selectors added to the leading electrolyte. The authors succeeded in resolving the enantiomers of pseudoephedrine, norpseudoephedrine, *O*-acetylpseudoephedrine, and *p*-hydroxynorpseudoephedrine. The first application of the capillary zone electrophoresis (CZE) for the chiral separation of drugs was described by Fanali [17]. Using DM-β-CD the authors resolved several sympathomimetics into their enantiomers. In the sequel, more than thousand papers appeared on applications of CDs for chiral separation of compounds of biological or pharmacological interest. Several parameters, such as concentration of the chiral selector, pH, the nature and ionic strength of the background electrolyte (BGE), and the addition of organic modifiers, have been found to have an important influence on the resolution. They are discussed in more detail in Chapter 4. Low pH is usually used to separate cationic compounds, while high pH is used for anionic analytes. The addition of organic modifiers has been found to influence efficiency and resolution. Theoretical models with respect to optimizing separation conditions were postulated by several authors [18–20]. A series of neutral CD derivatives is commercially available: heptakis-*O*-methyl-CD, heptakis(2,6-di-*O*-methyl) CD, heptakis(2,3,6-tri-*O*-methyl) CD, hydroxyethyl-CD (HE-CD), and hydroxypropyl-CD (HP-CD). Further neutral CD derivatives are cyanoethyl-β-CD [21], ethylcarbonate-β-CD [22], and mono-3-*O*-phenylcarbamoyl-β-CD [23]. Selectively methylated and acetylated CDs were synthesized by Miura et al. [24]. Matsunaga et al. [25] prepared a methylated glucuronyl glucosyl-β-CD and demonstrated its applicability for chiral separation by means of 16 drugs. A new β-CD derivative, 2-*O*-(2-hydroxybutyl)-β-CD was synthesized by Wei et al. [26] and successfully applied to the chiral separation of anisodamine, ketoconazole, propranolol, promethazine, adrenaline, and chlorphenamine. Recently, a highly water soluble CD derivative, 2-*O*-acetonyl-2-*O*-hydroxypropyl-β-CD (2-AHP-β-CD), was prepared by Lin et al. [27]. It represents a mixture of isomers with an average degree of substitution (DS) of about 1.0 for the acetonyl group and 3.8 for the hydroxypropyl group. This CD derivative showed improved resolution properties compared to other CDs for a series of basic and acidic drugs. Most of the commonly used CD derivatives are mixtures of isomers with different substitution patterns. Therefore reproducibility of the separations is rather bad. Recently, several groups started to synthesize selectively substituted CDs or single isomers. Schmitt et al. [28] prepared single isomers of heptakis(2,3,6-tri-*O*-methyl)-β-CD, heptakis(2,6-di-*O*-methyl)-β-CD, and heptakis(2,3-di-*O*-acetyl)-β-CD and compared the reproducibility of resolution with that of the corresponding randomly substituted CDs. It was shown that the batch-to-batch variation of the randomly substituted CDs is rather high. Cucinotta et al. [29] introduced a new type of CD derivatives, hemispherodextrins, in which a trehalose-capping moiety is bonded to β-CD (Figure 3.3). The authors demonstrated the applicability of these selectors for chiral separation by means of NSAIDs (Figure 3.4). Recently, Wistuba et al. [30] showed that δ-CD can also be used as a chiral selector. The authors applied this selector to the chiral separation of dansyl (Dns)-amino acids, fluorenylmethoxycarbonyl (Fmoc)-amino acids, dinitrophenyl (DNP)-amino

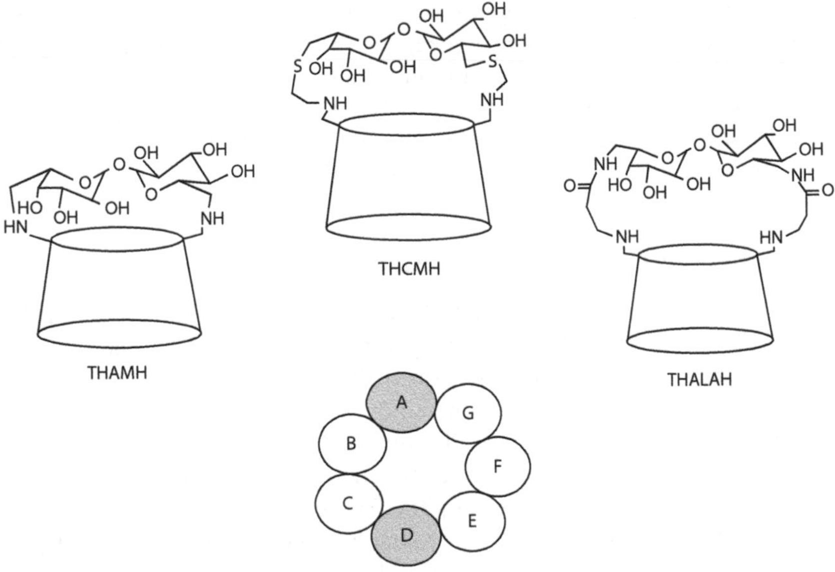

FIGURE 3.3 Schematic structures of hemispherodextrins. (From Cucinotta, V. et al., *J. Chromatogr. A*, 979, 137, 2002. With permission.)

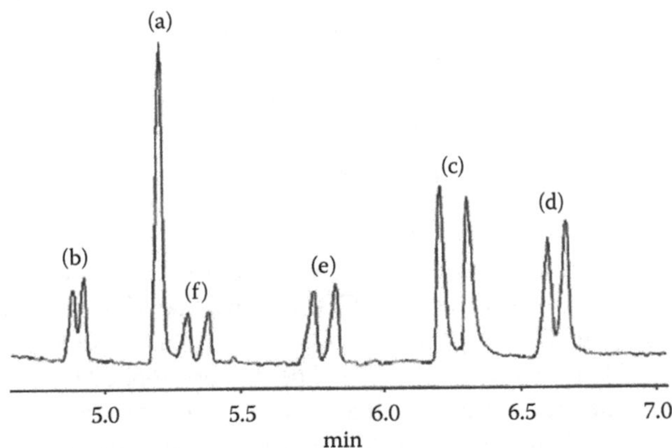

FIGURE 3.4 Chiral separation of six profen racemates. (From Cucinotta, V. et al., *J. Chromatogr. A*, 979, 137, 2002. With permission.)

acids, flavones, and some drugs. Jiang et al. [31] developed 3-hydroxypropyl-β-CD (3-HP-β-CD) and 2,3-dihydroxypropyl-β-CD (2,3-HP-β-CD) as new selectors and compared their chiral recognition properties with those of 2-HP-β-CD.

3.3 USE OF CHARGED CYCLODEXTRIN DERIVATIVES

Since with neutral CDs uncharged analytes cannot be resolved, charged CD derivatives have been developed. They can be used for both uncharged and charged analytes. In the ionized state, they migrate with their own electrophoretic mobility. Improved resolutions are obtained by making use of a countercurrent selector–analyte mobility. Furthermore, charged CDs show a higher solubility. Williams and Vigh [32] developed the charged resolving agent migration (CHARM) model, which is based on the consideration of simultaneous protonation and complexation equilibria. This model is helpful for selecting the appropriate separation conditions based on rational predictions (see Chapter 4).

3.3.1 Negatively Charged CD Derivatives

The most frequently used anionic CDs are carboxyfunctional CDs such as carboxymethyl-β-CD (CM-β-CD), carboxyethyl-β-CD (CE-β-CD), succinyl-β-CD [33–36], sulfated CDs, sulfobutyl-, and sulfoethyl ether-β-CD [37,38]. Highly sulfated CDs with a DS of 10 are commercially available from Beckman (Coulter, Fullerton, California, United States). Phosphated CD derivatives are only used in a limited number of applications [39]. A broad spectrum of drugs were screened using sulfated β-CD as a chiral selector [40]. The authors succeeded in resolving 37 out of 50 drugs investigated. A family of single isomers of sulfated β- and γ-CDs was introduced by Vigh's group [41–43]. They prepared derivatives completely sulfated in 6 position and completely substituted on their larger rims with hydrophilic groups, such as octakis(2,3-diacetyl-6-sulfato)–γ-CD [44], octakis-6-sulfato–γ-CD [45], (2-*N,N*-dimethylcarbamoyl)-β-CD [46], (2,3-*O*-dimethyl-6-*O*-sulfo-)–γ-CD [47], 2,3-di-*O*-acetyl-6-*O*-sulfo-α-CD [48], (2,3-dimethyl-6-*O*-sulfo-)–γ-CD [49], heptakis(2-*O*-methyl-3,6-di-*O*-sulfo-β-CD [50], and octa-(6-*O*-sulfo)–γ-CD [51]. The applicability of these derivatives for chiral separation has been demonstrated by means of a broad spectrum of drugs.

More recently, the same group prepared a new series of single isomers of sulfated CDs: hexakis(2,3-diacetyl-6-*O*-sulfo)-α-CD [52], hexakis(2,3-di-*O*-methyl-6-*O*-sulfo)-α-CD [53], hexakis(6-*O*-sulfo)-α-CD [54], the tetrabutylammonium salt of heptakis(2,3-*O*-diacetyl-6-*O*-sulfo)-β-CD, which was used in nonaqueous medium [55], heptakis(2-*O*-methyl-6-*O*-sulfo)-β-CD, and heptakis(2-*O*-methyl-3-*O*-acetyl-6-*O*-sulfo)-β-CD [56]. The latter two derivatives carried nonidentical substituents at all of the C2, C3, and C6 positions [57]. The authors applied these selectors to a broad spectrum of neutral, basic, as well as acidic and zwitterionic drugs.

A single isomer, 6-*O*-succinyl-β-CD (CDsuc6), was synthesized by Cucinotta et al. [58]. The authors demonstrated the chiral recognition ability of this selector by means of norephedrine, epinephrine, terbutaline, and norphenylephrine. Improved resolution of these compounds was observed compared to randomly succinylated CDs. Culha et al. [59] prepared a regioselective methylated CD derivative, 6-*O*-carboxymethyl-2,3-di-*O*-methyl-β-CD (CDM-β-CD). Besides the chiral separation of 4-hydroxylated polychlorbiphenyl derivatives, some catechins, and biphenyl compounds, the authors resolved the eight positional isomers of dihydroxynaphthalene.

3.3.2 Positively Charged CD Derivatives

The first cationic CDs were 6[(3-aminoethyl)amino]-6-deoxy-β-CD [60], 6^A-methylamino-β-CD, 6^A,6^D-dimethylamino-β-CD [61], a heptasubstituted methylamino-β-CD

[62], and mono(6-amino-6-deoxy)-β-CD [63] which found application in a broad spectrum of acidic and neutral compounds. Histamine-modified cationic β-CDs were prepared by Galaverna et al. [64] and applied to the chiral separation of Dns-amino acids, carboxylic acids, and hydroxy acids.

Haynes et al. [65] introduced a hepta-substituted single isomer cationic β-CD, heptakis(6-methoxyethylamine-6-deoxy)-β-CD, and demonstrated its chiral separation ability by means of NSAIDs and phenoxypropionic acid herbicides.

Ivanyi et al. [66] reported on the synthesis of a series of single isomer amino-β-CD derivatives (Figure 3.5) and applied them to the chiral separation of mandelic acid, *cis*-permethrinic acid, and *cis*-deltamethrinic acid. The authors found out that the substituent at the amino-*N*-atom has a significant influence on the enantioselectivity. Recently, a family of new single isomer cationic CD derivatives, 6-mono(alkylimidazolium)-β-CD [67], mono-6(A)-*N*-pentylammonium-6(A)-deoxy-β-CD chloride [68], mono-6(A)-butylammonium-6(A)-deoxy-β-CD tosylate [69], mono-6(A)-allyl ammonium-6(A)-deoxy-β-CD chloride [70], and mono-6(A)-propyl ammonium-6(A)-deoxy-β-CD chloride [71] were presented by Ng's group. These derivatives found application in a broad spectrum of acidic compounds of biological and pharmaceutical interest. Tang and Ng [72] prepared a series of single isomer 6-mono(3-alkylimidazolium)-β-CDs and applied them to the chiral separation of Dns-amino acids. The authors found out that the length of the alkyl chains plays an important role in chiral recognition. Recently, Cucinotta et al. [73] introduced two new CD derivatives, a β-CD derivative containing an ethylenediamine group in primary position (CDen) and a cysteamine-bridged hemispherodextrin

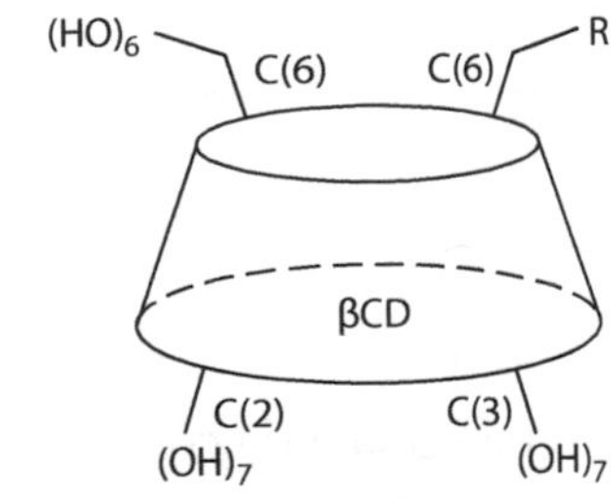

No.	Structure in 6^A position	pK_a(amino)
1	R = -OH = βCD	-
2	$-NH_2$	8.70 ± 0.2
3	$-NH-CH_2-CH_2-OH$	8.64 ± 0.2
4	$-NH-CH_2-CH(CH_3)-OH$	8.75 ± 0.2
5	$-NH-CH(CH_3)-CH_2-OH$	8.81 ± 0.2
6	$-NH-CH_2-CH_2-CH_2-OH$	8.73 ± 0.2
7	$-N(CH_2-CH_2-OH)_2$	8.22 ± 0.2
8	$-N[CH_2-CH(CH_3)-OH]_2$	8.31 ± 0.2

FIGURE 3.5 Simplified chemical structures and pK_a values of the amino-β-CD derivatives studied by Ivanyi et al. (From Ivanyi, R. et al., *Electrophoresis*, 25, 2675, 2004 With permission.)

(THCMH). The authors demonstrated the applicability of these selectors by means of the chiral separation of Dns-amino acids. Lin et al. [74] prepared a highly water soluble β-CD derivative, 2-*O*-(2-aminoethyl-imino-propyl)-β-*O*-hydroxypropyl-β-CD. On an acrylamide-coated capillary, the authors resolved the enantiomers of some acidic compounds such as hydroxy acids and NSAIDs.

CDs containing quaternary ammonium groups are charged over the whole pH range and therefore show a pH-independent electrophoretic mobility. Only very low selector concentrations are necessary. Furthermore, the electroosmotic flow (EOF) is reversed when using quaternary ammonium compounds. 2-Hydroxy-3-trimethylammonio-propyl-β-CD found application in the chiral separation of basic, neutral, and acidic compounds [75–79]. A quaternary ammonium compound of undefined structure (QA-β-CD) which is commercially available (CerestarUSA, Hammond, Indiana, United States) was applied to various acidic analytes [80–83]. This selector was also shown to be applicable in nonaqueous solvents such as formamide, *N*-methylformamide, methanol, and dimethyl sulfoxide [82].

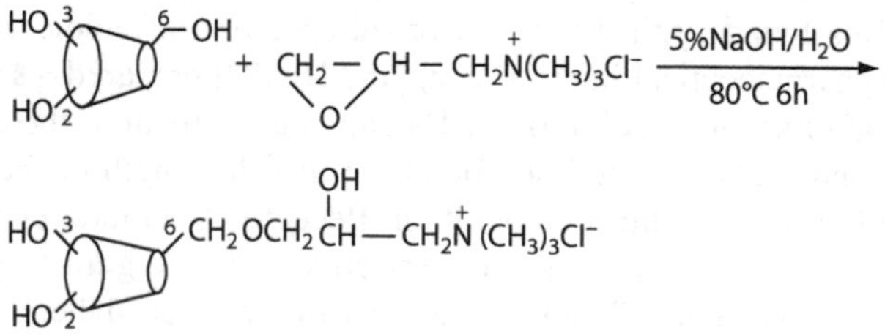

$$\text{HO}_3,\text{HO}_2\text{-CD-}_6\text{OH} + CH_2-CH-CH_2\overset{+}{N}(CH_3)_3Cl^- \text{ (epoxide)} \xrightarrow[80^\circ C\ 6h]{5\%NaOH/H_2O} \text{HO}_3,\text{HO}_2\text{-CD-}_6CH_2OCH_2CH(OH)-CH_2\overset{+}{N}(CH_3)_3Cl^-$$

FIGURE 3.6 Synthesis of 6-*O*-(2-hydroxy-3-trimethylammoniopropyl)-β-CD. (From Lin, X. et al., *Electrophoresis*, 27, 872, 2006. With permission.)

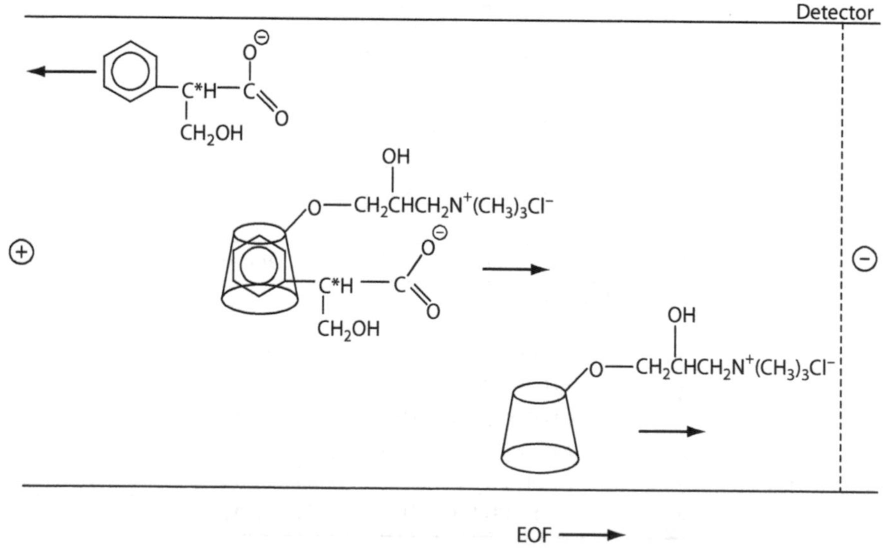

FIGURE 3.7 Flow diagram of the species on the uncoated capillary. (From Lin, X.L. et al., *Anal. Chim. Acta*, 517, 95, 2004. With permission.)

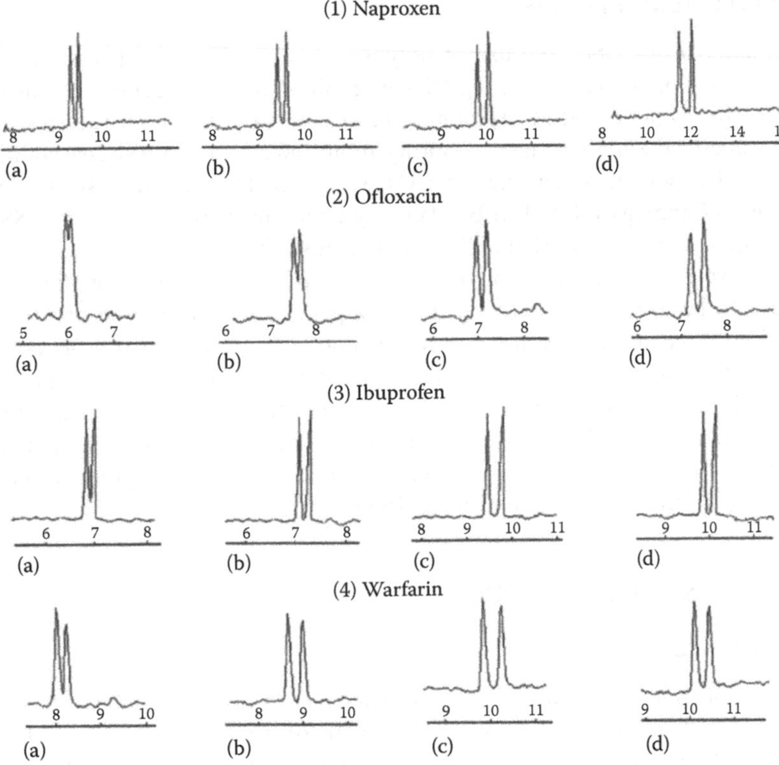

FIGURE 3.8 Electropherogram of the chiral separation of drug enantiomers at different 6-*O*-(2-hydroxy-3-trimethylammoniopropyl)-β-CD concentrations. (From Lin, X. et al., *Electrophoresis*, 27, 872, 2006. With permission.)

Recently, Lin et al. [84] synthesized a highly water soluble β-CD derivative, 6-*O*-(2-hydroxy-3-trimethylammoniopropyl)-β-CD (6-HPTMA-β-CD) (Figure 3.6) and compared its separation properties with that of QA-β-CD by means of a series of acidic analytes. A flow diagram showing the dierction of selectors and analytes is given in Figure 3.7. Figure 3.8 shows the effect of different concentrations of HPTMA-β-CD on the resolution of several drug enantiomers.

3.3.3 Amphoteric CD Derivatives

These selectors can be uncharged, positively, or negatively charged, depending on the pH [85]. Two amphoteric CDs, namely mono-(6-glutamylamino-6-deoxy-β-CD) [86] and AM-β-CD containing a 2-hydroxypropyltrimethyl-ammonium and a carboxyl group [81], were described. These selectors were shown to be applicable to the enantioseparation of neutral, acidic, and basic compounds.

3.4 POLYMERIZED CDS

Several authors described the use of polymerized CDs [87–93]. The synthesis is usually carried out by cross-linking CDs with bifunctional reagents, such as diepoxides or diisocyanates [94]. CD polymers have a more rigid structure and a reduced mobility due to the higher mass, resulting in an improved enantioseparation, and a much higher solubility. Uncharged CDs were used for the chiral separation of trimetoquinol analogs [88,89], β-blockers, sympathomimetics, anesthetic [88,90], amphetamine analogs [88,91], and 2-hydroxy acids [87].

A negatively charged β-CD polymer was prepared by Aturki and Fanali [95]. The polymer can be used either uncharged or charged, depending on the pH. It found application in the chiral separation of a selection of basic drugs. At a pH above 4, inversion of the mobility of the analytes was observed. Chiari et al. synthesized a vinylpyrrolidine-β-CD copolymer [96] by radical copolymerization of vinylpyrrolidone and methacroyl-β-CD. This polymer consists of an alkyl backbone and pendant β-CD units. The authors checked the enantioselectivity of this polymer selector by means of a mixture of sympathomimetic drugs.

3.5 DUAL SELECTOR SYSTEMS

3.5.1 Dual CD Systems

Overviews of this technique are given in specialized review articles [97,98] and in Chapter 4 of this book. A theoretical model for prediction of electrophoretic mobility differences using a dual CD system consisting of β-CD and DM-β-CD was presented by Nhujak et al. [99]. Matthijs et al. [100] developed a strategy for the optimization of the enantioseparation of drugs using mixtures of neutral and highly sulfated CDs in different concentrations and ratios.

Szökö et al. [101] applied a mixture of DM-β-CD and CM-β-CD for the quantitative analysis of *R*-(–)-deprenyl and its metabolites in rat urine. Beaufour et al. [102] succeeded in resolving a new antianginal drug, a benzoaxathiepin derivative, into the four stereoisomers, using a combination of DM-γ-CD and CM-β-CD (Figure 3.9). A dual CD system consisting of sulfated β-CD and M-β-CD was recently proposed by Chu et al. [103] for the chiral separation of the antiparkinson drug rotigotine.

3.5.2 CDs and Carbohydrates

Quan et al. [104] reported on the determination of D-serine as naphthalene-2,3-dicarboxaldehyde derivative using a combination of HP-γ-CD and sugars as enhancing chiral selectors. Among the sugars tested, such as β-lactose, sucrose, D-(–)-fructose, and D-(+)-glucose, the latter showed the best results. The authors applied this system using laser induced fluorescence (LIF) detection in the determination of D-serine in rat tissues. The combination of CM-γ-CD with the entangled polymer hydroxypropylmethylcellulose for the separation of citalopram enantiomers in human urine was described by Berzas-Nevado et al. [105].

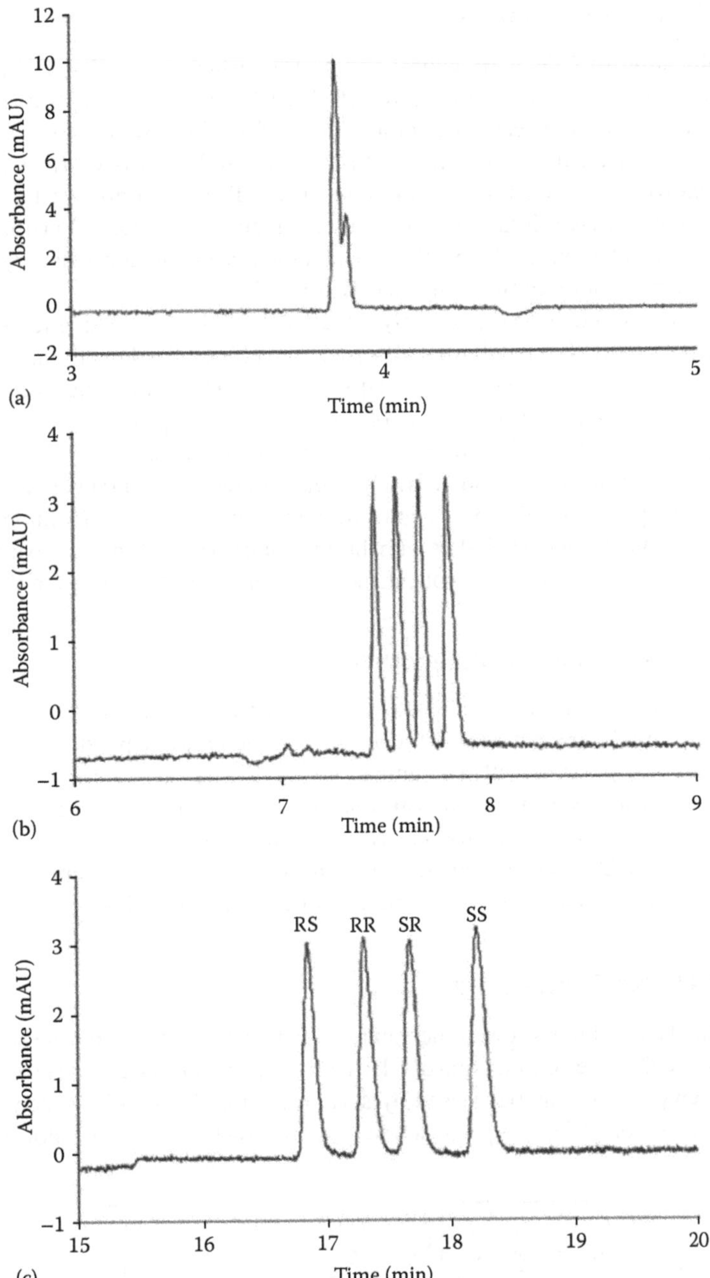

FIGURE 3.9 Effect of CM-β-CD concentration in a CM-β-CD/10 mM HP-γ-CD dual system on the separation of the benzoaxathiepin stereoisomers. (a) = 0 mM, (b) = 5 mM, (c) = 20 mM HP-γ-CD. (From Beaufour, M. et al., *J. Sep. Sci.*, 28, 529, 2005. With permission.)

3.5.3 CDs and Surfactants

The combination of CDs with achiral and chiral surfactants, called CD-mediated micellar electrokinetic chromatography (CD-MEKC) [106], represents a frequently practiced technique. For example, sodium dodecyl sulfate (SDS) serves as achiral surfactant and bile salts or polymeric surfactants [107,108] as chiral surfactants [109]. Negatively charged micelles migrate in the direction opposite to the EOF, while uncharged CDs migrate with the same velocity as the EOF. Partition of the analytes between the bulk solution, the CD, and the micelle pseudostationary phase is responsible for retention of the analytes (Figure 3.10).

Mertzman and Foley [110] recently developed a CD-modified microemulsion technique (CD-MEEKC) combining HP-β-CD or sulfated β-CD with the surfactant dodecoxycarbonyl valine, 1-butanol as a co-surfactant and ethyl acetate as a lipophilic core. The authors demonstrated the applicability of this approach by means of the chiral separation of sympathomimetics and some other drugs.

The use of chiral ionic liquids (Ils) represents a new trend. Francois et al. [111] combined DM-β-CD and TM-β-CD with ethyl- or phenylcholine bis(trifluoromethyl-sulfonyl)imide and resolved NSAIDs with this technique. These Ils did not exhibit chiral recognition ability by themselves but enhanced the separation properties of the CDs.

3.5.4 CDs and Ion-Pairing Reagents

Ion-pairing reagents have been shown to exhibit a supporting effect on the separation abilities of CDs. Bunke et al. [112] found out that the chiral separation of the drug cyclodrine is only possible with a combination of β-CD and (+)-camphor-10-sulfonic acid. No separation was obtained with one of these components. Jira et al. [113] showed that both chiral and achiral ion-pairing reagents show a synergistic effect on resolution using CDs. The combination of CDs and ion-pairing reagents was shown to be effective both in aqueous [113] and nonaqueous medium [114,115].

3.5.5 CDs and Crown Ethers

It has been observed that crown ethers exhibit a synergistic effect on the separation properties of CDs. The combination of CDs with the chiral crown ether, (+)-18-crown-6-tetracarboxylic acid has been used by several authors [116–119]. A new class of chiral crown ethers of the type of tetraoxadiaza-crown derivatives showed chiral

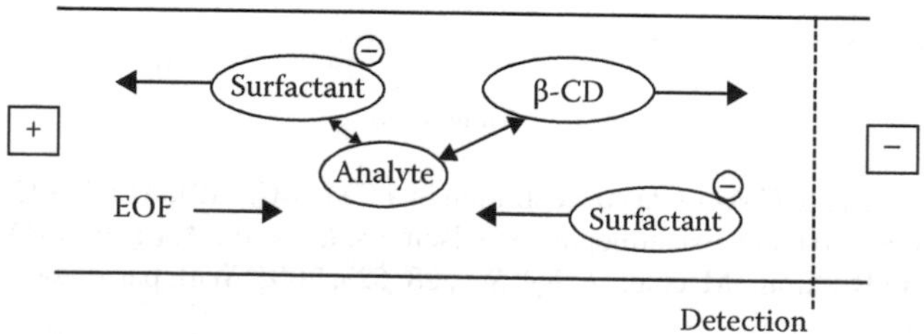

FIGURE 3.10 Scheme of MEKC.

resolution only in the presence of CDs [120,121]. It has been observed that achiral crown ethers also show a synergistic effect on the chiral recognition ability of CDs [122]. Armstrong et al. [123] discussed possible mechanisms and postulated the formation of "three body complexes" between the chiral amine, the CD, and the crown ether. A crown(3)L-alanine capped β-CD was prepared by Corradini et al. [124] and applied to the chiral separation of Dns-amino acids.

3.5.6 CDs and Macrocyclic Antibiotics

Recently, Dai et al. [125] made an attempt to combine the chiral recognition ability of CDs with that of antibiotics. The authors synthesized a water soluble β-CD derivative of erythromycin using 1-oxy-2,3-epoxypropane as binding reagent. It turned out that this selector exhibited significantly improved enantioselectivity compared to single β-CD and erythromycin. The authors used chlorpheniramine, salsolinal, and propranolol as model analytes (Figure 3.11).

3.5.7 Inclusion Complexation and Ligand-Exchange

The combination of host–guest interaction with the ligand-exchange principle has been investigated by several groups. Horimai et al. combined γ-CD with a

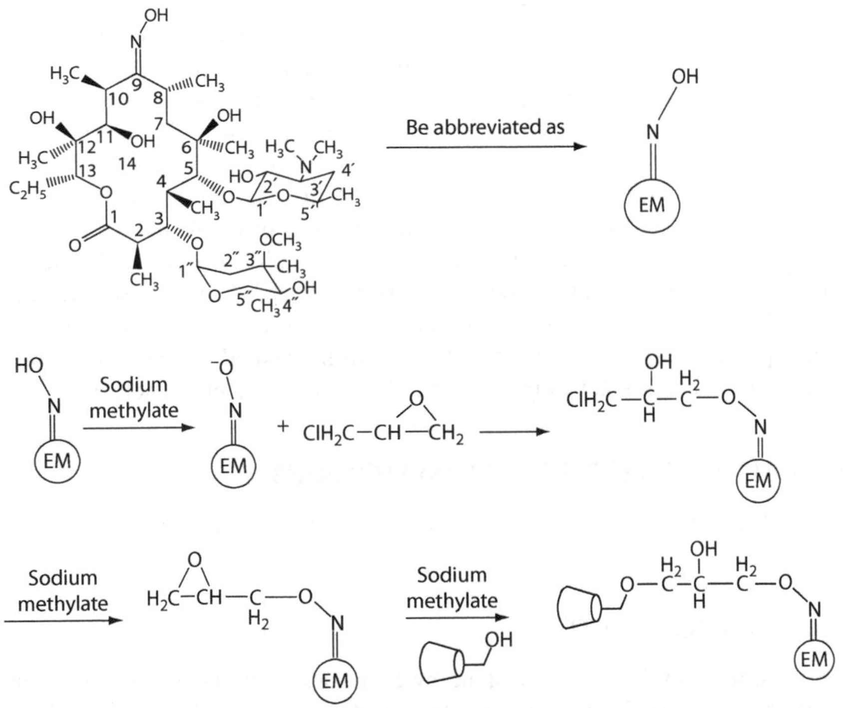

FIGURE 3.11 Synthesis of β-CD-derivatized erythromycin. (From Dai, R. et al., *Electrophoresis*, 28, 2566, 2007. With permission.)

Zn(II)-D-phenylalanine complex for the chiral separation of quinolone derivatives [126]. The use of a histamine-modified β-CD in the presence of Cu(II) ions for the chiral separation of amino acids based on a mixed inclusion/chelate complexation mechanism was reported by Cucinotta et al. [127]. Recently, the same group [128] prepared five new CD derivatives containing chelate complexing side chains. Using Cu(II) as complexing agent the authors checked the chiral separation ability of this approach by means of three aromatic amino acids. Wu et al. [129,130] synthesized a series of amino-substituted CD derivatives. Based on mixed inclusion/ligand-exchange mechanism the authors resolved aromatic amino acids and Dns-amino acids. The authors observed that the addition of PEG 20000 or *tert*-butyl alcohol improved resolution.

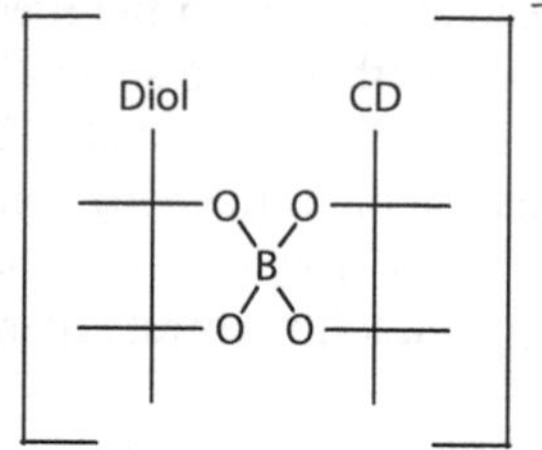

FIGURE 3.12 Scheme of mixed CD-borate-diol complexation.

A separation principle related to ligand exchange represents the use of mixed CD-borate-diol complexation (Figure 3.12). This approach was first described by Stefansson and Novotny [131] for the chiral separation of fluorescent-labeled sugars using β-CD in the presence of borate. Jira et al. [132] applied this technique for the chiral separation of various diol compounds including quinazoline derivatives of potential pharmaceutical interest that contain diol structure. Schmid and coworkers [132] studied different CD derivatives for their resolving ability for different *cis*-diols, among them hydrobenzoin. A dual chiral recognition mechanism can be assumed based on inclusion of the aromatic moiety into the cavity of the CD and interaction of the borate with the hydroxy groups at C2 and C3 at the mouth of the CD. The authors obtained good resolutions with β-CD, Succ-β-CD, M-β-CD, and HE-β-CD. No separations were obtained with 2,6-DM-β-CD and 2,3,6-TM-β-CD confirming the prerequisite of the availability of the hydroxy groups at C2 and C3. Noroski et al. [133] applied this principle to the chiral separation of a cholesterol-reducing drug containing a diol group. Again, negative results were obtained when 2,6-DM-β-CD or triacetyl-β-CD was used. Recently, Lin et al. [134] made use of the mixed CD-borate-diol complexation for the chiral separation of hydrobenzoin and structurally related compounds using the single isomer heptakis(2,3-dihydroxy-6-sulfo)-β-CD as a chiral selector.

3.6 CAPILLARY ELECTROCHROMATOGRAPHY

A detailed description of capillary electrochromatography (CEC) is given in Chapter 14 of this book.

3.6.1 Open-Tubular CEC

Pioneering work in this field was done by Schurig's group [135,136]. The authors attached permethylated β-CD via an octamethylene spacer to dimethylpolysiloxane and coated the product onto the capillary wall. Schurig demonstrated the principle of unified chromatography using the same capillary for CEC, nano-HPLC (high

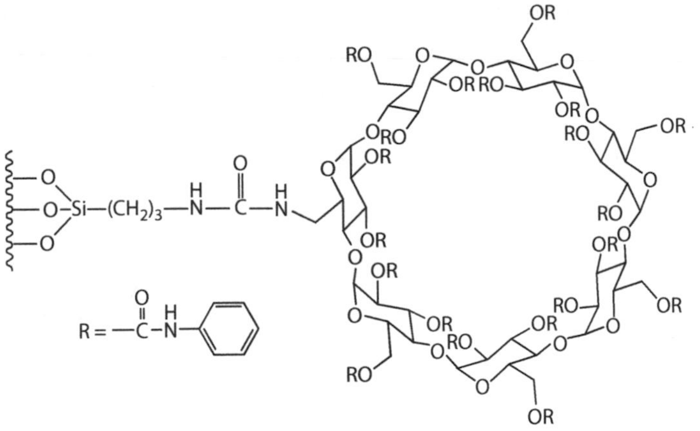

FIGURE 3.13 Structure of perphenylcarbamoylated-β-cyclodextrin-silica. (From Lin, B. et al., *Electrophoresis*, 27, 3057, 2006. With permission.)

performance liquid chromatography), gas chromatography (GC), and supercritical fluid chromatography (SFC) by means of a broad spectrum of analytes [137].

Lu and Ou [138] prepared an open-tubular (OT) capillary by bonding allyl-permethyl-β-CD and vinylsulfonic acid to polysiloxane.

3.6.2 Packed CEC (P-CEC)

Schurig and coworkers [139] prepared a silica-based stationary CEC phase by bonding permethylated β-CD to (mercaptopropyl)methyl silica. As an alternative, the same group used a thermal immobilization procedure for attaching a polysiloxane-linked permethyl-β-CD on silica [140]. The authors used mephobarbital, hexobarbital, and gluthetimide as analytes. Recently, Lin et al. [141] prepared a perphenylcarbamoylated β-CD phase and applied it to the chiral separation of a series of betablockers and sympathomimetics (Figures 3.13 and 3.14).

3.6.3 Monolithic Phases

A disadvantage of packed capillaries is the complicated packing procedure and the need of the preparation of frits by sintering a zone of the packing material. Such frits break easily and are sources of air bubbles. To circumvent these shortcomings, the preparation of monolithic phases became more and more popular. Two major types of monolithic phases have been prepared: monoliths on silica basis and monoliths on organic polymeric basis.

3.6.3.1 Silica-Based Monoliths

Monoliths on silica basis are prepared by in situ polycondensation of alkoxysilanes using a *sol–gel* process. Chiral selectors can be introduced simply by physical

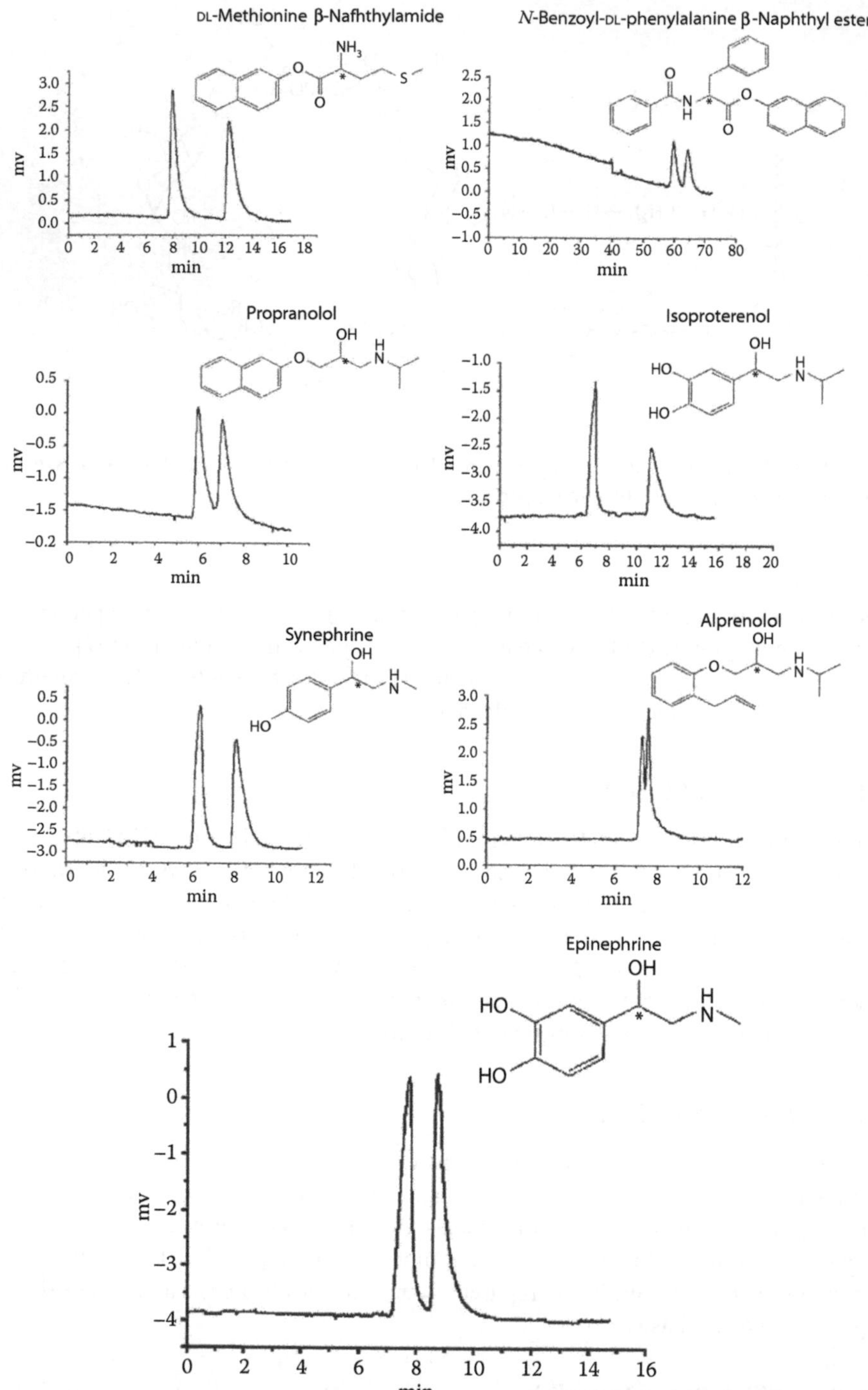

FIGURE 3.14 Enantioseparation by CEC on capillaries packed with perphenylcarbamoylated-β-CD-silica. (From Lin, B. et al., *Electrophoresis*, 27, 3057, 2006. With permission.)

adsorption, encapsulation, or on-column derivatization. Schurig's group prepared a monolithic phase by coating a silica-based monolith with a chiral polymer consisting of permethyl-β-CD grafted to polymethylsiloxane by an octamethylene spacer [142]. The monolithic phase was checked for its separation properties by means of hexobarbital, mephobarbital, benzoin, and carprofen.

Chen et al. [143] reported on the preparation of CD-containing monoliths by on-column derivatization of a silica-based monolith with β- or γ-CD via 3-glycidoxypropyltrimethoxy silane. The authors resolved Dns-amino acids and the positional isomers of *o*-, *m*-, and *p*-cresol on these phases.

Wistuba and Schurig [144] prepared a particle-fixed monolith by sintering a silica-based bed at 380°C after conventionally packing the capillary. Afterward the bed was coated with permethylated β-CD. Recently, the same group [145] developed a monolithic phase based on sol–gel glued CD-modified silica. The phase was prepared by the fusion of permethyl-β-CD-silica and linking them to the inner capillary wall. The authors demonstrated the chiral separation power of this monolithic phase by the chiral resolution of a broad spectrum of drugs, among them several barbiturates and NSAIDs, using pressure-supported CEC and capillary HPLC.

3.6.3.2 Polymer-Based Monoliths

The technique of preparation of organic monolithic phases (continuous beds) by in situ polymerization in the column has been introduced by Hjertén already in the 1980s [146].

Generally, the components are a monomer, a cross linker, a charge providing agent, a porogen, and a starter. An allylated chiral selector may be added to be copolymerized. To fix the polymer at the capillary wall, the capillary is first treated with γ-methacryloxypropyltrimethoxysilane. Koide and Ueno [147] prepared monolithic chiral stationary phases (CSPs) by incorporating CD polymers, such as poly-β-CD and CM-β-CD, in a polyacrylamide gel. Later, the same group [148,149] prepared a positively charged monolithic CSP, by allowing it to react with allyl carbamoylated-β-CD.

Végvári et al. [150] prepared a homogeneous gel on polyacrylamide basis by copolymerization of 2-hydroxy-3-allyloxy-propyl-β-CD with acrylamide, *N,N′*-methylene-bisacrylamide, and 2-acrylamido-2-methylpropane sulfonic acid as a negative charge providing agent or dimethyl allyl ammonium chloride for positively charged gels. The authors demonstrated the applicability of these phases by means of the chiral separation of a broad spectrum of acidic, basic, and neutral drugs. Pumera et al. [151] prepared neutral and negatively charged CD monoliths by either physically adsorbing *tert*-butyl-β-CD or copolymerizing peracetyl-2′-*O*-β-CD. The authors resolved ephedrine and ibuprofen on these phases. Sinner and Buchmeiser [152] developed a metal-catalyzed ring-opening metathesis polymerization process (ROMP) using a norbornene derivative of β-CD. The authors resolved the enantiomers of proglumide as a model compound on this monolithic phase. Kornysova et al. [153] reported on the use of a polyrotaxane approach for the preparation of monolithic phases based on the addition of cationic or anionic β-CD derivatives to the solution of a neutral acrylic acid monomer and a cross linker prior to polymerization. The chiral separation ability was demonstrated by means of metoprolol.

3.7 MICROFLUID DEVICES

The first microchip system for chiral separation was developed by Hutt et al. [154]. The authors presented a CD-MEKC system with γ-CD in combination with SDS and resolved fluorescein isothiocyanate (FITC)-labeled amino acids using LIF detection. Similar systems were described by Rodriguez et al. [155] and Wang et al. [156]. Skelley and Mathies [157] substituted FITC by fluorescamine and resolved a selection of amino acids using HP-β-CD as a chiral selector. As a detector, the authors used a confocal fluorescence microscope with a 404 nm blue diode laser excitation source. The authors demonstrated that this technique has potential for determination of the enantiomeric ratio of amino acids for extraterrestrial life detection. Wallenborg et al. [158] developed a microchip system in combination with a LIF detector for the chiral separation of the NBD-derivatives of ephedrine, norephedrine, cathinone, pseudoephedrine, methcathinone, amphetamine, and methylamphetamine using a mixture of HP-β-CD and sulfobutyl ether (SBE)-β-CD or highly sulfated CDs. No separations were obtained with neutral CDs. The authors applied this method for forensic analysis.

Schwarz and Hauser [159] resolved the enantiomers of adrenaline, noradrenaline, ephedrine, and pseudoephedrine on a microchip device using a two-electrode amperometric detection technique. CM-β-CD with and without addition of 18-crown-6 was used as a chiral selector. Later, the same group [160] resolved the enantiomers of small amines by conventional and on-chip CE using a combination of DM-β-CD and (+)-18-crown-6-tetracarboxylic acid as chiral selectors. Since the authors separated non-UV-absorbing amines, detection was carried out by contactless conductivity measurement.

Liu et al. [161] designed a microchip CE system combined with on-chip chemiluminescence detection. Making use of HP-β-CD as a chiral selector and the peroxyoxalate/hydrogen peroxide chemiluminescence detection system, the authors resolved Dns-amino acids with high sensitivity. Ludwig et al. [162] used a commercially available microchip electrophoresis system (MCE 2010, Shimadzu, Kyoto, Japan) containing a linear imaging UV detection system. Since the detection is carried out on the whole separation channel, the separation can be monitored continuously. Using highly sulfated α-, β-, or γ-CD as chiral selectors, 19 structurally different compounds were resolved. The same chip system was applied by Kitagawa et al. [163] for the separation of 1-aminoindan enantiomers by CD-mediated MEKC. Another commercially available microchip electrophoresis system (Microni, Enschede, the Netherlands) was tested by Belder et al. [164]. The chips were made from borofloat glass containing channels manufactured by powder-blasting technology. The authors resolved FITC-labeled 1-cyclohexylethylamine using HP-γ-CD as a chiral selector comparing chips containing channels coated with polyvinylalcohol and noncoated channels. The coated channels gave significantly better results. The fastest separation, reported so far, was described by Piehl et al. [165]. The authors resolved Dns-amino acids within 720 ms using highly sulfated CDs as chiral selectors in a microchip system with fluorescence detection employing a separation channel of 7 mm only in length. Gao et al. [166] reported on the development of multichannel microfluid devices for high throughput screening (Figure 3.15). The authors demonstrated the applicability of this device by means of FITC-labeled baclofen, norfenefrine, and tocainamide using γ-CD and DM-β-CD as chiral selectors in different channels. With this device, it was possible to find the best selector for one compound and to screen one selector with different

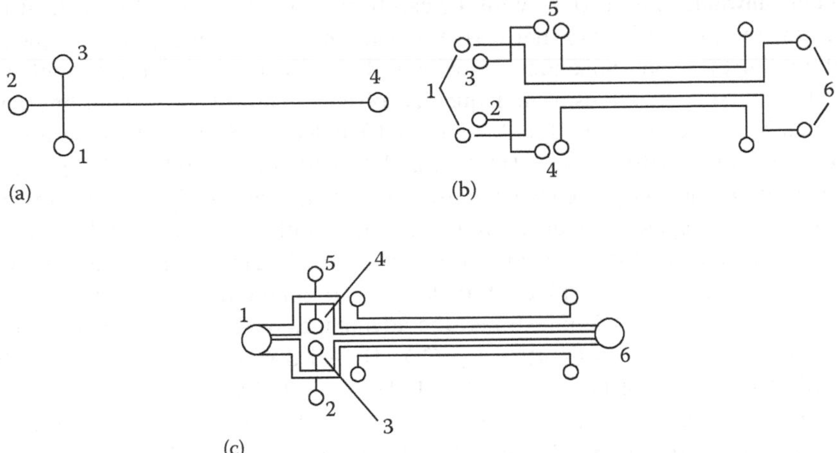

FIGURE 3.15 Schematics of the microfluidic chip channel structures: (a) mono-channel chip, (b) two-channel chip, and (c) four-channel chip. 1–6: sequence number of electrodes. (From Gao, Y. et al., *Electrophoresis*, 26, 4774, 2005. With permission.)

analytes. Recently, Belder et al. [167] presented a combined microfluid reactor/separation unit in one UV-transparent fused-silica microchip. The authors applied this system for the online study of the cleavage rate of the substrate glycidyl phenyl ether by different enzymes. Using heptakis-6-sulfato-β-CD as a chiral selector, the enantiomers of the substrate and the reaction product were separated online in the separation channel.

Ölvecka et al. [168] first applied the principle of ITP on a polymethylmethacrylate chip in combination with conductivity detection. Using DL-tryptophan as a model compound, the authors developed three different modes: single channel ITP, tandem-coupled channel ITP, and concentration cascade tandem-coupled ITP. In the first approach only one separation channel is used, in the second approach there are two separation channels used in series, and in the third approach a second leading electrolyte with a lower concentration than the first leading electrolyte is used in the second separation channel resulting in a concentration cascade. However, best results were obtained with the latter approach.

An electrochromatographic approach was developed by Zeng et al. [169] using a monolithic phase. The authors synthesized a polyacrylamide gel in situ in the microchannel using allyl-β-CD acting both as a cross linker and as a chiral selector. The wall of the channel was pretreated with 3-(trimethoxysilyl) propyl methacrylate for fixing the polymer in the channel. The authors applied this system to the chiral separation of FITC amino acids.

For more detailed information, the reader is referred to specialized review articles [170,171] and Chapter 16.

3.8 NONAQUEOUS MEDIUM

Comprehensive review articles focus on the use of nonaqueous solvents for chiral separation in electromigration techniques [172,173]. The reasons for using nonaqueous solvents are the insolubility of chiral selectors or analytes in water, for reducing unwanted interactions with the capillary wall, and reduced Joule heating. Weak interactions which are ineffective in water may be enhanced in nonaqueous medium. Furthermore, nonaqueous solvents show a better compatibility with MS detection. First applications of nonaqueous solvents in chiral CE separations were reported by Valko et al. [174] and Wang and Khaledi [175] using CDs in formamide (FA), dimethylformamide (DMF), and *N*-methylformamide (NMF) as solvents. Neutral and charged CDs were used in nonaqueous CE (NACE) for several applications. Comparative studies on the separation of basic compounds in aqueous and nonaqueous systems using sulfated β-CD (S-β-CD) in formamide showed that a significant reduction of band broadening is observed in nonaqueous medium [176]. Vincent and Vigh demonstrated the advantages of the use of the single isomer heptakis(2,3-diacetyl-6-sulfato)-β-CD (HDAS-β-CD) in pure methanol for the chiral separation of basic drugs [177]. The use of QA-β-CD in pure organic solvents for the chiral separation of amino acids derivatives and some profens was reported by Wang and Khaledi [82]. Karbaum and Jira [178] investigated a series of solvents for their suitability in NACE and discussed the advantages of NACE. Recently, Mol et al. [179] coupled CD-MEKC with electrospray ionization-mass spectrometry (ESI-MS). The authors used single isomer anionic CDs, such as heptakis(2,3-di-*O*-methyl-6-*O*-sulfo)-β-CD and heptakis(2,3-di-*O*-acetyl-6-*O*-sulfo)-β-CD, for the chiral separation of positively charged analytes and made the observation that the addition of camphorsulfonate as a counter ion to the BGE improved resolution. Later, the same group [180] developed a nonaqueous electrokinetic chromatography approach using cationic single isomer CD derivatives in combination with MS detection. The authors employed ESI-MS in the negative-ion mode using a sheath–liquid interface. The detection limit for acidic drugs was in the range of 50–400 ng/mL.

3.9 MISCELLANEOUS

Reversal of the enantiomer migration order (EMO) is often required, for example, in connection with enantiomeric purity checks for drugs. The distomer present in traces in a sample of the eutomer should always appear as the first peak; otherwise, it could be covered by the tailing of the eutomer. One way would be to change to a selector possessing opposite chirality. This is, however, not possible with CDs. In CE, reversal of the EMO can sometimes be obtained by changing from a neutral to a charged CD, by changing the mobility of the analyte or the selector by varying the pH. A further means is the reversal of the direction of the EOF by adding quaternary ammonium compounds. Moreover, addition of a surfactant can reverse the migration order. A survey of different possibilities for reversing the EMO in CE has been given by Chankvetadze et al. [181]. The influence of the BGE components on resolution was studied by Servais et al. [182]. Using heptakis(2,3-di-*O*-methyl-6-*O*-sulfo)-β-CD (HDMS-β-CD) as a chiral selector, the influence of cationic (sodium, ammonium or potassium) and anionic (chloride, formate, methanesulfonate, or camphorsulfonate) BGE additives was investigated. The best results were obtained with ammonium formate or camphorsulfonate in methanolic solution acidified with formic acid using three β-blockers (atenolol, celoprolol, and propranolol) and three local anesthetics (bupivacaine, mepivacaine, and prilocaine) as model analytes. Na et al. [183] observed an improvement in resolution when adding polystyrene nanoparticles to the electrolyte containing HP-β-CD as a chiral selector. Propranolol was used as a model analyte. The authors assume that the CD is adsorbed on the polystyrene particles forming a pseudostationary phase.

Kubacak et al. [184] developed an ITP system for the chiral separation of the antihistamines, dimethindene, and pheniramine. The authors used CE-β-CD in potassium acetate, adjusted to pH 4.8 by acetic acid, as the leading electrolyte with methylhydroxyethylcellulose for suppressing the EOF and L-β-alanine as the terminating electrolyte. Online coupling of capillary ITP and CZE for enantiomer separation has been described by Dankova et al. [185]. The authors demonstrated the high potential of ITP for sample clean up and preconcentration of analytes. The enantiomers of tryptophan and 2,4-dinitrophenyl norleucine were applied in different enantiomer ratios as model analytes and α-CD was used in both systems as a chiral selector. More recently, Kaniansky's group presented an ITP–CZE coupling system for sample preconcentration and clean up in chiral pharmaceutical analysis. This technique allows the application of high sample volumes whereby the zones are narrowed by ITP before entering the CZE part. The authors demonstrated the applicability of this approach by chiral analysis of pheniramine and analogs using CE-β-CD as a chiral selector for purity tests in pharmaceutical preparations and enantioselective metabolism studies [186,187] (Figure 3.16). A limit of detection of 2.5 ng/mL for *levo*-brompheniramine was achieved which permits the detection of

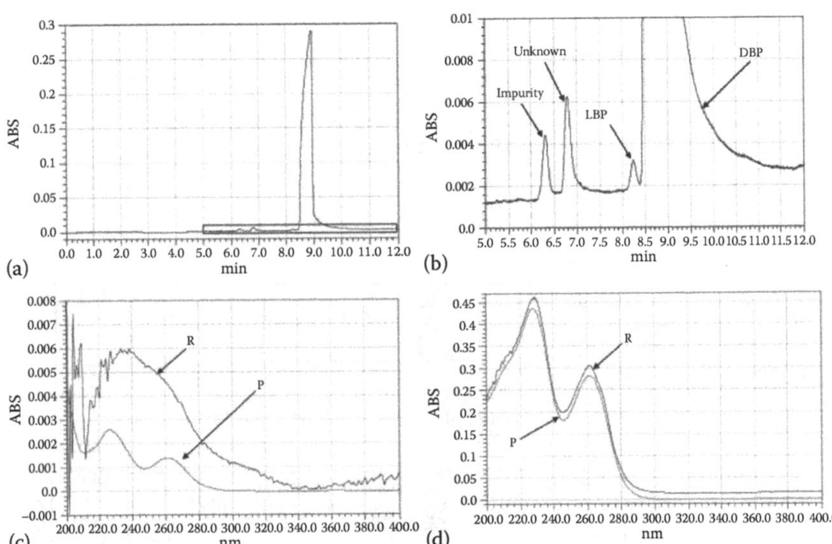

FIGURE 3.16 ITP-CZE analysis of 30 μL dexbrompheniramine (DBP) standard. The whole electropherogram obtained in CZE step taken at 261 nm wavelength (a). Its part given in the rectangle is shown in (b). Raw (R) (unprocessed) and processed (P) spectra of levobrompheniramine (LBP) impurity (c) and DBP (d), respectively. (From Marak, J. et al., *J. Pharm. Biomed. Anal.*, 46, 870, 2008 and Mikus, P. et al., *J. Chromatogr. A*, 1179, 9, 2008. With permission.)

traces of impurities in samples of *dextro*-brompheniramine. Zhong and Yeung [188] designed a capillary array system for combinatorial chiral separations. The system consisted of 96 capillaries, whereby the outlet ends of every eight capillaries were bundled together. Thereby in 12 bundles, 8 compounds can be tested in the same run at 12 different separation conditions. Using neutral and sulfated CDs as chiral selectors with different buffers, for 49 out of 54 compounds tested, the optimal separation conditions could be found within short time. Male and Luong [189] designed a CE system equipped with an array of microfabricated interdigitated platinum electrodes. The system was applied to the simultaneous chiral analysis of epinephrine, norephedrine, and isoproterenol, using heptakis(2,6-di-*O*-methyl)-β-cyclodextrin as a chiral selector. The interdigitated electrode chip served as an amplification/detection system and consisted of an array of seven electrodes at oxidizing potential to oxidize the analytes and a detector electrode set at reducing potential. Thereby fouling of the detection electrode is avoided. Henley et al. [190] developed a system making use of a counterflow produced by a pump in the direction opposite to the electrokinetic migration of the analyte. Using β-CD as a chiral selector, an improvement in resolution was obtained in this counterflow system for a series of drugs. Lee et al. [191]

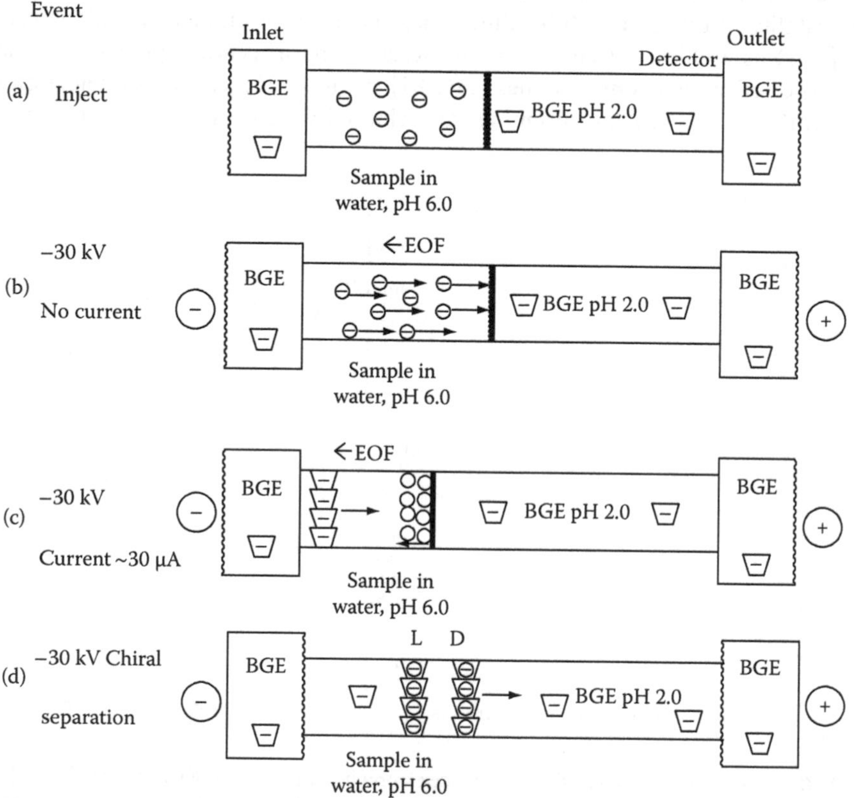

FIGURE 3.17 Schematic diagram of the pH-mediated stacking/sweeping-*S*-β-CD of racemic CBI-amino acids. (From Kirschner, D.L. et al., *Anal. Chem.*, 79, 736, 2007. With permission.)

designed a system consisting of chiral polyaniline nanobundles (C-PANI-NB) which are interconnected with CD sulfate units. The authors proposed a potential applicability of these enantioselective nanobundles as chiral selectors in CE. Sazelova et al. [192] developed a device for applying a radical field to control electrokinetic potential and EOF. The device consists of three high-voltage power supplies, which are used to form a radial electric field across the fused-silica capillary wall. The authors demonstrated the applicability of this device in a chiral separation system for terbutaline. Kirschner et al. [193] reported on the development of an online sample preconcentration approach based on field-amplified migration, pH-mediated stacking, and sweeping by the highly anionic sulfated β-CD. Using a CE system in reverse polarity mode in combination with LIF detection, the authors achieved baseline separation for cyanobenz[f]isoindole amino acids with detection limits in the range of 10^{-9} M (Figure 3.17).

REFERENCES

1. Chankvetadze, B. 2004. Combined approach using capillary electrophoresis and NMR spectroscopy for an understanding of enantioselective recognition mechanisms by cyclodextrins. *Chem Soc Rev* 33: 337–347.
2. Dodziuk, H, Kozminski, W, Ejchart, A. 2004. NMR studies of chiral recognition by cyclodextrins. *Chirality* 16: 90–105.
3. Bikadi, Z, Ivanyi, R, Szente, L, Ilisz, I, Hazai, E. 2007. Cyclodextrin complexes: Chiral recognition and complexation behaviour. *Curr Drug Discov Technol* 4: 282–294.
4. Zhang, G, Hong, Z, Chai, Y, Zhu, Z, Song, Y, Liu, C, Ji, S, Yin, X, Wu, Y. 2007. A study on the chiral recognition mechanism of enantioseparation of adrenaline and its analogues using capillary electrophoresis. *Chem Pharm Bull (Tokyo)* 55: 324–327.
5. Fanali, S, Bocek, P. 1996. A practical procedure for the determination of association constants of the analyte-chiral selector equilibria by capillary zone electrophoresis. *Electrophoresis* 17: 1921–1924.
6. Tokes, B, Ferencz, L, Buchwald, P, Donath-Nagy, G, Vancea, S, Santa, N, Kis, EL. 2008. Structural studies on the chiral selector capacity of cyclodextrin derivatives. *J Biochem Biophys Methods* 70: 1276–1282.
7. Fanali, S. 1996. Identification of chiral drug isomers by capillary electrophoresis. *J Chromatogr A* 735: 77–121.
8. Fanali, S. 1997. Controlling enantioselectivity in chiral capillary electrophoresis with inclusion-complexation. *J Chromatogr A* 792: 227–267.
9. Gübitz, G, Schmid, MG. 1997. Chiral separation principles in CE. *J Chromatogr A* 792: 179–225.
10. Blaschke, G, Chankvetadze, B. 2000. Enantiomer separation of drugs by capillary electromigration techniques. *J Chromatogr A* 875: 3–25.
11. Chankvetadze, B. 2001. Enantioseparation of chiral drugs and current status of electromigration techniques in this field. *J Sep Sci* 24: 691–705.
12. Scriba, GKE. 2003. Pharmaceutical and biomedical applications of chiral capillary electrophoresis and capillary electrochromatography: An update. *Electrophoresis* 24: 2409–2421.
13. Ward, TJ, Hamburg, DM. 2004. Chiral separations. *Anal Chem* 76: 4635–4644.
14. Natishan, TK. 2005. Recent progress in the analysis of pharmaceuticals by capillary electrophoresis. *J Liq Chromatogr* 28: 1115–1160.
15. Van Eeckhaut, A, Michotte, Y. 2006. Chiral separations by capillary electrophoresis: Recent developments and applications. *Electrophoresis* 27: 2880–2895.

16. Snopek, J, Jelinek, I, Smolkova-Keulemansova, E. 1988. Use of cyclodextrins in isotachophoresis: IV. The influence of cyclodextrins on the chiral resolution of ephedrine alkaloid enantiomers. *J Chromatogr* 438: 211–218.
17. Fanali, S. 1989. Separation of optical isomers by capillary zone electrophoresis based on host-guest complexation with cyclodextrins. *J Chromatogr A* 474: 441–446.
18. Wren, SA, Rowe, RC. 1993. Theoretical aspects of chiral separation in capillary electrophoresis. III. Application to beta-blockers. *J Chromatogr* 635: 113–118.
19. Rawjee, YY, Williams, RL, Vigh, G. 1994. Efficiency optimization in capillary electrophoretic chiral separations using dynamic mobility matching. *Anal Chem* 66: 3777–3781.
20. Surapaneni, S, Ruterbories, K, Lindstrom, T. 1997. Chiral separation of neutral species by capillary electrophoresis—Evaluation of a theoretical model. *J Chromatogr A* 761: 249–257.
21. Aturki, Z, Desiderio, C, Mannina, L, Fanali, S. 1998. Chiral separations by capillary zone electrophoresis with the use of cyanoethylated-β-cyclodextrin as chiral selector. *J Chromatogr A* 817: 91–104.
22. Zerbinati, O, Trotta, F, Giovannoli, C, Baggiani, C, Giraudi, G, Vanni, A. 1998. New derivatives of cyclodextrins as chiral selectors for the capillary electrophoretic separation of dichlorprop enantiomers. *J Chromatogr A* 810: 193–200.
23. Li, GB, Lin, XL, Zhu, CF, Hao, AY, Guan, YF. 2000. New derivative of beta-cyclodextrin as chiral selectors for the capillary electrophoretic separation of chiral drugs. *Anal Chim Acta* 421: 27–34.
24. Miura, M, Kawamoto, K, Funazo, K, Tanaka, M. 1998. Chiral separation of several amino acid derivatives by capillary zone electrophoresis with selectively acetylated beta-cyclodextrin derivatives. *Anal Chim Acta* 373: 47–56.
25. Matsunaga, H, Tanimoto, T, Haginaka, J. 2002. Separation of basic drug enantiomers by capillary electrophoresis using methylated glucuronyl glucosyl-cyclodextrin as a chiral selector. *J Sep Sci* 25: 1175–1182.
26. Wei, YH, Li, J, Zhu, CF, Hao, AY, Zhao, MG. 2005. 2-O-(2-hydroxybutyl)-beta-cyclodextrin as a chiral selector for the capillary electrophoretic separation of chiral drugs. *Anal Sci* 21: 959–962.
27. Lin, XL, Zhu, C, Hao, A. 2004. Evaluation of newly synthesized derivative of cyclodextrin for the capillary electrophoretic separation. *J Chromatogr A* 1059: 181–189.
28. Schmitt, U, Ertan, M, Holzgrabe, U. 2004. Chiral capillary electrophoresis: Facts and fiction on the reproducibility of resolution with randomly substituted cyclodextrins. *Electrophoresis* 25: 2801–2807.
29. Cucinotta, V, Giuffrida, A, Grasso, G, Maccarrone, G, Messina, M. 2002. Simultaneous separation of different enantiomeric pairs in capillary electrophoresis by mixing different hemispherodextrins, a very versatile class of receptors. *J Chromatogr A* 979: 137–145.
30. Wistuba, D, Bogdanski, A, Larsen, KL, Schurig, V. 2006. Delta-cyclodextrin as novel chiral probe for enantiomeric separation by electromigration methods. *Electrophoresis* 27: 4359–4363.
31. Jiang, Z, Thorogate, R, Smith, NW. 2008. Highlighting the role of the hydroxyl position on the alkyl spacer of hydroxypropyl-beta-cyclodextrin for enantioseparation in capillary electrophoresis. *J Sep Sci* 31: 177–187.
32. Williams, BA, Vigh, G. 1997. Dry look at the CHARM (charged resolving agent migration) model of enantiomer separations by capillary electrophoresis. *J Chromatogr A* 777: 295–309.
33. Chen, YY, Wang, W, Yang, WP, Zhang, ZJ. 2004. Carboxymethyl-beta-cyclodextrin for chiral separation of amino acids derivatized with fluorescene-5-isothiocyanate by capillary electrophoresis and laser-induced fluorescence detection. *Chinese Chem Lett* 15: 112–114.

34. Mandrioli, R, Pucci, V, Sabbioni, C, Bartoletti, C, Fanali, S, Raggi, MA. 2004. Enantioselective determination of the novel antidepressant mirtazapine and its active demethylated metabolite in human plasma by means of capillary electrophoresis. *J Chromatogr A* 1051: 253–260.
35. Van Eeckhaut, A, Detaevernier, MR, Michotte, Y. 2004. Separation of neutral dihydropyridines and their enantiomers using electrokinetic chromatography. *J Pharm Biomed Anal* 36: 799–805.
36. Mikus, P, Kubacak, P, Valaskova, I, Havranek, E. 2005. Separation of phenothiazines applying capillary electrophoresis and micromolar concentration of carboxyethyl-beta-cyclodextrin. *Chemia Analityczna* 50: 1031–1041.
37. Ha, PTT, Van Schepdael, A, Van Vaeck, L, Augustijns, P, Hoogmartens, J. 2004. Chiral capillary electrophoretic method for quantification of apomorphine. *J Chromatogr A* 1049: 195–203.
38. Groom, CA, Halasz, A, Paquet, L, Thiboutot, S, Ampleman, G, Hawari, J. 2005. Detection of nitroaromatic and cyclic nitramine compounds by cyclodextrin assisted capillary electrophoresis quadrupole ion trap mass spectrometry. *J Chromatogr A* 1072: 73–82.
39. Juvancz, Z, Jicsinszky, L, Markides, KE. 1997. Phosphated cyclodextrins as new acidic chiral additives for capillary electrophoresis. *J Microcol Sep* 9: 581–589.
40. Yang, GS, Chen, DM, Yang, Y, Tang, B, Gao, JJ, Aboul-Enein, HY, Koppenhoefer, B. 2005. Enantioseparation of some clinically used drugs by capillary electrophoresis using sulfated beta-cyclodextrin as a chiral selector. *Chromatographia* 62: 441–445.
41. Vincent, JB, Kirby, DM, Nguyen, TV, Vigh, G. 1997. A family of single-isomer chiral resolving agents for capillary electrophoresis.2. Hepta-6-sulfato-beta-cyclodextrin. *Anal Chem* 69: 4419–4428.
42. Vincent, JB, Sokolowski, AD, Nguyen, TV, Vigh, G. 1997. A family of single-isomer chiral resolving agents for capillary electrophoresis.1. Heptakis(2,3-diacetyl-6-sulfato)-beta-cyclodextrin. *Anal Chem* 69: 4226–4233.
43. Cai, H, Nguyen, TV, Vigh, G. 1998. A family of single-isomer chiral resolving agents for capillary electrophoresis.3. Heptakis(2,3-dimethyl-6-sulfato)-beta-cyclodextrin. *Anal Chem* 70: 580–589.
44. Zhu, W, Vigh, G. 2000. Enantiomer separations by nonaqueous capillary electrophoresis using octakis(2,3-diacetyl-6-sulfato)-gamma-cyclodextrin. *J Chromatogr A* 892: 499–507.
45. Zhu, W, Vigh, G. 2001. Separation selectivity patterns in the capillary electrophoretic separation of anionic enantiomers by octakis-6-sulfato-gamma-cyclodextrin. *Electrophoresis* 22: 1394–1398.
46. Christians, T, Holzgrabe, U. 2001. New single-isomer chiral selector for capillary electrophoresis: The highly water-soluble heptakis(2-N,N-dimethylcarbamoyl)-beta-cyclodextrin. *J Chromatogr A* 911: 249–257.
47. Busby, MB, Maldonado, O, Vigh, G. 2002. Nonaqueous capillary electrophoretic separation of basic enantiomers using octakis(2,3-O-dimethyl-6-O-sulfo)-gamma-cyclodextrin, a new, single-isomer chiral resolving agent. *Electrophoresis* 23: 456–461.
48. Li, S, Vigh, G. 2003. Synthesis, analytical characterization and initial capillary electrophoretic use in acidic background electrolytes of a new, single-isomer chiral resolving agent: hexakis(2,3-di-O-acetyl-6-O-sulfo)-alpha-cyclodextrin. *Electrophoresis* 24: 2487–2498.
49. Busby, MB, Maldonado, O, Vigh, G. 2003. Electrophoretic enantiomer separations at high pH using the new, single-isomer octakis(2,3-dimethyl-6-O-sulfo)-gamma-cyclodextrin as chiral resolving agent. *J Chromatogr A* 990: 63–73.
50. Maynard, DK, Vigh, G. 2001. Heptakis(2-O-methyl-3,6-di-O-sulfo)-beta-cyclodextrin: A single isomer, 14-sulfated beta-cyclodextrin for use as a chiral resolving agent in capillary electrophoresis. *Electrophoresis* 22: 3152–3162.

51. Zhu, W, Vigh, G. 2003. A family of single-isomer, sulfated gamma-cyclodextrin chiral resolving agents for capillary electrophoresis: octa(6-O-sulfo)-gamma-cyclodextrin. *Electrophoresis* 24: 130–138.
52. Li, SL, Vigh, G. 2004. Use of the new, single-isomer hexakis(2,3-diacetyl-6-O-sulfo)-alpha-cyclodextrin in acidic methanol background electrolytes for nonaqueous capillary electrophoretic enantiomer separations. *J Chromatogr A* 1051: 95–101.
53. Li, SL, Vigh, G. 2004. Single-isomer sulfated alpha-cyclodextrins for capillary electrophoresis: Hexakis(2,3-di-O-methyl-6-O-sulfo)-alpha-cyclodextrin, synthesis, analytical characterization, and initial screening tests. *Electrophoresis* 25: 2657–2670.
54. Li, SL, Vigh, G. 2004. Single-isomer sulfated alpha-cyclodextrins for capillary electrophoresis. Part 2. Hexakis(6-O-sulfo)-alpha-cyclodextrin: Synthesis, analytical characterization, and initial screening tests. *Electrophoresis* 25: 1201–1210.
55. Sanchez-Vindas, S, Vigh, G. 2005. Non-aqueous capillary electrophoretic enantiomer separations using the tetrabutylammonium salt of heptakis(2,3-O-diacetyl-6-O-sulfo)-cyclomaltoheptaose, a single-isomer sulfated beta-cyclodextrin highly-soluble in organic solvents. *J Chromatogr A* 1068: 151–158.
56. Busby, MB, Vigh, G. 2005. Synthesis of a single-isomer sulfated beta-cyclodextrin carrying nonidentical substituents at all of the C2, C3, and C6 positions and its use for the electrophoretic separation of enantiomers in acidic aqueous and methanolic background electrolytes, Part 2: Heptakis(2-O-methyl-6-O-sulfo) cyclomaltoheptaose. *Electrophoresis* 26: 3849–3860.
57. Busby, MB, Vigh, G. 2005. Synthesis of heptakis(2-O-methyl-3-O-acetyl-6-O-sulfo)-cyclomaltoheptaose, a single-isomer, sulfated beta-cyclodextrin carrying nonidentical substituents at all the C2, C3, and C6 positions and its use for the capillary electrophoretic separation of enantiomers in acidic aqueous and methanolic background electrolytes. *Electrophoresis* 26: 1978–1987.
58. Cucinotta, V, Giuffrida, A, Maccarrone, G, Messina, M, Puglisi, A, Torrisi, A, Vecchio, G. 2005. The 6-derivative of beta-cyclodextrin with succinic acid: A new chiral selector for CD-EKC. *J Pharm Biomed Anal* 37: 1009–1014.
59. Culha, M, Schell, FM, Fox, S, Green, T, Betts, T, Sepaniak, MJ. 2004. Evaluation of newly synthesized and commercially available charged cyclomaltooligosaccharides (cyclodextrins) for capillary electrokinetic chromatography. *Carbohyd Res* 339: 241–249.
60. Terabe, S. 1989. Electrokinetic chromatography—an interface between electrophoresis and chromatography. *Trac-Trends in Anal Chem* 8: 129–134.
61. Nardi, A, Eliseev, A, Bocek, P, Fanali, S. 1993. Use of charged and neutral cyclodextrins in capillary zone electrophoresis—Enantiomeric resolution of some 2-hydroxy acids. *J Chromatogr* 638: 247–253.
62. Fanali, S, Desiderio, C, Aturki, Z. 1997. Enantiomeric resolution study by capillary electrophoresis—Selection of the appropriate chiral selector. *J Chromatogr A* 772: 185–194.
63. Lelievre, F, Gareil, P, Jardy, A. 1997. Selectivity in capillary electrophoresis: Application to chiral separations with cyclodextrins. *Anal Chem* 69: 385–392.
64. Galaverna, G, Corradini, R, Dossena, A, Marchelli, R. 1999. Histamine-modified cationic beta-cyclodextrins as chiral selectors for the enantiomeric separation of hydroxy acids and carboxylic acids by capillary electrophoresis. *Electrophoresis* 20: 2619–2629.
65. Haynes, JL 3rd, Shamsi, SA, O'Keefe, F, Darcey, R, Warner, IM. 1998. Cationic beta-cyclodextrin derivative for chiral separations. *J Chromatogr A* 803: 261–271.
66. Ivanyi, R, Jicsinszky, L, Juvancz, Z, Roos, N, Otta, K, Szejtli, J. 2004. Influence of (hydroxy)alkylamino substituents on enantioseparation ability of single-isomer amino-beta-cyclodextrin derivatives in chiral capillary electrophoresis. *Electrophoresis* 25: 2675–2686.

67. Ong, TT, Tang, W, Muderawan, W, Ng, SC, Chan, HSO. 2005. Synthesis and application of single-isomer 6-mono(alkylimidazolium)-beta-cyclodextrins as chiral selectors in chiral capillary electrophoresis. *Electrophoresis* 26: 3839–3848.
68. Tang, W, Muderawan, IW, Ong, TT, Ng, SC. 2005. Enantioseparation of acidic enantiomers in capillary electrophoresis using a novel single-isomer of positively charged beta-cyclodextrin: Mono-6(A)N-pentylammonium-6(A)-deoxy-beta-cyclodextrin chloride. *J Chromatogr A* 1091: 152–157.
69. Tang, WH, Muderawan, IW, Ong, TT, Ng, SC. 2005. A family of single-isomer positively charged cyclodextrins as chiral selectors for capillary electrophoresis: Mono-6(A)-butylammonium-6(A)-deoxy-beta-cyclodextrin tosylate. *Electrophoresis* 26: 3125–3133.
70. Tang, WH, Muderawan, IW, Ong, TT, Ng, SC. 2005. Enantioseparation of dansyl amino acids by a novel permanently positively charged single-isomer cyclodextrin: Mono-6-N-allylammonium-6-deoxy-beta-cyclodextrin chloride by capillary electrophoresis. *Anal Chim Acta* 546: 119–125.
71. Tang, WH, Muderawan, IW, Ng, SC, Chan, HSO. 2006. Enantioselective separation in capillary electrophoresis using a novel mono-6(A)-propylammonium-beta-cyclodextrin as chiral selector. *Anal Chim Acta* 555: 63–67.
72. Tang, W, Ng, SC. 2007. Synthesis of cationic single-isomer cyclodextrins for the chiral separation of amino acids and anionic pharmaceuticals. *Nat Protoc* 2: 3195–3200.
73. Cucinotta, V, Giuffrida, A, Grasso, G, Maccarrone, G, Messina, M, Vecchio, G. 2007. High selectivity in new chiral separations of dansyl amino acids by cyclodextrin derivatives in electrokinetic chromatography. *J Chromatogr A* 1155: 172–179.
74. Lin, XL, Zhu, CF, Hao, AY. 2004. Enantiomeric separations of some acidic compounds with cationic cyclodextrin by capillary electrophoresis. *Anal Chim Acta* 517: 95–101.
75. Roussel, C, Favrou, A. 1995. Cationic beta-cyclodextrin—a new versatile chiral additive for separation of drug enantiomers by high-performance liquid-chromatography. *J Chromatogr A* 704: 67–74.
76. Bunke, A, Jira, T. 1996. Chiral capillary electrophoresis using a cationic cyclodextrin. *Pharmazie* 51: 672–673.
77. Jakubetz, H, Juza, M, Schurig, V. 1997. Electrokinetic chromatography employing an anionic and a cationic beta-cyclodextrin derivative. *Electrophoresis* 18: 897–904.
78. Schulte, G, Chankvetadze, B, Blaschke, G. 1997. Enantioseparation in capillary electrophoresis using 2-hydroxypropyltrimethylammonium salt of beta-cyclodextrin as a chiral selector. *J Chromatogr A* 771: 259–266.
79. Bunke, A, Jira, T. 1998. Use of cationic cyclodextrin for enantioseparation by capillary electrophoresis. *J Chromatogr A* 798: 275–280.
80. Chankvetadze, B, Endresz, G, Schulte, G, Bergenthal, D, Blaschke, G. 1996. Capillary electrophoresis and H-1 NMR studies on chiral recognition of atropisomeric binaphthyl derivatives by cyclodextrin hosts. *J Chromatogr A* 732: 143–150.
81. Tanaka, Y, Terabe, S. 1997. Enantiomer separation of acidic racemates by capillary electrophoresis using cationic and amphoteric beta-cyclodextrins as chiral selectors. *J Chromatogr A* 781: 151–160.
82. Wang, F, Khaledi, MG. 1998. Nonaqueous capillary electrophoresis chiral separations with quaternary ammonium beta-cyclodextrin. *J Chromatogr A* 817: 121–128.
83. Wang, F, Khaledi, MG. 1998. Capillary electrophoresis chiral separations of basic compounds using cationic cyclodextrin. *Electrophoresis* 19: 2095–2100.
84. Lin, X, Zhao, M, Qi, X, Zhu, C, Hao, A. 2006. Capillary zone electrophoretic chiral discrimination using 6-O-(2-hydroxy-3-trimethylammoniopropyl)-beta-cyclodextrin as a chiral selector. *Electrophoresis* 27: 872–879.

85. Chankvetadze, B, Blaschke, G. 2001. Enantioseparations in capillary electromigration techniques: Recent developments and future trends. *J Chromatogr A* 906: 309–363.
86. Lelievre, F, Gueit, C, Gareil, P, Bahaddi, Y, Galons, H. 1997. Use of a zwitterionic cyclodextrin as a chiral agent for the separation of enantiomers by capillary electrophoresis. *Electrophoresis* 18: 891–896.
87. Fanali, S, Aturki, Z. 1995. Further study on the use of uncharged beta-cyclodextrin polymer in capillary electrophoresis—enantiomeric separation of some alpha-hydroxy acids. *Electrophoresis* 16: 1505–1509.
88. Ingelse, BA, Everaerts, FM, Sevcik, J, Stransky, Z, Fanali, S. 1995. Further study on the chiral separation power of a soluble neutral beta-cyclodextrin polymer. *J High Res Chromatogr* 18: 348–352.
89. Nishi, H, Nakamura, K, Nakai, H, Terabe, S. 1995. Enantiomeric separation of trimetoquinol, denopamine and timepidium by capillary electrophoresis and HPLC and the application of capillary electrophoresis to the optical purity testing of the drugs. *Chromatographia* 40: 638–644.
90. Ingelse, BA, Everaerts, FM, Desiderio, C, Fanali, S. 1995. Enantiomeric separation by capillary electrophoresis using a soluble neutral beta-cyclodextrin polymer. *J Chromatogr A* 709: 89–98.
91. Sevcik, J, Stransky, Z, Ingelse, BA, Lemr, K. 1996. Capillary electrophoretic enantioseparation of selegiline, methamphetamine and ephedrine using a neutral beta-cyclodextrin epichlorhydrin polymer. *J Pharm Biomed Anal* 14: 1089–1094.
92. Vespalec, R, Bocek, P. 1999. Chiral separations in capillary electrophoresis. *Electrophoresis* 20: 2579–2591.
93. Amini, A. 2001. Recent developments in chiral capillary electrophoresis and applications of this technique to pharmaceutical and biomedical analysis. *Electrophoresis* 22: 3107–3130.
94. Fenyvesi, E. 1988. Cyclodextrin polymers in the Pharmaceutical-Industry. *J Incl Phenom* 6: 537–545.
95. Aturki, Z, Fanali, S. 1994. Use of beta-cyclodextrin polymer as a chiral selector in capillary electrophoresis. *J Chromatogr A* 680: 137–146.
96. Chiari, M, Desperati, V, Cretich, M, Crini, G, Janus, L, Morcellet, M. 1999. Vinylpyrrolidine-beta-cyclodextrin copolymer: A novel chiral selector for capillary electrophoresis. *Electrophoresis* 20: 2614–2618.
97. Lurie, IS. 1997. Separation selectivity in chiral and achiral capillary electrophoresis with mixed cyclodextrins. *J Chromatogr A* 792: 297–307.
98. Fillet, M, Hubert, P, Crommen, J. 2000. Enantiomeric separations of drugs using mixtures of charged and neutral cyclodextrins. *J Chromatogr A* 875: 123–134.
99. Nhujak, T, Sastravaha, C, Palanuvej, C, Petsom, A. 2005. Chiral separation in capillary electrophoresis using dual neutral cyclodextrins: Theoretical models of electrophoretic mobility difference and separation selectivity. *Electrophoresis* 26: 3814–3823.
100. Matthijs, N, Van Hemelryck, S, Maftouh, M, Massart, DL, Vander Heyden, Y. 2004. Electrophoretic separation strategy for chiral pharmaceuticals using highly-sulfated and neutral cyclodextrins based dual selector systems. *Anal Chim Acta* 525: 247–263.
101. Szoko, E, Tabi, T, Halasz, AS, Palfi, M, Kalasz, H, Magyar, K. 2004. Chiral characterization and quantification of deprenyl-N-oxide and other deprenyl metabolites in rat urine by capillary electrophoresis. *Chromatographia* 60: S245–S251.
102. Beaufour, M, Morin, P, Ribet, JP. 2005. Chiral separation of the four stereoisomers of a novel antianginal agent using a dual cyclodextrin system in capillary electrophoresis. *J Sep Sci* 28: 529–533.
103. Chu, BL, Guo, B, Zuo, H, Wang, Z, Lin, JM. 2007. Simultaneous enantioseparation of antiparkinsonian medication rotigotine and related chiral impurities by capillary zone electrophoresis using dual cyclodextrin system. *J Pharm Biomed Anal* 56: 854–859.

104. Quan, Z, Song, Y, Feng, YZ, LeBlanc, MH, Liu, YM. 2005. Detection of D-serine in neural samples by saccharide enhanced chiral capillary electrophoresis. *Anal Chim Acta* 528: 101–106.
105. Berzas-Nevado, JJ, Villasenor-Llerena, MJ, Guiberteau-Cabanillas, C, Rodriguez-Robledo, V. 2006. Enantiomeric screening of racemic citalopram and metabolites in human urine by entangled polymer solution capillary electrophoresis: An innovatory robustness/ruggedness study. *Electrophoresis* 27: 905–917.
106. Terabe, S, Miyashita, Y, Shibata, O, Barnhart, ER, Alexander, LR, Patterson, DG, Karger, BL, Hosoya, K, Tanaka, N. 1990. Separation of highly hydrophobic compounds by cyclodextrin-modified micellar electrokinetic chromatography. *J Chromatogr A* 516: 23–31.
107. Wang, J, Warner, IM. 1995. Combined polymerized chiral micelle and gamma-cyclodextrin for chiral separation in capillary electrophoresis. *J Chromatogr A* 711: 297–304.
108. Valle, BC, Billiot, FH, Shamsi, SA, Zhu, XF, Powe, AM, Warner, IM. 2004. Combination of cyclodextrins and polymeric surfactants for chiral separations. *Electrophoresis* 25: 743–752.
109. Okafo, GN, Camilleri, P. 1993. Direct chiral resolution of amino-acid derivatives by capillary electrophoresis. *J Microcol Sep* 5: 149–153.
110. Mertzman, MD, Foley, JP. 2004. Chiral cyclodextrin-modified microemulsion electrokinetic chromatography. *Electrophoresis* 25: 1188–1200.
111. Francois, Y, Varenne, A, Juillerat, E, Villemin, D, Gareil, P. 2007. Evaluation of chiral ionic liquids as additives to cyclodextrins for enantiomeric separations by capillary electrophoresis. *J Chromatogr A* 1155: 134–141.
112. Bunke, A, Jira, T, Gübitz, G. 1995. Chiral separation of cyclodrine by means of capillary electrophoresis. *Pharmazie* 50: 570–571.
113. Jira, T, Bunke, A, Karbaum, A. 1998. Use of chiral and achiral ion-pairing reagents in combination with cyclodextrins in capillary electrophoresis. *J Chromatogr A* 798: 281–288.
114. Servais, AC, Fillet, M, Abushoffa, AM, Hubert, P, Crommen, J. 2003. Synergistic effects of ion-pairing in the enantiomeric separation of basic compounds with cyclodextrin derivatives in nonaqueous capillary electrophoresis. *Electrophoresis* 24: 363–369.
115. Servais, AC, Fillet, M, Chiap, P, Dewe, W, Hubert, P, Crommen, J. 2004. Enantiomeric separation of basic compounds using heptakis(2,3-di-O-methyl-6-O-sulfo)-beta-cyclodextrin in combination with potassium camphorsulfonate in nonaqueous capillary electrophoresis: Optimization by means of an experimental design. *Electrophoresis* 25: 2701–2710.
116. Huang, WX, Xu, H, Fazio, SD, Vivilecchia, RV. 1997. Chiral separation of primary amino compounds using a non-chiral crown ether with beta-cyclodextrin by capillary electrophoresis. *J Chromatogr B* 695: 157–162.
117. Kuhn, R. 1999. Enantiomeric separation by capillary electrophoresis using a crown ether as chiral selector. *Electrophoresis* 20: 2605–2613.
118. Verleysen, K, Sandra, P. 1999. Experimentation with chiral CE: The application of two chiral selectors and partial filling with chiral selector. *J High Res Chromatogr* 22: 33–38.
119. Salami, M, Jira, T, Otto, HH. 2005. Capillary electrophoretic separation of enantiomers of amino acids and amino acid derivatives using crown ether and cyclodextrin. *Pharmazie* 60: 181–185.
120. Ivanyi, T, Pal, K, Lazar, I, Massart, DL, Vander Heyden, Y. 2004. Application of tetraoxadiaza-crown ether derivatives as chiral selector modifiers in capillary electrophoresis. *J Chromatogr A* 1028: 325–332.
121. Elek, J, Mangelings, D, Ivanyi, T, Lazar, I, Vander Heyden, Y. 2005. Enantioselective capillary electrophoretic separation of tryptophane- and tyrosine-methylesters in a dual system with a tetra-oxadiaza-crown-ether derivative and a cyclodextrin. *J Pharm Biomed Anal* 38: 601–608.

122. Huang, WX, Xu, H, Fazio, SD, Vivilecchia, RV. 2000. Enhancement of chiral recognition by formation of a sandwiched complex in capillary electrophoresis. *J Chromatogr A* 875: 361–369.
123. Armstrong, DW, Chang, LW, Chang, SSC. 1998. Mechanism of capillary electrophoresis enantioseparations using a combination of an achiral crown ether plus cyclodextrins. *J Chromatogr A* 793: 115–134.
124. Corradini, R, Buccella, G, Galaverna, G, Dossena, A, Marchelli, R. 1999. Synthesis and chiral recognition properties of L-Ala-crown(3)-L-Ala capped beta-cyclodextrin. *Tetrahedron Lett* 40: 3025–3028.
125. Dai, R, Nie, X, Li, H, Saeed, MK, Deng, Y, Yao, G. 2007. Investigation of beta-CD-derivatized erythromycin as chiral selector in CE. *Electrophoresis* 28: 2566–2572.
126. Horimai, T, Ohara, M, Ichinose, M. 1997. Optical resolution of new quinolone drugs by capillary electrophoresis with ligand-exchange and host-guest interactions. *J Chromatogr A* 760: 235–244.
127. Cucinotta, V, Giuffrida, A, Grasso, G, Maccarrone, G, Vecchio, G. 2003. Ligand exchange chiral separations by cyclodextrin derivatives in capillary electrophoresis. *Analyst* 128: 134–136.
128. Cucinotta, V, Giuffrida, A, Maccarrone, G, Messina, M, Vecchio, G. 2006. Ligand exchange capillary electrophoresis by cyclodextrin derivatives, a powerful tool for enantiomeric separations. *Electrophoresis* 27: 1471–1480.
129. Wu, BD, Wang, QQ, Liu, Q, Xie, JW, Yun, LH. 2005. Capillary electrophoresis direct enantioseparation of aromatic amino acids based on mixed chelate-inclusion complexation of aminoethylamino-beta-cyclodextrin. *Electrophoresis* 26: 1013–1017.
130. Wu, BD, Wang, QQ, Guo, L, Shen, R, Xie, JW, Yun, LH, Zhong, BH. 2006. Amino-substituted beta-cyclodextrin copper(II) complexes for the electrophoretic enantioseparation of dansyl amino acids: Role of dual chelate-inclusion interaction and mechanism. *Anal Chim Acta* 558: 80–85.
131. Stefansson, M, Novotny, M. 1993. Electrophoretic resolution of monosaccharide enantiomers in borate oligosaccharide complexation media. *J Am Chem Soc* 115: 11573–11580.
132. Jira, T, Bunke, A, Schmid, MG, Gübitz, G. 1997. Chiral resolution of diols by capillary electrophoresis using borate-cyclodextrin complexation. *J Chromatogr A* 761: 269–275.
133. Noroski, JE, Mayo, DJ, Moran, M. 1995. Determination of the enantiomer of a cholesterol-lowering drug by cyclodextrin-modified micellar electrokinetic chromatography. *J Pharm Biomed Anal* 13: 45–52.
134. Lin, CE, Lin, SL, Fang, IJ, Liao, WS, Chen, CC. 2004. Enantioseparations of hydrobenzoin and structurally related compounds in capillary zone electrophoresis using heptakis(2,3-dihydroxy-6-O-sulfo)-beta-cyclodextrin as chiral selector and enantiomer migration reversal of hydrobenzoin with a dual cyclodextrin system in the presence of borate complexation. *Electrophoresis* 25: 2786–2794.
135. Mayer, S, Schurig, V. 1993. Enantiomer separation by electrochromatography in open tubular columns coated with Chirasil-Dex. *J Liq Chromatogr* 16: 915–931.
136. Mayer, S, Schurig, V. 1994. Enantiomer separation using mobile and immobile cyclodextrin derivatives with electromigration. *Electrophoresis* 15: 835–841.
137. Schurig, V, Jung, M, Mayer, S, Fluck, M, Negura, S, Jakubetz, H. 1995. Unified enantioselective capillary chromatography on a Chirasil-Dex stationary phase—Advantages of column miniaturization. *J Chromatogr A* 694: 119–128.
138. Lu, HJ, Ou, QY. 2002. Preparation and evaluation of open tubular electrochromatography column with derivated polysiloxane as stationary phase. *Chem J Chinese U* 23: 30–33.

139. Wistuba, D, Czesla, H, Roeder, M, Schurig, V. 1998. Enantiomer separation by pressure-supported electrochromatography using capillaries packed with a permethyl-beta-cyclodextrin stationary phase. *J Chromatogr A* 815: 183–188.
140. Wistuba, D, Schurig, V. 1999. Enantiomer separation by pressure-supported electrochromatography using capillaries packed with Chirasil-Dex polymer-coated silica. *Electrophoresis* 20: 2779–2785.
141. Lin, B, Shi, ZG, Zhang, HJ, Ng, SC, Feng, YQ. 2006. Perphenylcarbamoylated beta-cyclodextrin bonded-silica particles as chiral stationary phase for enantioseparation by pressure-assisted capillary electrochromatography. *Electrophoresis* 27: 3057–3065.
142. Kang, JW, Wistuba, D, Schurig, V. 2002. A silica monolithic column prepared by the sol-gel process for enantiomeric separation by capillary electrochromatography. *Electrophoresis* 23: 1116–1120.
143. Chen, ZL, Ozawa, H, Uchiyama, K, Hobo, T. 2003. Cyclodextrin-modified monolithic columns for resolving dansyl amino acid enantiomers and positional isomers by capillary electrochromatography. *Electrophoresis* 24: 2550–2558.
144. Wistuba, D, Schurig, V. 2000. Enantiomer separation by capillary electrochromatography on a cyclodextrin-modified monolith. *Electrophoresis* 21: 3152–3159.
145. Wistuba, D, Banspach, L, Schurig, V. 2005. Enantiomeric separation by capillary electrochromatography using monolithic capillaries with sol-gel-glued cyclodextrin-modified silica particles. *Electrophoresis* 26: 2019–2026.
146. Hjertén, S, Liao, J-L, Zhang, R. 1989. High-performance liquid chromatography on continuous polymer beds. *J Chromatogr A* 473: 273–275.
147. Koide, T, Ueno, K. 1998. Enantiomeric separations of cationic and neutral compounds by capillary electrochromatography with charged polyacrylamide gels incorporating chiral selectors. *Anal Sci* 14: 1021–1023.
148. Koide, T, Ueno, K. 1999. Enantiomeric separations of cationic and neutral compounds by capillary electrochromatography with beta-cyclodextrin-bonded charged polyacrylamide gels. *Anal Sci* 15: 791–794.
149. Koide, T, Ueno, K. 2000. Enantiomeric separations of acidic and neutral compounds by capillary electrochromatography with beta-cyclodextrin-bonded positively charged polyacrylamide gels. *J High Res Chromatogr* 23: 59–66.
150. Vegvari, A, Foldesi, A, Hetenyi, C, Kocnegarova, O, Schmid, MG, Kudirkaite, V, Hjertén, S. 2000. A new easy-to-prepare homogeneous continuous electrochromatographic bed for enantiomer recognition. *Electrophoresis* 21: 3116–3125.
151. Pumera, M, Jelinek, I, Jindrich, J, Benada, O. 2002. Beta-cyclodextrin-modified monolithic stationary phases for capillary electrochromatography and nano-HPLC chiral analysis of ephedrine and ibuprofen. *J Liq Chromatogr* 25: 2473–2484.
152. Sinner, FM, Buchmeiser, MR. 2000. Ring-opening metathesis polymerization: Access to a new class of functionalized, monolithic stationary phases for liquid chromatography. *Angew Chem Int Ed Engl* 39: 1433–1436.
153. Kornysova, O, Surna, R, Snitka, V, Pyell, U, Maruska, A. 2002. Polyrotaxane approach for synthesis of continuous beds for capillary electrochromatography. *J Chromatogr A* 971: 225–235.
154. Hutt, LD, Glavin, DP, Bada, JL, Mathies, RA. 1999. Microfabricated capillary electrophoresis amino acid chirality analyzer for extraterrestrial exploration. *Anal Chem* 71: 4000–4006.
155. Rodriguez, I, Jin, LJ, Li, SFY. 2000. High-speed chiral separations on microchip electrophoresis devices. *Electrophoresis* 21: 211–219.
156. Wang, H, Dai, ZP, Wang, L, Bai, JL, Lin, BC. 2002. Enantiomer separation of amino acids on microchip-based electrophoresis. *Chinese J Anal Chem* 30: 665–669.
157. Skelley, AM, Mathies, RA. 2003. Chiral separation of fluorescamine-labeled amino acids using microfabricated capillary electrophoresis devices for extraterrestrial exploration. *J Chromatogr A* 1021: 191–199.

158. Wallenborg, SR, Lurie, IS, Arnold, DW, Bailey, CG. 2000. On-chip chiral and achiral separation of amphetamine and related compounds labeled with 4-fluoro-7-nitrobenzofurazane. *Electrophoresis* 21: 3257–3263.
159. Schwarz, MA, Hauser, PC. 2003. Chiral on-chip separations of neurotransmitters. *Anal Chem* 75: 4691–4695.
160. Gong, XY, Hauser, PC. 2006. Enantiomeric separation of underivatized small amines in conventional and on-chip capillary electrophoresis with contactless conductivity detection. *Electrophoresis* 27: 4375–4382.
161. Liu, BF, Ozaki, M, Utsumi, Y, Hattori, T, Terabe, S. 2003. Chemiluminescence detection for a microchip capillary electrophoresis system fabricated in poly(dimethylsiloxane). *Anal Chem* 75: 36–41.
162. Ludwig, M, Kohler, F, Belder, D. 2003. High-speed chiral separations on a microchip with UV detection. *Electrophoresis* 24: 3233–3238.
163. Kitagawa, F, Aizawa, S, Otsuka, K. 2005. Rapid enantioseparation of 1-aminoindan by microchip electrophoresis with linear-imaging UV detection. *Anal Sci* 21: 61–65.
164. Belder, D, Kohler, F, Ludwig, M, Tolba, K, Piehl, N. 2006. Coating of powder-blasted channels for high-performance microchip electrophoresis. *Electrophoresis* 27: 3277–3283.
165. Piehl, N, Ludwig, M, Belder, D. 2004. Subsecond chiral separations on a microchip. *Electrophoresis* 25: 3848–3852.
166. Gao, Y, Shen, Z, Wang, H, Dai, ZP, Lin, BC. 2005. Chiral separations on multichannel microfluidic chips. *Electrophoresis* 26: 4774–4779.
167. Belder, D, Ludwig, M, Wang, LW, Reetz, MT. 2006. Enantioselective catalysis and analysis on a chip. *Angew Chem Int Ed Engl* 45: 2463–2466.
168. Olvecka, E, Masar, M, Kaniansky, D, Johnck, M, Stanislawski, B. 2001. Isotachophoresis separations of enantiomers on a planar chip with coupled separation channels. *Electrophoresis* 22: 3347–3353.
169. Zeng, HL, Li, HF, Wang, X, Lin, JM. 2006. Development of a gel monolithic column polydimethylsiloxane microfluidic device for rapid electrophoresis separation. *Talanta* 69: 226–231.
170. Belder, D, Ludwig, M. 2003. Microchip electrophoresis for chiral separations. *Electrophoresis* 24: 2422–2430.
171. Mangelings, D, Vander Heyden, Y. 2007. High-throughput screening and optimization approaches for chiral compounds by means of microfluidic devices. *Comb Chem High Throughput Screen* 10: 317–325.
172. Wang, F, Khaledi, MG. 2000. Enantiomeric separations by nonaqueous capillary electrophoresis. *J Chromatogr A* 875: 277–293.
173. Lammerhofer, M. 2005. Chiral separations by capillary electromigration techniques in nonaqueous media I. Enantioselective nonaqueous capillary electrophoresis. *J Chromatogr A* 1068: 3–30.
174. Valko, IE, Siren, H, Riekkola, ML. 1996. Chiral separation of dansyl amino acids by capillary electrophoresis: Comparison of formamide and N-methylformamide as background electrolytes. *Chromatographia* 43: 242–246.
175. Wang, F, Khaledi, MG. 1996. Chiral separations by nonaqueous capillary electrophoresis. *Anal Chem* 68: 3460–3467.
176. Wang, F, Khaledi, MG. 1999. Non-aqueous capillary electrophoresis chiral separations with sulfated beta-cyclodextrin. *J Chromatogr B* 731: 187–197.
177. Vincent, JB, Vigh, G. 1998. Nonaqueous capillary electrophoretic separation of enantiomers using the single-isomer heptakis(2,3-diacetyl-6-sulfato)-beta-cyclodextrin as chiral resolving agent. *J Chromatogr A* 816: 233–241.
178. Karbaum, A, Jira, T. 1999. Nonaqueous capillary electrophoresis: Application possibilities and suitability of various solvents for the separation of basic analytes. *Electrophoresis* 20: 3396–3401.

179. Mol, R, Servais, AC, Fillet, M, Crommen, J, de Jong, GJ, Somsen, GW. 2007. Nonaqueous electrokinetic chromatography-electrospray ionization mass spectrometry using anionic cyclodextrins. *J Chromatogr A* 1159: 51–57.
180. Mol, R, de Jong, GJ, Somsen, GW. 2008. Coupling of non-aqueous electrokinetic chromatography using cationic cyclodextrins with electrospray ionization mass spectrometry. *Rapid Commun Mass Spectrom* 22: 790–796.
181. Chankvetadze, B, Schulte, G, Blaschke, G. 1997. Nature and design of enantiomer migration order in chiral capillary electrophoresis. *Enantiomer* 2: 157–179.
182. Servais, AC, Fillet, M, Chiap, P, Dewe, W, Hubert, P, Crommen, J. 2005. Influence of the nature of the electrolyte on the chiral separation of basic compounds in nonaqueous capillary electrophoresis using heptakis(2,3-di-O-methyl-6-O-sulfo)-beta-cyclodextrin. *J Chromatogr A* 1068: 143–150.
183. Na, N, Hu, YP, Ouyang, J, Baeyens, WRG, Delanghe, JR, De Beer, T. 2004. Use of polystyrene nanoparticles to enhance enantiomeric separation of propranolol by capillary electrophoresis with HP-beta-CD as chiral selector. *Anal Chim Acta* 527: 139–147.
184. Kubacak, P, Mikus, P, Valaskova, I, Havranek, E. 2007. Chiral separation of alkylamine antihistamines in pharmaceuticals by capillary isotachophoresis with charged cyclodextrin. *Drug Dev Ind Pharm* 33: 1199–1204.
185. Dankova, M, Kaniansky, D, Fanali, S, Ivanyi, F. 1999. Capillary zone electrophoresis separations of enantiomers present in complex ionic matrices with on-line isotachophoretic sample pretreatment. *J Chromatogr A* 838: 31–43.
186. Marak, J, Mikus, P, Marakova, K, Kaniansky, D, Valaskova, I, Havranek, E. 2008. Potentialities of ITP-CZE method with diode array detection for enantiomeric purity control of dexbrompheniramine in pharmaceuticals. *J Pharm Biomed Anal* 46: 870–876.
187. Mikus, P, Marakova, K, Marak, J, Kaniansky, D, Valaskova, I, Havranek, E. 2008. Possibilities of column coupling electrophoresis provided with a fiber-based diode array detection in enantioselective analysis of drugs in pharmaceutical and clinical samples. *J Chromatogr A* 1179: 9–16.
188. Zhong, WW, Yeung, ES. 2002. Combinatorial enantiomeric separation of diverse compounds using capillary array electrophoresis. *Electrophoresis* 23: 2996–3005.
189. Male, KB, Luong, JHT. 2003. Chiral analysis of neurotransmitters using cyclodextrin-modified capillary electrophoresis equipped with microfabricated interdigitated electrodes. *J Chromatogr A* 1003: 167–178.
190. Henley, TH, Wilburn, RT, Crouch, AM, Jorgenson, JW. 2005. Flow counterbalanced capillary electrophoresis using packed capillary columns: Resolution of enantiomers and isotopomers. *Anal Chem* 77: 7024–7031.
191. Lee, KP, Gopalan, AI, Lee, SH, Kim, MS. 2006. Polyaniline and cyclodextrin based chiral nanobundles—functional materials having size and enantioselectivity. *Nanotechnology* 17: 375–380.
192. Sazelova, P, Kasicka, V, Koval, D, Prusik, Z, Fanali, S, Aturki, Z. 2007. Control of EOF in CE by different ways of application of radial electric field. *Electrophoresis* 28: 756–766.
193. Kirschner, DL, Jaramillo, M, Green, TK. 2007. Enantioseparation and stacking of Cyanobenz[f]isoindole-amino acids by reverse polarity capillary electrophoresis and sulfated beta-cyclodextrin. *Anal Chem* 79: 736–743.
194. Sanger-van de Griend, CE, Ek, AG, Widahl-Nasman, ME, Andersson, EK. 2006. Method development for the enantiomeric purity determination of low concentrations of adrenaline in local anaesthetic solutions by capillary electrophoresis. *J Pharm Biomed Anal* 41: 77–83.
195. Musenga, A, Mandrioli, R, Morganti, E, Fanali, S, Raggi, MA. 2008. Enantioselective analysis of amisulpride in pharmaceutical formulations by means of capillary electrophoresis. *J Pharm Biomed Anal* 46: 966–970.

196. Lee, WS, Chan, MF, Tam, WM, Hung, MY. 2007. The application of capillary electrophoresis for enantiomeric separation of N,N-dimethylamphetamine and its related analogs: intelligence study on N,N-dimethylamphetamine samples in crystalline and tablet forms. *Forensic Sci Int* 165: 71–77.
197. Du, GH, Zhang, SZ, Xie, JW, Zhong, BH, Liu, KL. 2005. Chiral separation of anticholinergic drug enantiomers in nonaqueous capillary electrophoresis. *J Chromatogr A* 1074: 195–200.
198. Matthijs, N, Vander Heyden, Y. 2006. Enantiomeric impurity determination in capillary electrophoresis using a highly-sulfated cyclodextrins-based method. *Biomed Chromatogr* 20: 696–709.
199. Castro-Puyana, M, Crego, AL, Marina, ML, Garcia-Ruiz, C. 2007. Enantioselective separation of azole compounds by EKC. Reversal of migration order of enantiomers with CD concentration. *Electrophoresis* 28: 2667–2674.
200. Vaccher, MP, Lipka, E, Bonte, JP, Foulon, C, Goossens, JF, Vaccher, C. 2005. Enantiomeric analysis of baclofen analogs by capillary zone electrophoresis, using highly sulfated cyclodextrins: Inclusion ionization constant pK(a) determination. *Electrophoresis* 26: 2974–2983.
201. Kavran-Belin, G, Rudaz, S, Veuthey, JL. 2005. Enantioseparation of baclofen with highly sulfated beta-cyclodextrin by capillary electrophoresis with laser-induced fluorescence detection. *J Sep Sci* 28: 2187–2192.
202. Long, PH, Trung, TQ, Oh, JW, Kim, KH. 2006. Chiral purity test of bevantolol by capillary electrophoresis and high performance liquid chromatography. *Arch Pharm Res* 29: 808–813.
203. Wei, S, Guo, H, Lin, JM. 2006. Chiral separation of salbutamol and bupivacaine by capillary electrophoresis using dual neutral cyclodextrins as selectors and its application to pharmaceutical preparations and rat blood samples assay. *J Chromatogr B* 832: 90–96.
204. Kvasnicka, F, Biba, B, Cvak, L, Kratka, J, Voldrich, M. 2005. Separation of enantiomers of butorphanol and cycloamine by capillary zone electrophoresis. *J Chromatogr A* 1081: 87–91.
205. Denola, NL, Quiming, NS, Catabay, AP, Saito, Y, Jinno, K. 2006. Optimization of capillary electrophoretic enantioseparation for basic drugs with native beta-CD as a chiral selector. *Electrophoresis* 27: 2367–2375.
206. Gagyi, L, Gyeresi, A, Kilar, F. 2008. Role of chemical structure in stereoselective recognition of beta-blockers by cyclodextrins in capillary zone electrophoresis. *J Biochem Biophys Methods* 70: 1268–1275.
207. Denola, NL, Quiming, NS, Saito, Y, Jinno, K. 2007. Simultaneous enantioseparation and sensitivity enhancement of basic drugs using large-volume sample stacking. *Electrophoresis* 28: 3542–3552.
208. Wei, SL, Song, GQ, Lin, JM. 2005. Separation and determination of norepinephrine, epinephrine and isoprinaline enantiomers by capillary electrophoresis in pharmaceutical formulation and human serum. *J Chromatogr A* 1098: 166–171.
209. Lin, CE, Cheng, HT, Fang, IJ, Liu, YC, Kuo, CM, Lin, WY, Lin, CH. 2006. Strategies for enantioseparations of catecholamines and structurally related compounds by capillary zone electrophoresis using sulfated beta-cyclodextrins as chiral selectors. *Electrophoresis* 27: 3443–3451.
210. Liu, H, Yu, A, Liu, F, Shi, Y, Han, L, Chen, Y. 2006. Chiral separation of cefadroxil by capillary electrochromatography. *J Pharm Biomed Anal* 41: 1376–1379.
211. Van Eeckhaut, A, Michotte, Y. 2006. Chiral separation of cetirizine by capillary electrophoresis. *Electrophoresis* 27: 2376–2385.
212. Mikus, P, Valaskova, I, Havranek, E. 2005. Enantioselective analysis of cetirizine in pharmaceuticals by cyclodextrin-mediated capillary electrophoresis. *J Sep Sci* 28: 1278–1284.

213. Chou, YW, Huang, WS, Ko, CC, Chen, SH. 2008. Enantioseparation of cetirizine by sulfated-beta-cyclodextrin-mediated capillary electrophoresis. *J Sep Sci* 31: 845–852.
214. Tsimachidis, D, Cesla, P, Hajek, T, Theodoridis, G, Jandera, P. 2008. Capillary electrophoretic chiral separation of Cinchona alkaloids using a cyclodextrin selector. *J Sep Sci* 31: 1130–1136.
215. Nevado, JJB, Cabanillas, CG, Llerena, MJV, Robledo, VR. 2005. Enantiomeric determination, validation and robustness studies of racemic citalopram in pharmaceutical formulations by capillary electrophoresis. *J Chromatogr A* 1072: 249–257.
216. Zhao, SL, Song, YR, Liu, YM. 2005. A novel capillary electrophoresis method for the determination of D-serine in neural samples. *Talanta* 67: 212–216.
217. Kubacak, P, Mikus, P, Valaskova, I, Havranek, E. 2006. Separation of dimetinden enantiomers in drugs by means of capillary isotachophoresis. *Ceska Slov Farm* 55: 32–35.
218. Lu, YH, Zhang, M, Meng, Q, Zhang, ZX. 2006. Separation and determination of donepezil hydrochloride enantiomers in plasma by capillary electrophoresis. *Yao Xue Xue Bao* 41: 471–475.
219. Cheng, Y, Yang, X, Yu, Y. 2007. Chiral separation of doxazosin and its intermediate by capillary electrophoresis using tetramethylammonium hydroxide to control electroosmotic flow. *Se Pu* 25: 478–481.
220. Zhao, Y, Yang, XB, Jiang, R, Sun, XL, Li, XY, Liu, WM, Zhang, SY. 2006. Analysis of optical purity and impurity of synthetic D-phenylalanine products using sulfated beta-cyclodextrin as chiral selector by reversed-polarity capillary electrophoresis. *Chirality* 18: 84–90.
221. Zhao, SL, Yuan, HY, Xiao, D. 2005. Detection of D-serine in rat brain by capillary electrophoresis with laser induced fluorescence detection. *J Chromatogr B* 822: 334–338.
222. Liu, LB, Zheng, ZX, Lin, JM. 2005. Application of dimethyl-beta-cyclodextrin as a chiral selector in capillary electrophoresis for enantiomer separation of ephedrine and related compounds in some drugs. *Biomed Chromatogr* 19: 447–453.
223. de Pablos, RR, Garcia-Ruiz, C, Crego, AL, Marina, ML. 2005. Separation of etodolac enantiomers by capillary electrophoresis. Validation and application of the chiral method to the analysis of commercial formulations. *Electrophoresis* 26: 1106–1113.
224. Hammitzsch, M, Rao, RN, Scriba, GK. 2006. Development and validation of a robust capillary electrophoresis method for impurity profiling of etomidate including the determination of chiral purity using a dual cyclodextrin system. *Electrophoresis* 27: 4334–4344.
225. Zhou, S, Ouyang, J, Baeyens, WR, Zhao, H, Yang, Y. 2006. Chiral separation of four fluoroquinolone compounds using capillary electrophoresis with hydroxypropyl-beta-cyclodextrin as chiral selector. *J Chromatogr A* 1130: 296–301.
226. Khan, M, Viswanathan, B, Rao, DS, Reddy, GS. 2006. A validated chiral CE method for frovatriptan, using cyclodextrin as chiral selector. *J Pharm Biomed Anal* 41: 1447–1452.
227. Ammazzalorso, A, Amoroso, R, Bettoni, G, Chiarini, M, De Filippis, B, Fantacuzzi, M, Giampietro, L, Maccallini, C, Tricca, ML. 2005. Enantiomeric separation of gemfibrozil chiral analogues by capillary electrophoresis with heptakis(2,3,6-tri-O-methyl)-beta-cyclodextrin as chiral selector. *J Chromatogr A* 1088: 110–120.
228. de Oliveira, AR, Cardoso, CD, Bonato, PS. 2007. Stereoselective determination of hydroxychloroquine and its metabolites in human urine by liquid-phase microextraction and CE. *Electrophoresis* 28: 1081–1091.
229. Cardoso, CD, Jabor, VAP, Bonato, PS. 2006. Capillary electrophoretic chiral separation of hydroxychloroquine and its metabolites in the microsomal fraction of liver homogenates. *Electrophoresis* 27: 1248–1254.
230. Glowka, FK, Karazniewicz, M. 2005. High performance capillary electrophoresis method for determination of ibuprofen enantiomers in human serum and urine. *Anal Chim Acta* 540: 95–102.

231. Glowka, F, Karazniewicz, M. 2007. Enantioselective CE method for pharmacokinetic studies on ibuprofen and its chiral metabolites with reference to genetic polymorphism. *Electrophoresis* 28: 2726–2737.
232. Yi, F, Guo, B, Peng, Z, Li, H, Marriott, P, Lin, JM. 2007. Study of the enantioseparation of imazaquin and enantioselective degradation in field soils by CZE. *Electrophoresis* 28: 2710–2716.
233. Castro-Puyana, M, Crego, AL, Marina, ML. 2006. Separation and quantitation of the four stereoisomers of itraconazole in pharmaceutical formulations by electrokinetic chromatography. *Electrophoresis* 27: 887–895.
234. Theurillat, R, Knobloch, M, Schmitz, A, Lassahn, PG, Mevissen, M, Thormann, W. 2007. Enantioselective analysis of ketamine and its metabolites in equine plasma and urine by CE with multiple isomer sulfated beta-CD. *Electrophoresis* 28: 2748–2757.
235. Tan, L, Wang, Y, Liu, XQ, Ju, HX, Li, H. 2005. Simultaneous determination of L- and D-lactic acid in plasma by capillary electrophoresis. *J Chromatogr B* 814: 393–398.
236. Baldacci, A, Thormann, W. 2006. Analysis of lorazepam and its 30-glucuronide in human urine by capillary electrophoresis: Evidence for the formation of two distinct diastereoisomeric glucuronides. *J Sep Sci* 29: 153–163.
237. Santilio, A, D'Amato, M, Cataldi, L, Sorbo, A, Dommarco, R. 2006. Chiral separation of metalaxyl by capillary zone electrophoresis using cyclodextrins. *J Capill Electrophor Microchip Technol* 9: 79–84.
238. Martins, LF, Yegles, M, Wennig, R. 2008. Simultaneous enantioselective quantification of methadone and of 2-ethylidene-1,5-dimethyl-3,3-diphenyl-pyrrolidine in oral fluid using capillary electrophoresis. *J Chromatogr B* 862: 79–85.
239. Iio, R, Chinaka, S, Takayama, N, Hayakawa, K. 2005. Simultaneous chiral analysis of methamphetamine and related compounds by capillary electrophoresis/mass spectrometry using anionic cyclodextrin. *Anal Sci* 21: 15–19.
240. Hefnawy, MM. 2005. Optimization of the chiral resolution of metyrosine by capillary electrophoresis and/or micellar electrokinetic capillary chromatography. *J Liq Chromatogr* 28: 439–452.
241. Cruz, LA, Hall, R. 2005. Enantiomeric purity assay of moxifloxacin hydrochloride by capillary electrophoresis. *J Pharm Biomed Anal* 38: 8–13.
242. Fradi, I, Servais, AC, Pedrini, M, Chiap, P, Ivanyi, R, Crommen, J, Fillet, M. 2006. Enantiomeric separation of acidic compounds using single-isomer amino cyclodextrin derivatives in nonaqueous capillary electrophoresis. *Electrophoresis* 27: 3434–3442.
243. Nevado, JJB, Penalvo, GC, Dorado, RMR. 2005. Method development and validation for the separation and determination of omeprazole enantiomers in pharmaceutical preparations by capillary electrophoresis. *Anal Chim Acta* 533: 127–133.
244. Olsson, J, Stegander, F, Marlin, N, Wan, H, Blomberg, LG. 2006. Enantiomeric separation of omeprazole and its metabolite 5-hydroxyomeprazole using non-aqueous capillary electrophoresis. *J Chromatogr A* 1129: 291–295.
245. Mikus, P, Valaskova, I, Havranek, E. 2005. Enantioselective determination of pheniramine in pharmaceuticals by capillary electrophoresis with charged cyclodextrin. *J Pharm Biomed Anal* 38: 442–448.
246. Wang, W, Chen, YY, Yang, WP, Zhang, ZJ. 2005. Chiral separation and laser-induced fluorescence detection of propranolol enantiomers by capillary electrophoresis. *Chinese J Anal Chem* 33: 1113–1115.
247. Lucangioli, SE, Tripodi, V, Masrian, E, Scioscia, SL, Carducci, CN, Kenndler, E. 2005. Enantioselective separation of rivastigmine by capillary electrophoresis with cyclodextrines. *J Chromatogr A* 1081: 31–35.
248. Servais, AC, Fillet, M, Mol, R, Somsen, GW, Chiap, P, de Jong, GJ, Crommen, J. 2006. On-line coupling of cyclodextrin mediated nonaqueous capillary electrophoresis to mass spectrometry for the determination of salbutamol enantiomers in urine. *J Pharm Biomed Anal* 40: 752–757.

249. Yuan, P, Lin, L, Fan, Q, Zeng, L. 2006. Separation of sitafloxacin epimers by capillary electrophoresis. *Se Pu* 24: 513–515.
250. Maier, V, Horakova, J, Petr, J, Tesarova, E, Coufal, P, Sevcik, J. 2005. Chiral separation of tamsulosin by capillary electrophoresis. *J Pharm Biomed Anal* 39: 691–696.
251. Marini, RD, Rozet, E, Vander Heyden, Y, Ziemons, E, Boulanger, B, Bouklouze, A, Servais, AC, Fillet, M, Crommen, J, Hubert, P. 2007. Robustness testing of a chiral NACE method for R-timolol determination in S-timolol maleate and uncertainty assessment from quantitative data. *J Pharm Biomed Anal* 44: 640–651.
252. Bitar, Y, Holzgrabe, U. 2007. Enantioseparation of chiral tropa alkaloids by means of cyclodextrin-modified microemulsion electrokinetic chromatography. *Electrophoresis* 28: 2693–2700.

4 Factors Influencing Cyclodextrin-Mediated Chiral Separations

Anne-Catherine Servais, Jacques Crommen, and Marianne Fillet

CONTENTS

4.1 Type of CD 87
4.2 Influence of the CD Concentration 91
4.3 BGE pH and Ionic Strength 93
4.4 Addition of Organic Solvents 95
4.5 Injection Mode 96
4.6 Temperature 96
4.7 Dual Systems 97
4.7.1 Achiral Additive-CD Dual Systems 97
4.7.2 Dual CD Systems 98
4.7.2.1 General Considerations 98
4.7.2.2 Recent Applications of Dual CD Systems 100
Acknowledgments 103
References 104

In this chapter, several factors likely to influence cyclodextrin (CD)-mediated enantioseparations, such as CD type and concentration, pH and ionic strength of the background electrolyte (BGE), addition of organic solvents, injection mode as well as temperature are discussed from a practical point of view and illustrated by examples. Finally, several types of dual systems including at least one CD are also presented.

4.1 TYPE OF CD

Native CDs and their derivatives have been extensively used for the enantioseparation of chiral compounds. Although the set of CDs is rather large, new derivatives are continuously developed, as demonstrated in Table 4.1 which presents the CD derivatives recently synthesized and applied for capillary electrophoresis (CE) enantioseparations.

TABLE 4.1
Applications of Recently Synthesized CD Derivatives

CD Derivative	BGE	Analytes	Ref.
2,3-Dihydroxypropyl-β-CD; 3-Hydroxypropyl-β-CD	100 mM Tris-phosphate buffer pH 2.5	Basic chiral drugs	[40]
Chloride salt of mono-(3-methyl-imidazolium)-β-CD	50 mM NaH_2PO_4 adjusted to pH 6–9.6 with NaOH or H_3PO_4	Dansyl amino acids	[41]
Chloride salts of 6-mono(3-ethylimidazolium)-β-CD, 6-mono(3-propylimidazolium)-β-CD and 6-mono(3-hexylimidazolium)-β-CD	50 mM Acetic acid adjusted to pH 5.0 and 6.0 with NaOH	Dansyl amino acids	[42]
Dichloride salt of mono-6-deoxy-6-*N*,*N*,*N*′,*N*′,*N*′-pentamethylethylenediammonio-cyclomaltoheptaose	25 mM Phosphoric acid solution titrated to pH 2.5 with LiOH and 25 mM ethanolamine solution titrated to pH 9.3	Anionic, weak acid, and nonionic analytes	[43]
δ-CD	Phosphate/borate buffer (25 mM, pH 9 or 100 mM, pH 2.5)	Negatively and positively charged drugs	[44]
6-*O*-(2-hydroxy-3-trimethylammoniopropyl)-β-CD	50 mM Phosphate buffer (pH 4–6)	Acidic compounds	[30]
Chloride salt of mono-6^A-propylammonium-6^A-deoxy-β-CD	50 mM Phosphate buffer (pH 6.5)	Hydroxyl, carboxylic acids, and amphoteric analytes	[45]
2-*O*-(2-hydroxybutyl)-β-CD	50 mM Tris-phosphate buffer (pH 2.5)	Anisodamine, ketoconazole, propranolol, promethazine, adrenaline, and chlorphenamine	[46]
	50 mM Phosphate buffer (pH 2.5)	Anisodamine, promethazine, adrenaline, verapamil, and salbutamol	[47]

Chloride salt of mono-6-*N*-allylammonium-6-deoxy-β-CD	20 mM Acetate buffer (5.0–6.0) and 20 mM phosphate buffer (6.5–9.3)	Amino acids and dansyl amino acids	[48]
Tosylate and chloride salts of mono-6-deoxy-6-(3-methylimidazolium)-β-CD, mono-6-deoxy-6-(3-butyl-imidazolium)-β-CD, mono-6-deoxy-6-(3-decyl-imidazolium)-β-CD, and mono-6-deoxy-6-(1,2-dimethyl-imidazolium)-β-CD	Phosphate buffer titrated to the required pH (6.5–9.6) with H_3PO_4 or acetate buffer; titrated to the required pH (4.0–6.0) with acetic acid	Dansyl amino acids	[49]
Sodium salt of heptakis(2-*O*-methyl-6-*O*-sulfo)-cyclomaltoheptaose	Aqueous BGE: 25 mM H_3PO_4 solution titrated to pH 2.5 with LiOH Methanolic BGE: 25 mM H_3PO_4 and 12.5 mM NaOH	Weak base analytes	[50]
Tosylate salt of mono-6^A-butylammonium-6^A-deoxy-β-CD	50 mM Phosphate buffer	α-Hydroxy acids, carboxylic acids, and ampholytic analytes	[51]
Sodium salt of heptakis(2-*O*-methyl-3-*O*-acetyl-6-*O*-sulfo)cyclomaltoheptaose	Aqueous BGE: 25 mM H_3PO_4 solution titrated to pH 2.5 with LiOH Methanolic BGE: 25 mM H_3PO_4 and 12.5 mM NaOH	Weak base analytes	[52]
Chloride salt of mono-6^A-pentylammonium-6^A-deoxy-β-CD	50 mM Phosphate buffer (pH 5–9)	Anionic and ampholytic analytes	[53]
6-*O*-succinil-β-CD	10 mM Acetate buffer (pH 5.6)	Norepinephrine, epinephrine, terbutaline, and norphenilephrine	[54]

Most of the substituted CDs are mixtures of several isomers which differ in the degree of substitution or in the position of the substituents. When such a mixture is used, the chiral recognition mechanism can hardly be studied. Moreover, problems related to batch-to-batch variability may occur, as recently illustrated by Theurillat et al. [1]. In this paper, the enantiomers of ketamine and norketamine were separated using a BGE made up of 50 mM Tris-phosphate (pH 2.5) and 10 mg/mL of sulfated β-CD (S-β-CD) in the reversed polarity mode. The sodium salt of S-β-CD used was from Aldrich and consists of a mixture of β-CD with 7–11 mol sulfate/mol β-CD, characterized by an optical rotation ranging from +45° to +85°. Figure 4.1A through C illustrates the enantioseparation obtained with three different batches of this S-β-CD. From these figures, it can be concluded that for this randomly substituted CD, the resolution of ketamine enantiomers and to a minor extent that of norketamine enantiomers strongly differ from one batch to another. HS-β-CD from Beckman Coulter, with an average of 12.2 sulfate groups per molecule, was also tested but was not able to separate the enantiomers of ketamine (cf. Figure 4.1D).

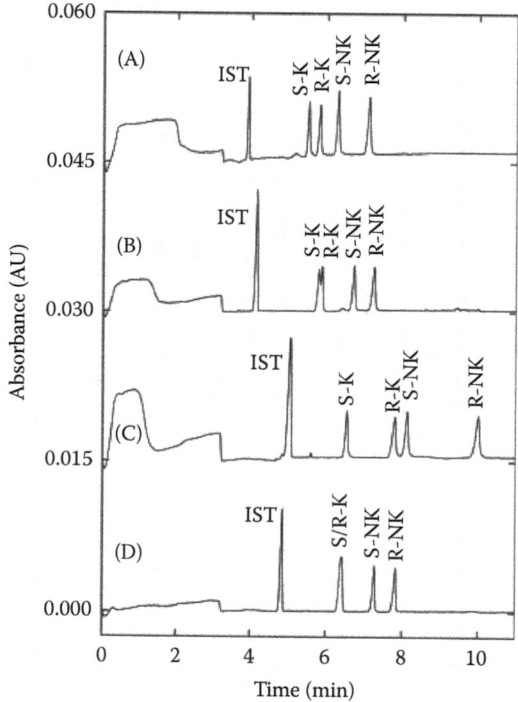

FIGURE 4.1 Influence of the batch and the number of sulfate groups per molecule of sulfated β-CD on the enantioresolution of ketamine and norketamine (about 0.5 μg/mL of each enantiomer). (A) Aldrich, batch 07222HO, (B) Aldrich, batch 13307MA, (C) Aldrich, batch 13112JD, (D) HS-β-CD from Beckman Coulter. CD concentration: 10 mg/mL. Peak identification: *S*-K, *S*-ketamine; *R*-K, *R*-ketamine; *S*-NK, *S*-norketamine; *R*-NK, *R*-norketamine; internal standard (IST), lamotrigine added as the IST. (Reproduced from Theurillat, R. et al., *Electrophoresis*, 28, 2748, 2007. With permission.)

In order to overcome these reproducibility problems and to investigate aspects related to the chiral recognition mechanism, well-defined single-isomer charged CD derivatives were developed and used in a great number of enantiomeric separations. It is worth noting that not only single-isomer anionic CDs derivatives were developed, but also single-isomer cationic CDs [2].

4.2 INFLUENCE OF THE CD CONCENTRATION

In the excellent review written by Rizzi, the theory of direct enantioseparation in CE is extensively developed [3]. Moreover, the equations required for the optimization of selectivity, resolution, and analysis time in capillary zone electrophoresis (CZE) are listed.

The influence of the CD concentration on the enantiomers' migration depends on the charge of both the selector and the analyte. The mobilities of neutral (or zwitterionic) analytes increase with the charged CD concentration. In the case of a neutral selector, the charged analytes are decelerated. When analytes and CD are charged, an increase in the mobilities is observed, if both species carry the same charge but a decrease or even a reversal of the migration order occurs if both exhibit opposite charges.

Different mathematical models such as the mobility difference model [4] and the charged resolving agent migration model (CHARM) [5] were developed to describe the mobilities and mobility differences as a function of experimental parameters such as selector concentration (mobility difference model and CHARM) and BGE pH (CHARM).

As the mobilities of the free enantiomers are equal ($\mu_R^{free} = \mu_S^{free} = \mu^{free}$) and assuming $\mu_R^{cplx} = \mu_S^{cplx}$, the optimal CD concentration with respect to selectivity is given by Equation 4.1 [3]

$$[CD]_\alpha^{opt} = \frac{1}{\sqrt{K_R K_S}} \sqrt{\frac{\mu^{free}}{\mu^{cplx}}} \tag{4.1}$$

K_R and K_S are the complexation constants.

The optimal CD concentration will be affected by a buffer pH value close to the pK_a of the weakly acidic or basic groups of the analyte or the selector. Moreover, if an organic solvent or any other additive is added to the BGE and if this addition has an impact on the binding strength, the optimal CD concentration will also be modified.

The CHARM model, which is described in detail in Ref. [5], is especially suited in the prediction of enantioselectivity and enantioresolution for neutral and ionized analytes when a charged chiral selector is used. Besides the effect of the concentration and charge of the chiral selector, the influence of the buffer pH, the charge of the analyte, and the electroosmotic flow (EOF) is also taken into account. Among other things, this model clearly shows that charged chiral selectors, which have the unique advantage to be useful in the separation of both uncharged and ionized analyte enantiomers, always lead in principle to higher resolution values and faster separations than neutral selectors with similar chiral discrimination ability. The model can be considerably simplified if the EOF is suppressed (e.g., in neutral-coated capillaries) and the selector (e.g., a CD derivative) has strongly acidic or basic functional groups so that it is fully charged in the whole pH range considered. Under these conditions, the BGE pH will only affect the charge of the analyte, which makes method development much easier. For neutral and permanently charged analytes, resolution optimization is very simple since it depends essentially on the charged chiral selector concentration. For analytes which are weak electrolytes, it can be deduced from the model that efficient method development requires the use of only two BGEs: one at low pH (e.g., pH 2.2) and one at high pH (e.g., pH 9.5). Resolution will be high at low pH for the "desionoselective" separation (enantioselective only for the analyte in uncharged form) of weak acids and the "ionoselective" separation (enantioselective only for the analyte in charged form) of weak bases while resolution will be high at high pH for the ionoselective separation of weak acids and the desionoselective separation of weak bases. On the other hand, resolution will be high both at low and high pH for the "duoselective" separation (enantioselective for both charged and uncharged forms of the analyte) of weak acids and bases. It is also worth noting that for chiral selectors with opposite charge to the analytes (i.e., the most common case), either cathodic or anodic migration can occur depending on the magnitude of the complexation constants.

The CHARM model was applied by Michaslka et al., in order to separate *S*-linezolid, the first available oxazolidinone antibacterial agent, from its enantiomeric impurity [6]. The studied analyte was found to be protonated under pH 3 and uncharged above pH 4, while both forms are present at pH 4. The resolution of the enantiomers was optimized according to the CHARM model for neutral and monoprotic analytes [5]. Therefore, two BGE pH values (low and high) were investigated while varying the concentration of the charged resolving agent, namely heptakis(2,3-di-*O*-acetyl-6-*O*-sulfo)-β-CD (HDAS-β-CD) in the present study. At pH 9 (50 mM borate buffer), the best enantioseparation was observed using 27.5 mM HDAS-β-CD under normal polarity mode in a bare-fused silica capillary (R_s value: 3.7). It is worth noting that, under these conditions, the enantiomeric impurity migrates after the *S*-enantiomer. In such a case, the quantification of the impurity may be hampered by the first-migrating large peak of the main isomer. Therefore, several strategies were investigated in order to reverse the enantiomer migration order, such as variations of the pH and the CD concentration, the use of a dual selector system as well as the reversal of the EOF. It was necessary to use a coated capillary (with linear polyacrylamide in which the EOF was suppressed) and a reversed polarity mode in order to reverse the migration order. Under these conditions, the use of 18.75 mM HDAS-β-CD in a 50 mM borate buffer pH 8 led to a high enantioresolution of linezolid (R_s value close to 7). At a low BGE pH (50 mM phosphate buffer, pH 2.2) and using a very low HDAS-β-CD concentration (1 mM), the linezolid enantiomers were also completely resolved in an uncoated capillary but under these conditions, the migration times were very high (about 75 min) and further strongly increased with the CD concentration. Finally, the method developed in the basic pH conditions was validated with respect to sensitivity, linearity, precision, and accuracy.

As previously mentioned, several approaches can be applied in order to reverse the enantiomer migration order. Very interesting results were obtained

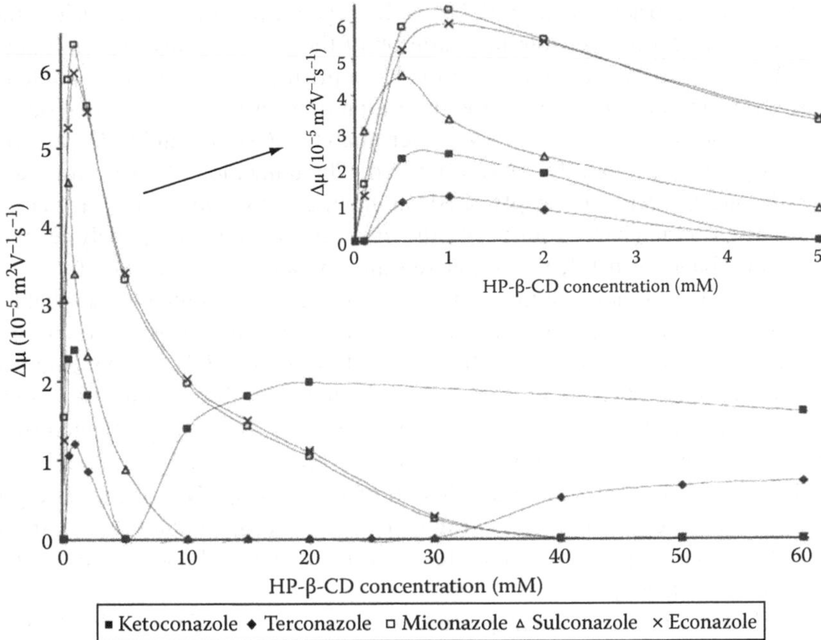

FIGURE 4.2 Influence of the HP-β-CD concentration on the mobility differences between the enantiomers of ketoconazole, terconazole, miconazole, sulconazole, and econazole. (Reproduced from Castro-Puyana, M. et al., *Electrophoresis*, 28, 2667, 2007. With permission.)

by Castro-Puyana et al., in a paper dealing with the chiral separation of six azole compounds using a neutral CD (β-CD, HP-β-CD, or heptakis-(2,3,6-tri-*O*-methyl)-β-CD) [7]. A reversal of the migration order of ketoconazole and terconazole enantiomers with HP-β-CD concentration was observed compared to the other CDs. As can be seen in Figure 4.2, the mobility difference of both compounds enantiomers reached a maximum at low HP-β-CD concentrations, decreased down to zero at 5 mM CD concentration, and then increased again at concentrations higher than 5 mM for ketoconazole and 30 mM for terconazole. Apparent binding constants for the azole enantiomers with the three investigated CDs were also determined.

4.3 BGE pH AND IONIC STRENGTH

In the case of ionizable CD derivatives, such as carboxymethyl- and carboxyethyl-β-CDs, it is obvious that the buffer pH influences the binding strength and the complex mobility. As the pH determines the effective net charge and the effective mobility of the complexes, it will affect the migration times, the shape, and the efficiency of the analytes' peaks, like in achiral CE systems [5].

Kirschner et al. developed a CE method with laser-induced fluorescence detection to resolve and determine the enantiomers of cyanobenz[*f*]isoindole (CBI) derivatives of amino acids using S-β-CD in the negative polarity mode [8]. The most critical factor affecting the enantioseparation of the studied analytes was found to be the BGE pH. As can be seen in Figure 4.3, a pH increase led to a complete loss of CBI-ser enantiomers resolution, the migration times were not strongly modified. A similar trend was observed for two other analytes (CBI-ala and CBI-glu) while for CBI-arg, the opposite effect was observed. The ionization of CBI-amino acids carboxylic groups depends on pH. CBI-ser having a pK_a value of 3, at pH 2, is mainly present in uncharged form and the enantiomers were completely resolved. On the other hand, at pH 3, the unionized and the anionic forms equally coexist and there was no enantioresolution. It is worth noting that, despite the fact that both the analyte and the chiral selector exhibit (at least partially) a negative charge at pH 3, CBI-ser strongly interacts with S-β-CD. Indeed, at pH 3, CBI-ser migration time in the absence of S-β-CD is much longer than that observed in the presence of 2% S-β-CD. Therefore, the complex analyte/S-β-CD is strongly affected by the analyte ionization state. The opposite trend observed for CBI-arg can be related to the solute charge. In contrast to the other studied analytes, CBI-arg carries a positive charge over the investigated pH range (due to the protonation of its guanidine group). Therefore, at pH 2, this analyte is positively charged and will be in zwitterionic form at higher pH.

The pH also determines the magnitude and the direction of the EOF which significantly influence the apparent selectivity and the enantioresolution. As illustrated in the review of Rizzi, the optimal selector concentration is shifted to slightly lower values with a comigrating EOF and to slightly higher values in the case of a weak countermigrating EOF [3]. A reversal of migration order and a shift of the maximum concentration to lower values are observed in the presence of a strong countermigrating EOF. If the value of the countermigrating EOF is comprised between the

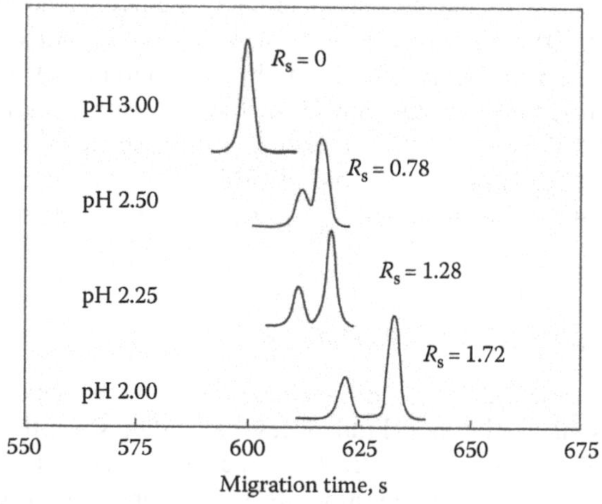

FIGURE 4.3 Influence of BGE pH on chiral resolution of cyanobenz[*f*]isoindole-ser. BGE: 2% S-β-CD, 25 mM phosphate. (Reproduced from Kirschner, D.L. et al., *Anal. Chem.*, 79, 736, 2007. With permission.)

mobilities of the free analyte and the complex, a so-called discontinuity is observed where very high selectivity values are observed. As the fine tuning of the EOF is not straightforward, a practical solution may consist in working with a constant countermigrating EOF, adjusting the mobility by changing the CD concentration or the binding strength [3].

For a given CD concentration, the effective selectivity (defined as the ratio of the effective mobilities) is influenced by all the components of the BGE which modify the binding strength. Besides its buffering capacity, the BGE determines the ionic strength of the electrophoretic medium. An increase in ionic strength reduces the electromigration dispersion, leading to a higher efficiency. However, high BGE concentrations lead to an increase in the current and possibly to excessive Joule heating. Moreover, when the analyte and the chiral selector are bound through electrostatic interactions, an increase in the ionic strength is generally unfavorable. Even if the ionic strength does not modify the ratio of the two complexation constants (i.e., the intrinsic enantioselectivity), it affects the EOF and the binding constants and therefore it may influence the apparent selectivity (i.e., the ratio of the two apparent mobilities) [5].

4.4 ADDITION OF ORGANIC SOLVENTS

With organic solvent proportions up to 30% v/v, a low impact on analytes and selectors pK_a values can be usually expected [9,10]. Nevertheless, the presence of an organic solvent in the BGE can strongly modify the binding constants, the EOF, the migration times, the BGE conductivity, and the solubility of both analyte and CD [11]. Even if the addition of an organic solvent does not modify the intrinsic selectivity, the effective selectivity may be affected, depending on the used CD concentration. For example, when the CD concentration is higher than the optimal value, the addition of an organic solvent is likely to improve the selectivity values since this addition decreases the complexation constants.

To illustrate the positive effect of the addition of an organic solvent on enantioseparation, the study carried out by Huang et al. is briefly summarized [12]. In this paper, six β-blockers enantiomers were simultaneously separated using carboxymethyl-β-CD (CM-β-CD). Four BGE systems containing the chiral selector were tested, namely phosphate, Tris-H_3PO_4, borate, and citrate. The best enantioseparation for all analytes was observed in a BGE made up of 20 mM pH 5.5 phosphate buffer and 1.5% w/v CM-β-CD, except for propranolol which was not completely enantioseparated. SDS and sodium cholate were then added to the BGE, but with no positive effect on the enantioresolution. The influence of the addition of several organic solvents was also evaluated. Among the different organic solvents tested, both methanol and ethanol were found to positively affect the enantiomeric resolution values and the best results were observed in the presence of a mixture of 5% ethanol and 5% methanol.

It is worth noting that percentages of organic solvent higher than 30% have a pronounced impact on the pK_a values.

In a review dedicated to enantioselective nonaqueous CE (NACE), the feasibility and potential of enantiomeric separations have been summarized [13]. Molecules poorly or not soluble in water or hydro-organic media can be analyzed in NACE. The limited solubility of some chiral selectors in aqueous buffers can prevent the use of their optimal concentrations. NA systems having a low dielectric constant constitute in principle a more favorable environment for chiral discrimination due to their ability to promote intermolecular interactions [3]. The introduction of NA electrolyte solutions in CE has expanded the range of the solvent parameters, offering new possibilities for optimization of mobilities, enantioselectivity, efficiency, resolution, and analysis time.

The enantiomeric purity determination of *R*-flurbiprofen was carried out in a methanolic BGE made up of ammonium acetate and ammonium camphorsulfonate using 6-monodeoxy-6-mono(2-hydroxy)propylamino-β-CD [14]. The method had to be carefully optimized in order to prevent the adsorption of the cationic CD onto the capillary wall which may impair efficiency and enantioresolution. With this aim in view, the addition of ammonium camphorsulfonate to the BGE was found to be particularly suitable to maintain high peak efficiency. Using the method of standard additions, the determination of 0.1% of *S*-flurbiprofen in *R*-flurbiprofen could be performed and a full validation of the NACE method was carried out by applying a novel strategy using accuracy profiles.

4.5 INJECTION MODE

It is well known that the choice of an appropriate injection mode is very important with respect to sensitivity. In the study carried out by Huang et al., an online sample stacking method, namely field-amplified sample injection, was applied in order to increase the method sensitivity [12]. After optimization, it was found that the injection of an aqueous sample solution by applying a voltage of 10 kV during 8 s gave rise to the best sensitivity compared to a hydrodynamic injection (0.5 psi during 8 s).

Besides its influence on method sensitivity, the injection mode may also have an impact on the enantioresolution. Castro-Puyana et al. evaluated the enantiomer recognition ability and pattern of three native and five substituted CDs toward an antiparkinsonian chiral drug [15]. As it has already been demonstrated previously [16], limit of quantitation and limit of detection were improved when the enantiomeric impurity, i.e., *S*-(+)-deprenyl, was migrating before the large peak of *R*-(−)-deprenyl. Three different sample injection techniques were compared. It was found that electrokinetic injection of the sample in the diluted buffer improved the enantioresolution as well as the method repeatability compared to hydrodynamic injection and electrokinetic injection of the sample in the BGE.

4.6 TEMPERATURE

The increase in temperature generally gives rise to a lower enantioselectivity [11]. Westall et al. have reported an unusual temperature effect on the enantioseparation of basic and neutral analytes using highly sulfated-β-CD in the reversed-polarity mode [17]. They suggested that the chiral selector concentration generally used is higher than the concentration leading to the maximal mobility difference, in order to reduce the analysis time. Therefore, in such a system, an increase in temperature results in an improvement of enantioselectivity combined with shorter migration times.

It is worth noting that Lin et al. also observed an unusual temperature effect on enantioselectivity for three basic analytes using randomly sulfate-substituted-β-CD under reversed-polarity mode [18]. They demonstrated that the temperature differently affects the mobility of both enantiomers. Indeed, the mobility differences of the first-migrating (+)-enantiomers between 40°C and 25°C are higher than those of the (−)-enantiomers above a well-defined CD concentration depending on the studied analyte.

In order to explain the temperature effect on enantioselectivity for four basic analytes, Peng et al. have studied the thermodynamic process during the chiral separation using sulfated-β-CD in the reversed polarity mode [19]. An increase in temperature was unfavorable to enantioselectivity for ephedrine and norephedrine, but favorable for synephrine and epinephrine. The differences in enthalpy and entropy were calculated in order to determine if the driving force of the enantioseparation mechanism is either enthalpy or entropy. For ephedrine and norephedrine, the enantioseparations were an enthalpy-driven process, but an entropy-driven process for synephrine and epinephrine. Therefore, the authors concluded that these differences in the driving force explain the temperature effect on enantioselectivity. In addition, they related both temperature effects to the structure of the analytes and more precisely, to the presence of one (or two) hydroxyl group(s) on the phenyl ring.

4.7 DUAL SYSTEMS

Several types of dual systems including at least one CD can be applied in the separation of enantiomers. In this section, applications dealing with the addition of an achiral additive (calixarene or SDS) to a CD as well as dual CD mixtures are presented.

4.7.1 ACHIRAL ADDITIVE-CD DUAL SYSTEMS

Hashem et al. studied the effect of achiral *p*-sulfonatocalixarenes on the enantioseparations of propranolol and brompheniramine using a complete filling technique (CFT) of HP-β-CD combined with a partial FT (PFT) of calixarene/HP-β-CD [20]. The PFT implies that the capillary is filled with the BGE containing the selector only up to the detection window. The use of the PFT is very advantageous when the selector exhibits a strong UV-absorbance, as it is the case for calixarenes. In both techniques (CFT and PFT), the BGE used during the separation process in the inlet and outlet ports of the capillary did not contain the selector. The effect of two pH conditions (2.50 and 4.65) as well as the addition of an organic solvent (MeOH and ACN) was evaluated. The application of HP-β-CD (CFT) followed by calixarene/HP-β-CD (PFT) at pH 4.65 was found to give the best enantioselectivity although the second enantiomer peak always showed a strong tailing. The authors proposed several assumptions to explain the results but investigations with techniques such as NMR should be undertaken in order to elucidate the molecular interactions between the analyte, the CD derivative, and the calixarene.

Huang et al. reported the resolution of the enantiomers of four poorly water-soluble organophosphorus pesticides in CE using nonaqueous (MeOH/ACN) and aqueous-organic (MeOH/H_2O/ACN) systems [21]. Sodium cholate, which is a chiral selector, was used as an electrolyte. Since sodium cholate alone was not able to separate the enantiomers of pyraclofos in MeOH/ACN mixture, SDS was added. This combination led to the complete enantioresolution of this compound (R_s value: 1.97), while no enantioselectivity could be observed for the three other pesticides. The three native CDs (α-, β-, and γ-CDs) were then evaluated in the presence of sodium cholate. It was necessary to add water in this system in order to increase the solubility of the neutral CDs. The chiral separation of the three analytes could only be obtained with γ-CD. It is worth noting that, in contrast to what is observed in the purely nonaqueous system for pyraclofos enantiomers peaks, the efficiency in the aqueous-organic media is very low, most probably due to analyte adsorption onto the capillary wall. This explains the relatively low resolution values obtained (between 0.99 and 1.36). However, a NACE system can considerably reduce the interactions of the analytes with the capillary wall (R_s value for pyraclofos enantiomers: 1.97).

4.7.2 Dual CD Systems

4.7.2.1 General Considerations

The use of only one CD in the BGE sometimes fails to completely enantioseparate chiral compounds. Therefore, a combination of CDs in dual systems has been introduced in order to enhance both selectivity and resolution. A dual CD system is usually a combination of a neutral and a charged CD [16,18,22–31], more rarely two charged CDs [32,33], or two neutral CDs [34]. Such a combination can lead to higher resolution due to the differences in the complexation mechanisms of the two CDs with the enantiomers, regarding the stability of complexation, the chiral recognition pattern, and the effect on analyte mobility.

It is now well known that analyte-CD complexation can lead to enantiomeric resolution in CE provided that there are not only differences in stability between the complexes formed with the CD and the enantiomers ($K_1 \neq K_2$), but also differences in mobility between the free and complexed forms of the analytes ($\mu_f \neq \mu_c$) (cf. equations).

$$R_s = \left(\frac{V}{32D}\right)^{1/2}\left(\frac{\ell}{L}\right)^{1/2}\frac{\Delta\mu_{ep}}{\left(\overline{\mu_{ep}}+\mu_{eof}\right)^{1/2}}$$

$$\Delta\mu_{ep} = \mu_{ep1} - \mu_{ep2} = \frac{[CD](\mu_f - \mu_c)(K_2 - K_1)}{1+[CD](K_1+K_2)+K_1K_2[CD]^2}$$

In separation systems containing two CDs, both selectors can act in a cooperative way as well as counteract each other. Therefore a better understanding of the mechanisms contributing to the enantioseparation is required.

Crommen and coworkers have developed mathematical models for prediction of enantioselectivity in dual selector systems [27,32,33], so that the optimization of these systems with respect to resolution and analysis time is facilitated. These models are based on the assumptions that the two selectors only form 1:1 complexes with the analyte enantiomers and that they lead to independent complexation (absence of mixed complexes). Moreover, the two selectors, which are most often CD derivatives, are assumed to be single-isomer, well-characterized compounds, which is unfortunately rarely the case in practice. This model is more general than previous ones: It is applicable to the enantioseparation of neutral or charged analytes using neutral or charged selectors.

In order to develop a dual separation system, it is feasible to manipulate the affinity pattern (choosing the suitable CD) or the mobility terms (optimizing the concentration of an anionic, cationic, or neutral CD). The latter seems to be more flexible and predictable in CE. Using dual systems, the selectivity can be improved if the two CDs used in combination affect the analyte mobility in opposite or at least different way (one accelerates and the other one decelerates it or has no effect on analyte mobility) and if the enantiomers possess opposite affinity patterns for these two chiral selectors.

Dual CD systems can provide very good selectivity for the enantioseparation of various kinds of drugs but they seem to be especially useful in the case of neutral or ionizable solutes for which selective complexation mainly occurs with neutral CDs and when they are themselves present in uncharged form ("desionoselective" separation, such as, e.g., profens with trimethyl-β-CD) [25–27,32]. In such instances, the charged CD should preferably have either no enantioselectivity or a chiral recognition pattern opposite to that of the neutral CD [16,25–27].

Following the principle that the mobility difference between enantiomers would be maximal if a chiral selector can accelerate preferably one enantiomer and the other CD can have a stronger decelerating effect on the other enantiomer, the use of combinations of an anionic CD derivative and a cationic CD derivative with opposite affinity patterns can also offer excellent possibilities for an enhancement of selectivity, resolution, and peak efficiency (i.e., for some profens at low pH) [33]. Under these conditions, an unlimited increase in selectivity and resolution can be obtained in principle but at the expense of analysis time. In such dual systems, the two CD derivatives have an opposite effect on the analyte mobility besides their opposite affinity patterns, which is favorable for an enhancement of selectivity and resolution.

On the other hand, the use of dual CD systems cannot be recommended when the two CDs have a similar effect on the analyte mobility. This is the case, for example, for dual systems where combinations of neutral CDs are employed for the enantioseparation of weak acids or bases in charged form. According to the models, no selectivity enhancement can be expected, in principle, using these kinds of dual systems. On the contrary, a competition effect for complexation with the analytes will take place between the two CDs, leading to a loss of selectivity compared to that obtained in a single system containing the most selective CD at its optimal concentration. This was demonstrated, in practice, in the case of aminoglutethimide. The latter was enantioseparated in cationic form using dual systems containing mixtures of α- and γ-CD (same affinity pattern) or of β- and γ-CD (opposite affinity pattern). In both dual systems, selectivity and resolution were lower than those obtained with single systems containing only γ-CD [35].

However, most applications described so far have been based on a trial and error approach, but recently Danel et al. used an experimental design (central composite design) in order to optimize concentrations of both CDs [24].

Besides, other theoretical models have been developed recently to determine thermodynamic parameters of the interconversion [36,37]. Furthermore, studies involving other instrumental techniques (NMR spectroscopy, MS, x-ray, crystallography,...) can also be useful for a better understanding of the intermolecular interactions involved in CE enantiomeric separations based on the use of mixtures of CDs [22].

4.7.2.2 Recent Applications of Dual CD Systems

The interest of using dual CD systems has been demonstrated in several studies in which different combinations of charged and neutral CDs have been employed [16,18,22–31]. Some major applications from the last 2 years are presented and discussed in this part of the chapter.

In a recent paper, enantioseparations of phenothiazines with dual CD systems made of single isomer sulfate-substituted β-CD (SI-S-β-CD) and a neutral CD (HP-β-CD, β-CD or γ-CD) were reported by Liao et al. [28]. In citrate BGE buffer of pH 3, the binding strength of phenothiazines to γ-CD is much weaker than to β- or HP-β-CD, allowing the use of a much higher CD concentration to obtain satisfactory enantioseparation. Figure 4.4A shows the influence of γ-CD concentration on

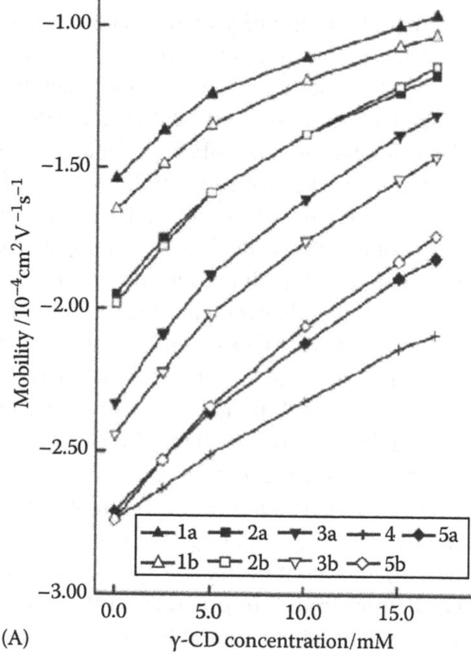

FIGURE 4.4 (A) Variations in the electrophoretic mobility of phenothiazines as a function of γ-CD concentration in the range of 0–17 mM using a dual CD system consisting of γ-CD and SI-S-β-CD (0.75%w/v) in a citrate buffer (100 mM) at pH 3.0. Sample concentration, 10 μg/mL; detection wavelength, 254 nm; other conditions, 20 kV, 25°C. Curve identification: 1, promethazine (▲, △); 2, ethopropazine (■, □); 3, trimeprazine (▼, ▽); 4, methotrimeprazine (+); 5, thioridazine (◆, ◇).

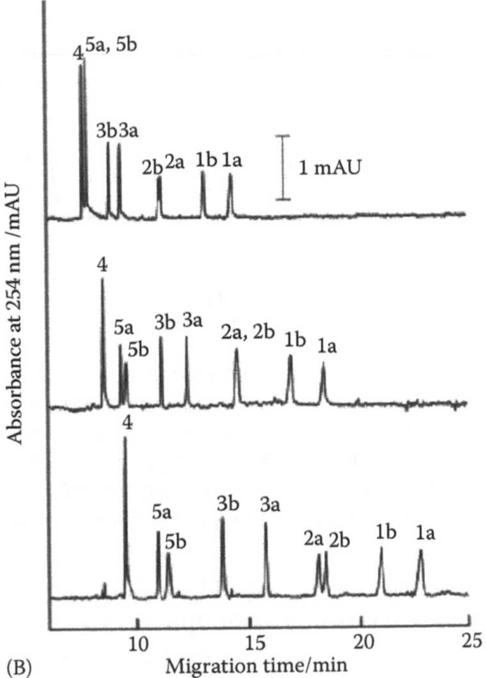

FIGURE 4.4 (continued) (B) Electropherogram of phenothiazines obtained with addition of γ-CD at different concentrations (2, 10, 17 mM) in a dual CD system consisting of γ-CD and SI-S-β-CD (0.75% w/v) using 100 mM citrate buffer at pH 3.0. Peak identification: 1, promethazine; 2, ethopropazine; 3, trimeprazine; 4, methotrimeprazine; 5, thioridazine. (Reproduced from Liao, W.S. et al., *Electrophoresis*, 28, 3922, 2007. With permission.)

the electrophoretic mobility of phenothiazines in the 0–17 mM concentration range in the presence of 0.75% SI-S-β-CD. Interestingly, the authors observed a reversal of the enantiomer migration order of both ethopropazine (2) and thioridazine (5) with increasing γ-CD concentration. The reversal of the migration order of the enantiomers may result from opposite effects of charged and neutral CDs on the mobility of the enantiomers (SI-S-β-CD has an accelerating effect, although γ-CD shows a decelerating effect), despite the fact that the chiral recognition is the same. As can be seen in Figure 4.4B, this dual CD system gave excellent enantioseparation of phenothiazines, except for methotrimeprazine.

An important application in the field of chiral separation is the enantiomeric purity determination of drugs, notably due to the increase of new chiral pharmaceuticals in the market. Furthermore, it has been largely proved that wide differences may exist between the pharmacological activity and the toxicity of two enantiomers. In this context, Lorin et al. chose CE CD systems to quantify very low amounts of the enantiomeric impurity of efaroxan, a highly selective α2-adrenoreceptor antagonist [31]. The test of several single CD systems gave rise to a resolution value of 3.4, which was insufficient to quantify the enantiomeric impurity at 0.1% level. The addition of an anionic CD to the pH 3 buffer already containing a neutral CD was found

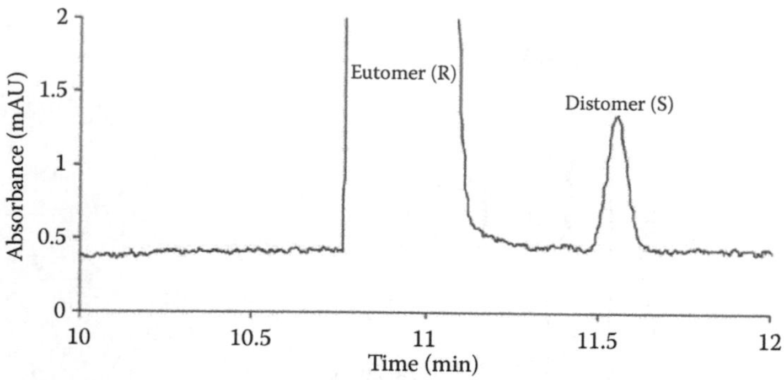

FIGURE 4.5 Quantification of the enantiomeric impurity at the 0.1% level. Conditions: fused-silica capillary, 50 cm total length (41.5 cm effective length) × 50 μm, i.d.; electrolyte, 100 mM H_3PO_4/triethanolamine (pH 3) + 7.5 mM DM-β-CD + 3 mM CM-β-CD; voltage, +25 kV; temperature, 25°C; detection, 214 nm; hydrodynamic injection 20 s at 50 mbar preceded by a water-plug preinjection for 5 s at 30 mbar; [*R*]: 200 μg/L; [*S*]: 0.2 μg/L. (Reprinted from Lorin, M. et al., *Anal. Chim. Acta*, 592, 139, 2007. With permission.)

to strongly increase the enantioselectivity. The careful optimization of the nature and concentration of both CDs, as well as the applied voltage and the injection time, allowed the authors to separate ($R_s = 7.2$) in short analysis time and quantify successfully 0.05% of enantiomeric impurity, even if the impurity was migrating after the major enantiomer (cf. Figure 4.5).

Since many parameters may influence the enantioselectivity, Danel et al. used a chemometric approach to find optimal conditions for the enantioselective analysis of an antipsychotic and its main metabolite [24]. They used a central composite design through the response surface methodology to maximize enantioresolution, finding optimal CDs (sulfated-α-CD and hydroxypropyl-β-CD) and phosphate buffer concentrations. The linear and quadratic coefficients for the three factors were all significant but no interactions were found. This methodology was successfully applied and the method was validated according to the acceptance criteria of ICH. The perspectives would be to undertake the metabolism study of this antipsychotic in biological media.

Recently Chu et al. validated successfully the assay of a chiral antiparkinsonian drug (rotigotine) and its chiral impurities [38]. They first tried single CD systems (β-CD, S-β-CD or methyl-β-CD), but they could not separate rotigonine and related impurities simultaneously. However, they observed a synergistic effect and complete separation using a dual CD system made of S-β-CD and methyl-β-CD in an acidic BGE (pH = 2.5). The optimal concentration for both S-β-CD and methyl-β-CD was found to be 2% (w/v), taking enantioresolution and migration times into consideration (total analysis time: 15 min). They also observed that the increase in the BGE ionic strength is favorable to enantioresolution due to the reduction of the EOF and the increase in buffer viscosity. Then method validation was performed evaluating specificity, precision, linearity, sensitivity, and accuracy. Finally, they applied the method to estimate the impurities of five batches of rotigotine.

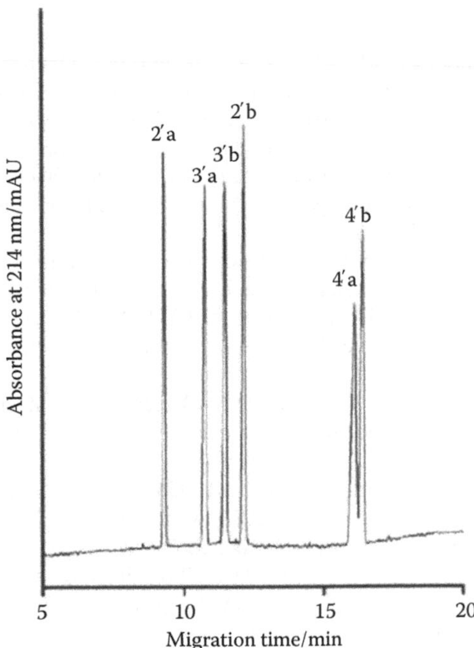

FIGURE 4.6 Electropherogram of hydroxyflavanones obtained with dual CD systems consisting of 2.0% (w/v) S-β-CD and 2 mM γ-CD with the addition of 2 mM SDS monomers in a phosphate buffer (50 mM) at pH 3.0. Capillary, 50.2 cm × 50 μm, i.d.; sample concentration, 30 μg/mL; detection wavelength, 214 nm. Other operating conditions: −21 kV, 25°C. Peak identification, 2′ = 2′-hydroxyflavanone, 3′ = 3′-hydroxyflavanone, 4′ = 4′-hydroxyflavanone. (Reprinted from Lin, C.H. et al., *J. Chromatogr. A*, 1188, 301, 2008. With permission.)

Another recent study by Lin et al. reported the interest of the addition of SDS monomers into a dual CD system for the enantioseparation of uncharged hydroxyflavanones [29]. Indeed, it has been shown that SDS as monomers or micelles strongly interacts with CDs and thus may affect enantioselectivity [39]. Figure 4.6 illustrates the complete enantioseparation of three hydroxyflavanones obtained with dual CD systems consisting of 2% of randomly sulfate-substituted β-CD (S-β-CD) (w/v) and 2 mM γ-CD with 2 mM SDS monomers in a phosphate buffer of pH 3. Under these conditions, SDS monomers act as a complexing agent increasing the effective mobility of the analytes toward the anode and modulate selectivity for the separation of hydroxyflavanones. Moreover, a synergistic effect of dual CD systems on the enantioselectivity of hydroxyflavanone was demonstrated. In correlation with the molecular structures, the authors suggested that both inclusion complexation and hydrogen bonding interactions play an important role in the chiral recognition of hydroxyflavanones.

ACKNOWLEDGMENTS

Research grants from the Belgium National Fund for Scientific Research (FNRS) to two of us (A.-C. S. and M. F.) are gratefully acknowledged.

REFERENCES

1. Theurillat, R, Knobloch, M, Schmitz, A, Lassahn, PG, Mevissen, M, Thormann, W. 2007. Enantioselective analysis of ketamine and its metabolites in equine plasma and urine by CE with multiple isomer sulfated beta-CD. *Electrophoresis* 28: 2748–2757.
2. Scriba, GKE. 2008. Cyclodextrins in capillary electrophoresis enantioseparations—Recent developments and applications. *J Sep Sci* 31: 1991–2011.
3. Rizzi, A. 2001. Fundamental aspects of chiral separations by capillary electrophoresis. *Electrophoresis* 22: 3079–3106.
4. Wren, SAC, Rowe, RC. 1992. Theoretical aspects of chiral separation in capillary electrophoresis 1. Initial evaluation of a model. *J Chromatogr* 603: 235–241.
5. Williams, BA, Vigh, G. 1997. Dry look at the CHARM (charged resolving agent migration) model of enantiomer separations by capillary electrophoresis. *J Chromatogr A* 777: 295–309.
6. Michalska, K, Pajchel, G, Tyski, S. 2008. Determination of enantiomeric impurity of linezolid by capillary electrophoresis using heptakis-(2,3-diacetyl-6-sulfato)-beta-cyclodextrin. *J Chromatogr A* 1180: 179–186.
7. Castro-Puyana, M, Crego, AL, Marina, ML, Garcia-Ruiz, C. 2007. Enantioselective separation of azole compounds by EKC. Reversal of migration order of enantiomers with CD concentration. *Electrophoresis* 28: 2667–2674.
8. Kirschner, DL, Jaramillo, M, Green, TK. 2007. Enantioseparation and stacking of cyanobenz[f]isoindole-amino acids by reverse polarity capillary electrophoresis and sulfated beta-cyclodextrin. *Anal Chem* 79: 736–743.
9. Sarmini, K, Kenndler, E. 1998. Capillary zone electrophoresis in mixed aqueous-organic media: Effect of organic solvents on actual ionic mobilities, acidity constants and separation selectivity of substituted aromatic acids. I. Methanol. *J Chromatogr A* 806: 325–335.
10. Sarmini, K, Kenndler, E. 1999. Capillary zone electrophoresis in mixed aqueous-organic media: Effect of organic solvents on actual ionic mobilities and acidity constants of substituted aromatic acids IV. Acetonitrile. *J Chromatogr A* 833: 245–259.
11. Fanali, S. 2000. Enantioselective determination by capillary electrophoresis with cyclodextrins as chiral selectors. *J Chromatogr A* 875: 89–122.
12. Huang, L, Lin, JM, Yu, LS, Xu, LJ, Chen, GN. 2008. Field-amplified on-line sample stacking for simultaneous enantioseparation and determination of some beta-blockers using capillary electrophoresis. *Electrophoresis* 29: 3588–3594.
13. Lammerhofer, M. 2005. Chiral separations by capillary electromigration techniques in nonaqueous media I. Enantioselective nonaqueous capillary electrophoresis. *J Chromatogr A* 1068: 3–30.
14. Rousseau, A, Chiap, P, Ivanyi, R, Crommen, J, Fillet, M, Servais, AC. 2008. Validation of a nonaqueous capillary electrophoretic method for the enantiomeric purity determination of R-flurbiprofen using a single-isomer amino cyclodextrin derivative. *J Chromatogr A* 1204: 219–225.
15. Castro-Puyana, M, Lomsadze, K, Crego, AL, Marina, ML, Chankvetadze, B. 2007. Separation of enantiomers of deprenyl with various CDs in CE and the effect of enantiomer migration order on enantiomeric impurity determination of selegiline in active ingredients and tablets. *Electrophoresis* 28: 388–394.
16. Fillet, M, Bechet, I, Chiap, P, Hubert, P, Crommen, J. 1995. Enantiomeric purity determination of propranolol by cyclodextrin-modified capillary electrophoresis. *J Chromatogr A* 717: 203–209.
17. Westall, A, Mallmstrom, T, Petersson, P. 2006. An observation of unusual temperature effects for enantioselective CZE employing highly sulfated-beta-cyclodextrin. *Electrophoresis* 27: 859–864.

18. Lin, CH, Kuo, CM, Liu, YC, Cheng, HT, Lin, WY, Wu, JC, Wang, LF, Lin, CE. 2006. Enantioselectivity of basic analytes in CZE enantioseparation under reversed-polarity mode using sulfated beta-cyclodextrins as chiral selectors: An unusual temperature effect. *Electrophoresis* 27: 4345–4350.
19. Peng, ZL, Yi, F, Guo, BY, Lin, JM. 2007. Temperature effects on the enantioselectivity of basic analytes in capillary EKC using sulfated beta-CDs as chiral selectors. *Electrophoresis* 28: 3753–3758.
20. Hashem, H, Kinzig, M, Jira, T. 2008. The effect of achiral calixarenes on chiral separation of propranolol-HCl and brompheniramine maleate in capillary electrophoresis using cyclodextrin as chiral selector. *Pharmazie* 63: 256–262.
21. Huang, L, Lin, JM, Xu, LJ, Chen, GN. 2007. Nonaqueous and aqueous-organic media for the enantiomeric separations of neutral organophosphorus pesticides by CE. *Electrophoresis* 28: 2758–2764.
22. Chankvetadze, B, Fillet, M, Burjanadze, N, Bergenthal, D, Bergander, C, Luftmann, H, Crommen, J, Blaschke, G. 2000. Enantioseparation of aminoglutethimide with cyclodextrins in capillary electrophoresis and studies of selector-selectand interactions using NMR spectroscopy and electrospray ionization mass spectrometry. *Enantiomer* 5: 313–322.
23. Chankvetadze, B, Lomsadze, K, Bergenthal, D, Breitkreutz, J, Bergander, K, Blaschke, G. 2001. Mechanistic study on the opposite migration order of clenbuterol enantiomers in capillary electrophoresis with beta-cyclodextrin and single-isomer heptakis(2,3-diacetyl-6-sulfo)-beta-cyclodextrin. *Electrophoresis* 22: 3178–3184.
24. Danel, C, Chaminade, P, Odou, P, Bartelemy, C, Azarzar, D, Bonte, JP, Vaccher, C. 2007. Enantioselective analysis of the antipsychotic 9-hydroxyrisperidone, main metabolite of risperidone, by chiral capillary EKC using dual CDs. *Electrophoresis* 28: 2683–2692.
25. Fillet, M, Hubert, P, Crommen, J. 1997. Enantioseparation of nonsteroidal anti-inflammatory drugs by capillary electrophoresis using mixtures of anionic and uncharged beta-cyclodextrins as chiral additives. *Electrophoresis* 18: 1013–1018.
26. Fillet, M, Hubert, P, Crommen, J. 1998. Method development strategies for the enantioseparation of drugs by capillary electrophoresis using cyclodextrins as chiral additives. *Electrophoresis* 19: 2834–2840.
27. Fillet, M, Hubert, P, Crommen, J. 2000. Enantiomeric separations of drugs using mixtures of charged and neutral cyclodextrins. *J Chromatogr A* 875: 123–134.
28. Liao, WS, Lin, CH, Chen, CY, Kuo, CM, Liu, YC, Wu, JC, Lin, CE. 2007. Enantioseparation of phenothiazines in CD-modified CZE using single isomer sulfated CD as a chiral selector. *Electrophoresis* 28: 3922–3929.
29. Lin, CH, Fang, WR, Kuo, CM, Chang, WY, Liu, YC, Lin, WY, Wu, JC, Lin, CE. 2008. Chiral separation of hydroxyflavanones in cyclodextrin-modified capillary zone electrophoresis using sulfated cyclodextrins as chiral selectors. *J Chromatogr A* 1188: 301–307.
30. Lin, XL, Zhao, MG, Qi, XY, Zhu, CF, Hao, AY. 2006. Capillary zone electrophoretic chiral discrimination using 6-O-(2-hydroxy-3-trimethylammoniopropyl)-beta-cyclodextrin as a chiral selector. *Electrophoresis* 27: 872–879.
31. Lorin, M, Delepee, R, Morin, P, Ribet, JP. 2007. Quantification of very low enantiomeric impurity of efaroxan using a dual cyclodextrin system by capillary electrophoresis. *Anal Chim Acta* 592: 139–145.
32. Abushoffa, AM, Fillet, M, Hubert, P, Crommen, J. 2002. Prediction of selectivity for enantiomeric separations of uncharged compounds by capillary electrophoresis involving dual cyclodextrin systems. *J Chromatogr A* 948: 321–329.
33. Abushoffa, AM, Fillet, M, Servais, AC, Hubert, P, Crommen, J. 2003. Enhancement of selectivity and resolution in the enantioseparation of uncharged compounds using mixtures of oppositely charged cyclodextrins in capillary electrophoresis. *Electrophoresis* 24: 343–350.

34. Nhujak, T, Sastravaha, C, Palanuvej, C, Petsom, A. 2005. Chiral separation in capillary electrophoresis using dual neutral cyclodextrins: Theoretical models of electrophoretic mobility difference and separation selectivity. *Electrophoresis* 26: 3814–3823.
35. Abushoffa, AM, Fillet, M, Marini, RD, Hubert, P, Crommen, J. 2003. Enantiomeric separation of aminoglutethimide by capillary electrophoresis using native cyclodextrins in single and dual systems. *J Sep Sci* 26: 536–542.
36. Dubsky, P, Svobodova, J, Gas, B. 2008. Model of CE enantioseparation systems with a mixture of chiral selectors. Part I. Theory of migration and interconversion. *J Chromatogr B Analyt Technol Biomed Life Sci* 875: 30–34.
37. Dubsky, P, Svobodova, J, Tesarova, E, Gas, B. 2008. Model of CE enantioseparation systems with a mixture of chiral selectors. Part II. Determination of thermodynamic parameters of the interconversion in chiral and achiral environments separately. *J Chromatogr B Analyt Technol Biomed Life Sci* 875: 35–41.
38. Chu, BL, Guo, B, Zuo, H, Wang, Z, Lin, JM. 2008. Simultaneous enantioseparation of antiparkinsonian medication Rotigotine and related chiral impurities by capillary zone electrophoresis using dual cyclodextrin system. *J Pharm Biomed Anal* 46: 854–859.
39. Yunus, WMZW, Taylor, J, Bloor, DM, Hall, DG, Wynjones, E. 1992. Electrochemical measurements on the binding of sodium dodecyl-sulfate and dodecyltrimethylammonium bromide with alpha-cyclodextrin and beta-cyclodextrins. *J Phys Chem* 96: 8979–8982.
40. Jiang, Z, Thorogate, R, Smith, NW. 2008. Highlighting the role of the hydroxyl position on the alkyl spacer of hydroxypropyl-beta-cyclodextrin for enantioseparation in capillary electrophoresis. *J Sep Sci* 31: 177–187.
41. Tang, WH, Ong, TT, Ng, SC. 2007. Chiral separation of dansyl amino acids in capillary electrophoresis using mono-(3-methylimidazolium)-beta-cyclodextrin chloride as selector. *J Sep Sci* 30: 1343–1349.
42. Tang, WH, Ong, TT, Muderawan, IW, Ng, SC. 2007. Effect of alkylimidazolium substituents on enantioseparation ability of single-isomer alkylimidazolium-beta-cyclodextrin derivatives in capillary electrophoresis. *Anal Chim Acta* 585: 227–233.
43. Nzeadibe, K, Vigh, G. 2007. Synthesis of mono-6-deoxy-6-*N*,*N*,*N*′,*N*′,*N*′-pentamethylethylenediammonio-cyclomaltoheptaose, a single-isomer, monosubstituted, permanently dicationic beta-CD and its use for enantiomer separations by CE. *Electrophoresis* 28: 2589–2605.
44. Wistuba, D, Bogdanski, A, Larsen, KL, Schurig, V. 2006. Delta-cyclodextrin as novel chiral probe for enantiomeric separation by electromigration methods. *Electrophoresis* 27: 4359–4363.
45. Tang, WH, Muderawan, IW, Ng, SC, Chan, HSO. 2006. Enantioselective separation in capillary electrophoresis using a novel mono-6(A)-propylammonium-beta-cyclodextrin as chiral selector. *Anal Chim Acta* 555: 63–67.
46. Wei, YH, Li, J, Zhu, CF, Hao, AY, Zhao, MG. 2005. 2-*O*-(2-hydroxybutyl)-beta-cyclodextrin as a chiral selector for the capillary electrophoretic separation of chiral drugs. *Anal Sci* 21: 959–962.
47. Zhao, MG, Hao, AY, Li, J, Wei, YH, Guo, P. 2005. New cyclomaltoheptaose (beta-cyclodextrin) derivative 2-*O*-(2-hydroxybutyl)cyclomaltoheptaose: Preparation and its application for the separation of enantiomers of drugs by capillary electrophoresis. *Carbohydr Res* 340: 1563–1565.
48. Tang, WH, Muderawan, IW, Ong, TT, Ng, SC. 2005. Enantio separation of dansyl amino acids by a novel permanently positively charged single-isomer cyclodextrin: Mono-6-N-allylammonium-6-deoxy-beta-cyclodextrin chloride by capillary electrophoresis. *Anal Chim Acta* 546: 119–125.

49. Ong, TT, Tang, W, Muderawan, W, Ng, SC, Chan, HSO. 2005. Synthesis and application of single-isomer 6-mono(alkylimidazolium)-beta-cyclodextrins as chiral selectors in chiral capillary electrophoresis. *Electrophoresis* 26: 3839–3848.
50. Busby, MB, Vigh, G. 2005. Synthesis of a single-isomer sulfated beta-cyclodextrin carrying nonidentical substituents at all of the C2, C3, and C6 positions and its use for the electrophoretic separation of enantiomers in acidic aqueous and methanolic background electrolytes. Part 2: Heptakis(2-O-methyl-6-O-sulfo) cyclomaltoheptaose. *Electrophoresis* 26: 3849–3860.
51. Tang, WH, Muderawan, IW, Ong, TT, Ng, SC. 2005. A family of single-isomer positively charged cyclodextrins as chiral selectors for capillary electrophoresis: Mono-6(A)-butylammonium-6(A)-deoxy-beta-cyclodextrin tosylate. *Electrophoresis* 26: 3125–3133.
52. Busby, MB, Vigh, G. 2005. Synthesis of heptakis(2-*O*-methyl-3-*O*-acetyl-6-*O*-sulfo)-cyclomaltoheptaose, a single-isomer, sulfated beta-cyclodextrin carrying nonidentical substitutents at all the C2, C3, and C6 positions and its use for the capillary electrophoretic separation of enantiomers in acidic aqueous and methanolic background electrolytes. *Electrophoresis* 26: 1978–1987.
53. Tang, W, Muderawan, IW, Ong, TT, Ng, SC. 2005. Enantioseparation of acidic enantiomers in capillary electrophoresis using a novel single-isomer of positively charged beta-cyclodextrin: Mono-6(A)N-pentylammonium-6(A)-deoxy-beta-cyclodextrin chloride. *J Chromatogr A* 1091: 152–157.
54. Cucinotta, V, Giuffrida, A, Maccarrone, G, Messina, M, Puglisi, A, Torrisi, A, Vecchio, G. 2005. The 6-derivative of beta-cyclodextrin with succinic acid: A new chiral selector for CD-EKC. *J Pharm Biomed Anal* 37: 1009–1014.

5 Macrocyclic Antibiotics as Chiral Selectors

Salvatore Fanali, Zeineb Aturki, Giovanni D'Orazio, and Anna Rocco

CONTENTS

5.1 Introduction 109
5.2 Macrocyclic Antibiotics: Properties and Enantioresolution Mechanism 110
5.2.1 Macrocyclic Antibiotic's Properties 110
5.2.2 Enantioresolution Mechanism 111
5.2.2.1 Enantioresolution of Uncharged Compounds 113
5.3 Detection of Enantiomers When Macrocyclic Antibiotics Are Used as Mobile Phase Additives 114
5.4 Electromigration Separation Modes Employing Macrocyclic Antibiotics 116
5.4.1 Capillary Zone Electrophoresis 116
5.4.2 Capillary Electrochromatography 122
5.5 "Less Common" Used Chiral Selectors in Capillary Electromigration Methods 125
5.6 Conclusions and Future Trends 130
References 131

5.1 INTRODUCTION

The availability of analytical separation methods suitable for discriminating enantiomeric compounds is of great importance in several research and/or application fields such as environmental, agricultural, industrial, biomedical, and pharmaceutical ones. Those methods have to offer the possibility to carry out qualitative and quantitative determinations in short time, with low costs, high efficiency, and high resolution.

In the last decade great attention was given to the development of new methodologies to be applied to chiral analysis, examining thoroughly the theory of chiral selector–analyte interactions, method optimization, "new" chiral selectors, applications, etc. Among the recently developed analytical methods, electromigration methods (EMs) such as capillary zone electrophoresis (CZE), micellar electrokinetic chromatography (MEKC), capillary electrochromatography (CEC), and capillary isotachophoresis (CITP) were widely investigated for chiral separations. For that purpose, a wide number of chiral selectors were used. They were mainly either added to the background electrolyte (BGE) or bonded to various media such as (1) capillary wall, (2) silica stationary phase, and (3) polymers. The chiral selector interacts with the two enantiomers during the electrophoretic run modifying their chemical–physical properties with the formation of labile diastereoisomers and consequent separation of optical antipodes. The chiral process, previously described, is called direct resolution method. A second approach, called indirect resolution method, has also been proposed. Stable diastereoisomeric complexes are formed by chemical reaction between enantiomers and a chiral derivatizing agent before the analysis. The complexes exhibit different properties and can be easily separated [1,2].

Among the large number and type of chiral selectors successfully employed in electrodriven methods macrocyclic antibiotics exhibit very high enantioselectivity toward a large number of enantiomeric compounds. These chiral selectors were first introduced by Armstrong and coworkers in high performance liquid chromatography (HPLC) and later on in EMs [3,4]. The use of macrocyclic antibiotics as chiral selector in EMs is documented by several review papers [5–11].

The list of these chiral selectors includes rifamycin B and SV [4,12], vancomycin [13–15], ristocetin A [16], teicoplanin [17], fradiomycin, kanamycin, streptomycin [18], balhimycin [19], Hepta-Tyr (teicoplanin derivative) [20], actaplanin [21], and avoparcin [22].

In spite of the high enantiomeric resolution capability of the above mentioned chiral selectors toward a wide numbers of racemic analytes, it is noteworthy to mention the drawbacks related to their use in separation science mainly due to the lack of sensitivity because of their strong UV absorption at the commonly employed wavelengths in capillary electrophoresis (CE) [7].

The aim of this chapter is to illustrate the state of the art related to the use of macrocyclic antibiotics in EMs. The properties of the most common employed macrocyclic antibiotic, the enantioresolution mechanism that takes place in the enantiorecognition process, and the perspectives in separation science will be considered. Furthermore some selected applications in the field of the chiral separations performed by using both CZE and CEC are also presented, updating our last review dealing with this topic [7].

5.2 MACROCYCLIC ANTIBIOTICS: PROPERTIES AND ENANTIORESOLUTION MECHANISM

5.2.1 Macrocyclic Antibiotic's Properties

Antibiotics are organic compounds with chemotherapeutic properties inhibiting the growth of bacteria. These compounds are typically produced by different species of microorganisms; however, nowadays synthetic or semisynthetic antibiotics are also available. Chemical modifications have been introduced with the aim to reduce toxicity, enhance antibacterial activity, and increase solubility.

Several classes of antibiotics have been studied and used in medical applications, among them are macrocyclic antibiotics. In their chemical structure, several

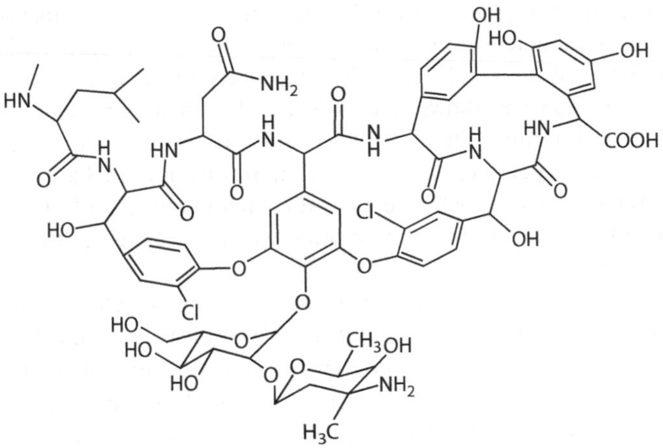

FIGURE 5.1 Chemical structure of vancomycin.

asymmetric centers, aromatic rings, sugar moieties, amino as well as carboxylic groups can occur. The presence of asymmetric centers ranging from 9 to 38 for rifamycin B or SV and ristocetin A, respectively, grants those compounds very high enantiorecognition capabilities. Due to the only just aforesaid properties, macrocyclic antibiotics have been employed in analytical chemistry, particularly in HPLC and CE for enantiomeric separations [3,5–11,23–25].

Macrocyclic antibiotics show similar chemical–physical behavior and enantiorecognition mechanism. All these properties will be discussed and examined in detail mainly considering vancomycin, one of the most used chiral selectors in CE, as a model compound.

This glycopeptidic chiral selector is characterized by the presence of 18 asymmetric centers and a large number of functional groups such as hydroxyl, amino, amide, carboxylic functions, and aromatic rings (for its chemical structure see Figure 5.1) in its structure. Vancomycin may exist in either charged (positively or negatively) or uncharged form depending on the pH of the medium in which this compound is dissolved. This is clearly due to the presence of two amino groups and one carboxylic group in its chemical structure. The antibiotic can be easily dissolved in aqueous solutions while it is less soluble in organic solvents with a lower polarity [13]. Vancomycin solutions are stable at low temperature and at acidic pH in the pH range 3–6.

5.2.2 Enantioresolution Mechanism

Resolution mechanisms so far exploited for chiral separations by EMs include ligand exchange, inclusion complexation, optical micelles solubilization, affinity, and ion-pairing interactions [26,27].

Macrocyclic antibiotics represent a class of chiral selectors capable of resolving a wide number of enantiomers offering very high selectivity and good efficiency.

As above reported, they possess several stereogenic centers and functional groups involved in nonstereo/stereoselective interactions based on hydrogen–hydrogen, π–π, hydrophobic, or repulsion bonds, etc. The aglycone pocket, which is characteristic of numerous antibiotics and allows inclusion complex formation, is therefore usually mentioned as a high-affinity site [28]. Chargeable groups, which determine positive or negative charges depending on the pH of environmental media, promote one of the primary interactions, namely, electrostatic interaction. Finally the enantiorecognition capability of macrocyclic antibiotics was also related to their aptitude to undergo dimerization [19,29].

Considering vancomycin, it is a very effective chiral selector for the resolution of anionic compounds (with carboxylic groups) when it is used in CZE with BGEs at pH range of 4–6. At these pH values, in fact, electrostatic interactions between vancomycin (positively charged) and anionic compounds are favored.

In the chiral recognition mechanism, the free secondary amino group present in the aglycone basket of vancomycin is also very important for interactions between chiral selector and analytes. This has been demonstrated by Nair et al. [30] comparing the enantioselectivity of native vancomycin to that of copper–vancomycin complex by using CZE. In this chapter, the secondary amino group belonging to *N*-methyl-leucine in the antibiotic structure was complexed with copper ions. Therefore this group was not available for electrostatic interactions with enantiomers. As a consequence enantiorecognition with acidic compounds was strongly reduced or eliminated.

Recently Schurig's group [19] studied a "new" macrocyclic antibiotic belonging to the vancomycin family, namely, balhimycin for the chiral resolution of eleven dansyl amino acids and six 2-arylpropionic acids by CE. In this study, the authors examined the antibiotic and two of its derivatives (dechlorobalhimycin and bromobalhimycin) in order to verify the role of chlorine atoms (present in two aromatic rings of balhimycin) in the enantiorecognition mechanism. They confirmed that the main interaction site is the aglycone cavity and observed higher enantioresolution when chlorine substituents were in the chemical structure. It was concluded that the presence of halide atoms promotes dimerization of the antibiotic improving its enantioresolution capability. The same research group previously published another mechanistic study of enantiomeric separation comparing vancomycin and balhimycin. The two chiral selectors have the same aglycone but different type and linking position of sugar moieties. The amino group on the sugar plays an important role in the enantioselectivity of balhimycin. In fact blocking the amino group of this antibiotic, by N-carbamoylation reaction, caused a decrease in enantioselectivity as a consequence of a lessening of the dimerization phenomenon [29].

Teicoplanin as well exhibits excellent enantiorecognition capabilities toward acidic compounds. In our study on this antibiotic, we carried out CZE experiments utilizing BGEs at pH range between 4 and 7. Best enantiomeric resolutions were observed at pH 5 where the macrocyclic compound was positively charged while analytes possessed a negative charge. The enantioselectivity of both teicoplanin and vancomycin toward anionic compounds is documented by the separation of a wide number of enantiomers, including herbicides, drugs, hydroxyacids, etc. [31–33].

The chemical structure of analytes also plays a very important role in the chiral recognition mechanism with macrocyclic antibiotics. The presence of carboxylic groups on the molecule has been demonstrated to be necessary to achieve chiral resolution in HPLC employing teicoplanin (chirobiotic T), vancomycin (chirobiotic V), and ristocetin (chirobiotic R) stationary phases. Hui et al. [34] studied a wide number of racemic compounds (tryptophan derivatives and indoles) possessing both carboxylic and amino groups in their structure. Modifying the chemical structure of those analytes, e.g., esterifying the carboxylic group, resulted in poor or no enantiomeric resolution. Tryptophan methyl ester or α-methyl-tryptamine was not resolved at all by using any of the studied chiral stationary phases. It was concluded that the carboxylic group of the analytes is the main interaction site while the amino substituent in the analyte's structure may be responsible for secondary interactions.

In the same study, experiments have also been carried out using CZE with the chiral selector simply added to the BGE. The authors remarked that most of the studied compounds, before being resolved into their enantiomers with chirobiotic T stationary phase, were partially resolved or unresolved by CZE. However they found some difference; e.g., abscisic acid was resolved with teicoplanin only in HPLC while in CZE baseline resolution was recorded employing all three studied chiral selectors. Additionally *N*-*t*-Boc-tryptophan enantiomers were separated in HPLC with only Chirobiotic R while in CZE all chiral selectors worked well. The influence of the analytes' structure on the chiral resolution mechanism was also discussed by Wan and Blomberg [35] analyzing amino acid derivatives in CZE.

9-fluorenylmethyl chloroformate-alanine (FMOC-Ala) was resolved with both vancomycin and teicoplanin while 9-fluorenyl-ethyl chloroformate-glycine (FLEC-Gly) was not resolved using both antibiotics. Those results were ascribed to the sterical hindrance effect of the second compound.

The enantiomeric separation of amino acids (free or derivatized form) has also been investigated by Cavazzini et al. by using a teicoplanin stationary phase in HPLC with the aim of elucidating the enantioresolution mechanism. In this study, the authors also concluded that the presence of carboxylic groups in the structure of analytes was of paramount importance in the chiral resolution. Furthermore enantiomers capable of fitting the aglycone basket may form several hydrogen bonds and hydrophobic interactions stabilizing the diastereoisomeric complex with the aglycone [25].

A theoretical approach for studying vancomycin-herbicide enantiomer binding by CE was proposed by André and Guillaume [36]. The model was applied to the determination of binding constants of herbicides, such as flamprop, mecoprop, fenoprop, etc., with vancomycin. Special attention was given to the optimization of the salt additive and the chiral selector concentration.

5.2.2.1 Enantioresolution of Uncharged Compounds

In 1984, there was a great revolution in CE with the introduction of micelles for the separation of uncharged compounds by Terabe et al. [37]. Later on this approach was also applied for the separation of uncharged chiral compounds [38]. More information concerning this topic can be found in Chapter 8 of this book.

The effect of micelles addition, such as sodium dodecyl sulfate (SDS), on chiral resolution of uncharged pesticides or drugs using vancomycin or teicoplanin as chiral selector was also investigated. Experiments carried out with a BGE supplemented with vancomycin and SDS revealed that the enantiomeric migration order of studied amino acid derivatives was reversed in comparison with data obtained in the absence of SDS. This effect was due to the different electrophoretic mobilities of vancomycin and vancomycin–SDS complex. The enantiomers interacted with bulk solution, SDS micelles, and SDS–vancomycin complex during the electrophoretic run [39].

5.3 DETECTION OF ENANTIOMERS WHEN MACROCYCLIC ANTIBIOTICS ARE USED AS MOBILE PHASE ADDITIVES

As can be observed in Figure 5.1, vancomycin has several aromatic rings responsible for strong absorption in the UV wavelengths commonly employed in analytical chemistry. Clearly this can also be observed for other macrocyclic antibiotics.

Although this was a drawback, it was possible to use glycopeptide antibiotics like vancomycin, ristocetin A, and teicoplanin in CE at low concentrations (1–5 mM) [6]. Therefore, concentrations of chiral selectors higher than 5 mM are useless for practical analysis because of the very low sensitivity. This problem was resolved by applying the partial-filling countercurrent method illustrated in Figure 5.2.

In this procedure, the capillary is first flushed with the BGE without the absorbing chiral selector, and then the BGE, containing the antibiotic, is injected. In the last procedure, the injected zone is stopped before the detector. After sample injection, an appropriate electric field is applied for the electrophoretic separation. The partial-filling countercurrent method works well, if analytes and chiral selector migrate in opposite directions. Consequently it is necessary that they have different charges. Taking vancomycin into account, it can be employed to separate acidic compounds

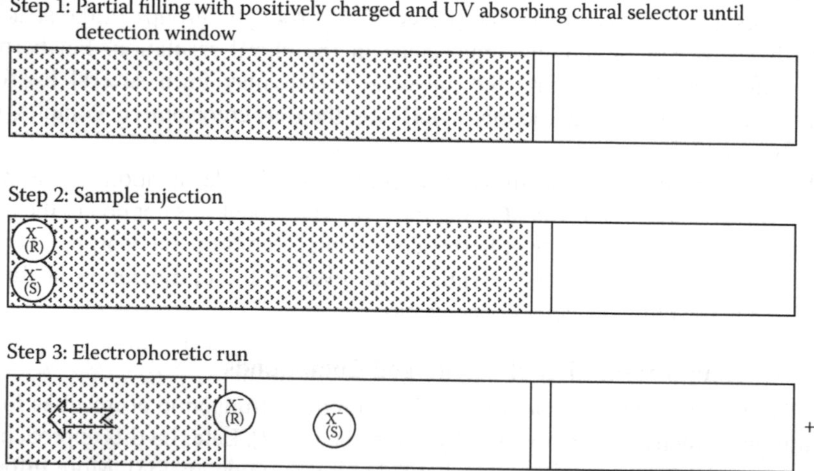

FIGURE 5.2 Scheme of partial-filling countercurrent method for the analysis of acidic enantiomers and using CZE with antibiotics added to the BGE.

selecting a suitable pH (4–7), where analytes are negatively charged, while the macrocyclic antibiotic is positively charged. Moreover, in order to achieve good results, e.g., high sensitivity, good resolution, high efficiency, and short analysis time, the chiral selector should not be present in the electrolyte vessels during run and the electroosmotic flow (EOF) should be reduced [31–33].

As an example of the good performance of the applied method, Figure 5.3 shows the enantiomeric separation of loxiglumide enantiomers where the L-isomer was present at a concentration of 0.2%.

Ward et al. utilized a similar technique to separate some nonsteroidal anti-inflammatory drugs. They filled the whole capillary. Consequently, when the electrophoretic run started, for a certain time a high detector signal was recorded until the antibiotic, migrating in the opposite direction of the analytes, disappeared out of the detector. Obviously those enantiomers moving with high mobility could not be detected [40]. We also applied the countercurrent method to combine CZE with mass spectrometry (MS). In this manner, the MS signal was not influenced by the antibiotic and the enantiomeric separation of some drugs was carried out determining the mass spectra of each enantiomer [41]. Experiments were performed in polyacrylamide-coated capillaries in order to minimize the EOF and therefore the chiral selector moved mainly for its electrophoretic mobility. The use of coated capillaries was helpful in reducing the antibiotic adsorption onto the capillary wall. This problem was also pointed out by Wan and Blomberg [35] when vancomycin or teicoplanin was used for the separation of dipeptide enantiomers by CZE. In this work several solutions were proposed, e.g., selection of a pH close to the pI of the antibiotic, use of SDS, decrease of cationic character, etc. Kang et al. [42] analyzed amino acid derivatives and acidic drugs by CZE in the presence of vancomycin using a co-EOF. The EOF was reversed

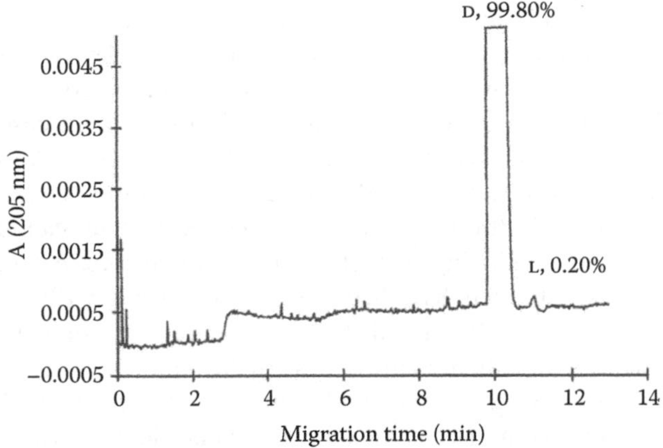

FIGURE 5.3 Electropherogram of the enantiomeric separation of loxiglumide by capillary zone electrophoresis. Experimental conditions: capillary (polyacrylamide coated) 50 cm × 50 μm ID; background electrolyte (BGE), 50 mM phosphate buffer pH 6; 3 mM of vancomycin was dissolved in the BGE, this solution was pumped for 90 s. Applied voltage, 18 kV. Sample: D- and L-Loxiglumide 99.80% and 0.20% (w/w). (Reproduced from Fanali, S. and Desiderio, C., *J. High Resolut. Chromatogr.*, 19, 322, 1996. With permission.)

by adding hexadimethrine bromide (HDB) to the BGE and fast separations were achieved. The use of HDB was helpful in both reducing analysis time and minimizing vancomycin adsorption. A similar approach was done by Schurig's group which was able to separate dansyl amino acid enantiomers in about 50 s [43]. Armstrong et al., using rifamycin B (belonging to the family of ansamycin) for the separation of several amino alcohols by CZE, recorded analytes as negative peaks (indirect UV detection) due to the strong absorption of the BGE [4]. We also adopted indirect detection of UV nonabsorbing compounds by CZE employing vancomycin for the chiral analysis. Aspartic and glutamic acid enantiomers were separated in a polyacrylamide-coated capillary. A BGE containing also sorbic acid/histidine in the pH range 4.5–6.5 was used. The optimized method was successfully applied to the analysis of teeth dentine and beer [44]. The same experimental conditions were used by us for the enantiomeric resolution of carboxymethylcysteine and its derivatives [45].

5.4 ELECTROMIGRATION SEPARATION MODES EMPLOYING MACROCYCLIC ANTIBIOTICS

One of the most important characteristics of EMs is represented by the different CE modes available for the improvement of analytes' separation. Among them CZE and CEC may offer high enantioresolution capability by employing several chiral selectors including macrocyclic antibiotics.

5.4.1 CAPILLARY ZONE ELECTROPHORESIS

In Section 5.2, we already described the separation of chiral compounds by using CZE where the chiral selector is simply added to the BGE for the electrophoretic run. Here the macrocyclic antibiotic is generally moving in the opposite direction of analytes altering in a different way the effective mobility of enantiomers. As already illustrated acidic racemic compounds such as amino acid derivatives, herbicides, drugs, peptide derivatives were resolved into their enantiomers by using vancomycin, teicoplanin, ristocetin, and teicoplanin derivatives. In all cases, negatively charged compounds have been analyzed [13,16,17,30,33,41,44–49]. Although several macrocyclic antibiotics have been used in CZE for the chiral separation of negatively charged compounds, this does not mean that basic analytes cannot be resolved into their enantiomers employing such class of chiral selectors. A large number of positively charged enantiomers have been separated by using CZE in the presence of rifamycin B or rifamycin SV and this fact has been reported in literature. The list includes β-blockers, sympathomimetic drugs, amphetamines, or their analogs, etc. [4]. The two antibiotics are *ansa* compounds with small differences in their chemical structure. Rifamycin B contains a carboxylic group outside the cavity while rifamycin SV has a phenolic function. Depending on the buffer pH (conditions usually used in CZE), rifamycin B is negatively charged while SV is uncharged.

It has been reported that in the chiral resolution of basic compounds with rifamycin B, electrostatic interaction played a very important role in the enantiorecognition mechanism [4]. Some review papers dealing with chiral separations by CE using antibiotics as chiral selectors appeared in the last decade and have documented the work done in this field [6–8,50–58]. In Table 5.1, the main studies concerning the

TABLE 5.1
Use of Antibiotics as Chiral Selectors in Capillary Zone Electrophoresis

Compounds	Chiral Selector	BGE	Ref.
Ala-Ala, Ala-Gly, Ala-Gly-Gly, Ala-Leu, Gly-Ala, Gly-Asn, Gly-Asp, Gly-Leu, Gly-Met, Gly-Phe, Gly-Val, Leu-Ala, Leu-Gly, Leu-Gly-Gly Leu-Leu	Vancomycin (1 mM) Teicoplanin (1 mM)	50 mM phosphate buffer pH 7.5, 10% isopropanol 12.5 mM phosphate + 12.5 mM acetate-tris buffer, 40% ACN	[35]
Carprofen, flurbiprofen, ketoprofen, naproxen, ibuprofen, 2′-hydroxyibuprofen, carboxy-ibuprofen, etodolac, etodolacglucuronide, 7-hydroxyetodolac, 7-hydroxyetodolac-glucuronide	Vancomycin (5–12 mM) (partial filling)	50 mM acetic acid–ammonium acetate buffer pH 4.8 CE-MS	[41]
FMOC-(Ala, Asn, Asp, Glu, Isoleu, Leu, Met, Norleu, Norval, Phe, Pro, Ser, Val), ketoprofen, dimetoxy-5,6,5′6′-bismethylenedioxybiphenyl- 2,3′-dicarboxylic acid (DBDA)	Vancomycin (2 mM)	50 mM Tris-phosphate buffer pH 6.0, 0.002% (w/v) HDB (co-electroosmotic flow)	[59]
Carprofen, cicloprofen, fenoprofen, flurbiprofen, ibuprofen, ketoprofen, indoprofen, naproxen, suprofen, loxiglumide, acenocoumarol, warfarin dichlorprop, diclofop acid, fenoprop, fenoxaprop acid, flamprop acid, fluazifop acid, haloxyfop acid, mecoprop	(Hepta-Tyr) (1–3 mg/mL) (partial filling)	50 mM Britton Robinson buffer pH 5, 20% ACN (polyacrylamide-coated capillary)	[47]
Carprofen, fenoprofen, flurbiprofen, indoprofen, ketoprofen, suprofen	Actaplanin A (0.5 mM)	40 mM phosphate buffer pH 6, 15%–30% 2-methoxyethanol	[21]
N-3,5-Dinitrobenzoyl-alanine, 2-aminoadipic acid, α-amino-*n*-butyric acid, β-amino-*n*-caprylic acid, α-amino-*n*-butyric acid, 3-amino-3-phenylpropionic acid, α-amino-1-thiopheneacetic acid, aspartic acid, ethionine, *o*-fluorophenylglycine, glutamic acid, homoserine, isoleucine, leucine, methionine, norleucine, norvaline, serine, valine	Avoparcin (0.4 mM) Ristocetin A (0.4 mM) Teicoplanin (0.4 mM) Vancomycin (0.4 mM)	50 mM phosphate buffer pH 6	[22]

(*continued*)

TABLE 5.1 (continued)
Use of Antibiotics as Chiral Selectors in Capillary Zone Electrophoresis

Compounds	Chiral Selector	BGE	Ref.
Dansyl-phenylalanine, *N*-3,5-dinitrobenzoyl-(alanyl-glycine, α-amino-*n*-caprylic acid, β-amino-*n*-butyric acid, α-aminopimelic acid, α-amino-2-thiopheneacetic acid), *N*-3,5-dinitrobenzoyl-(α-amino-3-thiopheneacetic acid, arginine, asparagine, 5-benzyloxy-tryptophan, *p*-bromophenylalanine) *N*-3,5-dinitrobenzoyl-[*N*-carbamyl-norleucine], *N*-3,5-dinitrobenzoyl-[*N*-carbamyl-valine], *N*-3,5-dinitrobenzoyl-(*p*-chlorophenylalanine, cysteine, DOPA, *o*-fluorophenylalanine, *m*-fluorophenylalanine, *p*-fluorophenylalanine, *m*-fluorotyrosine, glutamine), *N*-3,5-dinitrobenzoyl-(glycyl-alanine, glycyl-phenylalanine, 5-hydroxytryptophan, iopanoic acid, 5-methyltryptophan, 6-methyltryptophan, 7-methyltryptophan), *N*-3,5-dinitrobenzoyl-(α-methyltyrosine, *p*-nitrophenylalanine, phenylalanine, α-phenylglycine, theanine, threonine, tryptophan, *o*-tyrosine, *m*-tyrosine, tyrosine), fenoprofen, indoprofen, ketoprofen, naproxen, suprofen	Avoparcin (0.4 mM)	50 mM phosphate buffer pH 6	[22]
Dansyl aspartic acid, dansyl glutamic acid, dansylphenylalanine, dansylvaline, carprofen, fenoprofen, indoprofen, ketoprofen, *o*-methoxymandelic, warfarin, DNP-norleucine, DNP-alanine	Ristocetin A (2 mM)	100 mM phosphate buffer pH 6 (polyacrylamide-coated capillary) countercurrent process	[60]
Fenoprop, fluazifop, haloxyfop, mecoprop	Vancomycin (8 mM) partial filling	75 mM Britton Robinson buffer pH 5, 10 mM γ-CD (polyacrylamide-coated capillary)	[61]

Mandelic acid, *m*-hydroxymandelic acid, *p*-hydroxymandelic acid, 3,4 dihydroxymandelic acid, 4 Cl-mandelic acid, 2-phenyllactic acid, 3-phenyllactic acid	Hepta-Tyr (1–4 mg/mL) partial filling	75 mM Britton Robinson buffer pH 4, 20% ACN (polyacrylamide-coated capillary)	[48]
Dansyl (aspartic acid, aminocaprylic acid, amino-*n*-butyric acid, glutamic acid, leucine, methionine, norleucine, norvaline, phenylalanine, serine, threonine, tryptophan, valine)	A35512B (5 mM)	40 mM phosphate buffer pH 7, (10%–30% 2-methoxyethanol)	[62]
DNP-glutamic acid	Vancomycin (2 mM) partial filling	20 mM phosphate buffer pH 7, [poly(AEG) covalent coating, poly(DMA) adsorbed coating, epoxy poly(AG-AA) adsorbed coating]	[63]
N-tert-butyloxycarbonyl-(arginine, phenylalanine, tyrosine, trypthophan)	Teicoplanin (0.5 mM)	60 mM phosphate buffer pH 4.3, 10% ACN	[64]
Seleno-cystine, selenoethionine, selenomethionine	Vancomycin (2 mM)	20 mM MOPS/tris buffer pH 7	[65]
Aspartic acid, glutamic acid	Vancomycin (10 mM) partial filling	10 mM sorbic acid/histidine buffer, pH 5 (polyacrylamide-coated capillary)	[44]
Trans,trans-abscisic acid, *N*-acetyl-tryptophan, *N-t*-Boc-tryptophan, Indole-3-lactic acid, *cis,trans*-abscisic acid, *N*-CBZ-tryptophan, Indoline-2-carboxylic acid, *N*-(3-indolylacetyl)-aspartic acid	Teicoplanin (5 mM) (10%–30% ACN) Vancomycin (2 mM) Ristocetin A (2 mM)	100 mM phosphate buffer pH 6	[34]
N-Acetamido-*S*-carboxymethylcysteine, *S*-carboxymethylcysteine	Vancomycin (1–10 mM) partial filling	10 mM sorbic acid/histidine buffer, pH 5 (polyacrylamide-coated capillary)	[45]
2-[(5′-Benzoil-2′-hydroxy)phenyl]-propionic acid (DF-1738y), 2-[(4′-benzoiloxy-2′-hydroxy)phenyl]-propionic acid (DF-1770Y)	Vancomycin (1–5 mM) partial filling	50 mM Britton Robinson buffer pH 5 (polyacrylamide-coated capillary)	[66]

(continued)

TABLE 5.1 (continued)
Use of Antibiotics as Chiral Selectors in Capillary Zone Electrophoresis

Compounds	Chiral Selector	BGE	Ref.
Atrolactic acid, 1,1-binaphtyl-2,2-hydrogen diylphosphate, carprofen, 2-(3-chlorophenoxy)propionic acid, fenoprofen, *p*-hydroxymandelic acid, mandelic acid, 2-(4-nitrophenyl) propionic acid, 2-phenoxypropionic acid	Vancomycin (4 mM) Ristocetin A (4 mM) (synergistic chiral separations vanco/risto)	100 mM phosphate buffer pH 6 (polyacrylamide-coated capillary)	[67]
2-[(4′-Benzoiloxy-2′-hydroxy)phenyl]-propionic acid (DF-1770Y)	Vancomycin (7 mM) partial filling	50 mM Britton Robinson buffer pH 6.4 (polyacrylamide-coated capillary)	[68]
N-Acetyl-cysteine, *N*-acetyl-glutamic acid, *N*-acetyl-proline, *N*-acetyl-serine, *N*-acetyl-tyrosine	Vancomycin (2.5 mM) partial filling	20 mM ammonium acetate buffer pH 5 (polyacrylamide-coated capillary)	[49]
Dansyl-(α-amino-*n*-butyric acid, glutamic acid, leucine, methionine, norleucine, phenylalanine, serine, threonine, valine), ketoprofen	Vancomycin (2 mM) partial filling	50 mM Tris/phosphate buffer pH 6.2, 0.001% HDB (co-EOF)	[43]
Dansyl-(glutamic acid, serine, valine, threonine), ketoprofen, suprofen	Vancomycin (1 mM) A82846B (1 mM) (analogue of vancomycin)	50 mM phosphate buffer pH 6, 0.002% (w/v) HDB (co-EOF)	[69]
Biphenyldimethylester derivatives	Erythromycin (20 mM)	50 mM phosphate buffer pH 6, 50% MeOH	[70]
Haloxyfop, fluazifop, fenoxafrop, flamprop, diclofop, mecoprop, dichlorprop, fenoprop, 2-phenoxypropionic acid	Vancomycin (1–10 mM) (enantiomer binding)	10–60 mM sodium acetate buffer pH 6.5	[36]
Dansyl-Norleu, dansyl-Met, dansyl-Leu, dansyl-ABA	Balhimycin (2 mM) Vancomycin (2 mM)	50 mM Tris-phosphate buffer pH 6.0, 0.001% (w/v) HDB 40 mM sodium phosphate buffer pH 6.0, 0.001% (w/v) HDB	[29]

Dopa, carbidopa, leucovorin, warfarin, indoprofen, ketoprofen, propiomazine, econazole, miconazole, sulconazole, isoprenaline, bamethane, bupivacaine, verapamil, chlorpheniramine, fenfluramine, betaxolol, metoprolol, oxprenolol, pindolol, propranolol	Erythromycin A, erythromycin A *N*-oxide, anhydroerythromycin A, anhydroerythromycin A *N*-oxide, erythralosamine, erythralosamine *N*-oxide (0.1–10 mM)	Sodium phosphate pHs 3.0 and 7.0 sodium borate pH 9.2	[71]
Dansyl-(norvaline, phenylalanine, serine, leucine, norleucine, methionine, glutamic acid, valine, ABA, threonine, aspartic acid), ketoprofen, 2,4,5-TP, pirprofen, tiaprofenic acid, flurbiprofen	Balhimycin (2 mM) Bromobalhimycin (2 mM) (partial filling)	50 mM Tris-phosphate buffer pH 6.0, 0.001% (w/v) HDB	[72]
Dansyl-(leucine, valine, threonine, serine, tryptophan, glutamic acid, norleucine, methionine, aspartic acid, phenylalanine, norvaline) ketoprofen, suprofen, fenoprofen, flurbiprofen, pirprofen, tiaprofenic acid	Balhimicin (2.5 mM) Dechlorobalhimicin (2.5 mM)	50 mM Tris-phosphate buffer pH 6.0, 0.001% (w/v) HDB	[19]
Analogues of antihepatitis drug dimethyl diphenyl bicarboxylate (DDB)	Vancomycin (6 mM)	40 mM Tris-phosphate buffer pH 6.0, 0.001% HDB	[73]
Demethylated analogues of clofibric acid	Vancomycin (10 mM)	25 mM ammonium acetate buffer pH 5.0	[74]
FMOC-(valine, phenylalanine, alanine, methionine, serine, leucine, isoleucine), ketoprofen, fenoprofen	Vancomycin (2 mM) partial filling	50 mM Tris-phosphate pH 6.0 poly(dimethylacrylamide) PDMA-coated capillary	[75]
DL-lactic acid	Vancomycin (10 mM)	5 mM oxalate L-histidinium pH 4.5	[76]
Chlorpheniramine, salsolinal, propranolol	β-CD-derivatized erythromycin (15 mM)	20 mM phosphate buffer pHs 3.01, 4.98	[77]

use of macrocyclic antibiotics in CZE for the resolution of enantiomers, published since 1998, are reported.

5.4.2 Capillary Electrochromatography

Another valuable way to perform CE with improved selectivity is given by CEC. This method has been used for the separation of a large number of compounds in either uncharged or charged form [78–83]. CEC is a modern separation method which uses the stationary phase under electrophoretic experimental conditions. Mobile phase and analytes are moved toward the detector by an EOF generated by applying a high electric field. Due to the flat flow profile of the EOF, peak efficiency is very high compared to that usually achieved in HPLC. Therefore the possibility to combine high efficiency of CE with high selectivity and high load capacity of liquid chromatography offers a powerful tool that can be advantageously applied in analytical chemistry including chiral analysis (see also Chapter 14). Several chiral selectors have been employed in CEC either bonded to the capillary wall (open-tubular mode) or to silica particles or to polymeric material (monoliths) [84–87] (see also Chapter 15). Macrocyclic antibiotics were also studied achieving excellent results [88–91].

The first report concerning the use of vancomycin modified silica stationary phase in CEC was published in 1998 by Dermaux et al. [88]. Warfarin and hexobarbital enantiomers were separated in a 100 μm ID capillary packed with vancomycin stationary phase. Enantiomers were eluted using a mixture of triethylamine solutions at different pH values and acetonitrile (80:20 or 90:10, v/v) to achieve baseline resolution. Tryptophan and dinitrobenzoyl (DNB) leucine were separated into their enantiomers by using CEC in a capillary column packed with commercially available teicoplanin silica phase of 5 μm particles. The macrocyclic antibiotic exhibited good interaction with analytes possessing both amino and carboxylic groups [92]. Later on Karlsson et al. [93] demonstrated that a larger number of racemic compounds could be separated by CEC employing a teicoplanin based stationary phase. Aqueous and nonaqueous mobile phases were applied for the enantiomeric resolution of compounds of pharmaceutical interest. Basic, neutral, and acidic enantiomers were successfully separated. A larger number of racemic compounds were separated by CEC using a capillary column packed with diol silica with in situ immobilized vancomycin. Good enantioresolution was obtained for compounds containing amino groups with the exception of thalidomide [94]. Vancomycin chiral stationary phase was evaluated in CEC using both polar organic and reversed-phase mode by several groups, either using commercially available CSP or CSP synthesized in the laboratory [89,95,96].

We synthesized a vancomycin silica stationary phase (Van-SP) for packing 75 or 100 μm ID capillaries. Since the chemical reaction was done using diol silica particles, native silica was mixed with Van-SP in order to have silanol groups in the stationary phase for the generation of a strong EOF [96]. The influence of mobile phase composition (pH and buffer concentration, organic modifier type, and concentration) on EOF, chiral resolution, and enantioselectivity was studied. Several basic compounds such as atenolol, oxprenolol, propranolol, venlafaxine, terbutaline, acebutolol, metoprolol, pindolol, and mefluoquine were baseline resolved in their enantiomers. The packed Van-SP was also studied by using a polar organic mobile phase. Optimum conditions for the enantiomeric resolution of basic compounds were obtained using ACN-MeOH with 13 mM ammonium acetate. Modifying the organic solvent ratio, the elution order of *R*-venlafaxine and one of its metabolite enantiomer was reversed [97]. The capillary columns packed with Van-SP were used for some applications in biomedical analysis, e.g., venlafaxine and *O*-desmethylvenlafaxine enantiomers in human plasma [89], fluoxetine and norfluoxetine enantiomers in human plasma [90], mirtazapine, and 8-hydroxymirtazapine and *N*-desmethylmirtazapine enantiomers in human urine samples [91].

As an example of the applicability of CEC to the analysis of chiral compounds in biological samples, Figure 5.4 shows the analysis of extracted plasma samples from patients treated with venlafaxine. As can be observed, different enantiomeric ratios for both patients were obtained [89].

Grobuschek et al. used a teicoplanin aglycone stationary phase for the enantiomeric resolution of several amino acids and drugs [98]. The chiral selector was bonded to 3.5 μm particles and the CEC experiments were pressure assisted probably due to the low EOF. The same chiral stationary phase was used by the same group

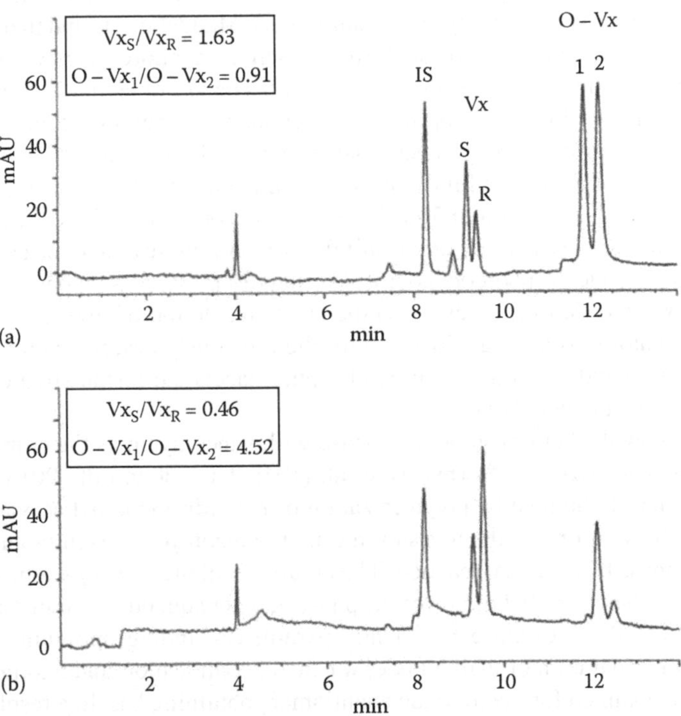

FIGURE 5.4 Analysis of extracted plasma samples (a) and (b) from different patients under therapy with venlafaxine. Experimental conditions: capillary 75 μm ID × 35 cm total length; 23 cm was the packed segment with vancomycin stationary phase; mobile phase, 100 mM ammonium acetate pH 6/water/acetonitrile (5:5:90, v/v); sample injection, electrokinetic 10 kV × 30 s. (Reproduced from Fanali, S. et al., *J. Chromatogr. A*, 919, 195, 2001. With permission.)

for the enantioresolution of some dipeptide enantiomers and the results achieved in CEC were compared to those obtained in microbore HPLC [99]. Better results concerning efficiency were obtained in CEC. Teicoplanin stationary phase (Te-SP) synthesized in our laboratory was studied by our group for the separation of several basic compounds by CEC [100]. Particles of different size (3.5 or 5 μm) were used in the presence or absence of native silica. The best results, regarding resolution and efficiency, were obtained both when Te-SP was used alone and when it was used with 3.5 μm particles.

As can be noticed, vancomycin and teicoplanin were the macrocyclic antibiotics used and with those chiral stationary phases mainly basic compounds were separated in their enantiomers. The search for new chiral selectors propelled our group to find a macrocyclic antibiotic that can be used for the enantioseparation of acidic compounds by CEC. Hepta-Tyr, a semisynthetic antibiotic belonging to the family of teicoplanin, was bonded to silica particles and packed into 75 μm ID capillaries. Amino silica material was mixed with the chiral stationary phase in the ratio 1:3 (w/w) in order to increase the EOF. In fact, the presence of amino groups only, as chargeable sites, in the antibiotic molecule was responsible for a reversed EOF. The newly packed capillary was studied using an aqueous reversed-phase mode for the separation of several hydroxyl acid enantiomers. Mandelic acid and their hydroxyl derivatives, 2- and 3-phenyllactic acid, were baseline enantioresolved while racemic mandelic acid methyl ester was not resolved [20]. The use of Hepta-Tyr in CEC was further investigated by our group trying to reduce the separation time of selected enantiomers. Optimizing experimental parameters such as organic modifier, buffer, pH, and the use of a short end injection we obtained the enantiomeric separation of racemic mandelic acid in about 70 s. Those results could be easily obtained due to the very high enantioselectivity of the chiral stationary phase and to the combination of high EOF and electrophoretic mobility of analytes moving as anions to the detector [101]. We extended the study to (1) the chiral resolution of several nonsteroidal anti-inflammatory drugs or analogous with the previously described chiral stationary phase [102] and (2) to the analysis of a pharmaceutical formulation containing loxiglumide enantiomers [103].

A continuous-bed chiral stationary phase with vancomycin as the chiral selector was in situ synthesized by Kornysova et al. [104]. The column of 100 μm ID contained a material obtained by polymerization of *N*-(hydroxymethyl)acrylamide and piperazine diacrylamide with vinyl sulfonic acid. Vancomycin was immobilized on a polymeric phase by chemical reaction. The column exhibited a very strong EOF useful for CEC separations in both reversed-phase and nonaqueous polar organic mode. The authors studied several experimental parameters finding optimum conditions for the chiral separation of thalidomide, warfarin, coumachlor, and felodipine. Best results were obtained for thalidomide enantiomers obtaining baseline resolution and high efficiency (120,000 N/m). Later on the same group presented a different approach to prepare polymeric material with the aim to simplify the synthesis procedure of the capillary containing nonparticulate vancomycin bed. *N-N′*-diallyltartardiamide monomer containing a diol group was copolymerized and reacted, after oxidation to aldehyde, with vancomycin. The prepared chiral stationary phase showed good enantioselectivity toward thalidomide, bupivacaine, and warfarin [105].

The use of continuous bed or monolithic columns in CEC can be advantageous because several problems related to packed capillaries are eliminated. Monolithic material does not contain frits, generally responsible for bubble formation, has no limitation for the column's length, and so forth. Just for the briefly mentioned advantages other groups investigated the possibility to use monolithic columns modified with macrocyclic antibiotics for the enantiomeric separation by CEC. Gatschelhofer et al. [106] separated several glycyl-dipeptides by CEC using monoliths prepared by ring-opening metathesis polymerization (ROMP) with teicoplanin aglycone. A sol-gel process was used by Dong et al. [107] to prepare a silica monolithic capillary with immobilized vancomycin. Good enantioseparation was obtained for eight racemic compounds by using both nonaqueous polar organic or aqueous mobile phases. Column efficiencies as high as 217,400 N/m were obtained.

Table 5.2 reports a summary of the studies done with CEC using stationary phases containing macrocyclic antibiotics as chiral selectors.

5.5 "LESS COMMON" USED CHIRAL SELECTORS IN CAPILLARY ELECTROMIGRATION METHODS

Based on the data reported till now in literature, it is evident that macrocyclic antibiotics exhibit very high enantioselectivity toward a large number of compounds due to many stereogenic centers in their structure. However different separation methods, experimental conditions, and most of all, new chiral selectors are nevertheless required to obtain better results for discriminating aged and new chiral drugs.

Actaplanin A ($Mr = 1970.27$) is a macrocyclic antibiotic belonging to the family of ristocetin and ristomycin. An amino sugar and several neutral sugars are bonded to a peptide group [21]. Nonsteroidal anti-inflammatory drugs, namely, carprofen, fenoprofen, flurbiprofen, ketoprofen, indoprofen, and suprofen were resolved into their enantiomers by CZE using this antibiotic. The chiral resolution of the analytes was investigated in the pH range 5–7 and in the presence of organic modifiers in the BGE. All compounds were resolved in their enantiomers using a phosphate buffer at pH 6 containing 0.5 mM of Actaplanin A and 2- methoxyethanol in the concentration range 15%–30%.

In 1998 Avoparcin, a new glycopeptide antibiotic, was analyzed by HPLC and two main structurally related compounds were characterized [114]: α-Avoparcin (*Mr* 1909) and β-avoparcin (*Mr* 1944). Several asymmetric centers, seven aromatic groups, 16 hydroxyl, one carboxylic, and three amino groups were studied and confirmed by nuclear magnetic resonance (NMR) [22]. The macrocyclic antibiotic was studied for the enantioseparation of a large number of compounds by CZE by changing the buffer pH, the concentration of avoparcin etc. Interesting results were obtained by modifying the content of organic modifier. The authors analyzed selected racemic compounds, namely, *N*-3,5-DNB-*p*-fluoro-phenylalanine, *N*-3,5-DNB-leucine, and *N*-3,5-DNB-methionine. The presence of low concentrations of organic modifiers (<10% methanol, acetonitrile, isopropanol) in the BGE containing avoparcin increased the enantioresolution of studied compounds. This effect was ascribed to the fact that the presence of organic modifier in the BGE destroyed the aggregation of the antibiotic and therefore increased resolution.

TABLE 5.2
Use of Antibiotics as Chiral Selectors in Capillary Electrochromatography

Compounds	Stationary Phase	Mobile Phase	Ref.
Warfarin, hexobarbital	Vancomycin immobilized 5 µm spherical silica gel	Reversed-phase mode ACN/0.1% TEA-acetic acid pH 5 (80:20, v/v) pH 6.6 (90:10, v/v)	[88]
Tryptophan, dinitrobenzoyl leucine	5 µm chirobiotic T	Reversed-phase mode ACN/2 mM phosphate, pH 7 (70:30, v/v) ACN/5 mM phosphate, pH 2.3 (70:30, v/v)	[92]
Alprenolol, atenolol, bupivacaine, ephedrine, isoprenaline, ketamine, metoprolol, phenylamine, practolol, thalidomide	Vancomycin-modified 5 µm LiChrospher diol silica	Polar organic mode MeOH/ACN/acetic acid/TEA (80:20:0.1:0.2; 80:20:0.2:0.2; 50:50:0.2:0.2; 20:80:0.1:0.2; 20:80:0.3:0.1, v/v/v/v)	[94]
Benzoin, bupivacaine, coumachlor, 5,5-diphenylhydantoin, felodipine, 5-(4-hydroxyphenyl)-5-phenylhydantoin, ibuprofen, 5-(4-methylphenyl)-5-phenylhydantoin, *N*-CBZ-L-glutamic acid, terbutaline, tryptophan, warfarin	5 µm chirobiotic T	Reversed-phase mode ACN/0.1% TEA-acetic acid, pH 4 (40:60, v/v) ACN/0.2% TEA-acetic acid, pH 4 (20:80, 50:50 v/v)	[93]
Alprenolol, atenolol, bupivacaine, fenoterol, β-hydroxy-phenetyl-amine, labetalol, metoprolol, phenylpropanol-amine, pindolol, propranolol, sotalol, terbutaline, verapamil	5 µm chirobiotic T	Polar organic mode MeOH/ACN/TEA/acetic acid (80:20:0.1:0.1; 95:5:0.05:0.3; 95:5:0.3:0.3, v/v/v/v)	[93]
Alprenolol, atenolol, bupivacaine, fenoterol, ketamine, labetalol, metoprolol, pindolol, propranolol, sotalol, terbutaline, thalidomide, verapamil	5 µm chirobiotic V	Polar organic mode ACN/MeOH/TEA/acetic acid (20:80:0.1:0.1; 15:85:0.1:0.1, v/v/v/v)	[95]

Binaphtol, coumachlor, felodipine, thalidomide, warfarin	5 μm chirobiotic V	Reversed-phase mode ACN/0.1% TEA-acetic acid pH 4 (30:70, v/v) pH 5 (30:70, 40:60, v/v)	[95]
Alprenolol, atenolol, fenoterol, metoprolol, pindolol, propranolol, sotalol	5 μm chirobiotic T	Polar organic mode MeOH/ACN/TEA/acetic acid (95:5:0.05:0.05, v/v/v/v)	[108]
Acebutolol, atenolol, clenbuterol, mefloquine, metoprolol, mianserin, oxprenolol, pindolol, propranolol, terbutaline, tolperisone, venlafaxine	Vancomycin-modified 5 μm LiChrospher diol silica	Reversed-phase mode ACN/100 mM ammonium acetate pH 6/H_2O (90:5:5, v/v/v)	[96]
Coumachlor, felodipine, thalidomide, warfarin	Vancomycin immobilized Acrylamide continuous bed	Reversed-phase mode ACN/triethylamine acetate buffer pH 5, 6.5 (15:85, v/v)	[104]
Venlafaxine, *O*-desmethylvenlafaxine	Vancomycin-modified 5 μm LiChrospher diol silica	Reversed-phase mode ACN/100 mM ammonium acetate pH 6/water (90:5:5, v/v/v)	[89]
Alprenolol, clenbuterol, mianserin, oxprenolol, propranolol, salbutamol, tolperisone, venlafaxine, *O*-desmethylvenlafaxine	Vancomycin-modified 5 μm LiChrospher diol silica	Polar organic mode ACN/MeOH/13 mM ammonium acetate (40:60, v/v)	[97]
Fluoxetine, norfluoxetine	Vancomycin-modified 5 μm LiChrospher diol silica	Reversed-phase mode ACN/MeOH/100 mM ammonium acetate pH 6/water (55:35:5:5, v/v/v/v)	[90]
DOPA, methyl-DOPA, *o*-tyrosine, phenylalanine, phenylserine, threo-3,4-dihydroxyphenylserine, 2,4,5-trihydroxyphenylalanine, tryptophan, *p*-tyrosine, 2-methyl-3-(3,4-dihydroxyphenyl)alanine, methyltryptophan, histidine	Teicoplanin aglycone 3.5 μm immobilized silica	Reversed-phase mode MeOH/0.2%TEA-acetic acid pH 4.1 (40:60, v/v)	[98]

(*continued*)

TABLE 5.2 (continued)
Use of Antibiotics as Chiral Selectors in Capillary Electrochromatography

Compounds	Stationary Phase	Mobile Phase	Ref.
Acebutolol, alprenolol, atenolol, bamethane, befunolol, betaxolol, bisoprol, bopindolol, buphenine, bupranol, carazolol, carvediol, celiprolol, chlortalidone, dipivefrine, epinephrine, etilephrine, fendilin, fenoterol, gallopamil, indoprofen, isoproterenol, mepindolol, metanephrine, methoxamine, metipranolol, metoprolol, norephedrine, normetanephrine, octopamine, orciprenaline, oxprenolol, pindolol, promethazine, propranolol, synephrine, sotalol, suprofen, thioridazine, tolamolol, warfarin	Teicoplanin aglycone 3.5 µm immobilized silica	Polar organic mode MeOH/ACN/TEA/acetic acid (80:20:0.2:0.2; 90:10:0.2:0.2, v/v/v/v)	[98]
Glycyl-(alanine, asparagine, aspartate, leucine, methionine, norleucine, norvaline, phenylalanine, serine, threonine, tryptophan, valine)	Teicoplanin aglycone 3.5 µm immobilized silica	Reversed-phase mode MeOH/0.2%TEA-acetic acid pH 4.1 (20:80, v/v) MeOH/ACN/0.2%TEA-acetic acid pH 4.1 (20:30:50, v/v/v)	[99]
Mandelic acid, *m*-, *p*-hydroxymandelic acid, 3,4-dihydroxymandelic acid, 3-hydroxy-4-methoxymandelic acid 4-chloromandelic acid, 2-phenyllactic acid, 3-phenyllactic acid, mandelic ethyl esther	(Hepta-Tyr)-modified 5 µm LiChrospher diol silica mixed with 5 µm amino silica	Reversed-phase mode ACN/50 mM ammonium acetate pH 6/water (5:1:4, v/v/v)	[20]
Ala-Ala, Ala-Met, Ala-Phe, Ala-Ser, Ala-Val, Leu-Ala, Leu-Leu, Leu-Phe, Leu-Val, Ala-Gly-Gly, Ala-Leu-Gly, Gly-Leu-Ala, Leu-Gly-Gly, Leu-Gly-Phe	Teicoplanin aglycone 3.5 µm immobilized silica	Reversed-phase mode MeOH/ACN/0.2%TEA-acetic acid pH 4.1 (20:30:50, v/v/v)	[109]
Acebutolol, alprenolol, atenolol, fluoxetine, metoprolol, nor-fluoxetine, oxprenolol, pindolol, propranolol, salbutamolol	Teicoplanin 3.5 and 5 µm Immobilized silica gel	Polar organic mode MeOH/ACN/ammonium acetate(60:40, v/v and 0.05% w/v)	[100]

Carprofen, cicloprofen, fenoprofen, ketoprofen, indoprofen, suprofen, 2-[(5′-benzoyl-2′hydroxy)phenyl]-propionic acid, 2-[(4′-benzoyloxy-2′-hydroxy)phenyl]propionic acid, 2-(4′-isobutylphenyl)-3-methylbutanoic acid, 2-(4′-isobutylphenyl)-butanoic acid, 2-(4′-isobutylphenyl) ciclopentylacetic acid	Vancomycin-modified 5 µm LiChrospher diol silica	Reversed-phase mode ACN/100 mM ammonium formate pH 3.5/water (90:5:5, v/v/v)	[102]
	(Hepta-Tyr)-modified 5 µm LiChrospher diol silica mixed with 5 µm amino silica (reversed EOF)	Reversed-phase mode ACN/100 mM sodium phosphate pH 6/water (50:5:35, v/v/v)	
Albuterol, carteolol, celiprolol, promethazine	5 µm chirobiotic V (p-CEC)	Polar organic phase MeOH/AcOH/TEA (100:0.1:0.1, v/v/v)	[110]
D, L loxiglumide	(Hepta-Tyr)-modified 5 µm LiChrospher diol silica mixed with 5 µm amino silica (reversed EOF)	Reversed-phase mode ACN/100 mM sodium phosphate pH 6 (50:50, v/v)	[103]
Mianserin	5 µm chirobiotic V sample focusing	9.4 mM TEAA pH 4.8/ACN (25:75, v/v)	[111]
Bupivacaine, thalidomide, warfarin	Vancomycin-based nonparticulate bed (monolithic)	Aqueous 0.15% TEAA pH 4.6/20% ACN	[105]
Oxprenolol, alprenolol, pindolol, metoprolol, propranolol, talinolol, atenolol, carteolol	3 µm vancomycin and teicoplanin CSPs	MeOH/ACN/AcOH/TEA (70:30:1.6:0.2, v/v/v/v) CEC-UV and CEC-MS	[112]
Gly-Ala, Gly-Asn, Gly-Asp, Gly-Leu, Gly-Met, Gly-Nle, Gly-Nva, Gly-Phe, Gly-Ser, Gly-Thr, Gly-Trp, Gly-Val	3 µm teicoplanin aglycone immobilized on particle-loaded monoliths by ring-opening metathesis polymerization (ROMP)	50% aq. TEAA solution pH 4.1 (adjusted with 0.2% AcOH)/50% ACN 40% aq. TEAA solution pH 4.1 (adjusted with 0.2% AcOH)/40% MeOH/20% ACN	[106]
Mirtazapine, 8-hydroxymirtazapine, *N*-desmethylmirtazapine	Vancomycin-modified 5 µm LiChrospher diol silica	100 mM ammonium acetate pH 6/H_2O/MeOH/ACN (5:15:30:50, v/v/v/v)	[91]
Propranolol, celiprolol, esmolol, bisoprolol, atenolol, metoprolol, carteolol, carvedilol, clenbuterol, bambuterol, terbutaline, isoprenaline, salbutamol, phenylpropanolamine, methylephedrine	5 µm vancomycin CSP	0.025% TEAA (pH 7)/MeOH (90/10, v/v) 0.05% TEAA (pH 7)/MeOH (90/10, v/v)	[113]
Propranolol, pindolol, metoprolol, atenolol, terbutaline, alprenolol, thalidomide, benzoin	Vancomycin immobilized silica monolith	MeOH/ACN/AcOH/TEA (80:20:0.1:0.1, v/v/v/v) 10 mM TEA phosphate buffer pH 6.5/20%ACN	[107]

Some of the *new* macrocyclic antibiotics used in CE were already introduced in Section 5.2.2. We studied Hepta-Tyr antibiotic, belonging to the family of teicoplanin, and found that it achieved good results in both CZE and CEC (as reported before). This compound is a semisynthetic glycopeptide related to the amides class of 34-de(acetylglucosaminyl)-34-deoxy teicoplanin and exhibited antibiotic activity against *Enterococcus fecalis* [47]. Because of its potential use in chiral separation, we studied the use of Hepta-Tyr as a BGE additive by CZE. Coated capillaries were used for experiments in order to minimize the antibiotic's adsorption and the EOF. Countercurrent partial-filling method was applied. 20 acidic compounds belonging to different classes of drugs and herbicides were studied by changing the BGE composition, the organic modifier, the antibiotic concentration, and the capillary temperature [47]. The same chiral selector was also applied to the chiral resolution of several hydroxyl acid derivatives by CZE [48]. Other *less commonly used* macrocyclic antibiotics employed in CE include A35512B [62], balhimycin, and bromobalhimycin [19,29,72].

Recently Petrusevska et al. reported a study where a new glycopeptide antibiotic, namely, eremomycin was bonded to the stationary phase and used for the separation of several amino acid enantiomers. This compound is quite similar to vancomycin but it differs in the trinucleus amino acid fragment by only one chlorine atom. Additionally eremomycin has two amino sugars (in vancomycin only one is present) [115]. The authors separated enantiomers of several α-amino acids that were not resolved by vancomycin stationary phase. The described macrocyclic antibiotic has not been used in CZE or with other electromigration methods.

The search for new chiral selectors to be used in the field of electromigration methods aimed at the study of a β-cyclodextrin derivatized with erythromycin by Dai et al. [77]. Chlorpheniramine, salsolinal, and propranolol were resolved into their enantiomers by optimizing some experimental parameters such as pH, concentration of the chiral selector, etc. β-cyclodextrin and erythromycin alone did not give good enantiomeric resolution of the studied compounds. On the other hand, it was already demonstrated that erythromycin alone did not resolve any enantiomers among the 21 racemic mixtures studied [71]. However, the authors demonstrated that the compounds interacted with the antibiotic added to the BGE modifying their effective mobility. The lack of enantioselectivity was ascribed to the size of the antibiotic aglycone ring, that is, only half as big as the β-cyclodextrin cavity.

5.6 CONCLUSIONS AND FUTURE TRENDS

CE, in its two separation modes, namely, CZE and CEC, is a recently developed technique successfully applied to the separation of chiral compounds. Enantiomers have been resolved employing macrocyclic antibiotics either added to the BGE (CZE) or immobilized into the capillary (CEC). The two CE modes offer different possibilities for improving the enantioseparation capability. CZE and CEC can certainly be considered complementary to each other. In fact the same glycopeptide antibiotic, e.g., vancomycin may exhibit a different enantiorecognition capability depending on the considered method. In CZE the chiral selector mainly resolved acidic enantiomeric compounds while in CEC basic isomeric analytes were easily separated. These results were obtained because in CZE vancomycin may offer two amino groups and one carboxylic group for interactions with analytes, while in CEC only one amino and one carboxylic function are available. As previously illustrated, the use of macrocyclic antibiotics in CZE presents some drawbacks such as poor sensitivity due to strong absorption. To resolve this problem, the countercurrent partial-filling method has been proposed.

Based on the presented data, it has been demonstrated that CE is a powerful electrodriven method capable of resolving a wide number of enantiomers with good sensitivity, high resolution capability, low cost, and short analysis time.

The employment of macrocyclic antibiotics for the analysis of enantiomers by using EMs coupled with MS has been demonstrated. This topic needs further studies considering the enormous potential of MS in determining the analytes' mass and structure, useful especially for the analysis of real samples.

Future development includes search of new macrocyclic antibiotics capable of separating positively, negatively charged, and neutral enantiomers in the same run.

REFERENCES

1. Chankvetadze, B. 1997. *Capillary Electrophoresis in Chiral Analysis*. Chichester-New York-Weinheim-Brisbane-Singapore-Toronto: John Wiley & Sons.
2. Gubitz, G, Schmid, MG. 2004. Chiral separation principles: An introduction. In *Methods in Molecular Biology, vol 243: Chiral Separations: Methods and Protocols*, Gubitz, G and Schmid, MG (Eds.), pp. 1–28. Totowa, New Jersey: Humana Press.
3. Armstrong, DW, Tang, YB, Chen, SS, Zhou, YW, Bagwill, C, Chen, JR. 1994. Macrocyclic antibiotics as a new class of chiral selectors for liquid chromatography. *Anal Chem* 66: 1473–1484.
4. Armstrong, DW, Rundlett, K, Reid III, GL. 1994. Use of a macrocyclic antibiotic, rifamycin B, and indirect detection for the resolution of racemic amino alcohols by CE. *Anal Chem* 66: 1690–1695.
5. Ward, TJ. 1996. Macrocyclic antibiotics—the newest class of chiral selectors. *LC-GC Int* 9: 428–435.
6. Ward, TJ, Oswald, TM. 1997. Enantioselectivity in capillary electrophoresis using the macrocyclic antibiotics. *J Chromatogr A* 792: 309–325.
7. Desiderio, C, Fanali, S. 1998. Chiral analysis by capillary electrophoresis using antibiotics as chiral selector. *J Chromatogr A* 807: 37–56.
8. Armstrong, DW, Nair, UB. 1997. Capillary electrophoresis enantioseparations using macrocyclic antibiotics as chiral selector. *Electrophoresis* 18: 2331–2342.
9. Fanali, S, Aturki, Z, Desiderio, C. 1999. Enantioresolution of pharmaceutical compounds by capillary electrophoresis. Use of cyclodextrins and antibiotics. *Enantiomer* 4: 229–241.
10. Hui, E, Caude, M. 1999. Enantioseparation in CE using macrocyclic antibiotics as chiral selectors. *Analysis* 27: 131–138.
11. Aboul-Enein, HY, Ali, I. 2001. Macrocyclic antibiotics as effective chiral selectors for enantiomeric resolution by liquid chromatography and capillary electrophoresis. *Chromatographia* 52: 679–691.
12. Ward, TJ, Dann III, C, Blaylock, A. 1995. Enantiomeric resolution using the macrocyclic antibiotics rifamycin B and rifamycin SV as chiral selectors for capillary electrophoresis. *J Chromatogr* 715: 337–344.
13. Armstrong, DW, Rundlett, KL, Chen, JR. 1994. Evaluation of the macrocyclic antibiotic vancomycin as a chiral selector for capillary electrophoresis. *Chirality* 6: 496–509.

14. Vespalec, R, Billiet, HAH, Frank, J, Bocek, P. 1996. Vancomycin as a chiral selector in capillary electrophoresis: An appraisal of advantages and limitations. *Electrophoresis* 17: 1214–1221.
15. Arai, T, Nimura, N, Kinoshita, T. 1996. Investigation of enantioselective separation of quinolonecarboxylic acids by capillary zone electrophoresis using vancomycin as a chiral selector. *J Chromatogr A* 736: 303–311.
16. Armstrong, DW, Gasper, MP, Rundlett, KL. 1995. Highly enantioselective capillary electrophoretic separations with dilute solutions of the macrocyclic antibiotic ristocetin A. *J Chromatogr* 689: 285–304.
17. Rundlett, KL, Gasper, MP, Zhou, EY, Armstrong, DW. 1996. Capillary electrophoretic enantiomeric separations using the glycopeptide antibiotic, teicoplanin. *Chirality* 8: 88–107.
18. Nishi, H, Nakamura, K, Nakai, H, Sato, T. 1996. Enantiomer separation by capillary electrophoresis using DEAE-dextran and aminoglycosidic antibiotics. *Chromatographia* 43: 426–430.
19. Jiang, Z, Bertazzo, M, Sussmuth, RD, Yang, Z, Smith, NW, Schurig, V. 2006. Highlighting the role of the chlorine substituents in the glycopeptide antibiotic balhimicin for chiral recognition in capillary electrophoresis. *Electrophoresis* 27: 1154–1162.
20. Fanali, S, Catarcini, P, Presutti, C, Quaglia, MG, Rigetti, PG. 2003. A glycopeptide antibiotic chiral stationary phase for the enantiomer resolution of hydroxy acid derivatives by capillary electrochromatography. *Electrophoresis* 24: 904–912.
21. Trelli-Seifert, LA, Risley, DS. 1998. Capillary electrophoretic enantiomeric separations of nonsteroidal anti-inflammatory compounds using the macrocyclic antibiotic actaplanin a and 2-methoxyethanol. *J Liq Chromatogr Relat Tech* 21: 299–313.
22. Ekborg-Ott, KH, Zientara, GA, Schneiderheinze, JM, Gahm, K, Armstrong, DW. 1999. Avoparcin, a new macrocyclic antibiotic chiral run buffer additive for capillary electrophoresis. *Electrophoresis* 20: 2438–2457.
23. Ekborg-Ott, KH, Wang, X, Armstrong, DW. 1999. Effect of selector coverage and mobile phase composition on enantiomeric separations with ristocetin a chiral stationary phases. *Microchem J* 62: 26–49.
24. Peter, A, Torok, G, Armstrong, DW, Toth, G, Tourwe, D. 2001. High-performance liquid chromatographic separation of enantiomers of synthetic amino acids on a ristocetin A chiral stationary phase. *J Chromatogr A* 904: 1–15.
25. Cavazzini, A, Nadalini, G, Dondi, F, Gasparrini, F, Ciogli, A, Villani, C. 2004. Study of mechanism of chiral discrimination of amino acids and their derivatives on a teicoplanin-based chiral stationary phase. *J Chromatogr A* 1031: 143–158.
26. Terabe, S, Otsuka, K, Nishi, H. 1994. Separation of enantiomers by capillary electrophoretic techniques. *J Chromatogr A* 666: 295–319.
27. Fanali, S. 1996. Identification of chiral drug isomers by capillary electrophoresis. *J Chromatogr A* 735: 77–121.
28. Loukili, B, Dufresne, C, Jourdan, E, Grosset, C, Ravel, A, Peyrin, E. 2003. Study of tryptophan enantiomer binding to a teicoplanin-based stationary phase using the perturbation technique: Investigation of the role of sodium perchlorate in solute retention and enantioselectivity. *J Chromatogr A* 986: 45–53.
29. Kang, J, Bischoff, D, Jiang, Z, Bister, B, Sussmuth, RD, Schurig, V. 2004. A mechanistic study of enantiomeric separation with vancomycin and balhimycin as chiral selectors by capillary electrophoresis. Dimerization and enantioselectivity. *Anal Chem* 76: 2387–2392.
30. Nair, UB, Chang, SSC, Armstrong, DW, Rawjee, YY, Eggleston, DS, McArdle JV. 1996. Elucidation of vancomycin's enantioselective binding site using its copper complex. *Chirality* 8: 590–595.

31. Desiderio, C, Porcaro, CM, Padiglioni, P, Fanali, S. 1997. Enantiomeric separation of acidic herbicides by capillary electrophoresis using vancomycin as chiral selector. *J Chromatogr A* 781: 503–513.
32. Fanali, S, Desiderio, C, Aturki, Z. 1997. Enantiomeric resolution study by capillary electrophoresis—selection of the appropriate chiral selector. *J Chromatogr A* 772: 185–194.
33. Fanali, S, Desiderio, C. 1996. Use of vancomycin as chiral selector in capillary electrophoresis. optimization and quantitation. *J High Resolut Chromatogr* 19: 322–326.
34. Hui, F, Ekborg-Ott, KH, Armstrong, DW. 2001. High-performance liquid chromatographic and capillary electrophoretic enantioseparation of plant growth regulators and related indole compounds using macrocyclic antibiotics as chiral selectors. *J Chromatogr A* 906: 91–103.
35. Wan, H, Blomberg, LG. 1997. Enantiomeric separation of small chiral peptides by capillary electrophoresis. *J Chromatogr A* 792: 393–400.
36. André, C, Guillaume, YC. 2003. Salt dependence on vancomycin-herbicide enantiomer binding: Capillary electrophoresis study and theoretical approach. *Electrophoresis* 24: 1620–1626.
37. Terabe, S, Otsuka, K, Ichikawa, K, Tsuchiya, A, Ando, T. 1984. Electrokinetic separations with micellar solution and open-tubular capillaries. *Anal Chem* 56: 111–113.
38. Terabe, S, Shibata, H, Miyashita, Y. 1989. Chiral separation by electrokinetic chromatography with bile salt micelles. *J Chromatogr* 480: 403–411.
39. Rundlett, KL, Armstrong, DW. 1995. Effect of micelles and mixed micelles on efficiency and selectivity of antibiotic-based capillary electrophoretic enantioseparations. *Anal Chem* 67: 2088–2095.
40. Ward, TJ, Dann III, C, Brown, AP. 1996. Separation of enantiomers using vancomycin in a countercurrent process by suppression of electoosmosis. *Chirality* 8: 77–83.
41. Fanali, S, Desiderio, C, Schulte, G, Heitmeier, S, Strickmann, D, Chankvetadze, B, Blaschke, G. 1998. Chiral capillary electrophoresis-electrospray mass spectrometry coupling using vancomycin as chiral selector. *J Chromatogr A* 800: 69–76.
42. Kang, JW, Yang, YT, You, JM, Ou, QY. 1998. Fast chiral separation of amino acid derivatives and acidic drugs by co-electroosmotic flow capillary electrophoresis with vancomycin as chiral selector. *J Chromatogr A* 825: 81–87.
43. Kang, J, Wistuba, D, Schurig, V. 2003. Fast enantiomeric separation with vancomycin as chiral additive by co-electroosmotic flow capillary electrophoresis: Increase of the detection sensitivity by the partial filling technique. *Electrophoresis* 24: 2674–2679.
44. Bednar, P, Aturki, Z, Stransky, Z, Fanali, S. 2001. Chiral analysis of UV nonabsorbing compounds by capillary electrophoresis using macrocyclic antibiotics: 1. Separation of aspartic and glutamic acid enantiomers. *Electrophoresis* 22: 2129–2135.
45. Fanali, S, Cartoni, C, Aturki, Z. 2001. Enantioseparation of S-carboxymethylcysteine and N-acetamidocarboxymethylcysteine by capillary electrophoresis using vancomycin. *J Sep Sci* 24: 789–794.
46. Vespalec, R, Corstjens, H, Billiet, HAH, Frank, J, Luyben, KCAM. 1995. Enantiomeric separation of sulfur- and selenium-containing amino acids by capillary electrophoresis using vancomycin as a chiral selector. *Anal Chem* 67: 3223–3228.
47. Fanali, S, Aturki, Z, Desiderio, C, Bossi, A, Rigetti, PG. 1998. Use of Hepta-tyr glycopeptide antibiotic as chiral selector in capillary electrophoresis. *Electrophoresis* 19: 1742–1751.
48. Fanali, S, Aturki, Z, Desiderio, C, Righetti, PG. 1999. Use of MDL 63,246 (Hepta-Tyr) antibiotic in capillary zone electrophoresis - ii. chiral resolution of alpha-hydroxy acids. *J Chromatogr A* 838: 223–235.

49. Fanali, S, Crucianelli, M, De Angelis, F, Presutti, C. 2002. Enantioseparation of amino acid derivatives by capillary zone electrophoresis using vancomycin as chiral selector. *Electrophoresis* 23: 3035–3040.
50. Desiderio, C, Fanali, S. 1998. Chiral analysis by capillary electrophoresis using antibiotics as chiral selector. *J Chromatogr A* 818: 281–282.
51. Hui, F, Caude, M. 1999. Enantioseparation in CE using macrocyclic antibiotics as chiral selectors. *Analusis* 27: 131–137.
52. Wan, H, Blomberg, LG. 2000. Chiral separation of amino acids and peptides by capillary electrophoresis. *J Chromatogr A* 875: 43–88.
53. Ward, TJ, Farris III, AB. 2001. Chiral separations using the macrocyclic antibiotics: A review. *J Chromatogr A* 906: 73–89.
54. Simo, C, Barbas, C, Cifuentes, A. 2003. Chiral electromigration methods in food analysis. *Electrophoresis* 24: 2431–2441.
55. Patel, BK, Hanna-Brown, M, Hadley, MR, Hutt, AJ. 2004. Enantiomeric resolution of 2-arylpropionic acid nonsteroidal anti-inflammatory drugs by capillary electrophoresis: Methods and applications. *Electrophoresis* 25: 2625–2656.
56. Ward, TJ. 2006. Chiral Separations. *Anal Chem* 78: 3947–3956.
57. Van Eeckaut, A, Michotte, Y. 2006. Chiral separations by capillary electrophoresis: Recent developments and applications. *Electrophoresis* 27: 2880–2895.
58. Ha, PTT, Hoogmartens, J, Van Schepdael, A. 2006. Recent advances in pharmaceutical applications of chiral capillary electrophoresis. *J Pharm Biomed Anal* 41: 1–11.
59. Kang, J, Yang, Y, You, J, Qu, Q. 1998. Fast chiral separation of amino acid derivatives and acidic drugs by co-electroosmotic flow capillary electrophoresis with vancomycin as chiral selector. *J Chromatogr A* 825: 81–87.
60. Oswald, TM, Ward, TJ. 1999. Enantioseparations with the macrocyclic antibiotic ristocetin A using a countercurrent process in CE. *Chirality* 11: 663–668.
61. Polcaro, CM, Marra, C, Desiderio, C, Fanali, S. 1999. Stereoselective analysis of acid herbicides in natural waters by capillary electrophoresis. *Electrophoresis* 20: 2420–2424.
62. Risley, DS, Trelli-Seifert, L, McKenzie, QJ. 1999. Enantiomeric separations of dansyl amino acids using the macrocyclic antibiotic A35512B as a chiral selector in capillary electrophoresis. *Electrophoresis* 20: 2749–2753.
63. Chiari, M, Cretich, M, Disperati, V, Marinzi, C, Galbusera, C, De Lorenzi, E. 2000. Evaluation of new adsorbed coatings in chiral capillary electrophoresis and the partial filling technique. *Electrophoresis* 21: 2343–2351.
64. Tesarová, E, Bosáková, Z, Zusková, I. 2000. Enantioseparation of selected N-tert-butyloxycarbonyl amino acids in high-performance liquid chromatography and capillary electrophoresis with a teicoplanin chiral selector. *J Chromatogr A* 879: 147–156.
65. Sutton, KL, Sutton, RMC, Stalcup, AM, Caruso, JA. 2000. A comparison of vancomycin and sulfated beta-cyclodextrin as chiral selectors for enantiomeric separations of selenoamino acids using capillary electrophoresis with UV absorbance detection. *Analyst* 125: 231–234.
66. Fanali, S, Cartoni, C, Desiderio, C. 2001. Chiral separation of newly synthesized arylpropionic acids by capillary electrophoresis using cyclodextrins or a glycopeptide antibiotic as chiral selectors. *Chromatographia* 54: 87–92.
67. Ward, TJ, Farris, AB, Woodling, K. 2001. Synergistic chiral separations using the glycopeptides ristocetin A and vancomycin. *J Biochem Biophys Meth* 48: 163–174.
68. Ficarra, R, Cutroneo, P, Aturki, Z, Tommasini, S, Calabrò, ML, Phan-Tan-Luu, R, Fanali, S, Ficarra, P. 2002. An experimental design methodology applied to the enantioseparation of a non-steroidal anti-inflammatory drug candidate. *J Pharm Biomed Anal* 29: 989–997.

69. Reilly, J, Sanchez-Felix, M, Smith, NW. 2003. Link between biological signaling and increased enantioseparations of acids using glycopeptide antibiotics. *Chirality* 15: 731–742.
70. Hou, JG, He, TX, Mao, XF, Du, XZ, Deng, HL, Gao, JZ. 2003. Enantiomeric separation using erythromycin as a new capillary zone electrophoresis chiral selector. *Anal Letters* 36: 1437–1449.
71. Ha, PTT, Van Schepdael, A, Roets, E, Hoogmartens, J. 2004. Investigating the potential of erythromycin and derivatives as chiral selector in capillary electrophoresis. *J Pharm Biomed Anal* 34: 861–870.
72. Jiang, Z, Kang, J, Bischoff, D, Bister, B, Sussmuth, RD, Schurig, V. 2004. Evaluation of balhimycin as a chiral selector for enantioresolution by capillary electrophoresis. *Electrophoresis* 25: 2687–2692.
73. Gao, W, Kang, J. 2006. Separation of atropisomers of anti-hepatitis drug dimethyl diphenyl bicarboxylate analogues by capillary electrophoresis with vancomycin as the chiral selector. *J Chromatogr A* 1108: 145–148.
74. Fantacuzzi, M, Bettoni, G, D'Orazio, G, Fanali, S. 2006. Enantiometric separation of some demethylated analogues of clofibric acid by capillary zone electrophoresis and nano-liquid chromatography. *Electrophoresis* 27: 1227–1236.
75. Wang, Z, Wang, J, Hu, Z, Kang, J. 2007. Enantioseparation by CE with vancomycin as chiral selector: Improving the separation performance by dynamic coating of the capillary with poly(dimethylacrylamide). *Electrophoresis* 28: 938–943.
76. Maier, V, Petr, J, Knob, R, Horakova, J, Sevcik, J. 2007. Electrokinetic partial filling technique as a powerful tool for enantiomeric separation of DL-lactic acid by CE with contactless conductivity detection. *Electrophoresis* 28: 1815–1822.
77. Dai, R, Nie, X, Li, H, Saeed, MK, Deng, Y, Yao, G. 2007. Investigation of β-CD-derivatized erythromycin as chiral selector in CE. *Electrophoresis* 28: 2566–2572.
78. Altria, KD. 1999. Overview of capillary electrophoresis and capillary electrochromatography. *J Chromatogr A* 856: 443–463.
79. Angus, PDA, Demarest, CW, Catalano, T, Stobaugh, JF. 2000. Aspects of column fabrication for packed capillary electrochromatography. *J Chromatogr A* 887: 347–365.
80. Bartle, KD, Carney, RA, Cavazza, A, Cikalo, MG, Myers, P, Robson, MM, Roulin, SCP, Sealey, K. 2000. Capillary electrochromatography on silica columns: Factors influencing performance. *J Chromatogr A* 892: 279–290.
81. Breadmore, MC, Hilder, EF, Macka, M, Haddad, PR. 2001. Determination of inorganic anions by capillary electrochromatography. *TRAC-Trends in Anal Chem* 20: 355–364.
82. Dittmann, MM, Masuch, K, Rozing, GP. 2000. Separation of basic solutes by reversed-phase capillary electrochromatography. *J Chromatogr A* 887 (1–2): 209–221.
83. Dorsey, JG. 1999. Electrochromatography: The hope and the promise. *Microchem J* 61: 6–11.
84. Chankvetadze, B. 1999. Recent trends in enantioseparations using capillary electromigration techniques. *TRAC-Trends in Anal Chem* 18: 485–498.
85. Girod, M, Chankvetadze, B, Blaschke, G. 2000. Enantioseparations in non-aqueous capillary electrochromatography using polysaccharide type chiral stationary phases. *J Chromatogr A* 887: 439–455.
86. Fanali, S, Catarcini, P, Blaschke, G, Chankvetadze, B. 2001. Enantioseparations by capillary electrochromatography. *Electrophoresis* 22: 3131–3151.
87. Preinerstorfer, B, Lämmerhofer, M. 2007. Recent accomplishments in the field of enantiomer separation by CEC. *Electrophoresis* 28: 2527–2565.
88. Dermaux, A, Lynen, F, Sandra, P. 1998. Chiral capillary electrochromatography on a vancomycin stationary phase. *J High Resolut Chromatogr* 21: 575–576.

89. Fanali, S, Rudaz, S, Veuthey, JL, Desiderio, C. 2001. Use of vancomycin silica stationary phase in packed capillary electrochromatography II. Enantiomer separation of venlafaxine and O-desmethylvenlafaxine in human plasma. *J Chromatogr A* 919: 195–203.
90. Desiderio, C, Rudaz, S, Veuthey, JL, Raggi, MA, Fanali, S. 2002. Use of vancomycin silica stationary phase in packed capillary electrochromatography: IV. Enantiomer separation of fluoxetine and norfluoxetine employing UV high sensitivity detection cell. *J Sep Sci* 25: 1291–1296.
91. Aturki, Z, Scotti, V, D'Orazio, G, Rocco, A, Raggi, MA, Fanali, S. 2007. Enantioselective separation of the novel antidepressant mirtazapine and its main metabolites by CEC. *Electrophoresis* 28: 2717–2725.
92. Carter-Finch, AS, Smith, NW. 1999. Enantiomeric separations by capillary electrochromatography using a macrocyclic antibiotic chiral stationary phase. *J Chromatogr A* 848: 375–385.
93. Karlsson, C, Wikstrom, H, Armstrong, DW, Owens, PK. 2000. Enantioselective reversed-phase and non-aqueous capillary electrochromatography using a teicoplanin chiral stationary phase. *J Chromatogr A* 897: 349–363.
94. Wikstrom, H, Svensson, LA, Torstensson, A, Owens, PK. 2000. Immobilisation and evaluation of a vancomycin chiral stationary phase for capillary electrochromatography. *J Chromatogr A* 869: 395–409.
95. Karlsson, C, Karlsson, L, Armstrong, DW, Owens, PK. 2000. Evaluation of a vancomycin chiral stationary phase in capillary electrochromatography using polar organic and reversed phase modes. *Anal Chem* 72: 4394–4401.
96. Desiderio, C, Aturki, Z, Fanali, S. 2001. Use of vancomycin silica stationary phase in packed capillary electrochromatography I. Enantiomer separation of basic compounds. *Electrophoresis* 22: 535–543.
97. Fanali, S, Catarcini, P, Quaglia, MG. 2002. Use of vancomycin silica stationary phase in packed capillary electrochromatography: III. Enantiomeric separation of basic compounds with the polar organic mobile phase. *Electrophoresis* 23: 477–485.
98. Grobuschek, N, Schmid, MG, Koldl, J, Gübitz, G. 2002. Enantioseparation of amino acids and drugs by CEC, pressure supported CEC, and micro-HPLC using a teicoplanin aglycone stationary phase. *J Sep Sci* 25: 1297–1302.
99. Schmid, MG, Grobuschek, N, Pessenhofer, V, Klostius, A, Gübitz, G. 2003. Enantioseparation of dipeptides by capillary electrochromatography on a teicoplanin aglycone chiral stationary phase. *J Chromatogr A* 990: 83–90.
100. Catarcini, P, Fanali, S, Presutti, C, D'Acquarica, I, Gasparrini, F. 2003. Evaluation of teicoplanin chiral stationary phases of 3.5 and 5 μm inside diameter silica microparticles by polar-organic mode capillary electrochromatography. *Electrophoresis* 24: 3000–3005.
101. Fanali, S, Catarcini, P, Presutti, C, Stancanelli, R, Quaglia, MG. 2003. Use of short-end injection capillary packed with a glycopeptide antibiotic stationary phase in electrochromatography and capillary liquid chromatography for the enantiomeric separation of hydroxy acids. *J Chromatogr A* 990: 143–151.
102. Fanali, S, Catarcini, P, Presutti, C. 2003. Enantiomeric separation of acidic compounds of pharmaceutical interest by capillary electrochromatography employing glycopeptide antibiotic stationary phases. *J Chromatogr A* 994: 227–232.
103. Fanali, S, D'Orazio, G, Quaglia, MG, Rocco, A. 2004. Use of a Hepta-Tyr antibiotic modified silica stationary phase for the enantiomeric resolution of d,l-loxiglumide by electrochromatography and nano-liquid chromatography. *J Chromatogr A* 1051: 247–252.
104. Kornysova, O, Owens, PK, Maruska, A. 2001. Continuous beds with vancomycin as chiral stationary phase for capillary electrochromatography. *Electrophoresis* 22: 3335–3338.

105. Kornysova, O, Jarmalaviciene, R, Maruška, A. 2004. A simplified synthesis of polymeric nonparticulate stationary phases with macrocyclic antibiotic as chiral selector for capillary electrochromatography. *Electrophoresis* 25: 2825–2829.
106. Gatschelhofer, C, Schmid, MG, Schreiner, K, Pieber, TR, Sinner, FM, Gübitz, G. 2006. Enantioseparation of glycyl-dipeptides by CEC using particle-loaded monoliths prepared by ring-opening metathesis polymerization (ROMP). *J Biochem Biophys Methods* 69: 67–77.
107. Dong, X, Dong, J, Ou, J, Zhu, Y, Zou, H. 2007. Preparation and evaluation of a vancomycin-immobilized silica monolith as chiral stationary phase for CEC. *Electrophoresis* 28: 2606–2612.
108. Carlsson, E, Wikström, H, Owens, PK. 2001. Validation of a chiral capillary electrochromatographic method for metoprolol on a teicoplanin stationary phase. *Chromatographia* 53: 419–424.
109. Schmid, MG, Grobuschek, N, Pessenhofer, V, Klostius, A, Gübitz, G. 2003. Chiral resolution of diastereomeric di- and tripeptides on a teicoplanin aglycone phase by capillary electrochromatography. *Electrophoresis* 24: 2543–2549.
110. Yao, C, Tang, S, Gao, R, Jiang, C, Yan, C. 2004. Enantiomer separations on a vancomycin stationary phase and retention mechanism of pressurized capillary electrochromatography. *J Sep Sci* 27: 1109–1114.
111. Enlund, AM, Andersson, ME, Hagman, G. 2004. Improved quantification limits in chiral capillary electrochromatography by peak compression effects. *J Chromatogr A* 1028: 333–338.
112. Zheng, J, Shamsi, SA. 2006. Simultaneous enantioseparation and sensitive detection of eight beta-blockers using capillary electrochromatography-electrospray ionization-mass spectrometry. *Electrophoresis* 27: 2139–2151.
113. Chen, Z, Zeng, S, Yao, T. 2007. Separation of beta-receptor blockers and analogs by capillary liquid chromatography (CLC) and pressurized capillary electrochromatography (pCEC) using a vancomycin chiral stationary phase column. *Pharmazie* 62: 585–592.
114. Ekborg-Ott, KH, Kullman, JP, Wang, X, Gahm, K, He, L, Armstrong, DW. 1998. Evaluation of the macrocyclic antibiotic avoparcin as a new chiral selector for HPLC. *Chirality* 10: 627–660.
115. Petrusevska, K, Kuznetsov, MA, Gedicke, K, Meshko, V, Satroverov, SM, Siedel-Morgenstern, A. 2006. Chromatographic enantioseparation of amino acids using a new chiral stationary phase based on a macrocyclic glycopeptides antibiotic. *J Sep Sci* 29: 1447–1457.

6 Chiral Separations Using Proteins and Peptides as Chiral Selectors

Jun Haginaka

CONTENTS

6.1 Introduction 139
6.2 Chiral Recognition Mechanism 140
6.3 Overview of the Different Proteins and Peptides Used 143
6.4 Modes of Use of Protein and Peptide Selectors in CE 143
6.4.1 Affinity Capillary Electrophoresis 143
6.4.1.1 Adsorption of a Protein or Peptide to the Capillary Wall 143
6.4.1.2 Interference of UV Detection 144
6.4.1.3 Protein Selectors 145
6.4.1.4 Peptide Selectors 148
6.4.2 Affinity Capillary Electrochromatography 151
6.4.2.1 Protein Selectors 151
6.4.2.2 Peptide Selectors 154
6.5 New Developments 154
6.6 Conclusions 156
List of Abbreviations 157
References 158

6.1 INTRODUCTION

Drug enantiomers can be different in potency, toxicity, and behavior in biological systems [1]. Thus, a lot of analytical methods have been developed for the discrimination of drug enantiomers. Among those, chiral high-performance liquid chromatography (HPLC) methods are widely employed for the assays of drug enantiomers in pharmaceutical preparations and biological fluids [2–4]. Recently, capillary electrophoresis (CE) methods have been developed for the separation of drug enantiomers by using chiral selectors, such as chiral ligand exchange, cyclodextrins, crown ethers, chiral micelles, polysaccharides, proteins, peptides, macrocyclic antibiotics, and molecularly imprinted polymers, as the running buffer additives or immobilized ligands [3]. Among those, proteins and peptides are of special interest as chiral selectors because of their unique properties of stereoselectivity and/or their suitability for separating a wide range of enantiomeric mixtures [3,5,6].

This chapter deals with chiral separations using proteins and peptides as chiral selectors in CE: chiral recognition mechanism based on proteins and peptides, overview of the different proteins and peptides used, modes of the use of protein and peptide selectors and new developments in CE.

6.2 CHIRAL RECOGNITION MECHANISM

A protein has enantioselectivity for a wide range of compounds because of multiple binding site(s) on its surface and/or multiple binding interaction(s) between a protein and a ligand [2,4]. Chiral recognition mechanism on a protein has been elucidated using spectroscopy, molecular modeling and ligand docking [7], or x-ray crystallography [8].

The chiral recognition mechanism of the third domain of ovomucoid from turkey egg white (OMTKY3) was elucidated by using NMR measurements, and molecular modeling and ligand docking [7]. On the surface of OMTKY3, there are two distinct binding sites (nonselective and enantioselective binding sites): in the former, hydrophobic interactions mainly work for the binding, while in the latter, hydrophobic, ionic, and hydrogen bonding interactions play important roles. The enantioselective binding model for (*R*)- and (*S*)-U-80413, which is a 2-arylpropionic acid derivative, with OMTKY3 is shown in Figure 6.1 [7]. One can see similarities and differences in orientation and intermolecular interactions between *(R)*- and *(S)*-U-80413. The carboxyl groups of each enantiomer engage in ionic interactions with the positive charge on Arg21. The carbonyl group on U-80413's central ring shares a hydrogen bond with NH_3^+ group of Lys34. The distinguishing difference between the enantiomers is the proximity of the phenyl group of (*R*)-U-80413 and Phe53.

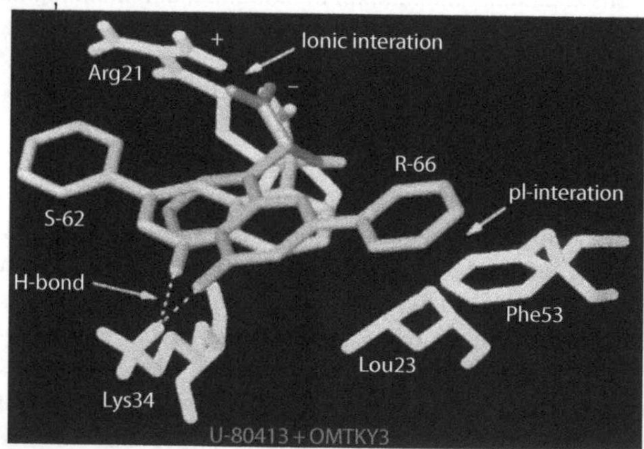

FIGURE 6.1 U-80413 enantiomers in the enantioselective binding site of OMTKY3. (From Pinkerton, T.C. et al., *Anal. Chem.*, 67, 2366, 1995. With permission.)

The chiral recognition mechanism on penicillin G-acylase (PGA) was investigated using molecular modeling and docking studies [9]. As illustrated in Figure 6.2A, the associated binding model for (*S*)-2-(4-chlorophenyl)-2-phenoxyacetic acid is characterized by the presence of numerous hydrogen bonding interactions involving the carboxyl group with both the backbone NH and the OH group of SerB386 [9]. An additional hydrogen bond is established between the ether oxygen of the ligand and the ArgB263 side chain. Moreover, an ionic interaction is formed between the negatively charged carboxyl group and the positively charged *N*-terminal SerB1. A set of charge–transfer interactions is also found between the ligand phenoxy and phenyl moieties and the PheB24, PheA146, PheB71, and PheB256 aromatic rings. Regarding (*R*)-2-(4-chlorophenyl)-2-phenoxyacetic acid, the calculated posing is similar to the one found for the (*S*)-isomer, even if the hydrogen bonding interaction between the ether oxygen and ArgB263 side chain is absent and the phenoxy moiety is placed in the same position occupied by the phenyl ring in the (*S*)-enantiomer (Figure 6.2B).

It is well known that stereoselective binding of drugs to human serum albumin (HSA) occurs principally at two major binding sites; warfarin-azapropazone site (site I) and indole-benzodiazepine site (site II), and that the binding sites I and II are located in hydrophobic cavities in subdomains IIA and IIIA, respectively [10]. The crystal structure of HSA-myristate complexed with (*R*)- and (*S*)-warfarin was determined [8]. The structures confirm that warfarin binds to subdomain IIA in the presence of fatty acids and reveal the molecular details of the protein–drug interaction. The two enantiomers of warfarin adopt very similar conformations when bound to the protein and make many of the same specific contacts with amino acid side chains at the binding site, thus accounting for the relative lack of stereospecificity of the HSA–warfarin interaction [8].

In the case of peptides, similar interactions described as in proteins could participate in enantioselective bindings. It should be noted that the secondary structure becomes more pronounced with an increase in the length of peptide. Secondary structure introduces more conformational rigidity to the selector, which may result in increased enantioselectivity.

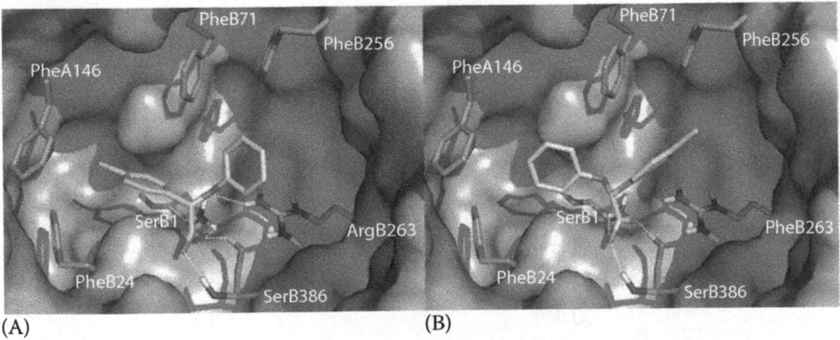

FIGURE 6.2 Binding mode of compound (*S*)-2-(4-chlorophenyl)-2-phenoxyacetic acid (A) and (*R*)-2-(4-chlorophenyl)-2-phenoxyacetic acid (B) within PGA. (From Lavecchia, A. et al., *J. Mol. Graph Model*, 25, 777, 2007. With permission.)

TABLE 6.1
Physical Properties of Proteins or Glycoproteins used as Chiral Selectors in CE

Protein	Molecular Mass	Carbohydrate Content (%)	Isoelectric Point	Origin
Albumins				
Bovine serum albumin (BSA)	66,000	—[a]	4.7	Bovine serum
Human serum albumin (HSA)	65,000	—	4.7	Human serum
Glycoproteins				
α_1-Acid glycoprotein (AGP)	41,000–43,000	45	2.7–3.8	Human serum
	33,000	35	2.7–3.8	
Ovomucoid (OMCHI)	28,000	30	4.1	Egg white
Ovoglycoprotein (OGCHI) (chicken α_1-acid glycoprotein (cAGP))	30,000	25	4.1	Egg white
Avidin	66,000	7	10	Egg white
Riboflavin binding protein (RfBP)	32,000–36,000	14	4	Egg white
Enzymes				
Cellulase				Fungus
Cellobiohydrolase I (CBH I, Cel7A)	65,000	6	3.9	
Lysozyme	14,300	—	10.5	Egg white
Pepsin	34,600	—	<1	Porcine stomach
Penicillin G-acylase (PGA)	85,000	—	8.1	Bacterium
Antibody (Immunoglobulin G)	150,000	2–3	6–8	Vertebrate
Others				
Ovotransferrin (conalbumin)	77,000	2.6	6.1	Egg white
Human serum transferring	78,000	5.5	5.2–5.6	Human serum
Cytochrome *c*	12,000	—	10.0–10.5	
Streptavidin	53,000	—	5	Bacterium
Casein[b]				Goat milk
α-	26,200		4.7	
β-	24,400		4.0–4.5	
γ-	30,000		5.8–6.0	

[a] No sugar moieties.
[b] A mixture of α-, β-, and γ-casein (70%, 27%, and 3%, respectively).

6.3 OVERVIEW OF THE DIFFERENT PROTEINS AND PEPTIDES USED

A lot of proteins and peptides have been used as chiral selectors in CE. Those proteins include albumins such as bovine serum albumin (BSA) [11], human serum albumin (HSA) [12], and serum albumin from other species [13]; glycoproteins such as α_1-acid glycoprotein (AGP) [14], ovomucoid from chicken egg whites (OMCHI) [15], ovoglycoprotein from chicken egg whites (OGCHI) [16] (now, it is termed chicken AGP (cAGP) [17]), avidin [18], and riboflavin binding protein (RfBP) [19]; enzymes such as cellobiohydrolase (CBH I, Cel7A) [20], lysozyme [21], pepsin [22], and PGA [23]; antibody [24]; and other proteins such as ovotransferrin (or conalbumin) [25], human serum transferrin (HST) [26], cytochrome *c* [21], streptavidin [27], and casein [28]. Physical properties of proteins used in chiral CE are shown in Table 6.1.

Peptides used as chiral selectors in CE include polymerized dipeptide anionic chiral surfactants [29] such as poly(sodium undecanoyl-L-valylleucinate) [poly(L-SUVL)], and poly(sodium undecanoyl-L-leucylvalinate) [poly(L-SULV)], poly(sodium undecanoyl-L-leucylleucinate) [poly(L-SULL)] and poly(sodium undecanoyl-L-valylvalinate) [poly(L-SUVV)], cyclohexapetides [30] such as cyclo(Arg-Lys-Tyr-Pro-Tyr-β-Ala), and basic peptides [21] such as Lys-Tyr and Lys-Ser-Tyr.

6.4 MODES OF USE OF PROTEIN AND PEPTIDE SELECTORS IN CE

There are two modes for chiral CE using a protein or peptide as a chiral selector. One is to dissolve the protein or peptide in the running buffers, and the other is to immobilize or adsorb it onto the capillary, or to pack protein-immobilized materials into the capillary. The former is termed as affinity CE (ACE) and the latter is termed affinity capillary electrochromatography (ACEC). The former is analogous to electrokinetic chromatography (EKC) using micelles or other kind of selectors. Thus, affinity EKC is also used to describe this technique.

The advantages of ACE based on a protein or peptide are that no immobilization of a protein or peptide to capillary walls and packing materials is required, and that packing procedures, needed for ACEC with a packed capillary, are not required. Furthermore, since binding properties of an immobilized protein are rather different from those of the native protein, it is favorable to use soluble proteins. The disadvantages of the ACE method include use of a larger amount of a protein or peptide, adsorption of a protein or peptide to the capillary wall, and interference of UV detection.

6.4.1 Affinity Capillary Electrophoresis

6.4.1.1 Adsorption of a Protein or Peptide to the Capillary Wall

The adsorption of proteins or peptides on the capillary wall will cause changes in the electroosmotic flow (EOF), which can affect the reproducibility of migration times and peak areas [31]. When uncoated capillaries are used, it is important to wash the capillaries between runs by sodium hydroxide or sodium dodecylsulfate (SDS) in order to remove the adsorbed proteins completely [5]. Two approaches to avoid the adsorption of proteins or peptides to a capillary wall are the use of chemically coated capillaries and the use of additives to minimize their interactions with the capillary wall. The permanent-coatings by linear polyacrylamide are most frequently used [32]. In addition, polyethylene glycol [33]-, methylcellulose [5]-, linear poly(dimethylacrylamide) [34]-, poly(acrylamide-*co*-allyl-β-D-glucopyranoside-*co*-allylglycidyl ether) [34]-, poly(dimethylacrylamide-*co*-allylglycidyl ether) [34]-, and pullulan [23]-coated capillaries are used. The additives to minimize the protein–wall interactions include hydroxypropylcellulose (HPC), dextran, *o*-phosphorylethanolamine (PEA), 2-(cyclohexylamino)ethanesulfonic acid (CHES), and 3-[(3-chloramidopropyl)dimethylammonio]-2-hydroxy-1-propanesulfonate (CHAPSO) [5].

6.4.1.2 Interference of UV Detection

When a protein or peptide is added in the running buffer, the background signal due to the protein or peptide interferes with detection of an analyte. To overcome this problem, a partial filling technique was first introduced by Valtcheva et al. [20] for the separation of β-blockers using CBH I (Cel7A) as a chiral selector. The technique was run automatically using a commercial CE instrument [25] with a slight modification. In the technique, the capillary was partially filled with a solution containing a protein and the protein was not in the detector cell when the analyte reached the cell. Figure 6.3 schematically illustrates the operating principle of the partial filling technique.

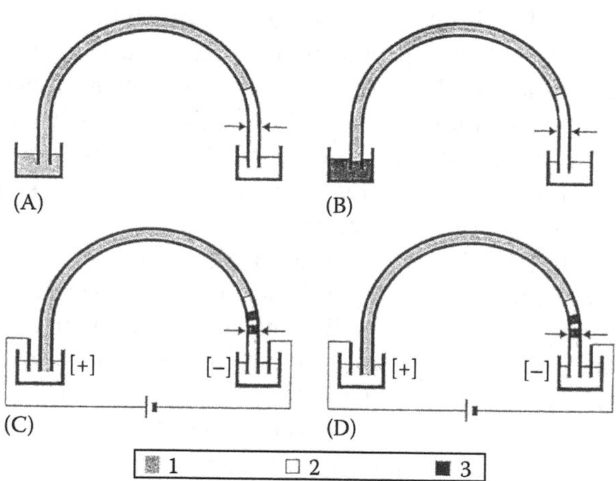

FIGURE 6.3 Schematic illustration of the partial filling technique. 1 = separation zone; 2 = running buffer; 3 = sample solution; arrows indicate detection window. (A) The separation zone is introduced from the injection end to a point short of the detector cell, (B) the sample solution is introduced into the capillary; (C) a high voltage is applied between both ends of the capillary after both ends are dipped into the running buffer and the analytes migrate toward the detector; (D) a separated zone reaches the detector cell but the separation zone does not reach this cell. (From Tanaka, Y. and Terabe, S., *J. Chromatogr. A*, 694, 279, 1995. With permission.)

At the beginning of the separation, the capillary is partially filled with a separation solution containing an acidic protein such as BSA or AGP (Figure 6.3A). A sample solution of a mixture of cationic analytes is introduced at the end of a capillary filled with the separation solution (Figure 6.3B). A mixture of cationic analytes migrates toward the cathode, while an acidic protein migrates in the opposite direction. Since in this example, a coated capillary is used to eliminate the EOF, the separation zone or protein does not migrate significantly during the run. In the separation zone, enantiomer separations are attained (Figure 6.3C), while the enantiomers migrate at identical velocities outside the separation zone and are detected in the absence of a protein (Figure 6.3D). The partial filling technique was applied for proteins such as BSA [25], crude OMCHI [25], AGP [25], avidin [18], CBH I (Cel7A) [20], pepsin [22], HST [26] and ovotransferrin [25], and PGA [23]. The technique gave improved detection sensitivity and comparable reproducibilities of migration times and peak areas, compared to the conventional technique where the protein was completely filled in the separation capillary [25].

6.4.1.3 Protein Selectors

BSA [11,35], HSA [12,36] and other serum albumins from rabbit, chicken, horse, guinea pig, dog, and goat [13] have been used as chiral selectors in ACE. Since the running buffer pH values used were near neutral, both BSA and HSA were negatively charged. Thus, under these conditions, cationic and uncharged solutes could be easily resolved because the analyte and protein migrated in the opposite direction. However, since anionic solutes migrated with the protein in the same direction, it was sometimes difficult to separate them. Resolution of ibuprofen, an anionic compound, was difficult using only BSA as a running buffer additive. The enantioseparation of ibuprofen was attained by addition of dextran, which modifies the protein mobility [37].

A lot of cationic solutes were enantioseparated using AGP as the chiral selector in ACE using the partial filling technique on linear polyacrylamide-coated capillaries [25]. Twenty-nine cationic racemates were separated by optimizing the separation conditions such as protein concentration, running buffer pH, and organic modifier [25]. Around 50–1000 μM AGP was used for the separation of those racemates.

The first trial of OMCHI as a chiral selector in ACE resulted in no enantioseparation of cationic, anionic, or uncharged compounds tested [15]. Later, a lot of compounds could be enantioseparated using OMCHI as chiral selector in ACE [25,28]. The enantioseparation of 2,4-dinitrophenyl (DNP)-amino acids was attained using OMCHI as the chiral selector in ACE with uncoated capillaries [28]. In addition, optical resolution of some drugs was achieved by optimizing the concentrations of both OMCHI and modifier (organic modifier, PEA) using dynamically coated capillaries by the addition of 0.25% HPC or PEG-coated capillaries. Many cationic compounds were enantioseparated using a high concentration of OMCHI (500 μM) in combination with the partial filling technique and linear polyacrylamide-coated capillaries [25]. The use of a high concentration of OMCHI with an additive such as an amphoteric surfactant or an organic modifier gave better resolution than the use of the low concentration. The results described above are easily understood, taking into account that OMCHI used in previous studies was crude [38]. A glycoprotein from chicken egg whites was isolated and termed OGCHI. It was found that about 10% OGCHI was included in crude OMCHI preparations, that chiral recognition ability of OMCHI reported previously [39] came from OGCHI and that pure OMCHI had no chiral recognition ability [38]. Furthermore, a cDNA clone encoding OGCHI was isolated and the amino acid sequence of OGCHI was clarified [17]. OGCHI consisted of 203 amino-acids including a predictable signal peptide of 20 amino acids. Mature OGCHI showed 31%–32% identities to rabbit and human AGPs. Thus, OGCHI should be chicken AGP (cAGP), a member of the lipocalin family.

cAGP was utilized as chiral selector in ACE for the separation of cationic enantiomers [16]. Figure 6.4, parts A, B, and C, shows the separation of tolperisone enantiomers using crude OMCHI, pure OMCHI, and cAGP, respectively, as chiral selectors in ACE, employing linear polyacrylamide-coated capillaries. The obtained results were consistent with those described above: tolperisone enantiomers were not resolved by using pure OMCHI, slightly resolved by crude OMCHI, and completely resolved by cAGP. Only the use of 50 μM cAGP as chiral selector resulted in good enantioseparation of tolperisone enantiomers. Separations of various drug enantiomers could be achieved by optimizing the concentrations of cAGP and type and concentration of modifier [16].

Avidin, succinylated avidin, and streptavidin were utilized as chiral selectors in ACE by a partial filling technique [27]. Avidin is a basic, glycosylated protein with an isoelectric point (p*I*) of 10.5, while succinylated avidin, which has a p*I* of 3.5–6.0, is chemically modified avidin, and streptavidin is a neutral nonglycosylated protein with a p*I* of ca. 7.0. Basic avidin was useful for the enantioseparation of acidic analytes such as 2-arylpropionic acid derivatives, vanilmandelic acid, and leucovorin, while acidic succinylated avidin was useful for the enantioseparation of basic analytes. Streptavidin was useful for both acidic and basic enantiomer separations. Since it is well known that avidin binds tightly to biotin, the authors investigated the effect of the addition of biotin on the enantioselectivity [27]. Not only the enantioselectivity but also the interaction with racemates was significantly lost by the formation of an avidin–biotin complex. Although it was already reported that a structural change of avidin occurs by the formation of an avidin–biotin complex [40], this possible change in the three-dimensional structure was not studied. Therefore, the authors could not conclude that the enantiorecognition arised from the interaction of the biotin binding sites on avidin with the analytes.

The use of quail RfBP [19] as a chiral selector in ACE was carried out with uncoated capillaries at running buffer pH 6.5. Uncharged and cationic compounds were enantioseparated, but anionic compounds such as indoprofen and warfarin were not because their mobilities were similar to that of the acidic protein. On the other hand, chicken RfBP [41] was used as the chiral selector at running buffer pH 4.4–6.8 on PEG-coated capillaries. Under these conditions, anionic and cationic compounds were enantioseparated. Among the running buffer pHs tested, the highest enantioselectivity was obtained for ketoprofen, an anionic compound, at pH 4.4.

CBH I from *Trichoderma reesei* (Cel7A) was used as a chiral selector in ACE [20,42]. The enantiomers of β-blockers were separated by a partial filling technique with linear polyacrylamide-coated capillaries [20] under unusual conditions; a high concentration of CBH I (625 μM), a high ionic strength (0.4 M sodium phosphate

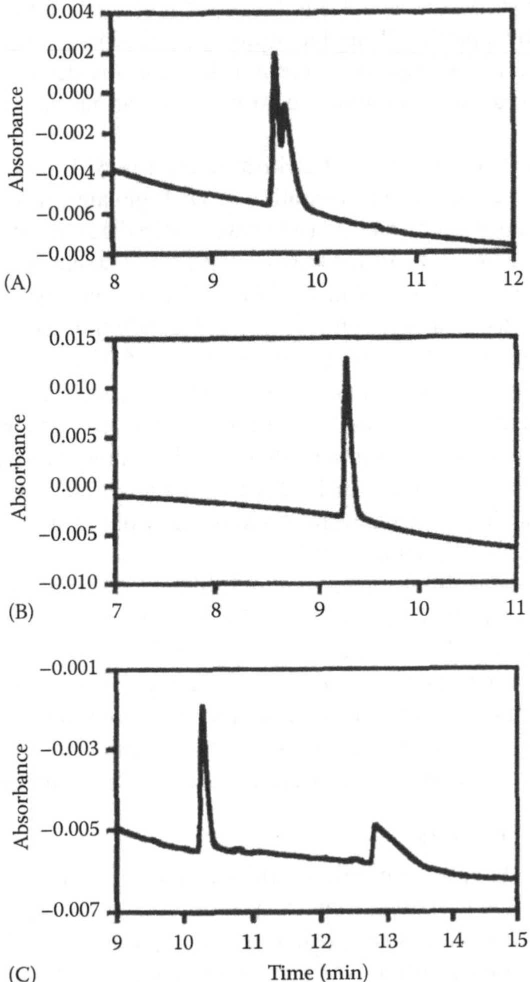

FIGURE 6.4 Separation of tolperisone enantiomers with crude OMCHI, OMCHI or cAGP as a chiral selector on a linear polyacrylamide-coated capillary in ACE. Conditions: capillary, 75 μm I.D., effective length 30 cm; running buffer, 50 mM phosphate buffer (pH 5.0)-2-propanol (95:5, v/v); separation solution, 50 mM phosphate buffer (pH 5.0)-2-propanol (95:5, v/v) containing 50 μM crude OMCHI (A), OMCHI (B) or cAGP (C); sample, 0.1 mg/mL racemic tolperisone hydrochloride; applied voltage, +12 kV; detection, 254 nm. (From Haginaka, J. and Kanasugi, N., *J. Chromatogr. A*, 782, 284, 1997. With permission.)

buffer), a high concentration of organic solvent (up to 30%), and a relatively low voltage. Since the running buffer pH was 5.1, the cationic β-blockers and the protein migrated in the opposite direction, and baseline resolution of the enantiomers was attained. CBH I was used as short plugs of 1.0–10 cm by a partial filling technique with a linear polyacrylamide-coated capillary of an effective length of 20 cm [42]. The enantiomers of oxprenolol and propranolol could be completely resolved with a selector plug of only 1.0 cm at the running buffer pH of 5.0.

Pepsin, with a p*I*<1, was used as a chiral selector in ACE on a linear-polyacrylamide-coated capillary by a partial filling technique [22]. Cationic compounds and chiral selectors migrated in the opposite direction, keeping the detector cell free of the strong absorbing protein. Enantiomer separations of various basic compounds were achieved.

Casein from goat milk was used as a chiral selector in ACE, where a concentration close to 1 mM allowed the enantioseparation of DNP-glutamic acid and DNP-proline [28]. Since a mixture of α-, β-, and γ-casein was used in the above study, a better chiral resolution may be achieved by using one of α-, β-, and γ-casein as the chiral selector.

As a chiral selector, 0.64–4.6 mM HST in 2-*N*-(*N*-morpholino)ethanesulfonic acid (pH 6.0) was used in ACE on a linear polyacrylamide-coated capillary by a partial filling technique [26,43]. HST, with a p*I* of ca. 6.0, did not migrate or slowly migrated, but the cationic compounds, tryptophan esters, and some drug enantiomers, migrated to the anodic end. Optimal conditions include ca. 1–15 cm selector plugs in a 30–45 cm capillary. As iron-saturated (differic) HST showed no enantioseparations for the compounds tested, iron-free HST was used as the selector. Ovotransferrin, a family of iron-binding proteins from chicken egg whites, was only applied for the enantioseparation of trimetoquinol [25].

PGA was used as a chiral selector in ACE [23]. The capillaries were silanized with glycidoxypropyltrimethoxysilane and the subsequent coupling of the hydroxyl groups of pullulan onto the silanized capillary. The EOF suppression obtained with coated capillaries allowed easy fulfillment of the typical countercurrent mode of a partial filling technique, where the analyte and chiral selector migrate in opposite directions. Enantioseparations of a series of acidic compounds such as ketoprofen, flubiprofen, fenoprofen, and suprofen were attained under the optimized conditions.

6.4.1.4 Peptide Selectors

The polymerized dipeptide anionic chiral surfactant, poly(L-SUVV), was synthesized and used as pseudo-stationary phases for chiral separations of basic, acidic, and neutral enantiomers [44]. For the separations of basic and acidic enantiomers, chiral recognition was significantly enhanced with poly(L-SUVV) as compared with a polymerized single amino acid anionic chiral surfactant [poly(sodium undecanoyl-L-valinate), poly(L-SUV)] [44]. To clarify the importance of the order of amino acids in the dipeptide surfactants in terms of chiral recognition and separations, four polymerized dipeptide anionic chiral surfactants, poly(L-SUVL), poly(L-SULV), poly(L-SULL), and poly(L-SUVV), were compared [29]. Poly(L-SULV) provides the best enantioselectivity among the four dipeptide surfactants for the separation of (±)-1,1′-bi-2-naphthol and (±)-1,1′-bi-2-naphthyl-2,2′-diyl hydrogen phosphate. Furthermore, the polymerized dipeptide anionic chiral surfactants investigated included all possible dipeptide combinations of the L-form of alanine, valine, leucine, and the achiral amino acid glycine (except glycine–glycine) as well as the single chiral center dipeptide surfactant poly(sodium undecanoyl-L-leucylβ-alaninate), poly(L-SULβA) were investigated for the separations of a lot of enantiomers [45]. The results indicate that the position and number of chiral centers and amino acid order play an important role in chiral recognition and separations, and that poly(L-SULV) is a broadly

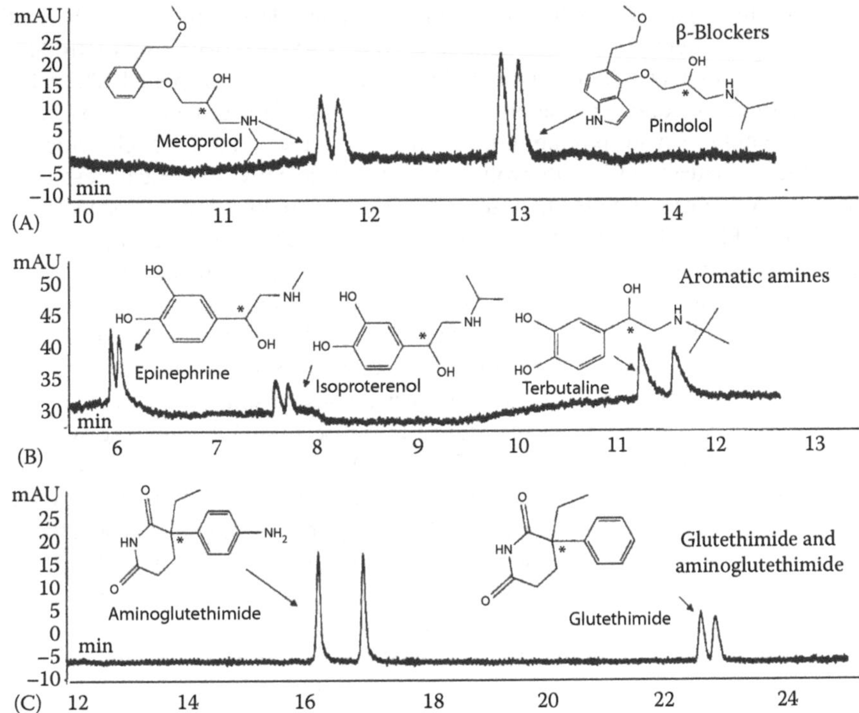

FIGURE 6.5 (A) Simultaneous separation and enantioseparation of several β-blockers. ACE conditions: 75 mM poly(L-SULV), 14°C, pH 6.5, 50 mM Na_2HPO_4/NaH_2PO_4. Pressure injection, 30 mbar s, +30 kV applied for separation; UV detection at 220 nm. (B) Simultaneous separation and enantioseparation of aromatic amines. ACE conditions: same as Figure 6.5A except UV detection was performed at 214 nm. (C) Simultaneous separation and enantioseparation of 0.5 mg/mL each of aminoglutethimide and glutethimide. ACE conditions: 80 mM poly(L-SULV), 12°C, pH 8.5, 25 mM Tris/Borate. Pressure injection, 90 mbar s, +30 kV applied for separation; UV detection at 254 nm. (From Shamsi, S.A. et al., *Anal. Chem.*, 75, 387, 2003. With permission.)

applicable chiral selector for ACE [46]. Figure 6.5, parts A, B, and C, shows the simultaneous separation and enantioseparation of β-blockers, aromatic amines, and aminoglutethimide and glutethimide, respectively, using poly(L-SULV) as a chiral selector. A total of 58 out of 75 racemic compounds could be resolved after choosing an appropriate concentration of poly(L-SULV) [46].

In ACE, highly complex hexapeptide libraries synthesized by combinatory approaches have been used as chiral selectors. Three randomly chosen sublibraries, Arg-Lys-X-X-X-D-Ala, Arg-Phe-X-X-X-D-Ala, and Arg-Met-X-X-X-D-Ala, where X represents natural amino acids except for cysteine and tryptophan were added to the background electrolytes in CE. These peptide libraries consist of $18^3 = 5832$ individual cyclopeptides which are synthesized as a mixture. The sublibrary Arg-Lys-X-X-X-D-Ala gave the best results in terms of separation factor and resolution of *N*-derivatized amino acid enantiomers among the sublibraries tested [30].

In order to identify the individual peptides responsible for the observed enantioselectivity, the slightly modified sublibrary, Arg-Lys-X-X-X-β-Ala, was introduced and a deconvolution procedure based on CE separation data was applied. The role of proline in three randomized positions on the separation of the test analytes was clarified. The sublibrary with proline in position 5 did not separate DNP-glutamic acid enantiomers whereas those with proline in 3 and especially 4 provided resolution; therefore, proline was assigned to position 4. Next, the sublibraries were subjected to preparative HPLC. The HPLC fractionation demonstrated in a simple way that some or all of the six amino acids, Val, Met, Ile, Leu Tyr, and Phe, were deeply involved in the chiral recognition of the analytes. Therefore, six sublibraries of general structure cyclo(Arg-Lys-O-Pro-X-β-Ala), with O equal to one of the six above-mentioned amino acids and X equal to a randomized position containing 18 natural amino acids, were synthesized and checked in CE. Each sublibrary possessed similar separation capacities for a set of racemates. Tyr was assigned to the position 3. Finally, the synthesis of six cyclohexapeptides, cyclo(Arg-Lys-Tyr-Pro-O-β-Ala), with O equal to Val, Met, Ile, Leu, Tyr, or Phe was performed. In this case, the presence of an aromatic ring on the amino acid lateral chain is essential for the cyclohexapeptide to be an efficient chiral selector. Only the cyclopeptides with Tyr or Phe in position 5 showed a marked enantioselectivity and provided excellent resolution of the test sample. The cyclohexapeptide, cyclo(Arg-Lys-Tyr-Pro-Tyr-β-Ala), was identified as the best selector. As shown in Figure 6.6, enantioseparations of DNP-D,L-amino acids were attained using cyclo(Arg-Lys-Tyr-Pro-Tyr-β-Ala) as the chiral selector [30].

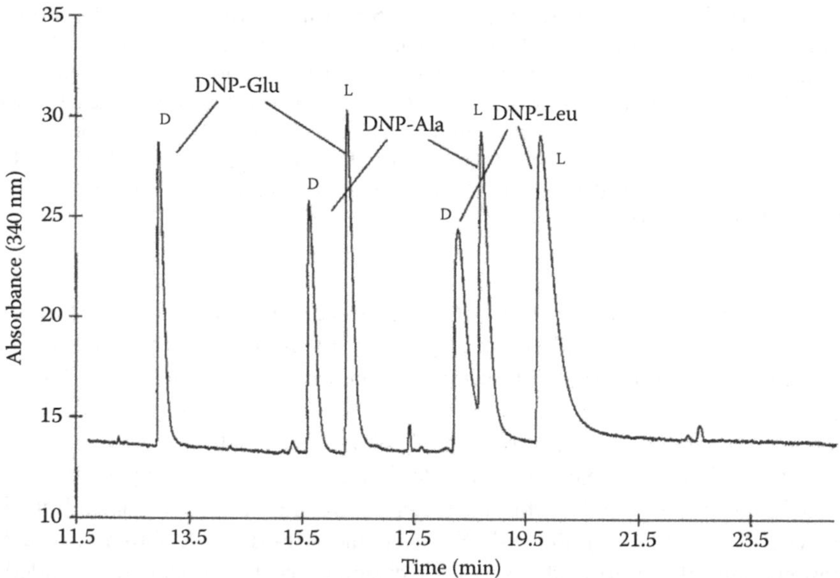

FIGURE 6.6 Separation of DNP-D,L-amino acids in a running electrolyte containing cyclo(Arg-Lys-Tyr-Pro-Tyr-β-Ala). Conditions: 10 mM cyclopeptide in 20 mM sodium phosphate buffer, pH 7.0; capillary, 50 μm I.D., 65 cm total length, 57 cm to the window; applied voltage, −20 kV; electrokinetic injection, −10 kV 3 s; detection at 340 nm, 15°C. (From Chiari, M. et al., *Anal. Chem.*, 70, 4972, 1998. With permission.)

Several cyclohexa- and cycloheptapeptides identified by a deconvolution approach were examined as chiral selectors in CE, where a coated capillary with poly(acrylamide-*co*-allyl amide of D-gluconic acid-*co*-allylglycidyl ether) is used to suppress the EOF. However, cycloheptapeptides could not separate *N*-derivatized amino acids. These findings suggest that the size and rigidity of the cyclopeptide system is important for ensuring chiral discrimination [47].

6.4.2 Affinity Capillary Electrochromatography

For chiral separations in protein- and peptide-based CE, proteins and peptides were immobilized or adsorbed onto the capillary, or protein-immobilized silica gels were packed into the capillary. The applied electric fields result in solvent and solute flow through the system. The enantioseparation occurs by differences in interactions with an immobilized or adsorbed protein selector between the enantiomers. This system is very similar to a HPLC-chiral stationary phase (CSP) system, which is operated by pressure-driven flow. Thus, this technique is termed ACEC. In addition, the term affinity capillary gel electrophoresis is used when a cross-linked-gel is used. A more detailed description of CEC can be found in Chapter 14.

6.4.2.1 Protein Selectors

There are several techniques for ACEC. The first one is that capillaries filled with gels consisting of BSA cross-linked with glutaraldehyde are prepared for the resolution of tryptophan enantiomers [11]. In this case, the detection window of the capillary was free from the cross-linked-BSA gel in order to avoid its UV absorbance. This was attained by pumping a mixture of BSA and glutaraldehyde till just before the detection window and by allowing the mixture to gel. The method could have the potential to be applicable for all types of affinity-based separations that include cross-linkable specific binding partners. The applicability of the method was demonstrated by making a mixed protein gel based on CBH I and BSA [48]. Since it was difficult to obtain a stable gel with CBH I alone, it was copolymerized with BSA. Another advantage of the mixed protein gel is that they could enantioseparate a wider range of compounds because of two protein ligands. However, since the EOF is eliminated or negligible with these capillaries, the technique is not applicable to the enantioseparation of uncharged compounds.

The second way for an ACEC technique is to use capillaries packed with protein-immobilized silica particles. HSA- [49] and AGP-immobilized [14] HPLC silica gels were packed into fused-silica capillaries. This work has been performed using commercially available silica packing materials of 5–7 μm diameter. In CEC, the magnitude of the EOF is important, since it determines whether uncharged compounds can be separated or not. EOF is generally weaker in these packed capillaries compared to open capillaries [14]. Furthermore, it is governed by the protein used and background electrolyte composition. Since both AGP and HSA are acidic proteins, EOF in these capillaries at neutral pH is in the same direction as that in open tubular capillaries. If basic proteins were used, the direction of the EOF is reversed. Figure 6.7 shows electrochromatograms showing enantioseparations of disopyramide, pentobarbital, hexobarbital, cyclophosphamide, and benzoin by ACEC with an AGP-Immobilized silical gel packed capillary [14].

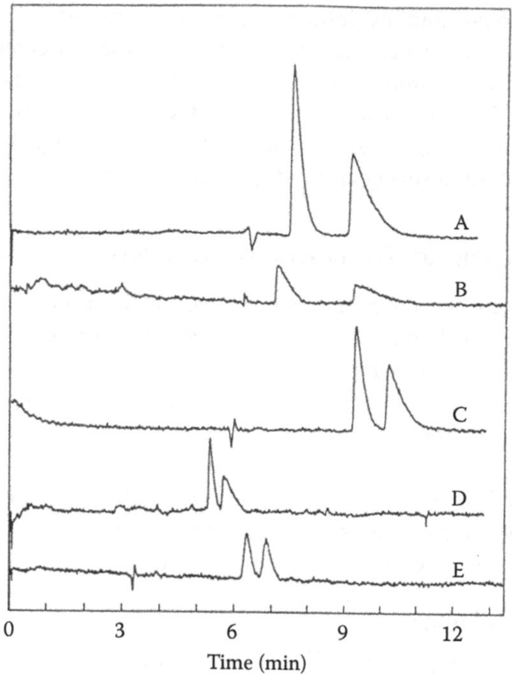

FIGURE 6.7 Electrochromatograms showing the enantiomeric separations of disopyramide (A), pentobarbital (B), hexobarbital (C), cyclophosphamide (D), and benzoin (E). Conditions: (A) 15% 2-propanol/4 mM phosphate, pH 6.8, applied voltage 12 kV, current 2 μA; (B) 2% 2-propanol/2 mM phosphate, pH 5.5, applied voltage 20 kV, current 2 μA; (C) 2% 2-propanol/2 mM phosphate, pH 5.5, applied voltage 18 kV, current 2 μA; (D) 3% 2-propanol/2 mM phosphate, pH 6.5, applied voltage 25 kV, current 2 μA; (E) 5% 1-propanol/5 mM phosphate, pH 6.5, applied voltage 15 kV, current 3 μA. (From Li, S. and Lloyd, D.K., *Anal. Chem.*, 65, 3688, 1993. With permission.)

A third method employed is to dynamically coat a protein onto the capillary. However, the protein coating slowly desorbs in the presence of an electric field. The desorbed protein can automatically be replaced by adding a small amount of soluble protein to the running buffer. The advantages of the method are that it does not require the use of any packing material or the immobilization of a protein onto the capillary, and that the same capillary can be used for work with additional proteins. When HSA was coated onto the capillary, about 0.7 monolayer of HSA adsorbed onto a typical uncoated-silica capillary at pH 7.4 [50]. Enantioseparation of warfarin showing strong bindings to HSA ($Ka \geq 10^5\ M^{-1}$) was attained using the above method, but no enantioseparation of tryptophan showing weaker bindings to HSA ($Ka \leq 10^4\ M^{-1}$) was attained. This is due to the low effective protein concentration on the capillary wall. Lysozyme was adsorbed onto the capillary wall [21] and enantioseparations of four amino acids and mephenytoin were attained. Since lysozyme is a basic protein, with a p*I* of 11.4, it might be bound tightly to the capillary by electrostatic interaction in the neutral running buffer (pH 7.2). Similarly, avidin, which is a basic protein, was adsorbed onto the capillary wall and applied to enantioseparations of acidic compounds [open-tubular CEC (OTCEC)] [51]. Furthermore, the physical adsorption method of avidin was successfully applied to a monolithic silica [52]. The method proved to be comparable in phase ratio to the chemical bonding methods used in HPLC. Due to its larger phase ratio, the resulting column showed more powerful separation capability as compared to OTCEC, as shown in Figure 6.8 [52].

An alternative way of coating a protein onto a capillary wall has been developed; stepwise coating of the capillary wall with a multilayer assembly of exfoliated γ-zyrconium phosphate and lysozyme [53]. OTCEC with this layer-by-layer coating was applied to the chiral separation of D- and L-tryptophan. Another way is to use a phospholipid coating with lysozyme [54], avidin [55], or BSA [56] as a chiral selector permeated into the phospholipid bilayer membrane. 1-(4-Iodobutyl)-1,4-dimethylpiperazin-1-ium iodide was applied as a first coating layer in the capillary to effectively suppress the EOF and to stabilize the phospholipid coating and lysozyme or BSA immobilization in the capillary. In addition, immobilized avidin in phospholipid-membrane coated fused-silica capillaries were prepared using biotinylated phospholipids and the protein was employed as a chiral selector for the enantioseparation of tryptophan and amino acids derivatives in ACEC [55].

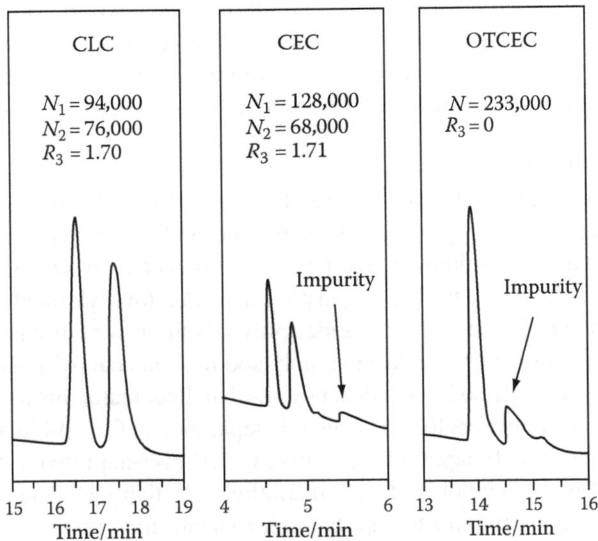

FIGURE 6.8 Separations of racemic 4-fluoromandelic acid by CLC, CEC, and OTCEC with physically adsorbed avidin stationary phase. Conditions: column, 27 cm (20 cm effective) × 50 μm I.D. monolithic silica capillary for CLC and CEC and 57 cm (50 cm effective) × 50 μm I.D. for OTCEC; mobile phase, 10 mM phosphate buffer containing 15% v/v methanol (pH 5.95) for CLC and CEC and 10 mM phosphate buffer (pH 5.95) for OTCEC; temperature, 20°C; pressure for CLC, 138 kPa; applied voltage, −15 kV for CEC and −10 kV for OTCEC. (From Liu, Z. et al., *Electrophoresis*, 23, 2977, 2002. With permission.)

A fourth way for the ACEC technique is to chemically immobilize a protein onto the surface of fused-silica capillaries. BSA was bound to the capillary wall [57]. The method includes etching the capillary wall with sodium hydroxide, epoxy-diol coating with 3-glycidoxypropyltrimethoxysilane followed by hydrolysis with hydrochloric acid, and activation with tresyl chloride and BSA coupling. Avidin was bound to an aminopropyl-silylated fused-silica surface using glutaraldehyde by a Schiff base formation reaction [58].

A fifth way for the ACEC technique is to encapsulate a protein using a sol–gel method. Two proteins, BSA and OMCHI, were encapsulated in tetramethoxysilane-based hydrogel and their enantioselectivity was evaluated for the separation of tryptophan, benzoin, eperisone, and chlorpheniramine enantiomers [59]. Furthermore, a sol–gel/organic hybrid composite material using gelatin or chitosan with tetramethoxysilane was prepared for the BSA-encapsulated monolithic column for ACEC [60]. The monolithic column prepared from chitosan with tetramethoxysilane showed a higher enantioselectivity for tryptophan enantiomers and the composite materials exhibited a higher stability compared to the silica sol–gel column.

Another approach is the use of protein–dextran polymer networks for ACE application. BSA was covalently linked to a high-molecular-mass dextran (M_r 2,000,000) using cyanogen bromide, and applied to the enantioseparation of leucovorin enantiomers [61]. Since the capillary was coated with linear polyacrylamide, an EOF would be small. Thus, the polymer would not flow out of the capillary during the course of several hours of the run. It was found that the amount of protein in the polymer network could be varied by dilution with nonderivatized dextran. This can be useful for optimizing the enantioseparation of a solute that adsorbs strongly to the ligand. The BSA–dextran polymer network can be removed and replaced by means of a syringe or by applying a fresh polymer mixture to the capillary.

6.4.2.2 Peptide Selectors

Basic peptides such as Lys-Tyr and Lys-Ser-Tyr were adsorbed onto the capillary wall and used for separations of tyrosine, phenylalanine, and fenoprofen enantiomers [21].

The polyelectrolyte multilayer coating was constructed in situ with alternating rinses of positively and negatively charged polymers. Poly(diallyldimethylammonium-chloride) and poly(L-lysine) hydrobromide [poly(L-lysine)] were used as the cationic polymers, and the polymeric surfactants poly(sodium undecanoyl-L-leucylalaninate) [poly(L-SULA)] and poly(sodium undecanoyl-L-alanylleucinate) [poly(L-SUAL)] were used as the anionic polymers [62]. The enantioseparations of the β-blockers, labetalol and sotalol, and the binaphthyl derivatives, 1,1′-bi-2-naphthyl-2,2′-dihydrogen phosphate, 1,1′-bi-2-naphthol and 1,1-binaphthyl-2,2′-diamine were attained with high run-to-run and capillary-to-capillary reproducibilities using a polyelectrolyte multilayer coating by poly(L-lysine) and poly(L-SULA).

6.5 NEW DEVELOPMENTS

The separation of (±)-*cis*- and (±)-*trans*-benzo[*a*]pyrene tetrols (BPTs) can be accomplished by the flow-through partial-filling ACE (FTPFACE) using a cross-reactive monoclonal antibody (mAb) against a polycyclic aromatic hydrocarbon [24]. In the

FTPFACE approach, the capillary is partially filled with a plug of mAb followed by a plug of a sample containing ligands. Upon application of a voltage gradient, and due to the high difference in the electrophoretic mobility induced by the difference in effective charge of ligand molecules and mAbs, the ligand and mAb zones temporarily overlap during the course of the electrophoretic run. The differences in the binding strength of the (±)-*cis*- and (±)-*trans*-BPTs with the antipolycyclic aromatic hydrocarbon mAb resulted in the separation of diastereomers and enantiomers as shown in Figure 6.9 [24]. Furthermore, it was demonstrated that (+)-enantiomers are more strongly immunocomplexed than their (–)-antipodes. FTPFACE approach allows for the use of much smaller amounts (pmol to fmol) of antibodies compared

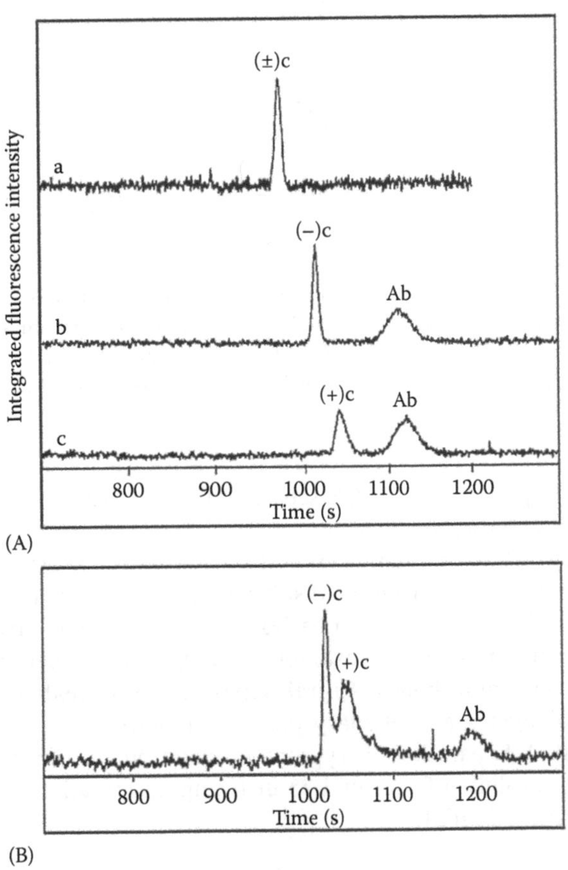

FIGURE 6.9 (A) FTPFACE electropherograms of enantiomeric *cis*-BPT; (a) nonchiral electrophoresis of a racemic mixture of (+)*cis*-BPT and (–)*cis*-BPT; (b) FTPFACE run of a (–)*cis*-BPT plug; (c) FTPFACE run of a (+)*cis*-BPT plug. (B) FTPFACE run of a racemic mixture of (+)*cis*-BPT and (–)*cis*-BPT. Fifty mM of sodium phosphate buffer at pH 8.8 was used as the CE running buffer. A 257 nm laser excitation source and a CCD camera were used for LIF detection. (From Grubor, N.M. et al., *Electrophoresis*, 26, 1080, 2006. With permission.)

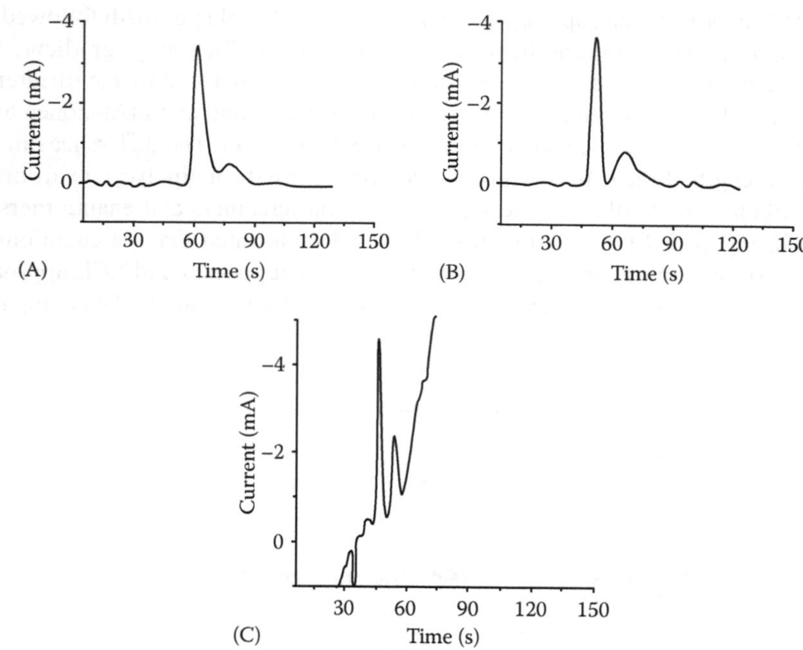

FIGURE 6.10 Effect of the separation field strength on the separation of 24 mM tryptophan enantiomers. Electrical field strength: (A) 187, (B) 218, and (C) 250 V/cm. Other conditions: effective separation length, 32 mm; electrolyte, 20 mM phosphate buffer; sample, 24 mM tryptophan; electrical detection, 0.6 V (vs. Ag/AgCl). (From Weng, X. et al., *Electrophoresis*, 27, 3135, 2006. With permission.)

to other modes of protein-based ACE separations (i.e., conventional partial-filling techniques, running buffer modifier, etc.).

Single-walled carbon nanotubes (SWNTs) were cut into short pipes by chemical oxidation in a mixture of concentrated sulfuric and nitric acids. BSA was bound to the shortened SWNTs, which bear carboxyl groups on their surfaces, by using water-soluble carbodiimide (for example, 1-ethyl-3-(3′-dimethylaminopropyl) carbodiimide) via amide bonds. Chiral separation was realized via utilizing SWNT–BSA conjugates as stationary phase immobilized in microchannel for microchip CE [63]. Separation of tryptophan enantiomers was achieved in less than 70 s with a resolution factor of 1.35 utilizing a separation length of 32 mm as shown in Figure 6.10 [63].

6.6 CONCLUSIONS

Chiral separations based on proteins or peptides in CE were carried out in two modes: one is that proteins or peptides are dissolved in the running buffer (ACE mode), and the other is that proteins or peptides are immobilized or adsorbed onto the capillary, or protein-immobilized silica gels are packed into the capillary (ACEC mode). The former is simpler and easier to operate than the latter. However, the ACE method has disadvantages of adsorption of a protein or peptide to the capillary wall, and absorption of UV light at the detection wavelength. Adsorption of a protein or peptide was overcome by using wall-coated capillaries or additives to minimize its interactions with the wall, and absorption of UV light at the detection wavelength was overcome by partial filling techniques.

Recently, FTPFACE using a small amount of cross-reactive mAb against a polycyclic aromatic hydrocarbon was applied for enantioseparation of BPTs. Furthermore, SWNT–BSA conjugates were utilized for enantioseparation of tryptophan in chip CE. These techniques could be promising for fast and efficient separation of enantiomers.

LIST OF ABBREVIATIONS

ACE	affinity capillary electrophoresis
ACEC	affinity capillary electrochromatography
AGP	α_1-acid glycoprotein
BPT	benzo[*a*]pyrene tetrol
BSA	bovine serum albumin
cAGP	chicken α_1-acid glycoprotein
CBH I, Cel7A	cellobiohydrolase I from *Trichoderma reesei*
CE	capillary electrophoresis
CEC	capillary electrochromatography
CHAPSO	3-[(3-chloramidopropyl)dimethylammonio]-2-hydroxy-1-propanesulfonate
CHES	2-(cyclohexylamino)ethanesulfonic acid
CLC	column liquid chromatography
CSP	chiral stationary phase
DNP	2,4-dinitrophenyl
EKC	electrokinetic chromatography
EOF	electroosmotic flow
FTPFACE	flow-through partial-filling affinity capillary electrophoresis
HPC	hydroxypropylcellulose
HPLC	high-performance liquid chromatography
HSA	human serum albumin
HST	human serum transferrin
mAb	monoclonal antibody
NMR	nuclear magnetic resonance
OMCHI	ovomucoid from chicken egg whites
OMTKY3	the third domain of ovomucoid from turkey egg white
OGCHI	ovoglycoprotein from chicken egg whites
OTCEC	open-tubular capillary electrochromatography
PEA	*o*-phosphorylethanolamine
PEG	polyethylene glycol
PGA	penicillin G-acylase
p*I*	isoelectric point
poly(L-SUAL)	poly(sodium undecanoyl-L-alanylleucinate)
poly(L-SULA)	poly(sodium undecanoyl-L-leucylalaninate)

poly(L-SULβA)	poly(sodium undecanoyl-L-leucylβ-alaninate)
poly(L-SULL)	poly(sodium undecylenyl-L-leucyleucinate)
poly(L-SULV)	poly(sodium undecanoyl-L-leucylvalinate)
poly(L-SUV)	poly(sodium undecanoyl-L-valinate)
poly(L-SUVL)	poly(sodium undecanoyl-L-valylleucinate)
poly(L-SUVV)	poly(sodium undecanoyl-L-valylvalinate)
RfBP	riboflavin binding protein
SDS	sodium dodecylsulfate
SWNT	single-walled carbon nanotube
UV	ultraviolet

REFERENCES

1. Allenmark, S. 1991. *Chromatographic Enantioseparation: Methods and Applications*, 2nd edn. New York: Ellis Horwood.
2. Haginaka, J. 2001. Protein-based chiral stationary phases for high-performance liquid chromatography enantioseparations. *J Chromatogr A* 906: 253–273.
3. Gubitz, G, Schmid, MG. (eds.) 2004. *Chiral Separations: Methods and Protocols*. Totowa: Humana Press.
4. Haginaka, J. 2008. Recent progresses in protein-based chiral stationary phases for enantioseparations in liquid chromatography. *J Chromatogr B* 875: 12–19.
5. Haginaka, J. 2000. Enantiomer separation of drugs by capillary electrophoresis using proteins as chiral selectors. *J Chromatogr A* 875: 235–254.
6. Millot, MC. 2003. Separation of drug enantiomers by liquid chromatography and capillary electrophoresis, using immobilized proteins as chiral selectors. *J Chromatogr B* 797: 131–159.
7. Pinkerton, TC, Howe, WJ, Urlich, EL, Comiskey, JP, Haginaka, J, Murashima, T, Walkenhorst, WF, Westler, WM, Markley, JL. 1995. Protein-binding chiral discrimination of HPLC stationary phases made with whole, fragmented and third domain turkey ovomucoid. *Anal Chem* 67: 2354–2367.
8. Petitpas, I, Bhattacharya, AB, Twine, S, East, M, Curry, S. 2001. Crystal structure analysis of warfarin binding to human serum albumin. *J Biol Chem* 276: 22804–22809.
9. Lavecchia, A, Cosconati, S, Novellino, E, Calleri, E, Temporini, C, Massolini, G, Carbonara, G, Fracchiolla, G, Loiodice, F. 2007. Exploring the molecular basis of the enantioselective binding of penicillin G acylase towards a series of 2-aryloxyalkanoic acids: A docking and molecular dynamics study. *J Mol Graph Mod* 25: 773–783.
10. He, XM, Carter, DC. 1992. Atomic structure and chemistry of human serum albumin. *Nature* 358: 209–215.
11. Birnbaum, S, Nilsson, S. 1992. Protein-based capillary affinity gel electrophoresis for the separation of optical isomers. *Anal Chem* 64: 2872–2874.
12. Vespalec, R, Šustáček, V, Boček, P. 1993. Prospects of dissolved albumin as a chiral selector in capillary electrophoresis. *J Chromatogr* 638: 255–261.
13. Arai, T, Nimura, N, Kinoshita, T. 1995. Investigation of enantioselective ofloxacin-albumin binding and displacement interactions using capillary affinity zone electrophoresis. *Biomed Chromatogr* 9: 68–74.
14. Li, S, Lloyd, DK. 1993. Direct chiral separations by capillary electrophoresis using capillaries packed with an α_1-acid glycoprotein chiral stationary phase. *Anal Chem* 65: 3684–3690.

15. Busch, S, Kraak, JC, Poppe, H. 1993. Chiral separations by complexation with proteins in capillary zone electrophoresis. *J Chromatogr* 635: 119–126.
16. Haginaka, J, Kanasugi, N. 1997. Separation of basic drug enantiomers by capillary zone electrophoresis using ovoglycoprotein as a chiral selector. *J Chromatogr A* 782: 281–288.
17. Sadakane, Y, Matsunaga, H, Nakagomi, K, Hatanaka, Y, Haginaka, J. 2002. Protein domain of chicken α_1-acid glycoprotein is responsible for chiral recognition ability. *Biochem Biophys Res Commun* 295: 587–590.
18. Tanaka, Y, Matsubara, N, Terabe, S. 1994. Separation of enantiomers by affinity elecrokinetic chromatography using avidin. *Electrophoresis* 15: 848–853.
19. De Lorenzi, E, Massolini, G, Lloyd, DK, Monaco, HL, Galbusera, C, Caccialanza, GJ. 1997. Evaluation of quail egg white riboflavin binding protein as a chiral selector in high-performance liquid chromatography and capillary electrophoresis. *J Chromatogr A* 790: 47–64.
20. Valtcheva, L, Mohammad, J, Pettersson, G, Hjertén, S. 1993. Chiral separation of β-blockers by high-performance capillary electrophoresis based on non-immobilized cellulase as enantioselective protein. *J Chromatogr* 638: 263–267.
21. Liu, Z, Zou, H, Ye, M, Ni, J, Zhang, Y. 1999. Study of physically adsorbed stationary phases for open tubular capillary electrochromatography. *Electrophoresis* 20: 2891–2897.
22. Fanali, S, Caponecchi, G, Aturki, Z. 1997. Enantiomeric resolution by capillary zone electrophoresis: Use of pepsin for separation of chiral compounds of pharmaceutical interest. *J Microcol Sep* 9: 9–14.
23. Gotti, R, Calleri, E, Massolini, G, Furlanetto, S, Cavrini, V. 2007. Penicillin G acylase as chiral selector in CE using a pullulan-coated capillary. *Electrophoresis* 27: 4746–4754.
24. Grubor, NM, Armstrong, DW, Jankowiak, R. 2006. Flow-through partial-filling affinity capillary electrophoresis using a crossreactive antibody for enantiomeric separations. *Electrophoresis* 26: 1078–1083.
25. Tanaka, Y, Terabe, S. 1995. Partial separation zone technique for the separation of enantiomers by affinity electrokinetic chromatography with proteins as pseudo-stationary phases. *J Chromatogr A* 694: 277–284.
26. Kilar, F, Fanali, S. 1995. Separation of tryptophan-derivative enantiomers with iron-free human serum transferring by capillary electrophoresis. *Electrophoresis* 16: 1510–1518.
27. Tanaka, Y, Terabe, S. 1999. Studies on enantioselectivities of avidin, avidin-biotin complex and streptavidin by affinity capillary electrophoresis. *Chromatographia* 49: 489–495.
28. Wistuba, D, Diebold, H, Schurig, V. 1995. Enantiomer separation of DNP-amino acids by capillary electrophoresis using chiral buffer additives. *J Microcol Sep* 7: 17–22.
29. Billiot, E, Macossay, J, Thibodeaux, S, Shamsi, SA, Warner IM. 1998. Chiral separations using dipeptide polymerized surfactants: Effect of amino acid order. *Anal Chem* 70: 1375–1381.
30. Chiari, M, Desperati, V, Manera, E, Longhi, R. 1998. Combinatorial synthesis of highly selective cyclohexapeptides for separation of amino acid enantiomers by capillary electrophoresis. *Anal Chem* 70: 4967–4973.
31. Lloyd, DK, Aubry, A-F, De Lorenzi, E. 1995. Selectivity in capillary electrophoresis: the use of proteins. *J Chromatogr A* 792: 349–369.
32. Hjertén, S. 1985. High-performance electrophoresis elimination of electroendosmosis and solute adsorption. *J Chromatogr* 347: 191–198.
33. Barker, GE, Russo, P, Hartwick, RA. 1992. Chiral separation of leucovorin with bovine serum albumin using affinity capillary electrophoresis. *Anal Chem* 64: 3024–3028.

34. Chiari, M, Cretich, M, Desperati, V, Marinzi, C, Galbusera, C, De Lorenzi, E. 2000. Evaluation of new adsorbed coatings in chiral capillary electrophoresis and partial filling technique. *Electrophoresis* 21: 2343–2351.
35. Eberle, D, Hummel, RP, Kuhn, R. 1997. Chiral resolution of pantoprazole sodium and related sulfoxides by complex formation with bovine serum albumin in capillary electrophoresis. *J Chromatogr A* 759: 185–192.
36. Ahmed, A, Lloyd, DK. 1997. Effect of organic modifiers on retention and enantiomeric separations by capillary electrophoresis with human serum albumin as a chiral selector in solution. *J Chromatogr A* 766: 237–244.
37. Sun, P, Wu, N, Barker, G, Hartwick, RA. 1993. Chiral separations using dextran and bovine serum albumin as run buffer additives in affinity capillary electrophoresis. *J Chromatogr* 648: 475–480.
38. Haginaka, J, Seyama, C, Kanasugi, N. 1995. Absence of chiral recognition ability in ovomucoid: Ovoglycoprotein-bonded HPLC stationary phases for chiral recognition. *Anal Chem* 67: 2539–2547.
39. Miwa, T, Ichikawa, M, Tsuno, M, Hattori, T, Miyakawa, T, Kayano, M, Miyake, Y. 1987. Direct liquid chromatographic resolution of racemic compounds. Use of ovomucoid as a column ligand. *Chem Pharm Bull* 35: 682–686.
40. Miwa, T, Miyakawa, T, Miyake, Y. 1988. Characteristics of an avidin-conjugated column in direct liquid chromatographic resolution of racemic compounds. *J Chromatogr* 457: 227–233.
41. Mano, N, Oda, Y, Ishihama, Y, Katayama, H, Asakawa, N. 1998. Investigation of interactions between drug enantiomers and flavoprotein as chiral selector by affinity capillary electrophoresis. *J Liq Chromatogr Rel Technol* 21: 1311–1332.
42. Hedeland, M, Isaksson, R, Pettersson, C. 1998. Cellobiohydrolase I as a chiral additive in capillary electrophoresis and liquid chromatography. *J Chromatogr A* 807: 297–305.
43. Schmid, MG, Gubitz, G, Kilar, F. 1998. Stereoselective interaction of drug enantiomers with human serum transferring in capillary zone electrophoresis (II). *Electrophoresis* 19: 282–287.
44. Shamsi, SA, Macossay, J, Warner, IM. 1997. Improved chiral separations using a polymerized dipeptide anionic chiral surfactant in electrokinetic chromatography: Separations of basic, acidic, and neutral racemates. *Anal Chem* 69: 2980–2987.
45. Billiot, E, Warner, IM. 2000. Examination of structural changes of polymeric amino acid-based surfactants on enantioselectivity: Effect of amino acid order, steric factors, and number and position of chiral centers. *Anal Chem* 72: 1740–1748.
46. Shamsi, SA, Valle, BC, Billiot, F, Warner, IM. 2003. Polysodium N-undecanoyl-L-leucylvalinate: A versatile chiral selector for micellar electrokinetic chromatography. *Anal Chem* 75: 379–387.
47. De Lorenzi, E, Massolini, G, Galbusera, C, Longhi, R, Marinzi, C, Consonni, R, Chiari, M. 2001. Chiral capillary electrophoresis and nuclear magnetic resonance investigation on the structure-enantioselectivity relationship in synthetic cyclopeptides as chiral selectors. *Electrophoresis* 22: 1373–1384.
48. Nilsson, S, Schweitz, L, Petersson, M. 1997. Three approaches to enantiomer separation of β-adrenergic antagonists by capillary electrochromatography. *Electrophoresis* 18: 884–890.
49. Lloyd, DK, Li, S, Ryan, P. 1995. Protein chiral selectors in free-solution capillary electrophoresis and packed-capillary electrochromatography. *J Chromatogr A* 694: 285–296.
50. Yang, J, Hage, DS. 1994. Chiral separation in capillary electrophoresis using human serum albumin as buffer additive. *Anal Chem* 66: 2719–2725.
51. Liu, Z, Otsuka, K, Terabe, S. 2001. Chiral separation by open tubular capillary electrochromatography. *J Sep Sci* 24: 17–26.

52. Liu, Z, Otsuka, K, Terabe, S, Motokawa, M, Tanaka, N. 2002. Physically adsorbed chiral stationary phase of avidin on monolithic silica column for capillary electrochromatography and capillary liquid chromatography. *Electrophoresis* 23: 2973–2981.
53. Geng, L, Bo, T, Liu, H, Li, N, Liu, F, Li, K, Gu, J, Fu, R. 2004. Capillary coated layer-by-layer assembly of γ-zirconium phosphate/lysozyme nanocomposite film for open tubular capillary electrochromatography. *Chromatographia* 59: 65–70.
54. Bo, T, Wiedmer, SK, Riekkola, M-L. 2004. Phopholipid-lysozyme coating for chiral separation in capillary electrophoresis. *Electrophoresis* 25: 1784–1791.
55. Han, NY, Hautala, JT, Wiedmer, SK, Riekkola, M-L. 2006. Immobilization of phopholipid-avidin on fused-silica capillaries for chiral separation in open-tubular capillary electrochromatography. *Electrophoresis* 27: 1502–1509.
56. Wiedmer, SK, Bo, T, Riekkola, M-L. 2008. Phospholipid-protein coatings for chiral capillary electrochromatography. *Anal Biochem* 373: 26–33.
57. Hofstetter, H, Hofstetter, O, Schurig, V. 1998. Enantiomer separation using BSA as chiral stationary phase in affinity OTEC and OTLC. *J Microcol Sep* 10: 287–291.
58. Kitagawa, F, Inoue, K, Hasegawa, T, Kamiya, M, Okamoto, Y, Kawase, M, Otsuka, K. 2006. Chiral separation of acidic drug components by open tubular electrochromatography using avidin immobilized capillaries. *J Chromatogr A* 1130: 219–226.
59. Kato, M, Kato-Sakai, K, Matsumoto, N, Toyo'oka, T. 2002. A protein-encapsulation technique by the sol-gel method for the preparation of monolithic columns for capillary electrochromatography. *Anal Chem* 74: 1915–1921.
60. Kato, M, Saruwatari, H, Kato-Sakai, K, Toyo'oka, T. 2004. Silica sol–gel/organic hybrid material for protein encapsulated column of capillary electrochromatography. *J Chromatogr A* 1044: 267–270.
61. Sun, P, Barker, GE, Hartwick, RA, Grinberg, N, Kaliszan, R. 1993. Chiral separations using an immobilized protein-dextran polymer network in affinity capillary electrophoresis. *J Chromatogr A* 652: 247–252.
62. Kamande, MW, Zhu, X, Kapnissi-Christodoulou, C, Warner, IM. 2004. Chiral separations using a polypeptide and polymeric dipeptide surfactant polyelectrolyte multilayer coating in open-tubular capillary electrochromatography. *Anal Chem* 76: 6681–6692.
63. Weng, X, Bi, H, Liu, B, Kong, J. 2006. On-chip chiral separation based on bovine serum albumin-conjugated carbon nanotube as stationary phase in a microchannel. *Electrophoresis* 27: 3129–3315.

7 Chiral Ligand Exchange Capillary Electrophoresis

Vincenzo Cucinotta and Alessandro Giuffrida

CONTENTS

7.1 Introduction 163
7.1.1 General Remarks 163
7.1.2 Theoretical Considerations on CLECE 165
7.2 Historical Developments 168
7.2.1 "All in Solution" CLECE 168
7.2.2 Cyclodextrins in CLECE 172
7.2.3 LECE in the Presence of Micelles 179
7.2.4 Ligand Exchange in CEC 185
7.3 Future Perspectives 188
References 190

7.1 INTRODUCTION

7.1.1 General Remarks

As known, metal ions can behave as electron pair acceptors. This property permits the bonding of a huge variety of both inorganic and organic molecules to a metal ion: the only condition that must be fulfilled is that this molecule has one or more electron pairs available for such bonding. The species formed following this bonding is called metal coordination compound, and the molecule or ion able to coordinate the metal ion is called ligand.

It is just this specific ability of metal ions that is exploited in the techniques based on the ligand exchange process: the ligand exchange is what occurs on the binding sites of the metal ion and corresponds to the substitution of a ligand with one or more other ligands.

This short recalling of elementary inorganic chemistry permits to focus on some decisive aspects of the separation techniques that exploit the ligand exchange process, including what gives the title to this chapter: chiral ligand exchange capillary electrophoresis (CLECE).

The first aspect is that we are speaking of an exchange process, a chemical reaction, which proceeds with a specific velocity: it has its own kinetics, which is decisive for the performance of the technique. We can obviously anticipate that the substitution reaction should occur quickly, and therefore the metal coordination compound must be labile. In such kind of complexes, the bond(s) between the ligand and the metal ions are continuously formed and broken: in this situation, the thermodynamic stability of the complex can be conveniently expressed as the mean time spent by the ligand molecules bound to the metal ion. Consequently, we can appreciate the importance of the stability of the complex and the degree of affinity of the ligand toward the metal ion. If the affinity is too low, regardless of the lability of the complex, the ligand will spend a negligible fraction of time bonded to the metal ion. Therefore, this process, though occurring, will have only a slight effect on the ligand behavior. The complex stability is conveniently expressed by the formation constant of the complex.

While the value of the complex stability depends on both the ligand and the metal ion, the lability of the complex depends almost exclusively on the metal ion: in the choice of the metal ion that we will use in these techniques, we will be forced to exclude all the metal ions which give rise to inert complexes, typically chromium(III) or cobalt(III).

The ligand exchange process was originally exploited in liquid chromatography, starting from the pioneering work by Helfferich as far back as 1961 [1]. In this first version of ligand exchange chromatography (LEC), the metal ion was bound to a cationic exchanger, and the analyte to be separated, interacting with this metal ion, could reside on the stationary phase. By adding a ligand containing eluent, a differential retention of the analytes occurred. By using a purpose-made chiral ion exchanger, Davankov et al. obtained discrimination of enantiomers of α-amino acids [2]. In this first developed procedure, through the ligand exchange, the affinity of a specific molecule for the metal ion became a sort of affinity for the stationary phase. Thus, LEC was a classical chromatographic separation, a two sites technique, the free ligand in the mobile phase and the complexed ligand in the stationary phase. Later on, modifications were introduced, and the methods can be conveniently classified as

- Chiral stationary phase (CSP)
- Chiral coated stationary phase (CCP)
- Chiral mobile phase (CMP)

While details about all these techniques can be easily found in excellent reviews [3], here it is interesting to give a glance at the CMP techniques, obtained by the addition of a metal ion complex in the mobile phase. From the point of view of the procedure, the CMP methods show some advantages, since a preliminary preparation of the stationary phase is not required: a commercially available reverse phase resin appears more than adequate for this task. Thus, only the preparation of a suitable mobile phase is required in this case. What soon appears new in this technique, is the possibility to exploit the homogeneous equilibria occurring in the mobile phase. An accurate study of such equilibria can be afforded by a plurality of techniques, e.g., spectrophotometry and, especially, pH-metric potentiometry. Unfortunately, what goes on in the mobile phase does not represent the totality of the processes, since, as always in chromatographic techniques, we should not forget the presence of the stationary phase. All the species in solution undergo an additional heterogeneous "quasi-equilibrium," distributing themselves between the mobile and the stationary phase. Since what happens in solution does not exhaust what is going on during the chromatographic run, the results obtained are a complex function not only of the equilibria occurring in solution, but also of these heterogeneous processes. From this point of view, we have two distinct possibilities: if one species in solution is neutral, it is the only one that will be able to interact with the reversed phase. However, if two or more species are neutral, any prediction of the success of the separation will be very hard, since we have to consider the differential affinity of the different neutral species for the reversed phase.

While most of the ligand exchange processes were exploited in column liquid chromatography, several applications in gas-chromatography [4], in supercritical chromatography [5] and in planar chromatography (for a review, see [6]) can also be found in literature.

In 1985, the ligand exchange process was exploited for the first time in capillary electrophoresis by Gassman et al. [7]. In this first report, the copper(II) ion, complexed by two histidine molecules, was added to an ammonium acetate solution to give a chiral background electrolyte (BGE), thus separating some α-amino acid racemates. By changing the ligand bonded to copper(II) in the complex added to BGE, other separations were obtained by the same authors [8]. With these two papers, a technique was introduced to obtain the separation of racemates which hardly could be separated before. These developments will be the subject of this chapter. Later on, we will come back to the historical developments of CLECE, which widened the same meaning of this term.

No exhaustive review of literature in this field will be given. The aim of this chapter is to show how CLECE was developed and which problems arise when trying to explain the results obtained. Thus, the cited literature will only have an exemplificative function. As excellent reviews, the papers by Schmid et al. [9] and Davankov [3] are acknowledged.

7.1.2 Theoretical Considerations on CLECE

We have seen in the previous section that the occurrence of complexation equilibria in solution can be exploited for the separation of analytes that show ligand properties toward a selected metal ion. We have also seen that, although the ligand exchange can be exploited for any kind of separation, the first reported examples of separation in the context of capillary electrophoresis concerned a chiral separation. Further, the CLECE separations were the first chiral separation obtained in capillary electrophoresis. One might reflect about this fact: is there any specific reason why the ligand exchange in capillary electrophoresis was from the beginning applied just to chiral separations or is it simply a casual occurring? Our opinion is that a specific reason indeed exists: the formation of metal complexes with the analytes induces a fine-tuning of the separation, permitting to achieve selectivity hardly achievable otherwise. In the field of the molecular recognition, though only recently, it is nowadays acquired that when a molecule must coordinate a metal ion, this corresponds to an additional recognition site, and, what is more important, this site involves very strict steric needs. A very significant example may be offered by the enzymatic catalysis which often involves metalloenzymes: the metal site is often the active site of the enzyme.

The enantiomeric separation is a difficult matter, since two molecules, otherwise equal, must be separated just on the basis of a slight structural difference, the spatial disposition of the same groups. The conditions to be fulfilled in capillary electrophoresis in order to obtain successful separation of the investigated enantiomeric analytes are

- The analytes give rise to a complex in solution.
- The formation rate of the complexes is high (labile complexes).
- The formation degree of the complexes is significantly different between the two enantiomeric analytes.
- The complex and the free ligand have different electrophoretic mobilities.

If all these conditions are fulfilled, the chiral separation has a fair probability to be successful. Anyhow, while the first two conditions must be necessarily fulfilled, the last two conditions both contribute to the separation and must be considered in greater detail.

In CE, the observed mobility can be easily deducted from the electropherograms and by the experimental parameters of separation by the formula:

$$\mu(\text{obs}) = Ll/(t_m \Delta V)\ (\text{m}^2\,\text{s}^{-1}\,\text{V}^{-1}) \tag{7.1}$$

where
L is the total length of the capillary
l is the length from the injection vial to the detection window
t_m is the migration time of the analyte
ΔV is the applied voltage

By applying Equation 7.1 to a marker of the electroosmotic flow (EOF), we obtain

$$\mu(\text{eo}) = Ll/(t(\text{eo}))\Delta V \tag{7.2}$$

and we can easily calculate the μ(eff):

$$\mu(\text{eff}) = \mu(\text{obs}) - \mu(\text{eo}) \tag{7.3}$$

Obviously, the μ(eff) is an algebraic quantity, its sign accounting for its either positive or negative charge. In LECE, this mobility, determined by Equation 7.3, can be correlated to the actual mobilities of the individual species present through the following formula:

$$\mu(\text{eff}) = \sum(\mu_I\, x_i) \tag{7.4}$$

where
μ_I is the actual mobility of the *i*-species
x_i is its molar fraction

Let us consider now the simplest case: the analyte forms one complex, and thus it distributes between only two species, namely, the free analyte and the complexed analyte. In this simplified case, considering that $x(\mathrm{f}) + x(\mathrm{c}) = 1$, by a simple substitution in Equation 7.4, we can write:

$$\mu(\mathrm{eff}) = (\mu(\mathrm{c}) - \mu(\mathrm{f})) \times x(\mathrm{c}) + \mu(\mathrm{f}) \tag{7.5}$$

for each enantiomer.

Thus, by subtracting enantiomer 2 from enantiomer 1, and considering that $\mu(\mathrm{f})$ must obviously be equal for both the enantiomers, we obtain:

$$\Delta(\mu(\mathrm{eff})) = \mu(\mathrm{c2}) \times x(\mathrm{c2}) - \mu(\mathrm{c1}) \times x(\mathrm{c1}) - (x(\mathrm{c2}) - x(\mathrm{c1}))\mu(\mathrm{f}) \tag{7.6}$$

The reported equations permit to correlate the observed mobilities with the actual mobilities and the molar fractions of the species. At the same time, it can be seen that, even in the simplest case of only two species, $\Delta(\mu(\mathrm{eff}))$ still depends on five different parameters. It clearly appears that a preliminary knowledge of the molar fractions, and therefore of the species distribution, is needed in order to calculate the actual mobilities of the involved species.

In this context, it can be interesting to consider the two extreme cases. In the case where no complexes are formed, by appropriate substitution in Equation 7.6, we simply obtain no difference in the mobilities between the enantiomers. It is an obvious result: no complex, no separation. We should, however, consider that, by using Equation 7.5, we can easily determine the actual mobility of the free analyte at a certain pH. The pH aspect needs to be underlined, since, generally, if an analyte has affinity for a metal ion, it will have an affinity for the proton, and thus it is absolutely necessary to control the pH value of BGE to have reproducible mobility values. In the other case, when both the enantiomers are fully complexed, by substituting in Equation 7.6, we obtain

$$\Delta(\mu(\mathrm{eff})) = \mu(\mathrm{c2}) - \mu(\mathrm{c1}) \tag{7.7}$$

Equation 7.7 shows that even in this case, a difference between the observed enantiomer mobilities can be observed, if their complexes have a different mobility. While this conclusion also appears quite obvious, it should be underlined that, by carrying out an electrophoretic separation in this condition, it is possible to get information on the actual mobilities of both the complexes.

These two extreme cases can exemplify how the knowledge of the species distribution in solution is the only way to get detailed information on the mechanism of ligand exchange actually occurring during the electrophoretic run. The conclusion is that the stereoselective formation of the analyte complexes gives rise to a difference in the mobilities of the two enantiomers. By measuring the EOF in the BGE, either by adding a marker in the sample, or by a separate electrophoretic run carried out with the same BGE, the effective mobilities can be calculated. At constant pH, a monotonic trend of the values of mobilities should be obtained when increasing the selector concentration. This behavior can represent an easy system suitability test. The mobilities of the individual enantiomers also permit to calculate the selectivity of the chiral selecting system. In this context, it is surprising that, in the related papers, only the resolution values are often reported. Though, obviously an optimal resolution is looked for, one should however not forget that the resolution is the combined effect, not only of selectivity, but also of efficiency, which depends on all the experimental details of the electrophoretic experiment. Our recommendation is to report both these values.

A last consideration concerns the way to express selectivity. In chromatography, the α values are reported, simply calculated by

$$\alpha = t_2'/t_1' \tag{7.8}$$

where t_1' and t_2' are the corrected retention times (2 refers to the slower enantiomer). In CLECE, on the contrary, the corrected retention times are substituted by the effective mobilities values.

However, this simple borrowing of the parameter α from chromatography is unadvisable, since during separation the two analytes can even show mobilities of different signs. In this case, a paradox occurs: the α values become negative. The point is that the reference value (t_m) in chromatography lies at one extreme of the chromatogram, at the beginning, while in CE the reference value (μ(eo)) can be situated anywhere in the electropherogram. From this point of view, the use of the selectivity factor *S* can be recommended, defined as

$$S = (\mu_2 - \mu_1)/(\mu_1 + \mu_2) \tag{7.9}$$

In Equation 7.9, the numerator is just the difference in the mobility values, while the denominator takes into account how long the separation process has operated, thus acting as a normalizing factor. This parameter does not suffer from any possible paradox and seems perfectly appropriate to express selectivity.

7.2 HISTORICAL DEVELOPMENTS

7.2.1 "All in Solution" CLECE

CLECE without the presence of any heterogeneous phase will be considered first. To underline that all the processes occur in solution, it is called "all in solution." As mentioned before, the first paper dealing with CLECE [7] has used the $[Cu(L\text{–}His)_2]$ complex as chiral selector. The separation of amino acids enantiomers was based on the formation of ternary complexes, by substitution of one histidine molecule with an amino acid molecule. The obtained complexes, having two chiral centers, were diastereoisomeric.

The procedure used in this work and in the 1990s [10–17] was to have a BGE containing copper(II) and the stereoselective ligand in a 1:2 concentration ratio, yielding the $[CuL_2]$ complex in solution. By exchanging one molecule of this ligand with one molecule of analyte, the ternary diastereoisomeric complex is formed. Thus, the presence in the BGE of a complex including two molecules of the ligand appeared to be the only way to obtain the separation. However, this conclusion seems rushed for two different reasons. First, the formation of metal complexes in solution is a quite complex process. Writing the equilibria for the formation of the complexes,

$$Cu(II) + L = [CuL] \tag{7.10}$$

$$[CuL] + L = [CuL_2] \tag{7.11}$$

we can soon appreciate that, due to the presence of the reactants involved in the concept of equilibrium, it is not possible to hypothesize the quantitative formation of the complexes. In particular, if this concept is applied to equilibrium (7.11), it becomes clear that in solution both these two complexes will be simultaneously present. The concentration of each is a complex function not only of the analytical concentrations, but also of the values of the formation constants. Further, we should consider that we are dealing with ligands that can be protonated. In solution there is a competition for the ligand between the proton and the metal ion. Thus, the situation is even more complex, with the obvious consequence that the concentration of the complexes depends on the pH of the solution. Generally, it is not possible to draw any quantitative or even semiquantitative conclusion, without the knowledge of the equilibria constants involved, and the accurate simulation by a suitable computer program. Second, as realized in 2000 [18], there is no need to have a bis-complex to exploit the ligand exchange mechanism, since the coordination sphere of the metal ion is completed by the necessary number of solvent (water) molecules. Thus, we are not forced to start with a complex containing two ligand molecules. We can simply start with a complex containing one ligand molecule, and this complex will form the wished ternary complex by simply substituting the necessary water molecules. Although ignored, this process of water substitution has fortunately permitted to obtain satisfactory chiral separations, since the hypothesized bis-complex between copper(II) and the stereoselective ligand presumably was present in the BGE at quite a low concentration, and the separation process was mainly accomplished by the unexpected presence of the mono-complex.

Following the pioneering work carried out by different authors, the year 2000 represents a milestone for this technique, as exemplified by the paper of Chen et al. [18]. In this paper, a correlation between the apparent mobility observed in the electrophoretic runs and the stability constants is established. However, any conclusion on the values of formation constants so found relies on the correctness of the proposed model.

An original approach was used by Karbaum and Jira [19], who coupled the ligand exchange to nonaqueous CE. The copper(II) complex with L-proline or with L-isoleucine, used as selector toward eight enantiomeric pairs of unmodified amino acids (AAs), was dissolved in methanol. Methanol is certainly a far weaker ligand than water. In this case, the ratio of the ligand exchange complex should be 1:2, since methanol would not be strong enough to give rise to the necessary ternary complex. Correspondingly, the 1:3 ratio between the analytical concentrations of copper(II) and the chiral selecting ligand results in the optimal chiral resolution.

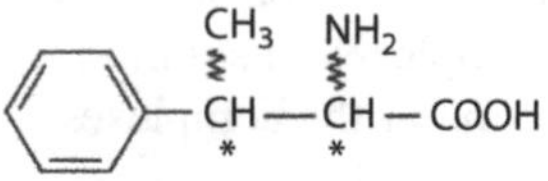

erythro-(2*R*,3*R*),(2*S*,3*S*)- and *threo*-(2*R*,3*S*),(2*S*,3*R*)-β-methlyphenylalanine (*e*/*t*-β-Me-Phe)

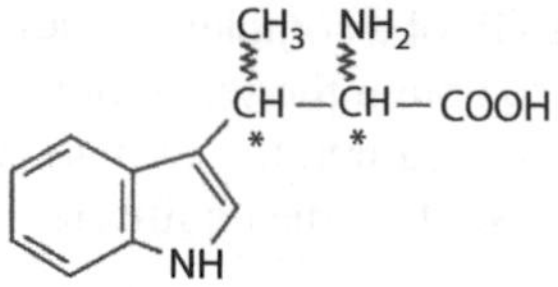

erythro-(2*R*,3*R*),(2*S*,3*S*)- and *threo*-(2*R*,3*S*),(2*S*,3*R*)-β-methyltryptophan (*e*/*t*-β-Me-Trp)

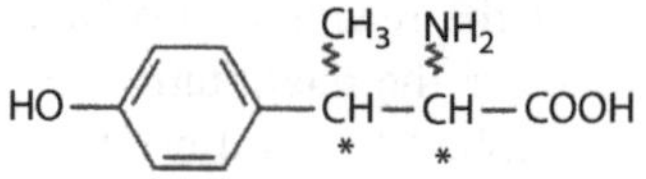

erythro-(2*R*,3*R*),(2*S*,3*S*)- and *threo*-(2*R*,3*S*),(2*S*,3*R*)-β-methyl-tyrosine (*e*/*t*-β-Me-Tyr)

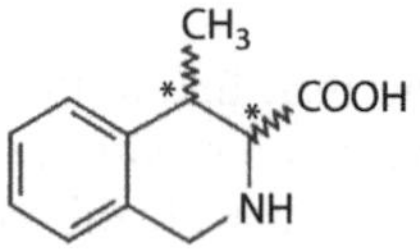

erythro-(2*R*,3*R*),(2*S*,3*S*)- and *threo*-(2*R*,3*S*),(2*S*,3*R*)-β-methyl-1,2,3,4-tetrahydroisoquinoline-3-carboxylic acid (*e*/*t*-β-Me-Tic)

FIGURE 7.1 Schematic formula of the stereoisomeric β-methyl-amino acids investigated in reference [22]. (From Grobuschek, N. et al., *J. Pharm. Biomed. Anal.*, 27, 599, 2002. With permission.)

Until 2000, only AAs were used as stereoselective ligand in CLECE experiments. In 2001, at last, a new class of compounds is proposed as chiral selector, the hydroxy acids [20,21], till that time only used as enantiomeric analytes to separate.

In 2002, Grobuschek et al. [22] reported an interesting study on the separation of the optical isomers of four β-methyl-amino acids. These compounds have two different chiral centers, and thus each of them can give rise to two enantiomeric pairs, whose schematic structure is shown in Figure 7.1. The electrophoretic separation was accomplished by using the copper(II) complex with 4-L-hydroxyproline, *N*-(2-hydroxypropyl)-4-L-hydroxyproline, and *N*-(2-hydroxyoctyl)- 4-L-hydroxyproline, respectively. Some electropherograms are shown in Figure 7.2. The same authors, in 2005 [23], by a similar mechanism and selector, separated a series of halogenated amino acids.

Yi Chen et al. introduced the use of a different ion in CLECE. They used zinc(II) instead of copper(II) [24,25]. Possible problems inherent to the use of zinc is related to its lesser coordinating ability with respect to copper(II), and the reported tendency of zinc to give rise to a tetrahedral coordination sphere. We should recall that zinc, being in the IIB group of the periodical table, is not a transition element (more relevantly, the zinc(II) ion is not a transition ion) and thus does not show ***d–d*** transitions. Thus, it is not so easy to characterize its coordination sphere, and its preference for a tetrahedral coordination is mainly extrapolated from structural studies at the solid state. In specific complexes used as selectors, the L-lysine [24] and the L-arginine [25] zinc(II) complexes respectively, the coordination is certainly not tetrahedral, since chiral resolution by the formation of ternary complexes with some AAs is observed. We should consider that, since the four coordination positions in a tetrahedral coordination are fully equivalent, the side chain position of

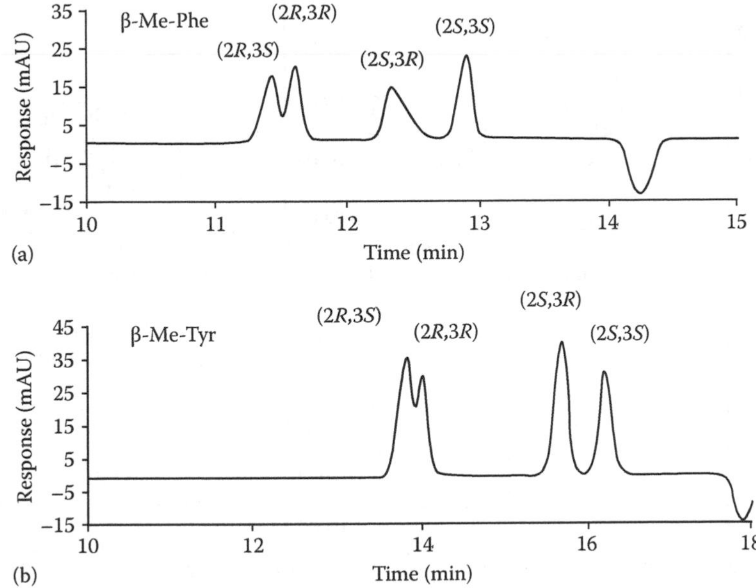

FIGURE 7.2 Electropherogram of the chiral separation of the stereoisomers of (a) *erythro*-(2*R*,3*R*),(2*S*,3*S*) and *threo*-(2*R*,3*S*),(2*S*,3*R*)-β-Me-Phe, (b) *erythro*-(2*R*,3*R*),(2*S*,3*S*) and *threo*-(2*R*,3*S*),(2*S*,3*R*)-β-Me-Tyr. Conditions: 20 mM HO-L-4-Hypro, 10 mM Cu(II), in 5 mM phosphoric acid solution adjusted to pH 4.3 by 5% ammonia. (From Grobuschek, N. et al., *J. Pharm. Biomed. Anal.*, 27, 599, 2002. With permission.)

one enantiomer could become fully equivalent to the position of the side chain in the other enantiomer complex, by simply inverting the two coordination positions. This inversion, in the case of a square planar or square planar based coordination (like the octahedral, more or less tetragonally distorted), should give rise to a different complex by the well known *cis/trans* isomerism. This subject will be considered again later on. The electropherograms, reported in Figure 7.3, show the fair results obtained, at quite low selector concentration. The second paper [25] is of particular interest since it reports the application to the analysis of a real food sample.

Besides the papers that have been briefly reviewed till now, describing the formation of the classical ternary first coordination sphere metal complexes, papers in which the diastereoisomeric species responsible for the chiral separation are obtained differently should also be mentioned. The first case is still based on metal ion complexes, but in this case an outer coordination sphere interaction is involved. Fanali et al. [26] proposed a very interesting application of CLECE, namely, the separation of enantiomeric cobalt(III) inert complexes. When a metal ion is coordinated by three molecules of a bidentate ligand, a chirality comes out, due to the two possible different enantiomers, named Λ and Δ, respectively. Such chirality can be observed for inert complexes only, since, in labile complexes, the breaking and the reforming of the metal–ligand bonds prevents the isolation of each enantiomer. The chiral complexes thus obtained were separated by using L-tartrate as chiral selector, leading to the formation of diastereoisomeric outer sphere complexes, which permitted the

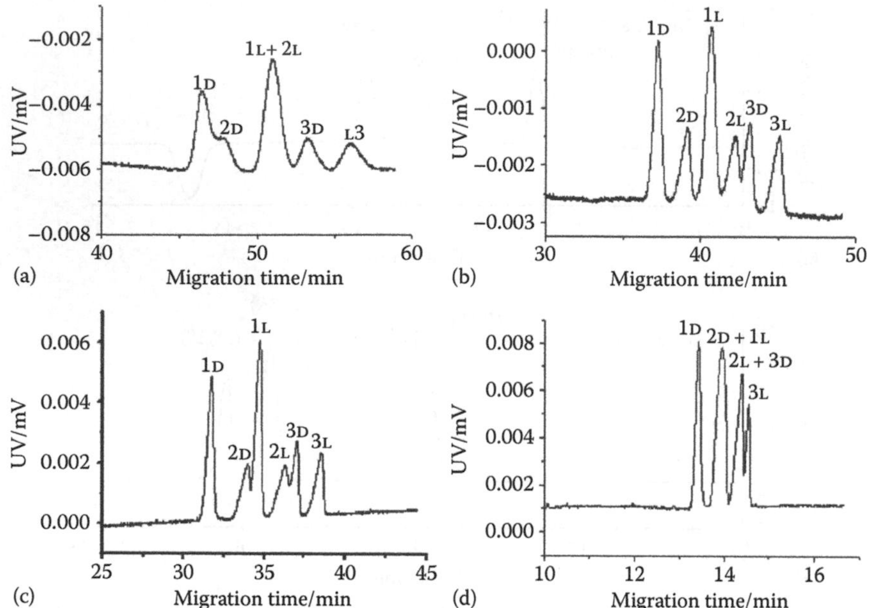

FIGURE 7.3 Electropherograms of mixed aromatic AAs carried out in a BGE containing 5 mM ammonium acetate, 3 mM zinc(II), 6 mM L-Lys and (a) 50, (b) 100, (c) 150, or (d) 200 mM boric acid, adjusted to pH 7.6 with Tris. Peak identity: 1D = D-Trp, 1L = L-Trp, 2D = D-Phe, 2L = L-Phe, 3D = D-Tyr, 3L = L-Tyr. (From Qi, L. et al., *Electrophoresis*, 28, 2629, 2007. With permission.)

chiral separation. The second case regards the use of a borate ion [27–29] as central moiety. As known, these ions show a strong affinity toward *cis*-1,2- and 1,3-diols. To separate D, L-panthotenic acid, Kodama et al. [27] used a BGE including the borate ion, together with the chiral diol *S*-3-amino-1,2-propanediol (SAP). They accurately investigated the dependence of the resolution and of the migration times on several parameters, like pH, analytical concentrations of both borate and SAP, temperature of the capillary, and percentage of methanol in the BGE. They could also ascertain the inversion of the enantiomer migration order (EMO) between the enantiomers of panthotenic acid when using, instead of SAP, its *R*-enantiomer (RAP). More recently, the same procedure was used by the same authors [28] to separate a series of monosaccharides, functionalized by bonding two 1-phenyl-3-methyl-5-pyrazolone units. A third paper by the same authors reports the addition of micelles in a similar system [29].

7.2.2 Cyclodextrins in CLECE

Cyclodextrins [30] are a family of naturally occurring cyclic oligosaccharides, characterized by the presence of a cavity in their molecular structure, which shows relatively hydrophobic characteristics. In virtue of this hydrophobic site, they behave in aqueous solution as receptors toward molecules having at least an apolar moiety. Being carbohydrates, they are water soluble, as well they are chiral molecules. These properties, all together, make them almost ideal candidates as chiral selectors in aqueous solution. Indeed, cyclodextrins have found a lot of different applications. No doubt that they represent by far the most used family of chiral selectors in CE, as also shown by the two chapters specifically devoted to them in this book (Chapters 3 and 4).

As concerns specifically CLECE, however, until 2003, their involvement was limited to the use of γ-cyclodextrin (the cyclo-malto-octaose), together with zinc(II) ion and D-phenylalanine [31]. In this interesting paper, some enantiomeric pairs of quinolone derivatives (NQs) were separated by exploiting their ligand ability toward the zinc(II) ion. This paper appears innovative in the choice of the metal ion. By adding the binary complex between zinc(II) and D-phenylalanine, NQs diastereoisomeric ternary complexes were presumably formed. However, no chiral separation was obtained in the absence of γ-cyclodextrin. The authors hypothesize that a differential inclusion between the two diastereoisomeric complexes occurred inside the cyclodextrin cavity. Therefore, in this system, the γ-cyclodextrin/metal ion interaction should be considered as a second sphere coordination, similar to what was discussed before concerning the paper of Fanali et al. [26].

The reason of the delayed use of cyclodextrins in LECE is that unmodified carbohydrates show poor coordination properties toward metal ions: only at the highest pH values, the metal ions win the competition with protons toward hydroxyl oxygen. Thus, if we want a direct covalent bond between the metal ion and cyclodextrin, we need to appropriately modify it. By exploiting our expertise in the synthesis and in the characterization of cyclodextrin derivatives [32–38], in 2003 we reported the first case of CLECE by using as chiral selector a copper(II) complex of a functionalized β-cyclodextrin [39]. We chose a histamine derivative 6-deoxy-6-[4-(2-aminoethyl) imidazolyl]cyclomaltoheptaose (CDmh), obtained by substituting the β-cyclodextrin on the primary position of one of its seven glucopyranose rings by a nitrogen atom of the imidazole ring. This compound had already shown its ability to separate the enantiomers of tryptophan in the presence of copper(II) ion in LEC chromatography [40]. The racemate to separate was also tryptophan in this case, and the electropherograms at selected concentrations of the selector are reported in Figure 7.4. In a previous paper, the equilibria of the ternary systems copper(II)/CDmh with both the enantiomers of tryptophan were investigated by pH-metric potentiometry [41]. Evidence of a thermodynamic stereoselectivity in these systems was shown by

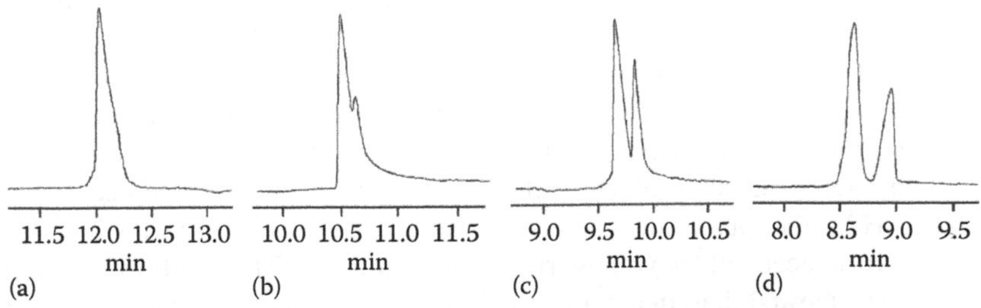

FIGURE 7.4 Examples of electropherograms in the presence of copper(II)–CDmh (concentration ratio 1:1.2). CDmh concentrations (mM): (a) 1.80, (b) 0.80, (c) 0.60, and (d) 0.25. (From Cucinotta, V. et al., *Analyst*, 128, 134, 2003. With permission.)

the formation constants of the complexes $[Cu(CDmh)Trp]^+$. A significant difference between the log β value for the complex of the L-enantiomer (18.99) and that for the analogous complex of the D-enantiomer (18.14) was obtained. In this case, the mechanism of chiral selection, as shown in the electropherograms reported in Figure 7.4, is straightforward. At the chosen pH (6.8), a high percentage of the monocationic complex is formed, while the free analyte is neutral. Thus, the higher is the fraction of complexed analyte, the shorter is its migration time. This is in full agreement with the EMO observed: the L-enantiomer is faster since its formation constant is higher. Another interesting aspect that differentiates this system from most of those already reported in literature, is the dependence of the selectivity on the selector concentration. While at the highest tested concentration no selectivity was observed, maximum selectivity is obtained for an intermediate concentration. This is not surprising, since not the formation of the two diastereoisomeric complexes, but rather the difference between their extent of formation needs to be optimized.

Lastly, it should be mentioned that the concentration used is in the 10^{-4} M range and the highest selectivity value is obtained for a 0.15 mM concentration. The very low concentration of the selector must be considered an important advantage, mainly because no interference from the BGE is observed in the optical detection of the analytes. In this case, since our basic selector at neutral pH is present as cationic species, a very low concentration minimizes its interaction with the capillary wall, and consequently phenomena of peak fronting and tailing occurs, while keeping the efficiency high.

A more complex system is obtained when using the 3-amino derivative of β-cyclodextrin (CD3NH2) as chiral selector [42]. In this case, the substitution occurs at a secondary position of the glucopyranose ring. As it is known, the acidity of the secondary hydroxyls is higher and, differently from the analogous primary groups. In the presence of a suitable metal ion and of a suitable steric disposition, they can be deprotonated even at quite low pH values. Thus, the strong donor properties of the amino nitrogen give rise to the bonding of the copper(II) ion. Favorable steric conditions occur that permit to the deprotonated 2-hydroxyl to bond the ion as well, as confirmed by the stability constants obtained by pH-metric potentiometry. The result is that in the ternary systems with each enantiomer of phenylalanine, both the unprotonated $[Cu(CD3NH2)(Phe)]^+$ and the deprotonated $[Cu(CD3NH2)(Phe)H_{-1}]$ ternary species are simultaneously present. It is important once more to carefully consider the electric charge of the complexes. Since the analyte is neutral at the BGE pH, only the unprotonated species, being cationic, can differentiate the migration times between the phenylalanine enantiomers. As shown in the distribution diagram in Figure 7.5, the L-enantiomer unprotonated complex is formed in higher percentage, due to its higher stability constant, and correspondingly, the L-enantiomer has a shorter migration time as observed in the electropherograms.

Another interesting study concerns the copper(II)/AB3NH2/tryptophan complex [43]. AB3NH2 is the cyclodextrin derivative, obtained by substituting the 3-hydroxyl groups of two adjacent glucopyranose rings with ammonia. The equilibrium study by potentiometry showed that, in contrast to alanine, tryptophan enantiomers give rise to a stereoselective effect. More precisely, this stereoselectivity specifically concerns the ternary unprotonated $[Cu(AB3NH2)(Trp)H_{-1}]$ species. The dicationic dimeric

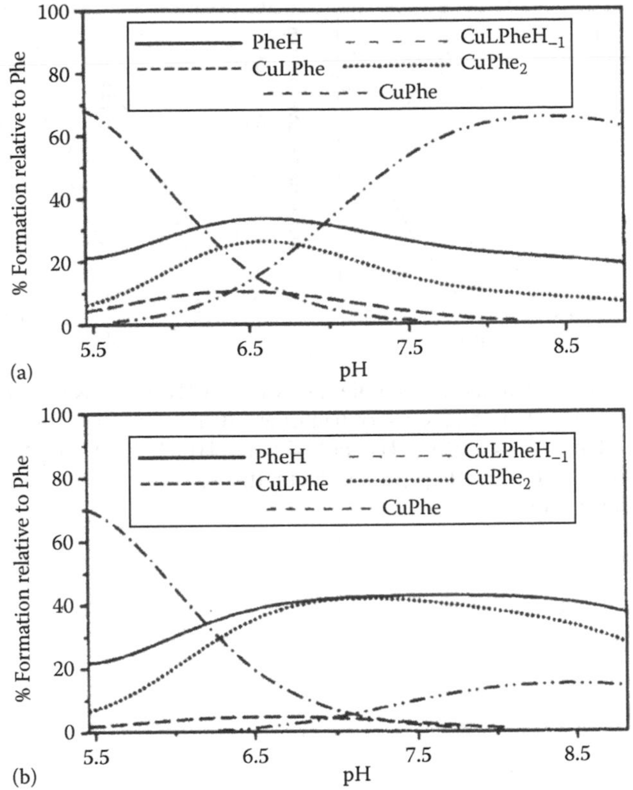

FIGURE 7.5 Distribution diagram for the Cu(II)–CD3NH2–AA systems: $[Cu^{2+}] = 7.5\ 10^{-3}$ M, [CD3NH2] = 9.0 10^{-3} M, [Phe] = 1.0 10^{-4} M: (a) L-Phe; (b) D-Phe. (From Cucinotta, V., *J. Chromatogr. B.*, 800, 127, 2004. With permission.)

species, which is the other species present at neutral pH, shows no enantioselectivity. The mechanism of separation is in this case indirect: the "active" (dicationic) dimeric species is formed in different percentages for the two enantiomers, as shown in the distribution diagrams in Figure 7.6. This is not caused by a significant difference in their stability constants, but rather by the difference in the stability constants between the unprotonated monomeric "inactive" (neutral) complexes, resulting in a different competition for the two enantiomers. In conclusion, all aspects involved should be carefully considered when defining the real mechanism of CLECE separation. In Figure 7.7, the electropherograms concerning the excellent separation of tyrosine are reported.

In a more recent paper [44], systems formed by the copper(II) ion with three AAs enantiomeric pairs and with five more different cyclodextrin derivatives used as chiral selectors were reported. The results were compared with the analogous systems of the already reported three derivatives, just discussed before. A summary of the electrophoretic data obtained are reported in Tables 7.1 through 7.3. In Figure 7.8, as an example, the electropherograms reported concern one of the studied systems,

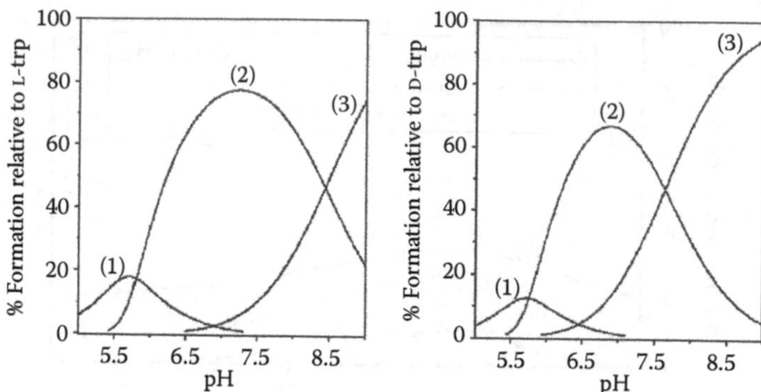

FIGURE 7.6 Amino acid distribution diagrams for Cu(II)–AB3NH2–L-Trp–D-Trp systems referred to L-Trp (left), and to D-Trp (right) at $C(\text{Cu}) = 7.5 \times 10^{-4}$ M; $C(\text{AB3NH2}) = 8.0 \times 10^{-4}$ M; $C(\text{Trp}) = 1.0 \times 10^{-4}$ M: (1) $[\text{Cu(AB3NH2)(Trp)H}]^{2+}$; (2) $[\text{Cu}_2(\text{AB3NH2})_2(\text{Trp})\text{H}_{-1}]^{2+}$; (3) $[\text{Cu(AB3NH2)(Trp)H}_{-1}]$. (From Cucinotta, V. et al., *Dalton Trans.*, 16, 2731, 2005. With permission.)

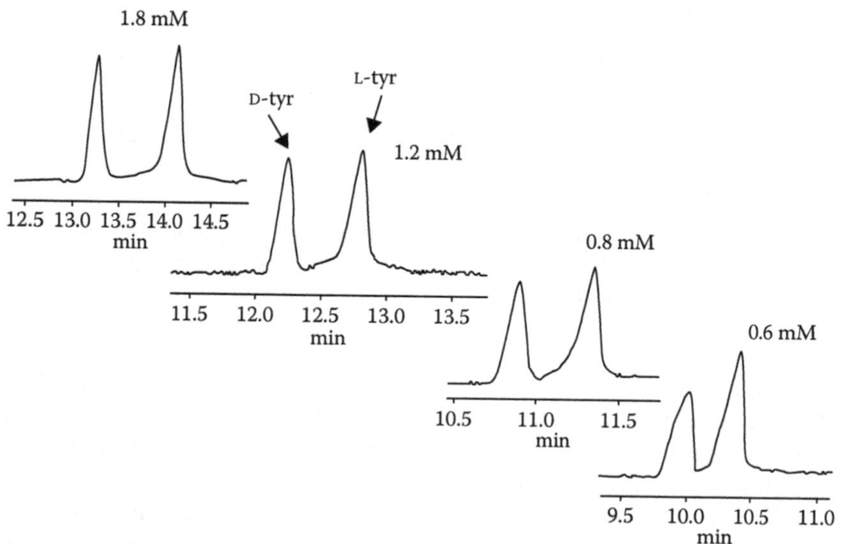

FIGURE 7.7 Electropherograms of tyrosine chiral separation by copper(II)/AB3NH2 complex. (From Cucinotta, V. et al., *Dalton Trans.*, 16, 2731, 2005. With permission.)

namely, the Cu(II)/Tyr systems with the 6-deoxy-6-[*N*-(2-methylamino)pyridine]-β-cyclodextrin (CDampy), which had previously been used in LEC chromatography [37]. Here, we will only underline the results of the 6-derivative with histamine, like the derivative already discussed (CDmh), but in this case the histamine is bound to the amino group (CDhm). Interestingly, inverting the disposition of the histamine moiety with respect to the cyclodextrin cavity, the EMO of the tryptophan

TABLE 7.1
Results of LECE Separations of Phenylalanine Enantiomers with the Investigated Selectors

Selector[a]	C (Selector) (mM)	t_L	t_D	S	R
CDampy[b]	1.8	10.62	10.53	0.009	[c]
	1.2	9.92	9.92	0	—
	0.8	8.78	8.65	0.015	[c]
	0.6	8.55	8.55	0	—
CDen[b]	1.8	10.56	10.56	0	—
	1.2	9.16	9.07	0.010	[c]
CD3NH2 [34]	1.8	10.62	10.78	0.015	1.45
	1.2	10.40	10.64	0.023	1.71
	0.8	9.15	9.26	0.012	0.54
	0.6	8.44	8.48	0.005	[c]
AB3NH2 [35]	1.8	13.46	13.10	0.027	0.86
	1.2	12.62	12.41	0.017	0.45
	0.8	11.48	11.30	0.040	0.41
	0.6	10.85	10.75	0.009	[c]

Source: Cucinotta, V. et al., *Electrophoresis*, 27, 1471, 2006. With permission.
Note: t_L, t_D are the apparent migration times (min) of two enantiomers. *S* is the selectivity factor, and *R* is the enantiomeric resolution.
[a] References in square brackets.
[b] This work.
[c] Low value.

enantiomers inverts too. The structural hypothesis is schematically proposed in Figure 7.9. It involves the presence of a *cis*-effect in these complexes: the tryptophan coordination has the steric constraint to give rise to the *cis*-disposition of the amino-groups of the two ligands. Obviously, such constraint would not exist in a tetrahedral coordination, as already underlined before when discussing the use of zinc(II)/chiral selector complex in CLECE.

The β-cyclodextrin selectors used in reference [44] also included a 6-ethylendiamine derivative (CDen) (see also reference [45] for equilibria data). This same selector was used by the only other group reporting CLECE studies using cyclodextrin derivatives [46–47]. In the first paper [46], CDen was the only selector used to simultaneously separate the enantiomeric pairs of four different unmodified AAs. The results of [44] and [46] are difficult to compare, since different concentrations of the complex were used, 1.8 and 20 mM, respectively. Interestingly, Wu et al. [46] used both PEG20000, a polymer interacting with the capillary wall, and *tert*-butyl alcohol. The authors succeeded in separating the tryptophan enantiomers only by adding this alcohol in the 1% v/v range. They propose that this alcohol competes with the indole moiety of tryptophan for the cyclodextrin cavity. This reasonable hypothesis once more confirms that

TABLE 7.2
Results of LECE Separations of Tryptophan Enantiomers with the Investigated Selectors

Selector[a]	C (Selector) (mM)	t_L	t_D	S	R
CDhm[b]	1.8	11.76	11.53	0.020	1.15
	1.2	10.19	10.00	0.019	1.27
	0.8	9.31	9.31	0	—
ABNH2[b]	1.8	12.11	11.99	0.010	0.65
	1.2	11.22	11.22	0	—
	0.8	9.32	9.32	0	—
	0.6	9.19	9.19	0	—
CDen[b]	1.8	10.46	10.52	0.006	[c]
	1.2	8.60	8.73	0.015	[c]
	1.8	10.52	10.52	0	—
	1.2	9.95	9.95	0	—
CD3en[b]	0.8	9.58	9.29	0.031	0.63
	0.6	9.15	8.97	0.020	[c]
CDmh [33]	1.8	15.9	15.9	0	—
	1.2	11.9	12.1	0.017	0.80
	0.8	10.5	10.6	0.014	0.53
	0.6	9.7	9.9	0.018	0.95
	0.4	10.4	10.7	0.029	1.76
	0.25	8.7	9.0	0.036	1.52
	0.15	8.4	8.8	0.044	1.07
	0.1	8.5	8.8	0.037	0.71
	0.07	8.5	8.8	0.029	0.62
AB3NH2 [35]	1.8	14.19	13.75	0.031	1.31
	1.2	12.85	12.48	0.029	0.95
	0.8	10.91	10.91	0	—
	0.6	9.84	9.84	0	—

Source: Cucinotta, V. et al., *Electrophoresis*, 27, 1471, 2006. With permission.
Note: t_L, t_D are the apparent migration times (min) of the two enantiomers. *S* is the selectivity factor, and *R* is the enantiomeric resolution.
[a] References in square brackets.
[b] This work.
[c] Low value.

we must not always try to favor the formation of the complexes: Sometimes we need, on the contrary, to disfavor their formation to increase what we really need, the difference in the formation degree between the two diastereoisomeric complexes.

In a more recent paper [47], the same authors used CDen along with three other cyclodextrin derivatives for the separation of the dansyl derivatives of five different AAs (DNS-AAs).

TABLE 7.3
Results of LECE Separations of Tyrosine Enantiomers with the Investigated Selectors

Selector[a]	C (Selector) (mM)	t_L	t_D	S	R
CDhm[b]	1.8	11.47	11.31	0.014	1.07
	1.2	10.46	10.39	0.007	[c]
	0.8	9.22	9.17	0.005	[c]
CDampy[b]	1.8	10.81	10.48	0.031	1.18
	1.2	9.92	9.69	0.023	0.92
	0.8	9.95	9.83	0.012	0.91
	0.6	8.72	8.51	0.024	0.88
ABNH2[b]	1.8	12.80	12.61	0.015	1.09
	1.2	11.79	11.63	0.014	0.80
	0.8	10.13	10.13	0	—
	0.6	10.03	10.03	0	—
CDen[b]	1.8	11.90	11.90	0	—
	1.2	10.74	11.20	0.042	1.26
CD3NH2 [34]	1.8	10.53	10.69	0.015	1.33
	1.2	9.97	10.06	0.009	0.47
	0.8	8.84	8.90	0.007	0.39
	0.6	8.07	8.07	0	—
AB3NH2 [35]	1.8	14.15	13.28	0.063	3.22
	1.2	12.84	12.26	0.046	2.42
	0.8	11.35	10.90	0.040	2.57
	0.6	10.41	10.02	0.039	1.70

Source: Cucinotta, V. et al., *Electrophoresis*, 27, 1471, 2006. With permission.
Note: t_L, t_D are the apparent migration times (min) of the two enantiomers. S is the selectivity factor, and R is the enantiomeric resolution.

[a] References in square brackets.
[b] This work.
[c] Low value.

7.2.3 LECE in the Presence of Micelles

The use of micelles in capillary electrophoresis dates back to 1984, to a pioneering work by Terabe et al. [48]. Chiral micelle-forming compounds were already explored with success as early as 1989 [49]. Different ways to obtain enantiomeric separations were investigated. Among them, there was the coupling of achiral micelles to the use of the ligand exchange mechanism. A paper reported the separation of a series of DNS-AAs racemates by *N,N*-didecyl-L-alanine copper(II) complexes in the presence of sodium dodecyl sulfate (SDS) [50]. In this case, the ligand is an amphiphilic molecule, which can give rise to micelles, but it was used at a concentration below the critical micelle concentration (CMC). However, on addition of

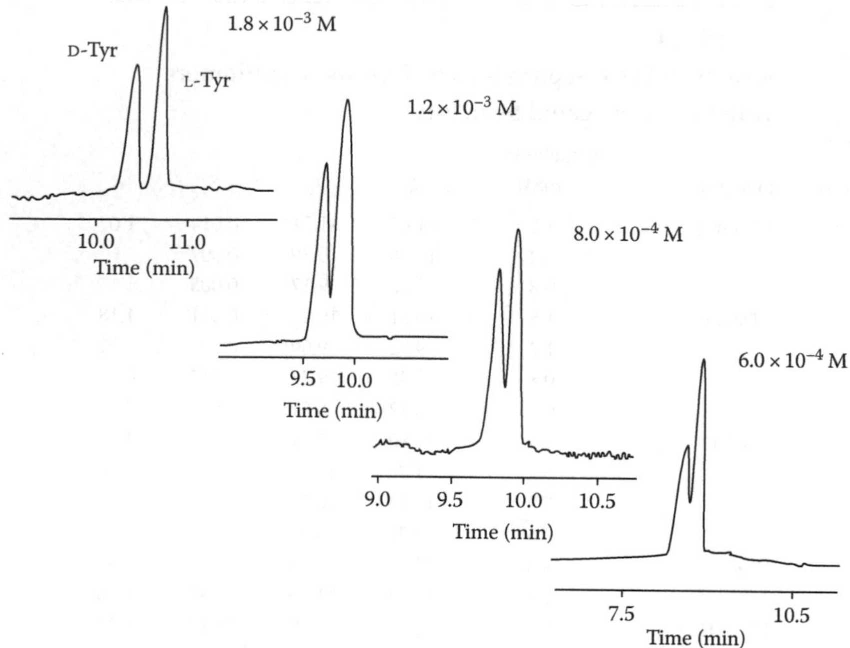

FIGURE 7.8 Electropherograms for the system Cu/CDampy/Tyr at different concentrations of CDampy in BGE ([Cu] = 5/6 [CDampy]). Migration times are uncorrected for EOF. Sample solution was 0.25 mM in racemate (0.05 mM in L-enantiomer excess). (From Cucinotta, V. et al., *Electrophoresis*, 27, 1471, 2006. With permission.)

the copper(II) complexes in the BGE, micelles of mixed composition are readily formed. The hydrophilic alanine moiety of the ligand exposes the copper(II) ion to the solution, permitting the bonding of the analyte to the metal ion. The differential affinities of two enantiomers of the dansyl amino acid gave rise to a different percentage of mixed complexes. Thus, each enantiomer was included in the micelles in a different amount and, by exploiting the different velocities of the free analyte and the micelles hosting the analyte ternary complex, chiral separations were obtained. By a similar selector system and a similar procedure, other DNS-amino acids were separated by other authors [51].

The approach usually followed in literature in the coupling of micelles to the ligand exchange process was simply to add the micelles to systems generally already studied in their absence, in order to understand how the micelles influenced their separation. In this context, we should recall that this kind of procedure is called micellar electrokinetic chromatography (MEKC). The name "chromatography" has been adopted, since the micelles represent a pseudostationary phase. They indeed represent another phase with respect to the running solution, and this second phase substitutes the stationary phase in chromatography. While in chromatography the stationary phase does not move during the runs, in MEKC it moves under the influence of the electrical field. Anyhow, it moves with its own velocity, and thus, if the micelles are electrically charged, a difference of velocity occurs. Since micelles

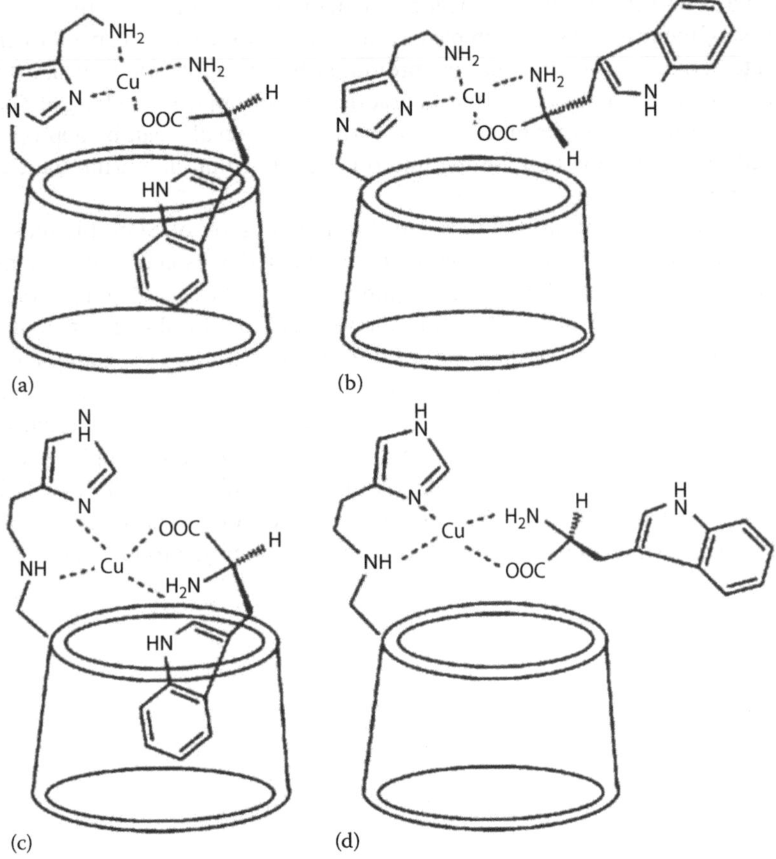

FIGURE 7.9 Schematic structures of the complexes (charges omitted): (a) [Cu(CDmh)(L-Trp)]; (b) [Cu(CDmh)(D-Trp)]; (c) [Cu(CDhm)(D-Trp)]; (d) [Cu(CDhm)(L-Trp)]. (From Cucinotta, V. et al., *Electrophoresis*, 27, 1471, 2006. With permission.)

are a different phase, their interaction with analytes cannot be regarded as a simple chemical interaction, similar to what is observed in solution. On the contrary, we must consider the associated physical phenomena of their adsorption on the micelle surface and of their diffusion inside the micelle, similar to what is observed on the stationary phases in liquid chromatography.

The consequence of this chromatographic aspect inherent to MEKC causes the same kind of uncertainty, of unpredictability, already mentioned in the introduction in the context of CMP chromatography when using a reversed phase as stationary phase. The similarity lies in the fact that in both techniques two distinct processes go on: from one side, the formation of the ternary diastereoisomeric complexes in solution, from the other side the "heterogeneous" interaction of the micelles with the different species. In addition, the micelles, representing the pseudostationary phase, are generally charged particles, either anionic or cationic. Therefore, they can be applied for the separation of neutral analytes. This charge, coupled to their apolar core, makes the behavior of the micelles particularly complex, resulting in a poor prediction of the result. As an example of practical relevance, it is difficult to predict if the interaction of a neutral species, promoted by the apolar core of the micelle will be stronger with respect to the interaction of an oppositely charged species. Nonetheless, once the experiments are performed, the results can be appropriately explained, throwing new light on the microprocesses going on during the electrophoretic run.

In this context, Lu et al. [52] reported that the addition of SDS, the most used anionic surfactant, induces not only an improvement in the resolution of the analytes, but also a reversal of the EMO of the two enantiomers, as shown in Figure 7.10. In this paper, a copper(II) 1:2 complex with L-lysine was used as chiral selector. To explain the shorter migration time of the AAs with respect to the EOF, the authors hypothesize the presence of a cationic species. As the pH is neutral, it can be excluded that the free amino acids have a net charge. Thus, the only cationic species present should be a copper(II) complex. However, it is not clear which specific complex should be present. The most obvious complex [Cu(Lys)(AA)] would be neutral. In order to explain the experimental data, the possible formation of the protonated $[Cu(Lys)(AA)H]^+$ complex should be considered. If the lysine ε-amino group bonding has not

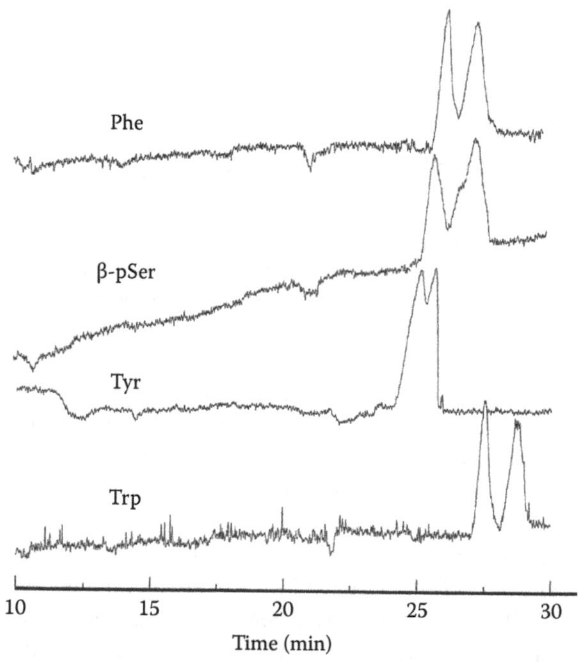

FIGURE 7.10 Typical electropherograms of the chiral separations of four amino acids by CZE. For all analytes, the D-form eluted before the L-from. Buffer: 10 mm $N4_4$AC, 6.67 mM cu(II), 13.33 mM L-Lys, at pH 7.0. (From Lu, X. et al., *J. Chromatogr. A.*, 945, 249, 2002. With permission.)

the suitable steric disposition to give rise to a chelate effect, it is possible that in this complex the lysine behaves as a bidentate ligand. The unbonded ε-amino group can therefore be protonated, as it happens in the free lysine. In this model, the shorter migration time of the D-enantiomer could be ascribed to the higher strength of its protonated complex with respect to the L-enantiomer analogous complex. However, this same result can be obtained from a lower strength of the D-enantiomer [Cu(Lys)(AA)] unprotonated complex. In this case, the lower competition could give rise to a higher formation degree of its protonated species. This latter hypothesis of the simultaneous presence of two different ternary species seems more likely, since a difference of eight log K units between the protonated and the unprotonated complex formation constants should also cause the presence of a significant percentage of the unprotonated species. Further, the steric needs, due to the third bond that unprotonated lysine could form with copper(II), should favor its enantioselectivity. Anyhow, only an accurate equilibrium study could give a definitive answer on this aspect.

The addition of SDS micelles makes the system even more complex, since all possible interactions of the species in solution with the micelles should be considered. Besides the interaction of the other ternary complex, whose presence seems likely, the free amino acids, being present mainly as zwitterions at neutral pH, can also interact with the micelles. The hydrophilic surface of the micelle can interact with the oppositely charged amino group. In addition, the hydrophobic side chain can reside inside the micelle, whereby the carboxylate group is projected toward the outside and fully solvated by the water molecules.

This case was discussed in depth to show how the exact mechanism can be easily oversimplified.

One of the most active groups in the field of LE-MEKC is that of Zilin Chen, who is also involved in CLECE without micelles (see previous section [18]). In a paper [53], they have described the simultaneous separation of the *o*-, *m*-, and *p*-isomers of fluoro-phenylalanine and of tyrosine. These are six different enantiomeric pairs of amino acids, with minimal structural differences. The chiral selectors were the complexes formed between copper(II) and L-4-hydroxy-proline (1:2 concentrations ratio). In Figure 7.11, it can be fully appreciated how the addition of 10 mM SDS dramatically improves the separation. The same group carried out a series of separations of tryptophan and phenylalanine at increasing concentrations of various anionic surfactants, by using as chiral selector the copper(II) complex with 4-hydroxyproline [54]. The observed inversion of EMO of the enantiomers along this series of experiments was ascribed to the formation of the micelles. When the surfactant is present as oligomers an interaction occurs, while, as soon as the micelles are formed, a different kind of interaction occurs. The selectivity should go through zero just at the CMC, thus permitting to obtain the CMC by extrapolating the linear trend observed at the higher concentration of the surfactant to the zero value of selectivity. Surprisingly, the authors have used the resolution values, and this procedure, does not seem advisable for the reasons already discussed in this chapter. Anyhow, the CMC data thus obtained appear in good agreement with those reported in literature, as determined by other techniques.

In a recent work [55], an accurate comparative study between LECE and LE-MEKC was carried out. A series of analytes racemates, including DNS-AAs and other

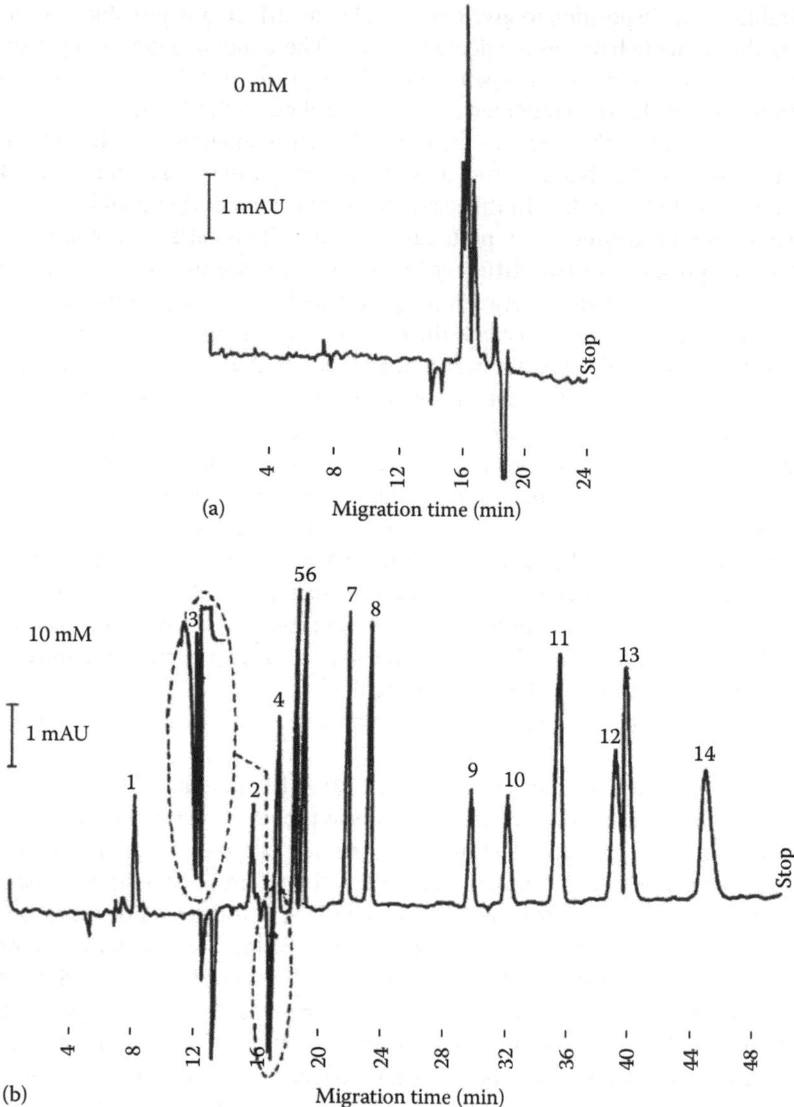

FIGURE 7.11 Electropherograms of 12 positional and optical isomers of amino acids in a BGE at pH 4.0, containing 25 mM Cu(II), 50 mM 4-OH-proline. (a) Without SDS; (b) with 10 mM SDS. The analytes peaks are labeled from 3 to 14. (From Chen, Z. et al., *J. Chromatogr. A.*, 813, 369, 1998. With permission.)

compounds, were separated by an ornithine complex of copper(II). For the DNS-AAs, the authors studied how the electrophoretic separation is influenced, besides by the pH and the selector concentration, by the presence of SDS, as well as by the coating of the capillary wall. We will shortly discuss both these two last aspects.

Using SDS, as observed previously, the EMO between the two enantiomers is inverted, while the selectivity in the separations worsens. This shows how the parameters of the separation in LE-MEKC are a complex function both of the equilibria in solution and of the partition of the different species between the solution and the micelles. Let us now discuss the inversion in the EMO. The selectors in this study are copper(II) complexes of an enantiomer of an amino acid. At pH 8 the best resolution was obtained. At this pH no protonated complex should be present and thus the complex should be neutral. Differently from the unmodified AAs, the dansyl AAs are however already anionic at neutral pH. Thus, with increasing degree of complex formation, the migration times should shorten, remaining always longer than EOF, since the analyte is exchanged between a neutral species (the complex) and an anionic species (the free analyte). This is what is actually observed in the CLECE experiments, and thus the shorter migration time of the L-enantiomer corresponds to a more stable complex formed compared to the D-enantiomer. Surprisingly, the authors ascribe the inversion of EMO to the interaction of the DNS-AAs with the micelles, but these are achiral and thus any preferential interaction of an enantiomer with them is not possible. It is likely that the complexed analyte interacts more strongly than the free analyte, since its negative charge is counterbalanced in the complex. If this is the case, the inversion can be simply explained by the preferential interaction of the complex with respect to the free analyte. Consequently, the higher the fraction of complexed analyte, the higher is the fraction of analyte in the micelles and the longer is its migration time in LE-MEKC.

In this same paper, the experiments done in a coated capillary appear particularly interesting [55]. Different results are obtained compared to the uncoated capillary, confirming the interaction with components in the solution. The authors hypothesize that the diprotonated free ornithine, in virtue of its double positive charge, interacts with the negatively charged (pH 8) capillary wall. The presence of a significant concentration of free ornithine can be easily understood, since, as previously said, the 1:2 ratio of the concentration of copper(II) and ornithine does not mean that ornithine is present in this species only. It is sufficient to consider that a significant concentration of the 1:1 complex will be present, resulting approximately in an equivalent concentration of free ligand. From this point of view, the hypothesized process occurring on the wall should be differently described as involving the diprotonated ornithine on the wall interacting with the required species in solution and there is no need to suppose that all the concerned species must be bonded together to the capillary wall.

A recent interesting paper proposes the application of CLECE in the presence of SDS in microchip electrophoresis (ME) [56]. This represents an obvious development of this technique toward miniaturization, a general tendency in CE. The selector is the copper(II) complex with L-prolinamide, while the chosen enantiomeric analytes were 4-fluoro-7-nitrobenzofurazan derivatives of amino acids (NBA-AA). This fluorescent moiety was added in order to decrease the detection limit, an important issue in microchip devices.

7.2.4 Ligand Exchange in CEC

Capillary electrochromatography (CEC) represents the borderline between chromatographic and electrophoretic techniques. Since it involves the distribution of the analytes between a stationary and a liquid mobile phase, similar to liquid chromatography and, analogously to capillary electrophoresis, it exploits an electrical field for the flow of the mobile phase along the column, even if a pressure can be applied to speed the flow and so the separation. This technique is explained in more detail in Chapter 14.

In the context of ligand exchange, chiral CEC was first introduced by Schmid et al. in 2000 [57] for the chiral separation of the racemates of a series of AAs. The stationary phase was obtained by an "in situ" copolymerization reaction, where in the reaction mixture the chiral unit *N*-(2-hydroxy-3-allyloxypropyl)-L-4-hydroxyproline was included, and thus this technique should be classified as CSP. A monolithic (continuous bed) column was obtained. As mobile phase, a 0.1 mM copper(II) solution at pH = 4.6 (sodium hydrogendiphosphate) was used. The copper(II) ion, eluting along the column, saturates the active sites of the stationary phase. When the analyte flows along the column, it interacts with these copper(II) ions by coordinating them, and thus a mechanism of retention occurs. Due to the chiral nature of the stationary phase, the interaction is different for the two enantiomers, thus resulting in their separation. As shown in Figure 7.12, the efficiency in CEC is much higher than in HPLC, and a decrease in the time necessary for the separation could be obtained either by applying a pressure or by using a shorter column (from the opposite end to the detection window). In 2001 the same authors reported the separation of a series of hydroxy acids by the same chiral phase [58].

In the meanwhile, Chen and Hobo [59] reported the synthesis of a new chiral silica-based monolithic column involving L-phenylalaninamide as chiral unit . They obtained the separation of twelve enantiomeric pairs of DNS-AAs. Interestingly, by using a copper(II) water/acetonitrile solution as mobile phase, buffered by ammonium acetate, they observed a negative EOF, that implies a positively charged stationary phase. However, the mechanism of retention that the authors describe is not in agreement with a negative EOF. The point is that, as already previously underlined in this chapter, many researchers see the ligand exchange mechanism as necessarily implying the substitution by the analyte of a molecule of selector from the 1:2 complex. Thus, also to describe the bonding of the copper(II) to the chiral phase, they show two phenylalaninamide groups occupying all the coordination planes of copper(II). Really, in this case, the phase would be neutral, since the amide group can bond a metal ion only if deprotonated. Consequently, the two positive charges of copper(II) would be counterbalanced by the two negative charges of the two amide groups involved in the coordination, and the stationary phase would be neutral. However, if we imagine the copper(II) anchored to the stationary phase by one phenylalaninamide group, with water molecules completing its coordination sphere, then a positive charge still exists, thus fully explaining the negative EOF. The ligand exchange, in this case, would correspond to the substitution of the required number of water molecules by the analyte molecule. In a second paper [60], the same authors used a similar CSP, but in this case the chiral unit was an L-prolinamide and was used to separate a series of dansyl-AAs and of hydroxy acids. Some electropherograms are reported in Figure 7.13, showing the good separations. More recently, a comparison has been made by the same group between "all in solution" CLECE, LE-MEKC, and LE-CEC [61].

The particle-loaded technique, first proposed by Kato et al. [62], is an alternative procedure for the preparation of monolithic columns. Very recently, Schmid et al. obtained partial or complete chiral separation of the enantiomeric pairs of 20 different AAs and of 14 different hydroxy acids using a short column of 6–8.5 cm [63].

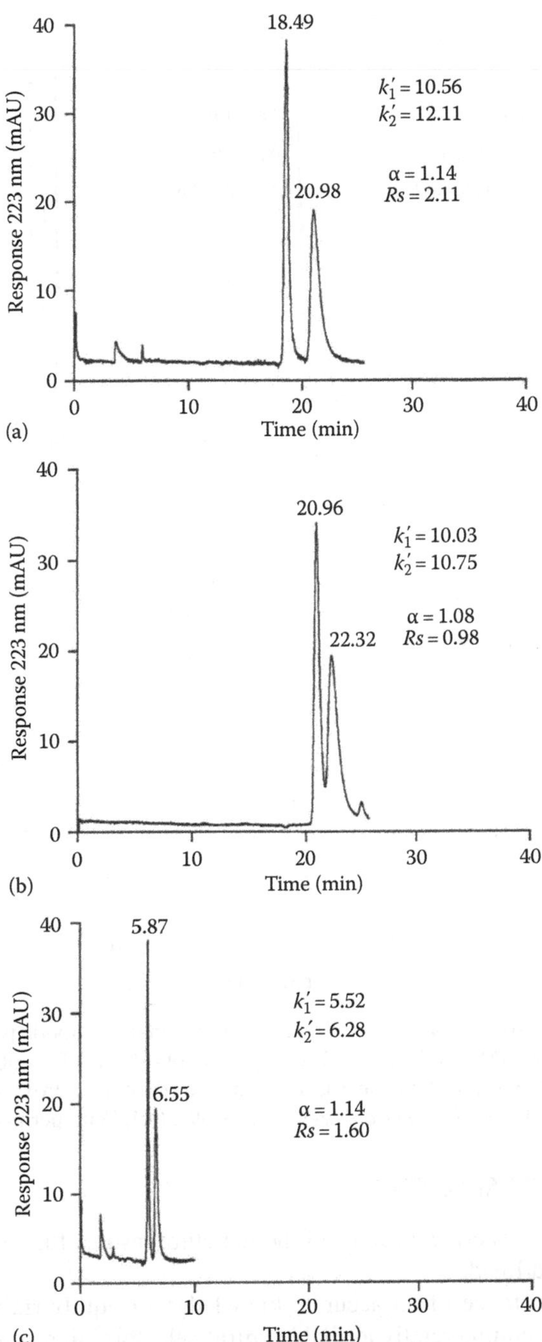

FIGURE 7.12 Chiral separation of DL-Phe, comparing (a) CEC, (b) nano-HPLC, and (c) pressure-supported CEC. Mobile phase: 50 mM sodium dihydrogenphosphate, 0.1 mM Cu(II). (From Schmid, M.G. et al., *Electrophoresis*, 21, 3141, 2000. With permission.)

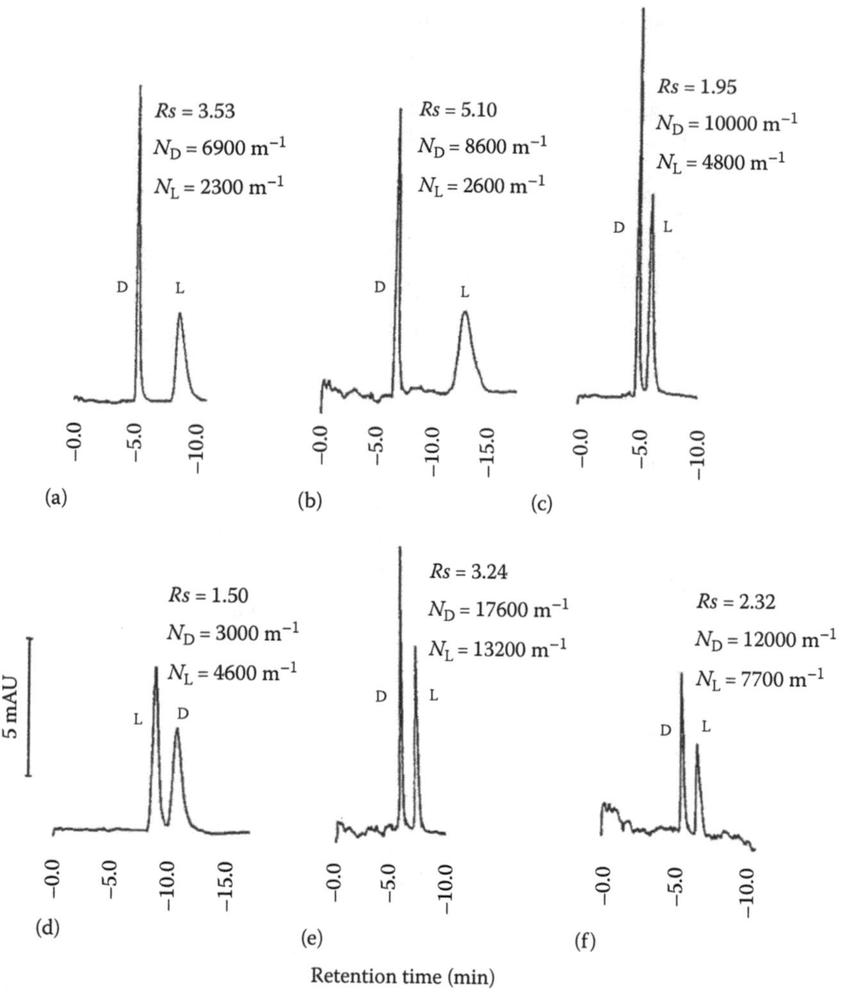

FIGURE 7.13 Electrochromatograms of representative samples. Mobile phase: acetonitrile/ 0.50 mM $Cu(Ac)_2$–50 mM NH_4Ac (7:3), pH 6.5. (a) Dns-DL-Phe, (b) Dns-DL-Trp, (c) Dns-DL-NorLeu, (d) Dns-DL-Ser, (e) DL-indole-3-lactic acid, and (f) DL-p-hydroxy phenyllactic acid. (From Chen, Z. and Hobo, T., *Electrophoresis*, 22, 3339, 2001. With permission.)

7.3 FUTURE PERSPECTIVES

Based on the previous critical survey of the literature results, future perspectives in CLECE will be outlined.

First, the importance of an accurate knowledge of equilibria in solution can be stressed. With commercially available chiral selectors, like AAs, the possibility to carry out an experimental study of the system under investigation should be taken into serious consideration. Preferably, this should be performed by pH-metric potentiometry. Once the equilibria involved in the system are characterized, it is possible to finely tune the experimental details of the electrophoretic runs, in terms of analytical concentrations and of pH. Of course, it is not easy to computationally reproduce the effective conditions inside the capillary. Some interesting suggestions can be found in a recent paper by Sanaie et al. [64], where the authors develop a model for liquid chromatography, but, in my opinion, also useful in capillary electrophoresis.

A theoretical question that is worth addressing is the relative contribution to the selectivity of the difference in the formation degree of the complexes, and of their different mobility.

If, however, the chiral selector to be used is not commercially available, a synthetic effort is also requested. This is the typical case of cyclodextrin derivatives, whose commercial availability is not appropriate, since they are commonly marketed as mixtures. Generally the individual components of these mixtures differ from one another not only in the relative position of the substituting groups (regioisomerism), but even in the extent of substitution. Indeed, the product is characterized by its average substitution degree. It is obvious that these mixed products prevent any possibility to clearly define what is going on in these systems. Thus, if the peculiar recognition characteristics of the cyclodextrin cavity have to be exploited, a demanding synthetic effort is also required.

A question could arise in this context: why should a researcher use cyclodextrin derivatives if they are so much demanding from the point of view of the requested experimental effort? The answer lies in their peculiar stereoselective recognition properties, which determined the huge number of papers in all the fields of chiral capillary electrophoresis published in the last 20 years. In the context of CLECE, the coupling of three different recognition sites, namely, the cyclodextrin cavity, the substituted moiety, and the metal coordination site, make them ideal molecules for this task. What we found in our CLECE investigations is that these selectors work at very low concentrations. It is sufficient to recall that the copper(II) complex of CDmh reaches its maximum chiral selectivity at a concentration as low as 0.15 mM, where commonly used concentrations are a hundred times higher.

We have seen that in CLECE the copper(II) ion is most commonly used as coordinated ion. The reported studies, using zinc(II) ion [24,25] and even borate ion [27–29], show that alternatives are available and their exploration should be encouraged. Also, the exploitation of the second sphere coordination [26,31], the complex-including mixed micelles [50], and, lastly, miniaturization [55] certainly open very promising paths for future perspectives.

There is no doubt that the "all in solution" CLECE is the best suitable technique for a rational prediction of the ideal experimental conditions for the chiral separation. When including micelles in the BGE, a heterogeneous process, and thus a sort of unpredictability, is introduced. The short review in this chapter showed that the addition of SDS can either improve or worsen the selectivity obtained. The accurate tuning of the electrophoretic run will inevitably need some blind trials before attaining the best experimental conditions. Similar considerations can be drawn for LE-CEC. However, even if the retention process is heterogeneous, and thus the kinetic aspect is quite unpredictable, the mechanism of retention itself in LE-CEC is quite simple. Therefore, the unpredictability should concern almost exclusively the efficiency and its contribution to the resolution.

As final comment, we would recommend researchers to carrying out, parallel to the ligand exchange experiments, experiments in the absence of the central ion. For example, we have carried out the chiral separation of DNS-AAs by CD-EKC [65,66]. Cyclodextrin derivatives were used as chiral selector at low concentrations, without the addition of copper(II) ions to the BGE. If we had added copper(II) ion in solution without any preliminary measurement, we could erroneously believe that we were carrying out a CLECE separation.

REFERENCES

1. Helfferich, F. 1961. "Ligand exchange": A novel separation technique. *Nature* 189: 1001–1002.
2. Davankov, VA, Rogozhin, SV. 1971. Ligand chromatography as a novel method for the investigation of mixed complexes: Stereoselective effects in a-amino acid copper(II) complexes. *J Chromatogr* 60: 280–283.
3. Davankov, VA. 2003. Enantioselective ligand exchange in modern separation techniques. *J Chromatogr A* 1000: 891–915 and references therein reported.
4. Schurig, V. 2001. Separation of enantiomers by gas chromatography. *J Chromatogr A* 906: 275–299.
5. Schurig, V, Fluck, M. 2000. Enantiomer separation by complexation SFC on immobilized Chirasil-nickel and Chirasil-zinc. *J Biochem Biophys Methods* 43: 223–240.
6. Guenther, K, Richter, P, Moeller, K. 2004. Separation of enantiomers by thin-layer chromatography: An overview. In *Methods in Molecular Biology*, Vol. 243: *Chiral Separations*, ed. G Gubitz and MG Schmid, 29–49. Totowa, NJ: Humana Press.
7. Gassman, E, Kuo, JE, Zare, RN. 1985. Electrokinetic separation of chiral compounds. *Science* 230: 813–814.
8. Gozel, P, Gassman, E, Michelsen, H, Zare, RN. 1987. Electrokinetic resolution of amino acid enantiomers with copper(II)-aspartame support electrolyte. *Anal Chem* 59: 44–49.
9. Schmid, MG, Grobuschek, N, Lecnik, OL, Gubitz, G. 2001. Chiral ligand-exchange capillary electrophoresis. *J Biochem Biophys Methods* 48: 143–154.
10. Desiderio, C, Aturki, Z, Fanali, S. 1994. Separation of a-hydroxy acid enantiomers by high performance capillary electrophoresis using copper(II)-L-amino acid and copper(II)–aspartame complexes as chiral selectors in the background electrolyte. *Electrophoresis* 15: 864–869.
11. Schmid, MG, Gubitz, G. 1996. Direct resolution of underivatized amino acids by capillary zone electrophoresis based on ligand exchange. *Enantiomer* 1: 23–27.
12. Yuan, ZB, Yang, LL, Zhangh, SS. 1999. Enantiomeric separation of amino acids by copper(II)-L-arginine ligand exchange capillary zone electrophoresis. *Electrophoresis* 20: 1842–1845.
13. Soontornniyomkij, B, Scandrett, K, Pietrzyk, DJ. 1998. Capillary zone electrophoresis separation of amino acid enantiomers as dansylated derivatives through control of electroosmotic flow. *J Liq Chromatogr Relat Technol* 21: 2245–2263.
14. Vegvari, A, Schmid, MG, Kilar, F, Gubitz, G. 1998. Chiral separation of alpha-amino acids by ligand-exchange capillary electrophoresis using *N*-2-hydroxy-octyl-L-4-hydroxyproline as a selector. *Electrophoresis* 19: 2109–2112.
15. Schmid, MG, Rinaldi, R, Dvereny, D, Gubitz, G. 1999. Enantioseparation of α-amino acids and dipeptides by ligand-exchange capillary electrophoresis of various L-4-hydroxyproline derivatives. *J Chromatogr A* 846: 157–163.
16. Schmid, MG, Laffranchini, M, Dreveny, D, Gubitz, G. 1999. Chiral separation of sympathomimetics by ligand exchange capillary electrophoresis. *Electrophoresis* 20: 2458–2461.

17. Schmid, MG, Lecnik, OL, Sitte, U, Gubitz, G. 2000. Application of ligand-exchange capillary electrophoresis to the chiral separation of a-hydroxy acids and β-lockers. *J Chromatogr A* 875: 307–314.
18. Chen, Z, Uchiayma, K, Hobo, T. 2000. Estimation of formation constants of ternary Cu(II) complexes with mixed amino acid enantiomers based on ligand exchange by capillary electrophoresis. *Anal Sci* 16: 837–841.
19. Karbaum, A, Jira, T. 2000. Chiral separation of unmodified amino acids with non-aqueous capillary electrophoresis based on the ligand-exchange principle. *J Chromatogr A* 874: 285–292.
20. Kodama, S, Yamamoto, A, Matsunaga, A, Hayakawa, K. 2001. Direct chiral resolution of tartaric acid in food products by ligand exchange capillary electrophoresis using copper(II)–D-quinic acid as a chiral selector. *J Chromatogr A* 932: 139–143.
21. Kodama, S, Yamamoto, A, Matsunaga, A, Soga, T, Hayakawa, K. 2001. Direct chiral resolution of malic acid in apple juice by ligand-exchange capillary electrophoresis using copper(II)-L-tartaric acid as a chiral selector. *Electrophoresis* 22: 3286–3290.
22. Grobuschek, N, Schmid, MG, Tuscher, C, Ivanova, M, Gubitz, G. 2002. Chiral separation of β-methyl-amino acids by ligand exchange using capillary electrophoresis and HPLC. *J Pharm Biomed Anal* 27: 599–605.
23. Koidl, J, Hödl, H, Schmid, MG, Pantcheva, S, Pajpanova, T, Gübitz, G. 2005. Chiral separation of halogenated amino acids by ligand-exchange capillary electrophoresis. *Electrophoresis* 26: 3878–3883.
24. Qi, L, Han, Y, Zuo, M, Chen, Y. 2007. Chiral CE of aromatic amino acids by ligand-exchange with zinc(II)–L-lysine complex. *Electrophoresis* 28: 2629–2634.
25. Qi, L, Liu, M, Guo, Z, Xie, M, Qiu, C, Chen, Y. 2007. Assay of aromatic amino acid enantiomers in rice-brewed suspensions by chiral ligand-exchange CE. *Electrophoresis* 28: 4150–4155.
26. Fanali, S, Ossicini, L, Foret, F, Bocek, P. 1989. Resolution of optical isomers by capillary zone electrophoresis: study of enantiomeric and distereoisomeric cobalt(III) complexes with ethylenediamine and amino acid ligands. *J Microcolumn Sep* 1: 190–194.
27. Kodama, S, Yamamoto, A, Iio, R, Sakamoto, K, Matsunaga, A, Hayakawa, K. 2004. Chiral ligand exchange capillary electrophoresis using borate anion as a central ion. *Analyst* 129: 1238–1242.
28. Kodama, S, Yamamoto, A, Iio, R, Aizawa, S, Nakagomi, K, Hayakawa, K. 2005. Chiral ligand exchange micellar electrokinetic chromatography using borate anion as a central ion. *Electrophoresis* 26: 3884–3889.
29. Kodama, S, Aizawa, S, Taga, A, Yamashita, T, Yamamoto, A. 2006. Chiral resolution of monosaccharides as 1-phenyl-3-methyl-5-pyrazolone derivatives by ligand-exchange CE using borate anion as a central ion of the chiral selector. *Electrophoresis* 27: 4730–4734.
30. As a general survey, refer to the special issue of *Chemical Reviews*, just devoted to cyclodextrins: Szejtli, J, ed. 1998, *Chem Rev* 98: 1741–2076.
31. Horimai, T, Ohara, M, Ichinose, M. 1997. Optical resolution of new quinolone drugs by capillary electrophoresis with ligand-exchange and host-guest interactions. *J Chromatogr A* 760: 235–244.
32. Cucinotta, V, D'Alessandro, F, Impellizzeri, G, Pappalardo, G, Rizzarelli, E, Vecchio, G. 1991. Cyclopeptide functionalized β-cyclodextrin. A new class of potentially enzyme mimicking compounds with two recognition sites. *J Chem Soc Chem Comm* 5: 293–294.
33. Bonomo, RP, Cucinotta, V, D'Alessandro, F, Impellizzeri, G, Maccarrone, G, Vecchio, G, Rizzarelli, E. 1991. Conformational features and coordination properties of functionalized cyclodextrins. Formation, stability, and structure of proton and copper(II) complexes of histamine- bearing β-cyclodextrin in aqueous solution. *Inorg Chem* 30: 2708–2713.

34. Cucinotta, V, D'Alessandro, F, Impellizzeri, G, Vecchio, G. 1992. Synthesis and conformation of dihistamine derivatives of cyclomalto-heptaose (β-cyclodextrin). *Carbohydr Res* 224: 95–102.
35. Di Blasio, B, Pavone, V, Nastri, F, Isernia, C, Saviano, M, Pedone, C, Cucinotta, V, Impellizzeri, G, Rizzarelli, E, Vecchio, G. 1992. Conformation for a β-cyclodextrin monofunctionalized with a cyclic dipeptide. *Proc Natl Acad Sci USA* 89: 7218–7221.
36. Cucinotta, V, D'Alessandro, F, Impellizzeri, G, Maccarrone, G, Rizzarelli, E, Vecchio, G. 1996. Functionalized β-cyclodextrins: Thermodynamic study and NMR titration on 6-diethylentriamine derivative. *J Chem Soc Perkin Trans II*: 1785–1788.
37. Bonomo, R, Cucinotta, V, D'Alessandro, F, Impellizzeri, G, Maccarrone, G, Rizzarelli, E, Vecchio, G, Carima, L, Corradini, R, Sartor, G, Marchelli, R. 1997. Chiral recognition by the copper(II) complex of 6-deoxy-6-N-(2-methylaminopyridine)-β-cyclodextrin. *Chirality* 9: 341–349.
38. Cucinotta, V, Grasso, G, Vecchio, G. 1998. From Capped to three-dimensional cyclodextrins: The first example of a new class of receptors by trehalose capping of β-cyclodextrin. *J Incl Phenomena*. 31: 43–55.
39. Cucinotta, V, Giuffrida, A, Grasso, G, Maccarrone, G, Vecchio, G. 2003. Ligand exchange chiral separation by cyclodextrin derivative in capillary electrophoresis. *Analyst* 128: 134–136.
40. Cucinotta, V, D'Alessandro, F, Impellizzeri, G, Vecchio, G. 1992. The copper(II) complex with the imidazole bound histamine derivative of ß-cyclodextrin as a powerful chiral discriminating agent. *J Chem Soc Chem Comm*:1743–1745.
41. Bonomo, RP, Cucinotta, V, Maccarrone, G, Rizzarelli, E, Vecchio, G. 2001. Thermodynamic stereoselectivity assisted by weak interactions in metal complexes. Chiral recognition of L/D-aminoacids by the copper(II) complex of 6-deoxy-6-[4-(2-aminoethyl) imidazolyl] – cyclomaltoheptaose. *J Chem Soc Dalton Trans* 8: 1366–1373.
42. Cucinotta, V, Giuffrida, A, La Mendola, D, Maccarrone, G, Puglisi, A, Rizzarelli, E, Vecchio, G. 2004. 3-Amino derivative of β-cyclodextrin: Thermodynamics of copper(II) complexes and exploitation of its enantioselectivity in the separation of amino acid racemates by LECE. *J Chromatogr B* 800: 127–133.
43. Cucinotta, V, Giuffrida, A, Maccarrone, G, Messina, M, Puglisi, A, Rizzarelli, E, Vecchio, G. 2005. Coordination properties of 3-functionalized β-cyclodextrins. Thermodynamic stereoselectivity of copper(II) complexes of the A,B-diamino derivative and its exploitation in LECE. *Dalton Trans* 16: 2731–2736.
44. Cucinotta, V, Giuffrida, A, Maccarrone, G, Messina, M, Vecchio, G. 2006. Ligand exchange capillary electrophoresis by cyclodextrin derivatives, a powerful tool for enantiomeric separations. *Electrophoresis* 27: 1471–1480.
45. Bonomo, RP, Cucinotta, V, D'Alessandro, F, Impellizzeri, G, Maccarrone, G, Rizzarelli, E, Vecchio, G. 1993. Coordination properties of 6-deoxy-6-[1-(2-amino)-ethylamino]-β-cyclodextrin and the ability of its copper(II) complex to recognize and separate amino acid enantiomeric pairs. *J Incl Phenomena* 15: 167–180.
46. Wu, B, Wang, Q, Liu, Q, Xie, J, Yun, L. 2005. Capillary electrophoresis direct enantioseparation of aromatic amino acids based on mixed chelate-inclusion complexation of aminoethylamino-β-cyclodextrin. *Electrophoresis* 26: 1013–1017.
47. Wu, B, Wang, Q, Xie, J, Yun, L. 2006. Amino-substituted β-cyclodextrin copper(II) complexes for the electrophoretic enantioseparation of dansyl amino acids: Role of dual chelate–inclusion interaction and mechanism. *Anal Chim Acta* 558: 80–85.
48. Terabe, S, Otsuka, K, Ichikawa, K, Tsuchiya, A, Ando, T. 1984. Electrokinetic separations with micellar solutions and open-tubular capillaries. *Anal Chem* 56: 111–113.
49. Terabe, S, Shibata, M, Miyashita, Y. 1989. Chiral separation by electrokinetic chromatography with bile salt micelles. *J Chromatogr* 480: 403–411.
50. Cohen, AS, Paulus, A, Karger, BL. 1987. High-performance capillary electrophoresis using open tubes and gels. *Chromatographia* 24: 15–24.

51. Sundin, NG, Dowling, TM, Grindberg, N, Bicker, G. 1996. Enantiomeric separation of dansyl amino acids using MECC with a ligand exchange mechanism. *J Microcolumn Sep* 8: 323–329.
52. Lu, X, Chen, Y, Guo, L, Yang, Y. 2002. Chiral separation of underivatized amino acids by ligand-exchange capillary electrophoresis using a copper(II)–L-lysine complex as selector. *J Chromatogr A* 945: 249–255.
53. Chen, Z, Lin, JM, Uchiyama, K, Hobo, T. 1998. Simultaneous separation of *o*-, *m*-, *p*-fluoro-DL-phenylalanine and *o*-, *m*-, *p*-DL-tyrosine by ligand-exchange micellar electrokinetic capillary chromatography. *J Chromatogr A* 813: 369–378.
54. Chen, Z, Lin, JM, Uchiyama, K, Hobo, T. 2000. Determination of critical micelle concentrations of anionic surfactants based on ligand exchange micellar electrokinetic chromatography. *Anal Chim Acta* 403: 173–178.
55. Zheng, Z, Wei, Y, Lin, J. 2004. Chiral separation based on ligand-exchange capillary electrophoresis using a copper(II)-L-ornithine ternary complex as selector. *Electrophoresis* 25: 1007–1012.
56. Nakajima, H, Kawata, K, Shen, H, Nakagama, T, Uchiyama, K. 2005. Chiral separation of NBD-Amino acids by ligand-exchange micro-channel electrophoresis. *Anal Sci* 21: 67–71.
57. Schmid, MG, Grobuschek, N, Tuscher, C, Gübitz, G, Vegvari, A, Machtejevas, E, Maruska, A, Hjerten, S. 2000. Chiral separation of amino acids by ligand-exchange capillary electrochromatography using continuous beds. *Electrophoresis* 21: 3141–3144.
58. Schmid, MG, Grobuschek, N, Lecnik, O, Gübitz, G, Vegvari, A, Hjerten, S. 2001. Enantioseparation of hydroxy acids on easy-to-prepare continuous beds for capillary electrochromatography. *Electrophoresis* 22: 2616–2619.
59. Chen, Z, Hobo, T. 2001. Chemically L-phenylalaninamide-modified monolithic silica column prepared by a sol-gel process for enantioseparation of dansyl amino acids by ligand exchange-capillary electrochromatography. *Anal Chem* 73: 3348–3357.
60. Chen, Z, Hobo, T. 2001. Chemically L-prolinamide-modified monolithic silica column for enantiomeric separation of dansyl amino acids and hydroxy acids by capillary electrochromatography and μ-high performance liquid chromatography. *Electrophoresis* 22: 3339–3346.
61. Chen, Z, Niitsuma, M, Uchiyama, K, Hobo, T. 2003. Comparison of enantioseparations using Cu(II) complexes with L-amino acid amides as chiral selectors or chiral stationary phases by capillary electrophoresis, capillary electrochromatography and micro liquid chromatography. *J Chromatogr A* 990: 75–82.
62. Kato, M, Dulay, MT, Bennett, B, Chen, JR, Zare, R. 2000. Enantiomeric separation of amino acids and nonprotein amino acids using a particle-loaded monolithic column. *Electrophoresis* 21: 3145–3151.
63. Schmid, MG, Koidl, J, Wank, P, Kargl, G, Zöhrer, H, Gübitz, G. 2007. Enantioseparation by ligand-exchange using particle-loaded monoliths: Capillary-LC versus capillary electrochromatography. *J Biochem Biophys Methods* 70: 77–85.
64. Sanaie, N, Haynes, CA. 2006. A multiple chemical equilibria approach to modeling and interpreting the separation of amino acid enantiomers by chiral ligand-exchange chromatography. *J Chromatogr A* 1132: 39–50.
65. Cucinotta, V, Giuffrida, A, Grasso, G, Maccarrone, G, Messina, M, Vecchio, G. 2007. High selectivity in new chiral separations of dansyl amino acids by cyclodextrin derivatives in electrokinetic chromatography. *J Chromatogr A* 1155: 172–179.
66. Cucinotta, V, Giuffrida, A, Maccarrone, G, Messina, M, Puglisi, A, Vecchio, G. 2007. Synthesis and NMR characterization of β-alanine-bridged hemispherodextrin, a very efficient chiral selector in EKC. *Electrophoresis* 28: 2580–2588.

8 Chiral Separations by Micellar Electrokinetic Chromatography

Joykrishna Dey and Arjun Ghosh

CONTENTS

8.1 Introduction ... 195
8.2 Self-Assembly of Surfactants ... 196
8.3 Separation Principle of MEKC ... 199
8.4 Mechanism of Chiral Separation ... 201
8.5 Surfactant-Based Chiral Selectors ... 202
8.5.1 Anionic Surfactants ... 202
8.5.1.1 Bile Salts ... 202
8.5.1.2 Alkylglucoside Surfactants ... 204
8.5.1.3 Amino Acid-Derived Surfactants ... 204
8.5.1.4 Vesicle-Forming Surfactants ... 206
8.5.1.5 Miscellaneous Surfactants ... 207
8.5.2 Cationic Surfactants ... 207
8.5.3 Zwitterionic Surfactants ... 208
8.5.4 Neutral Surfactants ... 209
8.5.5 Enantioseparations Using Achiral and Chiral Surfactant Mixtures ... 209
8.6 Cyclodextrin-Modified MEKC Separations ... 211
8.7 Enantioseparations Using Polymerized Chiral Surfactants ... 214
8.8 MEKC with Mass Spectrometric Detection ... 221
References ... 222

8.1 INTRODUCTION

Micellar electrokinetic chromatography (MEKC) is a powerful method for the separation of ionic as well as nonionic solutes [1,2] by capillary electrophoresis (CE). It was originally developed by Terabe et al. [3] in the early 1980s. In MEKC, solute molecules are partitioned between an ionic pseudostationary phase (PSP) and the electrophoretic run buffer, also called background electrolyte (BGE) and are separated according to their different affinities for the PSP. The separation principle is very similar to high-performance liquid chromatography (HPLC). However, since uncoated fused silica capillaries are used for separations in MEKC as a rule, the electroosmotic flow (EOF) is superimposed on the separation process. MEKC is found to have several advantages over HPLC, e.g., (1) high efficiency (large theoretical plate count), (2) short analysis time, (3) small sample size (~1 nL) and buffer quantities, and (4) ease of changing PSP and running media, which make method development easy. Though MEKC has advantages over HPLC in terms of simplicity, resolution, and cost, it suffers from low concentration sensitivity as a consequence of the limited sample volume and short path length for UV-vis detection. The aforementioned advantages, however, have made MEKC attractive for enantiomeric separations. In order to achieve enantiomeric separations, a chiral PSP or a chiral secondary additive must be employed. The importance of chiral separation led to an enormous number of papers based on chiral separation using electrokinetic chromatography (EKC) and other CE-based techniques in the last two decades. Many authors have reviewed the subject [4–45].

Among the various CE-based methods in chiral analysis, MEKC has been the most popular technique. Mainly two modes of chiral separations are used. One is MEKC, where the PSP typically consists of surfactant aggregates, such as micelles, vesicles (or liposomes), etc. The other is called cyclodextrin (CD) modified MEKC (CD-MEKC) which uses a mixture of nonderivatized or derivatized CD and an achiral ionic surfactant. The surfactants employed in MEKC should have some characteristics to provide optimum resolution [46]. First of all, the aggregate formed by the surfactant should (1) be stable under the analytical conditions to allow adjustment of retention factors, (2) be monodisperse to provide large theoretical plate counts, and (3) have high electrophoretic mobility to provide a wide migration time range. Second, to minimize Joule heating, the surfactant should have a low critical micelle concentration (CMC). Finally, surfactants should be such that simple chemical modification of the headgroup produces a wide variety of chemical structures enabling the modulation of chromatographic separation for a range of analytes. Also the separation condition should allow fast mass transfer for analytes between the PSP and the aqueous phase.

8.2 SELF-ASSEMBLY OF SURFACTANTS

Surfactant molecules with a polar (ionic or neutral) headgroup and long hydrocarbon tail are known to self-assemble in aqueous medium above a critical concentration called critical aggregation concentration (CAC) to form different types of aggregates. The term CAC is preferred to CMC because besides micelles, surfactants can also form various types of aggregates. The shape and size of the aggregates depend on the molecular architecture of the surfactant [47]. Shapes of the spontaneously formed surfactant aggregates can be predicted with considerable certainty using three nominal geometric parameters of the surfactant molecule: (1) the optimal headgroup area, a_0, (2) the volume of hydrocarbon tail, v, which is assumed to be fluid and incompressible, and (3) the critical chain length, l_c, which is the maximum effective length the chain can assume. Israelachvili defined a critical packing parameter or shape factor, P ($P = v/l_c a_0$) that can be used to predict which structures an amphiphile will assemble into [47]. The model predicts the formation of spherical micelle at $P < 1/3$, a cylindrical micelle at $1/3 < P < 1/2$, a vesicle or flexible bilayer at $1/2 < P < 1$, and inverted micelles at $P > 1$.

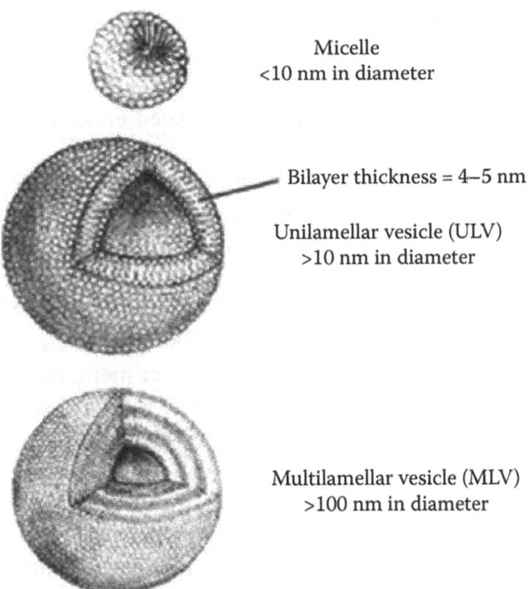

FIGURE 8.1 Shape of a micelle, a unilamellar vesicle (ULV), and a multilamellar vesicle (MLV). (Reproduced from Owen, R.L. et al., *Electrophoresis*, 26, 735, 2005. With permission.)

Micelles are the simplest and most studied ones [48–50] among the surfactant aggregates. Micellar structures of different shapes, such as spherical, ellipsoid, disk, rod, and wormlike have been reported in the literature. Spherical micelles are usually referred to as normal micelles or simply micelles (Figure 8.1). This type of aggregates has small sizes (hydrodynamic diameter, $d_H < 10\,\text{nm}$). Micelles are generally formed by cationic, anionic, zwitterionic as well as nonionic surfactants having short alkyl chains. In general, the ionic surfactants have higher CAC value compared to nonionic surfactants. The CAC in aqueous media decreases as the hydrophobic character of the surfactant increases. In the aqueous solution, the presence of electrolyte causes a decrease in the CAC value. The effect is more pronounced for anionic and cationic surfactants than for zwitterionic surfactants and more pronounced for zwitterionic than for nonionics. Water-soluble polar organic compounds, such as alcohols and amides, reduce the CAC to much lower concentrations. Shorter-chain alcohols are mainly solubilized in the water–micelle interfacial region (Stern layer). On the other hand, longer-chain compounds are solubilized in the outer portion of the micelle core, between the surfactant molecules. Additives that have more than one group capable of forming hydrogen bonds with water appear to produce greater reduction of the CAC value. However, additives like urea, formamide, *N*-methylacetamide, guanidinium salts, short-chain alcohols, ethylene glycol, and other polyhydric alcohols, such as fructose and xylose, increase the CAC to relatively higher concentrations by modifying the interaction of water with surfactant molecules. It should be noted that like the CAC value, the shape of a micelle formed in aqueous solution is also influenced by the nature of counterions, electrolyte concentration, organic additives, and temperature [48].

Considering the value of P, surfactants of long alkyl chains are expected to form rodlike micelles. For a homologous series of surfactants, the sphere-to-rod transition occurs as the chain length is increased. These micelles are normally formed from growth of spherical and disklike micelles at concentrations much higher than the CAC value of the surfactants. This is often manifested by the increase in solution viscosity. The growth of micelles can also be induced either by increasing the concentration or hydrophobicity of the counterion.

Unlike micelles, vesicles are spontaneously formed of closed bilayer structures (Figure 8.1). They are constructed of alternating layers of lipid or surfactant bilayers spaced by aqueous layers or compartments arranged in approximately concentric shells. As a result, the vesicles have two distinct domains: the lipophilic membrane and the interior aqueous cavity. Vesicles are classified in terms of number of lamellae and size. Vesicles with single bilayer membrane are called unilamellar vesicles (ULVs). On the other hand, vesicles having multilayer membranes are referred to as multilamellar vesicles (MLVs). Figure 8.1 presents a schematic view of ULV and MLV. The hydrodynamic diameter of vesicles is greater than 10 nm. Vesicles are normally formed by double-tailed surfactants with the alkyl chain containing more than ten carbon atoms [51]. Stable vesicles can also be produced from single-chain amphiphiles whose conformations have been restricted by the incorporation of rigid segments (e.g., biphenyl unit, azobenzene unit, diphenyl azomethine unit, etc.) or by intermolecular hydrogen-bonding interactions [51,52]. The simplest single-chain surfactants that form vesicles are the long chain alkyl carboxylates [53–61]. In recent literature, it has been reported that vesicles are also produced spontaneously in aqueous solutions of cationic and anionic surfactant mixtures [62–65]. The work on structural characterization and application of lipid vesicles in CE has been reviewed by Owen et al. [66].

Aggregation behavior of chiral surfactants has attracted tremendous attention because of their potential applications. Chiral surfactants are speciality molecules that are used in stereoselective synthesis [67]. Chiral recognition between chiral centers significantly increases the interaction between the hydrophobic alkyl chains.

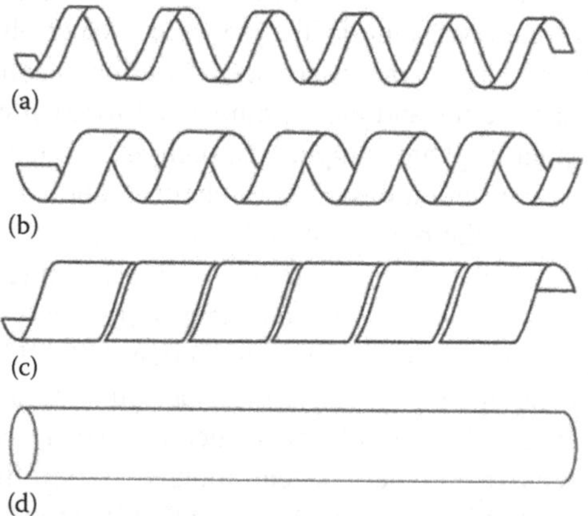

FIGURE 8.2 Self-assembly structures of chiral surfactants: (a) twisted ribbon, (b) helical ribbons, (c) helical ribbon closer to tubular structure, and (d) tubule. (Reproduced from Nakashima, N. et al., *J. Am. Chem. Soc.*, 107, 509, 1985. With permission.)

This results in a reduction of the CAC value of the surfactant [68–72]. Similarly assemblies formed from racemic mixtures have a different morphology from those formed by enantiomerically pure surfactants [73,74]. In addition to micellar and vesicular structures, many chiral amphiphilic molecules have been found to form long, twisted, or helical ribbons, and narrow tubules (Figure 8.2) [75–77]. These morphologies are usually observed with chiral amphiphilic molecules, such as carbohydrate [78,79] and *N*-acyl amino acid amphiphiles [80–84]. Vesicles formed by enantiomerically pure surfactants are found to be more stable compared to those formed by achiral or racemic mixtures under the same conditions [69,85].

8.3 SEPARATION PRINCIPLE OF MEKC

The separation principle of EKC involving surfactants is related to liquid chromatography rather than electrophoretic migration mechanisms [3]. The aggregates formed by the surfactant in the electrophoretic run buffer act as a PSP. The aggregates of ionic surfactants thus formed have electrophoretic self-mobility. Analytes are separated based on the differential partitioning between the PSP and the BGE. The EKC methods involving surfactants are usually called MEKC irrespective of the nature of the aggregate formed by the surfactant in the separation medium. However, the use of the term vesicular EKC (VEKC) can also be found in the literature when vesicles act as the PSP.

Traditionally an uncoated fused silica capillary and an anionic surfactant such as sodium dodecyl sulfate (SDS) are used for MEKC separation. A schematic representation of the separation principle of MEKC is depicted in Figure 8.3. Upon application of a high voltage to the capillary filled with BGE (pH > 2) containing one or more surfactants, an EOF that moves the entire liquid stream is generated in the direction

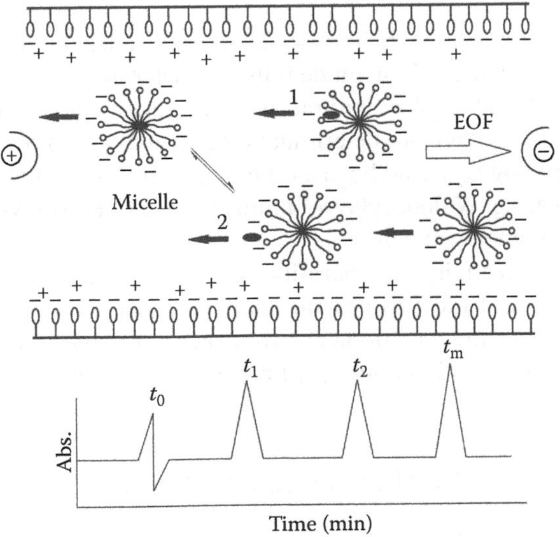

FIGURE 8.3 Separation principles in MEKC.

of the cathode (negative electrode). Anionic micelles owing to its surface negative charge will have mobility in solution toward the anode (positive electrode), which is opposite to the direction of EOF. However, at pH > 5, the EOF becomes much stronger than the mobility of the micelles and the latter migrate toward the cathode. This results in a fast-moving aqueous phase and a slow-moving micellar phase. The difference in time ($t_{mc} - t_o$) taken by an unretained neutral analyte (t_o) and an analyte that is totally retained (due to complete incorporation into the micelle) (t_{mc}) to reach the detector is called migration window. When neutral analytes are injected at the anode end of the capillary, a fraction is incorporated into the micelles and it migrates with the velocity of the micelle. The remaining fraction migrates with the velocity of the EOF. The distribution equilibrium is quickly established and the analytes are separated by the difference in distribution coefficients. The migration order for neutral analytes in MEKC is dictated by the hydrophobicity of the analyte. Under the condition as shown in Figure 8.3, all analytes including anionic, cationic, and neutral molecules injected at the positive end move toward the negative end. The migration time (t_r) of the neutral analyte is limited within the migration window.

In MEKC, the retention factor, k, is defined as $k = n_{mc}/n_{aq}$, where n_{mc} and n_{aq} are the number of analytes in the micelle and aqueous phases, respectively. For neutral analytes, k can be obtained from the equation [3,86]

$$k_N = (t_r - t_o)/(t_o - t_r t_o/t_{mc}) \tag{8.1}$$

where t_r is the migration time of the analyte. For anionic electrolytes, however, a different equation has been derived by Khaledi et al. [87]

$$k_A = (t_r - t_{ion})/(t_{ion} - t_r t_{ion}/t_{mc}) \tag{8.2}$$

where t_{ion} is the migration time in the absence of the micelle. Traditionally, methanol, acetonitrile, mesityl oxide, or formamide is used to determine t_o and t_{mc} is determined by use of Sudan III, Sudan IV, or anthracene. There are methods reported in the literature [88–90] for simultaneous determination of t_o and t_{mc}. However, an alternative method of determining t_{mc} would be to use an oppositely charged surfactant containing a chromophore, e.g., *N*-dodecylpyridinium chloride (DPC) or *N*-cetylpyridinium chloride (CPC) for anionic micelles.

In MEKC, the selectivity or separation factor (α) is calculated as the ratio of the retention factors of the two neutral analytes. That is, $\alpha_N = k_2/k_1$, where the subscript 2 stands for the enantiomer with larger retention time. The selectivity factors of anionic (α_A) and cationic (α_C) analytes can be given by the following equations [91]:

$$\alpha_A = (t_2 - t_o + t_{ion})/(t_1 - t_o + t_{ion}) \tag{8.3}$$

$$\alpha_C = (t_2 - t_o - t_{ion})/(t_1 - t_o - t_{ion}) \tag{8.4}$$

The resolution (R_S) for the neutral enantiomeric pair 1 and 2 in MEKC can be defined by the equation [3,92]:

$$R_S = N^{1/2}/4(\alpha - 1)/\alpha\left[k_2/(1+k_2)\right]\left[(1-t_o/t_{mc})/(1+t_ok_1/t_{mc})\right] \quad (8.5)$$

where N is the number of theoretical plates. This equation predicts the effects of N, α, k_2, and t_o/t_{mc} on the resolution. The migration time and hence retention factor can be altered by changing the buffer components, ionic strength, pH, capillary length, inside diameter, inside surface characteristics, and temperature. The optimum k_2 corresponding to the highest resolution is given by $k_2 = (t_{mc}/t_o)^{1/2}$. On the other hand, the ratio, t_o/t_{mc}, that is directly related to the migration time window is inversely proportional to R_S. Decrease of the ratio will thus increase the resolution, which can be achieved by the addition of organic solvents, such as methanol, acetonitrile, 2-propanol, or by lowering the pH to acidic. Selectivity in MEKC is easily manipulated through the addition of tetraalkylammonium salts, urea, and organic solvents as well as due to changing surfactants. The details of such manipulations are described elsewhere [93].

8.4 MECHANISM OF CHIRAL SEPARATION

The existence and the extent of chiral recognition in two-dimensional as well as in three-dimensional systems is not a simple consequence of molecular asymmetry. It is the result of intermolecular interactions and for separations on surfaces, molecule–substrate interactions specific to the chiral compound. It is generally accepted that a three-point-interaction including a short-range repulsive part that ensures that molecules do not overlap (steric hindrance), an attractive component due to van der Waals (π–π, ion–dipole, dipole–dipole, dipole–induced dipole), and a specific interaction (electrostatic, hydrogen bonding) is required for chiral discrimination. Local chiral structure formation occurs through preferential orientation of the molecules according to these interactions. For example, combination of steric interactions and dispersive forces that describe the topological arrangement of atoms in the molecule may promote alignment for like molecules but inhibit for unlike molecules. However, the opposite effect is also possible. On the other hand, hydrogen bonding, which is an electrostatic interaction, may promote a specific alignment of a set of molecules.

Surfaces can help in chiral resolution in a number of ways. Chiral resolution of racemates has been experimentally observed at a surface. Eckhardt et al. [94] showed that chiral amphiphilic tetracyclics separate in two mirror-image domains at the aqueous surface. On the other hand, chiral stationary phases have been used in chiral HPLC and gas chromatography (GC) for many years. In chiral chromatography, diastereomeric interactions of one of the enantiomers either with the chiral stationary phase or the chiral solvent (when a chiral solvent is used in combination with an achiral column) are generated. Chiral separations in MEKC are based on stereoselective interactions between the chiral analyte and chiral micelles, which act as a PSP. It is believed that as in HPLC, chiral separation in MEKC involves a temporary diastereomeric interaction between the enantiomers and the chiral surfactant.

The enantiomers are separated when their binding constants with the surfactant and/or the mobilities of their diastereomeric complexes are different. However, it should be remembered that diastereomeric interactions are dependent upon the partition coefficient of the analyte between the micelles and the bulk aqueous phase. In chiral MEKC, the chiral moiety of the surfactant is located close to the surfactant polar headgroup, which constitutes the Stern layer of the micelle. Thus, chiral recognition takes place near the surface of the micelle. In other words, the chiral interface of the surfactant micelles promotes chiral resolution. It is evident by the fact that enantiomeric separations are not obtained when the surfactant concentration is below the CAC value. For further understanding of the mechanism of chiral recognition and for improvement of MEKC separations of enantiomers, an analysis of the separation data for each chiral selector–analyte combination is essential.

8.5 SURFACTANT-BASED CHIRAL SELECTORS

Soon after the introduction of MEKC, Zare's group in 1985 first reported enantiomeric separations of dansyl-amino acids by CE using L-histidine/Cu(II) complex [95]. Since then much effort has been made to develop and characterize chiral PSPs for MEKC with enhanced performance and/or unique selectivity. A summary is given in Table 8.1. The chemical structures of some representative chiral surfactants are shown in Figure 8.4.

8.5.1 Anionic Surfactants

8.5.1.1 Bile Salts

Potential applications of MEKC in chiral analysis using various chiral anionic surfactants, for example, bile salts and sodium *N*-dodecanoyl-L-valinate (SDV) were first shown by Terabe's group [96]. Bile salts, such as sodium cholate (SC), were among the first chiral surfactants used in MEKC [7]. Bile salts are natural chiral amphiphiles with a helical structure and are able to interact with relatively planar and rigid analytes through electrostatic, hydrogen bonding, and hydrophobic interactions. SC micelles were also shown to effectively separate flavanone-7-*O*-glycoside epimers of neohesperidin and naringin [97]. Sodium taurocholate (STC) micelles in acetate buffer were used for the resolution of metyrosine enantiomers [98]. Tian et al. [99] also used SC micelles for the analysis of palonosetron hydrochloride stereoisomers. Commercially available anionic surfactants derived from bile acids, such as sodium deoxycholate (SDC) and sodium taurodeoxycholate (STDC), are the most popular micelle-forming chiral selectors [17,100]. They have been used to separate enantiomers of dansyl (DNS) derivatives of amino acids [7], 1-naphthylethylamine [101], various anionic and neutral compounds [102], as well as drugs, e.g., diltiazem and trimetoquinol [101,103–105], mephenytoin [106], and 3-hydroxy-1,4-benzodiazepins [107]. However, diltiazem and trimetoquinol could only be enantioseparated by STDC. In fact, STDC is considered to be the most effective chiral selector among bile salts [98,108,109]. Recently, Hebling et al. [110] have demonstrated that chiral resolution of (*R*,*S*)-1,1′-binaphthyl-2,2′-dihydrogen phosphate (BNP) is very sensitive to the aggregation state of the SC surfactant. It was shown that the NMR spectral resolution of the enantiomers strongly correlates with the chiral separation in MEKC. The H5-H7 edge

TABLE 8.1
Chiral Separation by MEKC Using Anionic, Cationic, Neutral, and Zwitterionic Surfactants

Surfactants	Separated Enantiomers
Anionic surfactants	
Bile salts	
Sodium cholate (SC)	BOH, oxazepam, temazepam, lorazepam, lormetazepam
Sodium deoxycholate (SDC)	BOH, BNP, BNC, biphenanthrene dihydroxide, diltiazem, trimetoquinol, carboline, 1-naphthylethylamine
Sodium taurocholate (STC)	Diltiazem, trimetoquinol, carboline, 1-naphthylethylamine
Sodium taurodeoxycholate (STDC)	DNS-AAs, diltiazem, trimetoquinol, carboline, 1-naphthylethylamine, BOH, tetrahydropapveroline, 2,2,2-trifluoro-1-(9-anthryl)-ethanol, laudanosoline, norlaudanosoline, laudanosine, BNC
Glucopyranoside-based surfactants	
n-Dodecyl-β-D-glucopyranoside monosulfate *n*-Hexadecyl-β-D-glucopyranoside monosulfate *n*-Dodecyl-β-D-glucopyranoside monophosphate	DNS-AAs, AA-carbamates, metoprolol, cromakilin, Troger's base, ephedrine, hexobarbital, phenobarbital, binaphthyl compounds, mephenytoin, fenoldopam, hydroxymephenytoin, silvex, mecoprop, dichloroprop, CPPAs, 2-PPA
Acylamino acid surfactants	
Sodium *N*-dodecanoyl-L-alaninate (SDA) Sodium *N*-dodecanoyl-L-valinate (SDV) Sodium *N*-dodecanoyl-L-serinate (SDSer) Sodium *N*-dodecanoyl-L-threoninate (SDT) Sodium *N*-dodecanoyl-L-glutamate (SDGlu) Sodium *N*-undecylenyl-L-valinate (SUV)	PTH-AAs, benzoin, warfarin, amino acid derivatives, (*N*-benzoyl-*O*-isopropyl, etc.)
Sodium *N*-(4-*n*-dodecyloxybenzoyl)-L-valinate (SDLV), *N*-[4-*n*dodecyloxybenzoyl]-L-leucinate (SDLL) and sodium *N*-[4-*n*-dodecyloxybenzoyl]-L-isoleucinate (SDLIL), Sodium *N*-(4-*n*-decyloxybenzoyl)-L-valinate (SDeLV), Sodium *N*-(4-*n*-octyloxybenzoyl)-L-valinate (SOLV)	BOH, BNP, BZN, TB
Other amino acid type surfactants	
(*R*)-*N*-Dodecoxycarbonylvaline (*S*)-*N*-Dodecoxycarbonylvaline (*S*)-2-[(Dodecoxycarbony)amino]-3(*S*)-methyl-l-sulfooxypentane	Atenolol, ketamine, benzoin, pindolol, bupivacaine, metoprolol, homatropin, terbutaline, octopamine. DNB-AA methyl esters, PMP derivatized aldose, norephedrine, pseudoephedrine, ephedrine, *N*-methylpseudoephedrine, norphenylephrine, Piperidine-2,6-dione

(*continued*)

TABLE 8.1 (continued)
Chiral Separation by MEKC Using Anionic, Cationic, Neutral, and Zwitterionic Surfactants

Surfactants	Separated Enantiomers
Cationic surfactants	
(1*R*,2*S*)-(−)- *N*-Dodecyl-*N*-methylephedrinium bromide (DMEB)	Profen type NSAIDs
3-(*N*-Dodecanoyl-L-valylamino) propyltrimethylammonium bromide, 3-(*N*-nonanoyl-L-valylamino)-hexyltrimethylammonium bromide	DNB-AAs, DNB-AA-isopropyl ester, DNB-AA-isobutyl ester, DNB-AA-*s*-pentyl ester, BzAla-isoprpyl ester, NB-Ala-isopropylester
Neutral surfactants	
n-Heptyl-β-D-thioglucopyranoside *n*-Octyl-β-D-glucopyranoside *n*-Octyl-β-D-thioglucopyranoside *n*-Dodecyl-β-D-glucopyranoside Nonyl-β-D-glucopyranoside Octyl-β-D-maltopyranoside	DNS-AAs, AA-carbamates, TB, metoprolol, cromakilin, ephedrine, binaphthyl compounds, hexobarbital, phenobarbital mephenytoin, CPPAs, 2-PPA, fenoldopam, silvex, mecoprop, dichloroprop, hydroxymephenytoin
L-α-Palmitoyllysophosphatidylcholine	DNS-AAs
Zwitterionic surfactants	
N-(2-Hydroxy-*n*-dodecyl)-L-threonine (2-HDT)	BOH, BNP, BZN
3-[(3-Cholamidopropyl)dimethylammonio]-1-propane sulfonate (CHAPS)	BNP, TB
3-[(3-Cholamidopropyl)dimethylammonio]-2-hydroxy-1-propane sulfonate (CHAPSO)	

Note: AA, amino acid; BNC, binaphthyldicarboxylic acid; Bz, benzoyl; CPPA, chlorophenoxypropionic acid; NB, *N*-benzoyl; 2-PPA, 2-phenoxypropionic acid; PMP, 1-phenyl-3-methyl-5-pyrazolone.

of BNP was shown to be sensitive to chirally selective interactions with the primary micellar aggregate of SC.

8.5.1.2 Alkylglucoside Surfactants

Tickle et al. [111] introduced anionic alkyl glucoside surfactants, dodecyl-β-D-glucopyranoside monophosphate and dodecyl-β-D-glucopyranoside monosulfate. Enantioseparations of cromakalin, BNP, mephenytoin, hydroxy mephenytoin, and 3,4-dimethyl-5,7-dioxo-2phenylperhydro-1,4-oxazepine were obtained using dodecyl-β-D-glucopyranoside 4,6-hydrogen phosphate. On the other hand, hexadecyl-D-glucopyranoside 6-hydrogensulfate was employed for the enantiomeric separations of some phenylthiohydantoin (PTH) derivatives of amino acids and binaphthol (BOH) [112].

8.5.1.3 Amino Acid-Derived Surfactants

Many researchers compared various negatively charged micelles of chiral surfactants [20]. Synthetic anionic surfactants containing different amino acids as

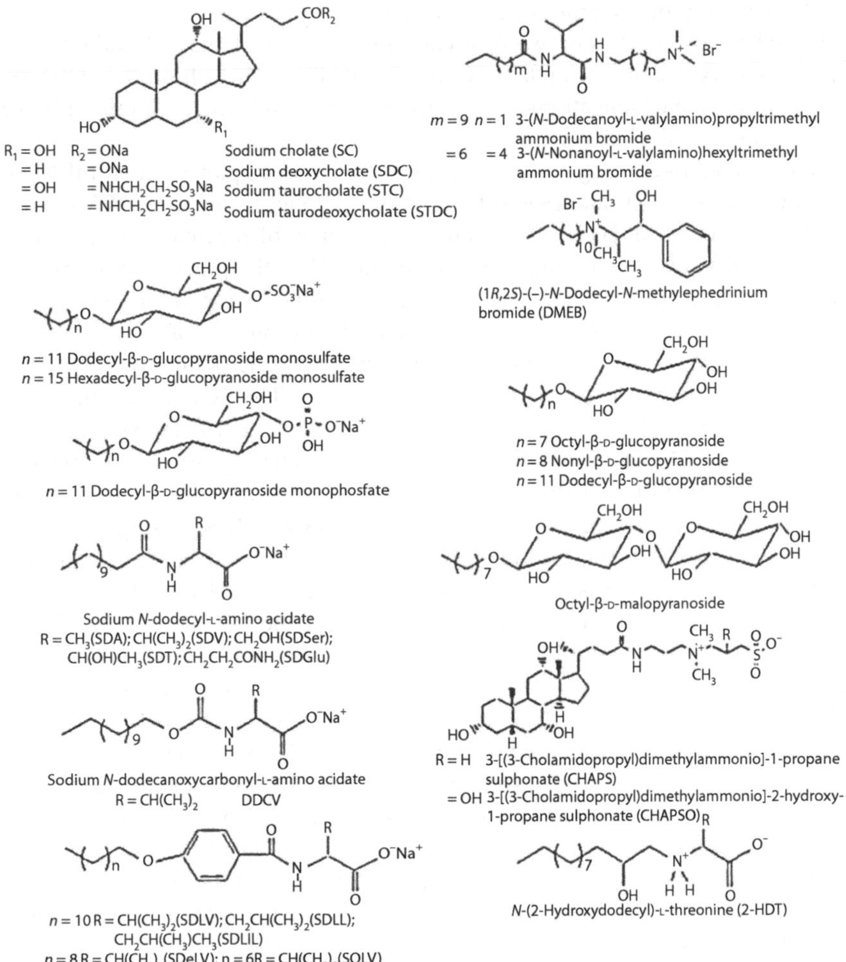

FIGURE 8.4 Chemical structures of anionic, cationic, neutral, and zwitterionic surfactants employed in chiral MEKC separations.

headgroup have been evaluated through enantiomeric separations of PTH and dinitrobenzoyl (DNB) derivatives of DL-amino acids and compounds of pharmaceutical interest [113–117]. Typically, *N*-acylamino acid surfactants, such as sodium dodecanoyl-L-valinate (SDV), sodium dodecanoyl-L-alaninate (SDA), sodium dodecanoyl-L-serinate (SDSer), sodium dodecanoyl-L-threoninate (SDT), and sodium dodecanoyl-L-glutamate (SDGlu), were employed. Otsuka and Terabe [116] investigated the effects of organic additives, e.g., methanol, and urea on the enantioselectivity and peak symmetry. The chiral recognition of *N*-acylamino acid micelles has been shown to be due to (1) hydrogen-bonding interaction of the enantiomeric solutes with the amide group buried in the micelle core, (2) the formation of a chiral barrier by the amino acid side chain near the Stern layer, and (3) differences in penetration depth of the solute molecules that perturbs the chiral barrier. Consequently, an increase in enantioselectivity has been observed with the increase in hydrophobicity of the amino acid side chain of the surfactant [117]. Warner's group used mixed micelles of two *N*-acylamino acid surfactants and evaluated their chiral selectivity in comparison to pure surfactant [118]. They showed that the resolution of atropisomers/enantiomers of benzoin, BNP, and aminoglutethimide were better in a mixed micellar system.

Another very interesting class of chiral surfactants employed in chiral MEKC are sodium salts of (*S*)- and (*R*)-*N*-dodecyloxycarbonyl-L-valine (DDCV) synthesized by Mazzeo et al. [119]. Enantiomers of a range of racemates, e.g., atenolol, metoprolol, pindolol, ephedrine, norephedrine, *N*-methylpseudoephedrine, norphenylephedrine, bupivacaine, homatropine, ketamine, octopamine, and terbutaline were successfully separated in phosphate/borate buffer (pH 8.8). Foley and coworkers [120] used (*R*)- and (*S*)-DDCV for enantiomeric separations of 20 basic drugs, most of which were β-agonists or β-antagonists. The BGE containing 25 mM DDCV, 100 mM 2-[*N*-cyclohexylamino]ethane sulfonic acid (CHES), and 10 mM triethylamine (TEA) was found to be most effective. The highest enantioselectivity of 1.25 was obtained for clenbuterol. The influence of surfactant counterions (Li^+, Na^+, K^+) on the enantioseparations of 15 basic drugs by MEKC using DDCV was also investigated [121]. Improvement in peak symmetry and hence enhancement of separation efficiency was observed using the DDCV with Li^+ as the counterion. Optical resolution of several *N*-benzoylamino acid methyl esters [122,123] and piperidine-2,6-dione [124] enantiomers, except glutethimide, were also successfully achieved by using DDCV. Swartz et al. [124] demonstrated resolution of glutethimide enantiomers using (*S*)-*N*-dodecyloxycarbonylproline (DDCP) surfactant. Enantioselectivities were observed to be better with DDCV in comparison to dodecanoylvaline (DV). Interestingly, sodium (*S*)-*N*-dodecyloxycarbonyl-L-valinate was found to be a better resolving agent compared to sodium (*R*)-*N*-dodecyloxycarbonyl-L-valinate at the same experimental conditions. Ding and Fritz [125] evaluated amino acid-derived carbamate-type surfactants with C4-C11 chain length using propanolol, atenolol, ketamine, laudanosine, nefopam, benzoin, and hydrobenzoin as model compounds.

8.5.1.4 Vesicle-Forming Surfactants

As discussed earlier, vesicles are either single bilayer or multilayer aggregates with bigger size compared to the micelles and have higher surface charge density. Due to high electrophoretic mobility, vesicles provide wider migration window and thus have the potential for separation of compounds with similar hydrophobicity. Vesicles formed from SDS and *n*-dodecyltrimethylammonium bromide (DTAB) surfactants were shown to provide about two times wider migration window, higher polar group selectivity, retention time, and efficiency as compared to the SDS micellar system [126]. Improvements in migration range and pH stability of vesicular systems formed by *n*-cetyltrimethylammonium bromide (CTAB) and DDCV mixtures were also observed by Pascoe et al. [127]. Chiral vesicles are a special class of supramolecules that are used as enantioselective analytical reagents [128,129]. However, among surfactant-based chiral PSPs, vesicles have not received much attention in chiral separations by EKC. There are only a few papers on chiral separations using VEKC [130–133]. Dey's group [130,131] have demonstrated atropisomeric/enantiomeric separations of BOH, BNP, binaphthyldiamine (BNA), benzoin, and Tröger's base (TB) using vesicles composed of sodium *N*-(4-*n*-dodecyloxybenzoyl)-L-valinate (SDLV) as chiral selector. Surfactant concentration of only 5 mM was required for the separations reported. They observed better chiral resolution compared to other micelle-forming surfactant monomers. A brief discussion on the aggregation properties and microstructures of the surfactant in aqueous buffer solutions was also included in the paper [131]. Chiral selectivity of the vesicles prepared from sodium *N*-(4-*n*-dodecyloxybenzoyl)-L-leucinate (SDLL) and sodium *N*-(4-*n*-dodecyloxybenzoyl)-L-isoleucinate (SDLIL) was evaluated by Mohanty and Dey [132] for the enantioseparations of the above mentioned analytes focusing more on the chiral recognition mechanism. Enantioselectivities of SDLL and SDLIL were compared with that of SDLV. It was concluded that hydrophobic interactions did not contribute to the selectivity, but steric hindrance near the chiral center played a major role in differentiation of enantiomers. However, the presence of two chiral centers in SDLIL had a negative effect on the enantioseparations of BOH and BZN. In a more recent study [133], enantioselectivity of SDLV was also compared with two surfactants with identical headgroup (L-valine), sodium *N*-(4-*n*-decyloxybenzoyl)-L-valinate (SDeLV) and sodium *N*-(4-*n*-octylyloxybenzoyl)-L-valinate (SOLV), which form micelles in the separation conditions employed. Nevertheless, successful enantiomeric separations were obtained for the analytes using both SDeLV and SOLV. Interestingly, no significant difference in enantioselectivities of vesicles and micelles was observed by the researchers. The type of aggregate, however, had an effect on other chromatographic parameters, e.g., analysis time, retention factor, and chiral resolution.

8.5.1.5 Miscellaneous Surfactants

Using an approach similar to *N*-acylamino acids, Dalton and coworkers synthesized (*R*,*R*)-tartaric acid [134,135] and employed it in MEKC for enantioseparations of compounds having single and fused polyaromatic rings. The enantiomers of the latter type resolved better than those with single aromatic ring. Sodium salt of maleopimaric acid (SMA), a new type of chiral surfactant, has been synthesized and employed for the enantioseparation of amino acids derivatized with naphthalene-2,3-dicarboxaldehyde (NDA) [136]. Enantiomeric separations of two compounds, NDA-DL-tryptophan and NDA-DL-kynurenine, were achieved under the conditions selected. Chiral and achiral MEKC in combination with laser-induced fluorescence (LIF) detection was employed in the work by Herrero et al. [137] to identify and quantify a group of fluorescein isothiocyanate (FITC) derivatized D- and L-amino acids in different microalgae samples. The contents of each pair of enantiomers of arginine (Arg), lysine (Lys), alanine (Ala), glutamic acid (Glu), and aspartic acid (Asp) in three samples were compared. In a review by Otsuka [138], the use of a sensitive detection scheme that includes online sample concentration and the use of thermal lens microscopy in MEKC using avidin as a chiral selector has been discussed.

8.5.2 Cationic Surfactants

Although enantiomeric separations by MEKC have been traditionally performed using chiral anionic surfactants, there are a few reports on enantiomeric separations using cationic surfactants. The use of cationic surfactants is normally avoided because of their adsorption onto the negatively charged capillary wall, which either completely eliminates or reduces the EOF and causes baseline distortion. Bunke et al. [139] evaluated the selectivity of (–)-*N*-dodecyl-*N*-methylephedrinium bromide (DMEB) (see Figure 8.4 for structure) surfactant for the chiral separation of the atropisomeric methaqualone. Using a high concentration of DMEB, the enantiomers of this compound were only separated with a lower separation factor. Dey and coworkers [140] also used DMEB for the enantioseparation of eight profen type (carprofen, fenoprofen, fluorbiprofen, ibuprofen, indoprofen, ketoprofen, naproxen, and suprofen) nonsteroidal anti-inflammatory drugs (NSAIDs) at a relatively lower concentration (6–10 mM). Enantiomeric separations of individual compounds as well as in mixtures of three or five compounds have been reported. Dobashi's group used trimethylammonium-terminated surfactants whose chiral moieties are incorporated into the hydrocarbon tail of the surfactants [141,142]. In these surfactants, the chiral valinediamide moiety is shielded by the surfactant headgroup (Figure 8.4). However, they showed effective chiral recognition of enantiomers inside hydrophobic micelles. The authors showed that the surfactant with longer spacer between the chiral moiety and the headgroup exhibits significantly greater selectivity compared to surfactants containing a shorter spacer in the case of DNB-amino acid isopropyl esters [143].

8.5.3 Zwitterionic Surfactants

Chemical structures of representative chiral zwitterionic (or amphoteric) surfactants that have been used for chiral analysis are shown in Figure 8.4. There are only a few studies that have examined the chiral zwitterionic surfactants as chiral selectors [144–147]. The advantage of zwitterionic surfactants over the anionic ones is that their aqueous solutions have very low or zero electrical conductivity, which allows their use at high concentrations and high voltages for efficient separations. This reduces Joule heating even when used at higher concentrations. However, most zwitterionic surfactants have poor aqueous solubility limiting their use as PSP in MEKC. It should be noted that since the micelles of electrically neutral surfactants do not migrate, separation methods based solely on such surfactants cannot be classified as MEKC.

Hadley et al. [148] have used two zwitterionic steroidal chiral selectors, e.g., the cholic acid derivative 3-[(3-cholamidopropyl)dimethylamino]-1-propane sulfonate (CHAPS) and the corresponding 2-hydroxy-1-propane sulfonate derivative (CHAPSO) as sole chiral selectors for the enantioseparation of racemic TB and BNA [149]. Both selectors were, however, found to adsorb onto the bare fused silica capillary wall shielding the basic analytes from interactions with the silanol groups and facilitated efficient separations. Nimura and coworkers [150] employed L-α-palmitoyllysophosphatidylcholine, a phospholipid, for the enantioseparations of DNS-amino acids. In 2008, Ghosh and Dey [151] reported enantioseparations of BOH, BNA, BNP, benzoin, and TB using a zwitterionic surfactant, *N*-(2-hydroxydodecyl)-L-threonine (HDT) (Figure 8.4) that forms tubular aggregates in aqueous solution. Successful separation of atropisomers/enantiomers was obtained at a very low concentration of the surfactant. The authors reported improvement of optical resolution in the presence of organic additives, such as methanol, acetonitrile, and 2-propanol.

8.5.4 Neutral Surfactants

In comparison to chiral zwitterionic surfactants more reports on chiral analysis by MEKC using neutral surfactants are available. Among neutral surfactants the *n*-alkyl-β-D-glucopyranosides were first employed to examine chiral discrimination using neutral micelles [152]. The glucopyranoside-derived surfactants with both L- and D-form are known to have very low CAC and high aqueous solubility. Since the neutral micelles migrate with the velocity of the EOF, they are employed for enantiomeric separations of only charged compounds. However, these chiral selectors (see Figure 8.4 for structures) are known to form complexes with borate ions in solution at higher pH forming in situ charged micelles and therefore, they do not require additional charged surfactant. Neutral alkyl-β-D-glucopyranoside surfactants with C7–C10 chain length have been used for the enantioseparations of amino acid derivatives, and phenoxy acid herbicides [153,154]. For the optical resolution of the phenoxy acid herbicide enantiomers, octylmaltopyranoside was also employed [155,156]. The glucopyranoside-derived surfactants were employed in the chiral separations of drugs, such as metoprolol, ephedrine, hexobarbital, phenobarbital, mephenytoin, hydroxymephenytoin, and fenoldopam [152,157]. Ju and El Rassi [158,159] synthesized cyclohexyl-alkyl-β-D-maltosides surfactants and evaluated their chiral selectivity using dansyl amino acids, dinitrophenyl amino acids, and tryptophan derivatives. Mechref and El Rassi [160], on the other hand, employed neutral steroidal glucoside surfactants, such as *N,N*-bis-(3-D-gluconamidopropyl)-cholamide (Big CHAP) and *N,N*-bis-(3-D-gluconamidopropyl)-deoxycholamide (deoxy Big CHAP) and obtained chiral separations of some DNB-amino acids, binaphthyl enantiomers, TB, and a phenoxy acid herbicide by MEKC in borate buffers.

8.5.5 Enantioseparations Using Achiral and Chiral Surfactant Mixtures

As discussed above micelles of neutral chiral surfactants alone cannot be employed in MEKC separations because they migrate with the velocity of EOF. However, their use as a chiral additive to an anionic achiral/chiral surfactant has been reported in the literature. For example, digitonin, which is a naturally occurring electrically neutral chiral surfactant, has been used as a chiral additive to an SDS solution at acidic pH for the enantioseparation of some PTH-amino acids [161]. Terabe and coworkers [112] have employed glycyrrhizic acid (GRA) and β-escin as chiral PSPs in MEKC. Some DNS-amino acids were enantioseparated using a mixture of 30 mM GRA, 50 mM octyl-β-D-glucopyranoside, and 10 mM SDS solution. However, the migration times were not reproducible. Similarly β-escin-SDS system provided optical resolution of some PTH-amino acids under acidic conditions.

The applicability of a mixed micelle system, such as SDS and SDV, for the enantioseparations of *N*-3,5-dinitrobenzoylated amino acid isopropyl esters was first demonstrated by Dobashi et al. [113,162]. The micropolarity of the mixed micelle was observed to be lower than that of pure SDS micelles. According to the authors, this provided a favorable ordered microenvironment for hydrogen-bonding interactions between the chiral surfactant and the enantiomers near the Stern layer. In this work, using a phosphate-borate buffer system the D-isomer was observed to elute first. Employing the same mixed micelle system in the presence of 10% methanol and 5 M urea six PTH-amino acids were enantioseparated simultaneously along with warfarin and benzoin [163,164]. It was also demonstrated that the addition of SDS, urea, methanol, and 2-propanol to each of the chiral surfactant solution of sodium *N*-dodecanoyl-L-serinate (SDSer), sodium *N*-dodecanoyl-L-aspartate (SDAsp), and sodium *N*-tetradecanoyl-L-glutamate (STGlu) was essential for the improvement of enantioselectivity and peak shapes in chiral analysis of PTH-amino acids [144,165].

The addition of bile salts to the aqueous solutions of polyoxyethylene ethers (PEs; $C_{12}E_4$, $C_{12}E_6$, $C_{10}E_8$) have been reported by Clothier et al. [166,167] to improve enantiomeric separations of verapamil and related compounds by MEKC. In one experiment, mixtures of SDC and three PEs were evaluated. In another investigation, chiral resolving abilities of micellar solutions of four different bile salts alone and in mixtures with $C_{12}E_4$ and methanol were studied. The addition of $C_{12}E_4$ to the solutions of SC and SDC with or without methanol improved chiral resolution for compounds containing longer hydrocarbon chain, there by separating the chiral center from the major functional groups. Following the same method, enantiomeric separations of three binaphthyl derivatives were performed using the combination of SC and SDC surfactants with SDS [168]. Recently, Park and Jung [169] performed enantioseparations of some chiral flavonoids by MEKC using SDS micelles in the presence of microbial cyclosophoraoses and their sulfated derivatives.

Zhao et al. [170] demonstrated the utility of abietic acid, a naturally occurring chiral derivatizing agent, through separations of diastereomers of 10

TABLE 8.2
Chiral Separations by MEKC Using Mixed Chiral and Achiral Surfactant Systems

Chiral Surfactant	Achiral Surfactant	Additives	Separated Enantiomers
SC or SDC	Brij 30	Methanol	Atenolol, verapamil, norverapamil, BOH, gallopamil, BAYK8644
SDC	Brij 30	Urea	Verapamil, gallopamil, norverapamil,
STDC	Brij 30		Bupivacaine, prilocaine mepivacaine
Octyl-β-D-glucoside	SDS, GRA		DNS-amino acids
β-Escin	SDS		PTH-amino acids
SDV	SDS	Methanol, urea	PTH-amino acids
SDV		Methanol, urea	PTH-amino acids
SDV	SDS		PTH-amino acids, benzoin, warfarin
SDGlu	SDS	Urea, methanol/isopropanol	PTH-amino acids, benzoin
SDSer	SDS	Methanol	PTH-amino acids
STGlu	SDS		PTH-amino acids

DL-amino acid enantiomers including Ala, valine (Val), leucine (Leu), phenylalanine (Phe), methionine (Met), serine (Ser), threonine (Thr), and tryptophan (Trp) in phosphate buffer containing 18 mM SDS and 25% (v/v) acetonitrile. In another study, they employed dehydroabietylisothiocyanate (DHAIC) as chiral derivatizing agent for enantioseparation of five amino acids and a chiral drug 3,4-dihydroxyphenylalanine (DOPA) using similar separation condition. A summary of chiral separations using mixed micellar systems can be found in Table 8.2.

8.6 CYCLODEXTRIN-MODIFIED MEKC SEPARATIONS

Addition of cyclodextrins (CDs) to achiral surfactants in a mixed micellar system has been proved to be an effective approach to improve chiral separations by MEKC [22]. Addition of CDs to aqueous solutions of the most commonly used achiral SDS surfactant has been shown to be very effective for chiral resolution of uncharged compounds. The partition equilibrium established between the micelle and the bulk aqueous phase causes a difference in the velocity of the analyte that moves with the EOF. The chiral separation is a consequence of the difference in CD-analyte binding constant. Competition between SDS and chiral analyte molecules for the CD cavity acts against chiral discrimination [171]. Since the separation principle involves partition as well as complexation equilibria, the method is often referred to as CD-MEKC. In this method, the chiral selectivity can be improved by the modification of the chemical structure of the CD and the surfactant, the concentration of the CD and the surfactant, the buffer type, the ionic strength, the pH, and the organic additives. CD-MEKC has been successfully used for chiral separations of a wide variety of compounds including amino acids, pharmaceuticals, pollutants, and biological extracts. The discrimination capability of micelles containing CDs for various classes of compounds has been discussed in a review by Fanali et al. [172]. A summary of CD-MEKC separations is given in Table 8.3.

CD-MEKC has been widely used for the enantioseparation of derivatized amino acids [173–176]. For chiral separations of amino acids, Tran and Kang [177] used a three-component system containing CD, 1-*S*-octyl-β-D-thioglucopyranoside, and SDS surfactants, which showed better separations than 1-*S*-octyl-β-D-thioglucopyranoside alone. Another three-component system utilized 3-((3-chloroamidopropyl)-dimethylamino)-1-propane sulfonate with SDS and CD for the separation of dansyl amino acids [145]. The enantioseparations of FITC derivatives of DL-Ala, DL-Glu, and DL-Asp were obtained by CD-MEKC using LIF detection [178]. LIF detection was used for both qualitative and quantitative analysis of L- and D-amino acids (DL-Ala, DL-proline, DL-Glu, and DL-Asp) in vinegars with high efficiency (720,000 plates/m) and good sensitivity (LOD < 16.6 nM), using 20 mM β-CD, 30 mM SDS in 100 mM borate buffer (pH 9.7) [179]. Herrero et al. [180] employed CD-MEKC coupled with LIF detection for the chiral analysis of FITC derivatized major DL-amino acids in conventional and transgenic maize using a BGE composed of 100 mM borate buffer (pH 10.0), 80 mM SDS, and 20 mM β-CD. Enantiomeric excess of several D-amino acids in one of the studied maize extractions were also determined in this work. Recently, chiral analysis of FITC derivatized DL-Ala, DL-Asn, DL-Arg, DL-Glu, and DL-Asp from extracts of conventional and transgenic yeasts by CD-MEKC coupled

TABLE 8.3
Chiral Separation by CD-MEKC Method

Cyclodextrin	Surfactant	Additives	Separated Enantiomer
Amino acid			
β-CD, γ-CD	SDS	Methanol	DNS-AAs
β-CD, γ-CD	SDS		CBI-AAs
β-CD	SDS	2-Propanol	FMOC-AAs
β-CD	SDS		FITC-amino acids (Arg, Pro, GABA, Ala, Glu, Asp)
β-CD	SDS		CBI-D, L-Asp
γ-CD	SDS		FA-D, L-amino acid
α-CD, β-CD, γ-CD, HP-β-CD, HE-β-CD, Me-β-CD	SDS		Uncharged derivatives of chiral amino acid C-protected by alkyl group and *N*-protected by cyclic acyl group
β-CD	SDS		The D- and L-forms of proline, Arg, Asn, Ala, Glu and Asp
α-,β-,γ-CDs, HP-α-,β-, γ-CDs	SDS		Naphthalene-2,3-dicarboxaldehyde derivatized DOPA
Chiral drug			
HP-β-CD	SDS		Propiconazole, DOPA
β-CD, γ-CD, HP-γ-CD	SDS		Aryl allenic acids
Neutral cyclosophoraoses (Cys)	SDS		Catechin
β-CD, γ-CD	SDS	Acetonitrile	Cicletanine
β-CD, DM-β-CD	SDS	Urea, methanol	Hexobarbital, fadrozole
DM-β-CD	SDS	Urea, methanol	Aminoglutethimide analogues
β-CD	SDS	Urea, methanol	Mephobarbital, glutethimide
γ-CD	SDS	Urea, methanol	Secobarbital
γ-CD	SDS	l-Menthoxyacetic acid D-Camphorsulfonic acid	Pentobarbital, thiopental, binaphthyl compounds, trifluoroanthrylethanol
α-CD + β-CD	SDS		Alprenolol, atenolol, propanolol
HP-β-CD	SDS		BMS-180431-09 (cholesterol-lowering drug)
β-CD	SDS	Urea	SM-8849, analogs (immunomodulating antirheumatic drug)
β-CD	SDS		Aminoglutethimide
Highly sulfated cyclosophoraoses (HS-Cys)	SDS		Isosakuranetin, neohesperidin
Carboxymethyl-beta-cyclodextrin (CM-β-CD)	SDS		Raltitrexed (RD)

TABLE 8.3 (continued)
Chiral Separation by CD-MEKC Method

Cyclodextrin	Surfactant	Additives	Separated Enantiomer
Mono-3-*O*-phenylcarbamyl-β-CD	SDS	Methanol	Primaquine, Pemoline
β-CD	SDS		Chiral methotrexate
γ-CD	SDS		Hesperidin, neohesperidin, naringenin, hesperetin, naringin, pinostrobin, isosakuranetin, eriodictyol, homoeriodictyol
HP-β-CD	SDS		Lorazepam
Pesticides and herbicides			
CM-γ-CD	HPMC		Citalopram derivative (CIT, DCIT, DDCIT, CIT-NO, and CIT-PA)
γ-CD	SDS		Doxorubicinol, doxo
HS-γ-CD	Brij 35		Oxazolidinone
HP-γ-CD	SDS	Urea, propanol	PCB
β-CD, γ-CD	SDS	Urea	PCB
α-,β-,γ-CD, hydroxypropyl-β-, dimethyl-β- and trimethyl-β-CD	SDS		Chiral pesticides—organophosphorus, DDT congeners and methyl esters of phenoxy acids
HP-γ-CD, HS-β-CD	SDS		Triadimenol
γ-CD	SDS	Urea, 2-methyl-2-propanol	Uniconazole
γ-CD, DM-β-CD	SDS	Urea, acetonitrile	Diniconazole
HP-β-CD	SDS		Porphyrin and phthalocyanines

Note: CBI, cyanobenzoisoindole; CIT, citalopram; CIT NO, citalopram *N*-oxide; CIT-PA, citalopram propionic acid; DCIT, *N*-desmethylcitalopram; DDCIT, *N,N*-didesmethylcitalopram; FA, fluorescamine; GABA, γ-aminobutyric acid; HPMC, hydroxypropylmethylcellulose.

with LIF detection with high efficiency (800,000 plates/m) and good sensitivity (LOD ~ 40 nM) has been reported [181]. Pumera et al. [182] showed the enantioseparation of *N*-fluorenyl-methoxycarbonyl-L-alanyl *N*-carboxyanhydride using γ-CD in the presence of SDS micelles with a detection limit of 0.2% of the minor enantiomer.

Enantioseparations of seven chiral aryl allenic acids using SDS micelles containing α-, β-, γ-CDs, and hydroxypropyl-CDs (HP-CD) as chiral selectors and 2-propanol as organic additive have been reported by Wang et al. [183]. Method validation including repeatability, linearity, limit of detection (LOD), and limit of quantitation (LOQ) was performed. Determination of enantiomeric excess (ee) as high as 99.65%, was reported for a nonracemic mixture. A CD-MEKC method for enantioseparation of nateglinide with a resolution of 1.68 was developed using HP-β-CD as chiral selector in SDS micellar solution containing 10% v/v *n*-propanol [184].

The applicability of CD-MEKC in enantioseparation of pharmaceutical compounds, e.g., thiopental and thiobarbital using SDS and γ-CD has been shown by Nishi et al. [185]. Enantioseparation of other pharmaceutical compounds was performed using CD-MEKC including cycletanine [186], diniconazol and uniconazol [187], thiazole derivatives [188], hexobarbital, mephobarbital, secobarbital, gluthetimide analogs, and fadrozole [189]. Addition of β-CD to the SDS micelles afforded the enantioseparation of mephenytoin, phenytoin, and their derivatives [190]. Pharmaceuticals, such as β-agonists, β-antagonists, phenylethylamine stimulants, and the antidepressant diclofensine were also enantioseparated by CD-MEKC using charged or uncharged β-CD in SDS micellar systems. Chiral separation of many cationic compounds [106] was obtained by addition of sodium sulfobutylether-β-CD to the SDS micelles. Improved enantioseparations of five phenothiazines obtained by CD-MEKC using β-CD, and HP-β-CD in citrate buffer containing *n*-tetradecyltrimethylammonium bromide (TTAB) as cationic surfactant at low pH have been reported [191]. Zhu et al. [192] obtained enantioseparation of pemoline by CD-MEKC. The enantioseparation was found to be dependent on the type of CDs used as chiral selectors. Enantiomeric separations of ephedrine and salbutamol by MEKC using β-CD (8 mM) and SDS (15 mM) in ammonium chloride-ammonia solution were reported by Zheng and MO [193]. In this work, the analytes were detected by electrochemical detection using a gold microelectrode. Simultaneous separation of methotrexate (MTX) enantiomers and its impurities such as 2,4-diamino-6-(hydroxymethyl)pteridine, aminopterine hydrate, 4-[*N*-(2-amino-4-hydroxy-6-pteridinylmethyl)-*N*-methylamino] benzoic acid, and 4-[*N*-(2,4-diamino- 6-pteridinylmethyl)-*N*-methylamino] benzoic acid using β-CD (45 mM) and SDS (100 mM) mixture in borate buffer (pH 9.3) containing 25% v/v methanol has been reported by Gotti et al. [194]. Both L- and D-MTX were quantified in pharmaceuticals. A CD-MEKC method was developed by Shen and Zhao [195] for the enantioseparation of the cyanobenzoisoindole (CBI) derivative of DOPA using α-, β-, γ-CDs, and HP-CDs in the presence of SDS micelles. The developed method was also employed to control the optical purity of levodopa. Enantioseparation of uniconazole by CD-MEKC method using β-CD (10 mM) and mono-3-*O*-phenylcarbamoyl-β-CD (10 mM) mixture in 50 mM SDS solution (in 50 mM borate buffer, pH 9.5) containing 5% v/v 1-propanol was reported by Zhang and Zhu [196].

Marina et al. [197] reported individual separations of twelve polychlorinated biphenyls (PCBs) into their sterically different forms and simultaneous separations of nine of these compounds within ca. 35 min by the use of a γ-CD-SDS mixed micelle system in 2-(*N*-cyclohexylamino)ethane sulfonic acid (CHES) buffer. A CD-MEKC method for the enantiomeric separation of chiral pesticides and other pollutants in environmental samples has been described recently [198].

8.7 ENANTIOSEPARATIONS USING POLYMERIZED CHIRAL SURFACTANTS

MEKC is characterized by a limited migration time range due to electrophoretic mobility of the PSP. This makes separation of hydrophobic compounds difficult as they have high partition coefficients and tend to have migration times close to t_{mc} with high retention factors. For optimum resolution, the retention factors of the analytes have to be adjusted within the limited migration time range by changing the analytical parameters. For example, addition of organic solvents can alter the retention factors. However, considering micellar stability it is not always possible to alter the retention factors by changing the analytical parameters. This has led many researchers to seek alternative PSPs in MEKC. Several alternative PSPs, such as polymers, proteins, charged cyclodextrins, dendrimers, and polymerized surfactants have been employed to overcome the limitation of surfactant micelles. A huge amount of work has been done in this area since its introduction. Numerous reviews [199–210] have compiled these works and applications of MEKC in related areas.

Among other alternative PSPs, polymerized surfactants (polymer micelles or micelle polymers) have been well studied. The use of polymer micelles overcomes the limitation of the dynamic nature of micelles and provides reasonably wide migration range and better resolution. Since the requirement of self-association at CAC is eliminated, polymerized surfactants with virtually any ionic headgroup structure can in principle be employed to improve separation efficiency and selectivity. The use of polymerized surfactants also minimizes or eliminates detection interferences and Joule heating that is very common with micellar systems. The primary covalent structure and concentration of polymeric PSPs are not affected by changes in the analytical conditions despite changes in solvation conditions, which, however, affect physical structure, conformation, or aggregation behavior of the polymer. This simplifies the optimization of separations because it allows the separation conditions to be selected and adjusted without concern for the stability of the PSP. Indeed, the use of polymerized surfactants has allowed use of BGEs containing high concentrations of organic modifiers for the separation of the hydrophobic compounds.

Polymer micelles have potential, but also real limitations. For example, lower separation efficiency compared to conventional micelles is often observed with polymers of higher polydispersity and/or slower mass transfer kinetics. Further the presence of impurities in polymeric PSPs, which is very common with synthetic polymers, can interfere with detection. Impurities in the polymeric PSPs pose a challenge with UV and fluorescence detection, where very low levels of UV-absorbing or fluorescent impurities would cause large background signals.

Despite limitations, polymer micelles have been successfully used in chiral analysis. Wang and Warner [211] and Dobashi et al. [117] are the first to show enantioseparation using polymerized surfactants. It is worth to mention here that the use of chiral polymerized surfactants, e.g., poly(sodium *N*-undecenoyl-L-valinate), poly-SUV, is an interesting approach in MEKC. Following these publications a significant fraction of effort over the past decade has been made for the development and characterization of chiral polymer micelles [3,27,29,42,43,201,212–222]. An overview of enantioseparations by MEKC using chiral polymerized surfactants is presented in Table 8.4. Chemical structures of representative polymers and their corresponding monomers are shown in Figure 8.5. Different terminologies, such as polymeric (or high-molecular-mass) surfactants, micelle polymers, and molecular micelles, are used in the literature to represent polymeric PSPs. However, a distinction among these should be made. The so-called polymeric surfactants are amphiphilic polymers

TABLE 8.4
Enantiomeric Separations by MEKC Using Molecular Micelles

Polymer	Additives	Separated Enantiomers
Amino acid-based polymers		
Poly-SUV		BNA, BOH, TB, warfarin, norlaudanosoline, laudanosoline, laudanosine, coumachlor, alprenolol, propanolol, GR 24, GR 7
Poly-SUV, poly-SUT		PTH-amino acids
Poly-L-SUL, poly-D-SUL, poly-L-SUA, poly-D-SUA, poly-L-SUV, poly-D-SUV	Hexanol, undecylenyl alcohol, β-CD, γ-CD	BOH, BNP, BNA, PTH-amino acids, (3,5-dinitrobenzoy1)amino acid isopropyl esters, laudanosine, benzoin, benzoin methyl ether (BME), warfarin, and coumachlor, verapamil, aminoglutethimide
Poly-SOLV		BNP, warfarin, coumachlor, benzoin methyl ether, benzoin, hydrobenzoin, temazepam
Poly-SOLV		BNP, BOH, BNA
Poly-SULV, poly-SUVL, poly-SUVV, poly-SUAL, poly-SUSL, poly-SUTL, poly-SUILV, poly-SULL, poly-SULG, poly-SUGL	Ionic liquids (EMIMBF4, EMIMPF6, BMIMBF$_4$)	BNP, BOH, BNA, TB, temazepam, oxazepam, lorazepam, laudanosoline, norlaudanosoline, laudanosine, oxprenolol, alprenolol, propanolol, aminoglutethimide, 2,2,2-trifluro-1-(9-anthryl)ethanol
Poly-SUE, poly-SUEE, poly-SUEM, poly-SUETB		
Alkenoxy amino acid-based polymer		
Poly-SUCL		BNP, BOH, BNA temazepam, oxazepam, lorazepam
Poly-SUCIL		
Copolymers		
Poly-(SUS-*co*-SUL)		BNP, BOH, BNA, atenolol, metoprolol, pindolol, oxprenolol, alprenolol, and propranolol, carteolol, talinolol
		BNP, BOH, BNA, flunitrazepam, nitrazepam, clonazepam, temazepam, diazepam, oxazepam, lorazepam

which have been synthesized in a conventional way and which show surface activity due to the formation of macromolecular aggregates in water. On the other hand, micelle polymers are synthesized in micellar form of the surfactants that maintain micellar structure after polymerization. The polymerized surfactants with carboxylate headgroup are also called polysoaps. It is well accepted that the polymerized surfactant adopts a micelle-like spherical shape in aqueous solution as shown in

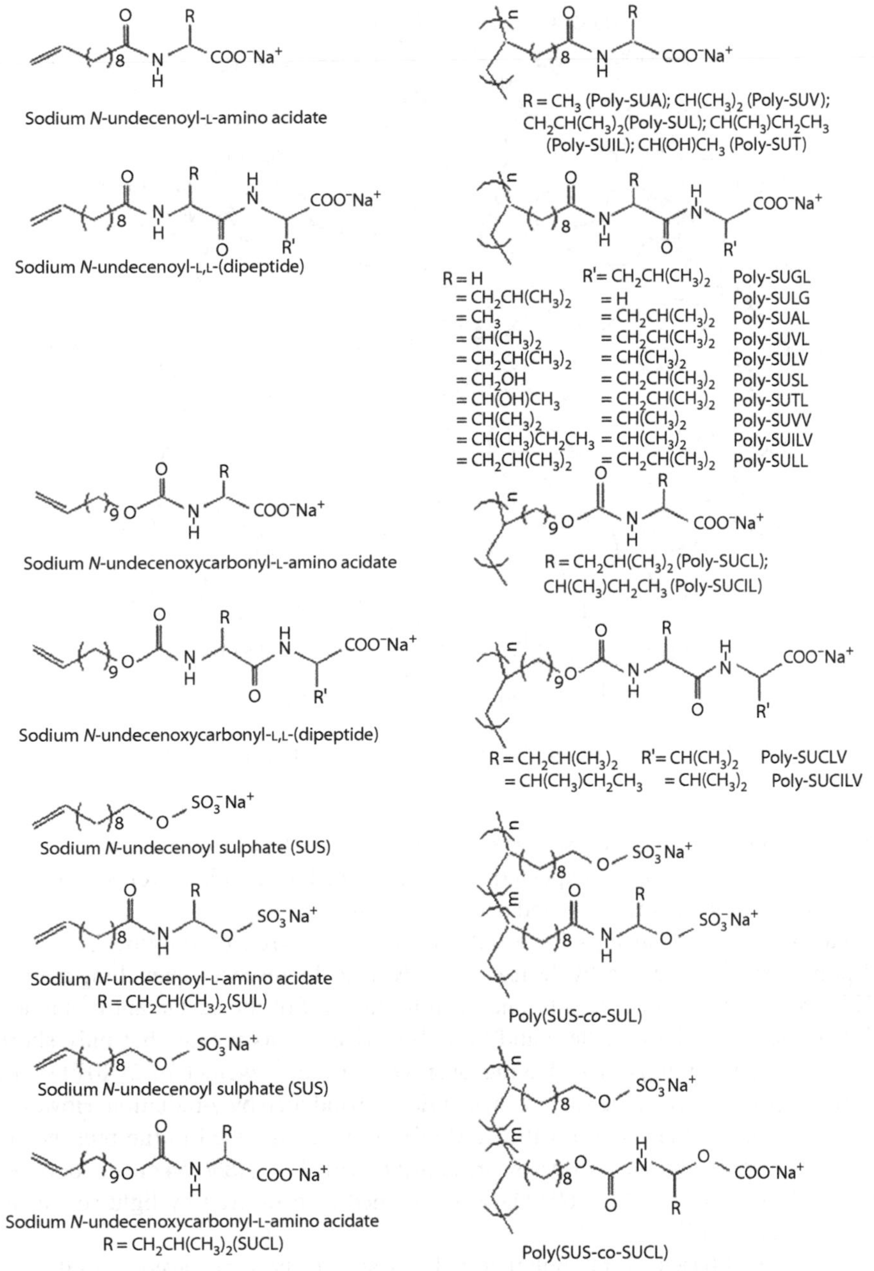

Sodium *N*-undecenoyl-L-amino acidate

$R = CH_3$ (Poly-SUA); $CH(CH_3)_2$ (Poly-SUV); $CH_2CH(CH_3)_2$(Poly-SUL); $CH(CH_3)CH_2CH_3$ (Poly-SUIL); $CH(OH)CH_3$ (Poly-SUT)

Sodium *N*-undecenoyl-L,L-(dipeptide)

R	R'	
$R = H$	$R' = CH_2CH(CH_3)_2$	Poly-SUGL
$= CH_2CH(CH_3)_2$	$= H$	Poly-SULG
$= CH_3$	$= CH_2CH(CH_3)_2$	Poly-SUAL
$= CH(CH_3)_2$	$= CH_2CH(CH_3)_2$	Poly-SUVL
$= CH_2CH(CH_3)_2$	$= CH(CH_3)_2$	Poly-SULV
$= CH_2OH$	$= CH_2CH(CH_3)_2$	Poly-SUSL
$= CH(OH)CH_3$	$= CH_2CH(CH_3)_2$	Poly-SUTL
$= CH(CH_3)_2$	$= CH(CH_3)_2$	Poly-SUVV
$= CH(CH_3)CH_2CH_3$	$= CH(CH_3)_2$	Poly-SUILV
$= CH_2CH(CH_3)_2$	$= CH_2CH(CH_3)_2$	Poly-SULL

Sodium *N*-undecenoxycarbonyl-L-amino acidate

$R = CH_2CH(CH_3)_2$ (Poly-SUCL); $CH(CH_3)CH_2CH_3$ (Poly-SUCIL)

Sodium *N*-undecenoxycarbonyl-L,L-(dipeptide)

R	R'	
$R = CH_2CH(CH_3)_2$	$R' = CH(CH_3)_2$	Poly-SUCLV
$= CH(CH_3)CH_2CH_3$	$= CH(CH_3)_2$	Poly-SUCILV

Sodium *N*-undecenoyl sulphate (SUS)

Sodium *N*-undecenoyl-L-amino acidate
$R = CH_2CH(CH_3)_2$(SUL)

Poly(SUS-*co*-SUL)

Sodium *N*-undecenoyl sulphate (SUS)

Sodium *N*-undecenoxycarbonyl-L-amino acidate
$R = CH_2CH(CH_3)_2$(SUCL)

Poly(SUS-co-SUCL)

FIGURE 8.5 Chemical structures of representative anionic surfactant monomers and corresponding homo- and copolymers employed in chiral MEKC separations.

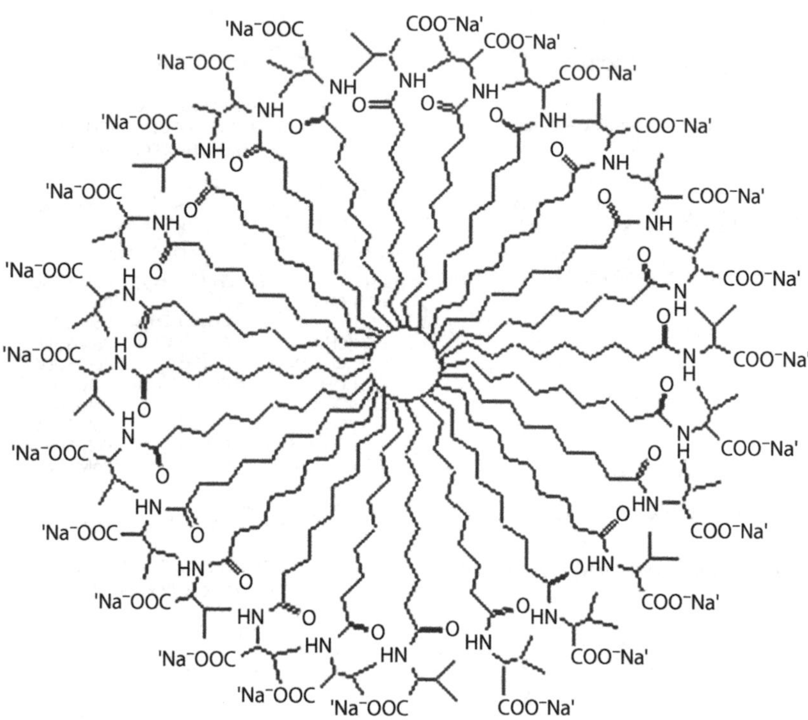

FIGURE 8.6 Structure of the molecular micelle formed by poly(sodium undecenoyl-L-valinate).

Figure 8.6 and therefore they are called "molecular micelles." However, such conformation of the polymer is not expected to be maintained in organic solvents or organic rich aqueous-organic mixtures and needs to be investigated.

Various research groups have determined the degree of polymerization of 15 polymerized surfactants by the use of steady-state fluorescence quenching method [223–230]. They found that only one third to one half of the surfactant monomers [231] constituting the micelle could be polymerized. This means that only short-chain polymers or oligomers having average molecular weight of 7–10 kDa are formed especially when the surfactant solution is irradiated by γ-radiation. However, when sodium *N*-undecenoyl-L-valinate (SUV) was polymerized in the presence of 25 mM phosphate-borate buffer by irradiation with UV (185/254 nm) a polymer (poly-SUV) of average $M_w \sim 69$ kDa was obtained as measured by light scattering measurements [117].

Amino acid-based chiral polymerized surfactants have received considerable attention over the past several years, and a number of studies using these polymers were reported in the literature [232,233]. These include studies to modify mainly the headgroup structure of polymerizable surfactants to improve selectivity. The effects of hydrophobicity and steric hindrance of the ionic headgroup as well as counter-ions have been investigated. Warner's group [233] have evaluated and compared chiral selectivities of 18 amino acid-based chiral surfactants and their polymeric counterparts through enantioseparations of lorazepam, temazepam, propanolol, and BOH by MEKC. The methyl, ethyl, and *t*-butyl esters of glutamic acid-derived polymer, poly-SUG, were synthesized to vary the hydrophobicity, steric hindrance, and orientation of the amino acid side chain [234]. Changes in the orientation or ionization of the headgroup were found to enhance chiral separation of most of the analytes employed. On the other hand, steric hindrance of the glutamic acid side chain and bulky *t*-butyl group either reduced or completely destroyed chiral resolution of some analytes. Interestingly, the *t*-butyl group was found to be essential for the chiral separation of the other compounds. Thibodeaux et al. [235] synthesized polymers with proline, valine, norvaline, leucine, norleucine, isoleucine, and *t*-butyl leucine as headgroups and used them for enantioseparations by MEKC. The authors noted lower chiral resolution due to the steric hindrance of the side chain of norleucine and *t*-butyl leucine, while side chains containing one or two carbons did not significantly affect the resolution. Polymers of alkenoxy amino acid-based surfactants, poly(sodium *N*-undecenoxycarbonyl-L-leucinate) (poly-SUCL) and poly(sodium *N*-undecenoxycarbonyl-L-isoleucinate) (poly-SUCIL) have been developed and employed for enantioseparations of binaphthyl derivatives [236], β-blockers [237], and β-blockers with two stereogenic centers [238] by Shamsi and coworkers. A difference in chiral recognition mechanism was noted. The polymer with isoleucine headgroup provided better resolution for β-blockers with two stereogenic centers. Chiral resolution of binaphthyl derivatives was nearly equivalent with both polymers while poly-SUCL provided overall better resolution and a wider migration window.

In order to examine the effects of steric hindrance of the surfactant headgroup, and number and position of chiral centers in the surfactant molecule on the separation performance, chiral separations of seven organic racemates (three benzoin derivatives, laudanosine, norlaudanosoline, laudanosoline, and chlorthalidone) using eighteen polymeric single amino acid and dipeptide surfactants were studied [239]. For six of the seven analytes, dipeptide-derived polymeric surfactants provided better chiral separation than polymeric single amino acid surfactants. Steric factors were also found to be important in determining the chiral separation performance. In general, chiral selectors with more sterically hindered environments provided better chiral selectivity for less sterically hindered analytes, while chiral selectors with less steric hindrance provided better selectivity for analytes with sterically hindered chiral centers. Mwongela et al. [240] synthesized a new chiral polymeric surfactant, poly(sodium *N*-oleyl-L-leucylvalinate) (poly-SOLV) for chiral separations of BOH, BNA, BNP, benzoin, benzoin derivatives, warfarin, and coumachlor. The selectivity of the polymer was compared with that of the corresponding monomeric surfactant. The higher hydrophobicity of the surfactant was found to provide better chiral recognition for hydrophobic compounds. The superior performance of the polymers with L-leucyl-valinate headgroup in chiral MEKC was demonstrated by an extensive study in which the resolution of 58 out of 75 neutral, anionic, and cationic racemates was achieved [241].

The effects of polymer polydispersity and alkyl chain length on the performance and chiral resolution were also studied. Tarus et al. [242] polymerized micellar solutions of sodium *N*-undecenoyl-L-leucinate (SUL) in the presence of alcohols, such as hexanol and undecenyl alcohol. The polymers thus obtained were found to be larger compared to those obtained by polymerization in pure aqueous medium. Both types of poly-SUL were employed for enantioseparations of BNP, and coumarinic and benzoin derivatives. The polymer obtained in the presence of alcohols was found to exhibit remarkable resolution for coumarinic derivatives. In another work, the authors [243] used poly-SUL in the presence of various concentrations of Triton X-102, which is a nonionic surfactant. The effects of molecular weight and polydispersity of poly-SUL on various properties and EKC separation of PTH-amino acids and coumarinic derivatives were studied by the use of three molecular weight fractions of poly-SUL [244]. The EKC performance of the individual fractions was observed to be better than that of the polydisperse unfractionated material, with the fractions often providing better separation efficiency. However, since the results are analyte dependent and since molecular weight often has a more dramatic effect on efficiency than polydispersity, the efficiency results do not demonstrate conclusively that polydispersity is the most important factor affecting efficiency.

In order to investigate chiral micelle–analyte interactions, Warner's group has performed fluorescence and NMR experiments in combination with CE [244,245]. They demonstrated that a strong correlation exists between the selectivity factor (α) and the ratio (r_R/r_S) of fluorescence anisotropy values of *R*- and *S*-isomer of the probe analyte. In a more recent work, using four diastereomeric molecular micelles of poly(sodium *N*-undecenoyl-L-leucylvalinate), poly-SULV as chiral selectors, Warner and coworkers showed that depending on the molecular structure of the analytes the primary site of interaction is either the leucine or valine moiety of the chiral headgroup of the molecular micelles [91].

Polymers of sodium *N*-alkenoxycarbonyl-L-leucinate surfactants with C8–C11 chains were evaluated as PSPs for the chiral separation of seven β-blockers of different hydrophobicity [246]. However, only the sodium *N*-octenoxycarbonyl-L-leucinate polymer provided good simultaneous chiral resolution of all seven β-blockers. The C11 monomer with a terminal triple bond was also polymerized to obtain a more rigid polymer containing unsaturation (double bond) in the polymer chain. The double bond containing polymer poly(sodium *N*-undecynoxycarbonyl-L-leucinate) exhibited slightly better or similar chiral resolution and efficiency than did poly(sodium *N*-undecenoxycarbonyl-L-leucine). It was demonstrated that the hydrophilic–lipophilic balance (HLB) of polymeric PSPs is an important factor in determining performance.

For enhancement of aqueous solubility in acidic pH, Akbey et al. [247,248] synthesized copolymers from sodium 10-undecenoyl sulfate and sodium *N*-undecenoyl-L-leucinate (SUL) at different molar ratios. Physicochemical characterization of the copolymers was also performed. The copolymers were employed for chiral as well as achiral solutes, e.g., alkylphenyl ketones, and benzodiazepines. Each copolymer was observed to have different selectivities for the three biphenyl derivatives employed. More recently, chiral selectivities of six L-leucine- and L-isoleucine-derived polymeric surfactants with carboxylate and sulfate headgroups were compared by Rizvi and Shamsi [249]. Enanioseparations of several groups of organic racemates (e.g., phenylethylamines (PEAs), β-blockers, benzoin derivatives, PTH-amino acids, and benzodiazepines) were obtained under neutral and basic pH conditions. For the three classes of PEAs, the polymers with carboxylate headgroup provided better separations compared to those with sulfate headgroup. However, increasing the number of hydroxyl groups in the phenyl ring of the PEAs caused deterioration of the resolution. Among carboxylate polymers poly(sodium undecenoyl-L-isoleucinate), poly-SUIL, showed the best enantioseparation capability. However, among polymers with sulfate headgroups both poly(sodium *N*-undecenoxycarbonyl-L-isoleucine sulfate) and poly(sodium *N*-undecenoxycarbonyl-L-valine sulfate) proved to be equally effective for enantioseparation. A report summarizing the chiral separations by MEKC using polymeric chiral surfactants has been published recently [250].

8.8 MEKC WITH MASS SPECTROMETRIC DETECTION

Several ways of coupling CE and electrospray ionization-mass spectrometry (ESI–MS) for analysis of chiral compounds are discussed in a review by Shamsi [251] and also in Chapter 13 of this book. EKC–MS is an important mode in which a charged chiral selector, such as sulfated β-CD or vancomycin, is employed. On the other hand, the MEKC–MS mode that uses conventional chiral surfactant molecules, which are highly surface active and nonvolatile, was found not to be suitable for chiral analysis. However, coupling of MEKC and MS became feasible with the advent of molecular micelles (i.e., polymeric surfactants) [252]. For (1) their high molecular weight compared to conventional surfactants, which eliminates interference in the mass region of interest, and (2) less interference with the ionization process than conventional surfactants, the polymeric PSPs can be used with mass spectrometric detection. Akbay et al. [253] reported the simultaneous enantioseparation and ESI–MS detection of eight weakly basic β-blockers using poly-SUCL as the PSP. The detection and separation parameters were systematically and sequentially optimized to yield stable, high-resolution separations, and good signal-to-noise ratio (S/N). Comparison of the optimal performance of SUCL and poly-SUCL showed that the enantioresolution, efficiency, and S/N are better with the polymeric surfactant. However, a different elution order for each pair of enantiomers was observed with the polymer. In another study, the same group investigated three undecenoxycarbonyl-amino acids and three sodium *N*-undecenoyl-amino acid-derived polymerized surfactants with one or two amino acids in the headgroup for the chiral EKC separation and ESI–MS detection of two benzodiazepines and one benzoxazocine [254]. The effect of organic modifier was investigated using poly-SUL. Though addition of methanol, acetonitrile, or 2-propanol to the BGE at 10%–20% improved the chiral resolution and the electrospray stability, baseline resolution was not achieved with all enantiomers. However, systematic method development using the dipeptide containing polymer (poly-SUCLV) resulted in sufficiently resolved peaks. An interesting observation that was made in this study was that various surfactant types were found to affect the S/N ratio with dipeptide-based PSPs generally giving higher S/N ratio for all analytes.

More recently, Rizvi et al. [255] employed polymerized sulfated amino acid surfactants in MEKC–MS for the chiral separation of a range of analytes including β-blockers, phenoxypropionic acids, benzodiazepines, benzoin derivatives, and PTH-amino acids. The sensitivity of MS detection was demonstrated with all the separations. In another report, Shamsi's group compared MEKC–UV separations with those obtained by MEKC–ESI–MS. The use of poly-SUCL in MEKC–MS provided significantly higher sensitivity for the stereoisomers of ephedrines and related compounds in comparison to MEKC–UV. In the same year, these authors also employed MEKC–MS for the separation, identification, and quantitation of 10 enantiomers of ephedrine and related compounds [256]. The separations were obtained using poly-SUCL in 30% v/v acetonitrile. Complete method validation was performed and the method was employed for three standard reference materials. The same authors [257] also achieved enantioseparations of warfarin and coumachlor drugs employing MEKC–MS using poly-SULV as chiral PSP. The method was also successfully used to determine the ratio of the enantiomers of warfarin in plasma samples of patients' undergoing warfarin therapy. A review [250] of their work on chiral analysis using MEKC–MS has been recently published.

REFERENCES

1. Cammileri, P. 1997. *Capillary Electrophoresis: Theory and Practice*, 2nd edition. Boca Raton, FL: CRC press.
2. Pyell, U. 2006. *Electrokinetic Chromatography. Theory, Instrumentation and Applications*. West Sussex, U.K.: John Wiley & Sons.
3. Terabe, S, Otsuka, K, Ichikawa, A, Tsuchiya, A, Ando, T. 1984. Electrokinetic separations with micellar solutions and open-tubular capillaries. *Anal Chem* 56: 111–113.
4. Terabe, S, Otsuka, K, Nishi, H. 1994. Separation of enantiomers by capillary electrophoretic techniques. *J Chromatogr A* 666: 295–319.
5. Nishi, H, Terabe, S. 1995. Optical resolution of drugs by capillary electrophoretic techniques. *J Chromatogr A* 694: 245–276.
6. Nishi, H, Terabe, S. 1996. Micellar electrokinetic chromatography: Perspectives in drug analysis. *J Chromatogr A* 735: 3–27.
7. Nishi, H. 1996. Enantiomer separation of drugs by electrokinetic chromatography. *J Chromatogr A* 735: 57–76.
8. Fanali, S. 1996. Identification of chiral drug isomers by capillary electrophoresis. *J Chromatogr A* 735: 77–121.
9. Bressolle, F, Audran, M, Pham, TN, Jean, JV. 1996. Cyclodextrins and enantiomeric separation of drugs by liquid chromatography and capillary electrophoresis: Basic principles and new developments. *J Chromatogr B* 687: 303–336.
10. Riekkola, ML, Wiedmer, SK, Valkó, IE, Sirén, H. 1997. Selectivity in capillary electrophoresis in the presence of micelles, chiral selectors and non-aqueous media. *J Chromatogr A* 792: 13–35.
11. Williams, CC, Shamsi, SA, Warner, IM. 1997. Chiral micelle polymers for chiral separations in capillary electrophoresis. *Adv Chromatogr* 37: 363–423.
12. Marina, ML, Crego, AL. 1997. Capillary electrophoresis: A good alternative for the separation of chiral compounds of environmental interest. *J Liq Chromatogr Rel Technol* 20: 1337–1365.
13. Shamsi, SA, Warner, IM. 1997. Central composite design in the chiral analysis of amphetamines by capillary electrophoresis. *Electrophoresis* 18: 933–944.
14. Gubitz, G, Schmid, MG. 1997. Chiral separation principles in capillary electrophoresis. *J Chromatogr A* 792: 179–225.
15. Waetzig, H, Degenhardt, M, Kunkel, A. 1998. Strategies for capillary electrophoresis: Method development and validation for pharmaceutical and biological applications. *Electrophoresis* 19: 2695–2752.
16. Verleysen, K, Sandra, P. 1998. Separation of chiral compounds by capillary electrophoresis. *Electrophoresis* 19: 2798–2833.

17. Chankvetadze, B. 1997. *Capillary Electrophoresis in Chiral Analysis*. Chichester, U.K.: Wiley.
18. Vespalec, R, Boček, P. 1999. Chiral separations in capillary electrophoresis. *Electrophoresis* 20: 2579–2591.
19. Chankvetadze, B, Blaschke, G. 1999. Selector-select and interactions in chiral capillary electrophoresis. *Electrophoresis* 20: 2592–2604.
20. Otsuka, K, Terabe, S. 2000. Enantiomer separation of drugs by micellar electrokinetic chromatography using chiral surfactants. *J Chromatogr A* 875: 163–178.
21. Wan, H, Blomberg, LG. 2000. Chiral separation of amino acids and peptides by capillary electrophoresis. *J Chromatogr A* 875: 43–88.
22. Ward, TJ. 2000. Chiral separations. *Anal Chem* 72: 4521–4528.
23. Santoro, MIRM, Prado, MSA, Steppe, M, Kedor-Hackmann, ERM. 2000. Capillary electrophoresis: Theory and applications in drug analysis [Eletroforese capilar: Teoria e aplicacoes na analise de medicamentos]. *Rev Bras Cienc Farm* 36: 97–110.
24. De Boer, T, De Zeeuw, RA, De Jong, GJ, Ensing, K. 2000. Recent innovations in the use of charged cyclodextrins in capillary electrophoresis for chiral separations in pharmaceutical analysis. *Electrophoresis* 21: 3220–3239.
25. Vespalec, R, Bocek, P. 2000. Chiral separations in capillary electrophoresis. *Chem Rev* 100: 3715–3753.
26. Haginaka, J. 2000. Enantiomer separation of drugs by capillary electrophoresis using proteins as chiral selectors. *J Chromatogr A* 875: 235–254.
27. Yarabe, HH, Billiot, E, Warner, IM. 2000. Enantiomeric separations by use of polymeric surfactant electrokinetic chromatography. *J Chromatogr A* 875: 179–206.
28. El Rassi, Z. 2000. Chiral glycosidic surfactants for enantiomeric separation in capillary electrophoresis. *J Chromatogr A* 875: 207–233.
29. Palmer, CP. 2000. Polymeric and polymer-supported pseudostationary phases in micellar electrokinetic chromatography: Performance and selectivity. *Electrophoresis* 21: 4054–4072.
30. Gübitz, G, Schmid, MG. 2000. Recent progress in chiral separation principles in capillary electrophoresis. *Electrophoresis* 21: 4112–4135.
31. Raggi, MA, Pucci, V. 2005. Analysis of antiepileptic drugs in biological fluids by means of electrokinetic chromatography. *Electrophoresis* 26: 767–782.
32. Amini, A. 2001. Recent developments in chiral capillary electrophoresis and applications of this technique to pharmaceutical and biomedical analysis. *Electrophoresis* 22: 3107–3130.
33. Rizzi, A. 2001. Fundamental aspects of chiral separations by capillary electrophoresis. *Electrophoresis* 22: 3079–3106.
34. Chankvetadze, B, Blaschke, G. 2001. Enantioseparations in capillary electromigration techniques: recent developments and future trends. *J Chromatogr A* 906: 309–363.
35. Stalcup, AM. 2001. In *Chiral Separation Techniques*, 2nd edition. Subramanian, G. (Ed.), pp. 287–298. Weinheim, Germany: Wiley-VCH.
36. Molina, M, Silva, M. 2002. Micellar electrokinetic chromatography: Current developments and future. *Electrophoresis* 23: 3907–3921.
37. Ward, TJ. 2002. Chiral separations. *Anal Chem* 74: 2863–2872.
38. Scriba, GKE. 2003. Pharmaceutical and biomedical applications of chiral capillary electrophoresis and capillary electrochromatography: An update. *Electrophoresis* 24: 2409–2421.
39. Patel, B, Hanna-Brown, M, Hadley, MR, Hutt, AJ. 2005. Enantiomeric resolution of 2-arylpropionic acid nonsteroidal anti-inflammatory drugs by capillary electrophoresis: Methods and applications. *Electrophoresis* 25: 2625–2656.
40. Gübitz, G, Schmid, MG. 2004. Recent advances in chiral separation principles in capillary electrophoresis and capillary electrochromatography. *Electrophoresis* 25: 3981–3996.
41. Ward, TJ, Hamburg, D-M. 2004. Chiral separations. *Anal Chem* 76: 4635–4644.

42. Palmer, CP, McCarney, JP. 2004. Recent progress in the use of soluble ionic polymers as pseudostationary phases for electrokinetic chromatography. *Electrophoresis* 25: 4086–4094.
43. Holland, LA, Gayto-Ely, M, Pappas, TJ. 2005. Recent advances in micellar electrokinetic chromatography. *Electrophoresis* 26: 719–734.
44. Poinsot, V, Lacroix, M, Maury, D, Chataigne, G, Feurer, B, Couderc, F. 2006. Recent advances in amino acid analysis by capillary electrophoresis. *Electrophoresis* 27: 176–194.
45. Palmer, CP. 2007. Recent progress in the use of ionic polymers as pseudostationary phases for EKC. *Electrophoresis* 28: 164–173.
46. Kahle, KA, Foley, JP. 2007. Review of aqueous chiral electrokinetic chromatography (EKC) with an emphasis on chiral microemulsion EKC. *Electrophoresis* 28: 2503–2526.
47. Israelachvili, J. 1991. *Intermolecular and Surface Forces*, 2nd edition. London: Academic Press.
48. Rosen, MJ. 2004. *Surfactants and Interfacial Phenomena*. Hoboken, NJ: Wiley-Interscience.
49. Laughlin, RG. 1994. *The Aqueous Phase Behavior of Surfactants*. London, U.K.: Academic Press.
50. Fendler, JH. 1982. *Membrane Mimetic Chemistry*. New York: Wiley.
51. Kunitake, T. 1992. Synthetic bilayer membranes: Molecular design, self-organization, and application. *Angew Chem Int Ed Engl* 31: 709–726.
52. Kunitake, T, Okahata, Y, Shinomura, M, Yasunami, S, Takarabe, K. 1981. Formation of stable bilayer assemblies in water from single-chain amphiphiles. Relationship between the amphiphile structure and the aggregate morphology. *J Am Chem Soc* 103: 5401–5413.
53. Bloechliger, E, Blocher, M, Walde, P, Luisi, PL. 1998. Matrix effect in the size distribution of fatty acid vesicles. *J Phys Chem B* 102: 10383–10390.
54. Zhang, Y-J, Jin, M, Lu, R, Song, Y, Jiang, L, Zhao, Y, Li, TJ. 2002. Interfacial-dependent morphologies in the self-organization system of chiral molecules observed by atomic force microscopy. *J Phys Chem B* 106: 1960–1967.
55. Boettcher, C, Schade, B, Fuhrhop, J-H. 2001. Comparative cryo-electron microscopy of noncovalent N-dodecanoyl-(D- and L-) serine assemblies in vitreous toluene and water. *Langmuir* 17: 873–877.
56. Borocci, S, Mancini, G, Cerichelli, G, Luchetti, L. 1999. Conformational behavior of aqueous micelles of sodium N-dodecanoyl-L-prolinate. *Langmuir* 15: 2627–2630.
57. Vollhardt, D, Gehlert, U. 2002. Chiral discrimination in 1-stearylamine-glycerol monolayers. *J Phys Chem B* 106: 4419–4423.
58. Kawasaki, H, Souda, M, Tanaka, S, Nemoto, N, Karlsson, G, Almgren, M, Maeda, H. 2002. Reversible vesicle formation by changing pH. *J Phys Chem B* 106: 1524–1527.
59. Gebicki, JM, Hicks, M. 1976. Preparation and properties of vesicles enclosed by fatty acid membranes. *Chem Phys Lipids* 16: 142–160.
60. Hargreaves, WR, Deamer, DW. 1978. Liposomes from ionic, single-chain amphiphiles. *Biochemistry* 17: 3759–3767.
61. Gebicki, JM, Hicks, M. 1973. Ufasomes are stable particles surrounded by unsaturated fatty acid membranes. *Nature* 243: 232–234.
62. Marques, EF. 2000. Size and stability of catanionic vesicles: effects of formation path, sonication, and aging. *Langmuir* 16: 4798–4807.
63. Bergström, M. 1996. Thermodynamics of vesicle formation from a mixture of anionic and cationic surfactants. *Langmuir* 12: 2454–2463.
64. Caillet, C, Hebrant, M, Tondre, C. 2000. Sodium octyl sulfate/cetyltrimethylammonium bromide catanionic vesicles: Aggregate composition and probe encapsulation. *Langmuir* 16: 9099–9102.
65. Safran, SA, Pincus, P, Andelman, D. 1990. Theory of spontaneous vesicle formation in surfactant mixtures. *Science* 248: 354–356.

66. Owen, RL, Strasters, JK, Breyer, ED. 2005. Lipid vesicles in capillary electrophoretic techniques: Characterization of structural properties and associated membrane-molecule interactions. *Electrophoresis* 26: 735–751.
67. (a) Diego-Castro, MJ, Hailes, HC. 1998. Novel application of chiral micellar media to the Diels-Alder reaction. *Chem Commun* 15: 1549–1540. (b) Zhang, Y, Sun, P. 1996. The asymmetric induction and catalysis of chiral reverse micelle: Asymmetric reduction of prochiral ketones. *Tetrahedron: Asymmetry* 7: 3055–3058.
68. Miyagishi, S, Nishida, M. 1978. Influence of chirality on micelle formation of sodium *N*-acylalanates and sodium *N*-lauroylvalinates. *J Colloid Interface Sci* 65: 380–386.
69. Miyagishi, S, Higashide, M, Asakawa, T, Nishida, M. 1991. Surface tensions and critical micelle concentrations in mixed solutions of potassium perfluorononanoylalaninate and potassium acylalaninates. *Langmuir* 7: 51–55.
70. Miyagishi, S, Asakawa, T, Nishida, M. 1989. Hydrophobicity and surface activities of sodium salts of *N*-dodecanoyl amino acids. *J Colloid Interface Sci* 131: 68–73.
71. Miyagishi, S, Ishibai, Y, Asakawa, T, Nishida, M. 1985. Critical micelle concentration in mixtures of *N*-acyl amino acid surfactants. *J Colloid Interface Sci* 103: 164–169.
72. Shinitsky, M, Haimovitz, R. 1993. Chiral surfaces in micelles of enantiomeric N-palmitoyl- and N- stearoylserine. *J Am Chem Soc* 115: 12545–12549.
73. Fuhrhop, JH, Schnieder, P, Rosenberg, J, Boekema, E. 1987. The chiral bilayer effect stabilizes micellar fibers. *J Am Chem Soc* 109: 3387–3390.
74. Zhang, L, Lu, Q, Liu, M. 2003. Fabrication of chiral Langmuir-Schaefer films from achiral TPPS and amphiphiles through the adsorption at the air/water interface. *J Phys Chem B* 107: 2565–2569.
75. Nakashima, N, Asakuma, S, Kunitake, T. 1985. Optical microscopic study of helical superstructures of chiral bilayer membranes. *J Am Chem Soc* 107: 509–510.
76. Blanzat, M, Massip, S, Spéziale, V, Perez, E, Rico-Lattes, I. 2001. *Langmuir* 17: 3512–3514.
77. Fuhrhop, J-H, Helfrich, W. 1993. Fluid and solid fibers made of lipid molecular bilayers. *Chem Rev* 93: 1565–1582.
78. Emmanouil, V, El Ghoul, M, Andre-Barres, C, Guidetti, B, Rico-Lattes, I, Lattes, A. 1998. Synthesis of new long-chain fluoroalkyl glycolipids: Relation of amphiphilic properties to morphology of supramolecular assemblies. *Langmuir* 14: 5389–5396.
79. Köning, J, Boettcher, C, Winkler, H, Zeitler, E, Talmon, Y, Furhop, J-H. 1993. Magic angle (54.7°) gradient and minimal surfaces in quadruple micellar helices. *J Am Chem Soc* 115: 693–700.
80. Mohanty, A, Dey, J. 2004. Spontaneous formation of vesicles and chiral self-assemblies of sodium N-(4-dodecyloxybenzoyl)-L-valinate in water. *Langmuir* 20: 8452–8459.
81. Roy,S,Dey,J.2005.SpontaneouslyformedvesiclesofsodiumN-(11-acrylamidoundecanoyl)-glycinate and L-alaninate in water. *Langmuir* 21: 10362–10369.
82. Roy, S, Dey, J. 2006. Self-organization properties and microstructures of sodium N-(11-acrylamidoundecanoyl)-L-valinate, and L-threoninate in water. *Bull Chem Soc Jpn* 79: 59–66.
83. Mohanty, A, Dey, J. 2007. Self-assembly formation of sodium N-(4-alkoxybenzoyl)-L-aminoacidates: Effects of chain length and headgroup structure. *Langmuir* 23: 1033–1040.
84. Roy, S, Dey, J. 2007. Effect of hydrogen-bonding interaction of the amino acid side chain on self-assembly formation of sodium N-(11-acrylamidoundecanoyl)-L-asparaginate, -L-glutaminate, and -L-serinate. *J Colloid and Interface Sci* 307: 229–234.
85. Morigaki, K, Dallavalle, S, Walde, P, Colonna, S, Luisi, PL. 1997. Autopoietic self-reproduction of chiral fatty acid vesicles. *J Am Chem Soc* 119: 292–301.
86. Otsuka, K, Terabe, S, Ando, T. 1985. Electrokinetic chromatography with micellar solutions: Retention behaviour and separation of chlorinated phenols. *J Chromatogr A* 348: 39–47.

87. Khaledi, MG, Smith, SC, Strasers, JK. 1991. Micellar electrokinetic capillary chromatography of acidic solutes: Migration behavior and optimization strategies. *Anal Chem* 63: 1820–1830.
88. Bushey, MM, Jorgenson, JW. 1989. Effects of methanol-modified mobile phase on the separation of isotopically substituted compounds by micellar electrokinetic capillary chromatography. *J Microcol Sep* 1: 125–130.
89. Vindevogel, J, Sandra, P. 1992. *Introduction to Micellar Electrokinetic Chromatography*. Heidelberg, Germany: Hüthig.
90. Kuzdal, SA, Hagen, JJ, Monnig, CA. 1995. Simultaneous determination of the electroosmotic flow and pseudo-stationary phase migration time in electrokinetic capillary chromatography. *J High Resolut Chromatogr* 18: 439–442.
91. Valle, BC, Morris, KF, Fetcher, KA, Fernand, V, Sword, DM, Eldridge, S, Larive, CK, Warner, IM. 2007. Understanding chiral molecular micellar separations using steady-state fluorescence anisotropy, capillary electrophoresis, and NMR. *Langmuir* 23: 425–435.
92. Terabe, S, Otsuka, K, Ando, T. 1985. Electrokinetic chromatography with micellar solution and open-tubular capillary. *Anal Chem* 57: 834–841.
93. Terabe, S. 1992. Selectivity manipulation in micellar electrokinetic chromatography. *J Pharm Biomed Anal* 10: 705–715.
94. Eckhardt, CJ, Peachey, NM, Swanson, DR, Takacs, JM, Khan, MA, Gong, X, Kim, J-H, Wang, J, Uphaus, RA. 1993. Separation of chiral phases in monolayer crystals of racemic amphiphiles. *Nature* 362: 614–616.
95. Gassmann, E, Kuo, JE, Zare, RN. 1985. Electrokinetic separation of chiral compounds. *Science* 230: 813–814.
96. Terabe, S, Shibata, M, Miyashita, Y. 1989. Chiral separation by electrokinetic chromatography with bile salt micelles. *J Chromatogr A* 480: 403–411.
97. Wistuba, D, Trapp, O, Gel-Moreto, N, Galensa, R, Schurig, V. 2006. Stereoisomeric separation of flavanones and flavanone-7-o-glycosides by capillary electrophoresis and determination of interconversion barriers. *Anal Chem* 78: 3424–3433.
98. Hefnawy, MM. 2005. Optimization of the chiral resolution of metyrosine by capillary electrophoresis and/or micellar electrokinetic capillary chromatography. *J Liq Chromatogr Relat Technol* 28: 439–452.
99. Tian, K, Chen, H, Tang, J, Chen, X, Hu, Z. 2006. Enantioseparation of palonosetron hydrochloride by micellar electrokinetic chromatography with sodium cholate as chiral selector. *J Chromatogr A* 1132: 333–336.
100. Wren, S. 2001. Other chiral selectors. *Chromatographia* 54 (Suppl. 1): 78–92.
101. Nishi, H, Fukuyama, T, Matsuo, M, Terabe, S. 1989. Chiral separation of optical isomeric drugs using micellar electrokinetic chromatography and bile salts. *J Microcol Sep* 1: 234–241.
102. Cole, RO, Sepaniak, MJ, Hinze, WL. 1990. Optimization of binaphthyl enantiomer separations by capillary zone electrophoresis using mobile phases containing bile salts and organic solvent. *J High Res Chromatogr* 13: 579–582.
103. Nishi, H, Fukuyama, T, Matsuo, M, Terabe, S. 1990. Chiral separation of trimetoquinol hydrochloride and related compounds by micellar electrokinetic chromatography using sodium taurodeoxycholate solutions and application to optical purity determination. *Anal Chim Acta* 236: 281–286.
104. Nishi, H, Fukuyama, T, Matsuo, M, Terabe, S. 1990. Chiral separation of diltiazem, trimetoquinol and related compounds by micellar electrokinetic chromatography with bile salts. *J Chromatogr A* 515: 233–243.
105. Amini, A, Ingegerd, I, Pettersson, C, Westerlund, D. 1996. Enantiomeric separation of local anaesthetic drugs by micellar electrokinetic capillary chromatography with taurodeoxycholate as chiral selector. *J Chromatogr A* 737: 301–313.

106. Aumatell, A, Wells, RJ. 1994. Enantiomeric differentiation of a wide range of pharmacologically active substances by cyclodextrin-modified micellar electrokinetic capillary chromatography using a bile salt. *J Chromatogr A* 688: 329–337.
107. Okafo, GN, Bintz, C, Clarke, SE, Camilleri, P. 1992. Micellar electrokinetic capillary chromatography in a mixture of taurodeoxycholic acid and β-cyclodextrin. *J Chem Soc Chem Commun* 17: 1189–1192.
108. Okafo, GN, Camilleri, P. 1993. Direct chiral resolution of amino acid derivatives by capillary electrophoresis. *J Microcol Sep* 5: 149–143.
109. Nishi, H. 1995. Separation of binaphthyl enantiomers by capillary zone electrophoresis and electrokinetic chromatography. *J High Res Chromatogr* 18: 659–664.
110. Hebling, CM, Thompson, LE, Eckenroad, KW, Manley, GA, Fry, RA, Mueller, KT, Strein, TG, Rovnyak, D. 2008. Sodium Cholate aggregation and chiral recognition of the probe molecule (R,S)-1,1′-binaphthyl-2,2′-dihydrogenphosphate (BNDHP) observed by ^{1}H and ^{31}P NMR spectroscopy. *Langmuir* 24: 13866–13874.
111. Tickle, DC, Okafo, GN, Camilleri, P, Jones, RFD, Kirby, AJ. 1994. Glucopyranoside-based surfactants as pseudostationary phases for chiral separations in capillary electrophoresis. *Anal Chem* 66: 4121–4126.
112. Otsuka, K, Sugimoto, M, Terabe, S, Oida, T, Nakamura, M. 1998. Separation of optical isomers by micellar electrokinetic chromatography using chiral surfactants. *Jpn J Electrophoresis* 42(Suppl 1): 23.
113. Dobashi, A, Ono, T, Hara, S, Yamaguchi, J. 1989. Optical resolution of enantiomers with chiral mixed micelles by electrokinetic chromatography. *Anal Chem* 61: 1984–1986.
114. Ishihama, Y, Terabe, S. 1993. Enantiomeric separation by micellar electrokinetic chromatography using saponins. *J Liq Chromatogr* 16: 933–944.
115. Otsuka, K, Kashihara, M, Kawaguchi, Y, Hisamitsu, T, Terabe, S. 1993. Optical resolution by high-performance capillary electrophoresis: Micellar electrokinetic chromatography with sodium N-dodecanoyl-L-glutamate and digitonin. *J Chromatogr* 652: 253–257.
116. Otsuka, K, Terabe, S. 1990. Effects of methanol and urea on optical resolution of phenylthiohydantion-DL-amino acids by micellar electrokinetic chromatography with sodium N-dodecanoyl-L-valinate. *Elelctrophoresis* 11: 982–984.
117. Dobashi, A, Hamada, M, Dobashi, Y. 1995. Enantiomeric separation with sodium dodecanoyl-L-amino acidate micelles and poly(sodium(10-undecenoyl)-L-valinate) by electrokinetic chromatography. *Anal Chem* 67: 3011–3017.
118. Williams, AA, Tarus, JJ, Agbari, RA, Warner, IM. 2002. Effect of mixed micelles in chiral separations using micellar electrokinetic chromatography. *Proc NOBCChE* 29: 111–117.
119. Mazzeo, JR, Grover, ER, Swartz, ME, Petersen, JS. 1994. Novel chiral surfactant for the separation of enantiomers by micellar electrokinetic capillary chromatography. *J Chromatogr* 680: 125–135.
120. Peterson, AG, Ahuja, ES, Foley, JP. 1996. Enantiomeric separations of basic pharmaceutical drugs by micellar electrokinetic chromatography using a chiral surfactant, N-dodecoxycarbonylvaline. *J Chromatogr B* 683:15–28.
121. Peterson, AG, Foley, JP. 1997. Influence of the inorganic counterion on the chiral micellar electrokinetic separation of basic drugs using the surfactant N-dodecoxycarbonylvaline. *J Chromatogr B* 695:131–145.
122. van Hove, E, Sandra, P. 1995. Considerations on the enantiomeric separation by MEKC of N-Bz-amino acids with N-dodecoxycarbonylvaline as chiral selector. *J Liq Chromatogr* 18: 3675–3683.
123. Swartz, ME, Mazzeo, JR, Grover, ER, Brown, PR. 1996. Validation of enantiomeric separations by micellar electrokinetic capillary chromatography using synthetic chiral surfactants. *J Chromatogr A* 735: 303–310.

124. Swartz, ME, Mazzeo, JR, Grover, ER, Brown, PR, Abul-Enein, HY. 1996. Separation of piperidine-2,6-dione drug enantiomers by micellar electrokinetic capillary chromatography using synthetic chiral surfactants. *J Chromatogr A* 724: 307–316.
125. Ding, W, Fritz, JS. 1999. Carbamate chiral surfactants for capillary electrophoresis. *J Chromatogr A* 831: 311–320.
126. Hong, M, Weekley, BS, Grieb, SJ, Foley, JP. 1998. Electrokinetic chromatography using thermodynamically stable vesicles and mixed micelles formed from oppositely charged surfactants. *Anal Chem* 70: 1394–1403.
127. Pascoe, RJ, Peterson, AG, Foley, JP. 2000. Investigation of the chiral surfactant N-dodecoxycarbonylvaline in electrokinetic chromatography: Improvements in elution range and pH stability via mixed micelles and vesicles, and the hydrophobicity determination of basic pharmaceutical drugs. *Electrophoresis* 21: 2033–2042.
128. Locascio-Brown, L, Plant, A, Horvath, V, Durst, RA. 1990. Liposome flow injection immunoassay: Implications for sensitivity, dynamic range, and antibody regeneration. *Anal Chem* 62: 2587–2593.
129. Plant, AL, Brizgys, MV, Locascio-Brown, L, Durst, RA. 1989. Generic liposome reagent for immunoassays. *Anal Biochem* 176: 420–426.
130. Mohanty, A, Dey, J. 2003. A giant vesicle forming single tailed chiral surfactant for enantioseparation by micellar electrokinetic chromatography. *Chem Commun* 12: 1384–1385.
131. Mohanty, A, Dey, J. 2005. Vesicles as pseudostationary phase for enantiomer separation by capillary electrophoresis. *J Chromatogr A* 1070: 185–192.
132. Mohanty, A, Dey, J. 2006. Enantioselectivity of vesicle-forming chiral surfactants in capillary electrophoresis. Role of the surfactant headgroup structure. *J Chromatogr A* 1128: 259–266.
133. Mohanty, A, Dey, J. 2007. Effect of hydrophobic chain length of the chiral surfactant on enantiomeric separations by electrokinetic chromatography: Comparison between micellar and vesicular pseudo-stationary phases. *Talanta* 71: 1211–1218.
134. Dalton, DD, Taylor, DR, Waters, DG. 1995. Synthesis and use of a novel chiral surfactant based on (R,R)-tartaric acid and its application to chiral separations in micellar electrokinetic capillary chromatography (MECC). *J Microcol Sep* 7: 513–520.
135. Dalton, DD, Taylor, DR, Waters, DG. 1995. Synthesis and use of novel chiral surfactants in micellar electrokinetic capillary chromatography. *J Chromatogr A* 712: 365–371.
136. Wang, H, Zhao, S, He, M, Zhao, Z, Pan, Y, Liang, Q. 2007. Sodium maleopimaric acid as pseudostationary phase for chiral separations of amino acid derivatives by capillary micellar electrokinetic chromatography. *J Sep Sci* 30: 2748–2753.
137. Herrero, M, Ibanez, E, Fanali, S, Cifuentes, A. 2007. Quantitation of chiral amino acids from microalgae by MEKC and LIF detection. *Electrophoresis* 28: 2701–2709.
138. Otsuka, K. 2007. Chiral separations using avidin as a chiral selector and highly sensitive detection using thermal lens microscopy in capillary electrophoresis. *Chromatography* 28: 1–7.
139. Bunke, A, Jira, TH, Beyrich, TH. 1997. (−)-N-Dodecyl-N-methylephedrinium bromide as chiral selector in capillary electrophoresis. *Pharmazie* 52: 762–764.
140. Dey, J, Mohanty, A, Roy, S, Khatua, D. 2004. Cationic vesicles as chiral selector for enantioseparations of nonsteroidal antiinflammatory drugs by micellar electrokinetic chromatography. *J Chromatogr A* 1048: 172–132.
141. Dobashi, A, Hamada, M. 1997. Molecular recognition with micellar and micelle-like aggregates in aqueous media. *J Chromatogr A* 780: 179–189.
142. Hara, S, Dobashi, A, Ono, T, Yamaguchi, J. 1991. Surfactant and their use as carrier in optical resolution and electrochromatography. *Jpn. Kokai Tokkyo Koho* A19911202, Heisei, P.6.
143. Dobashi, A, Hamada, M, Yamaguchi, J. 2001. Molecular recognition by chiral cationic micellar and micelle-like aggregates in electrokinetic capillary chromatography. *Electrophoresis* 22: 88–96.

144. Otsuka, K, Kawakami, H, Tamaki, W, Terabe, S. 1995. Optical resolution of amino acid derivatives by micellar electrokinetic chromatography with sodium N-tetradecanoyl-L-glutamate. *J Chromatogr A* 716: 319–322.
145. Tran, CD, Kang, J. 2003. Chiral separation of amino acids by capillary electrophoresis with 3-[(3-cholamidopropyl)-dimethylammonio]-1-propane sulfonate as chiral selector. *Chromatographia* 57: 81–86.
146. Tivesten, A, Lundqvist, A, Folestad, S. 1997. Selective chiral determination of aspartic and glutamic acid in biological samples by capillary electrophoresis. *Chromatographia* 44: 623–633.
147. Gilges, M, Hardley, M. 1997. Resolution of the diastereomers of a large synthetic peptide by capillary electrophoresis using nonionic surfactants. *Electrophoresis* 18: 2944–2949.
148. Hadley, MR, Harrison, MW, Hutt, AJ. 2003. Use of chiral zwitterionic surfactants for enantiomeric resolutions by capillary electrophoresis. *Electrophoresis* 24: 2508–2513.
149. Xu, RJ, Vidal-Madjar, C, Sebile, B. 1998. Capillary electrophoretic behavior of milk proteins in the presence of non-ionic surfactants. *J Chromatogr B* 706: 3–11.
150. Nimura, N, Itoh, H, Mitsumo, C, Knoshita, T. 1994. Chiral recognition modeling for exploring the origin of molecular chirality: Simulation of chiral discrimination between biologically–relevant compounds using capillary electrophoresis. *Proceedings of Separation Sciences'94*, Tokyo, Japan, 283–285.
151. Ghosh, A, Dey, J. 2008. Enantiomeric separations of binaphthyl derivatives by capillary electrophoresis using N-(2-hydroxydodecyl)-L-threonine as chiral selector: Effect of organic additives. *Electrophoresis* 29: 1540–1547.
152. Tickle, DC, George, A, Jennings, K, Cammileri, P, Kirgby, AJ. 1998. A study of the structure and chiral selectivity of micelles of two isomeric D-glucopyranoside-based surfactants. *J Chem Soc Perkin Trans* 2: 467–474.
153. Mechref, Y, El Rassi, Z. 1997. Capillary electrophoresis of herbicides II. Evaluation of alkylglucoside chiral surfactants in the enantiomeric separation of phenoxy acid herbicides. *J Chromatogr A* 757: 263–273.
154. Desbene, P, Fulchic, C. 1996. Utilization of n-alkyl-β-D-glucopyranosides in enantiomeric separation by micellar electrokinetic chromatography. *J Chromatogr A* 749: 257–269.
155. Mechref, Y, El Rassi, Z. 1996. Capillary electrophoresis of herbicides. III. Evaluation of octylmaltopyranoside chiral surfactant in the enantiomeric separation of phenoxy acid herbicides. *Chirality* 8: 518–524.
156. Mechref, Y, El Rassi, Z. 1997. Capillary electrophoresis of herbicides: IV. Evaluation of octylmaltopyranoside chiral surfactant in the enantiomeric separation of fluorescently labeled phenoxy acid herbicides and their laser-induced fluorescence detection. *Electrophoresis* 18: 220–226.
157. Horimai, T, Arai, T, Sato, Y. 2000. New amphiphilic aminosaccharide derivatives as chiral selectors in capillary electrophoresis. *J Chromatogr A* 875: 295–305.
158. Ju, M, El Rassi, Z. 1997. Enantioseparations by capillary electrophoresis using chiral glycosidic surfactants. I. Evaluation of cyclohexyl-pentyl-β-D-maltoside surfactant. *Electrophoresis* 20: 2766–2771.
159. Ju, M, El Rassi, Z. 2000. Enantioseparations by capillary electrophoresis using chiral glycosidic surfactants. II. Comparison of chiral cyclohexyl-alkyl-β-D-maltoside surfactants. *J Liq Chromatogr Rel Technol* 23: 35–45.
160. Mechref, Y, El Rassi, Z. 1996. Micellar electrokinetic capillary chromatography with in-situ charged micelles VI. Evaluation of novel chiral micelles consisting of steroidal-glycoside surfactant-borate complexes. *J Chromatogr A* 724: 285–296.
161. Otsuka, K, Terabe, S. 1990. Enantiomeric resolution by micellar electrokinetic chromatography with chiral surfactants. *J Chromatogr A* 515: 221–226.

162. Dobashi, A, Ono, T, Hara, S, Yamaguchi, J. 1989. Enantioselective hydrophobic entanglement of enantiomeric solutes with chiral functionalized micelles by electrokinetic chromatography. *J Chromatogr A* 480: 413–420.
163. Terabe, S, Ishihama, Y, Nishi, H, Fukuyama, T, Otsuka, K. 1991. Effect of urea addition in micellar electrokinetic chromatography. *J Chomatogr* 545: 359–368.
164. Otsuka, K, Kawahara, J, Tatekawa, K, Terabe, S. 1991. Chiral separations by micellar electrokinetic chromatography with sodium N-dodecanoyl-L-valinate. *J Chomatogr A* 559: 209–214.
165. Otsuka, K, Karuhaka, K, Higashimori, M, Terabe, S. 1994. Optical resolution of amino acid derivatives by micellar electrokinetic chromatography with N-dodecanoyl-L-serine. *J Chromatogr A* 680: 317–320.
166. Clothier, Jr. JG, Tomellini, SA. 1996. Chiral separation of verapamil and related compounds using micellar electrokinetic capillary chromatography with mixed micelles of bile salt and polyoxyethylene ethers. *J Chromatogr A* 723: 179–187.
167. Clothier, Jr. JG, Daley, LM, Tomellini, SA. 1996. Effects of bile salt structure on chiral separations with mixed micelles of bile salts and polyoxyethylene ethers using micellar electrokinetic capillary chromatography. *J Chromatogr B* 683: 37–45.
168. Penn, SG, Chiu, RW, Monning, CA. 1994. Separation and analysis of cyclodextrins by capillary electrophoresis with dynamic fluorescence labeling and detection. *J Chromatogr A* 680: 233–241.
169. Park, H, Jung, S. 2005. Separation of some chiral flavonoids by microbial cyclosophoraoses and their sulfated derivatives in micellar electrokinetic chromatography. *Electrophoresis* 26: 3833–3838.
170. Zhao, S, Wang, H, Zhang, R, Tang, L, Liu, Y-M. 2006. Degrading dehydroabietylisothiocyanate as a chiral derivatizing reagent for enantiomeric separations by capillary electrophoresis. *Electrophoresis* 27: 3428–3433.
171. Pedersen, C. 1967. Cyclic polyethers and their complexes with metal salts. *J Am Chem Soc* 89: 2495–2496.
172. Fanali, S, Cristalli, M, Vespalec, R, Boček, P. 1994. Chiral separations in capillary electrophoresis. *Adv Electrophoresis* 7: 1–88.
173. Salami, M, Otto, H-H, Jira, T. 2001. Chiral separation of amino acid esters by micellar electrokinetic chromatography. *Electrophoresis* 22: 3291–3296.
174. Miyashita, Y, Terabe, S. 1990. Application Data DS-767, Beckmann Instruments, Fullerton, CA.
175. Ueda, T, Kitamura, F, Mitchel, R, Metcalf, T, Kuwana, T, Nakamoto, A. 1991. Chiral separation of naphthalene-2,3-dicarboxaldehyde-labeled amino acid enantiomers by cyclodextrin-modified micellar electrokinetic chromatography with laser-induced fluorescence detection. *Anal Chem* 63: 2979–2981.
176. Ueda, T, Mitchel, R, Kitamura, F, Metcalf, T, Kuwana, T, Nakamoto, A. 1992. Separation of naphthalene-2,3-dicarboxaldehyde-labeled amino acids by high-performance capillary electrophoresis with laser-induced fluorescence detection. *J Chromatogr A* 593: 265–274.
177. Tran, CD, Kang, J. 2002. Chiral separation of amino acids by capillary electrophoresis with octyl-β-thioglucopyranoside as chiral selector. *J Chromatogr A* 978: 221–230.
178. Fu, Y, Yan, L, Luo, G, Chen, C, Wan, Y, Zhang, L. 2004. Chiral analysis of amino acids in biological samples by micellar electrokinetic chromatography with laser-induced fluorescence detection. *Fenxi Huaxue* 32: 1575–1579.
179. Carlavilla, D, Moreno-Arribas, MV, Fanali, S, Cifuentes, A. 2006. Chiral MEKC-LIF of amino acids in foods: Analysis of vinegars. *Electrophoresis* 27: 2551–2557.
180. Herrero, M, Ibanez, E, Martin-Alvarez, PJ, Cifuentes, A. 2007. Analysis of chiral amino acids in conventional and transgenic maize. *Anal Chem* 79: 5071–5077.
181. Giuffrida, A, Tabera, L, Gonzalez, R, Cucinotta, V, Cifuentes, A. 2008. Chiral analysis of amino acids from conventional and transgenic yeasts. *J Chromatogr B* 875: 243–247.

182. Pumera, M, Flegel, M, Jelinek, I. 2002. Chiral analysis of biogenic DL-amino acids derivatized by urethane - protected α-amino acid N-carboxyanhydride using capillary zone electrophoresis and micellar electrokinetic chromatography. *Electrophoresis* 23: 2449–2456.
183. Wang, Z, Tang, Z, Gu, Z, Hu, Z, Ma, S, Kang, J. 2005. Enantioseparation of chiral allenic acids by micellar electrokinetic chromatography with cyclodextrins as chiral selector. *Electrophoresis* 26: 1001–1006.
184. Huang, Y, Liu, L, Du, G, Xie, J. 2005. Enantioseparation of nateglinide by micellar electrokinetic chromatography. *Yaowu Fenxi Zazhi* 25: 852–855.
185. Nishi, H, Fukuyama, T, Terabe, S. 1991. Chiral separation by cyclodextrin-modified micellar electrokinetic chromatography. *J Chromatogr* 553: 503–516.
186. Prunonosa, J, Obach, R, Diez-Coscon, A, Gouesclou, L. 1992. Determination of cicletanine enantiomers in plasma by high-performance capillary electrophoresis. *J Chromatogr B* 574: 127–145.
187. Furuta, R, Doi, T. 1994. Chiral separation of diniconazole, uniconazole and structurally related compounds by cyclodextrin-modified micellar electrokinetic chromatography. *Electrophoresis* 15:1322–1325.
188. Furuta, R, Doi, T. 1995. Enantiomeric separation of a thiazole derivative by high-performance liquid chromatography and micellar electrokinetic chromatography. *J Chromatogr A* 708: 245–251.
189. Francotte, E, Cherkaoul, S, Faupel, M. 1993. Separation of the enantiomers of some racemic nonsteroidal aromatase inhibitors and barbiturates by capillary electrophoresis. *Chirality* 5: 516–526.
190. Desiderio, C, Fanali, S, Kupfer, A, Thormann, W. 1994. Analysis of mephenytoin, 4-hydroxymephenytoin and 4-hydroxyphenytoin enantiomers in human urine by cyclodextrin micellar electrokinetic capillary chromatography: Simple determination of a hydroxylation polymorphism in man. *Electrophoresis* 15: 87–93.
191. Lin, C-E, Chen, K-H, Hsiao, Y-Y, Liao, W-S, Chen, C-C. 2002. Enantioseparation of phenothiazines in cyclodextrin-modified micellar electrokinetic chromatography. *J Chromatogr A* 971: 261–266.
192. Zhu, C, Lin, X, Wei, Y. 2002. Chiral separation of pemoline enantiomers by cyclodextrin-modified micellar capillary chromatography. *J Pharm Biomed Anal* 30: 293–298.
193. Zheng, Y-P, Mo, J-Y. 2004. Enantiomeric separation of ephedrine and salbutamol by micellar electrokinetic chromatography using β-cyclodextrin as chiral additive. *Chinese J Chem* 22(8): 845–848.
194. Gotti, R, El-Hady, DA, Andrisano, V, Bertucci, C, El-Maali, NA, Cavrini, V. 2004. Determination of chiral and achiral related substances of tehotrexate by cyclodextrin-modified micellar electrokinetic chromatography. *Electrophoresis* 25(16): 2830–2837.
195. Shen, J, Zhao, S. 2004. Enantiomeric separation of naphthalene-2,3-dicarboxaldehyde derivatized DL-3,4-dihydroxyphenylalanine and optical purity analysis of L-3, 4-dihydroxyphenylalanine drug by cyclodextrin-modified micellar electrokinetic chromatography. *J Chromatogr A* 1059: 209–214.
196. Zhang, W, Zhu, C. 2004. Separation of uniconazole by enantiomers capillary electrophoresis with dual cyclodextrin systems. *Can J Sci Spectros* 49: 277–281.
197. Marina, ML, Benito, I, Diez-Masa, JC, Gonzáles, MJ. 1996. Chiral separation of polychlorinated biphenyls by micellar electrokinetic chromatography with gamma-cyclodextrin as modifier in the separation buffer. *Chromatographia* 42: 269–272.
198. Garrison, AW, Schmitt-Kopplin, P, Avants, JK. 2008. Analysis of the enantiomers of chiral pesticides and other pollutants in environmental samples by capillary electrophoresis. In *Methods in Molecular Biology, 384: Capillary Electrophoresis: Methods and Protocols*, Schmitt-Kopplin, Ph. (Ed.), pp. 157–170. Totowa, NJ: Humana Press.
199. Terabe, S. 2004. Micellar Electrokinetic Chromatography. *Anal Chem* 76: 240A–246A.

200. Pirogov, AV, Shpigun, OA. 2003. Application of water-soluble polymers as modifiers in electrophoretic analysis of phenols. *Electrophoresis* 24: 2099–2105.
201. Peric, I, Kenndler, E. 2003. Recent developments in capillary electrokinetic chromatography with replaceable charged pseudostationary phases or additives. *Electrophoresis* 24: 2924–2934.
202. Fritz, JS. 2003. The role of organic solvents in the separation of nonionic compounds by capillary electrophoresis. *Electrophoresis* 24: 1530–1536.
203. Urbánek, M, Křivánková, L, Boček, P. 2003. Stacking phenomena in electromigration: From basic principles to practical procedures. *Electrophoresis* 24: 466–485.
204. Welsch, T, Michalke, D. 2003. (Micellar) electrokinetic chromatography: An interesting solution for the liquid phase separation dilemma. *J Chromatogr A* 1000: 935–951.
205. Makino, K, Itoh, Y, Teshima, D, Oishi, R. 2004. Capillary electrophoresis analysis of a wide variety of seized drugs using the same capillary with dynamic coatings. *Electrophoresis* 25:1488–1495.
206. Trojanowicz, M, Pobozy, E, Gübitz, G. 2003. Speciation of oxidation states of elements by capillary electrophoresis. *J Sep Sci* 26: 983–995.
207. Boyce, MC, Haddad, PR. 2003. Tailoring the separation selectivity of metal complexes and organometallic compounds resolved by capillary electrophoresis using auxiliary separation processes. *Electrophoresis* 24: 2013–2022.
208. Dabek-Zlotorzynska, E, Chen, H, Ding, L. 2003. Recent advances in capillary electrophoresis and capillary electrochromatography of pollutants. *Electrophoresis* 24: 4128–4149.
209. Poole, SK, Poole, CF. 2003. Separation methods for estimating octanol–water partition coefficients. *J Chromatogr B* 797: 3–19.
210. Davankov, VA. 2003. Enantioselective ligand exchange in modern separation techniques. *J Chromatogr A* 1000: 891–915.
211. Wang, JA, Warner, IM. 1994. Chiral separations using micellar electrokinetic capillary chromatography and a polymerized chiral micelle. *Anal Chem* 66: 3773–3776.
212. Mammilery, P. 1997. Chiral surfactants in micellar electrokinetic capillary chromatography. *Electrophoresis* 18: 2322–2330.
213. Palmer, CP, Tanaka, N. 1997. Selectivity of polymeric and polymer-supported pseudostationary phases in micellar electrokinetic chromatography. *J Chromatogr A* 792: 105–124.
214. Palmer, CP. 1997. Micelle polymers, polymer surfactants and dendrimers as pseudostationary phases in micellar electrokinetic chromatography. *J Chromatogr A* 780: 75–92.
215. Palmer, CP. 2002. Recent progress in the development, characterization and application of polymeric pseudophases for electrokinetic chromatography. *Electrophoresis* 23: 3993–4004.
216. Palmer, CP, McCarney, JP. 2004. Developments in the use of soluble ionic polymers as pseudo-stationary phases for electrokinetic chromatography and stationary phases for electrochromatography. *J Chromatogr A* 1044: 159–176.
217. Nilsson, C, Nilsson, S. 2006. Nanoparticle-based pseudostationary phases in capillary electrochromatography. *Electrophoresis* 27: 76–83.
218. Maichel, B, Kenndler, E. 2000. Recent innovation in capillary electrokinetic chromatography with replaceable charged pseudostationary phases or additives. *Electrophoresis* 21: 3160–3173.
219. Schweitz, L, Spegel, P, Nilsson, S. 2001. Approaches to molecular imprinting based selectivity in capillary electrochromatography. *Electrophoresis* 22: 4053–4063.
220. Spegel, P, Schweitz, L, Nilsson, S. 2003. Molecularly imprinted polymers in capillary electrochromatography: Recent developments and future trends. *Electrophoresis* 24: 3892–3899.

221. Nilsson, J, Spegel, P, Nilsson, S. 2004. Molecularly imprinted polymer formats for capillary electrochromatography. *J Chromatogr B* 804: 3–12.
222. Shamsi, SA, Palmer, CP, Warner, IM. 2001. Molecular micelles: Novel pseudostationary phases for CE. *Anal Chem* 73: 140A–149A.
223. Norton, D, Shamsi, SA. 2003. Separation of methylated isomers of benzo[a]pyrene using micellar electrokinetic chromatography. *Anal Chim Acta* 496: 165–176.
224. Mwongela, SM, Numan, A, Gill, NL, Agbaria, RA, Warner, IM. 2003. Separation of achiral and chiral analytes using polymeric surfactants with ionic liquids as modifiers in micellar electrokinetic chromatography. *Anal Chem* 75: 6089–6096.
225. Akbay, C, Shamsi, SA. 2004. Polymeric sulfated surfactants with varied hydrocarbon tail: II. Chemical selectivity in micellar electrokinetic chromatography using linear solvation energy relationships study. *Electrophoresis* 25: 635–644.
226. Akbay, C, Shamsi, SA. 2004. Polymeric sulfated surfactants with varied hydrocarbon tail: I. Synthesis, characterization, and application in micellar electrokinetic chromatography. *Electrophoresis* 25: 622–634.
227. Schulte, S, Palmer, CP. 2003. Alkyl-modified siloxanes as pseudostationary phases for electrokinetic chromatography. *Electrophoresis* 24: 978–983.
228. Peterson, DS, Palmer, CP. 2002. Novel alkyl-modified anionic siloxanes as pseudostationary phases for electrokinetic chromatography: III. Performance in organic-modified buffers. *J Chromatogr A* 959: 255–261.
229. Peterson, DS, Palmer, CP. 2001. Novel alkyl-modified anionic siloxanes as pseudostationary phases for electrokinetic chromatography: II. Selectivity studied by linear solvation energy relationships. *Electrophoresis* 22: 3562–3566.
230. Pandey, S, Redden, RA, Hendricks, AE, Fletcher, KA, Palmer, CP. 2003. Characterization of the solvation environment provided by dilute aqueous solutions of novel siloxane polysoaps using the fluorescence probe pyrene. *J Colloid Interface Sci* 262: 579–587.
231. Billiot, FH, McCarroll, M, Billiot, EJ, Rugutt, JK, Morris, K, Warner, IM. 2002. Comparison of the aggregation behavior of 15 polymeric and monomeric dipeptide surfactants in aqueous solution. *Langmuir* 18: 2993–2997.
232. Hara, S, Dobashi, A. Japan, Patent # 149,205,192.
233. Billot, FH, Billot, EJ, Warner, IM. 2001. Comparison of monomeric and polymeric amino acid based surfactants for chiral separations. *J Chromatogr A* 922: 329–338.
234. Thibodeaux, SJ, Billiot, E, Torres, E, Valle, BC,Warner, IM. 2003. Enantiomeric separations using polymeric L-glutamate surfactant derivatives: Effect of increasing steric factors. *Electrophoresis* 24: 1077–1082.
235. Thibodeaux, SJ, Billiot, E, Warner, IM. 2002. Enantiomeric separations using poly (L-valine) and poly(L-leucine) surfactants: Investigation of steric factors near the chiral center. *J Chromatogr A* 966: 179–186.
236. Rizvi, SAA, Simons, DN, Shamsi, SA. 2004. Polymeric alkenoxy amino acid surfactants: III. Chiral separations of binaphthyl derivatives. *Electrophoresis* 25: 712–722.
237. Rizvi, SAA, Shamsi, SA. 2003. Polymeric alkenoxy amino acid surfactants: I. Highly selective class of molecular micelles for chiral separation of-β-blockers. *Electrophoresis* 24: 2514–2526.
238. Rizvi, SAA, Akbay, C, Shamsi, SA. 2004. Polymeric alkenoxy amino acid surfactants: II. Chiral separations of-β-blockers with multiple stereogenic centers. *Electrophoresis* 25: 853–860.
239. Billiot, FH, Billiot, EJ, Ng, YK, Warner, IM. 2006. Chiral separation of norlaudanosoline, laudanosoline, laudanosine, chlorthalidone, and three benzoin derivatives using amino acid based molecular micelles. *J Chromatogr Sci* 44: 64–69.
240. Mwongela, S, Akbey, C, Zhu, X, Collins, S, Warner, IM. 2003. Use of poly(sodium oleyl-L-leucylvalinate) surfactant for the separation of chiral compounds in micellar electrokinetic chromatography. *Electrophoresis* 24: 2940–2947.

241. Shamsi, SA, Valle, BC, Billiot, F, Warner, IM. 2003. Polysodium N-undecanoyl-L-leucylvalinate: A versatile chiral selector for micellar electrokinetic chromatography. *Anal Chem* 75: 379–387.
242. Tarus, J, Agbaria, RA, Morris, K, Billot, FH, Williams, AA, Chatman, T, Warner, IM. 2003. Enantioselectivity of alcohol-modified polymeric surfactants in micellar electrokinetic chromatography. *Electrophoresis* 24: 2499–2507.
243. Tarus, J, Jernigan, T, Morris, K, Warner, IM. 2004. Enantioselectivity of structurally modified poly(sodium undecenoyl-L-leucinate) by insertion of Triton X-102 surfactant molecules. *Electrophoresis* 25: 2720–2726.
244. Traus, J, Agbaria, RA, Morris, K, Mwongela, S, Numan, A, Simuli, L, Fletcher, KA, Warner, IM. 2004. Influence of the polydispersity of polymeric surfactants on the enantioselectivity of chiral compounds in micellar electrokinetic chromatography. *Langmuir* 20: 6887–6895.
245. McCarroll, ME, Billot, FH, Warner, IM. 2001. Fluorescence anisotropy as a measure of chiral recognition. *J Am Chem Soc* 123: 3173–3174.
246. Rizvi, SAA, Shamsi, SA. 2005. Polymeric alkenoxy amino acid surfactants: IV. Effects of hydrophobic chain length and degree of polymerization of molecular micelles on chiral separation of-β-blockers. *Electrophoresis* 26: 4172–4186.
247. Akbey, C, Gill, NL, Agberia, RA, Warner, IM. 2003. Copolymerized polymeric surfactants: Characterization and application in micellar electrokinetic chromatography. *Electrophoresis* 24: 4209–4220.
248. Akbey, C, Tarus, J, Gill, NL, Agberia, RA, Warner, IM. 2004. Novel anionic copolymerized surfactants of mixed achiral and chiral surfactants as pseudostationary phases for micellar electrokinetic chromatography. *Electrophoresis* 25: 758–765.
249. Rizvi, SAA, Shamsi, SA. 2007. Polymeric alkenoxy amino acid surfactants: V. Comparison of carboxylate and sulfate headgroup polymeric surfactants for enantioseparation in MEKC. *Electrophoresis* 28: 1762–1778.
250. Rizvi, SAA, Shamsi, SA. Chiral analysis using polymeric surfactants in micellar electrokinetic chromatography (MEKC) and MEKC coupled mass spectrometry. In *Chiral Separation Techniques,* 3rd edition. Ganapathy, S (Ed.), pp. 505–560. Weinheim, Germany: Wiley-VCH Verlag Gmbh & Co.
251. Shamsi, SA. 2002. Chiral capillary electrophoresis-mass spectrometry: Modes and applications. *Electrophoresis* 23: 4036–4051.
252. Hou, J, Zheng, J, Rizvi, SAA, Shamsi, SA. 2007. Simultaneous chiral separation and determination of ephedrine alkaloids by MEKC-ESI-MS using polymeric surfactant I: method development. *Electrophoresis* 28: 1352–1363.
253. Akbay, C, Rizvi, SAA, Shamsi, SA. 2005. Simultaneous enantioseparation and tandem UV-MS detection of eight β-blockers in micellar electrokinetic chromatography using a chiral molecular micelle. *Anal Chem* 77: 1672–1683.
254. Hou, J, Rizvi, SAA, Zheng, J, Shamsi, SA. 2006. Application of polymeric surfactants in micellar electrokinetic chromatography-electrospray ionization mass spectrometry of benzodiazepines and benzoxazocine chiral drugs. *Electrophoresis* 27: 1263–1275.
255. Rizvi, SAA, Zheng, J, Apkarian, RP, Dublin, SN, Shamsi, SA. 2007. Polymeric sulfated amino acid surfactants: A class of versatile chiral selectors for micellar electrokinetic chromatography (MEKC) and MEKC-MS. *Anal Chem* 79: 879–898.
256. Hou, J, Zheng, J, Shamsi, SA. 2007. Simultaneous chiral separation of ephedrine alkaloids by MEKC-ESI-MS using polymeric surfactant II: Application in dietary supplements. *Electrophoresis* 28: 1426–1434.
257. Hou, J, Zheng, J, Rizvi, SAA, Shamsi, SA. 2007. Separation and determination of warfarin enantiomers in human plasma using a novel polymeric surfactant for micellar electrokinetic chromatography-mass spectrometry. *J Chromatogr A* 1159: 208–216.

9 Chiral Microemulsion Electrokinetic Chromatography

Kimberly A. Kahle and Joe P. Foley

CONTENTS

9.1 Introduction 235
 9.1.1 General O/W Microemulsion Formulations for EKC 236
 9.1.2 Brief History of Microemulsions in EKC 238
9.2 Principles of Chiral EKC 238
 9.2.1 Mechanism of Chiral Separation 239
 9.2.2 Resolution in EKC 239
9.3 Principles of Chiral MEEKC 241
 9.3.1 Single-Chiral-Component Microemulsions 241
 9.3.2 Microemulsions Comprised of Two or More Chiral Components 241
9.4 Fundamental Studies and Findings in Chiral MEEKC 245
 9.4.1 Surfactant-Based Chiral Microemulsions 245
  9.4.1.1 Dodecoxycarbonylvaline as the Chiral Surfactant 245
  9.4.1.2 *N*-undecenoyl-D-valinate as the Chiral Surfactant 254
 9.4.2 Cosurfactant-Based Chiral Microemulsions 256
 9.4.3 Oil-Based Chiral Microemulsions 256
 9.4.4 Multiple-Chiral-Component Microemulsions 258
9.5 Applications and Concluding Remarks for Chiral MEEKC 265
 9.5.1 Analytes Separated 265
 9.5.2 Concluding Remarks on Chiral MEEKC 267
9.6 Future Directions 267
References 268

9.1 INTRODUCTION

Microemulsion electrokinetic chromatography (MEEKC) is the newest and least published of the electrokinetic chromatography (EKC) options, particularly in terms of chiral separations. Microemulsions are aggregates in the order of 1–5 nm that are created by combining and sonicating a surfactant, an oil, and usually a cosurfactant for stability. They are optically transparent and thermodynamically stable when the

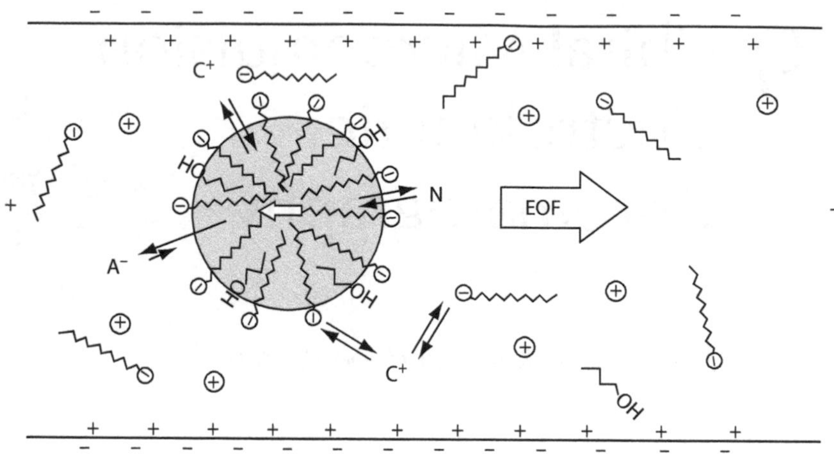

FIGURE 9.1 Schematic presentation of MEEKC. A typical running buffer consists of 0.8% *n*-octane + 3.3% SDS + 6.6% 1-butanol in borate buffer, pH 9.2. (From Hansen, S.H., *Electrophoresis*, 24, 3900, 2003. With permission.)

components are present in the proper ratio. In comparison to other surfactant pseudostationary phases (PSPs), microemulsions are less rigid [1], can solubilize more hydrophobic compounds, have a more tunable elution range (migration window) [2], and offer a wider variety of parameters to optimize. An important difference compared to other EKC modes is sample diluent. In MEEKC, it is best to dissolve the sample in the microemulsion solution instead of just the background electrolyte to obtain reproducible separations and satisfactory peak shape. The two main classes of microemulsions are water-in-oil and oil-in-water. Water-in-oil (w/o) systems contain a water core surrounded by a surfactant in an oil bulk phase and are not commonly used in EKC. Oil-in-water (o/w) microemulsions consist of an oil core (hydrocarbon or other water immiscible liquid) encased by a surfactant/cosurfactant (typically a short chain alcohol) outer shell; general representations of a MEEKC separation and an o/w aggregate are given in Figures 9.1 and 9.2, respectively. As shown in Figure 9.3, o/w microemulsions can be converted to w/o systems by increasing the oil concentration. When the water and oil are present in near equal amounts, a bicontinuous type of phase exists where a honeycomb-like structure is found instead of discrete droplets.

9.1.1 General O/W Microemulsion Formulations for EKC

Several MEEKC review articles have been published in recent years and provide more details on microemulsions and their separating ability in EKC [3–6]. Microemulsion formulating requires the consideration of all aggregate components and their concentrations. The surfactant impacts the size and charge of the PSP as well as the electroosmotic flow (EOF) and selectivity. Incorporation of a second surfactant to create a mixed surfactant microemulsion allows for greater variation in charge and selectivity. For achiral MEEKC, sodium dodecyl sulfate (SDS) is the most commonly used

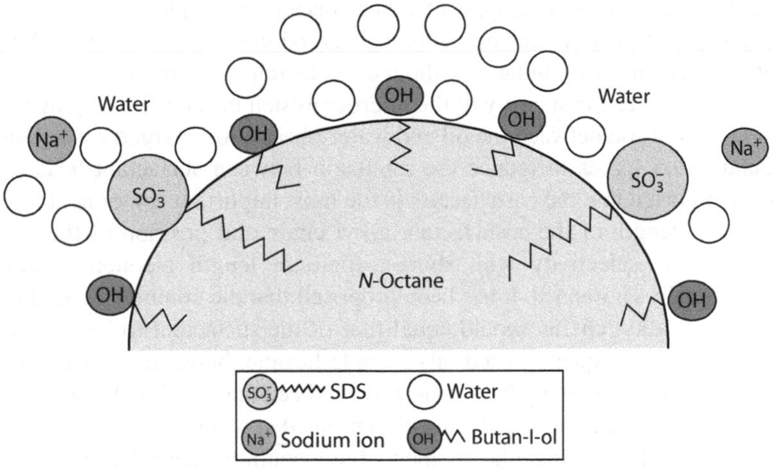

FIGURE 9.2 General representation of a microemulsion structure. (From Altria, K. *Chromatographia*, 52, 758, 2000. With permission.)

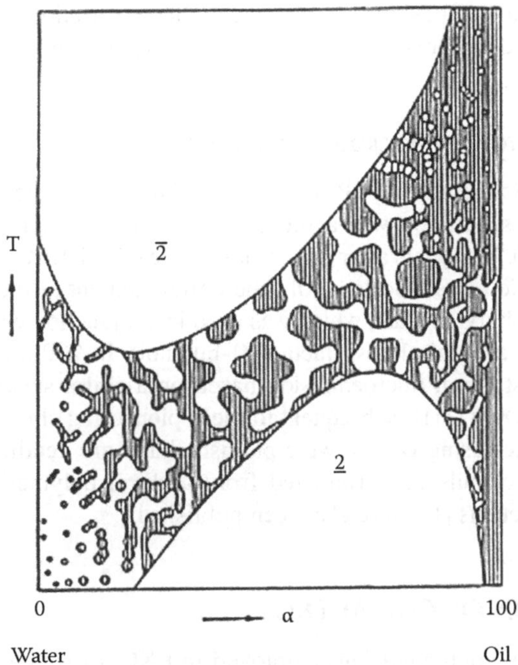

FIGURE 9.3 Schematic section of the phase prism of a ternary mixture of water, an organic liquid (oil), and a nonionic amphiphile. At constant amphiphile concentration, a region of isotropic single-phase solutions is observed extending from the water-rich to the oil-rich side of the phase prism. This single-phase region is surrounded by two two-phase regions in which the amphiphile is dissolved either in the aqueous bottom phase (2) or in the organic top phase (2). The hatching in the single-phase region illustrates the microstructure of the solutions. (From Schwuger, M., Stickdorn, K. and Schomaecker, R., *Chem Rev*, 95, 849, 1995. With permission.)

charged surfactant. Some examples of other surfactants employed include Tween, Triton, and Brij-35 (neutral) and bile salts such as sodium cholate (anionic). Although microemulsions can be formulated without cosurfactant, the presence of a cosurfactant greatly increases the stability of the microemulsion droplets by helping to lower the interfacial tension between the oil and water phases; for charged surfactants, the cosurfactant also serves to reduce the repulsion between surfactant head groups. It has been reported that the cosurfactant is the most important component for selectivity [5]. The length of the cosurfactant alkyl chain and position of the hydroxyl group affect the selectivity with short-to-medium length alcohols (butanol to hexanol) typically chosen [7]. It has been proposed that the combined lengths of the alcohol and oil alkyl chains should equal that of the surfactant to improve microemulsion stability. Frequently used oils include hexane, heptane, octane, and ethyl acetate. Changes in selectivity have generally not been observed with variation in oil identity. The concentration of each constituent needed to form a stable microemulsion is interdependent on the other elements but a common achiral system would be composed of 3.3% (w/v) SDS, 6.6% (v/v) butanol, and 0.8% (v/v) octane in a buffer. An article on non separation science uses of microemulsions described basic microemulsion properties including phase behaviour and inter facial tension [8]. Besides the microemulsion composition, buffer identity/concentration, pH, temperature, applied voltage, and buffer additives also affect the separation.

9.1.2 Brief History of Microemulsions in EKC

The first use of a microemulsion PSP in EKC was published by Watarai et al. in 1991 [9] where an achiral separation was achieved using an o/w aggregate (water/SDS/1-butanol/heptane). Two years after this initial article on MEEKC appeared, the first chiral microemulsion was reported [10]. The chiral element of the PSP was the oil core, (2*R*,3*R*)-di-*n*-butyl tartrate, which was used in conjunction with an achiral surfactant (SDS) and an achiral cosurfactant (1-butanol). In 2002, Pascoe and Foley introduced the first chiral microemulsion based on a chiral surfactant (dodecoxycarbonylvaline, DDCV) [11]. Subsequent to these pioneering chiral MEEKC works, several more studies using DDCV were published and are detailed in Section 9.4. Additionally, microemulsions formulated from a chiral polymeric surfactant [12] and chiral cosurfactants [13] have also been published.

9.2 PRINCIPLES OF CHIRAL EKC

In general, the o/w microemulsions employed in EKC utilize charged surfactants so that the PSP will have counter-electroosmotic migration. The charge composition of the microemulsion contributes to the overall electrical conductivity of the system. In particular, the interfacial tension of the oil greatly impacts the amount of surfactant needed to form a stable aggregate. If the oil has a high interfacial tension (high hydrophobicity), a larger amount of surfactant will be required to form a stable microemulsion, thus increasing conductivity (Joule heating) and vice versa for low-interfacial-tension oils.

9.2.1 Mechanism of Chiral Separation

To accomplish a chiral separation in EKC, a direct approach strategy is employed where a temporary diastereomeric interaction occurs between the enantiomers and the chiral selectand. The two enantiomers can be separated when either their binding constants with the chiral agent or the mobilities of their transient diastereomeric complexes are different. It is generally recognized that a three-point-interaction must take place for chiral discrimination [14]. The nature of these interactions can be electrostatic, hydrogen bonding, steric hindrance, π–π, ion–dipole, dipole–dipole, dipole-induced dipole, or van der Waals (in order of decreasing strength).

9.2.2 Resolution in EKC

Fundamental resolution equations for EKC and chiral EKC vary depending on the specific situation. For neutral molecules, the following equation developed by Terabe can be used [15]:

$$R_s = \frac{\sqrt{N}}{4}\left(\frac{\alpha-1}{\alpha}\right)\left(\frac{k_2}{1+k_2}\right)\left(\frac{1-t_0/t_{mc}}{1+(t_0/t_{mc})k_1}\right) \quad (9.1)$$

where
- R_s is the resolution
- N is the efficiency
- α is the selectivity (k_2/k_1)
- k_1 and k_2 are the capacity (retention) factors of neutral compounds
- t_0 is the migration time of an unretained compound
- t_{mc} is the migration time of the PSP (micellar PSP in this case)

When some or all of the analytes of interest are charged, the calculation of resolution becomes more complex due to the electrophoretic mobilities of the charged analytes.

For enantiomeric separations, an equation for the resolution of neutral and charged enantiomers was published by Foley [16,17]:

$$R_s = \frac{\sqrt{N}}{4}\left(\frac{\alpha-1}{\alpha}\right)\left(\frac{k_2}{1+k_{avg}}\right)\left(\frac{1+\mu_{ep,enant}/\mu_{eo}-t_0/t_{psp}}{1+\mu_{ep,enant}/\mu_{eo}+(t_0/t_{psp})k_{avg}}\right) \quad (9.2)$$

where
- R_s, N, α, and k have been identified in Equation 9.1 except that in Equation 9.2 they also pertain to charged compounds
- k_{avg} is $(k_1 + k_2)/2$
- t_{PSP} is the migration time of the PSP
- $\mu_{ep,enant}/\mu_{eo}$ is the electrophoretic mobility of the free enantiomer relative to the EOF

Equation 9.2 was experimentally verified for over 140 combinations of pairs of enantiomers and microemulsion formulations, with excellent agreement (<2% error) between theory and experiment [18]. Retention factors were calculated via

$$k = \frac{t_R(1+\mu_{ep,enant}/\mu_{eo}) - t_0}{t_0 - (1+\mu_{ep,PSP}/\mu_{eo})t_R} \tag{9.3}$$

where $\mu_{ep,PSP}$ is the electrophoretic mobility of the PSP (microemulsion).

Alternatively, the expression derived by Mazzeo can also be employed, although only a limited experimental verification was performed [19]:

$$R_s = \sqrt{\frac{VL_{det}}{16LD}}\left(\frac{|(x_1 - x_2)(\mu_{add,ep} - \mu_{fs,ep})|}{|(x_1 + x_2)(\mu_{add,ep} - \mu_{fs,ep}) + 2\mu_{eo} + 2\mu_{fs,ep}|^{1/2}}\right) \tag{9.4}$$

where

V is the applied voltage
L and L_{det} are the total and effective length of the capillary, respectively
D is the effective diffusion coefficient of the enantiomer
x_i is the fraction of ith enantiomer that is interacting with the chiral PSP at a given instant, i.e., $x_i = k_i/(1+k_i)$
$\mu_{add,ep}$ is the electrophoretic mobility of the chiral PSP
$\mu_{fs,ep}$ is the electrophoretic mobility of the free enantiomer

As shown in the above equations, there are several factors or figures of merit that influence resolution in MEEKC and include efficiency (N), selectivity (α or k_2/k_1), retention factor (k), and migration window or elution range (t_{PSP}/t_0). In order to achieve the maximum resolution obtainable for a given separation, the retention factor should be as close to its optimum as possible. An expression for the optimum retention factor, k_{opt}, is obtained by taking the first derivative of the last two terms on the right-hand side of Equation 9.2 [16,17]:

$$k_{opt} = \sqrt{\frac{t_{PSP}}{t_0}\left(1+\frac{\mu_{ep,enant}}{\mu_{eo}}\right)} = \sqrt{\frac{\mu_{eo}}{\mu_{me}}\left(1+\frac{\mu_{ep,enant}}{\mu_{eo}}\right)} \tag{9.5}$$

where μ_{me} is the net (apparent) mobility of the microemulsion and other parameters were identified below Equation 9.2.

In practice, resolution can be improved by increasing the difference in velocities between the free analytes and the PSP. General guidelines for the type of microemulsion to use with charged and/or neutral analytes are shown in Table 9.1.

Another aspect to consider for resolution is the migration window or elution range. The larger this value, the larger the resolution will be for a given separation. A main approach to increasing the migration window is the use of a PSP that has a large electrophoretic mobility that is opposite to the electroosmotic mobility. MEEKC differs

TABLE 9.1
Guidelines for Microemulsion Type Selection

Microemulsion Type	Analyte Type
Anionic	Neutral and/or cationic
Cationic	Neutral and/or anionic
Neutral	Cationic or anionic

from other forms of EKC in that the migration window is adjustable over a wider range by changing the charge of the aggregate, i.e., increasing/decreasing the charged surfactant concentration.

Once efficiency, retention factors, and the migration window have been optimized, enantioselectivity improvements should be sought. The main variable for enantioselectivity changes is the identity of the chiral microemulsion component. Depending on the specific microemulsion formulation, the other aggregate components may also contribute to enantioselectivity and as such their identity/concentration should be examined. Additionally, the background electrolyte can impact the enantioseparation quality not only from its contributions to conductivity, pH, and EOF but also by its interaction with the PSP.

9.3 PRINCIPLES OF CHIRAL MEEKC

9.3.1 Single-Chiral-Component Microemulsions

Theoretically, only one of the microemulsion elements (surfactant, cosurfactant, or oil) needs to be stereoselective in order to separate a pair of enantiomers, and this has been confirmed experimentally. However, the enantiomers must interact sufficiently with the chiral selector in order to be resolved. For example, a chiral oil may not be adequate for an analyte that is relatively hydrophilic or interacts electrostatically with the charged surfactant. It is important to choose the identity of the chiral aggregate agent to fit the enantiomers of interest. In some cases, better enantioselectivity can be achieved when more than one of the microemulsion constituents is chiral.

9.3.2 Microemulsions Comprised of Two or More Chiral Components

When more than one of the microemulsion elements (surfactant, cosurfactant, or oil) is chiral, the stereoselective situation is typically better, although more complex. The overall enantioselectivity provided by the chiral microemulsion is now a linear or nonlinear combination of logarithms of the enantioselectivities provided by the chiral components. If both stereochemical forms of $n - 1$ chiral microemulsion components are available, the enantioselectivity of an n-chiral-component microemulsion can always be made greater than the enantioselectivity of a single-chiral-component microemulsion by simply using the optimal stereochemical combination of the components (*vide infra*).

TABLE 9.2
Stereochemical Combinations of Microemulsion Components

Two chiral components		
RRX	RXR	XRR
RSX	RXS	XRS
SRX	*SXR*	*XSR*
SSX	*SXS*	*XSS*
Three chiral components		
	RRR	
	RRS	
	RSR	
	RSS	
	SRR	
	SRS	
	SSR	
	SSS	

Shown in Table 9.2 is an exhaustive list of the stereochemical combinations of microemulsion components when two or more of the components are chiral. In the three-letter abbreviations used to designate these combinations, the first, second, and third letters refer to the surfactant, cosurfactant, and oil, respectively. *R* and *S* indicate the absolute configurations and X represents an achiral or racemic component. Except for the migration order, which could be important when quantitating an enantiomeric impurity in the presence of the major enantiomer, half of the combinations in Table 9.2 are redundant in the sense that they would provide the same magnitude of enantioselectivity. These redundant combinations are italicized. The relative contribution to enantioselectivity provided by each microemulsion component will depend on both (1) the inherent stereoselective capability of each component and (2) the degree of stereoselective interaction of the stereogenic center(s) on the enantiomeric analytes with the stereoselective site(s) on the microemulsion components. The latter will depend in part on the physical proximity and chemical complementarity of the stereogenic centers of the enantiomeric analytes and microemulsion components.

The overall enantioselectivity provided by a multiple-chiral-component microemulsion will not usually be equal to that provided by the individual stereoselective microemulsion components, depending on whether the microemulsion components are working in concert or in opposition, i.e., on whether their stereoselective interactions with a given pair of enantiomers would result in the same or reversed migration order. If microemulsion components are found to be working in opposition in a given chiral microemulsion, a change of the stereochemical configuration of one of the components from *R* to *S* or vice versa is all that is needed to convert the chiral microemulsion to one in which the components are working in concert, thus increasing the overall enantioselectivity. This will be discussed in more detail later.

The overall enantioselectivity ($\alpha = k_2/k_1$, where k_1 and k_2 refer to the retention factors of the first and second enantiomers, respectively) provided by a chiral microemulsion is an easily measured parameter. However, it may also be predicted from the thermodynamic sum of the enantioselectivities provided by the individual microemulsion components (*vide infra*), whenever the interactions of these components with each of the enantiomers are independent of one another. Enantioselectivities provided by individual components are easily measured using single-chiral-component microemulsions that are formulated with the *R* or *S* form of one component and the racemic or achiral forms of the other components (e.g., *R* or *S* surfactant, racemic cosurfactant, racemic oil).

Although the independence of an enantiomer's interactions with different chiral components in the same microemulsion is perhaps a counterintuitive idea, we have nevertheless observed such behavior with several different multiple-chirality microemulsion formulations. Moreover, while such independence is not always observed, it is a convenient assumption to make for the purpose of enantioselectivity interpretation. Finally, though the independence of such enantiomer interactions with different chiral microemulsion components seems reasonably more plausible for those combinations (chiral surfactant/chiral oil or chiral cosurfactant/chiral oil) in which microemulsion components are in distinctly different regions of the microemulsion droplet (outer layer versus oil core, see Figure 9.2), we have occasionally observed independent, stereoselective enantiomer–microemulsion component interactions for the chiral surfactant/chiral cosurfactant combination, in which the microemulsion components are in the same region (outer layer) of the microemulsion.

Assuming for the moment that enantiomers do interact independently with each chiral microemulsion component, the overall enantioselectivity observed will be the thermodynamic sum of the enantioselectivities provided by individual microemulsion components, i.e.,

$$\alpha_{\text{microemulsion,predicted}} = \exp(\ln \alpha_{\text{surfactant}} + \ln \alpha_{\text{cosurfactant}} + \ln \alpha_{\text{oil}}) \tag{9.6}$$

Equation 9.6 was deduced from the well-known relationship $\Delta\Delta G = -RT \ln \alpha$. Since free energies for independent processes are additive, so are the logarithms of enantioselectivities.

Note that in Equation 9.6, it was assumed that the chiral microemulsion components are working in concert, i.e., the migration order of the enantiomers with each microemulsion component is the same. Obviously, when a second chiral microemulsion component is providing an enantioselectivity in opposition to the first, the thermodynamic "sum" involves a subtraction, i.e.,

$$\alpha_{\text{microemulsion,predicted}} = \exp(\ln \alpha_{\text{component \#1}} - \ln \alpha_{\text{component \#2}}) \tag{9.7}$$

since enantioselectivities are always defined to be greater than unity.

The concepts described in Equations 9.6 and 9.7 can be combined into a general equation that allows for the possibility of using more than one chiral surfactant, chiral cosurfactant, and/or chiral oil in a multiple-component chiral microemulsion:

$$\alpha_{\text{microemulsion,predicted}} = \exp\left(\sum \ln \alpha_i - \sum \ln \alpha_j\right) \quad (9.8)$$

where

α_i is the individual enantioselectivity provided by any chiral microemulsion component that results in a particular migration order for a pair of enantiomers

α_j is the enantioselectivity provided by any component that results in the opposite migration order

As mentioned earlier and to be detailed later, it is possible to maximize $\alpha_{\text{microemulsion,predicted}}$ in Equation 9.8 by optimizing the stereochemical combination of the microemulsion components (Table 9.2) so that they are all working in concert, thereby eliminating the second summation term that reduces the overall enantioselectivity.

If the chiral microemulsion components are not interacting independently with the enantiomers, i.e., when an enantiomer interacts simultaneously with two or more chiral microemulsion components, the overall enantioselectivity provided by the microemulsion will usually be higher or lower than that predicted by Equation 9.8 due to synergistic effects among the components. When $\alpha_{\text{microemulsion,observed}} > \alpha_{\text{microemulsion,predicted}}$, the synergy is termed beneficial and when $\alpha_{\text{microemulsion,observed}} < \alpha_{\text{microemulsion,predicted}}$, the synergy is said to be detrimental. Both types of synergies would appear to be equally probable, but we have observed a somewhat greater percentage of beneficial synergies.

Although synergistic stereoselective interactions have frequently been observed between different functionalities on a variety of chiral selectors employed in capillary electrophoresis (CE) and/or high-performance liquid chromatography (HPLC) (native and derivatized cyclodextrins, macrocyclic glycopeptides, polymer micelles, etc.) [20], the distinction between those and the synergies observed with chiral microemulsions is that the synergies observed previously were *intra*molecular, they occurred among different stereoselective sites on the same molecule, whereas the microemulsion synergies are *inter*molecular, they occur among different stereoselective sites on different molecules (surfactant, cosurfactant, and/or oil).

Shown in Table 9.3 are three examples of enantioselectivities achieved with two- or three-chiral-component microemulsions, each representing a different synergy scenario (beneficial synergy, no synergy, and detrimental synergy). Also reported is the stereochemical combination that provided the largest enantioselectivity for the specified analyte. Importantly, the data for pseudoephedrine show that a large beneficial synergy is possible between a surfactant (*R*-DDCV) and a cosurfactant (*S*-2-hexanol) even when the latter provides little to no enantioselectivity by itself. The large beneficial synergy that occurred caused this stereochemical combination (RSX) to have the highest enantioselectivity of all possible combinations. The data for synephrine showed that although no synergy was obtained with the first stereochemical combination, a different combination could be utilized that provided slightly higher enantioselectivity. Finally, the data for *N*-methylephedrine showed that when one stereochemical combination (RXR) resulted in low enantioselectivity due to a detrimental synergy, the enantioselectivity could be increased significantly by using a more optimal stereochemistry (RSX) for the microemulsion.

TABLE 9.3
Examples of Enantioselective Synergy (Beneficial, None, and Detrimental) for Two- and Three-Chiral-Component Microemulsions[a] and Comparison with the Enantioselectivity Obtained Using the Best Stereochemical Combination of Microemulsion Components

Oil Enantiomer	Ethyl acetate Pseudoephedrine		Dibutyl tartrate Synephrine		Dibutyl tartrate *N*-Methylephedrine	
	RXX	1.135	RXX	1.038	RXX	1.067
	XSX	<1.002	XXR	<1.002	XXR	1.009
Predicted α	RSX	1.135	RXR	1.038	RXR	1.077
Observed α	RSX	1.168	RXR	1.038	RXR	1.062
Synergy	Beneficial	0.033	None	0.000	Detrimental	–0.015
Best α	RSX	1.168	SXS	1.040	RXS	1.075

[a] Surfactant, dodecoxycarbonylvaline; cosurfactant, 2-hexanol.

While increasing the number of chiral microemulsion components can have a significant positive effect on enantioselectivity, particularly if the stereochemical combination of surfactant, cosurfactant, and/or oil is optimized, the effect on other chromatographic figures of merit (efficiency, retention factor, migration window) is usually only minimal.

9.4 FUNDAMENTAL STUDIES AND FINDINGS IN CHIRAL MEEKC

To date, only 14 articles on chiral MEEKC have been published. In addition to these original reports, review articles have also included chiral MEEKC [21,22]. The approaches and major findings from the MEEKC publications are presented in the following sections.

9.4.1 Surfactant-Based Chiral Microemulsions

The majority of chiral MEEKC studies have utilized chiral surfactants to achieve enantioresolution.

9.4.1.1 Dodecoxycarbonylvaline as the Chiral Surfactant

The first chiral MEEKC separation to utilize the chiral surfactant DDCV (see Figure 9.4) was published in 2002 by Pascoe and Foley [11]. The microemulsion contained 1.0% w/v surfactant, 1.2% v/v achiral cosurfactant (1-butanol), and 0.5% v/v achiral low-interfacial-tension oil (ethyl acetate). A zwitterionic buffer was employed to lower solution conductivity and allowed for a higher applied voltage. This chiral-surfactant-based aggregate was used to analyze nine pairs of pharmaceutical enantiomers (one neutral and eight basic); results were compared to those from chiral micellar

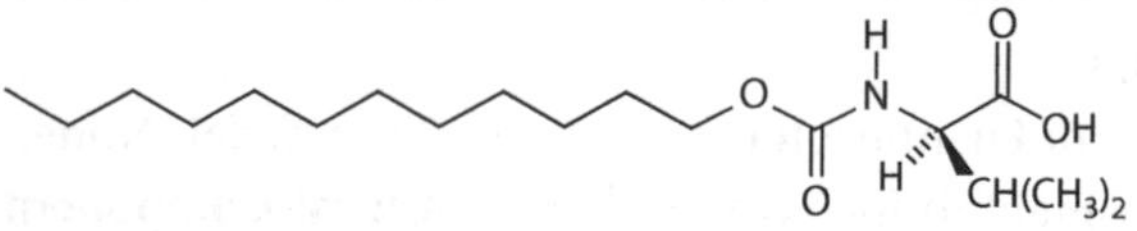

R-Dodecoxycarbonylvaline (*R*-DDCV)

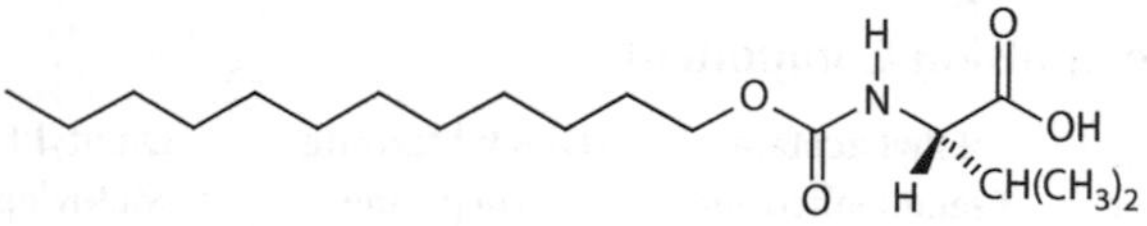

S-Dodecoxycarbonylvaline (*S*-DDCV)

FIGURE 9.4 Structures of the chiral surfactant dodecoxycarbonylvaline (DDCV).

electrokinetic chromatography (MEKC) using the same chiral surfactant. Importantly, the elution range with MEEKC was found to be approximately 2.5-fold greater than that in MEKC. MEEKC enantioselectivities were slightly larger than those with MEKC while efficiencies were lower, resolution remained essentially unchanged, and retention factors were more optimal in the MEEKC mode. Subsequent to this publication, it was determined that the lower efficiencies resulted from the buffer identity [23]. A dramatic decrease in analysis time with a microemulsion PSP instead of micelles was demonstrated for ephedrine and methylpseudoephedrine enantiomers (4 min runtime, three times shorter than MEKC, see Figure 9.5). Elution order reversal was easily accomplished by simply switching the stereochemical configuration of the surfactant as both forms are commercially available.

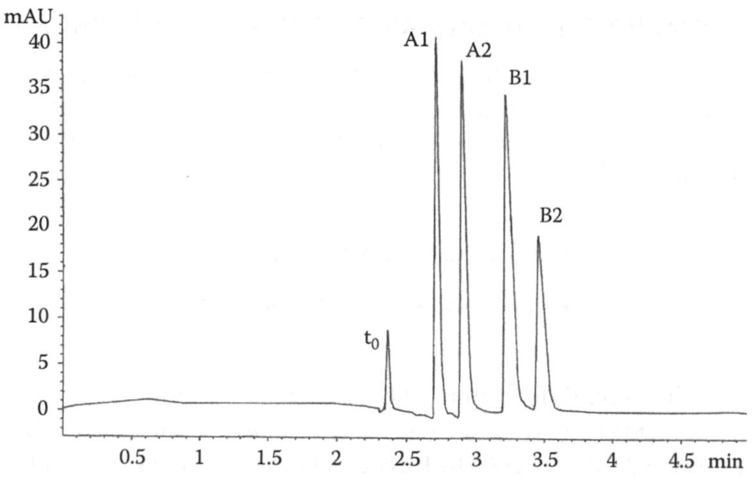

FIGURE 9.5 Separation of enantiomers of methylpseudoephedrine and ephedrine. (A1) *R*,*R*-methylpseudoephedrine, (A2) *S*,*S*-methylpseudoephedrine, (B1) *R*,*S*-ephedrine, (B2) *S*,*R*-ephedrine. Microemulsion system—1.0% (w/v) (*S*)-DDCV, 0.5% (v/v) ethyl acetate, 1.2% (v/v) 1-butanol, 50 mM ACES buffer, pH 7.0, and a separation voltage of 18 kV. (From Pascoe, R. and Foley, J.P., *Analyst*, 127, 710, 2002. With permission.)

DDCV was next used to examine the impact of oil identity and concentration [24]. Three new low-interfacial-tension oils (methyl formate, methyl acetate, and methyl propionate) were used to create chiral microemulsions (surfactant = 1.0% w/v DDCV, cosurfactant = 1.2% v/v 1-butanol). Equimolar (51 mM) and equal percentage (v/v, 0.5%) results were compared to the previously employed ethyl acetate aggregate for the analysis of 14 pharmaceutical compounds (one neutral and thirteen cations). The aims of this study were to further investigate how the oil affects chromatographic figures of merit and to see if a different low-interfacial-tension component would give better results than ethyl acetate. Microemulsions containing methyl formate exhibited short stability (one week) and problematic baselines. Among the remaining oils tested at constant volume percent, ethyl acetate gave the highest elution range, enantioselectivity, and was able to baseline separate the most pairs of enantiomers. Significant changes in efficiency were not observed with changes in oil identity. The data demonstrated that although the identity of the achiral oil does not drastically alter chiral MEEKC performance it can impact the overall quality of the separation. Ethyl acetate was deemed the best choice for an achiral oil under the conditions employed. Another aspect of this publication was selection of an appropriate PSP marker compound. Biphenyl, octanophenone, phenyldodecane, and Sudan III were tested with octanophenone giving the longest PSP retention time.

Dual-PSP experiments utilizing DDCV-based chiral microemulsions (1.0% w/v DDCV, 1.2% v/v 1-butanol, and 0.5% v/v ethyl acetate) and SDS-based achiral microemulsions (0.6% w/v SDS, 1.2% v/v 1-butanol, and 0.5% v/v ethyl acetate) in conjunction with derivatized cyclodextrins (HP-β-CD, s-β-CD, and HS-β-CD) were reported by Mertzman and Foley [23]. When more than one chiral selector is used, they will either work in a complementary or antagonistic fashion. Retention was shown to decrease when the neutral cyclodextrin HP-β-CD was added to either microemulsion, resulting in negligible enantioselectivity and resolution for the nine pairs of enantiomers. Combining charged cyclodextrins with charged microemulsions increases the conductivity of the solution and requires the applied voltage to be decreased to avoid excessive Joule heating. Systems containing DDCV-based aggregates and s-β-CD showed varying resolutions depending on the DDCV enantiomer, cyclodextrin concentration, and analyte identity. Compounds which were not resolved using DDCV-microemulsions (synephrine, propranolol, and chlorpheniramine) exhibited modest to moderate improvements in resolution with the second chiral agent (R_s values of 1.3, 0.5, and 1.7, respectively). The ephedrine-based analytes showed general trends of improved resolution with *S*-DDCV aggregates and increasing resolution with increased CD concentration. This indicates that a complementary interaction occurs between *S*-DDCV and s-β-CD. After the initial CD-MEEKC studies, further optimization was desired in terms of efficiency, peak shape, and baseline. A smaller set of samples was analyzed using the DDCV microemulsion, HS-β-CD, and phosphate buffer. The phosphate buffer was chosen over N-2-Acetamido-2-aminoethanesulfonic acid (ACES) and Tris due to its low UV absorbance and higher efficiency (Figure 9.6).

Another DDCV chiral MEEKC study focused on surfactant concentration and buffer identity [25]. As detailed above, the buffer composition was found to drastically affect the separation quality. Method robustness also improved with the phosphate buffer system. On the downside, the switch from ACES to phosphate caused

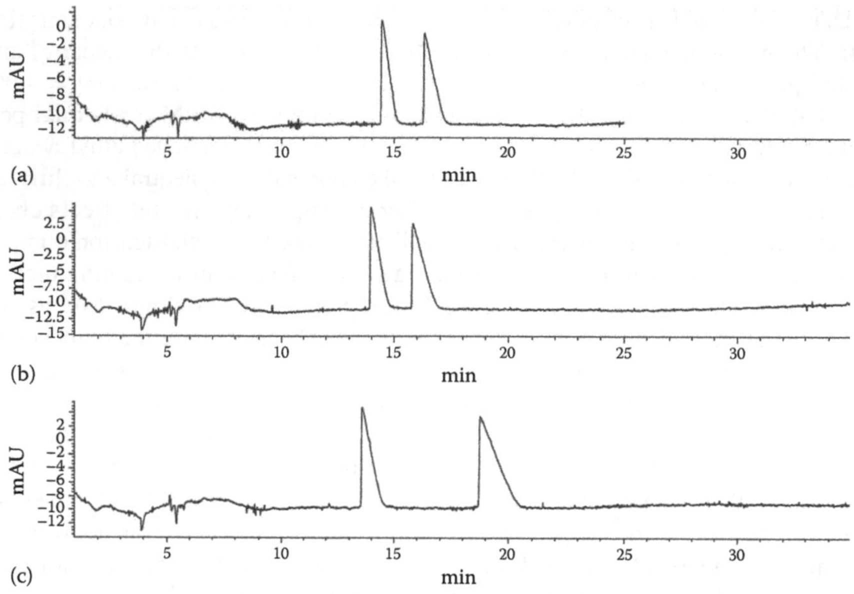

FIGURE 9.6 Representative chromatograms for enantiomeric separations using the optimized 1.5% HS-β-CD-(*S*)-DDCV microemulsion separation conditions with 50 mM phosphate buffer (pH 7.0). Voltage, 9 kV. (a) (±)-Ephedrine; (b) (±)-methylephedrine; (c) (±)-pseudoephedrine. (From Mertzman, M.D. and Foley, J.P., *Electrophoresis*, 25, 1188, 2004. With permission.)

reductions in resolution and elution range. To overcome these undesirable effects, the concentration of the surfactant was increased thereby increasing the charge of the aggregate and elution range. Microemulsions were prepared with 1%, 2%, 3%, and 4% w/v DDCV (1.2% v/v 1-butanol and 0.5% v/v ethyl acetate). The same general trend was observed for each figure of merit: increasing surfactant concentration increased efficiency, resolution, elution range, and retention (Figure 9.7). Possible explanations for this trend were a decrease in PSP overload via increased droplet size and increased mass transfer from smaller analyte penetration depths due to more tightly packed surfactant. A four times phase ratio microemulsion was also prepared in this study (4% w/v DDCV, 4.8% v/v 1-butanol, and 2.0% v/v ethyl acetate) and resulted in reduced retention, elution range, and resolution. The importance of surfactant concentration and buffer optimization in chiral MEEKC was demonstrated.

Following the surfactant concentration project, Mertzman and Foley used DDCV-based microemulsions to evaluate the influence of temperature [26]. Nine pairs of enantiomers were used as probes for van't Hoff thermodynamic analyses of 1% and 4% DDCV microemulsions (1.2% v/v 1-butanol and 0.5% ethyl acetate) over a temperature range of 15°C–35°C. As expected, both resolution and enantioselectivity decreased as temperature increased and the aggregate system containing more surfactant provided greater resolution and enantioselectivity than the lower surfactant concentration formulation. In order to conduct the desired thermodynamic calculations, the distribution coefficient (K_{eq}) and the phase ratio ($\beta = V_{PSP}/V_{aq}$) were calculated via the following equations:

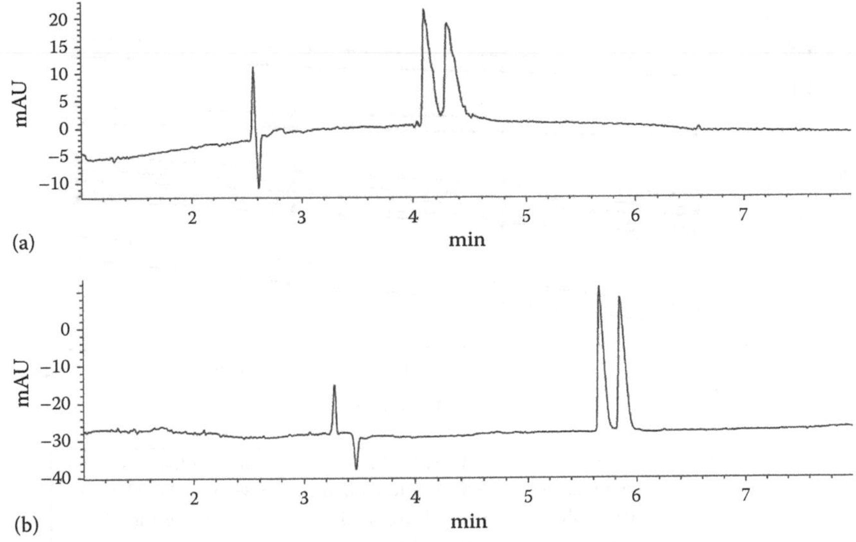

FIGURE 9.7 Representative chromatograms of the separation of (±)-norphenylephrine comparing the use of either 1% DDCV microemulsion in ACES buffer or 4% DDCV microemulsion in phosphate buffer. (a) (±)-Norphenylephrine employing 1% DDCV microemulsion in ACES buffer. Resolution, 1.26; average efficiency, 13,000; separation voltage, 17 kV. (b) (±)-Norphenylephrine employing 4% DDCV microemulsion in phosphate buffer. Resolution, 1.83; average efficiency, 50,000; voltage, 11.5 kV. Detection wavelength, 215 ± 5 nm; capillary dimensions, $L_d = 23.6$ cm, $L_t = 32$ cm, i.d. = 50 μm; injection, hydrodynamic (25 mbar × 2 s); separation voltage, 17 kV. (From Mertzman, M.D. and Foley, J.P., *Electrophoresis*, 25, 3247, 2004. With permission.)

$$k = K_{eq}\left(\frac{V_{PSP}}{V_{aq}}\right) \tag{9.9}$$

$$\beta = \frac{\sum_i^n \bar{V}_i (C_i - C_{crit,i})}{1 - \sum_i^n \bar{V}_i (C_i - C_{crit,i})} \tag{9.10}$$

where

$\bar{V}_i$ is the partial molar volume of the *i*th microemulsion component
C_i is the concentration of the *i*th microemulsion component
$C_{crit,i}$ is the critical aggregate concentration of the *i*th component (assumed to be zero for nonaggregating components, i.e., 1-butanol and ethyl acetate)

The phase ratios for the DDCV microemulsions were determined to be 0.028 and 0.060 for the 1% and 4% surfactant concentration, respectively. After values were acquired for the distribution coefficients, van't Hoff plots (ln K_{eq} versus 1/T) were

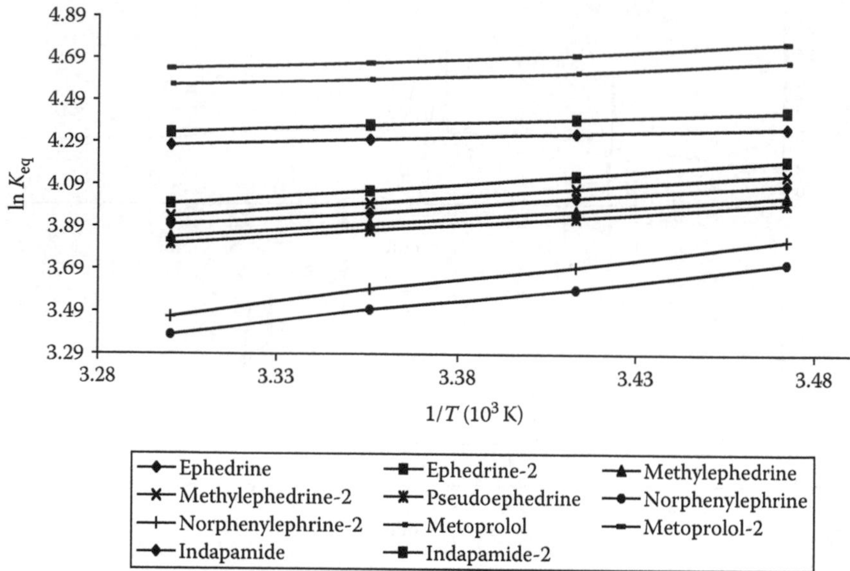

FIGURE 9.8 van't Hoff plot for moderate-to-highly retained solutes using 1% DDCV microemulsion over a temperature range of 15°C–30°C. Voltage 8.0 kV. Detection wavelength: 215 ± 5 nm, capillary dimensions: L_d = 23.6 cm, L_t = 32 cm, i.d. = 50 μm, injection: hydrodynamic (25 mbar × 2 s). (From Mertzman, M.D. and Foley, J.P., *J. Chromatogr. A*, 1073, 181, 2005. With permission.)

constructed. A representative plot is shown in Figure 9.8. From the graphs, the enthalpy (slope, $-\Delta H°/R$) and entropy (intercept, $\Delta S°/R$) of transfer were calculated. Linear plots indicate that a single mechanism occurs for the analyte–selectand interaction over the temperature range studied, with assumptions that no change in heat capacity occurs and the phase ratio does not change with temperature. Interestingly, the 4% DDCV microemulsion did not give linear van't Hoff plots as found for the 1% DDCV system and was therefore excluded from further thermodynamic examination. The lack of linearity for the more concentrated system was most likely due to a greater variation in the phase ratio or overall PSP structure with temperature. Values for enthalpy of transfer (for the transition from the aqueous buffer into the aggregate) were negative in all cases, thereby demonstrating the analyte's preference for partitioning into the microemulsion. In contrast, the entropies of transfer were inconsistent in terms of sign with positive values representing increased disorder for analyte partitioning into the PSP and vice versa for negative values. Although enthalpy and entropy results are informative for discerning separation mechanisms, they only give insight into achiral interactions. In order to reveal trends in chiral recognition mechanisms, the enthalpic and entropic contributions to changes in Gibb's free energy (proportional to the logarithm of enantioselectivity) were examined, with entropy contributing the most to enantioresolution. This finding is supported by the general 3-point-rule for chiral interaction: one enantiomer must interact with the chiral selectand via three interactions while the other enantiomer will have two, effectively boosting the difference in entropy for the enantiomer pair. Lastly for this

TABLE 9.4
Comparison of Micelles, Modified Micelles, and Microemulsion

	2% w/v DDCV Micelles	2% w/v Butanol-Modified DDCV Micelles	2% w/v DDCV Microemulsion	4% w/v DDCV Microemulsion
Resolution	1.8	1.5	1.5	1.7
Efficiency	26,000	35,000	35,000	68,000
Retention factor	3.2	2.6	2.8	2.9
Enantioselectivity	1.10	1.10	1.10	1.07
Elution range	7.8	5.3	4.7	6.5

study, the existence of a common retention mechanism for the analytes tested was shown via enthalpy/entropy compensation at a temperature of 227 K (linear plot of ΔH versus ΔS where the slope gives the compensation temperature).

To fairly compare and contrast the chiral separations obtainable with chiral micelles, solvent-modified chiral micelles and chiral microemulsions, Mertzman and Foley utilized the chiral surfactant DDCV [27]. Two levels of surfactant, 2% and 4% w/v, were employed with and without 1.2% v/v 1-butanol and/or 0.5% v/v ethyl acetate for the enantioseparation of eleven pharmaceutical enantiomers. At 4% DDCV, only a microemulsion system (1.2% 1-butanol and 0.5% ethyl acetate) provided stable baselines, indicating the instability of the micellar and modified-micellar (1.2% 1-butanol) solutions. Despite the erratic baselines, the elution range was largest for MEKC at both DDCV concentrations and lowest for MEEKC due to the electrophoretic mobility of the aggregate (influenced by charge and frictional drag, data for stable systems given in Table 9.4). In terms of chromatographic figures of merit, the largest and sometimes most nonoptimal retention factors occurred with micelles (modified micelles and microemulsions gave similar values, see Table 9.4). As shown in Table 9.4, the best resolution was also obtained with the micellar solution (2% w/v) and significant differences were not observed for enantioselectivity. The most dramatic differences among the DDCV aggregates were for efficiency at the 4% w/v level where microemulsion results were double those of butanol-modified micelles and more than twice those of micelles. Although baseline and stability issues may have contributed to this trend, larger retention with micelles most definitely would have increased longitudinal diffusion and hence broadened the peaks. At 2% DDCV, efficiency was comparable and slightly better with microemulsions and butanol-modified micelles than micelles (Table 9.4 and Figure 9.9). This effect likely stems from PSP fluidity where the 1-butanol could decrease the rigidity of the aggregate, thus allowing analytes to more easily partition in and out (increased mass transfer). The results from this study remind one that it is best to start with the simplest system possible but for difficult chiral separations, the potential benefits of MEEKC justify the added complexity.

The effect of achiral cosurfactant identity in DDCV-based microemulsions (2% w/v *R*-DDCV and 0.5% ethyl acetate as the oil) was examined for the analyses of six pairs of enantiomers [28]. The alcohols selected as cosurfactants included primary

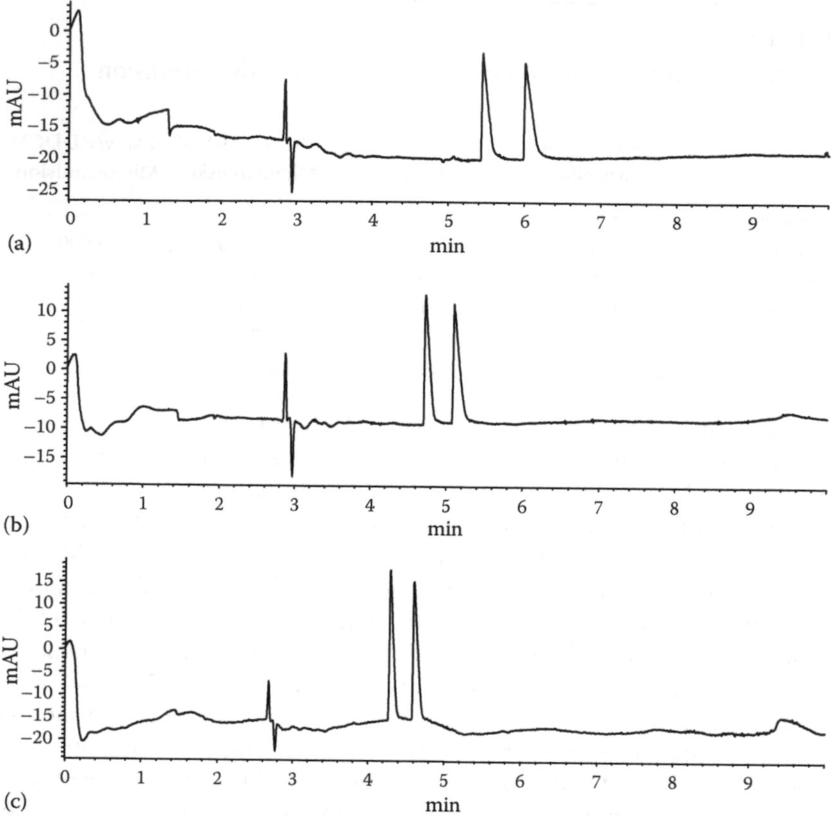

FIGURE 9.9 Representative chromatograms of the separation of (±)-pseudoephedrine illustrating the similarity in baseline stability provided by 2% DDCV (a) micelles, (b) butanol-modified micelles, or (c) microemulsion. Voltage: 12 kV. Detection wavelength: 215 ± 5 nm, capillary dimensions: $L_d = 23.6$ cm, $L_t = 32$ cm, i.d. = 50 μm, injection: hydrodynamic (25 mbar × 2 s). (From Mertzman, M.D. and Foley, J.P., *Electrophoresis*, 26, 4153, 2005. With permission.)

(1-butanol, 1-pentanol, and 1-hexanol), secondary (2-pentanol and 2-hexanol), and cyclic (cyclopentanol and cyclohexanol) structures, all at equimolar concentrations (131 mM). Particle size analysis of the aggregates showed that the cosurfactant structure did not impact the size of the microemulsion (range of 4.7–6.8 nm) and therefore the cosurfactants were incorporated into the structure in a similar manner. Only minor changes in elution order (based on retention factors) were observed for the ephedrine derivatives, indicating that the alcohol has some influence on selectivity [29]. Although the exact effects were analyte specific (see Figure 9.10 for example electropherograms), some general trends in chromatographic figures of merit were apparent: cyclic and short chain primary alcohols improved enantioselectivity, longer chain primary alcohols decreased enantioselectivity, best overall efficiency with 1-hexanol, and best overall resolution with 1-butanol. Enantioselectivity differences with cosurfactant were proposed to result from the alignment of the alcohol among the surfactant monomers: If the alcohol increased the distance between surfactant monomers (cyclic or secondary structures) enantiomers would have better chance to associate with the chiral selector. Another consequence of this postulate would be increased water content in the interior of the nanodroplet which could also improve the interaction with the chiral surfactant through hydrogen bonding. The hydrophobicity of the microemulsions was assessed via methylene selectivity and was found to increase in the order of cyclopentanol < 1-butanol ≤ cyclohexanol ≤ 2-pentanol < 1-pentanol < 2-hexanol < 1-hexanol, thereby confirming that secondary and cyclic alcohols align differently in the microemulsion and impact the amount of water that can penetrate into the PSP. Significant variations in efficiencies (almost 50%) were obtained with the cosurfactants tested. The inverse relationship between enantioselectivity and efficiency was explained to be due to the nature of the chiral three-point-interaction. Chiral associations will take place on a longer timescale (i.e., it is highly unlikely that all three chiral interactions occur precisely at the same time)

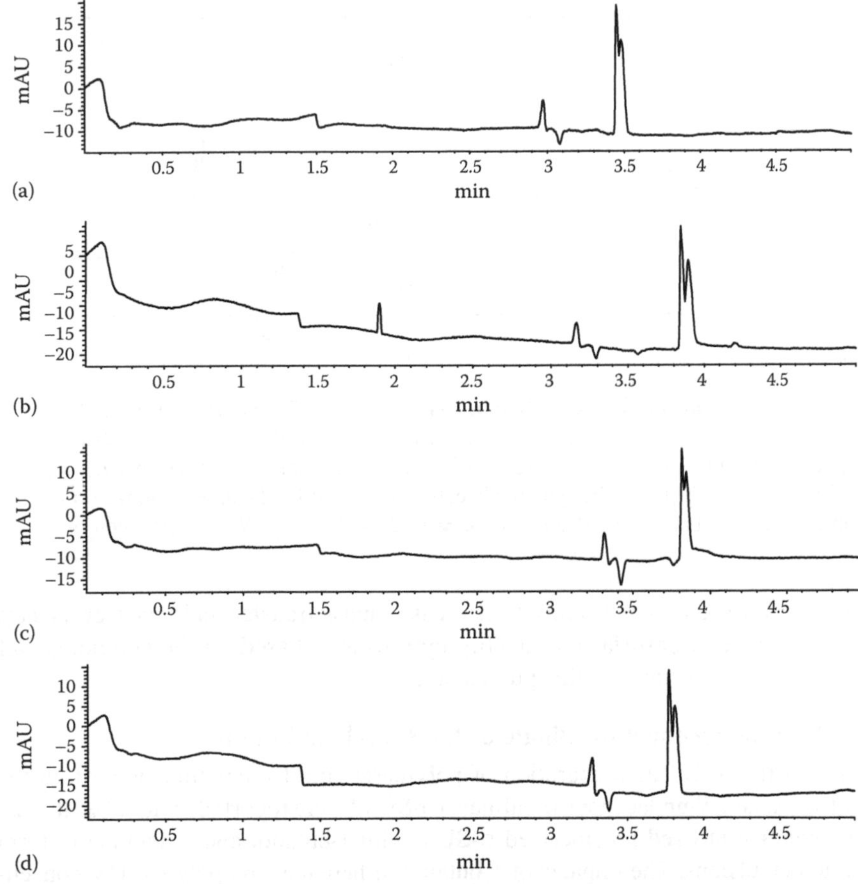

FIGURE 9.10 Effect of cosurfactant identity on the separation of atenolol enantiomers using chiral MEEKC. (a) 1-pentanol (1.40% v/v); (b) 2-pentanol (1.40% v/v); (c) 1-hexanol (1.65% v/v); (d) 2-hexanol (1.65% v/v);

(continued)

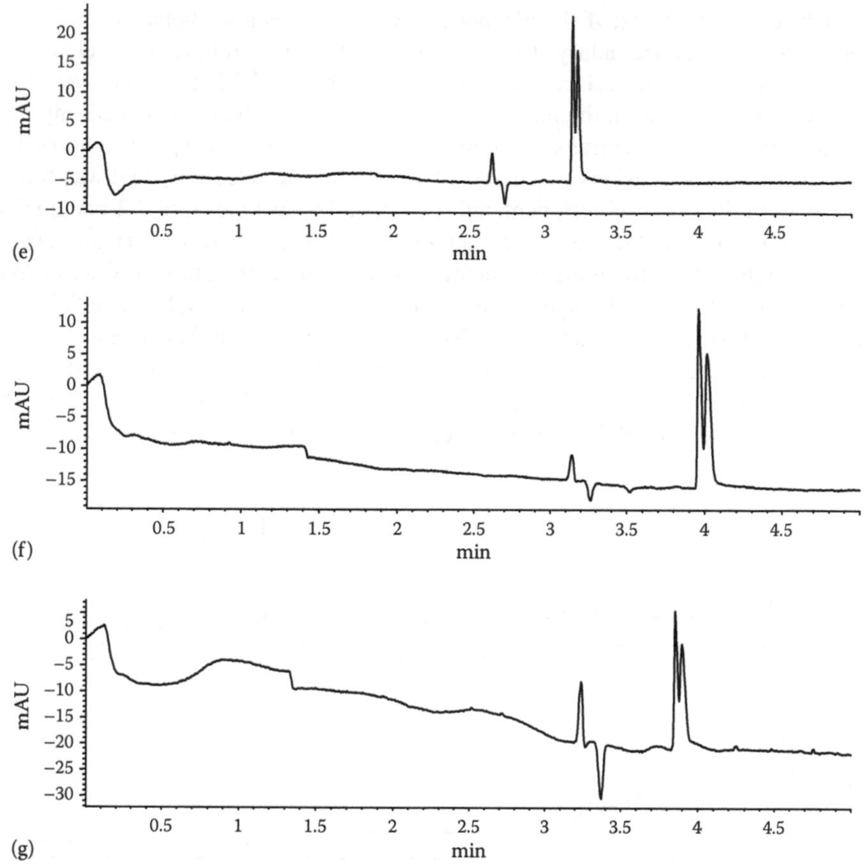

FIGURE 9.10 (continued) (e) 1-butanol (1.20% v/v); (f) cyclopentanol (1.19% v/v); (g) cyclohexanol (1.39% v/v). Other conditions: 2% w/v DDCV; 0.5% v/v ethyl acetate; 50 mM phosphate buffer, pH 7.0; detection wavelength = 215 ± 5 nm; capillary dimensions: L_{tot} = 32 cm, L_{eff} = 23.6 cm, i.d. = 50 μm; hydrodynamic injection = 25 mbar for 2 s; applied voltage = 11.5 kV. (From Kahle, K.A. and Foley, J.P., *Electrophoresis*, 27, 4321, 2006. With permission.)

than achiral associations, leading to decreased mass transfer and lower efficiency. The importance of cosurfactant identity optimization based on the compounds of interest was demonstrated in this publication.

9.4.1.2 *N*-undecenoyl-D-valinate as the Chiral Surfactant

Unique methods for the preparation of polymeric chiral microemulsions from the chiral surfactant *N*-undecenoyl-D-valinate (D-SUV) were reported by Iqbal et al. [12]. One technique utilized polymerized D-SUV with 1-butanol and *n*-heptane to form the microemulsions. The impacts of 1-butanol, *n*-heptane, and poly-D-SUV concentrations were examined using binaphthyl, barbiturate, and paveroline derivatives with different trends in resolution, retention, efficiency, and enantioselectivity found for the different analytes. For 1,1′-binaphthyl-2,2′-diyl hydrogen phosphate (BNP) no separation was obtained without 1-butanol and the optimal cosurfactant concentration was shown to be 3.5% (range of 1% to 7%). It was hypothesized that too much 1-butanol (3.5%–7.0% w/w) would fill the microemulsion so that analytes could not penetrate into the aggregate to interact with the chiral surfactant. The other binapthyl derivatives (1,1′-bi-2-naphthol (BOH) and 1,1′-binaphthyl-2,2′-diamine (BNA)) displayed highest resolution values with 1.0% w/w and 0.50% w/w 1-butanol, respectively. Barbiturate and paveroline derivatives had maximum resolutions with 3.5% and 1.0% cosurfactant, respectively. Oil phase concentrations between 0.21% and 1.60% w/w were studied with the best resolution of binaphthyl, barbiturate, and paverolines achieved with 0.82% w/w, 0.82% w/w, and 1.60% w/w, respectively. The changes observed with increasing oil concentration were explained to be due to increasing hydrophobicity of the droplet and decreasing space for analytes to penetrate. Overall, increasing the chiral surfactant concentration within the range of 0.25%–0.76% w/w resulted in resolution improvements. The chiral microemulsions prepared from chiral polymer were also compared to chiral MEKC (no butanol or *n*-heptane) and solvent-modified MEKC (butanol or *n*-heptane) with MEEKC generally performing better than the other EKC modes (Figure 9.11). The second technique for the preparation of chiral polymeric microemulsions started by combining surfactant (D-SUV, 3.00% w/w), cosurfactant (1-butanol, 0.0%–7.50% w/w), and oil (*n*-heptane, 0.82% w/w)

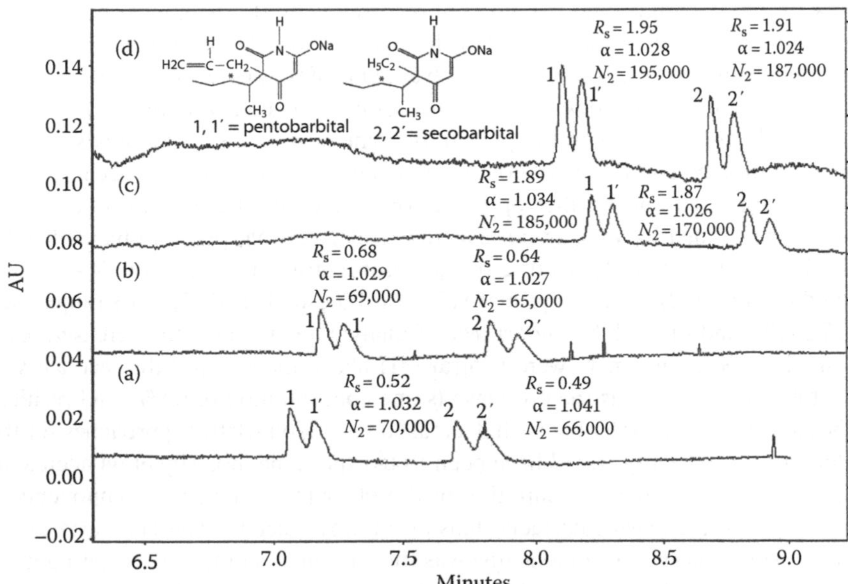

FIGURE 9.11 Comparison of MEEKC and MEKC for simultaneous separation and enantioseparation of 1,1′= (±) pentobarbital (*P*) and 2,2′= (±) secobarbital (*S*). Both MEEKC and MEKC contain 0.76% (w/w) poly-D-SUV, Na_2HPO_4/NaH_2PO_4 25 mM at pH 7.0. Enantioseparation is done using (a) MEKC without 1-butanol and *n*-heptane (b) solvent-modified MEKC with 0.82% (w/w) *n*-heptane (c) solvent-modified MEKC with 3.50% (w/w) 1-butanol, and (d) MEEKC 3.50% (w/w) with 1-butanol and 0.82% (w/w) *n*-heptane. (From Iqbal, R. et al., *J. Chromatogr. A*, 1043, 291, 2004. With permission.)

followed by sonication and polymerization. Residual 1-butanol and *n*-heptane not encapsulated within the microemulsion polymer (MP) were evaporated and the polymer was then freeze dried. The MP was dissolved in buffer before use. A smaller set of enantiomers was used to evaluate MP systems with better separations obtained with 3.5%, 1.50%, and 0.25% w/w 1-butanol for BNP, BOH, and the barbiturate derivatives, respectively. Compared to microemulsion formulations containing polymeric surfactant, improved resolution was obtained for BNP only although baseline separation could not be achieved.

9.4.2 Cosurfactant-Based Chiral Microemulsions

One chiral MEEKC publication demonstrated the ability of chiral 2-alkanols (cosurfactant) to resolve enantiomers [13]. Initially, *R*-(−)-2-butanol (6.6%), *R*-(−)-2-pentanol (5.0%), *R*-(−)-2-hexanol (5.0%), and *R*-(−)-2-heptanol (3.3%) were evaluated in microemulsion formulations containing SDS (3.5%) and *n*-octane (0.8%). The shortest chain cosurfactant, 2-butanol, did not provide any enantioresolution for the five pairs of enantiomers tested. This particular alcohol was deemed to have too short of an alkyl chain to effectively span the interfacial region of the nanodroplet. Separations with 2-pentanol were more successful for three compounds. When 2-hexanol was employed, better resolution was obtained for some of the analytes but two pairs of enantiomers remained unresolved. Chiral 2-heptanol enabled propranolol to be separated (unresolved with all other cosurfactants) but reduced the resolution for enantiomers previously resolved. One of the analytes, *N*-methyl ephedrine, could not be separated with any of the cosurfactants utilized. Structurally, this analyte lacks the β-amino proton present in some of the other compounds thereby indicating that the chiral recognition mechanism was reliant on hydrogen bonding with the chiral reagent. Further optimization of the chiral-2-hexanol microemulsion was performed by varying cosurfactant concentration, pH, oil phase identity, and oil concentration. In terms of 2-hexanol concentration, a range of 1%–6% was studied (Figure 9.12) with an optimal value of 5% established. The pH range used was 7.2–10.2 and showed the importance of analyte ionization with decreased resolution when the enantiomers were neutral (pH above their pK_a). Ethyl acetate was substituted for *n*-octane as the oil phase (same concentration of 0.8%) and resulted in the same migration order but with decreased resolution. Better approaches for the evaluation of oil identity would have been to use the same molarity ethyl acetate as *n*-octane or to adjust the concentration of the other microemulsion components to compensate for the lower interfacial tension of ethyl acetate. The last variation in the microemulsion system composition was the elimination of the oil, resulting in no resolution. The ability to reverse the elution order by simply switching the cosurfactant stereochemistry was demonstrated for the enantiomers of norephedrine.

9.4.3 Oil-Based Chiral Microemulsions

The first separation via chiral MEEKC was reported by Aiken and Huie in 1993 and combined the chiral oil (2*R*, 3*R*)-di-*n*-butyl tartrate (0.5% w/w, shown in Figure 9.13)

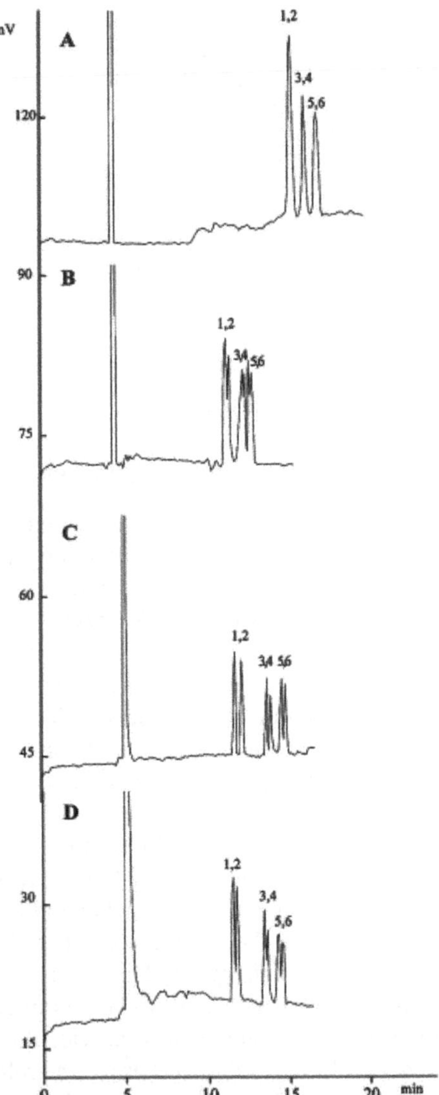

FIGURE 9.12 MEEKC separation of the enantiomers of norephedrine, ephedrine, and nadolol using different concentrations of *R*-(−)-2-hexanol: (A) 1%, (B) 3%, (C) 5%, and (D) 6%. Other microemulsion constituents composed of 0.8% *n*-octane and 3.5% SDS in 10 mM borate buffer (pH 9.2). Peak identification: 1, (1*R*,2*S*)-(−)-norephedrine; 2, (1*S*,2*R*)-(+)-norephedrine; 3, (1*R*,2*S*)-(−)-ephedrine; 4, (1*R*,2*S*)-(+)-ephedrine; 5 and 6, DL-nadolol (racemic). (From Zheng, Z.-X. et al., *Electrophoresis*, 25, 3263, 2004. With permission.)

with the achiral surfactant SDS (0.6% w/w) and the achiral cosurfactant 1-butanol (1.2% w/w) to achieve a selectivity of 2.6 for the enantiomers of ephedrine [10]. The presence of 1-butanol as the cosurfactant was shown to be essential: no enantioseparation could be obtained in its absence.

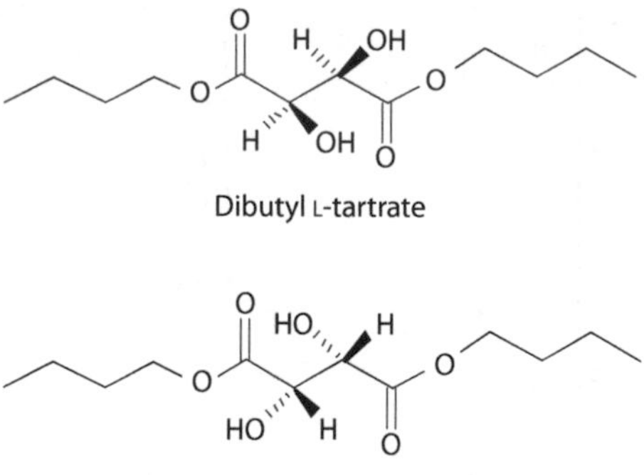

Dibutyl L-tartrate

Dibutyl D-tartrate

FIGURE 9.13 Structure of the chiral oil dibutyl tartrate.

9.4.4 Multiple-Chiral-Component Microemulsions

The first report that utilized two chiral reagents in a single microemulsion formulation was published by Kahle and Foley in 2006 [30]. Each enantiomer of the chiral surfactant DDCV (2% w/v) was combined with the chiral cosurfactant *S*-2-hexanol (1.65% v/v) and the achiral oil ethyl acetate (0.5% v/v). Initially, a lower concentration of alcohol was used (1.2% v/v) but the formulation containing both chiral selectors in the *S* configuration caused peak splitting indicating that this was not an optimized system. Increasing the cosurfactant concentration from 1.2% v/v to 1.65% v/v (millimolarity equal to that of previously employed 1-butanol formulations) alleviated the peak shape problems. Two pairs of enantiomers, *N*-methyl ephedrine and pseudoephedrine, were used to probe the performance of each dual-chirality microemulsion. The combination of *S*-DDCV and *S*-2-hexanol proved beneficial in terms of efficiency and resolution for both analytes. In Figure 9.14 representative electropherograms of pseudoephedrine are shown. No change in enantioselectivity was observed for *N*-methyl ephedrine but the other analyte, pseudoephedrine, exhibited a decrease with *S*-DDCV. Analysis of changes in Gibb's free energy allowed thermodynamic synergies (beneficial and detrimental) to be identified. These calculations required information on each chiral component's individual contribution to enantioselectivity. Racemic 2-hexanol was used to determine the role of the chiral surfactant and racemic surfactant was used for the chiral alcohol (no separation achieved for either enantiomer with this system). For pseudoephedrine and *N*-methyl ephedrine, employment of *S*-DDCV in conjunction with *S*-2-hexanol resulted in beneficial and detrimental synergisms, respectively. The cause of the synergistic effects was hypothesized to be either (1) an interaction between the chiral surfactant and chiral cosurfactant, or (2) a three-way interaction involving the chiral surfactant, the chiral cosurfactant, and the chiral analyte. Additionally, results from microemulsions containing racemic 2-hexanol and 1-butanol (data previously reported) were compared, with the former giving a significant improvement in efficiency (almost threefold).

The first chiral oil used in MEEKC, dibutyl tartrate (Figure 9.13), was combined with the chiral surfactant DDCV to further study dual-chirality microemulsions [31]. A series of four two-chiral-component aggregates containing *R*- or *S*-DDCV (2.0% w/v), racemic 2-hexanol (1.65% v/v), and *R*- or *S*-dibutyl tartrate (dibutyl-L-tartrate

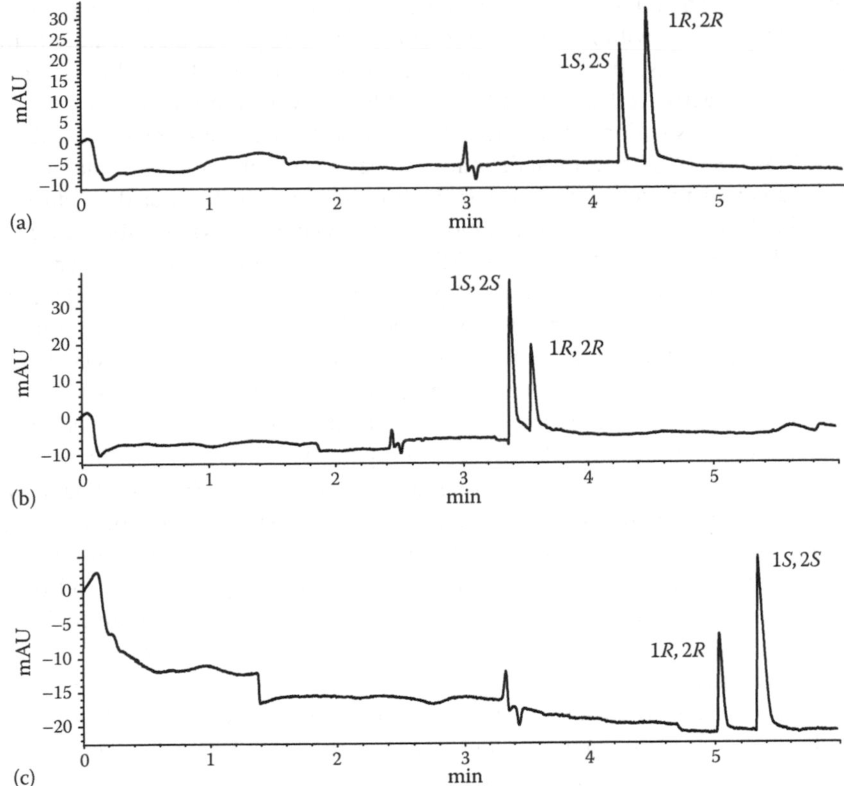

FIGURE 9.14 Effect of cosurfactant stereochemical configuration on the enantiomeric separations of pseudoephedrine. (a) *RX* microemulsion, (b) *RS* microemulsion, and (c) *SS* microemulsion. Peak identification: 1*S*,2*S* = (1*S*,2*S*)-pseudoephedrine; 1*R*,2*R* = (1*R*,2*R*)-pseudoephedrine. Surfactant concentration = 2% w/v; cosurfactant concentration = 1.65% v/v; 0.5% v/v ethyl acetate; 50 mM phosphate buffer, pH 7.0; detection wavelength = 215 ± 5 nm; capillary dimensions: L_{tot} = 32 cm, L_{eff} = 23.6 cm, i.d. = 50 μm; hydrodynamic injection = 25 mbar for 2 s; applied voltage = 11.5 kV. (From Kahle, K.A. and Foley, J.P., *Electrophoresis*, 27, 896, 2006. With permission.)

and dibutyl-D-tartrate, respectively, 1.23% v/v) were employed for the separation of six pairs of pharmaceutical enantiomers and results were statistically evaluated. In comparison to ethyl acetate, dibutyl tartrate proved to be slightly more difficult to use due to its higher hydrophobicity. It required longer sonication times for solubilization (microemulsions remained stable after formation for several months) and more rigorous column conditioning for stable baselines. Another observation was that one chiral analyte, atenolol that could be minimally resolved with ethyl acetate could not be resolved at all with the new oil. Enantioselectivity trends were mainly analyte specific but overall the system comprised of *R*-DDCV and *S*-dibutyl tartrate gave the highest value and dual-chirality aggregates provided slight improvements over single-chirality nanodroplets. The ephedrine derivatives displayed an enantioselective preference for the surfactant and the oil being present in opposite stereochemical configurations. The data for efficiency were clearly divided into two groups based on oil configuration, with racemic and *R*-dibutyl tartrate giving significantly better values. The root cause of this variation could not be discerned. The highest average efficiency, across all compounds, was obtained with the racemic oil system. With respect to resolution, on average the enantiomers were best resolved with the *R*-DDCV/*S*-dibutyl tartrate formulation. Changes in resolution were explained via the fundamental resolution equation (Equation 9.2). As in the first dual-chirality publication [30], the existence of enantioselective synergisms was established using a thermodynamic model. Two analyte-PSP combinations did not result in any synergy whereas five analyte-PSP cases displayed positive synergy (enantioselectivity higher than expected) and thirteen showed negative synergy. A lack of symmetry for synergy between oppositely configured microemulsions (i.e., *R*-DDCV/*S*-dibutyl tartrate vs. *S*-DDCV/*R*-dibutyl tartrate) was observed but the underlying source was not identified. The thermodynamic analysis enabled the authors to narrow down the potential interactions responsible for the synergy to a three-way interaction between chiral analyte and the two chiral microemulsion components. Synergy would only be observed for moderately hydrophobic compounds which would penetrate far enough into the aggregate to interact with the chiral oil as well as the surface-oriented surfactant. During the course of these experiments, single-chirality, oil-based microemulsions were tested with no resolution or enantioselectivity obtained. These findings contradicted those published by Aiken and Huie [10] and were surprising given the similarity in microemulsions.

Another chiral oil, diethyl tartrate (Figure 9.15), was simultaneously incorporated into chiral-DDCV-microemulsions and tested in the same manner as dibutyl tartrate (above) [32]. The oil concentration was adjusted to 0.88% v/v to keep the same molar concentration as with the ethyl acetate and dibutyl tartrate formulations. The absence of enantioselectivity was again observed for chiral-oil-only microemulsions. Variations in efficiency were analyte dependent (see Figure 9.16 for representative electropherograms of pseudoephedrine), with *R*-DDCV and *S*-DDCV containing

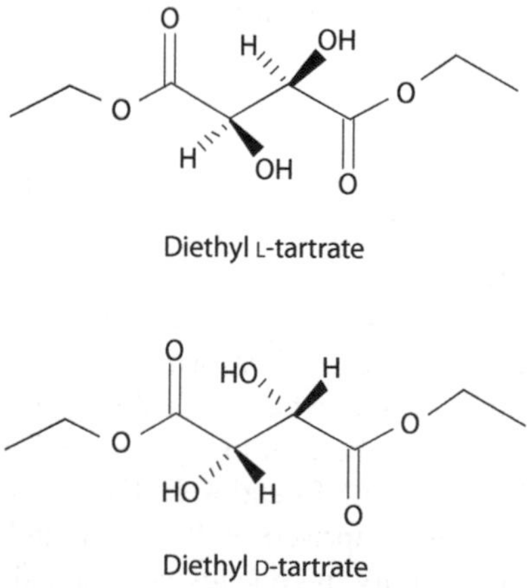

Diethyl L-tartrate

Diethyl D-tartrate

FIGURE 9.15 Structure of the chiral oil diethyl tartrate.

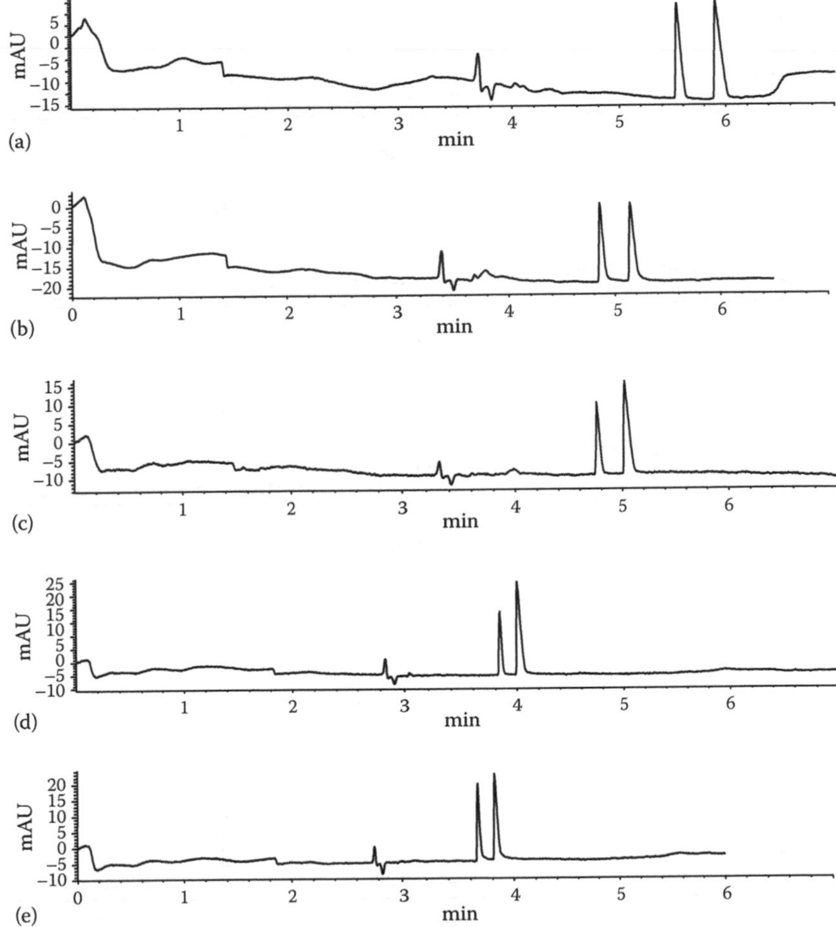

FIGURE 9.16 Effect of surfactant and oil stereochemical configurations on the efficiency and separation of pseudoephedrine enantiomers. (a) *SXS* microemulsion, (b) *RXR* microemulsion, (c) *RXX* microemulsion, (d) *RXS* microemulsion, and (e) *SXR* microemulsion. Surfactant (DDCV) concentration = 2.00% w/v; cosurfactant (2-hexanol) concentration = 1.65% v/v; oil (diethyl tartrate) concentration = 0.88% v/v; 50 mM phosphate buffer, pH 7.0. Capillary dimensions: L_{tot} = 32 cm, L_{eff} = 23.6 cm, i.d. = 50 μm. Detection wavelength was 215 ± 5 nm, injections were performed hydrodynamically for 2 s with 25 mbar pressure, and the applied voltage was 11.5 kV. (From Kahle, K.A. and Foley, J.P., *Electrophoresis*, 28, 2644, 2007. With permission.)

aggregates providing the best values and worst values, respectively. An in-depth description of EKC zone broadening contributors was given, with analyte-PSP mass transfer pinpointed as the most likely source of differing efficiencies. Specifically, the more polar microemulsion aggregates (determined from methylene selectivity) gave improvements in efficiency as a result of faster kinetics. As previously detailed, more hydrophilic microemulsions contain more water, are more fluid, and have greater spacing between surfactant monomers. Minimal changes in enantioselectivity

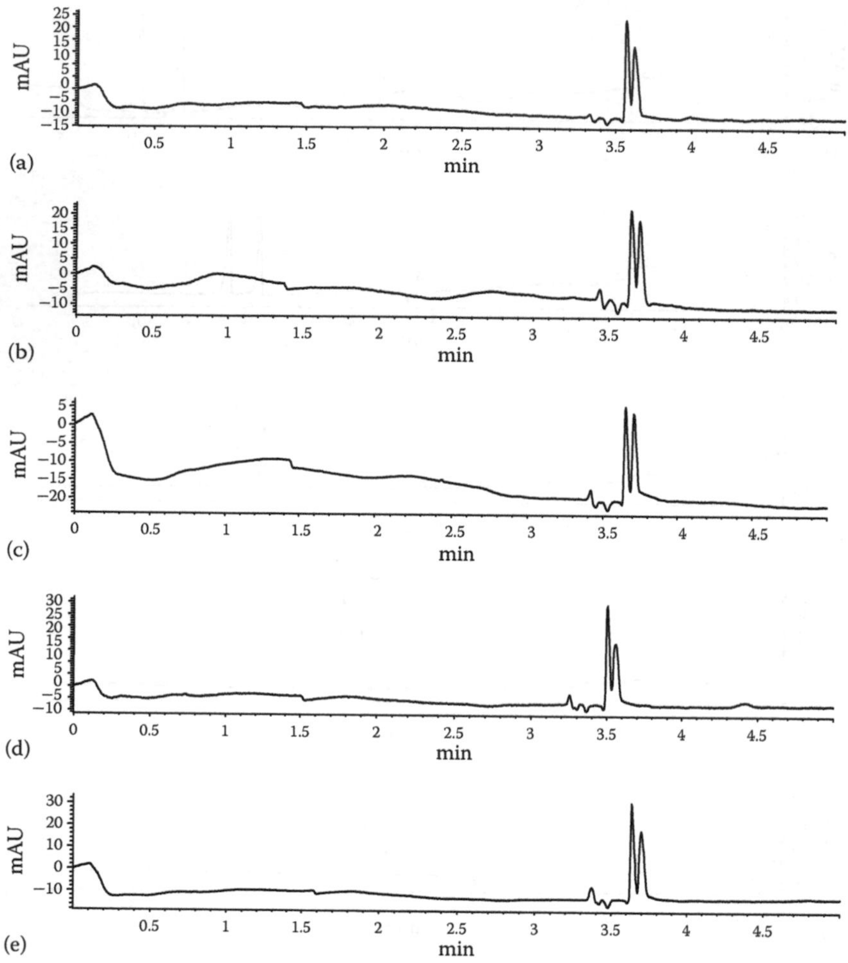

FIGURE 9.17 Effect of surfactant and oil stereochemical configurations on the α_{enant} and resolution of synephrine enantiomers. (a) *RXX* microemulsion, (b) *SXR* microemulsion, (c) *RXR* microemulsion, (d) *RXS* microemulsion, and (e) *SXS* microemulsion. Conditions as in Figure 9.16. (From Kahle, K.A. and Foley, J.P., *Electrophoresis*, 28, 2644, 2007. With permission.)

were observed and these depended on analyte identity. The effect of microemulsion stereochemical configuration on the separation of synephrine enantiomers is shown in Figure 9.17. Similarly, resolution was not drastically impacted by variation of aggregate chiral composition; the combination of *S*-DDCV with *S*-diethyl tartrate (diethyl-D-tartrate) resulted in a slightly higher average resolution for the six enantiomers tested. The fundamental resolution equation was used to determine which chromatographic figure of merit played the biggest role in resolution improvements. The data are shown in Table 9.5. Analysis of differences in Gibb's free energy changes revealed the existence of synergy in 20 out of 24 analyte-PSP cases, eleven beneficial and nine detrimental. Results from ethyl acetate, dibutyl tartrate, and

TABLE 9.5
Evaluation of Resolution Improvements via the Fundamental Resolution Equation[a]

Analyte	Best R_s^b	Rank of k_{avg}/k_{opt}	Rank of Efficiency	Rank of Enantioselectivity	Rank of Elution Range	Biggest Contributor to R_s
Ephedrine	*RR*	Worst	Best	Second best	Second best	Elution range
Pseudoephedrine	*RR*	Worst	Second worst	Second best	Second best	Elution range
N-Methyl ephedrine	*SR*	Second best	Second worst	Worst	Worst	Retention factor
Atenolol	*SS*	Second worst	Worst	Best	Best	Enantioselectivity
Synephrine	*SS*	Second best	Second worst	Best	Best	Enantioselectivity
Metoprolol	*SS*	Worst	Third best	Best	Best	Elution range

[a] Rankings based on a total of five microemulsions.
[b] Microemulsion abbreviations: *RR*, *R*-DDCV/*R*-diethyl tartrate; *SR*, *S*-DDCV/*R*-diethyl tartrate; *SS*, *S*-DDCV/*S*-diethyl tartrate.

these diethyl tartrate microemulsions were compared, with dibutyl tartrate providing the best average efficiency and most optimal retention factors and ethyl acetate providing the highest enantioselectivity and resolution. Ethyl acetate and diethyl tartrate aggregates had similar hydrophobicities that were lower than those for dibutyl tartrate systems. The studies on chiral surfactant/chiral oil microemulsions clearly demonstrated the ability of the oil to affect chiral separations.

Going one step further, Kahle and Foley explored microemulsions formulated with all chiral components (surfactant, cosurfactant, and oil) [33]. Four different systems containing DDCV (2.0% w/v, *R*- or *S*-), *S*-2-hexanol (1.65% v/v), and diethyl tartrate (0.88% v/v, *R*- or *S*-) were prepared and used to separate six pairs of pharmaceutical enantiomers. Aggregates formed from *R*-DDCV/*S*-2-hexanol/*R*-diethyl tartrate did not form true stable droplets but instead reverted back to individual layers. Considerable changes in enantioselectivity were not obtained with variations in microemulsion stereochemical composition. Three-chiral-component systems did provide higher values than the chiral-surfactant-only microemulsion, with the combination of *S*-DDCV/*S*-2-hexanol/*S*-diethyl tartrate performing the best for four out of six pairs of enantiomers. An example of the enantioselectivity differences is shown in Figure 9.18 for synephrine. Half the analytes experienced better efficiencies with triple- versus single-chirality nanodroplets and the *R*-DDCV/*S*-2-hexanol/*S*-diethyl tartrate combination afforded the best average efficiency among the baseline separated compounds. As seen in prior DDCV-MEEKC studies, higher efficiency was well correlated to lower enantioselectivity (mass transfer implications previously discussed). The novel triple-chirality PSPs surpassed the single-chirality version

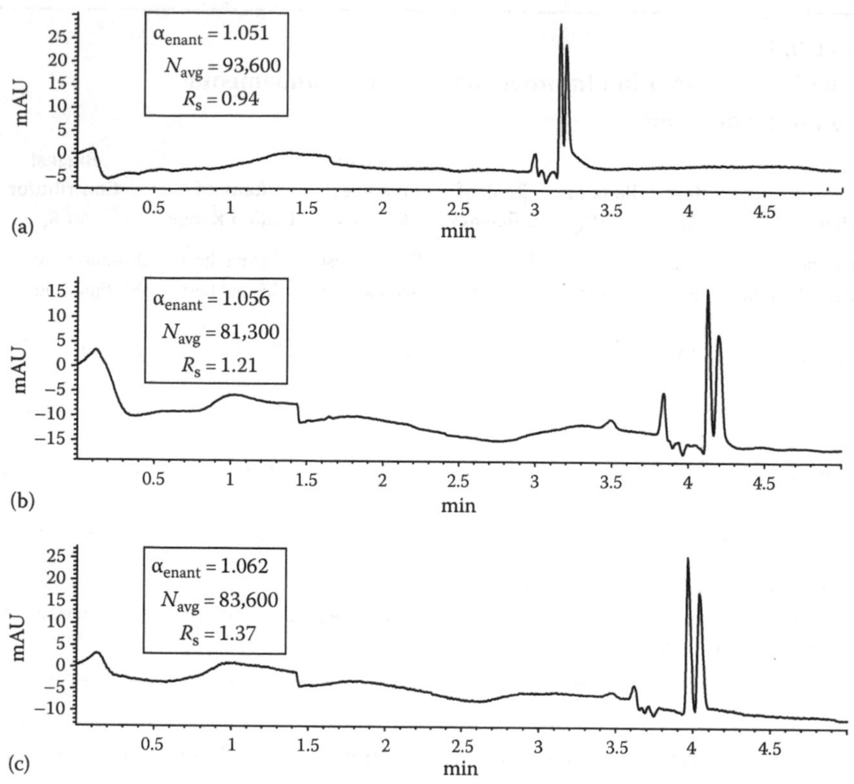

FIGURE 9.18 Effect of microemulsion stereochemical composition on synephrine enantioselectivity. (a) *RXX* microemulsion, (b) *SSR* microemulsion, and (c) *RSS* microemulsion. Surfactant concentration = 2.00% w/v; cosurfactant concentration = 1.65% v/v; oil concentration = 1.23% v/v; 50 mM phosphate buffer, pH 7.0. Capillary dimensions: L_{tot} = 32 cm, L_{eff} = 23.6 cm, i.d. = 50 μm. Detection wavelength was 215 ± 5 nm, injections were performed hydrodynamically for 2 s with 25 mbar pressure, and the applied voltage was 11.5 kV. (From Kahle, K.A. and Foley, J.P., *Electrophoresis*, 28, 3024, 2007. With permission.)

for resolution as a direct result of increased enantioselectivities (averages of 2.06 and 1.83 across all compounds, respectively). The largest improvement in resolution was found for the enantiomers of metoprolol as depicted in Figure 9.19. Synergies were determined as in earlier reports by Kahle and Foley and were all beneficial in nature. Dual-chirality microemulsion contributions to changes in Gibb's free energy were calculated during the thermodynamic analysis, giving new information on chromatographic figures of merit for chiral surfactant/chiral cosurfactant/racemic diethyl tartrate aggregates. These data allowed for a thorough comparison of two-chiral-component aggregates prepared in two ways: chiral surfactant/chiral cosurfactant and chiral surfactant/chiral oil (the combination of chiral cosurfactant with chiral oil was unable to resolve any of the enantiomers). The former gave slightly better enantioselectivity (both chiral selectors in close proximity and readily accessible) and much better resolution. Efficiencies were nearly equivalent for the two

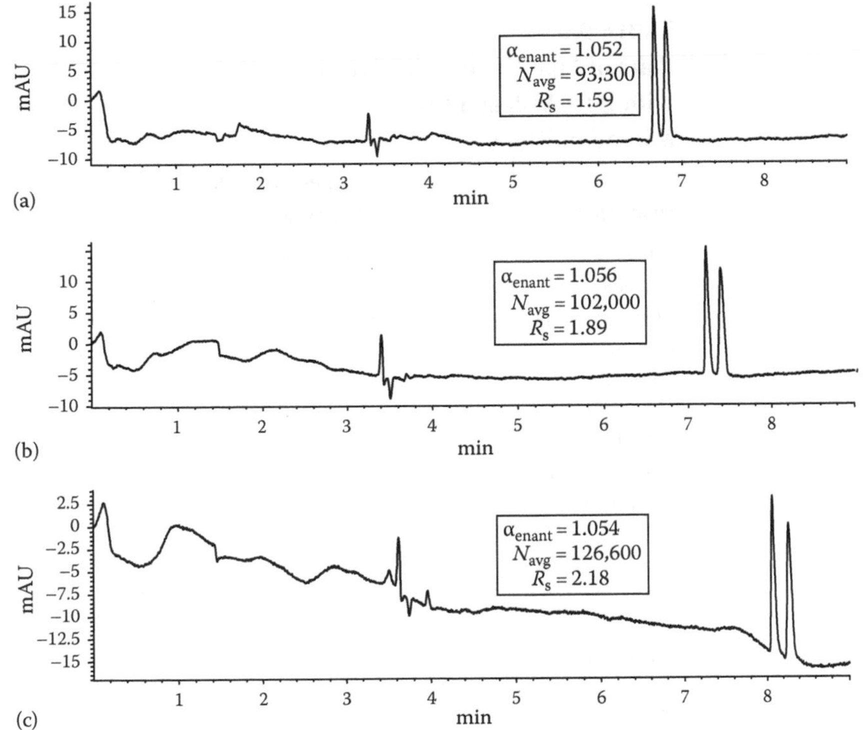

FIGURE 9.19 Effect of microemulsion stereochemical composition on metoprolol resolution. (a) *RXX* microemulsion, (b) *SSS* microemulsion, and (c) *RSS* microemulsion. Conditions as given in Figure 9.18. (From Kahle, K.A. and Foley, J.P., *Electrophoresis*, 28, 3024, 2007. With permission.)

types of dual-chirality PSPs. Methylene selectivity values were shown to be similar, confirming that unlike identity, component stereochemistry does not impact the microemulsion hydrophobicity. Findings from chiral surfactant/chiral cosurfactant formulations with different racemic oils (ethyl acetate and diethyl acetate) were also discussed. Values for enantioselectivity, efficiency, and resolution were examined with respect to the number of chiral reagents in the microemulsion formulation. In general, enantioselectivity and resolution improve as more chiral components are introduced whereas the opposite effect occurs for efficiency, demonstrating the ability to easily fine-tune MEEKC separations.

9.5 APPLICATIONS AND CONCLUDING REMARKS FOR CHIRAL MEEKC

9.5.1 Analytes Separated

A limited number of enantiomers have been separated in chiral MEEKC, as shown in Tables 9.6 and 9.7. Most are small pharmaceutical compounds with similar structures that are cationic under the analysis conditions and tested as purchased (i.e., not

TABLE 9.6
Pairs of Enantiomers Analyzed with DDCV-Based Chiral Microemulsions

Enantiomers	Charge under Analysis Conditions
Arterenol	Cationic
Atenolol	
Chlorpheniramine	
Ephedrine	
Epinephrine	
Metoprolol	
N-Methyl ephedrine	
Norphenylephrine	
Octopamine	
Propranolol	
Pseudoephedrine	
Synephrine	
Verapamil	
Indapamide	Neutral

Note: All compounds above Indapamide are cationic

TABLE 9.7
Pairs of Enantiomers Analyzed with Non-DDCV-Based Chiral Microemulsions

Enantiomers	Charge under Analysis Conditions
Ephedrine	Cationic
N-Methyl ephedrine	
Nadolol	
Norephedrine	
Propranolol	
Synephrine	
Laudanosoline	
Norlaudanosoline	
Secobarbital	Anionic
Pentobarbital	
1,1′-Binaphthyl-2,2′-diyl hydrogen phosphate (BNP)	
1,1′-Bi-2-naphthol (BOH)	
1,1′-Binaphthyl-2,2′-diamine (BNA)	

Note: All compounds above secobarbital are cationic

from real-world samples). The full power of chiral MEEKC has yet to be established for various analyte structures/charges.

9.5.2 Concluding Remarks on Chiral MEEKC

In stark contrast to the large body of literature on achiral MEEKC, research focusing on chiral MEEKC has been scarce. The existing publications have clearly demonstrated the importance of optimizing the identity and concentration of each microemulsion constituent in addition to the mobile phase for the analytes of interest. These options for fine-tuning a method make MEEKC a very attractive technique for the difficult task of enantioresolution. Even though some reports indicate that a chiral cosurfactant or chiral oil can elicit a chiral separation, experiments using DDCV and D-SUV have established the key role of the chiral surfactant in the chiral recognition mechanism. The cosurfactant has been shown to be the next most influential component in the aggregate with alterations in selectivity readily achieved via an identity change (chain length and hydroxyl group location both impact selectivity). Drastic alterations in electrochromatography were not realized with changes in the oil core of the droplet although one report directly contradicts this overall trend. The oil phase does play a minor role in varying the chromatographic figures of merit.

The systematic studies performed using DDCV-based microemulsions enable the proposal of some general formulating advice. Utilization of a short-chained primary alcohol such as 1-butanol or a more hydrophilic oil such as ethyl acetate can improve enantioselectivity and resolution. To increase efficiency and methylene selectivity, a longer-chained primary alcohol like 1-hexanol or more hydrophobic oil like dibutyl tartrate should be selected. Also, distinct enhancements can be achieved if racemic microemulsion constituents are switched to chiral ones. It appears to be more beneficial to combine a chiral surfactant with a chiral cosurfactant to boost enantioselectivity and resolution.

9.6 FUTURE DIRECTIONS

An abundance of research opportunities still exist in the field of chiral MEEKC. In order to effectively and efficiently develop selective methods, fundamental studies to probe the chiral recognition mechanism would be invaluable. Also, to broaden the current knowledge base on microemulsion formulations, more components (surfactant, cosurfactant, and oil), combinations, and ratios need to be evaluated. The introduction of more chiral surfactants with a wider variety of chiral centers and characteristics (i.e., chain length and charge) would allow for significant alterations in selectivity and expand the usefulness of chiral MEEKC. Employing this technique for the analysis of more diverse analytes, including mixture of chiral and achiral compounds, would undoubtedly provide further insights into solute–PSP interactions as well as aid in method development. Publications that demonstrate real-world chiral MEEKC applications and the benefits of this EKC subset would increase its popularity.

REFERENCES

1. Terabe, S, Matsubara, N, Ishihama, Y, Okada, Y. 1992. Microemulsion electrokinetic chromatography: Comparison with micellar electrokinetic chromatography. *J Chromatogr* 608: 23–29.
2. Pyell, U. 2004. Determination and regulation of the migration window in electrokinetic chromatography. *J Chromatogr A* 1037: 479–490.
3. Hansen, SH. 2003. Recent applications of microemulsion electrokinetic chromatography. *Electrophoresis* 24: 3900–3907.
4. Huie, CW. 2006. Recent applications of microemulsion electrokinetic chromatography. *Electrophoresis* 27: 60–75.
5. Marsh, A, Clark, B, Broderick, M, Power, J, Donegan, S, Altria, K. 2004. Recent advances in microemulsion electrokinetic chromatography. *Electrophoresis* 25: 3970–3980.
6. McEvoy, E, Marsh, A, Altria, K, Donegan, S, Power, J. 2007. Recent advances in the development and application of microemulsion EKC. *Electrophoresis* 28: 193–207.
7. Pomponio, R, Gotti, R, Luppi, B, Cavrini, V. 2003. Microemulsion electrokinetic chromatography for the analysis of green tea catechins: Effect of the cosurfactant on the separation selectivity. *Electrophoresis* 24: 1658–1667.
8. Schwuger, M, Stickdorn, K, Schomaecker, R. 1995. Microemulsions in technical processes. *Chem Rev* 95: 849–864.
9. Watarai, H, Ogawa, K, Abe, M, Monta, T, Takahashi, I. 1991. Capillary electrophoresis with O/W microemulsions of water/SDS/1-butanol/heptane. *Anal Sci* 7: 245–248.
10. Aiken, JH, Huie, CW. 1993. Use of a microemulsion system to incorporate a lipophilic chiral selector in electrokinetic capillary chromatography. *Chromatographia* 35: 448–450.
11. Pascoe, R, Foley, JP. 2002. Rapid separation of pharmaceutical enantiomers using electrokinetic chromatography with a novel chiral microemulsion. *Analyst* 127: 710–714.
12. Iqbal, R, Rizvi, SAA, Akbay, C, Shamsi, SA. 2004. Chiral separations in microemulsion electrokinetic chromatography. Use of micelle polymers and microemulsion polymers. *J Chromatogr A* 1043: 291–302.
13. Zheng, Z-X, Lin, J-M, Chan, W-H, Lee, AWM, Huie, CW. 2004. Separation of enantiomers in microemulsion electrokinetic chromatography using chiral alcohols as cosurfactants. *Electrophoresis* 25: 3263–3269.
14. Berthod, A. 2006. Chiral recognition mechanisms. *Anal Chem* 78: 2093–2099.
15. Terabe, S, Otsuka, K, Ando, T. 1985. Electrokinetic chromatography with micellar solution and open-tubular capillary. *Anal Chem* 57: 834–841.
16. Foley, JP, Ahuja, ES. 1996. Electrokinetic chromatography. In *Pharmaceutical and Biomedical Applications of Capillary Electrophoresis*, Lunte, SM and Radzik, DM (Eds.), pp. 81–178. Tarrytown, NY: Pergamon Press.
17. Nielsen, KR, Foley, JP. 1998. Micellar electrokinetic chromatography. In *Capillary Electrophoresis (2nd Edition)*, Camilleri, P. (Ed.), pp. 135–182. Boca Raton, FL: CRC Press.
18. Kahle, KA. 2007. Effect of identity and number of chiral microemulsion components in chiral microemulsion electrokinetic chromatography, PhD Thesis, Department of Chemistry, Drexel University, Philadelphia, PA.
19. Mazzeo, JR, Swartz, ME, Grover, ER. 1995. A resolution equation for elecrokinetic chromatography based on electrophoretic mobilities. *Anal Chem* 67: 2966–2973.
20. Stalcup, A. 2008. Personal communication, HPLC 2008, Philadelphia, PA.
21. Guebitz, G, Schmid, MG. 2007. Advances in chiral separation using capillary electromigration techniques. *Electrophoresis* 28: 114–126.
22. Kahle, KA, Foley, JP. 2007. Review of aqueous chiral electrokinetic chromatography (EKC) with an emphasis on chiral microemulsion EKC. *Electrophoresis* 28: 2503–2526.

23. Mertzman, MD, Foley, JP. 2004. Chiral cyclodextrin-modified microemulsion electrokinetic chromatography. *Electrophoresis* 25: 1188–1200.
24. Mertzman, MD, Foley, JP. 2004. Effect of oil substitution in chiral microemulsion electrokinetic chromatography. *Electrophoresis* 25: 723–732.
25. Mertzman, MD, Foley, JP. 2004. Effect of surfactant concentration and buffer selection on chromatographic figures of merit in chiral microemulsion electrokinetic chromatography. *Electrophoresis* 25: 3247–3256.
26. Mertzman, MD, Foley, JP. 2005. Temperature effects on chiral microemulsion electrokinetic chromatography employing the chiral surfactant dodecoxycarbonylvaline. *J Chromatogr A* 1073: 181–189.
27. Mertzman, MD, Foley, JP. 2005. Comparison of dodecoxycarbonylvaline microemulsion, solvent-modified micellar and micellar pseudostationary phases for the chiral analysis of pharmaceutical compounds. *Electrophoresis* 26: 4153–4163.
28. Kahle, KA, Foley, JP. 2006. Chiral microemulsion electrokinetic chromatography: Effect of cosurfactant identity on enantioselectivity, methylene selectivity, resolution, and other chromatographic figures of merit. *Electrophoresis* 27: 4321–4333.
29. Altria, KD, Clark, BJ, Mahuzier, PE. 2000. The effect of operating variables in microemulsion electrokinetic capillary chromatography. *Chromatographia* 52: 758–768.
30. Kahle, KA, Foley, JP. 2006. Chiral microemulsion electrokinetic chromatography with two chiral components: Improved separations via synergies between a chiral surfactant and a chiral cosurfactant. *Electrophoresis* 27: 896–904.
31. Kahle, KA, Foley, JP. 2007. Two-chiral-component microemulsion electrokinetic chromatography-chiral surfactant and chiral oil: part 1. Dibutyl tartrate. *Electrophoresis* 28: 1723–1734.
32. Kahle, KA, Foley, JP. 2007. Two-chiral component microemulsion EKC-chiral surfactant and chiral oil. Part 2: Diethyl tartrate. *Electrophoresis* 28: 2644–2657.
33. Kahle, KA, Foley, JP. 2007. Influence of microemulsion chirality on chromatographic figures of merit in EKC: Results with novel three-chiral-component microemulsions and comparison with one- and two-chiral-component microemulsions. *Electrophoresis* 28: 3024–3040.

10 Chiral Separations in Nonaqueous Media

Ylva Hedeland and Curt Pettersson

CONTENTS

10.1 Introduction 271
10.2 Organic Solvents and Electrolytes Used in Chiral NACE 277
10.3 Ion-Pair Selectors 281
  10.3.1 Electrolytes and Solvents in Chiral Ion-Pair Selector Systems 286
  10.3.2 Pharmaceutical Applications of Ion-Pair Selectors 289
10.4 Cyclodextrins 290
  10.4.1 Applications Using CDs Derivatives 297
10.5 Miscellaneous Chiral Selectors Used in NACE 298
10.6 Improvement of Detection Limit in Chiral NACE 299
10.7 Concluding Remarks 304
List of Abbreviations 304
References 306

10.1 INTRODUCTION

The use of nonaqueous solvents in capillary electrophoresis (CE) was introduced in 1984 by Walbroehl and Jorgensen [1]. One year later, the first chiral separation in aqueous CE was published by Gassmann et al. [2]. However, little or almost no attention was paid to the nonaqueous media in CE until the middle of the 1990s and the first application of chiral separation in a nonaqueous medium was presented as recently as 1996 [3–6]. Since then, approximately 50 articles concerning enantioseparations in nonaqueous capillary electrophoresis (NACE) have been published, and over 40 different chiral selectors have been described. This chapter covers articles published during the years 1996–2007. For reviews on chiral separation in CE and NACE, see the Refs. [7–13].

In CE, nonaqueous separation can be used as a complement to aqueous separation since it facilitates the use of chiral selectors with a low solubility in water [4,6,14]. Furthermore, it may improve the enantioseparation for selectors with a lack of or low enantioselectivity in aqueous buffers [15]. Another advantage of NACE is that the use of organic solvents enables poorly water-soluble substances to be analyzed [16]. The enhanced solubility of nonpolar substances is advantageous when injection of high concentrations is necessary to lower the limit of detection (LOD), e.g., for chiral impurity analysis [17–20]. Another way of improving the detection sensitivity in CE is the use of hyphenation with mass spectrometry (MS) [21,22]. Since the nonaqueous solvents are generally more volatile than water, they are preferable in these systems, because a higher volatility would facilitate ionization in the MS interface. An additional advantage of NACE is that some of the organic solvents used in the background electrolyte (BGE), e.g., formamide (FA), might also reduce the adsorption of the analytes to the capillary wall below the level found for aqueous BGEs [23,24].

The separation in CE is based on mobility differences of the analytes in an electrical field. For a pair of enantiomers, the mobilities of their uncomplexed forms are identical. Therefore, addition of a chiral selector to the BGE is required to achieve discrimination between them. Enantioselectivity can be obtained in CE by taking advantage of a difference in the degree of formation of the diastereomeric complexes between the enantiomers and the selector (i.e., $K_R \neq K_S^{\dagger}$) and/or formation of complexes with different mobilities (i.e., $\mu_{CR} \neq \mu_{CS}{}^{\dagger\dagger}$). A difference in the complexation constants between the enantiomers is the most common source of enantioselectivity, and this is also the only possible mechanism for differentiation when the diastereomeric complexes formed are uncharged, as for ion-pairs ($\mu_{CR} = \mu_{CS} = 0$). A mobility difference between the diastereomeric complexes only arises when they have a net charge (i.e., an effective mobility) and the diastereomers either differ in shape, which may occur in host–guest complexation where there is a difference in the "fitting" of the analyte into the cavity [25–27]. Furthermore, the authors claim that a difference in the conditional acid dissociation constants ($pK_a^{\dagger}$) for the enantiomers due to the complexation with the CD will give rise to a difference in the fraction charged of the enantiomers [26]. According to Wren and Rowe, the mobility difference between the enantiomers ($\Delta\mu$) can be described as a function of the free selector concentration [28]:

$$\Delta\mu = \frac{(\mu_f - \mu_c)\Delta K[C]}{(1 + K_1[C])(1 + K_2[C])} \tag{10.1}$$

where

μ_f is the mobility of the free solute
μ_c is the mobility of the complex
C is the selector concentration
K_1 and K_2 are the ion-pair formation constants for the first (fastest migrating) and second enantiomer, respectively
ΔK is the difference between them ($K_2 - K_1$)

This model assumes a selector–analyte ratio of 1:1 which can be simplified in the case of equal binding constants [27] or if the complex formed does not possess a net

† The association constant for the complex between the selector and the (*R*) and the (*S*) enantiomer respectively.

†† The mobilities of the complexes formed between the selector and the the (*R*) and the (*S*) enantiomer respectively.

mobility. The model has also been expanded further to include other stochiometric ratios [29]. Equation 10.1 has a maximum, corresponding to the optimal selector concentration at which $\Delta\mu_{max}$ is achieved [28]:

$$[C]_{opt} = \frac{1}{\sqrt{K_1 K_2}} \tag{10.2}$$

If the values of K_1 and K_2 decrease, the value of $\Delta\mu_{max}$ is shifted to higher concentrations and the slope around the maximum value of $\Delta\mu_{max}$ becomes flatter, as shown in Figure 10.1a and b [3], where curve A originates from a water-based BGE and has the highest values of *K*, and *C* originates from an FA-based system and has the lowest values of *K*.

The reversible formation of the diastereomeric complexes can be based on, e.g., (1) inclusion complexation (2) ion-pair formation, (3) a combination of (1) and (2), or (4) a ligand-exchange mechanism [11]. In order to obtain a difference between

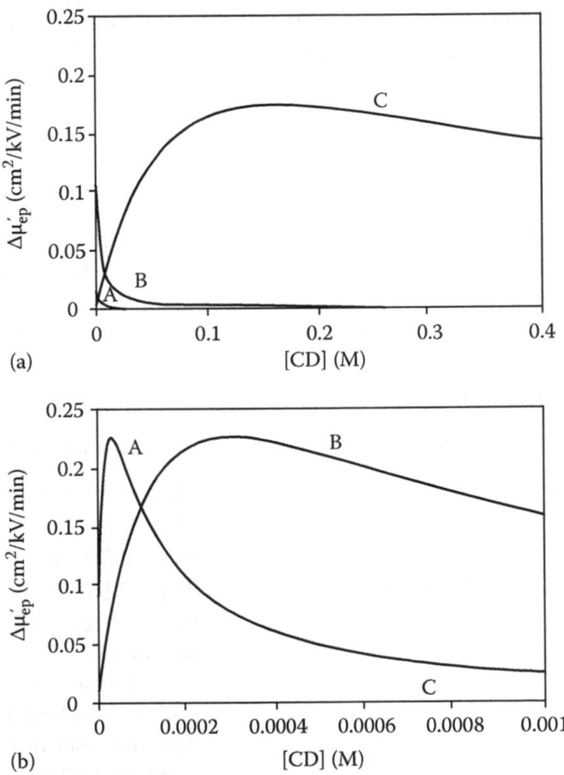

FIGURE 10.1 The relationship between $\Delta\mu$ and β-CD concentration for thioridazine in different BGE solvents. The BGEs contained β-CD, 50mM citric acid, and 25mM TRIS. (a) 0-0.4M β-CD; (b) 0-0.001M β-CD Curve: (A) aqueous buffer, (B) 6M urea in aqueous buffer, and (C) FA. K_1 was determined to be $2.91 \cdot 10^4$ in (A), 2.98×10^{-3} in (B) and 5.39 in (C). (From Wang, F. and Khaledi, M.G., *Anal. Chem.*, 68, 3460, 1996. With permission.)

the complex constants, it is generally assumed that the selector can enable a so-called three-point interaction or three contact points involving at least one of the enantiomers [30,31]. These interactions can be of an attractive (e.g., ionic interaction or hydrogen bonding) as well as of a repulsive nature (e.g., ionic repulsion or steric hindrance). It is preferable if the structure of the selector is rigid (nonflexible), to minimize the possibility of conformational changes. The chiral selectors that have been used in NACE are listed in Table 10.1. As can be distinguished in the table, ion-pair selectors based on small molecular weight selectors and cyclodextrin (CD) derivatives are most commonly used. Many of the selectors have only been used in organic solvents (for instance the majority of the ion-pair selectors), but some of them (e.g., β-cyclodextrin [β-CD] [4]) have also been applied in aqueous solutions.

In aqueous BGEs, changes in the pH, selector type and concentration, type of buffer, and ionic strength can alter the chiral resolution [32]. In nonaqueous media, there is also a possibility to change the enantioresolution through the exchange of one solvent for another or by changing the relative proportions of the solvents.

It should be stressed that the optimal enantioresolution (*Rs*) is not always found at the highest value of $\Delta\mu$, especially when the values of K_1 and K_2 are low [16]. In addition to $\Delta\mu$, the average of the observed mobilities of the enantiomers (μ_m), and the efficiency (*N*) also influence the enantioresolution obtained [33]:

$$Rs = \frac{\sqrt{N}}{4} \times \frac{\Delta\mu}{\mu_m} \tag{10.3}$$

According to Equation 10.3, the resolution will increase if the average of the observed mobilities (μ_m) is decreased (e.g., if the electroosmotic flow [EOF] is decreased) and if

TABLE 10.1
Chiral Selectors Used in NACE

Type of Interaction	Group of Selectors	Example of Selector
Inclusion complexation	Native CD	β-Cyclodextrin (β-CD) [4,3,104,109]
		γ-CD [3]
		methyl-β-CD [3]
	CD derivative	HP-β-CD [3,16]
		HE-β-CD [16]
		HxDAS [88]
		Mono-(6-*O*-norborn-2-ene-5-yl-carboxyl)-β-CD [104]
		Oligo-(6-*O*-norborn-2-ene-5-yl-carboxyl)-β-CD [104]
		Pentakis-(6-*O*-norborn-2-ene-5-ylmethyl-hydroxysiloxyl)-β-CD [104]
		Oligo (pentakis-(6-*O*-norborn-2-ene-5-ylmethyl-hydroxysiloxyl)-β-CD) [104]

TABLE 10.1 (continued)
Chiral Selectors Used in NACE

Type of Interaction	Group of Selectors	Example of Selector
Ion-pair formation	Chiral counterion (negatively charged)	(+) and (–)-camphorsulfonic acid [6] (1*S*,4*R*)-ketopinic acid [20] (–)-DIKGA [15,20,21] L-ZGP [70] *(S)*-and *(R)*-dinitrobenzoyl-leucine [51]
	Chiral counterion (positively charged)	Quinine [5,55,56] Quinine derivatives [14,56–60] Quinidine [55] Quinidine derivatives [14,57,61] Cinchonine [55] Cinchoine derivatives [55] Cinchonidine [55] Cinchonidine derivatives [55]
Inclusion complexation and ion-pair formation	Negatively charged CD derivative	Sulfated-β-CD [3,81,110] Sulfated-γ-CD [87,111], ODAS-γ-CD [87] ODMS-γ-CD [111] HDMS-β-CD [17,84,85] HDAS-β-CD [86] TBA$_7$HDAS-β-CD [90]
	Positively charged CD derivative	EA-β-CD [112] IPA-β-CD [112] PA-β-CD [112] QA-β-CD [24]
	Crown ether	(+)-18-Crown-6-tetracarboxylic acid [94]
Ligand-exchange mechanism		Complex of copper (II) and L-proline or L-isoleucine [95]

Abbreviations: HP-β-CD, hydroxypropyl-β-cyclodextrin; HE-β-CD, hydroxyethyl-β-cyclodextrin; HxDAS, hexakis(2,3-diacetyl-6-*O*-sulfo)-α-cyclodextrin; (–)-DIKGA, di-*O*-isopropylidene-keto-L-gulonic acid; L-ZGP, *N*-benzoxycarbonylglycyl-L-proline; ODAS-γ-CD, octakis(2,3-diacetyl.6.sulfato)-γ-cyclodextrin; ODMS-γ-CD, octakis(2,3-*O*-dimethyl-6-*O*-sulfo)-γ-cyclodextrin; HDMS-β-CD, heptakis(2,3-dimethyl-6-sulfato)-β-CD; HDAS-β-CD, heptakis(2,3-diacetyl-6-sulfato)-β-CD; TBA$_7$HDAS-β-CD, heptakis(2,3-di-*O*-acetyl-6-sulfo)-cyclomaltoheptanose tetrabutylammoniumsalt; EA-β-CD, 6-monodeoxy-6-mono(2-hydroxy)ethylamino-β-cyxlodextrin; IPA-β-CD, 6-monodeoxy-6-mono (2-hydroxy)propylamino-β-cyxlodextrin; PA-β-CD, 6-monodeoxy-6-mono (3-hydroxy) propylamino-β-cyxlodextrin; QA-β-CD, quartenary ammonium β-CD.

the efficiency (N) is subsequently preserved (or enhanced). Since $\Delta\mu$ is generally low in chiral NACE (with some exceptions, e.g., [34]), it is fundamental to the success of the separation that the EOF be minimized. A reduction of the EOF (μ_{EOF}) can be obtained, e.g., by the use of covalently or dynamically coated capillaries or by changing the solvent mixture in such a way that the $\varepsilon \cdot \zeta/\eta$ ratio in Equation 10.4 is reduced [35]:

$$\mu_{EOF} = -\frac{\varepsilon_{r,wall} \times \zeta_{wall}}{\eta_{wall}} \tag{10.4}$$

$\varepsilon_{r,wall}$ is the relative dielectric constant in the electrical double layer (EDL), η_{wall} is the viscosity in the EDL, and ζ is the potential between the stagnant and the diffuse layers of the EDL. For simplicity, and to obtain a rough estimate of the influence of different solvents, the values for $\varepsilon_{r,wall}$ and η_{wall} can be assumed to be the same as those in free solution. However, the ε and η values for solvent mixtures are often difficult to obtain, and furthermore, the ζ potential as well as the ε and η values (see Equation 10.4) change when the electrolytes are added to the solvent. The values of ε and η for the solvents that are commonly used in chiral NACE are presented in Table 10.2 together with the autoprotolysis constants and boiling points of the solvents. Water is included in the table as a reference. In nonaqueous media it is often necessary to minimize the EOF by, e.g., exchanging to a solvent with a lower ε value. In addition, the analytes of interest and/or the selector need to be charged to some extent to obtain mobility in the electric field. If ε in the solvent is too low, or if the solvent lacks acidic or basic properties, only negligible amounts of ions can exist.

TABLE 10.2
Properties of the Organic Solvents Used in Chiral NACE

Solvent	ε^a	η^b (kg/ms)	$K_{a(solvent)}$	t^c_{boil} (°C)
N-Methylformamide	182.4	1.99	$10^{0.04}$	180–185
Formamide	111.0[d]	3.764[d]	$10^{0.48}$	210.5
Water	78.36	1.137	$10^{-14.0}$	100.0
ACN	37.5	0.375	$10^{-26.5}$	81.60
N,N-Dimethylformamide	36.71	0.9243[d]	$10^{0.01}$	73.095
Methanol	32.70	0.5506[d]	$10^{-16.7}$	64.70
Ethanol	24.55	1.078[a]	$10^{-19.1}$	78.29
2-Propanol	19.92	2.859	N/A	82.26
1,2-Dichloroethane	10.36	0.887	N/A	83.48
Dichloromethane	8.93	0.449	N/A	39.75

Source: Riddick, J.A. et al., *Organic solvents-physical properties and methods of purification*, John Wiley & Sons, New York, 1986.

[a] 25°C.
[b] 15°C.
[c] 1 atm.
[d] 20°C.

10.2 ORGANIC SOLVENTS AND ELECTROLYTES USED IN CHIRAL NACE

The most commonly used organic solvents for the ion-pair selectors are methanol (MeOH) and ethanol (EtOH) and combinations thereof, but for the CDs, MeOH, and FA are the most frequently used (Table 10.3). There are examples in the

TABLE 10.3
Organic Solvents and Electrolytes Used in Chiral NACE Separations

Solvent	Electrolyte	Type of Selector	Refs.
Acetonitrile (ACN)	Acetic acid (HAc) and Tween 20	Ion-pair	[6]
ACN	Methanesulfonic acid and triethylamine (TEA)	CD	[90]
ACN:Methanol (MeOH) (70:30)	HAc and TEA	Ion-pair	[51]
Ethanol (EtOH) or MeOH	LiOH, KOH, NaOH, CsOH, or RbOH	Ion-pair	[40]
EtOH:MeOH (20:80) and (60:40)	TEA and HAc	Ion-pair	[58,60,61]
EtOH:MeOH (60:40)	Ammonia (NH_3) and octanoic acid (C-8 acid)	Ion-pair	[14,55,57]
EtOH:MeOH (60:40)	KOH	Ion-pair	[20]
EtOH:MeOH (60:40)	Octanoic acid and NH_3	Ion-pair	[14]
EtOH:MeOH (60:40)	TEA and octanoic acid	Ion-pair	[59]
Formamide (FA)	Ammonium acetate (NH_4Ac) and HAc	CD	[24]
FA	Citric acid, Tris, and TEA	CD	[114]
FA	Citric acid and Tris	CD	[3,81,114]
FA	HAc and NaCl	CD	[109]
FA and dimethylsulfoxide	Tetra-*N*-butylammoniumperchlorate	Crown ether	[94]
MeOH	NH_4Ac, HAc and $CuCl_2$*$2H_2O$	Ligand exchange	[95]
MeOH	NH_4Ac	Ion-pair, CD	[5,56,112]
MeOH	Formic acid, camphorsulfonic acid	CD + ion-pair	[54]
MeOH	NaOH	Ion-pair	[15]
MeOH	Phosphoric acid (H_3PO_4) and LiOH	CD	[88]
MeOH	H_3PO_4 and NaOH	CD	[17,84,85,87]
MeOH	Dichloro acetic acid and TEA	CD	[86]
MeOH:1,2-dichloroethane	NH_4Ac	Ion-pair	[70]
MeOH:2-PrOH (80:20) or 75:25	NH_4Ac	Ion-pair	[21,70]
MeOH:dichloromethane	NH_4Ac	Ion-pair	[70]
N,N-Dimethylformamide (*N*MF)	Citric acid and Tris	CD	[3]
NMF	NaCl	CD	[4]
NMF	$NaHCO_3$	CD	[104]

literature of other solvents that have only been used for achiral separations in nonaqueous media, such as *N*-,*N*-dimethylacetamide, butanol [36], nitromethane [37], and tetrahydrofurane in mixtures with MeOH [38] or a combination of MeOH and acetonitrile (ACN) [39]. These solvents might also prove to be useful for chiral separations.

Important properties for a BGE solvent are (1) low volatility, (2) high solubility of the electrolytes and the samples, (3) having a rather high self-dissociation constant ($K_{a,solvent}$), (4) having a reasonably low dielectric constant and (5) low viscosity. One difference between the achiral and the chiral separations in NACE is that solvents with a higher ε ($\varepsilon > 30$) are more commonly used in the achiral separations, especially in comparison with the solvents used for the ion-pair selectors [23]. The ion-pair formation is more pronounced in a solvent with a low ε, as a low ε often promotes the separation when ion-pair selectors are used. But, as already mentioned above, if the value of ε is too low, it will suppress the ionization of the species (the analyte and/or the selector) in the nonaqueous solvent, which results in the loss of enantioselectivity since uncharged molecules have no net velocity in an electrical field. The competing nonstereoselective ion-pair formation in the solvents with a low ε might be another complicating factor that cause difficulties in the design of separation conditions, since the proper choice of the other electrolytes in the BGE (in addition to the choice of the selector) is of major importance for the separation [40]. Commonly used electrolytes for chiral separations in nonaqueous media are listed in Table 10.3. The problem with nonstereoselective ion-pair formation in the BGE will be discussed further in Section 10.3.

For safety reasons, all flammable and toxic solvents should be avoided in the BGE, and if ultraviolet (UV) detection is used, the solvent should have as low a UV cutoff as possible. Of course, using a high purity solvent is also preferable. In this context, it should be stressed that the term "nonaqueous" is slightly misleading, since many of the organic solvents, e.g., EtOH, and some of the more common electrolytes, e.g., ammonia (NH_3), contains small amounts of water or produce water in the BGE by neutralization. Furthermore, some of the solvents, e.g., dichloromethane (DCM), and electrolytes, e.g., sodium hydroxide (NaOH), are hygroscopic, which needs to be taken into account since it might cause problems with the repeatability and reproducibility of the method if the water content changes during the analysis. Addition of small amounts of water may reduce these problems. Tjørnlund and Hansen [41] investigated the influence of water on an achiral separation by using four different organic solvents (MeOH, ACN, dimethylsulfoxide [DMSO], and *N*-methylformamide [NMF]). They found that the reproducibility increased with only a minor effect being observed on the selectivity, efficiency, and EOF when 0.5% water was added to the BGE. The addition of small amounts (i.e., 0.5%) of water has also shown to improve the repeatability of the method for a chiral separation of β-blockers in an EtOH-based BGE [40]. However, the addition of higher concentrations of water often decreases and even ruins the enantioresolution when ion-pair selectors are used [21]. The influence of water when ion-pair selectors are used is discussed further in Section 10.3.

Alterations in the organic solvent composition in the BGE give rise to simultaneous changes in several different parameters, including the pK_a^* value of the solutes, which is one of the parameters that alters the effective mobility. Bjørnsdottir and Hansen [42,43] observed selectivity changes for structurally similar amines when the organic solvent was exchanged. Cantu et al. [44] determined the pK_a^* for the same amines in MeOH and ACN and correlated the variation in their migration order to the change in their pK_a^* when exchanging one solvent for the other. A change in the migration order (and selectivity) between different solvents has also been observed for amphetamine related compounds when β-CD was used as the selector [45]. In addition, the environment for ion-pair formation and/or inclusion complexation will differ between solvents and solvent mixtures. Additionally, the exchange of the solvent also affects the EOF [46], which influences the enantioresolution (see Equation 10.3) [33]. Thus, there are several reasons why the use of nonaqueous solvents is not as straightforward as that of aqueous media. Some of the reasons have already been mentioned above. Another difference is that, in aqueous systems a well defined "acidity"-scale, based on the activity of H_3O^+ (pH), is used. The change in pH can easily be measured by commercially available electrodes and standard calibration solutions for the equipment are readily obtained. In organic solvents (that contain water), the pH*(or the apparent pH) is sometimes measured by a standard pH electrode. This can either be done by calibrating the electrode with an appropriate organic solvent mixture at a defined pH* (sometimes in combination with the use of an electrode filled with an electrolyte based on the same organic solvent or solvent mixture), or by calibration with standard aqueous solutions followed by measurements on the organic solvent mixture [47,48]. The first technique can be used with subsequent back calculation to the solvent mixture of interest, whereas the latter technique should only be used to measure approximate relative acidities in the solvents, and cannot be related to the true H_3O^+ activity in the solvent. It is worth mentioning that it has been pointed out that the use of an aqueous filling solution in the electrode may cause clogging, and permanent damage of the electrode, if the filling electrolyte or the salt bridge electrolyte has limited solubility in the nonaqueous solvent solution to be measured [48]. For a more detailed discussion on different strategies to be adopted for the measurement and control of the pH* in organic solvents, see the reviews by Porras and Kenndler [48–50] and Roses [47].

The reason for the interest in measuring the pH* in organic solvents is the desire to have an easy way to estimate the extent to which the solutes (and/or the selector) in the BGE have been charged in order to facilitate rational experimental design. However, it is often difficult to obtain information about the pK_a^* of the analyte and electrolytes in the actual solvent or solvent mixture. This often makes it necessary to determine the pK_a^* in the solvent [44], or to approximate the value by extrapolation from the pK_a obtained experimentally [40] or by theoretical calculation [51] in an aqueous solution. If the pK_a^*s of the electrolytes in the solvent/solvent mixture are known, the pH* in the solvent can be calculated using the Henderson–Hasselbalch equation [50]. But it should be stressed that, in this equation, the initial concentration is assumed to be the same as the equilibrium concentration, which is only true if the ion-pair formation between the different species in the electrolyte is negligible (which is not the case in many of the solvents that are used in chiral NACE, such as EtOH). In solvents with a high degree of ion-pairing, all ion-pair formation constants are required, and, based on this information, the actual pH* in the electrolyte can be calculated by iteration, using the formation constants for all equilibria in the actual BGE [40]. In organic solvents in the absence of significant amounts of water, the self-ionization of the solvent (which means, e.g., the activity of $MeOH_2^+$ in MeOH, instead of the activity of H_3O^+) might be the parameter of greatest interest since the presence of a hydrogen donor or a hydrogen acceptor influences the charged fraction of the analyte and the selector.

The preparation of a buffer in nonaqueous solvents is more complicated than in water. Knowledge of the pK_a^* values of the buffer components is, as mentioned above, frequently lacking. Furthermore, it is generally difficult to use a sufficiently high buffer concentration in the BGE to obtain a buffer of sufficient strength owing to the low solubility of many of the frequently used buffer components (such as NaOH) in organic solvents. An adequate ionic strength in the BGE would also decrease the electrokinetic dispersion that arises from conductivity mismatching between the BGE and the sample zone. However, a high buffer concentration will subsequently increase the nonenantioselective ion-pair formation between the selector and the buffer components, and between the analyte and the buffer components, which might decrease the difference in mobility (if the selector concentration is below or at $\Delta\mu_{max}$, Figure 10.1). This is probably one of the reasons why the type of buffer mixtures that are frequently used in aqueous CE are not often used in the NACE systems. As can be distinguished from Table 10.3, the kind of buffers that are most frequently used in aqueous media (e.g., phosphate buffer) are uncommon in NACE, but have been used in some CD systems (see Section 10.4).

To summarize, nonaqueous BGEs can be used as an alternative to aqueous ones for chiral separations in CE. Some of the advantages and disadvantages of using chiral NACE are listed below:

Advantages

- NACE facilitates the use of selectors with low solubility in water.
- NACE facilitates the use of selectors with a lack of or low enantioselectivity in water.
- NACE facilitates the analysis of solutes with low solubility in water.
- The higher volatility of the nonaqueous solvents facilitates the hyphenation to MS.
- The solvent or solvent mixtures in the BGE can be used to suppress or reverse the EOF.

Disadvantages

- The low solubility of many of the electrolytes in nonaqueous media and the competing nonchiral ion-pair formation makes it difficult to achieve a sufficiently high ionic strength in the BGE.
- It is difficult to easily measure the pH* in nonaqueous solvents and there is often a lack of pK_a^* data for the analytes and the electrolytes.
- It is difficult to predict the influence of a change of solvent or solvent composition.

10.3 ION-PAIR SELECTORS

The first chiral separations utilizing ion-pair selectors (chiral counterions) in CE were published in 1996 by Stalcup and Gahm [5] and Bjørnsdottir et al. [6] and were performed on quinine (QN) [5] and camphorsulfonic acid (CSS) [6] in MeOH and ACN-based BGEs. QN was used for the enantioseparation of amino acids and CSS was used for the enantioseparation of basic drugs. CSS has only been used in a few papers since 1996 [52–54]. One substance structurally related to CSS has also been described, the weaker acid ketopinic acid (KPA), where the sulfonic acid is exchanged for a carboxylic one [20]. Many studies have been performed to explore QN and the other cinchona alkaloids and derivatives thereof as chiral selectors [5,14,55–61]. The derivatization of the cinchona alkaloids has given a great enhancement in the enantioresolution. For example, the separation of dinitrobenzoyl-leucine (DNB-Leu) with QN as the chiral selector gave an Rs value of 1.8, but with the derivative *tert*-butylcarbamoylated quinine (tBuCQN) the Rs was increased to 56.2 (Table 10.4). However, the BGE differed between these two separations, but this trend has been noted when the same BGE was used for the majority of the derivatives [34]. A comparison between different cinchona alkaloid derivatives under similar conditions has been made by Piette et al. [34,55]. Examples of separations from published work with ion-pair selectors are tabulated in Table 10.4 together with information concerning the BGE used. By comparing the Rs values in Table 10.4 (ion-pair selectors) with those in Table 10.5, for CDs and their derivatives, and Table 10.6, for miscellaneous chiral selectors, it can be concluded that the cinchona alkaloid derivatives so far have given the greatest enantioresolution of all selectors used in NACE.

Chiral ion-pair selectors can be divided into two subgroups, the negatively charged and the positively charged ones (displayed respectively in Figures 10.2A and 10.2B). Separation with ion-pair selectors is based on the formation of neutral ion-pairs. A chiral discrimination, as mentioned previously in Section 10.1, also requires additional interactions between the chiral counterion and the enantiomeric solute. These can be attractive (e.g., hydrogen-bonding, dipole–dipole, or π–π interaction) and/or repulsive (e.g., steric hindrance, ion repulsion). Many of the ion-pair selectors are relatively small molecules with only one or a few chiral centra, which limit the number of types of molecules that can exhibit enantioselectivity with the same selector. On the other hand, these small molecules make the research on selector–selectand interactions much easier, since the number of possible contact points is limited. For CDs, it is sometimes possible to use the same type of selector for the separation of anionic, cationic, neutral compounds, and zwitter ions [62]. The mechanism for complex formation differs between ion-pairs and CDs, and the chiral recognition models for CDs are discussed further in Section 10.4.

The use of five different negatively charged and 17 positively charged ion-pair selectors has been published in NACE (Figure 10.2A and 10.2B). Three of the negatively charged ones have previously been used as chiral mobile phase additives (CMPA) in high performance liquid chromatography (HPLC) [63–65]. In addition, the positively charged QN and acetyl quinidine have already been used as CMPAs [66,67] prior to introduction in NACE. Their successful use as CMPAs had given indications that these selectors might be used as chiral stationary phases (CSPs) too,

TABLE 10.4
Selected Examples of Chiral Separations by Use of Ion-Pair Selectors

Substance	Selector	BGE	λ (nm)	*Rs*	Ref.
Atenolol	CSS	5–40 mM selector 1 M HAc and 0.1–0.2 mM Tween 20 in ACN	214	N/A	[6]
Bambuterol	KPA	100 mM selector, 40 mM KOH in MeOH:EtOH (2:3)	220	4.1	[20]
Derivatives of QN and QD	(*S*)-DNB-Leu	10 mM selector, 200 mM HAc, 12.5 mM TEA in ACN:MeOH (70:30)	216, 330	8.03	[51]
DNB-homoPhe	Quinine (QN)	2.4 mM selector and 13 mM NH_4Ac in MeOH	214	1.8	[5]
DNBLeu	tBuCQN	10 mM selector, 100 mM C-8 acid, 12.5 mM TEA in EtOH:MeOH 60:40. Plug length 30 cm (total 45.5 cm)	230, 250, 280, 360	17.73	[59]
DNB-Leu	QN	2.4 mM selector and 13 mM NH_4Ac in MeOH	214	1.8	[5]
DNB-Leu	tBuCQN	5 mM selector and 12.5 mM NH_3 and 100 mM C-8 acid in MeOH:EtOH (40:60)	214	56.2	[56]
DNB-Phe	cHex-bis-CQN	10 mM selector, 100 mM C-8 acid and 12.5 mM NH_3 in EtOH:MeOH (60:40)	214	77.2	[14]
DNB-Phe	cHex-bis-CQN-S-C_{12}	10 mM selector, 100 mM C-8 acid and 12.5 mM NH_3 in EtOH:MeOH (60:40)	214	78.3	[14]
DNB-Phe	HM-bis-CQN	10 mM selector, 100 mM C-8 acid and 12.5 mM NH_3 in EtOH:MeOH (60:40)	214	59.3	[14]
DNB-Phe	DCP-CQD	10 mM selector, 12.5 mM NH_3, 100 mM C-8 acid in EtOH:MeOH (60:40)	214	33.8	[57]
DNB-PheGly	tBuCQN	5 mM selector and 12.5 mM NH_3 and 100 mM C-8 acid in MeOH:EtOH (40:60)	214	37.9	[56]
DNB-PheGly	All-CDHQD	10 mM selector, 12.5 mM NH_3, 100 mM C-8 acid in EtOH:MeOH (60:40)	214	33.7	[57]
DNB-PheGly	chex-CCD	5 mM selector and 12.5 mM NH_3 and 100 mM C-8 acid in MeOH:EtOH (40:60)	214	59.6	[55]
DNB-PheGly	cHex-bis-CQN-SO-C_{12}	10 mM selector, 100 mM C-8 acid and 12.5 mM NH_3 in EtOH:MeOH (60:40)	214	71.7	[14]
DNB-PheGly	HM-bis-CQD	10 mM selector, 100 mM C-8 acid and 12.5 mM NH_3 in EtOH:MeOH (60:40)	214	79.5	[14]
DNB-PheGly	P-bis-CQN	10 mM selector, 100 mM C-8 acid and 12.5 mM NH_3 in EtOH:MeOH (60:40)	214	45.1	[14]
DNB-tLeu	1-Me-QN	10 mM selector, 12.5 mM NH_3, 100 mM C-8 acid in EtOH:MeOH (60:40)	214	3.1	[57]
DNB-tLeu	ADCQN	10 mM selector, 12.5 mM NH_3, 100 mM C-8 acid in EtOH:MeOH (60:40)	214	65.4	[57]
DNB-tLeu	All-CDHQD	10 mM selector, 12.5 mM NH_3, 100 mM C-8 acid in EtOH:MeOH (60:40)	214	52.5	[57]
DNB-tLeu	chex-CCD	5 mM selector and 12.5 mM NH_3 and 100 mM C-8 acid in MeOH:EtOH (40:60)	214	70.8	[55]
DNB-tLeu	DCP-CQD	10 mM selector, 12.5 mM NH_3, 100 mM C-8 acid in EtOH:MeOH (60:40)	214	35.0	[57]
DNB-tLeu	DNP-CCN	5 mM selector and 12.5 mM NH_3 and 100 mM C-8 acid in MeOH:EtOH (40:60)	214	24.3	[55]
DNB-tLeu	tBuCQN	5 mM selector and 12.5 mM NH_3 and 100 mM C-8 acid in MeOH:EtOH (40:60)	214	57.3	[55]
DNB-tLeu	tBuCQN	5 mM selector and 12.5 mM NH_3 and 100 mM C-8 acid in MeOH:EtOH (40:60)	214	71.5	[55]

DNB-tLeu	cHex-bis-CQN	10 mM selector, 100 mM C-8 acid and 12.5 mM NH_3 in EtOH:MeOH (60:40)	214	63.3	[14]
DNB-tLeu	HM-bis-CQN	10 mM selector, 100 mM C-8 acid and 12.5 mM NH_3 in EtOH:MeOH (60:40)	214	75.9	[14]
DNB-αAbu	QN	5 mM selector and 12.5 mM NH_3 and 100 mM C-8 acid in MeOH:EtOH (40:60)	214	4.8	[55]
DNB-αAbu	cHex-bis-CQN-S-C_{12}	10 mM selector, 100 mM C-8 acid and 12.5 mM NH_3 in EtOH:MeOH (60:40)	214	71.9	[14]
DNB-αAbu	HM-bis-CQD	10 mM selector, 100 mM C-8 acid and 12.5 mM NH_3 in EtOH:MeOH (60:40)	214	71.6	[14]
DNB-αAbu	P-bis-CQN	10 mM selector, 100 mM C-8 acid and 12.5 mM NH_3 in EtOH:MeOH (60:40)	214	52.5	[14]
DNB-αAbu	ADCQN	10 mM selector, 12.5 mM NH_3, 100 mM C-8 acid in EtOH:MeOH (60:40)	214	65.0	[57]
DNB-αAbu	All-CDHQD	10 mM selector, 12.5 mM NH_3, 100 mM C-8 acid in EtOH:MeOH (60:40)	214	71.3	[57]
DNB-αAbu	All-CDHQN	10 mM selector, 12.5 mM a NH_3, 100 mM C-8 acid in EtOH:MeOH (60:40)	214	62.7	[57]
DNB-αAbu	Cinchonidine	5 mM selector and 12.5 mM NH_3 and 100 mM C-8 acid in MeOH:EtOH (40:60)	214	4.1	[55]
DNB-αAbu	Cinchonine	5 mM selector and 12.5 mM NH_3 and 100 mM C-8 acid in MeOH:EtOH (40:60)	214	5.6	[55]
DNB-αAbu	Quinidine	5 mM selector and 12.5 mM NH_3 and 100 mM C-8 acid in MeOH:EtOH (40:60)	214	7.6	[55]
DNB-αAbu	tBuCQN	5 mM selector and 12.5 mM NH_3 and 100 mM C-8 acid in MeOH:EtOH (40:60)	214	61.7	[55]
DNB-αAbu	tBuCQN	5 mM selector and 12.5 mM NH_3 and 100 mM C-8 acid in MeOH:EtOH (40:60)	214	90.8	[55]
DNB-αAbu	1-Me-QN	10 mM selector, 12.5 mM NH_3, 100 mM C-8 acid in EtOH:MeOH (60:40)	214	17.0	[57]
DNB-αAbu	All-CDHQD	10 mM selector, 12.5 mM NH_3, 100 mM C-8 acid in EtOH:MeOH (60:40)	214	56.8	[57]
DNB-αAbu	All-CDHQN	10 mM selector, 12.5 mM NH_3, 100 mM C-8 acid in EtOH:MeOH (60:40)	214	42.1	[57]
DNB-αAbu	Cinchonidine	5 mM selector and 12.5 mM NH_3 and 100 mM C-8 acid in MeOH:EtOH (40:60)	214	6.6	[55]
DNB-αAbu	Cinchoine	5 mM selector and 12.5 mM NH_3 and 100 mM C-8 acid in MeOH:EtOH (40:60)	214	6.2	[55]
DNB-αAbu	Quinidine	5 mM selector and 12.5 mM NH_3 and 100 mM C-8 acid in MeOH:EtOH (40:60)	214	7.4	[55]
DNB-αAbu	QN	5 mM selector and 12.5 mM NH_3 and 100 mM C-8 acid in MeOH:EtOH (40:60)	214	5.8	[55]
DNB-βAbu	tBuCQN	10 mM selector, 100 mM C-8 acid, 12.5 mM TEA in EtOH:MeOH 60:40. Plug length 30 cm (total 45.5 cm)	230, 250, 280, 360	17.47	[59]
DNB-βAbu	cHex-bis-CQN-SO-C_{12}	10 mM selector, 100 mM C-8 acid and 12.5 mM NH_3 in EtOH:MeOH (60:40)	214	59.2	[14]
DNP-Ala-ala	tBuCQN	10 mM selector, 100 mM HAc,12.5 mM TEA in EtOH:MeOH (20:80)	254	11.14	[58]
DNZ-Ala-ala	tBuCQN	10 mM selector, 100 mM HAc,12.5 mM TEA in EtOH:MeOH (60:40)	254	9.40	[58]
DNZ-Phe	DNP-CCD	5 mM selector and 12.5 mM NH_3 and 100 mM C-8 acid in MeOH:EtOH (40:60)	214	16.3	[55]
Erythro-1-	tBuCQD	10 mM selector, 100 mM HAc, 12.5 mM TEA in EtOH:MeOH (60:40)	230, 250, 280, 360	6.88	[61]

(*continued*)

TABLE 10.4 (continued)
Selected Examples of Chiral Separations by Use of Ion-Pair Selectors

Substance	Selector	BGE	λ (nm)	*Rs*	Ref.
Isoprenaline	(–)-DIKGA	100 mM selector, 40 mM KOH in MeOH:EtOH (2:3)	220	11.3	[20]
Labetalol (RR/SS)	(–)-DIKGA	100 mM selector, 40 mM NaOH in MeOH	214	3.41	[15]
Mepivacaine	L-ZGP	125 mM selector, 50 mM NH_4Ac in MeOH:2-PrOH (80:20)	214, 254, 272	4.4	[70]
Metoprolol	CSS	5–40 mM selector 1 M HAc and 0.1–0.2 mM Tween 20 in ACN	214	N/A	[6]
Pronethalol	(–)-DIKGA	100 mM selector, 40 mM NaOH in MeOH	214	3.85	[15]
Pronethalol	(–)-DIKGA	100 mM selector, 40 mM NH_4Ac in MeOH:2-PrOH (75:25)	214	4.5	[21]
Sotalol	(–)-DIKGA	100 mM selector, 40 mM KOH in MeOH:EtOH (2:3)	220	17.5	[20]
Sotalol	(–)-DIKGA	100 mM selector, 40 mM NH_4Ac in MeOH:2-PrOH (75:25)	214	3.0	[21]
t-Bu-c-mefloquine	(*S*)-DNB-Leu	10 mM selector, 150 mM HAc,12.5 mM TEA in ACN:MeOH (70:30)	216, 330	0.90	[51]
Terodiline	L-ZGP	125 mM selector, 50 mM NH_4Ac in MeOH:2-PrOH (80:20)	214, 254, 272	1.7	[70]
Threo-2-	tBuCQN	10 mM selector, 100 mM HAc, 12.5 mM TEA in EtOH:MeOH (60:40)	230, 250, 280, 360	8.54	[61]
Timolol	KPA	100 mM selector, 40 mM KOH in MeOH:EtOH (2:3)	220	4.2	[20]

Note: For abbreviation of the most common BGE components, see Table 10.3. For abbreviation of the amino acids, see the abbreviation list.

Abbreviations: CSS, Camphersulfonic acid; KPA, (1*S*,4*R*)-ketopinic acid; (*S*)-DNB, 3,5-dinitrobenzoyl; tBuCQN, *O*-(tertbutylcarbamoyl) quinine; C-8 acid, octanoic acid; cHex-bis-CQN, *trans*-1,4-cyclohexylene(carbamoylated quinine); cHex-bis-CQN-S-C_{12}, *trans*-1,4-cyclohexylene-bis(carbamoylated-11-dodecylthio-dihydroquinine); HM-bis-CQN, 1,6-hexamethylene-bis(carbamoylated quinine); DCP-CQD, 3,4-dichlorophenyl carbamoylated quinidine; All-CDHQD, allyl carbamolylated dihydroquinidine; chex-CCD, cyclohexyl cinchonidine; cHex-bis-CQN-SO-C_{12}, *trans*-1,4-cyclohexylene-bis(carbamoylated-11-dodecylsulfinyl-dihydroquinine); HM-bis-CQD, 1,6-hexamethylene-bis(carbamoylated quinidine); P-bis-CQN, 1,3-phenylene-bis(carbamoylated quinine); ADCQN, 1-adamantyl carbamoylated quinine; DNP-CCN, dinitrophenyl cinchonine; All-CDHQN, allyl carbamoylated dihydroquinidine; DNP, 2,4-dinitrophenyl; DNZ, 3,5-dinitrobenzyloxycarbonyl; DNP-CCD, dinitrophenyl cinchoninine; Erythro-1-, erythro-1-amino-2-hydroxypropane phosphonic acid; tBuCQD, *tert*-butylquinidine; L-ZGP, *N*-benzoxycarbonylglycyl-L-proline; Threo-2-, Threo-2-amino-1-hydroxypropane phosphonic acid.

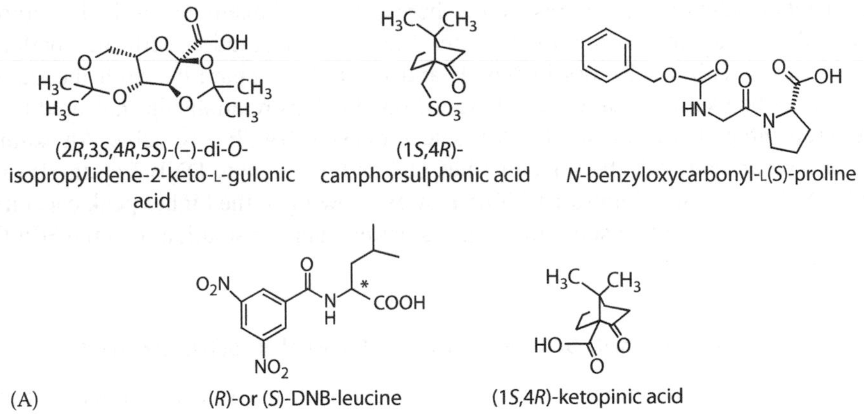

(2*R*,3*S*,4*R*,5*S*)-(−)-di-*O*-isopropylidene-2-keto-L-gulonic acid

(1*S*,4*R*)-camphorsulphonic acid

N-benzyloxycarbonyl-L(*S*)-proline

(A)

(*R*)-or (*S*)-DNB-leucine

(1*S*,4*R*)-ketopinic acid

(I) quinine (QN) cinchonine (CN) and derivatives

(II) quinidine (QD), cinchonidine (CD) and derivatives

(III) bis-CQN and bis-CQD derivatives

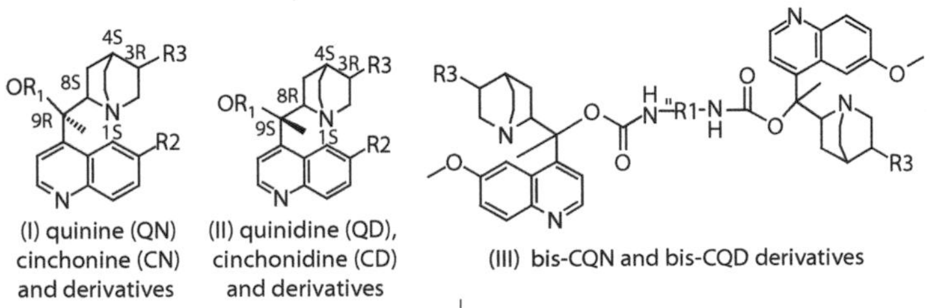

Structure (I)	R1	R2	R3	Structure (II)	R1	R2	R3
Quinine (QN)	H	OCH3	CH-CH2	Quinidine (QD)	H	OCH3	CH-CH2
Cinchonine	H	H	CH-CH2	Cinchonidine	H	H	CH-CH2
t-buCQN	CO-NH-C(CH3)3	OCH3	CH-CH2	*t*-buCQD	CO-NH-C(CH3)3	OCH3	CH-CH2
DNP-CQN	CO-NH-3,5-dinitrophenyl	OCH3	CH-CH2	DCP-CQD	CO-NH-3,4-dichlorophenyl	OCH3	CH-CH2
cHex-CQN	CO-NH-cyclohexyl	OCH3	CH-CH2	All-CNHQD	CO-NH-CH2-CH-CH2	OCH3	CH2-CH3
Ad-CQN	CO-NH-1-adamatyl	OCH3	CH-CH2	Structure (III)	R1	R3	
All-CNHQN	CO-NH-CH2-CH-CH2	OCH3	CH2-CH3	cHex-bisCQN	cyclohexyl	CH-CH2	
Structure (III)	R1	R3		cHex-bis-CQN-SO-C12	cyclohexyl	11-dodecylsulfinyl	
P-bis-CQN	phenyl	CH-CH2		cHex-bis-CQN-S-C12	cyclohexyl	11-dodecylthio	
HM-bis-CQD	hexamethylene	CH-CH2					

(B)

FIGURE 10.2 (A) Negatively charged ion-pair selectors; (B) positively charged ion-pair selectors.

provided that it is possible to attach them chemically to a solid support without loss of the chiral recognition site(s).

Many of the alkaloid derivatives exhibit a great enantioselectivity in NACE. An Rs value of 79.5 was, e.g., obtained for the dimeric form of a derivatized selector that originates from quinidine, namely 1,6-hexamethylene-bis(carbamoylated quinidine) (HMbisCQD) [14] (Table 10.4). This makes them of interest as CSPs, if the enantioresolution can be preserved. There are several examples in the literature where there is a good correlation between the enantioseparation with the alkaloid derivatives as chiral selectors in NACE and the HPLC separation with the same derivatives bonded to a stationary phase [61,68,69]. This relationship indicated that rapid screening of new potential HPLC selectors using NACE is possible. An example of a separation with the selector *t*-buCQN and the analyte DNB-Leu in HPLC and NACE is shown in Figure 10.3 [68]. However, owing to the higher peak capacity in CE, the NACE method sometimes gave a better enantioresolution than the HPLC one [61].

10.3.1 Electrolytes and Solvents in Chiral Ion-Pair Selector Systems

Since the values of ion-pair constants often are low, the selector concentration profile has a flatter curve which gives a broader interval where the value of $\Delta\mu$ is

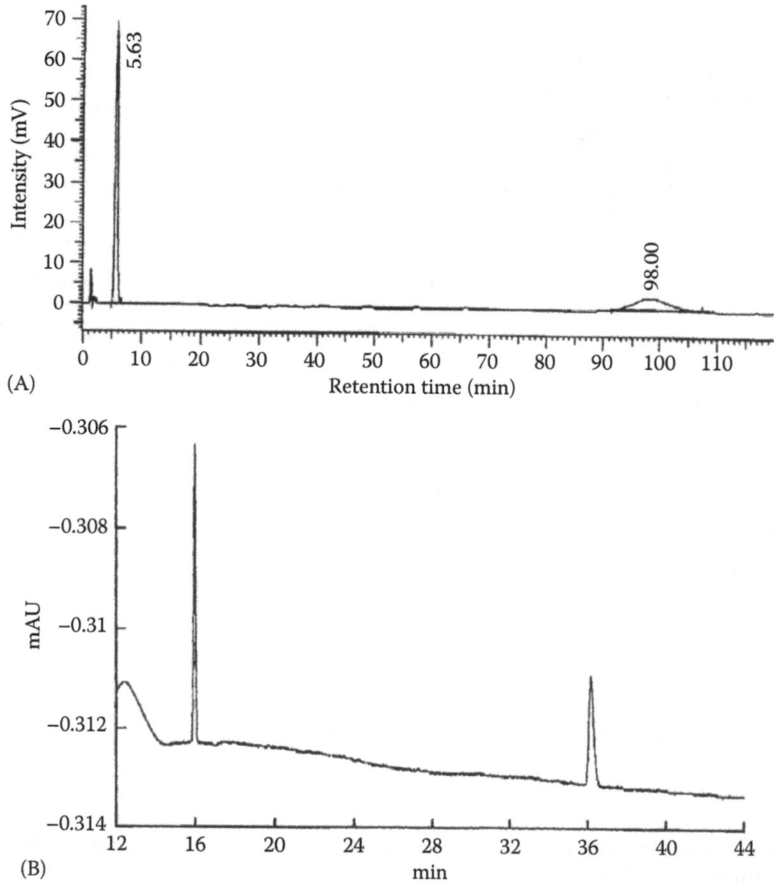

FIGURE 10.3 Enantiomer separation of DNB-Leu (A) by HPLC and (B) by CE. Mobile phase and background electrolyte: 100 mM octanoic acid and 12.5 mM ammonia in MeOH:EtOH (40:60). The BGE also contained 10 mM *tert*-butyl carbamoylated quinine. (From Piette, V. et al., *J. Chromatogr. A.*, 987, 421, 2003. With permission.)

near the $\Delta\mu_{max}$ value, resulting in a wider range over which $\Delta\mu$ is close to its maximum value (Figure 10.1). As discussed above, obtaining a minimal EOF is crucial in systems with a small mobility difference between the diastereomeric ion-pairs. The EOF in chiral NACE systems can be minimized in different ways, e.g., by applying covalent [15] or dynamic coatings [6], or by changing the BGE solvent. As discussed above, the decrease in EOF is often ascribed to a minimized ε/η ratio (Equation 10.4) resulting from a change of solvents [70]. However, it has also been revealed that the EOF can be minimized and even reversed without such an exchange, i.e., by minimizing (or reversal of) the ζ potential (Equation 10.4) through a change of the electrolyte [40,71]. The influence of different alkali metal hydroxides on the EOF has been studied with diisopropylidene ketogulonic acid (DIKGA) as the chiral selector [40]. These electrolytes were shown to influence the enantioresolution in more than one way. They formed competing nonstereoselective ion-pairs to a different extent with the chiral selector, which changed the amount of selector available for complexation with the enantiomers [40]. However, their greatest impact on the separation was their influence on the EOF. The cathodic EOF decreased with decreasing solvated radius of the alkali metal in MeOH and EtOH (i.e., the EOF decreased in the order Li^+, Na^+, K^+...) and even became anodic in some of the systems when KOH, RbOH, or CsOH was added to the BGE [40]. The separation of the enantiomers of isoprenaline is shown in Figure 10.4A–C. When KOH was added to an ethanolic BGE, the enantioseparation of isoprenaline with DIKGA as the chiral selector was performed within 60 s (No. 13 in Figure 10.4C), whereas the electropherogram where LiOH was added instead of KOH only displayed a partial separation (No. 11 in Figure 10.4C). The major reason for the difference in enantioresolution is the reversed EOF in the system containing KOH. A reversal of the EOF has also been observed when KPA was used as the chiral selector [20] (Figure 10.5). In this case, the cathodic EOF became anodic when the amount of EtOH was increased to 40% in a MeOH-based BGE. The examples mentioned above demonstrate that a proper choice of the electrolyte and solvents is crucial for a successful separation in NACE using ion-pair selectors. The influence of the BGE composition on separations in NACE has been reviewed by Fillet et al. [72].

For the ion-pair selectors, the majority gave decreased selectivity when water was added to the nonaqueous BGE and a lack of enantioselectivity in aqueous BGEs [7]. Many of the selectors, such as QN and quinidine (QD), have a limited solubility in water [14]. There are some exceptions, however, where enantioseparations with ion-pair selectors have been performed in the presence of higher amounts of water. The alkaloid derivative *t*-BuCQN was used in a BGE containing 20% water and 80% MeOH for a determination of the dissociation constants of its ion-pair with (*R*)-and (*S*)-DNB-Leu [73]. The authors found slightly lowered dissociation constants for the ion-pairs in this medium compared to a solely MeOH-based BGE, and the enantioresolution was lower too. This is, to the best of the author's knowledge, the highest amount of water that has been used with complete enantioseparation for ion-pair selectors. It should be noted that both QN [74] and CSS [75] have been used as ion-pairing reagents in combination with different CDs in water-based buffers. QN was used in chiral CE and CSS in CD modified chiral micellar electrokinetic

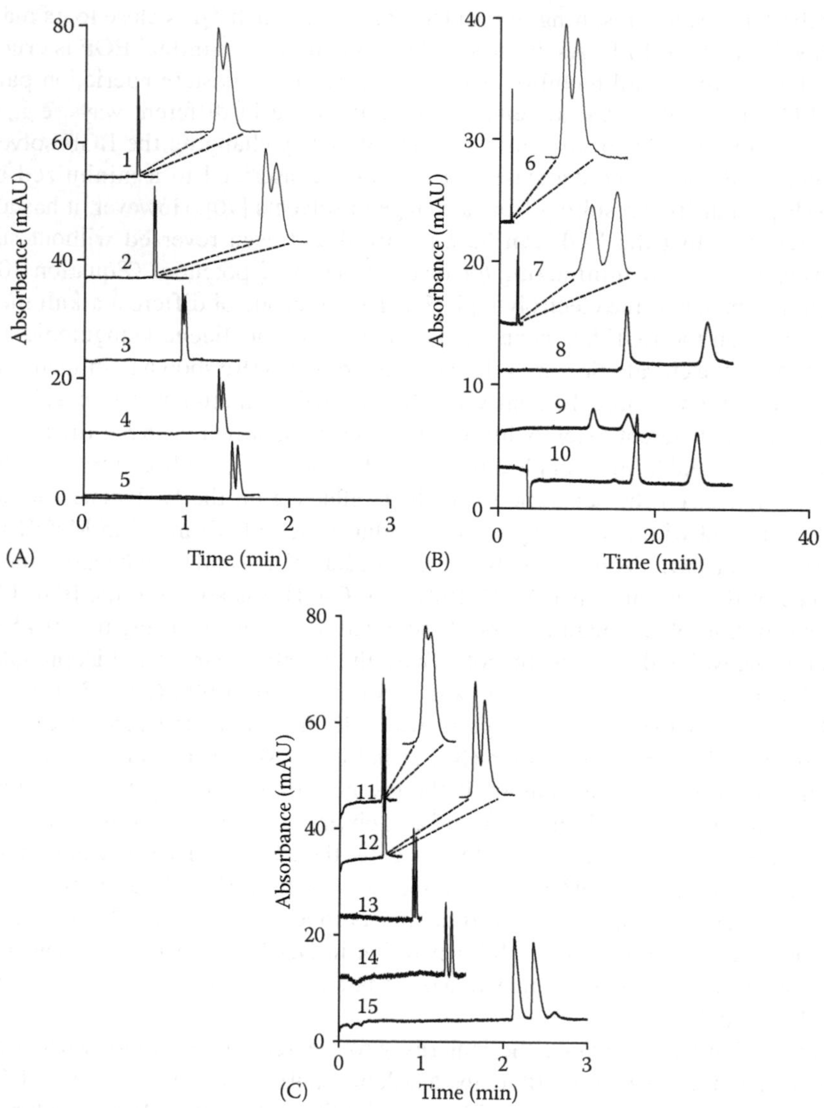

FIGURE 10.4 Influence of the alkali metal hydroxide and the organic solvent on the enantioseparation of isoprenaline. BGE: (A) 50 mM (–)-DIKGA and 20 mM alkali metal hydroxide in MeOH. 1: LiOH, 2: NaOH, 3: KOH, 4: RbOH, 5: CsOH, (B) 50 mM (–)-DIKGA and 20 mM alkali metal hydroxide in EtOH. 6: LiOH, 7: NaOH, 8: KOH, 9: RbOH, 10: CsOH. (C) 100 mM (–)-DIKGA and 40 mM alkali metal hydroxide in MeOH 11: LiOH, 12: NaOH, 13: KOH, 14: RbOH, 15: CsOH. L_{det} 8.5 cm, 30 kV. (From Hedeland, Y. et al., *Electrophoresis*, 27, 4469, 2006. With permission.)

chromatography (MEKC). However, this is not necessarily an enantioselective interaction between the analyte and the ion-pair selector in these applications, since achiral ion-pair reagents have been found to enhance the chiral separation in combination with CDs as well as chiral ones [74].

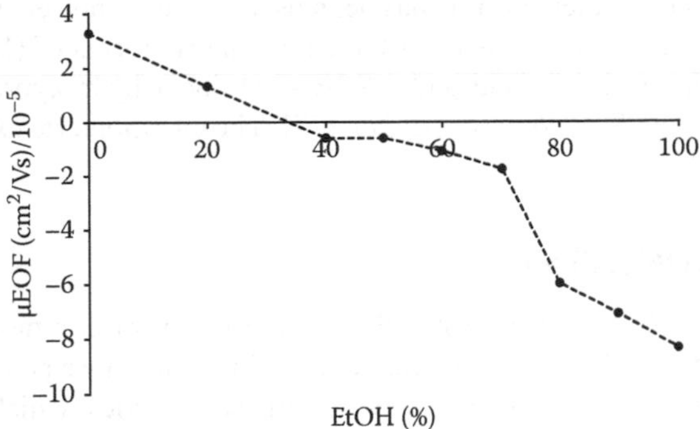

FIGURE 10.5 The influence of the ethanol concentration on the EOF. BGE: 100 mM (+)-KPA and 40 mM KOH in MeOH:EtOH (%, v/v), 30 kV, L_{det} 8.5 cm. EOF marker: 0.1% mesityl oxide (v/v) in MeOH. (From Hedeland, Y. et al., *J. Chromatogr. A.*, 1141, 287, 2007. With permission.)

10.3.2 Pharmaceutical Applications of Ion-Pair Selectors

The ion-pair selectors in NACE are rarely used for pharmaceutical applications, but the higher solubility of many drugs in nonaqueous solvents makes NACE an alternative to aqueous BGEs for, e.g., enantiomeric purity determination. The selector DIKGA has been used for a chiral purity determination of ephedrine in tablets [20]. The relatively high enantioresolution (Rs = 10.0) together with a high solubility of ephedrine in the selected solvent made it possible to inject 25 mM (1*R*,2*S*)-ephedrine with a minor impurity (0.071%) of (1*S*,2*R*)-ephedrine with maintained enantioresolution (Figure 10.6B). In accordance with the ICH-guidelines, no impurity in the tablet of Efedrin®† was detected above the reporting threshold (0.05%) for new drug substances [76] (Figure 10.6A), and the LOD for the method was determined to be 0.033% [20].

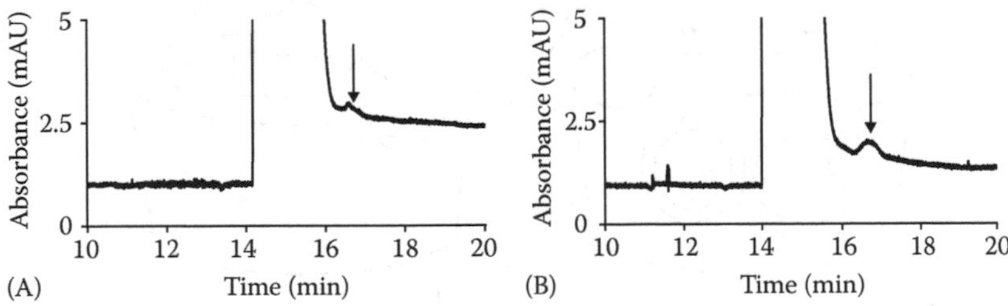

FIGURE 10.6 Enantiomeric purity of *(1S,2R)*-ephedrine. BGE: 100 mM (−)-DIKGA and 40 mM KOH in MeOH:EtOH (2:3 v/v), sample concentration 25 mM of *1S,2R*-ephedrine, 30 kV, L_{det} 23 cm. Polyacrylamide coated capillary. (A) Efedrin® (Registered trademark of Merck NM, Stockholm, Sweden) (batch 5A209A). (B) *(1S,2R)*-ephedrine spiked with 0.071% (w/w) *(1R,2S)*-ephedrine. (From Hedeland, Y. et al., *J. Chromatogr. A.*, 1141, 287, 2007. With permission.)

† Merck NM, Stockholm, Sweden

Tert-butylcarbamoylated quinine has been used as the chiral selector for determination of the enantiomeric excesses of the enantiomeric forms of (1*R*,2*S*)-1-amino-2-hydroxypropylphosphonic acid and (1*S*,2*R*)-2-amino-1-hydroxypropylphosphonic acid produced by the *Streptomyces fradiae* [77]. The enantiomeric excess was determined to be ≥98%.

10.4 CYCLODEXTRINS

Different types of CDs and derivates thereof are the most commonly used chiral selectors in aqueous CE [11]. They often exhibit high enantioselectivity for a broad range of substances [62]. The CDs are cyclic oligosaccharides, which are composed of D-glucose units, connected through α(1–4)-glycosidic bonds. The CD structure can be described as a truncated cone with a cavity in the center. The inside of the cavity is hydrophobic and the outside has a more hydrophilic character. The wider edge of the cavity of the native CDs consists of secondary hydroxyl groups and the narrower edge is comprised of primary ones. In NACE, native CDs are only rarely used and, so far, α-CD has not been successfully applied. One of the reasons for this is its limited solubility in many of the commonly used nonaqueous solvents, such as MeOH. γ-CD exhibits low solubility in MeOH (1 mM) compared to water (179 mM) [78]. In addition, β-CD is difficult to dissolve in MeOH [45] but has a higher solubility in NMF (>700 mM) [4] and in FA (where at least 150 mM was soluble) [45] than in water (18 mM or up to 60 mM have been reported) [4,45]. Another reason for their more frequent use in aqueous media may be associated with the type of complex formed with the solutes. For CDs, two different complexation mechanisms are possible [79]. In aqueous and hydroorganic solvents, the complexation between the selector and the solute is mainly driven by hydrophobic inclusion (Figure 10.7A) [79]. One prerequisite for this is that the hydrophobic group of the solute is small

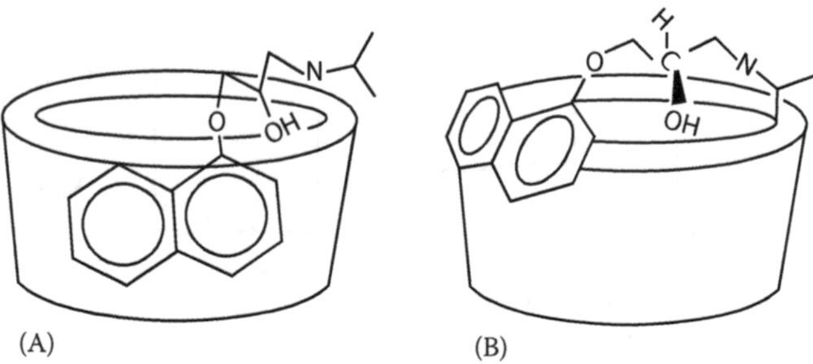

FIGURE 10.7 Schematic diagram of two different enantioselective mechanisms for CDs. Selector: β-cyclodextrin; solute: propranolol. (A) Hydrophobic inclusion complexation. The enantioselectivity also requires hydrogen-bonding and steric interactions at the mouth of the CD cavity. (B) The polar-organic mode where the solvent (ACN in this example) occupies the hydrophobic cavity and the complexation is a combination of hydrogen-bonding and dipolar interactions at the mouth of the CD. Steric interactions can also contribute to the chiral recognition. (From Armstrong, D.W. et al., *J. Liq. Chromatogr.*, 20, 3279, 1997. With permission.)

enough to fit into the cavity of the CD. The interaction with the CD is assumed to be a two-step process where the guest molecule first penetrates the host, then solvent is released from the guest and the host molecule [80]. The solute–selector interaction is assumed to be primarily van der Waals and hydrophobic interactions but hydrogen-bonding and/or steric interactions between the solute and selector may also be required to achieve enantioselectivity. In nonpolar solvents, the formation of the inclusion complex is suppressed, and the solute–selector complexation consists of hydrogen-bonding and dipole–dipole interactions in the vicinity of the edge of the CDs (Figure 10.7B) [79].

The binding constants for the interaction between the CD and the guest molecules are generally found to be lower in less polar solvents than in water [3]. As already discussed above, this results in a higher optimal selector concentration and a flatter curve than in aqueous BGE (Figure 10.1). Thus, the concentration of the CD is generally less critical for the chiral resolution in nonaqueous media than in aqueous ones [3,40,81]. In addition, it should be stressed that a lower binding constant does not necessarily mean that the enantioselectivity is lower, only that the optimal selector concentration is higher. This is demonstrated in Figure 10.8 for the separation of the enantiomers of trimipramine using β-CD in water (A) and FA (B), where the separation is enhanced in FA and the optimal concentration is increased from 0.2 mM to 100 mM [3].

The properties of the native CDs can be modified by derivatization. The attached groups can be neutral (e.g., methyl-), positive (e.g., propylamino-), or negative (e.g., 2,3-diacetyl-6-sulphato-). The derivatization changes the appearance of the edge of the truncated cone and opens up possibilities for additional types of interactions between the selector and the solutes. If the attached groups are charged, the CD derivative obtains electrophoretic mobility, which might increase the enantioresolution. A major drawback with the CD derivatives has been the batch-to-batch variation, since the substitution has been found to be random. In addition to the problem on interbatch variation in enantioseparations, it also made the selector–solute

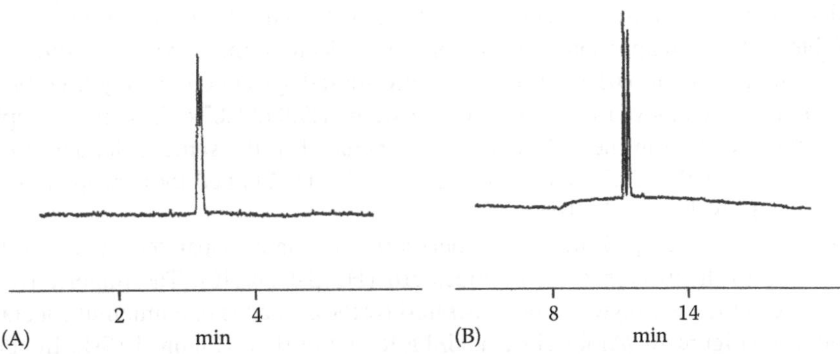

FIGURE 10.8 Chiral separation of trimipramine in aqueous and FA media. (A) 0.2 mM β-CD in 50 mM citric acid, 25 mM Tris in water, pH 3.02. (B) 100 mM β-CD in 150 mM citric acid, 100 mM Tris in FA, pH* 5.1. (From Wang, F. and Khaledi, M.G., *Anal. Chem.*, 68, 3460, 1996. With permission.)

interactions hard to evaluate. However, it is now possible to perform a selective synthesis which results in only one isomer of the CD derivative [82].

The use of charged derivatives is more limited in NACE than in aqueous systems, where one selector is often able to enantioseparate molecules regardless of their charge (positive, negative, or neural). In NACE systems, charged CD derivatives have been used primarily to separate solutes of opposite charge, e.g., sulfated ones have been used for cationic analytes and ammonium substituted ones for anions [7]. It has been suggested that the ionic-interaction compensates for the weakened inclusion complexation in the nonaqueous solvent. As mentioned above, the limited solubility (especially in alcohols) of the native CD and CD derivatives is one of the limitations with these selectors in NACE [83,84]. But the sulfated single CD-isomers have proved to be more soluble than the randomly substituted sulfated ones [11]. The single CD-isomers of heptakis(2,3-diacetyl-6-sulfato)-β-cyclodextrin (HDAS-β-CD), heptakis(2,3-methyl-6-sulfato)-β-cyclodextrin (HDMS-β-CD), octakis (2,3-diacetyl-6-sulfato)-γ-cyclodextrin (ODAS-γ-CD), and octakis(2,3-*O*-dimethyl-6-*O*-sulfo)-γ -cyclodextrin (ODMS-γ-CD) have a moderate solubility in, e.g., MeOH and have been utilized as selectors in this solvent [17,81,84–89]. However, the above mentioned single isomers, and the majority of the other single isomers that have been introduced are insoluble in ACN [90]. Sanchez-Vindas and Vigh [90] found that the low solubility of HDAS-β-CD in this solvent could be improved by the replacement of the sodium counterion by a more hydrophobic one, and applied the single isomer sulfated β-CD as its tetrabutylammonium salt. They used the chiral selector (TBA_7HDAS-β-CD) in an ACN-based BGE for the enantioseparation of 13 basic compounds.

The choice of solvent and electrolytes is of great importance for the enantioseparation in the CD systems. The influence of the solvent has already been mentioned above, and is primarily addressed at their hydrophobicity, which influences the host–guest complexation. Formamide often exhibits a better environment for the enantiomeric separation with the CDs than, e.g., MeOH which indicated that the solvophobic interaction between the CD and the solute is of major importance for the chiral recognition. Unless the major interactions in host–guest complexation are of hydrophobic character, the choice of electrolytes might also influence the separation. Typical BGEs for chiral separations with CDs as the chiral selectors are presented in Table 10.5. In comparison with the ion-pair selectors, these systems more often contain some kind of buffer, even the in water-based system commonly used phosphate buffer is employed in some of the systems [17,84,85,87,88]. Some examples of enantioresolution in the CD systems are included in the same table, and are in the range from 0.9 to 29.2, where the highest value is obtained for bupropion using ODAS-γ-CD as the chiral selector.

In Figure 10.9, a 1,3,4-thiadiazine derivative is enantioseparated by use of the chiral selector hydroxypropyl-β-cyclodextrin (HP-β-CD) [16]. The importance of the choice of the electrolyte can be illustrated by the exchange of ammonium acetate/acetic acid (Figure 10.9A) for citric acid/TRIS in the BGE (Figure 10.9B). In addition, the researchers found that MeOH was superior as a solvent for separations with HP-β-CD and with hydroxyethyl-β-cyclodextrin (HE-β-CD), but that FA was the solvent of choice for methyl-β-cyclodextrin (Me-β-CD). The influence of sodium, potassium, ammonium, chloride, formate, methaneformate, and camphorsulfonate

TABLE 10.5
Selected Examples of Chiral Separations with CD-Based Selectors

Substance	Selector	BGE	λ (nm)	*Rs*	Ref.
1,3,4-Selenadiazine derivatives	HE-β-CD	200 mM selector, 25 mM citric acid,12.5 mM TRIS in MeOH	340	N/A	[16]
1,3,4-Thiadiazine derivatives	HP-β-CD	200 mM selector, 25 mM citric acid,12.5 mM TRIS in MeOH	340	N/A	[16]
1,3,4-Thiadiazine derivatives	Methyl-β-CD	200 mM selector, 25 mM NH_4Ac, 1 M HAc in FA	340	N/A	[16]
P-chloroamphetamine	HDAS-β-CD	40 mM selector, 50 mM dichloroacetic acid and 25 mM TEA in MeOH	214	3.3	[86]
Pindolol	HDAS-β-CD	40 mM selector, 50 mM dichloroacetic acid and 25 mM TEA in MeOH	214	2.6	[86]
1-Aminoindan	HxDAS	5 mM selector, 25 mM H_3PO_4 and 12.5 mM LiOH in MeOH	214	8.9	[88]
2-Aminoindan	ODAS-γ-CD	45 mM selector, 25 mM H_3PO_4 and 12.5 mM NaOH in MeOH	214	20.6	[87]
4-Cl-amphetamine	HDMS-β-CD	40 mM selector, 0.02 M H_3PO_4 and 0.01 M NaOH in MeOH	214	6.8	[17]
Bepridil	β-CD	100 mM selector, 100 mM Tris,150 mM citric acid and 5% TEA in FA (pH* 5.1)	254	0.9	[114]
Bupropion	ODAS-γ-CD	30 mM selector, 25 mM H_3PO_4 and 12.5 mM NaOH in MeOH	214	29.2	[87]
Chlophedianol	TBA_7HDAS-β-CD	10.0 mM selector, 50 mM methanesulfonic acid, 21 mM TEA in ACN	210	4.4	[90]
Dansylalanine	β-CD	200 mM selector, 10 mM NaCl in NMF	254	100[1]	[4]
Dansylasparagine	β-CD	200 mM selector, 10 mM NaCl in NMF	254	100[1]	[4]
Dansylthreonine	QA-β-CD	20 mM selector, 20 mM NH_4Ac, 1% HAc in FA	254	6.36	[24]
Dansylvaline	QA-β-CD	20 mM selector, 20 mM NH_4Ac, 1% HAc in FA	254	9.60	[24]
Demethylbencynonatine	HDMS-β-CD	10 mM selector, 20 mM H_3PO_4, 10 mM NaOH in MeOH	254, 200	3.18	[84]
DNS-aspartic acid	Mono-β-CD no 1	4% selector, 35 mM $NaHCO_3$ in NMF	254, 285	1.39	[104]
DNS-aspartic acid	Oligo-β-CD no 1	4% selector, 35 mM $NaHCO_3$ in NMF	254, 285	2.03	[104]
DNS-aspartic acid	Mono-β-CD no 3	4% selector, 35 mM $NaHCO_3$ in NMF	254, 285	1.96	[104]

(continued)

TABLE 10.5 (continued)
Selected Examples of Chiral Separations with CD-Based Selectors

Substance	Selector	BGE	λ (nm)	*Rs*	Ref.
DNS-aspartic acid	β-CD	4% selector, 35 mM $NaHCO_3$ in NMF	254, 285	1.91	[104]
DNS-glutamic acid	Mono-β-CD no 1	4% selector, 35 mM $NaHCO_3$ in NMF	254, 285	1.28	[104]
DNS-glutamic acid	Oligo-β-CD no 3	4% selector, 35 mM $NaHCO_3$ in NMF	254, 285	2.07	[104]
DNS-glutamic acid	Oligo-β-CD no 1	4% selector, 35 mM $NaHCO_3$ in NMF	254, 285	1.72	[104]
DNS-glutamic acid	β-CD	4% selector, 35 mM $NaHCO_3$ in NMF	254, 285	1.85	[104]
DNS-Threonine	Oligo-β-CD no 3	4% selector, 35 mM $NaHCO_3$ in NMF	254, 285	2.10	[104]
DNS-Threonine	Mono-β-CD no 3	4% selector, 35 mM $NaHCO_3$ in NMF	254, 285	1.97	[104]
Epinephrine	HDMS-β-CD	40 mM selector, 25 mM H_3PO_4, 12.5 nM NaOH in MeOH	214	3.3	[85]
Fluoxetine	HxDAS	30 mM selector, 25 mM H_3PO_4 and 12.5 mM LiOH in MeOH	214	3.8	[88]
Homoatropine	HDMS-β-CD	40 mM selector, 0.02 M H_3PO_4 and 0.01 M NaOH in MeOH	214	7.3	[17]
Ibuprofen	EA-β-CD	10 mM selector, 20 mM NH_4Ac in MeOH	230, 250, 260, 265	7.3	[112]
Ibuprofen	IPA-β-CD	10 mM selector, 20 mM NH_4Ac in MeOH	230, 250, 260, 265	7.7	[112]
Ibuprofen	PA-β-CD	10 mM selector, 20 mM NH_4Ac in MeOH	230, 250, 260, 265	8.2	[112]
Mianserin	HP-β-CD	100 mM selector, 150 mM citric acid, 100 mM Tris in FA	254	1.61	[3]
Mianserin	β-CD	100 mM selector, 150 mM citric acid, 100 mM Tris in FA	254	1.88	[3]
N-benzoyl phe-ala methylester	β-CD	1.51 M selector, 0.06 M NaCl and 10% HAc in FA	260	1.5	[109]

Nefopam	HP-β-CD	100 mM selector, 150 mM citric acid, 100 mM Tris in FA	254	1.82	[3]
Nefopam	Methyl-β-CD	100 mM selector, 150 mM citric acid, 100 mM Tris in FA	254	0.97	[3]
Nefopam	γ-CD	100 mM selector, 150 mM citric acid, 100 mM Tris in FA	254	0.97	[3]
Ondanseron	β-CD	100 mM selector, 100 mM Tris, 150 mM citric acid and 10% TEA in FA (pH* 5.1)	254	3.3	[114]
Pinacidil	β-CD	100 mM selector, 100 mM Tris, 150 mM citric acid and 15% TEA in FA (pH* 5.1)	254	≈1.5	[114]
Propranolol	HDMS-β-CD	40 mM selector, 25 mM H_3PO_4, 12.5 nM NaOH in MeOH	214	5.2	[85]
Suprofen	EA-β-CD	10 mM selector, 20 mM NH_4Ac in MeOH	230, 250, 260, 265	7.2	[112]
Suprofen	IPA-β-CD	10 mM selector, 20 mM NH_4Ac in MeOH	230, 250, 260, 265	7.4	[112]
Suprofen	PA-β-CD	10 mM selector, 20 mM NH_4Ac in MeOH	230, 250, 260, 265	9.1	[112]
Timolol	HDMS-β-CD	30 mM CD, 30 mM K^+ camphor SO_3^- and 0.75 M formic acid in MeOH	295	9.2	[18]
Trimeprazine	β-CD-(SO_4^-)$_4$	3.85% selector, 100 mM Tris and 150 mM citric acid (pH* 5.1) in FA	Variable λ	3.19	[110]
Trimipramine	β-CD-(SO_4^-)$_4$	3.85% selector, 100 mM Tris and 150 mM citric acid (pH* 5.1) in FA	Variable λ	3.64	[110]
Trimipramine	γ-CD	100 mM Selector, 150 mM citric acid,100 mM Tris in FA	254	1.50	[3]
Trohexylphenidyl	Methyl-β-CD	100 mM Selector, 150 mM citric acid,100 mM Tris in FA	254	1.94	[3]
β-Bencynonatine	HDMS-β-CD	10 mM selector, 20 mM H_3PO_4, 10 mM NaOH in MeOH	254, 200	2.45	[84]
Hemicholium-15 Br	TBA_7HDAS	10.0 mM selector, 50 mM methanesulfonic acid, 21 mM TEA in CAN	210	4.5	[90]

Note: For abbreviation of the most common BGE components, see Table 10.3. For abbreviation of the chiral selectors, see the footnote in Table 10.1.

Abbreviations: Mono-β-CD no 1, mono-(6-*O*-norborn-2-ene-5-yl-carboxyl)-β-CD; Oligo-β-CD no 1, oligo-(6-*O*-norborn-2-ene-5-yl-carboxyl)-β-CD; Mono-β-CD no 3, pentakis-(6-*O*-norborn-2-ene-5-ylmethyl-hydroxysiloxyl)-β-CD; Oligo-β-CD no 3, oligo (pentakis-(6-*O*-norborn-2-ene-5-ylmethyl-hydroxysiloxyl)-β-CD).

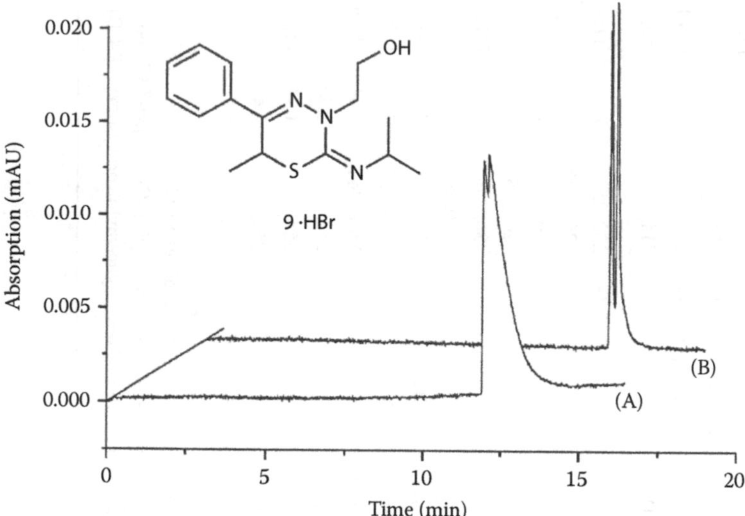

FIGURE 10.9 Influence of the electrolyte on the separations of a 1,3,4-thiadiazine derivative. Conditions: 200 mM HP-β-CD in MeOH and (A) 25 mM ammonium acetate, 1 M acetic acid pH* 5.0, (B) 25 mM citric acid, 12.5 mM TRIS pH* 4.9. (From Karbaum, A. and Jira, T., *J. Biochem. Biophys. Methods*, 48, 155, 2001. With permission.)

have been investigated for the separation using HDMS-β-CD as the selector [53]. The authors found that both the cationic and the anionic electrolytes had a great impact on the enantioresolution. In this study, ammonium formate and potassium camphorsulfonate seemed to be the best choice [53]. The synergistic effects of mixing CDs with ion-pair selectors had also been reported earlier by the same group [52], and had previously been observed in HPLC [91]. In the NACE study [52], a low enantioselectivity for trimipramine or none at all were observed when the HDMS-β-CD was used alone in the BGE (Figure 10.10B), whereas a complete enantioresolution was obtained when the camphorsulfonate was added to the BGE (Figure 10.10A). No data for the separation of trimipramine with CSS alone was presented in the study. However, CSS had not shown selectivity for the enantiomers of trimipramine in an earlier study performed in an ACN-based BGE [6]. A combination of the potassium salt of the camphorsulfonate and a high concentration of acetic acid was used to suppress the EOF in the later study [52]. The authors stressed that the effect of the ion-pair selector can not only be attributed to a difference in the formation of the diastereomeric ion-pairs with CSS, since the addition of achiral ion-pair selectors also give an improved enantioresolution with CDs. Thus, the effect seems to be related, at least partly, to a difference in interaction of the ion-pair complex formed (solute-CSS) with the CD. Furthermore, the use of ion-pair selectors in combination with CDs seems to be more favorable for an analyte with a high affinity to the CD [53]. The combined system with HDMS-β-CD and camphorsulfonate has been optimized further by experimental design and the method is published in a later paper [54].

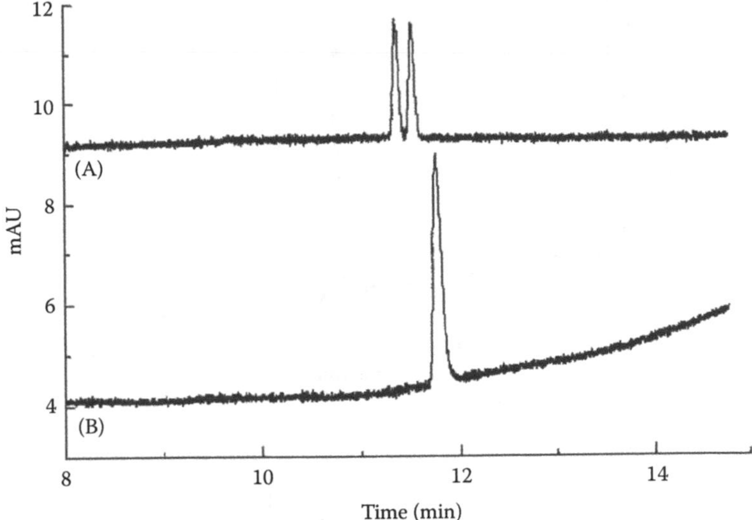

FIGURE 10.10 Enantioseparation of trimipramine. BGE: (A) 20 mM HDMS-β-CD, 1 M formic acid and 40 mM potassiumcamphorsulfonate in MeOH; (B) 20 mM HDMS-β-CD, 1 M formic acid and 40 mM NH_4Cl in MeOH. (From Servais, A.-C. et al., *Electrophoresis*, 24, 363, 2003. With permission.)

10.4.1 Applications Using CDs Derivatives

No applications using native CDs have been performed in nonaqueous media, but five studies have been carried out with CD derivatives. The CD derivative HDAS-β-CD has been used as a selector for the determination of the enantiomers of salbutamol in human urine [83]. The eluate from the off-line solid phase extraction was evaporated under nitrogen and reconstituted in MeOH prior to injection into the CE system. The LOD, when UV detection was used, was determined to be 125 ng/mL for each of the enantiomers. The LOD was further improved to 8 ng/mL for the fastest migrating enantiomer and to 14 ng/mL for the second one by hyphenation of the CE system with MS detection (Figure 10.11) [22].

Another biomedical application with HDMS-β-CD in NACE has been developed for omeprazole and one of its metabolites [92]. For omeprazole, nonaqueous media is preferable since the stability of the molecule is superior in MeOH than in aqueous solutions. However, the method has not been used on real samples since the LOD is too high (50 μM), but the authors expect that a MS hyphenation would increase the sensitivity.

In addition, HDMS-β-CD has been used in a chiral purity determination of (*S*)-timolol [18]. This method could be utilized for the separation of impurities of (*S*)-timolol (Figure 10.12) [18]. The LOD for *(R)*-timolol was determined to be 0.03% of the main peak and the LOQ was reported to be 0.1% [18]. The LOQ is in the same range as other recently published chiral purity determinations with CE, where the LOQ values varied between 0.05 and 0.5% [12]. A comprehensive

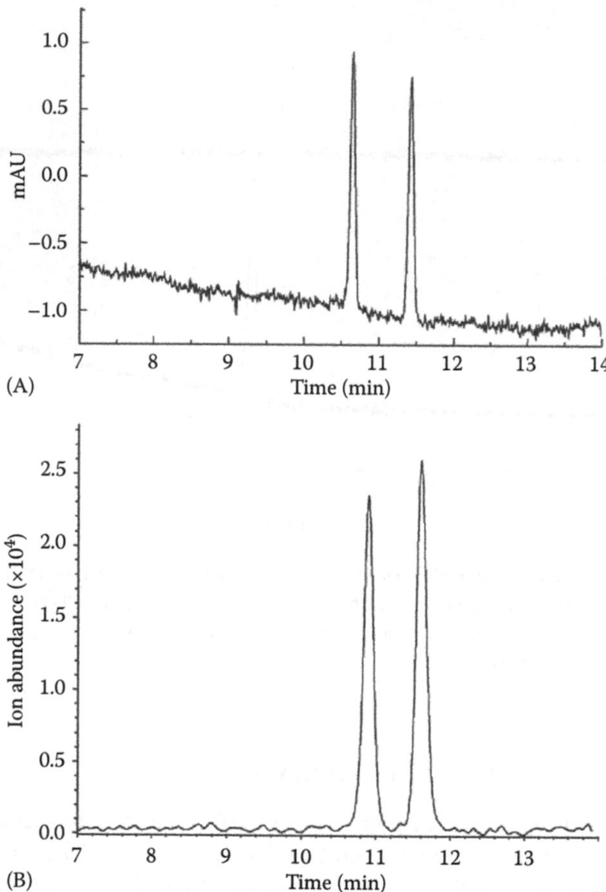

FIGURE 10.11 Enantioseparation of salbutamol by the use of UV and MS-detection. BGE: 10 mM ammonium formate and 15 mM HDAS-β-CD in methanol acidified with 0.75 M formic acid. (A) NACE-UV at 230 nm, 1.5 μg/mL of each enantiomer. (B) NACE ESI-MS (extracted ion chromatogram at m/z 240.4), 30 ng/mL of each enantiomer. Sheet liquid: ACN-H_2O containing 0.1% formic acid (75:25), sheet flow rate 2.5 μL/min. (From Servais, A.-C. et al., *J. Pharm. Biomed. Anal.*, 40, 752, 2006. With permission.)

inter-laboratory study of the above mentioned method for (*S*)-timolol has been performed between laboratories in Europe and Canada [19]. HDMS-β-CD has also been used before for a chiral purity determination of a hydrophobic intermediate of mitomycin, where the minor enantiomeric impurity corresponded to 4.2% of the main peak [17].

10.5 MISCELLANEOUS CHIRAL SELECTORS USED IN NACE

There are certain families of selectors that can be used in aqueous solutions, which have not so far been applied as chiral selectors in nonaqueous CE (e.g., proteins,

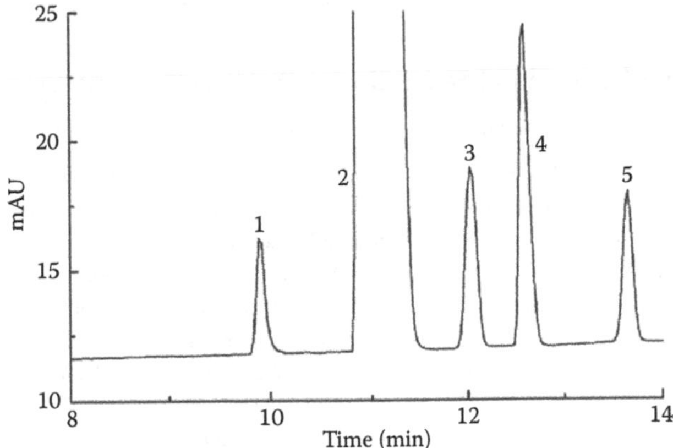

FIGURE 10.12 Typical electropherogram of a mixture solution containing the main impurities of timolol. Selector: HDMS-β-CD, internal standard: pyridoxine (peak 1). Peaks and concentrations: (1) pyridoxine (10 μg/mL), (2) (*S*)-timolol (2 mg/mL), (3) (*R*)-timolol (1.0% or 20 μg/mL), (4) isotimolol (1.09% or 21.8 μg/mL), and (5) dimer maleate (0.51% or 10.3 μg/mL). (From Marini, R.D. et al., *J. Chromatogr. A.*, 1120, 102, 2006. With permission.)

macrocyclic antibiotics, and chiral calixarenes) [12]. Some of these would probably also work in the nonaqueous mode, and, to provide an example, vancomycin has already been successfully used for the enantioseparation in nonaqueous capillary electrochromatography [93].

Mori et al. [94] used the macrocyclic polyether (+)-18-crown-6-tetracarboxylic acid as the chiral selector for separation of aromatic amines, amino acids, and amino alcohols. In total, Mori and coworkers investigated ten different solvents and successful enantioseparation was obtained in DMSO and FA. They also found that addition of tetra-*N*-butylammonium perchlorate to the FA-based BGE improved the separation efficiency.

The use of a selector based on a ligand-exchange mechanism has also been performed by Karbaum and Jira [95] who used copper(II) and L-proline or L-isoleucine in MeOH for the enantioseparation of eight amino acids, and investigated the influence of pH,* the concentration, and the ratio between the complexing agents. The stoichiometric ratio between Cu^{2+} and the ligand was found to be of utmost importance for the separation but it differed from the optimal ratio in aqueous media. The pH* was adjusted by NaOH and formic acid and measured by a conventional pH-electrode. For L-proline, a distinct maximum for the separation factor was found at pH* 3.9. Some typical separations using copper(II) and L-proline or the crown ether as chiral selectors are presented in Table 10.6.

10.6 IMPROVEMENT OF DETECTION LIMIT IN CHIRAL NACE

The detection limit is often higher for CE than for other separation techniques such as LC, because of the lower loading capacity and the use of on-column detection in CE, which gives a shorter path length. But the use of a preconcentration step or

TABLE 10.6
Selected Examples of Chiral Separation with Miscellaneous Chiral Selectors

Substance	Selector	BGE	λ (nm)	α[a]	Refs.
1-Amino-1, 2-diphenylethanol	(+)-18-Crown-6-tetracarboxylic acid	10 mM selector, 40 mM tetra-*N*-butylammoniumperchlorate in FA	260, 280 or 300 nm	1.21	[94]
1-Naphtylethylamine	(+)-18-Crown-6-tetracarboxylic acid	10 mM selector, 100 mM tetra-*N*-butylammoniumperchlorate in FA	260, 280 or 300 nm	1.24	[94]
Phenylalanine	Cu (II): L-proline	25 mM NH_4Ac, 1 M HAc, 5.86 mM $CuCl_2^*2H_2O$, 16 mM L-proline in MeOH (pH* 3.9)	214 nm	1.12	[95]
α-Methyltyrosine	Cu (II): L-proline	25 mM NH_4Ac, 1 M HAc, 5 mM $CuCl_2^*2H_2O$, 15 mM L-proline in MeOH (pH* 3.92)	214 nm	1.14	[95]

[a] Defined as $\alpha = t_{m1}/t_{m2}$ in the Ref. [97] and as $\alpha = \mu_{eff1}/\mu_{eff2}$ in the Ref. [98].

a more sensitive detector can improve the detection limit. In addition, the use of the partial filling technique might improve the sensitivity when the chiral selector increases the detection limit, e.g., when the selector shows absorbance in the UV region (when UV detection is used) or is nonvolatile (when MS detection is applied). The most commonly used detection method in chiral NACE is not only UV spectroscopy, but also MS [21,22] and Fourier transform infrared (FTIR) spectroscopy [96]. As well as providing quantitative information, the latter technique gives qualitative information about the chiral complexes. A spectral shift in the mid-IR region was observed when DNB-leucine was complexed with the selector *t*-BuCQN compared to the uncomplexed analyte [96]. Interestingly, the authors noticed a slight difference between the enantiomeric pairs.

Preconcentration can be achieved by manipulating the electrophoretic velocity of the analyte in the injection zone or by partitioning the analyte to a pseudostationary phase. In NACE, preconcentration techniques like field amplified stacking (FASS) [97–99], transient isotachophoresis (*t*-ITP) [100], low temperature zones [101], low and high conductivity zones [102], and online solid-phase extraction [103] have been utilized. For a recent review on different strategies for enhanced sensitivity in chiral CE, see the reference by García-Ruiz and Marina [13] and Chapter 12. However, it should be stressed that the use of a stacking technique to improve the detection limit subsequently increases the complexity of the system, which also might affect the robustness of the method. Furthermore, the often lower conductivity in nonaqueous solvents than in aqueous ones decreases the possibility of zone sharpening by, e.g., FASS.

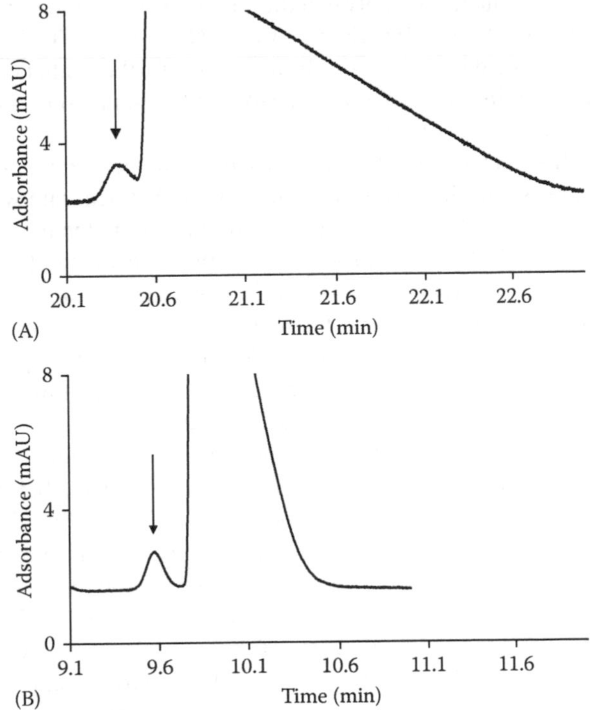

FIGURE 10.13 Peak sharpening of (*S*)-timolol. BGE: 100 mM (+)-KPA and 40 mM KOH in MeOH:EtOH (3:2 v/v). 30 kV, L_{det} 23 cm Sample: 2 mM (*S*)-timolol and 0.05 mM (*R*)-timolol (2.4%) dissolved in MeOH. Normal injection (A), transient ITP (B). Leading electrolyte: 100 mM NaAc in MeOH (25 mbar in 1 s). The capillary was filled with the selector solution, and the terminating electrolyte (200 mM of triethanolamine in MeOH) was used as the anodic vial. The migration time of the (*R*)-enantiomer is marked by the arrow. (From Hedeland, Y. et al., *J. Chromatogr. A.*, 1141, 287, 2007. With permission.)

The stacking technique *t*-ITP has been applied in combination with the chiral separation of timolol in nonaqueous media (Figure 10.13) [20]. A zone sharpening was observed for (*S*)-timolol when *t*-ITP was used (Figure 10.13B) which facilitated the injection of a higher sample concentration, thereby decreasing the LOD of the enantiomeric impurity from 2.5% to 0.2%. In conjunction with this, the migration time decreased from 23 to 10.5 min in the *t*-ITP system [20]. The decreased migration time was primarily due to the cathodic EOF caused by the terminating electrolyte (in the selector BGE alone, the EOF was anodic). If the sample amount is in the mL range, an off-line preconcentration, (e.g., SPE) is often preferable to *t*-ITP since the NACE system will probably be more straightforward to use. One example of this was already presented for the enantioseparation of salbutamol in urine (Figure 10.11) [83].

The use of matching conductivities between the sample and BGE zone is of importance for the peak efficiency, since mismatch leads to peak broadening and asymmetry. Eder et al. [104] increased the mobility of the analytes and received a closer matching of conductivities between the sample zone and the BGE by addition of up to 5% of water saturated bicarbonate to the NMF-based BGE. This caused a subsequent decrease of the EOF in the BGE, which further enhanced the separation of the D- and L-amino acids. Norbornene-β-CD-based monomers and oligomers were used as chiral selectors [104].

The partial filling technique [105] can be used to enhance detection sensitivity in BGEs with UV absorbing chiral selectors, such as *tert*-butylcarbamoyl quinine [59]. An example of an electropherogram where the partial filling technique is employed is displayed in Figure 10.14 (unpublished figure). In this example, the selector zone has roughly the same absorbance, measured in mAU, as the sample, which would increase the LOD. When the partial filling technique is employed, only a part of the capillary is filled with the selector solution [105]. This technique requires separation conditions where the enantiomers are separated in and migrate through the selector zone, into the selector free solution, and reach the detector before the selector [21]. An alternative, probably less common than the previous condition, is that the selector zone migrates through the sample and reaches the detector before the sample [54]. The experimental conditions need to be optimized, including the plug length, the selector concentration, and the composition of the selector free zone. Zarbl et al. [51] used PVA coated capillaries to avoid an early breakthrough of the selector zone into the detection window. Zarbl and coworkers found that the EOF in the system was lowered, and the repeatability, in terms of the migration times, was comparable to what they had observed previously in uncoated capillaries using the partial filling technique [59]. The agreement in the magnitude of the EOF between the two zones (with and without selector) is significant to reduce the extra peak-broadening

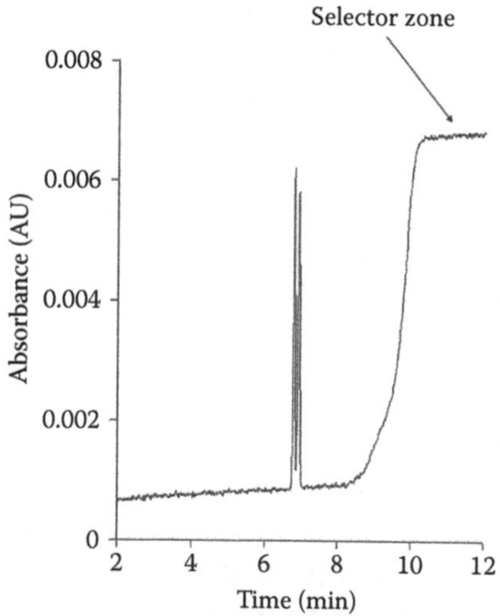

FIGURE 10.14 The use of the partial filling technique. BGE in selector zone: 100 mM (−)-DIKGA and 40 mM NH_4Ac in MeOH, selector free zone: 100 mM HAC and 40 mM NH_4Ac in MeOH. Solute: 0.1 mM pronethalol, λ 214 nm, L_{det} 40 cm, fused silica capillary (unpublished figure).

that occurs in the interface [21]. The magnitude of the EOF in the selector zone determines the lowest effective mobility for an analyte that can be detected without interference with the selector zone. In the ideal separation, no EOF exists in any of the zones. One major drawback with the partial filling technique is that the enantioresolution decreases, since only a part of the capillary is filled with the selector solution. This technique has been used for determination of enantiomeric excess of synthesized L-phosphaserine and L-phosphaisoserine to avoid the UV-absorbing selector (*t*-BuCQN) to reach the detector [60].

The partial filling technique can also be used in conjunction with MS detection to avoid contamination of the interface by nonvolatile buffer components, such as the chiral selectors [21]. Although the flow from the CE is low (in the nL/min range), the introduction of nonvolatile substances, e.g., chiral selectors, into the MS would probably require repetitive cleaning of the interface if high sensitivity is to be preserved. Thus, the use of volatile buffers in nonaqueous solvents would also improve the performance of the CE-MS hyphenation [106]. The partial filling technique has been applied for the separation of pronethalol using MS detection [21]. In this study, the selector plug was composed of 80 mM of the selector DIKGA and 20 mM NH_4Ac in MeOH: 2-propanol (2-PrOH) 75:25. The selector free zone consisted of 20 mM NH_4Ac and 20 mM HAc in MeOH:2-PrOH (75:25). The addition of acetic acid to the selector free zone facilitated the detection of more slowly migrating analytes or the use of a longer selector plug. The sheath liquid used was a MeOH:H_2O mixture (50:50) containing 0.25% acetic acid. The sheath flow was set to 1 μL/min and a counter pressure of 15 mbar was applied to avoid suction from the sheath flow in the electrospray interface [21]. However, the use of partial filling is not necessary if the chiral selector has a net migration to the anode. That has been shown by Servais and coworkers [22] for a chiral separation of salbutamol (Figure 10.11). This CD is negatively charged in the NACE-system and had a net migration to the anode. No ion suppression was observed, which might be expected by an introduction of the nonvolatile cyclodextrin derivative (HDAS-β-CD) into the MS [22].

The use of a sheath flow interface promotes the ionization process in the electrospray interface and facilitates the use of more non-volatile BGEs. However, since the flow from the CE capillary is low, the use of sheath flow will dilute the sample considerably. Furthermore, the sheath liquid may also influence the chiral separation, since the oppositely charged ions in the sheath liquid will be attracted by the potential at the inlet vial (i.e., the anions will be attracted if the inlet is the anode) and, as a consequence, will migrate into the capillary. This is called “the liquid sheath effect” [107] and can be minimized by the use of the same counterion in both the BGE and the sheath liquid. If the EOF is low or in opposite direction compared to the MS interface, this effect can be advantageous, since it facilitates the use of nonvolatile buffer components in the BGE, such as phosphate, given that they will be replaced by sheath liquid counterions before the BGE enters the MS interface [107].

One way to avoid the above mentioned problems with the sheath liquid is to use a sheathless electrospray interface. When a sheathless interface is used, the EOF needs to be in the direction of the MS interface. This makes the use of volatile buffers and solvents crucial, and the use of partial filling would probably be advantageous in order to ensure an efficient ionization and detection without interference from the

BGE containing the chiral selector. For a recent review on NACE-MS hyphenation see the reference [106].

10.7 CONCLUDING REMARKS

Since the majority of the chiral drugs that are introduced into the market today are produced in an enantiomerically pure form [108], there is a need for accurate and rapid determination of chiral impurity. The high peak capacity of CE, makes it a promising technique for the conduction of such determinations, since it is highly efficient and has a sufficient LOD for this type of applications [12]. The use of non-aqueous media in CE is preferable especially for purity determination of solutes with a low solubility in water. Chiral NACE has also proved its usefulness for biological applications, particularly in combination with the hyphenation to MS detection [22], e.g., for studies of chiral metabolism or chiral kinetics. Chiral CE is especially valuable for studies where the sample amount is limited, e.g., in samples from small children or laboratory animals, when CE could be used as an alternative to HPLC.

Nonaqueous solvents facilitate the use of ion-pair selectors for chiral separations in CE, and different selectivities compared to aqueous systems are obtained for the CDs and the CD derivatives. Although, the separation mechanism in HPLC (using a CSP) and in CE differs, the use of chiral NACE has been demonstrated to facilitate rapid screening of new potential chiral selectors for HPLC.

LIST OF ABBREVIATIONS

(–)-DIKGA	(–)-2,3:4,6-Di-*O*-isopropylidene-2-keto-L-gulonic acid
ACN	acetonitrile
Ad-CQN	1-adamantyl carbamoylated quinine
Ala–ala	alanine–alanine
All-CNHQD	allyl carbamolylated dihydroquinidine
ALL-CNHQN	allyl carbamoylated dihydroquinine
BGE	background electrolyte
C8-acid	octanoic acid
CD	cyclodextrin
CE	capillary electrophoresis
cHex-bisCQN	*trans*-1,4-cyclohexylene(carbamoylated quinine)
cHex-bis-CQN-S-C12	*trans*-1,4-cyclohexylene-bis(carbamoylated-11-dodecylthio-dihydroquinine)
cHex-bis-CQN-SO-C12	*trans*-1,4-cyclohexylene-bis(carbamoylated-11-dodecylsulfinyl-dihydroquinine)
cHex-CCD	cyclohexyl cinchonidine
CMPA	chiral mobile phase additive
CSP	chiral stationary phase
CSS	camphersulfonic acid
DCM	dichloromethane
DCP-CQD	3,4-dichlorophenyl carbamoylated quinidine

DMSO	dimethylsulfoxide
DNB-	3,5-dinitrobenzoyl
DNP-	2,4-dinitrophenyl
DNP-CCD	dinitrophenyl cinchoninine
DNP-CCN	dinitrophenyl cinchonine
DNZ	3,5-dinitrobenzyloxycarbonyl
EA-β-CD	6-Monodeoxy-6-mono (2-hydroxy)ethylamino-β-cyxlodextrin
EDL	electrical double layer
EOF	electro-osmotic flow
EtOH	ethanol
FA	formamide
HDAS-β-CD	heptakis(2,3-diacetyl-6-sulfato)-β-CD
HDMS-β-CD	heptakis-(2,3-dimethyl-6-sulfato)-β-CD
HM-bis-CQD	1,6-hexamethylene-bis(carbamoylated quinidine)
HMbisCQN	1,6-hexamethylene-bis(carbamoylated quinine)
HomoPhe	homophenylalanine
HP-β-CD	hydroxypropyl-β-cyclodextrin
HxDAS	hexakis(2,3-diacetyl-6-*O*-sulfo)-α-cyclodextrin
IPA-β-CD	6-monodeoxy-6-mono (2-hydroxy)propylamino-β-cyxlodextrin
KPA	ketopinic acid
Leu	leucine
L-ZGP	*N*-benzyloxycarbonyl-L-(*S*)-proline
MEKC	micellar electrokinetic chromatograpy
MeOH	methanol
MS	mass spectrometry
NACE	nonaqueous capillary electrophoresis
NMF	*N*-methylformamide
ODAS-γ-CD	octakis(2,3-diacetyl-6-sulfato)-γ-cyclodextrin
ODMS-γ-CD	octakis(2,3-*O*-dimethyl-6-*O*-sulfo)-γ-cyclodextrin
PA-β-CD	6-monodeoxy-6-mono (3-hydroxy)propylamino-β-cyclodextrin
P-bis-CQN	1,3-phenylene-bis(carbamoylated quinine)
Phe	phenylalanine
Phegly	phenylglycine
QA-β-CD	quartenary ammonium β-CD
QD	quinidine
QN	quinine
Rs	resolution (refers in this chapter to the *enantioresolution*)
TBA_7HDAS-β-CD	heptakis(2,3-di-*O*-acetyl-6-sulfo)-β-cyclodextrin cyclomaltoheptanose tetrabutylammoniumsalt
t-buCQD	*O*-(tertbutylcarbamoyl) quinidine
t-buCQN	*O*-(tertbutylcarbamoyl) quinine
*t*Leu	*tert*-leucine
UV	ultra violet
HE–β–CD	hydroxyethyl-β-cyclodextrin
α-Abu	α-aminobutyric acid

βAbu	β-aminobutyric acid
β-CD	β-cyclodextrin

REFERENCES

1. Walbroehl, Y, Jorgensen, J. 1984. On-column UV absorption detector for open tubular capillary electrophoresis. *J Chromatogr* 315: 135–143.
2. Gassmann, E, Kuo, JE, Zare, RN. 1985. Electrokinetic separation of chiral compounds. *Science* 230: 813–814.
3. Wang, F, Khaledi, MG. 1996. Chiral separations by nonaqueous capillary electrophoresis. *Anal Chem* 68: 3460–3467.
4. Valko, IE, Siren, H, Riekkola, M-L. 1996. Chiral separation of dansyl-amino acids in a nonaqueous medium by capillary electrophoresis. *J Chromatogr A* 737: 263–272.
5. Stalcup, AM, Gahm, KH. 1996. Quinine as a chiral additive in nonaqueous capillary zone electrophoresis. *J Microcol Sep* 8: 145–150.
6. Bjørnsdottir, I, Hansen, SH, Terabe, S. 1996. Chiral separation in non-aqueous media by capillary electrophoresis using the ion-pair principle. *J Chromatogr A* 745: 37–44.
7. Wang, F, Khaledi, MG. 2000. Enantiomeric separations by nonaqueous capillary electrophoresis. *J Chromatogr A* 875: 277–293.
8. Riekkola, M-L, Jussila, M, Porras, SP, Valko, IE. 2000. Non-aqueous capillary electrophoresis. *J Chromatogr A* 892: 155–170.
9. Rizzi, A. 2001. Fundamental aspects of chiral separations by capillary electrophoresis. *Electrophoresis* 22: 3079–3106.
10. Amini, A. 2001. Recent developments in chiral capillary electrophoresis and applications of this technique to pharmaceutical and biomedical analysis. *Electrophoresis* 22: 3107–3130.
11. Lämmerhofer, M. 2005. Chiral separations by capillary electromigration techniques in nonaqueous media. I. Enantioselective nonaqueous capillary electrophoresis. *J Chromatogr A* 1068: 3–30.
12. Van Eeckhaut, A, Michotte, Y. 2006. Chiral separations by capillary electrophoresis: Recent developments and applications. *Electrophoresis* 27: 2880–2895.
13. Garcia-Ruiz, C, Marina, ML. 2006. Sensitive chiral analysis by capillary electrophoresis. *Electrophoresis* 27: 195–212.
14. Piette, V, Lindner, W, Crommen, J. 2002. Enantiomeric separation of *N*-protected amino acids by non-aqueous capillary electrophoresis with dimeric forms of quinine and quinidine derivatives serving as chiral selectors. *J Chromatogr A* 948: 295–302.
15. Carlsson, Y, Hedeland, M, Bondesson, U, Pettersson, C. 2001. Non-aqueous capillary electrophoretic separation of enantiomeric amines with (−)-2,3:4,6-di-*O*-isopropylidene-2-keto-*L*-gulonic acid as chiral counter ion. *J Chromatogr A* 922: 303–311.
16. Karbaum, A, Jira, T. 2001. Chiral separations of 1,3,4-thia- and 1,3,4-selenadiazine derivatives by use of non-aqueous capillary electrophoresis. *J Biochem Biophys Methods* 48: 155–162.
17. Tacker, M, Glukhovskiy, P, Cai, H, Vigh, G. 1999. Nonaqueous capillary electrophoretic separation of basic enantiomers using heptakis(2,3-dimethyl-6-sulfato)-β-cyclodextrin. *Electrophoresis* 20: 2794–2798.
18. Marini, RD, Servais, A-C, Rozet, E, Chiap, P, Boulanger, B, Rudaz, S, Crommen, J, Hubert, P, Fillet, M. 2006. Nonaqueous capillary electrophoresis method for the enantiomeric purity determination of *S*-timolol using heptakis(2,3-di-*O*-methyl-6-*O*-sulfo)-β-cyclodextrin: Validation using the accuracy profile strategy and estimation of uncertainty. *J Chromatogr A* 1120: 102–111.

19. Marini, RD, Groom, C, Doucet, FR, Hawari, J, Bitar, Y, Holzgrabe, U, Gotti, R, Schappler, J, Rudaz, S, Veuthey, J-L, Mol, R, Somsen, GW, de Jong, GJ, Ha, PTT, Zhang, J, Van Schepdael, A, Hoogmartens, J, Briône, W, Ceccato, A, Boulanger, B, Mangelings, D, Vander Heyden, Y, Van Ael, W, Jimidar, I, Pedrini, M, Servais, A-C, Fillet, M, Crommen, J, Rozet, E, Hubert, P. 2006. Interlaboratory study of a NACE method for the determination of *R*-timolol content in *S*-timolol maleate: Assessment of uncertainty. *Electrophoresis* 27: 2386–2399.
20. Hedeland, Y, Lehtinen, J, Pettersson, C. 2007. Ketopinic acid and diisoproylideneketogulonic acid as chiral ion-pair selectors in capillary electrophoresis. Enantiomeric impurity analysis of *S*-timolol and *1R,2S*-ephedrine. *J Chromatogr A* 1141: 287–294.
21. Lodén, H, Hedeland, Y, Hedeland, M, Bondesson, U, Pettersson, C. 2003. Development of a chiral non-aqueous capillary electrophoretic system using the partial filling technique with UV and mass spectrometric detection. *J Chromatogr A* 986: 143–152.
22. Servais, A-C, Fillet, M, Mol, R, Somsen, GW, Chiap, P, de Jong, GJ, Crommen, J. 2006. On-line coupling of cyclodextrin mediated nonaqueous capillary electrophoresis to mass spectrometry for the determination of salbutamol enantiomers in urine. *J Pharm Biomed Anal* 40: 752–757.
23. Riekkola, M-L. 2002. Recent advances in nonaqueous capillary electrophoresis. *Electrophoresis* 23: 3865–3883.
24. Wang, F, Khaledi, MG. 1998. Nonaqueous capillary electrophoresis chiral separations with quaternary ammonium β-cyclodextrin. *J Chromatogr A* 817: 121–128.
25. Chankvetadze, B. 1997. Separation selectivity in chiral capillary electrophoresis with charged selectors. *J Chromatogr A* 792: 269–295.
26. Rizzi, AM, Kremser, L. 1999. pKa shift-associated effects in enantioseparations by cyclodextrin-mediated capillary zone electrophoresis. *Electrophoresis* 20: 2715–2722.
27. Chankvetadze, B, Lindner, W, Scriba, GKE. 2004. Enantiomer separations in capillary electrophoresis in the case of equal binding constants of the enantiomers with a chiral selector: Commentary on the feasibility of the concept. *Anal Chem* 76: 4256–4260.
28. Wren, SAC, Rowe, RC. 1992. Theoretical aspects of chiral separation in capillary electrophoresis: I. Initial evaluation of a model. *J Chromatogr A* 603: 235–241.
29. Sänger-van de Griend, CE, Groningsson, K, Arvidsson, T. 1997. Enantiomeric separation of a tetrapeptide with cyclodextrin extension of the model for chiral capillary electrophoresis by complex formation of one enantiomer molecule with more than one chiral selector molecules. *J Chromatogr A* 782: 271–279.
30. Dalgliesh, CE. 1952. The optical resolution of aromatic amino-acids on paper chromatograms. *J Chem Soc* 3: 3940–3942.
31. Davankov, VA. 1997. The nature of chiral recognition: Is it a three-point interaction? *Chirality* 9: 99–102.
32. Vespalec, R, Bocek, P. 2000. Chiral separations in capillary electrophoresis. *Chem Reviews (Washington, DC)* 100: 3715–3753.
33. Giddings, JC. 1969. Generation of variance, "theoretical plates", resolution, and peak capacity in electrophoresis and sedimentation. *J Sep Sci* 4: 181–189.
34. Piette, V, Fillet, M, Lindner, W, Crommen, J. 2000. Enantiomeric separation of amino acid derivatives by non-aqueous capillary electrophoresis using quinine and related compounds as chiral additives. *Biomed Chromatogr* 14: 19–21.
35. Hunter, RJ. 1981. *Zeta Potential in Colloid Science. Principles and Applications.* London: Academic Press Inc.
36. Palonen, S, Jussila, M, Porras, SP, Hyötyläinen, T, Riekkola, M-L. 2002. Nonaqueous capillary electrophoresis with alcoholic background electrolytes: Separation efficiency under high electrical field strengths. *Electrophoresis* 23: 393–399.

37. Subirats, X, Porras, SP, Roses, M, Kenndler, E. 2005. Nitromethane as solvent in capillary electrophoresis. *J Chromatogr A* 1079: 246–253.
38. Cottet, H, Struijk, MP, Van Dongen, JLJ, Claessens, HA, Cramers, CA. 2001. Non-aqueous capillary electrophoresis using non-dissociating solvents: Application to the separation of highly hydrophobic oligomers. *J Chromatogr A* 915: 241–251.
39. Thibon, VRA, Bartle, KD, Abbott, DJ, Mccormack, KA. 1999. Analysis of zinc dialkyl dithiophosphates by nonaqueous capillary electrophoresis and application to lubricants. *J Microcol Sep* 11: 71–80.
40. Hedeland, Y, Haglöf, J, Beronius, P, Pettersson, C. 2006. Effect of alkali metal hydroxides on the enantioseparation of amines using di-*O*-isopropylidene-keto-*L*-gulonic acid as the selector in NACE. *Electrophoresis* 27: 4469–4479.
41. Tjørnlund, J, Hansen, SH. 1997. The effect of water on separations in non-aqueous capillary electrophoresis systems. *Chromatographia* 44: 5–9.
42. Bjørnsdottir, I, Hansen, SH. 1995. Comparison of separation selectivity in aqueous and non-aqueous capillary electrophoresis. *J Chromatogr A* 711: 313–322.
43. Bjørnsdottir, I, Hansen, SH. 1995. Determination of opium alkaloids in crude opium using non-aqueous capillary electrophoresis. *J Pharm Biomed Anal* 13: 1473–1481.
44. Cantu, MD, Hillebrand, S, Carrilho, E. 2005. Determination of the dissociation constants (p*K*a) of secondary and tertiary amines in organic media by capillary electrophoresis and their role in the electrophoretic mobility order inversion. *J Chromatogr A* 1068: 99–105.
45. Huang, Y-S, Tsai, C-C, Liu, J-T, Lin, C-H. 2005. Comparison of the use of aqueous and nonaqueous buffers in association with cyclodextrin for the chiral separation of 3,4-methylenedioxymethamphetamine and related compounds. *Electrophoresis* 26: 3904–3909.
46. Valko, IE, Siren, H, Riekkola, M-L. 1999. Characteristics of electroosmotic flow in capillary electrophoresis in water and in organic solvents without added ionic species. *J Microcol Sep* 11: 199–208.
47. Roses, M. 2004. Determination of the pH of binary mobile phases for reversed-phase liquid chromatography. *J Chromatogr A* 1037: 283–298.
48. Porras, SP, Kenndler, E. 2004. Capillary zone electrophoresis in non-aqueous solutions: pH of the background electrolyte. *J Chromatogr A* 1037: 455–465.
49. Porras, SP, Riekkola, ML, Kenndler, E. 2001. Capillary zone electrophoresis of basic analytes in methanol as non-aqueous solvent. Mobility and ionization constant. *J Chromatogr A* 905: 259–268.
50. Porras, SP, Riekkola, M-L, Kenndler, E. 2003. The principles of migration and dispersion in capillary zone electrophoresis in nonaqueous solvents. *Electrophoresis* 24: 1485–1498.
51. Zarbl, E, Lämmerhofer, M, Franco, P, Petracs, M, Lindner, W. 2001. Development of stereoselective nonaqueous capillary electrophoresis system for the resolution of cationic and amphoteric analytes. *Electrophoresis* 22: 3297–3307.
52. Servais, A-C, Fillet, M, Abushoffa, AM, Hubert, P, Crommen, J. 2003. Synergistic effects of ion-pairing in the enantiomeric separation of basic compounds with cyclodextrin derivatives in nonaqueous capillary electrophoresis. *Electrophoresis* 24: 363–369.
53. Servais, A-C, Fillet, M, Chiap, P, Dewe, W, Hubert, P, Crommen, J. 2005. Influence of the nature of the electrolyte on the chiral separation of basic compounds in nonaqueous capillary electrophoresis using heptakis(2,3-di-*O*-methyl-6-*O*-sulfo)-β-cyclodextrin. *J Chromatogr A* 1068: 143–150.
54. Servais, A-C, Fillet, M, Chiap, P, Dewe, W, Hubert, P, Crommen, J. 2004. Enantiomeric separation of basic compounds using heptakis(2,3-di-*O*-methyl-6-*O*-sulfo)-β-cyclodextrin in combination with potassium camphorsulfonate in nonaqueous capillary electrophoresis: Optimization by means of an experimental design. *Electrophoresis* 25: 2701–2710.

55. Piette, V, Fillet, M, Lindner, W, Crommen, J. 2000. Non-aqueous capillary electrophoretic enantioseparation of *N*-derivatized amino acids using cinchona alkaloids and derivatives as chiral counter-ions. *J Chromatogr A* 875: 353–360.
56. Piette, V, Lämmerhofer, M, Lindner, W, Crommen, J. 1999. Enantiomeric separation of *N*-protected amino acids by non-aqueous capillary electrophoresis using quinine or tert-butyl carbamoylated quinine as chiral additive. *Chirality* 11: 622–630.
57. Piette, V, Lindner, W, Crommen, J. 2000. Enantioseparation of anionic analytes by non-aqueous capillary electrophoresis using quinine and quinidine derivatives as chiral counter-ions. *J Chromatogr A* 894: 63–71.
58. Czerwenka, C, Lämmerhofer, M, Lindner, W. 2002. Electrolyte and additive effects on enantiomer separation of peptides by nonaqueous ion-pair capillary electrophoresis using *tert*-butylcarbamoylquinine as chiral counterion. *Electrophoresis* 23: 1887–1899.
59. Lämmerhofer, M, Zarbl, E, Lindner, W. 2000. *tert.*-Butylcarbamoylquinine as chiral ion-pair agent in non-aqueous enantioselective capillary electrophoresis applying the partial filling technique. *J Chromatogr A* 892: 509–521.
60. Hammerschmidt, F, Lindner, W, Wuggenig, F, Zarbl, E. 2000. Enzymes in organic chemistry. Part 10: Chemo-enzymatic synthesis of *L*-phosphaserine and *L*-phosphaisoserine and enantioseparation of amino-hydroxyethylphosphonic acids by non-aqueous capillary electrophoresis with quinine carbamate as chiral ion pair agent. *Tetrahedron: Asymmetry* 11: 2955–2964.
61. Lämmerhofer, M, Zarbl, E, Lindner, W, Simov, BP, Hammerschmidt, F. 2001. Simultaneous separation of the stereoisomers of 1-amino-2-hydroxy and 2-amino-1-hydroxypropane phosphonic acids by stereoselective capillary electrophoresis employing a quinine carbamate type chiral selector. *Electrophoresis* 22: 1182–1187.
62. Zhu, W, Vigh, G. 2003. Capillary electrophoretic separation of enantiomers in a high-pH background electrolyte by means of the single-isomer chiral resolving agent octa(6-*O*-sulfo)-γ-cyclodextrin. *J Chromatogr A* 987: 459–466.
63. Pettersson, C, Schill, G. 1981. Separation of enantiomeric amines by ion-pair chromatography. *J Chromatogr* 204: 179–183.
64. Karlsson, A, Pettersson, C. 1991. Enantiomeric separation of amines using *N*-benzoxycarbonylglycyl-*L*-proline as chiral additive and porous graphitic carbon as solid phase. *J Chromatogr* 543: 287–297.
65. Pettersson, C, Gioeli, C. 1993. Chiral separation of amines using reversed-phased ion-pair chromatography. *Chirality* 5: 241–245.
66. Pettersson, C. 1984. Chromatographic separation of enantiomers of acids with quinine as chiral counder ion. *J Chromatogr* 316: 553–567.
67. Pettersson, C, Gioeli, C. 1988. Improved resolution of enantiomers of naproxen by the simultaneous use of a chiral stationary phase and a chiral additive in the mobile phase. *J Chromatogr* 435: 225–228.
68. Piette, V, Lammerhofer, M, Lindner, W, Crommen, J. 2003. Enantiomer separation of *N*-protected amino acids by non-aqueous capillary electrophoresis and high-performance liquid chromatography with *tert.*-butyl carbamoylated quinine in either the background electrolyte or the stationary phase. *J Chromatogr A* 987: 421–427.
69. Lämmerhofer, M, Zarbl, E, Piette, V, Crommen, J, Lindner, W. 2001. Evaluation of enantioselective nonaqueous ion-pair capillary electrophoresis as screening assay in the development of new ion exchange type chiral stationary phases. *J Sep Sci* 24: 706–716.
70. Hedeland, Y, Hedeland, M, Bondesson, U, Pettersson, C. 2003. Chiral separation of amines with *N*-benzoxycarbonylglycyl-*L*-proline as selector in non-aqueous capillary electrophoresis using methanol and 1,2-dichloroethane in the background electrolyte. *J Chromatogr A* 984: 261–271.

71. Salimi-Moosavi, H, Cassidy, RM. 1996. Control of separation selectivity and electroosmotic flow in nonaqueous capillary electrophoretic separations of alkali and alkaline earth metal ions. *J Chromatogr A* 749: 279–286.
72. Fillet, M, Servais, A-C, Crommen, J. 2003. Effects of background electrolyte composition and addition of selectors on separation selectivity in nonaqueous capillary electrophoresis. *Electrophoresis* 24: 1499–1507.
73. Bartak, P, Bednar, P, Kubacek, L, Lammerhofer, M, Lindner, W, Stransky, Z. 2004. Advanced statistical evaluation of the complex formation constants from electrophoretic data II: Diastereomeric ion-pairs of *(R,S)*-*N*-(3,5-dinitrobenzoyl)leucine and *tert*-butylcarbamoylquinine. *Anal Chim Acta* 506: 105–113.
74. Jira, T, Bunke, A, Karbaum, A. 1998. Use of chiral and achiral ion-pairing reagents in combination with cyclodextrins in capillary electrophoresis. *J Chromatogr A* 798: 281–288.
75. Nishi, H, Fukuyama, T, Terabe, S. 1991. Chiral separation by cyclodextrin-modified micellar electrokinetic chromatography. *J Chromatogr A* 553: 503–516.
76. ICH, 2003. Q3A Impurities in New Drug Substances.
77. Simov, BP, Wuggenig, F, Lämmerhofer, M, Lindner, W, Zarbl, E, Hammerschmidt, F. 2002. Indirect evidence for the biosynthesis of *(1S,2S)*-1,2-epoxypropylphosphonic acid as a co-metabolite of fosfomycin [*(1R,2S)*-1,2-epoxypropylphosphonic acid] by *Streptomyces fradiae*. *Eur J Org Chem* 2002: 1139–1142.
78. Huang, L, Lin, C-H, Xu, L, Chen, G. 2007. Nonaqueous and aqueous-organic media for the enantiomeric separations of neutral organophosphorus pesticides by CE. *Electrophoresis* 28: 2758–2764.
79. Armstrong, DW, Chang, LW, Chang, SC, Wang, X, Ibrahim, H, Reid, GR, Beesley, TE. 1997. Comparison of the enantioselectivity of β-cyclodextrin vs. heptakis-2,3-*O*-dimethyl-β-cyclodextrin LC stationary phases. *J Liq Chromatogr* 20: 3279–3295.
80. Rekharsky, MV, Inoue, Y. 1998. Complexation thermodynamics of cyclodextrins. *Chem Rev* 98: 1875–1918.
81. Wang, F, Khaledi, MG. 1999. Capillary electrophoresis chiral separation of basic pharmaceutical enantiomers with different charges using sulfated β-cyclodextrin. *J Microcol Sep* 11: 11–21.
82. Vincent, JB, Sokolowski, AD, Nguyen, TV, Vigh, G. 1997. A family of single-isomer chiral resolving agents for capillary electrophoresis. 1. Heptakis(2,3-diacetyl-6-sulfato)-β-cyclodextrin. *Anal Chem* 69: 4226–4233.
83. Servais, A-C, Chiap, P, Hubert, P, Crommen, J, Fillet, M. 2004. Determination of salbutamol enantiomers in human urine using heptakis(2,3-di-*O*-acetyl-6-*O*-sulfo)-β-cyclodextrin in nonaqueous capillary electrophoresis. *Electrophoresis* 25: 1632–1640.
84. Du, G, Zhang, S, Xie, J, Zhong, B, Liu, K. 2005. Chiral separation of anticholinergic drug enantiomers in nonaqueous capillary electrophoresis. *J Chromatogr A* 1074: 195–200.
85. Cai, H, Vigh, G. 1998. Capillary electrophoretic separation of weak base enantiomers using the single-isomer heptakis-(2,3-dimethyl-6-sulfato)-β-cyclodextrin as resolving agent and methanol as background electrolyte solvent. *J Pharm Biomed Anal* 18: 615–621.
86. Vincent, JB, Vigh, G. 1998. Nonaqueous capillary electrophoretic separation of enantiomers using the single-isomer heptakis(2,3-diacetyl-6-sulfato)-β-cyclodextrin as chiral resolving agent. *J Chromatogr A* 816: 233–241.
87. Zhu, W, Vigh, G. 2000. Enantiomer separations by nonaqueous capillary electrophoresis using octakis(2,3-diacetyl-6-sulfato)-γ-cyclodextrin. *J Chromatogr A* 892: 499–507.
88. Li, S, Vigh, G. 2004. Use of the new, single-isomer, hexakis(2,3-diacetyl-6-*O*-sulfo)-α-cyclodextrin in acidic methanol background electrolytes for nonaqueous capillary electrophoretic enantiomer separations. *J Chromatogr A* 1051: 95–101.

89. Du, G, Qin, L, Xie, J, Yun, L, Zhong, B, Liu, K. 2005. Nonaqueous capillary electrophoretic chiral separations of anticholinergic drugs with heptakis(2,3-di-*O*-methyl-6-*O*-sulfato)-β-cyclodextrin. *Chromatographia* 61: 527–531.
90. Sanchez-Vindas, S, Vigh, G. 2005. Non-aqueous capillary electrophoretic enantiomer separations using the tetrabutylammonium salt of heptakis(2,3-*O*-diacetyl-6-*O*-sulfo)-cyclomaltoheptaose, a single-isomer sulfated β-cyclodextrin highly-soluble in organic solvents. *J Chromatogr A* 1068: 151–158.
91. Szepesi, G, Gazdag, M. 1988. Enantiomeric separations and their application in pharmaceutical analysis using chiral eluents. *J Pharm Biomed Anal* 6: 623–639.
92. Olsson, J, Stegander, F, Marlin, N, Wan, H, Blomberg, LG. 2006. Enantiomeric separation of omeprazole and its metabolite 5-hydroxyomeprazole using non-aqueous capillary electrophoresis. *J Chromatogr A* 1129: 291–295.
93. Enlund, AM, Andersson, ME, Hagman, G. 2004. Improved quantification limits in chiral capillary electrochromatography by peak compression effects. *J Chromatogr A* 1028: 333–338.
94. Mori, Y, Ueno, K, Umeda, T. 1997. Enantiomeric separations of primary amino compounds by non-aqueous capillary zone electrophoresis with a chiral crown ether. *J Chromatogr A* 757: 328–332.
95. Karbaum, A, Jira, T. 2000. Chiral separation of unmodified amino acids with non-aqueous capillary electrophoresis based on the ligand-exchange principle. *J Chromatogr A* 874: 285–292.
96. Hinsmann, P, Arce, L, Svasek, P, Lämmerhofer, M, Lendl, B. 2004. Separation and online distinction of enantiomers: A non-aqueous capillary electrophoresis Fourier transform infrared spectroscopy study. *Appl Spectr* 58: 662–666.
97. Morales, S, Cela, R. 1999. Capillary electrophoresis and sample stacking in non-aqueous media for the analysis of priority pollutant phenols. *J Chromatogr A* 846: 401–411.
98. Morales, S, Cela, R. 2002. Field-amplified sample stacking and nonaqueous capillary electrophoresis determination of complex mixtures of polar aromatic sulfonates. *Electrophoresis* 23: 408–413.
99. Gao, W, Lin, S, Chen, Y, Chen, A, Li, Y, Chen, X, Hu, Z. 2005. Nonaqueous capillary electrophoresis for rapid and sensitive determination of fangchinoline and tetrandrine in Radix *Stephaniae tetrandrae* and its medicinal preparations. *J. Sep. Sci.* 28: 639–646.
100. Shihabi, ZK. 2002. Stacking for nonaqueous capillary electrophoresis. *Electrophoresis* 23: 1628–1632.
101. Tsai, C-H, Fang, C, Liu, J-T, Lin, C-H. 2004. Stacking and low-temperature technique in nonaqueous capillary electrophoresis for the analysis of 3,4-methylenedioxymethamphetamine. *Electrophoresis* 25: 1601–1606.
102. Tsai, C-H, Tsai, C-C, Liu, J-T, Lin, C-H. 2005. Sample-stacking techniques in nonaqueous capillary electrophoresis. *J Chromatogr A* 1068: 115–121.
103. Veraart, JR, Reinders, MC, Lingeman, H, Brinkman, UAT. 2000. Non-aqueous capillary electrophoresis of biological samples after at-line solid-phase extraction. *Chromatographia* 52: 408–412.
104. Eder, K, Sinner, F, Mupa, M, Huber, CG, Buchmeiser, MR. 2001. Evaluation of norbornene-β-cyclodextrin-based monomers and oligomers as chiral selectors by means of nonaqueous capillary electrophoresis. *Electrophoresis* 22: 109–116.
105. Valtcheva, L, Mohammad, J, Pettersson, G, Hjerten, S. 1993. Chiral separation of β-blockers by high-performance capillary electrophoresis based on non-immobilized cellulase as enantioselective protein. *J Chromatogr* 638: 263–267.
106. Scriba, GKE. 2007. Nonaqueous capillary electrophoresis-mass spectrometry. *J Chromatogr A* 1159: 28–41.
107. Foret, F, Thompson, TJ, Vouros, P, Karger, BL, Gebauer, P, Bocek, P. 1994. Liquid sheath effects on the separation of proteins in capillary electrophoresis/electrospray mass spectrometry. *Anal Chem* 66: 4450–4458.

108. Caner, H, Groner, E, Levy, L, Agranat, I. 2004. Trends in the development of chiral drugs. *Drug Discov Today* 9: 105–110.
109. Li, Y, Xie, LJ, Liu, HW, Hua, WT. 1999. Chiral separation of *N*-benzoyl phenylalanine methyl ester by nonaqueous capillary electrophoresis. *Chin Chem Lett* 10: 303–306.
110. Wang, F, Khaledi, MG. 1999. Non-aqueous capillary electrophoresis chiral separations with sulfated β-cyclodextrin. *J Chromatogr B* 731: 187–197.
111. Busby, MB, Maldonado, O, Vigh, G. 2002. Nonaqueous capillary electrophoretic separation of basic enantiomers using octakis(2,3-*O*-dimethyl-6-*O*-sulfo)-γ-cyclodextrin, a new, single-isomer chiral resolving agent. *Electrophoresis* 23: 456–461.
112. Fradi, I, Servais, A-C, Pedrini, M, Chiap, P, Ivanyi, R, Crommen, J, Fillet, M. 2006. Enantiomeric separation of acidic compounds using single-isomer amino cyclodextrin derivatives in nonaqueous capillary electrophoresis. *Electrophoresis* 27: 3434–3442.
113. Riddick, JA, Bunger, WB, Sakano, TK. 1986. *Organic Solvents-Physical Properties and Methods of Purification*. New York: John Wiley & Sons.
114. Ren, X, Huang, A, Wang, T, Sun, Y, Sun, Z. 1999. Enantiomeric separation of three chiral drugs by nonaqueous capillary electrophoresis with triethylamine as additive. *Chromatographia* 50: 625–628.

11 Quantitative Analysis in Pharmaceutical Analysis

Ulrike Holzgrabe, Claudia Borst, Christine Büttner, and Yaser Bitar

CONTENTS

11.1 Introduction 313
11.2 Advantages and Disadvantages of Chiral Capillary Electrophoresis in Comparison with Other Techniques 314
11.2.1 Sensitivity 315
11.2.2 Cyclodextrin Derivatives 315
11.2.3 Chiral CE versus HPLC 316
11.2.4 Other Techniques 316
11.3 Validation of Methods 317
11.3.1 Specificity 317
11.3.2 Accuracy 318
11.3.2.1 Assay 318
11.3.2.2 Impurities (Quantification) 318
11.3.3 Precision 318
11.3.4 Limit of Detection and Limit of Quantitation 319
11.3.5 Range and Linearity 320
11.3.6 Robustness 320
11.4 Content Assays and Enantiomeric Purity Testing 322
11.5 Examples of the Validation of Chiral CD-Modified CE Methods 323
11.5.1 Levodopa 323
11.5.2 Tropa Alkaloids 326
11.5.3 Aziridine Derivatives 330
11.6 Conclusion 335
References 338

11.1 INTRODUCTION

The different enantiomers of chiral drug substances often exhibit different pharmacological and toxicological properties as well as different pharmacokinetic behavior. Thus, the safety and efficacy of chiral pharmaceuticals are critically related to their enantiomeric purity, and therefore, the development and optimization of chiral analytical methods is a very important field for quality control in the pharmaceutical industry [1]. Hence, the demand for quantitative analytical techniques assessing the exact composition of a racemate or determining the enantiomeric excess (ee) of an enantiomeric pure substance is growing.

With the increasing globalization of the pharmaceutical market, the need for generally accepted requirements for pharmaceutical products also rises. The necessity for international regulatory guidance was already recognized in the 1980s. Europe, Japan, and the United States had begun discussions on the possibilities of harmonization, until in conclusion, the *International Conference on Harmonisation of Technical Requirements* for the Registration of Pharmaceuticals for Human Use, was born in 1990 [2].

Under the auspices of the *International Conference on Harmonisation* (*ICH*), various international guidelines have been developed to assimilate global requirements of pharmaceuticals [3]. Of particular relevance to the quality of chiral drug substances are the guidelines on analytical validation (Topic Q2 (R1)), impurities (Topic Q3A (R2)), specifications (Topic Q6A), and stability testing for chemicals in drug substances and drug products (Topics Q1A (R2) –Q1F), as well as the Good Manufacturing Practice Guide for Active Pharmaceutical Ingredients (API) (Topic Q7) [4].

The ICH guidelines echo bit by bit in the wording of the laws of the parties of the conference. The guideline on validation of analytical procedures is in line with part 3 (analytical validation) of the Technical Guide for the Elaboration of Monographs of the *European Pharmacopoeia* (*Ph. Eur.*), respectively. Though there are practical instructions given in the Technical Guide, which describe the accomplishment of the investigations [5].

Since regulatory authorities have recognized the significance of chirality in drug action, chiral capillary electrophoresis (CE) becomes a very important technique for chiral separations. For example, a chiral CZE-method for the determination of the enantiomeric purity of galantamine hydrobromide is suggested for the 6th edition of the *Ph. Eur.*

11.2 ADVANTAGES AND DISADVANTAGES OF CHIRAL CAPILLARY ELECTROPHORESIS IN COMPARISON WITH OTHER TECHNIQUES

Due to the serious advantages of CE in chiral analysis over other techniques, the relevance of CE methods is still increasing. Short analysis times, quick and simple method development, low consumption of sample, solvents and chiral selectors turns chiral CE into a valuable alternative to other analytical procedures, as long as financial and environmental aspects are concerned. Anyway, chiral CE also features high resolving power and high flexibility in choosing and changing types of selectors, which leads to enormous variability of applications [6].

Thus, nearly every separation problem can be solved quickly by modification of published chiral CE methods. Since about 90% of all drugs are chargeable, they can be analyzed by the simplest CE technique, the capillary zone electrophoresis (CZE). However, unchargeable molecules can also be separated and determined by adding a surfactant like sodium dodecylsulfate (SDS) to the background electrolyte (BGE) in micellar electrokinetic chromatography (MEKC). In this case, separation is carried out by differences in polarity (see also Chapter 8). By combining these two principles of separation, the performance can escalate in a way that even complicated separations can be accomplished easily. Based on similar conditions, microemulsion electrokinetic chromatography (MEEKC) is an attractive technique in analytical separation. Commonly used oil-in-water microemulsions contain nanometer-sized oil droplets suspended in an aqueous buffer. This oil core is able to solubilize many hydrophobic analytes so that chromatographic separation can be obtained by partition between the oil droplets and the buffer solution. This technique is described in more detail in Chapter 9.

11.2.1 Sensitivity

The two main reasons for not using CE, namely, lack of sensitivity and reproducibility, are no longer generally true. For long times, chiral CE methods suffered from the low sensitivity, caused by small injection volumes (few nanoliters) and a very short detection pathway (50–100 µm). Nevertheless the continuous further development of CE equipment, detector performance, and injection techniques, e.g., different kinds of "sample stacking," improved the situation [7,8]. The method based on field strength differences between the sample zone and the running buffer is called stacking. Concentration of the sample is obtained when conductivity of the sample-solvent is significantly lower than that of the running buffer. On applying the voltage, a proportionally greater field will develop across the sample zone causing the ions to migrate faster, until they reach the running buffer. Here the field decreases and the ions migrate more slowly. This effect leads to concentration at the boundary. If the conductivity of the sample is too high, it is possible to inject a short plug of water before the sample [3]. Also a good facility for sensitivity enhancement is the preconcentration of samples, e.g., by solid-phase extraction (SPE), liquid–liquid extraction, liquid-phase microextraction or microdialysis [9].

Nowadays, the required sensitivity of 0.1%, for the determination of the enantiomeric impurity of optically pure substances can be achieved in most cases, using UV or diode array detector (DAD) detection. There are numerous examples, where chiral CE methods surpass high performance liquid chromatography (HPLC) methods with regard to reproducibility, costs, expenditure of time, and resolution power [10–12]. Particularly since generic strategies of method development in chiral CE are published, the expenditure of time, concerning method development, should decrease [13–15].

11.2.2 Cyclodextrin Derivatives

When a validated quantitative method has to be performed, care must be taken by using randomly substituted cyclodextrin (CD) derivatives, because they show serious differences from batch to batch. Validation of separation is difficult when employing such CD derivatives [16].

As described in Section 11.3.6, the differences between single isomer CDs and randomly substituted CDs are enormous as well as the batch-to-batch variations. As long as randomly substituted CDs are concerned, the degree and the locus of substitution can influence the separation power and also the migration times and migration order of the compounds.

11.2.3 Chiral CE versus HPLC

The problem of poor robustness has to be faced by meticulous working, since CE methods are affected even by small changes in, e.g., pH value of the BGE. In contrast, resolution in HPLC is not as sensitive to changes of the mobile phase as is CZE. Although chiral stationary phases for HPLC are very expensive and the resolving power and robustness are often poor, the most often applied method in chiral analysis is HPLC [3]. That fact is particularly caused by the disregard of the often high-output CE methods by international pharmacopoeias [11].

A further disadvantage of chiral HPLC analysis is the high consumption of organic solvents, which is also an environmental problem. However, the high sensitivity of HPLC methods is a good reason for using this prevalent technique.

11.2.4 Other Techniques

The *Ph. Eur.* [17] and most other international pharmacopoeias make frequent use of optical rotation. This method is neither selective nor sensitive, especially in cases where [α] values are small, although it may be adequate for identification purposes. For example, the content of dextronorgestrel in levonorgestrel can be determined with a precision of ±1°; the limits given in the pharmacopoeia are −30° to −35°, which corresponds to ±5% [3]; this variation is too high for compliance with regulatory guidelines, nevertheless this method is still prescribed in the *Ph. Eur.*

Another analytical procedure to determine the content of enantiomers is chiral quantitative nuclear magnetic resonance (qNMR) spectroscopy. The main difference of the chiral NMR method compared to the methods described earlier, is the fact that the signals of the enantiomers are resolved without physically separating the compounds. The huge disadvantage of this technique is the extremely high purchasing costs of a NMR apparatus, though it is a suitable alternative with easy and less time consuming method development, short analysis times, and easy to validate [18,19].

Also chiral gas chromatography (GC) is described in current literature and a couple of published methods, often with hyphenation to mass spectrometry (MS), can be found. Established advantages of GC are the high efficiency at high speed, an enormous separation power, sensitive and straightforward detection formats, and multicolumn operations. However, only volatile and thermostable substances can be determined by GC, so this technique is restricted in applications. It is a wide-spread technique in fragrance and flavor industry [20].

It becomes apparent that there are two techniques in chiral analysis, which outclass all other techniques, namely, the already mentioned chiral CE and chiral HPLC. HPLC excels in terms of sensitivity with the trouble of high costs of consumable material and waste disposal. On the other hand, CE methods score with high resolving power and very low costs with the lack of robustness and sensitivity, even if these are already improved. However, since both techniques show high efficacy and variability of application, nearly all pharmaceutical substances can be assessed with them. Furthermore, CE methods can examine validity of HPLC methods and vice versa, because they are orthogonal and complementary techniques.

11.3 VALIDATION OF METHODS

Validation of analytical methods always follows the same schema; the instructions are given in the ICH guideline on validation [21] or in the analogous text of a law. There is no difference in validation procedure of enantioselective assessments in comparison to other analytical determinations, even though CE methods take some more experience for validation. CE relies on more complicated physical effects than, e.g., HPLC, which leads to a high number of parameters that have to be considered when validating a CE method. Thus, some CE specific aspects of validation have to be taken into account (see Table 11.1).

Many of the most common analytical procedures have to be validated, i.e., identification tests, quantitative content of impurities, limit tests for impurities or any other quantitative test for the determination of the content of the API or other ingredients of a drug product. Typical validation characteristics, which have to be taken into consideration, are listed and discussed in the following [21].

11.3.1 Specificity

Specificity is the ability to assess unequivocally the analyte in the presence of components, which may be expected to be present. This includes related substances, degradation products, matrix, and impurities caused by the synthesis process of pharmaceutical compounds or by the degradation of the API. It is also termed selectivity [21].

TABLE 11.1
Specific CE Validation Activities According to Ref. [25]

Requirement	Comment
Capillary variation	Lot-to-lot, different supplier
Reagent source	Pre-prepared electrolytes, additive purity
Electrolyte stability	May be several months
Instrument transfer	May require some revalidation for different settings
Response factor	Not necessarily the same as in HPLC
Long-term injection precision	Buffer depletion effects, test for expected sequence duration
Operator training	Alignment of capillaries
Robustness	Interactions more likely than HPLC-experimental designs
Indirect UV detection	Assess robustness fully
Capillary rinsing	Validate and specify in method

Source: Altria, K.D. and Rudd, D.R., *Chromatographia*, 41, 325, 1995. With permission.

Specificity in chiral analysis might be a problem, because both enantiomers differ neither in chemical nor in physical properties; thus, they can be only separated by chiral selectors, which have different affinity to the enantiomers. For an assay method, specificity is assessed by injecting a test solution spiked with degradation products, synthesis intermediates, excipients from formulation, and other possible impurities at their expected level.

Another way to validate specificity is to assess peak homogeneity or peak purity. This can be achieved using a DAD. Logically this method cannot be applied as long as enantiomers are concerned, since they exhibit identical absorption spectra, but for other impurities it is a feasible technique [22].

11.3.2 Accuracy

Accuracy expresses the closeness of agreement between the value which is accepted either as a conventional true value or an accepted reference value and the value measured [21].

11.3.2.1 Assay

The proofs of a method providing accurate outcomes can be brought forward by assessing reference substances with known content, expressed as % recovery, or by comparing the findings to those achieved by an orthogonal and validated method. Often a method is regarded as providing accurate outcomes, if specificity, linearity, and precision are proven.

If certain substances, also optical pure substances, are not available in the purity needed for accuracy determinations, one has to manage with a method called standard addition to determine the accurate content of the main component. Using the standard addition method, one adds certain amounts of a standard solution with known content of the analyte to the unknown solution, so any influence of the impurities is accounted for in the calibration. In this way, a calibration curve can be achieved and the operator can extrapolate and determine the initial concentration [23].

11.3.2.2 Impurities (Quantification)

Accuracy should be assessed on samples spiked with known amounts of impurities. In cases where it is impossible to obtain samples of certain impurities and/or degradation products, it is acceptable to compare results obtained by an independent procedure. The response factor of the drug substance can be used [5].

11.3.3 Precision

Precision expresses the closeness of agreement between a series of measurements obtained from multiple sampling of the same homogeneous sample under the prescribed conditions. A distinction is drawn between repeatability, intermediate precision, and reproducibility. Repeatability expresses the precision under the same operating conditions over a short interval of time; it is also termed intra-assay precision [21].

For evaluating repeatability, the relative standard deviation (RSD) values are to be assessed. The number of repeating measurements is at least 9 covering the specified range, e.g., 3 concentrations with 3 replicates each or 6 determination at 100% of the test concentration [5]. Intermediate precision expresses within-laboratories variations like different days, different analysts, different equipment, even different batches of substances needed for the analytical procedure. Special attention has to be paid to the variabilities of different batches of CDs. Intermediate precision can be tested by repetitive analysis on five separate days with different analysts, equipment, etc. For each day, the sample and standard solutions should be prepared fresh in order to include errors, e.g., from weighing and diluting. From a statistical point of view, precision can be described best using the confidence interval of the analytical result [24]. Data should be analyzed with respect to repeatability and intermediate precision. Reproducibility means the precision between laboratories. It is a measure for the applicableness and transferability of an analytical method from one plant of a pharmaceutical company to another. Reproducibility is assessed by means of an interlaboratory trial. It should be considered in case of the standardization of an analytical procedure, for instance, for inclusion of procedures in pharmacopoeias [5].

11.3.4 Limit of Detection and Limit of Quantitation

The limit of detection (LOD) expresses the lowest amount of analyte in a sample, which can be detected but not necessarily quantitated as an exact value [21].

Often the LOD is estimated by the signal-to-noise-ratio (s/n-ratio). If the s/n-ratio takes a value of at least 3, it is possible to distinguish unambiguously between the signal of the analyte and the background noise. Another estimation of the LOD is based on the standard deviation of the response and the slope of calibration curve.

The LOD may be expressed as

$$\text{LOD} = \frac{3.3\sigma}{S}$$

where

σ is the standard deviation of the response
S is the slope of the calibration curve

The slope S may be estimated from the calibration curve of the analyte. The estimate of σ may be carried out in a variety of ways. One way, for example, is based on the standard deviation of the blank, which is determined by the measurement of the magnitude of analytical background response obtained by analyzing an appropriate number of blank samples, or based on the calibration curve. For this purpose, a specific calibration curve should be studied using samples containing the analyte in the estimative range of LOD. The residual standard deviation of a regression line or the standard deviation of y-intercepts of regression lines may be used as the standard deviation.

The limit of quantitation (LOQ) expresses the lowest amount of analyte in a sample, which can be quantitatively determined with suitable precision and accuracy [21]. The facilities of determining the LOQ are in principle the same as for determining the LOD. Based on the s/n-ratio: LOQ is defined as the concentration, yielding a s/n-ratio of 10. Using the standard deviation of the response and the slope of calibration curve a calibration curve should be studied in the estimative range of the LOQ.

The LOQ may be expressed as

$$LOQ = \frac{10\sigma}{S}$$

For impurities known to be unusually potent or to produce toxic or unexpected pharmacological effects, the detection/quantitation limit has to be commensurate with the level at which they must be controlled [5].

11.3.5 Range and Linearity

The range of an analytical procedure is the interval between the upper and the lower concentration (including these concentrations) for which the analytical procedure has been demonstrated to have a suitable precision, accuracy, and linearity [21]. It is normally derived from linearity studies and depends on the intended application of the procedure.

Linearity is the ability of an analytical procedure (within a given range) to obtain test results which are directly proportional to the concentration (amount) of analyte in the sample [21]. Linearity should be established by visual evaluation of a plot of signals as a function of analyte concentration or content. If there is a linear relationship, test results should be evaluated by appropriate statistical methods, for example, by calculation of a regression line by the method of least squares. The correlation coefficient, y-intercept, slope of the regression line, and the deviation of the actual data points from the regression line should be drawn on linearity investigations. Main component assay linearity is assessed by analyzing five standards, typically placebo samples spiked with known amount of analyte, covering concentrations over the range 80%–120% of the nominal assay concentration. For impurity determinations, it is necessary to demonstrate detector linearity over the intended operating range for the method. Linearity has to be accounted for the range from LOQ or from 50% of the specification of each impurity, whichever is greater, to 120% of the specification [5]. Acceptable correlation coefficients (ideally 0.997 or better) and an intercept close to the origin should be achieved [25].

11.3.6 Robustness

Robustness of an analytical procedure is a measure of its capacity to remain unaffected by small, but deliberate variations in method parameters and provides an indication of its reliability during normal usage [21].

The investigation of robustness has to be achieved by accomplishment of measurements by different analysts, with different batches of reagents, with different capillaries from different manufacturers, and with different CE instruments from different manufactures, preferentially. Also variations in temperature, applied voltage or current, concentration of buffer additives, e.g., organic modifiers or chiral selectors, injection time, detection wavelength, etc., have to be taken into consideration. Also the stability of solutions may play a critical role and has to be investigated. If measurements are regarded to be susceptible to variations in analytical conditions, those should be suitably controlled or a precautionary statement should be included in an analytical procedure. One consequence of the evaluation of robustness should be that a series of system suitability parameters, e.g., resolution tests, or peak-to-valley ratios, is established to ensure that the validity of the analytical procedure is maintained whenever used [5].

Reproducibility and robustness are tender spots of CE methods; often published methods are not transferable from one laboratory to another. This is due to the fact, that CE methods, in contrast to HPLC methods, are easily affected even by small changes in parameters, even though the available instruments can be considered to be robust. Thus, the authors of a published method should give instructions about what can be done in case of failure of the analytical procedure. Also small varieties in different batches of CDs or other chiral selectors lead to differences in resolution power and migration times.

As aforementioned the kind of CDs used for chiral analysis plays an important role for robustness of an analytical procedure. The resolution of racemates, for example, depends strongly on the substitution pattern and position as well as the purity of the CD used. Commercially available are single isomer CD derivatives as well as randomly substituted CDs, which are mostly characterized by their molecular degree of substitution (DS). The DS indicates how many of the functional groups in the CD are substituted in average, the number starts with 0 for a totally unsubstituted CD up to 1 for a CD whose hydroxy groups are completely modified. The DS is an average number normally determined for a whole batch. As described by Chankvetadze et al. [26], even the migration order can change by switching from a single isomer CD to a randomly substituted one. The resolution of the enantiomers using a single isomer CD or a randomly substituted CD varies in a wide range, but also the differences between two batches of one CD can also be of high importance [16].

Beside the choice of chiral selector, an analyst should also investigate what kind of instrumental adjustments can be done to achieve accurate and precise outcomes and to get distinct separations, without fundamentally modifying the methods. These adjustments include the applied voltage, the capillary temperature, the pH limits, or composition of the BGE. Generic instructions on what can be done, if a certain separation problem appears are described in a review from Wätzig et al. [24].

In the *Ph. Eur.* such instructions can be found in the system suitability tests of the monographs. By following the system suitability tests, the analysts can check up the eligibility of their own equipment to carry out a certain analytical measurement. In the monographs of the *Ph. Eur.*, which describes the recommended analytical procedures for investigating the identity, the content, and the purity of an individual drug substance, system suitability testing instructions are given, e.g., for liquid chromatography (LC) methods and for CE methods in the monographs of individual drug substances. By means of chromatographic parameters, like resolution of two critically separated peaks, it is possible to screen the applied method. Unfortunately generic instructions in the CE chapter of the *Ph. Eur.* (Chapter 2.2.47) are almost completely missing, in contrast to the chapter of Chromatographic Separation Methods (Chapter 2.2.46). An example from the *Ph. Eur.* 6 [17] is given to clarify the approach: the system suitability test of the CE method for the investigation of related substances of the monograph of the tripeptide glutathione.

System suitability:

- Resolution: Minimum 1.5 between the peaks due to the internal standard and impurity B in the chromatogram obtained with reference solution (c); if necessary increase the pH with dilute sodium hydroxide solution R.
- Peak-to-valley ratio: Minimum 2.5, where H_p is the height above the baseline of the peak due to impurity D and H_v is the height above the baseline of the lowest point of the curve separating this peak from the peak due to glutathione in the chromatogram obtained with reference solution (c); if necessary lower pH with phosphoric acid R.
- Check that in the electropherogram obtained with test solution (a) there is no peak with the same migration time as the internal standard (in such case correct the area of the phenylalanine peak) [27].

Also critical to robustness is the rinsing performance. New capillaries have to be rinsed with 0.1 M NaOH and with water for 15–30 min each, to get rid of contaminations, which maybe caused by the manufacturing process. Furthermore there should be similar rinsing procedures before daily use and between each run. Without accomplishing similar rinsing procedure before each measurement no robust outcomes will be achieved. Since buffer components and other substances may remain in the capillary, each capillary should be dedicated to one application only. Also buffer depletion influences the migration times of the compounds. Thus, it has to be examined for each method, how many runs can be accomplished from the same buffer vial, before the vials have to be replaced [9].

11.4 CONTENT ASSAYS AND ENANTIOMERIC PURITY TESTING

For quantitative analysis of enantiomers only chromatographic, electrophoretic, and NMR spectroscopic techniques can be taken into account. Traditional titrations and UV spectroscopic methods cannot be used, because the enantiomers differ neither in chemical nor in physical properties.

There are two different kinds of quantitative determinations in enantioselective analysis, namely, the determination of both enantiomers in a racemate and the determination of the ee in an optically pure substance. Before quantifying the individual components of a mixture of substances they have to be separated from each other. Because of the similarity of the two analytes, the requirements on separating techniques for enantioselective assessments are higher than in case of other separations. After separating the individual components of a mixture they can be identified, usually based on comparison of times between standard and samples or by spiking the samples with a reference substance. By integration of the peaks areas the components can be quantified as well. Hence, identification, purity and content assays can be accomplished in one go. Whereas it is mostly simple to determine the amount of enantiomers in a racemate (should be 50%), it is more difficult to evaluate the ee, because the minor enantiomer has to be unequivocally evaluated in the presence of the major enantiomer, whose content is 100 or even 1000 times higher. The requirements on linearity, sensitivity, and limit of quantification, concerning the validation of the method, are extremely high.

The only feasible traditional method for determining an enantiomeric impurity is the suboptimal technique of optical rotation, which is often used in pharmacopoeias (see *Ph. Eur.* Chapter 2.4). As aforementioned, the sensitivity of this technique does not comply with regulatory requirements on purity assays. Because the ICH guidelines on impurities included the enantiomeric impurities in 2006, they are now handled like the other impurities by the regulatory authorities and validated methods are necessary to determine the enantiomeric excess. The guideline also contents a decision tree for the identification and qualification of impurities, because, if the content of a certain impurity exceeds a certain threshold, the impurity has to be reported (threshold 0.05%),* identified (threshold 0.10%),* or even qualified (threshold 0.15%).* If it is not possible to limit an impurity to the corresponding content, the prescribed approach is authoritative. Qualification, for example, includes toxicological studies to estimate the risks for human use [28].

Evidently chiral analytical methods do not yet comply with these numerical limits for enantiomeric impurities, but the principles and approach of identification and qualification still apply, particularly where the drug concerned is not a single enantiomer of a previously marketed racemate [3].

For the control of impurities the *Ph. Eur.* recommends the use of an internal standard to achieve appropriate precision.

11.5 EXAMPLES OF THE VALIDATION OF CHIRAL CD-MODIFIED CE METHODS

11.5.1 Levodopa

Recently the monograph levodopa (Figure 11.1) in the *Ph. Eur.* [17] was revised in order to replace the "Related substances" thin layer chromatography (TLC) test with a better HPLC method and the poor optical rotation test (angle of optical rotation −1.27 to −1.34) with a more sensitive method. Although a couple of CD-modified CE

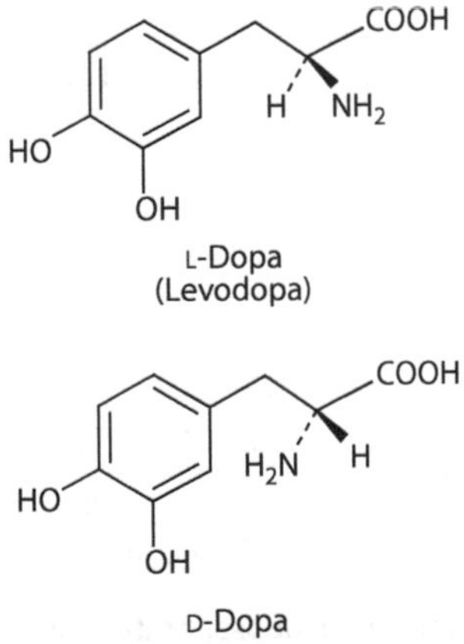

COOH
H NH₂
HO
OH
L-Dopa
(Levodopa)

COOH
H₂N H
HO
OH
D-Dopa

FIGURE 11.1 Structural formulae of L- and D-dopa.

* For drugs with maximum daily intake ≤2 g/day.

methods employing negatively charged CDs were already described [29,30] and all of them were capable of separating the enantiomers of D,L-dopa and limiting D-dopa to at least less than 0.2%, the *Ph. Eur.* has decided to use a chiral HPLC. Using an achiral RP-18 column, the resolution of the enantiomers ($R_s > 5$) can be achieved by a mobile phase composed of copper acetate and *N,N*-dimethyl-L-phenylalanine in a water–methanol mixture. The method is able to limit the D-dopa to less than 0.5% only and the peak shape of the major peak is rather poor which makes the determination of the minor component, i.e., the enantiomeric impurity, uncertain (data not shown).

In contrast, it was possible to fully validate a slightly modified CE method reported by Hoogmartens and coworkers [30]. Using an uncoated fused-silica capillary and a mobile phase composed of 10 mM sulfated β-CD in 20 mM sodium phosphate buffer pH 2.5 in a reversed polarity mode, the D-enantiomer migrates in front of the levodopa with a resolution higher than 6. According to the ICH guidelines [21], the LOD of D-dopa amounted to 0.015% and the limit of quantification to 0.04% of the major peak (Figure 11.2) [11]. Both values are slightly better than the data obtained by the group of Hoogmartens [30] and far below the limit given in the *Ph. Eur.* monograph [31].

In order to check linearity, a calibration was performed using eight concentration levels in the range of 0.4–6.0 µg/mL which corresponds to a 0.04%–0.60% content of D-dopa in a levodopa sample of 1.0 mg/mL. Each standard was injected three times randomly and the mean corrected peak area of each enantiomer was evaluated.

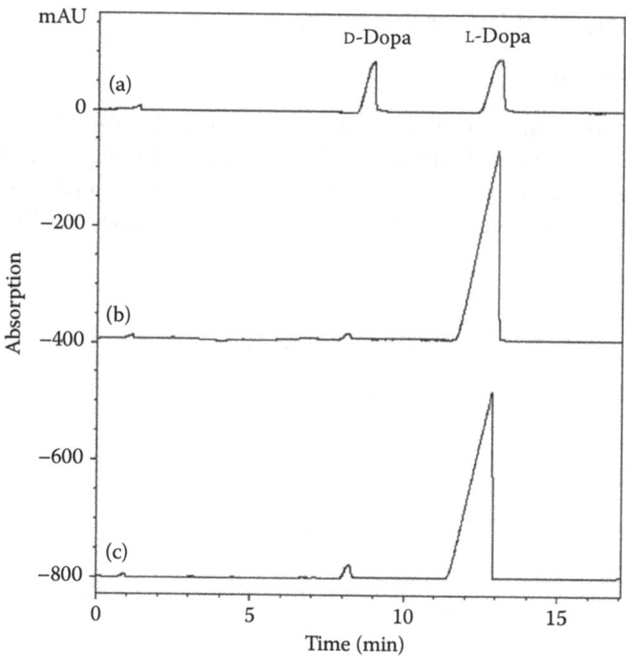

FIGURE 11.2 Electropherograms of enantioseparation of (a) racemic D,L-dopa, (b) 0.10%, and (c) 0.50% D-dopa as impurity of levodopa under optimized conditions: 31.0/22.5 cm fused-silica capillary, 10 mM sulfated-β-cyclodextrin dissolved in 20 mM phosphate buffer pH 2.5 as BGE, 20°C, −15 kV (I = 82.0 µA), and 200 nm detection wavelength (UV).

The calibration curve was obtained by plotting the concentration of D-dopa in the test solution against the peak-area ratios of both enantiomers. The good calibration data are demonstrated by the linear equation $y = 0.1095x + 0.0911$ and showed a linearity confirmed by the coefficient of determination $r^2 = 0.9984$ and the RSD (y-mean) between 0.29% and 3.82%.

The precision and the accuracy of the method were investigated in a similar way with respect to repeatability and intermediate precision or accuracy. Two different solutions of 1.2 and 4.0 µg/mL of D-dopa corresponding to the 0.12% and 0.40%, respectively, in 1.0 mg/mL levodopa were prepared. Each solution was injected six times on one day (intraday precision or accuracy) and on two different days, respectively (interday precision or accuracy). The corrected peak area and the relative migration times (RMT) as well as the accuracy values at the two concentration levels of D-dopa are summarized in Table 11.2. The acceptable RSDs with respect to the RMTs were found to be between 0.75% and 0.84% for intraday precision and between 0.53% and 0.90% for interday precision in 2 days. In the case of the corrected peak area-ratio, RSDs for the intraday precision were found to be in a range of 0.85% and 3.91% and for interday precision between 1.75% and 3.77%. Thus, the method proved to be precise.

The data, being summarized in Table 11.2, also demonstrate acceptable accuracy of the method. The highest RSD values for the interday and intraday accuracy were found to be 3.77% and 3.91%, respectively, and, thus, within the required range of ±10% of the theoretical values of the concentration of minor component D-dopa.

The robustness of the method was checked with respect to the most critical parameters such as variation of the applied voltage between −14 and −16 kV; the capillary temperature ranging from 18°C to 22°C; the pH of the BGE between 2.4 and 2.6; the buffer concentration in a range of 18 and 22 mM; the chiral selector concentration between 9.0 and 11.0 mM, considering corrected peak-area ratio, the RMT, and the resolution factor (R_s) of both enantiomers in the electropherogram. The solution containing 1.0 mg/mL of Levodopa spiked with 0.1% of D-dopa was analyzed

TABLE 11.2
Determination of D-Dopa in Levodopa: Intra- and Interday Precision of the Corrected Peak Area Ratio and Relative Migration Times as well as the Accuracy of the Method Tested at Two Different Concentration Levels

Day	Conc. D-Dopa (µg/mL)	Peak Area Ratio	RSD (%)	RMT	RSD (%)	Accuracy (µg/mL)	Accuracy (%)	RSD (%)
		Interday precision				Interday accuracy		
1	1.2	0.221	3.18	0.638	0.90	1.20	100.08	3.18
	4.0	0.503	1.75	0.638	0.53	4.02	100.62	1.75
2	1.2	0.223	2.52	0.629	0.78	1.21	101.15	2.52
	4.0	0.510	3.77	0.639	0.55	4.08	101.99	3.77
		Intraday precision				Intraday accuracy		
	1.2	0.218	0.85	0.623	0.84	1.18	98.64	0.85
	4.0	0.505	3.91	0.632	0.75	4.04	101.06	3.91

TABLE 11.3
Robustness Assessment of the Method for Determining D-Dopa in Levodopa Based on the Variation of the Corrected Peak Area Ratio and the Relative Migration Time (RMT) as well as the Resolution (R_s) between Enantiomers

Variation Conditions	Variation	Peak Area Ratio	RMT	R_s
	Standard conditions[a]	0.215	0.647	8.24
Buffer concentration	18 mM	0.217	0.629	8.64
	22 mM	0.217	0.657	7.50
Buffer pH	2.4	0.217	0.660	7.82
	2.6	0.217	0.643	8.70
Voltage	−14 kV	0.217	0.651	7.99
	−16 kV	0.216	0.629	8.57
Temperature	18°C	0.214	0.650	8.21
	22°C	0.219	0.626	8.43
Sulf-β-CD concentration	9 mM	0.216	0.637	7.72
	11 mM	0.209	0.640	8.95

[a] Standard conditions: Applied voltage 15 kV (reversed polarity), capillary temperature 20°C and BGE composed of 20 mM phosphate buffer, pH 2.5, and 10 mM Sulf-CD as chiral selector.

three times under either condition. The data of robustness are summarized in Table 11.3 and displayed in Figure 11.3.

The corrected peak area ratio and RMT varied between 0.209 and 0.219 and 0.626 and 0.660, respectively, corresponding to relative variations of −2.79% to +1.86% and −3.25% to +2.01%, respectively, in comparison to the standard conditions. The resolution between both enantiomers varied more significantly between 7.50 and 8.95 corresponding to the relative variation −8.98% to +8.62%. However, none of these changes significantly affected the enantiomeric impurity result.

Hence, acceptable relative variations of the corrected peak area ratio, the RMT and resolution values (R_s) were found and the method was considered to be very robust against small variations of the standard conditions. Taken together, the CE method described here is easy to perform and far more sensitive than the HPLC method described in the *Ph. Eur.* In addition, it seems to be more precise and robust.

11.5.2 Tropa Alkaloids

Chiral cyclodextrin-modified MEEKC (CD-MEEKC) can be applied for the enantiomeric separation of tropa alkaloids, namely, atropine, scopolamine, ipratropium, and homatropine (Figure 11.4) [32]. The standard oil-in-water microemulsion BGE solution consisted of 0.8% w/w octane, 6.6% w/w 1-butanol, 2.0% w/w SDS, and 90.6% w/w 10 mM sodium tetraborate buffer pH 9.2. The type and concentration of the CD is an important parameter affecting enantioseparation. In fact, the combined use

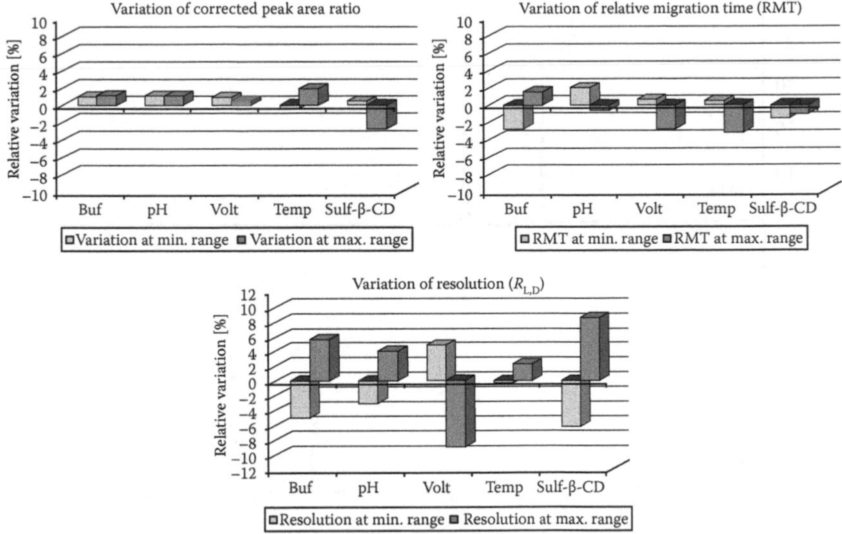

FIGURE 11.3 Determination of D-dopa in levodopa: relative variation of the corrected peak area ratio, relative migration time (RMT), and resolution factor (R_s) at small variations of the standard conditions. The bars are marked with Buf, buffer concentration; pH, buffer pH; Volt, voltage; Temp, temperature; and Sulf-ß-CD, Sulf-CD concentration. ($R_{L,D}$ is equal to R_s).

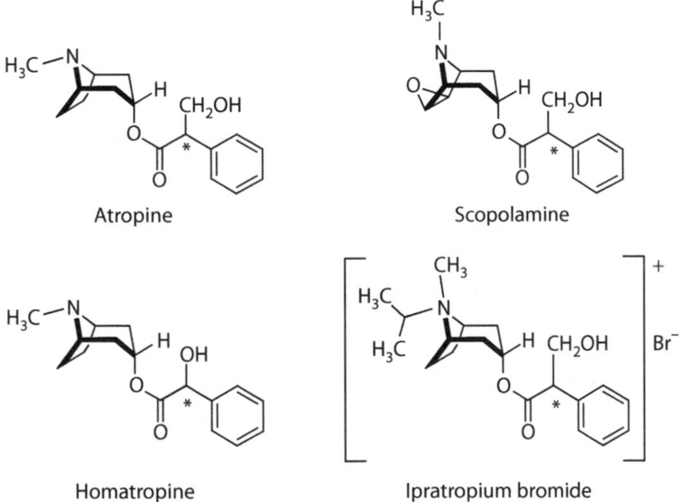

FIGURE 11.4 Structural formulae of the racemic tropa alkaloids.

of negatively charged β-CDs as a stereoselective pseudostationary phase with achiral SDS-microemulsion allowed successful separations. However, only by the addition of small concentration of heptakis-(2,3-*O*-dimethyl-6-*O*-sulfo)-β-CD (HDMS-β-CD) or sulfated β-CD (Sulf-β-CD) to this MEEKC a high resolution and selectivity as well as rather short analysis time were achieved (see Figures 11.5 and 11.6).

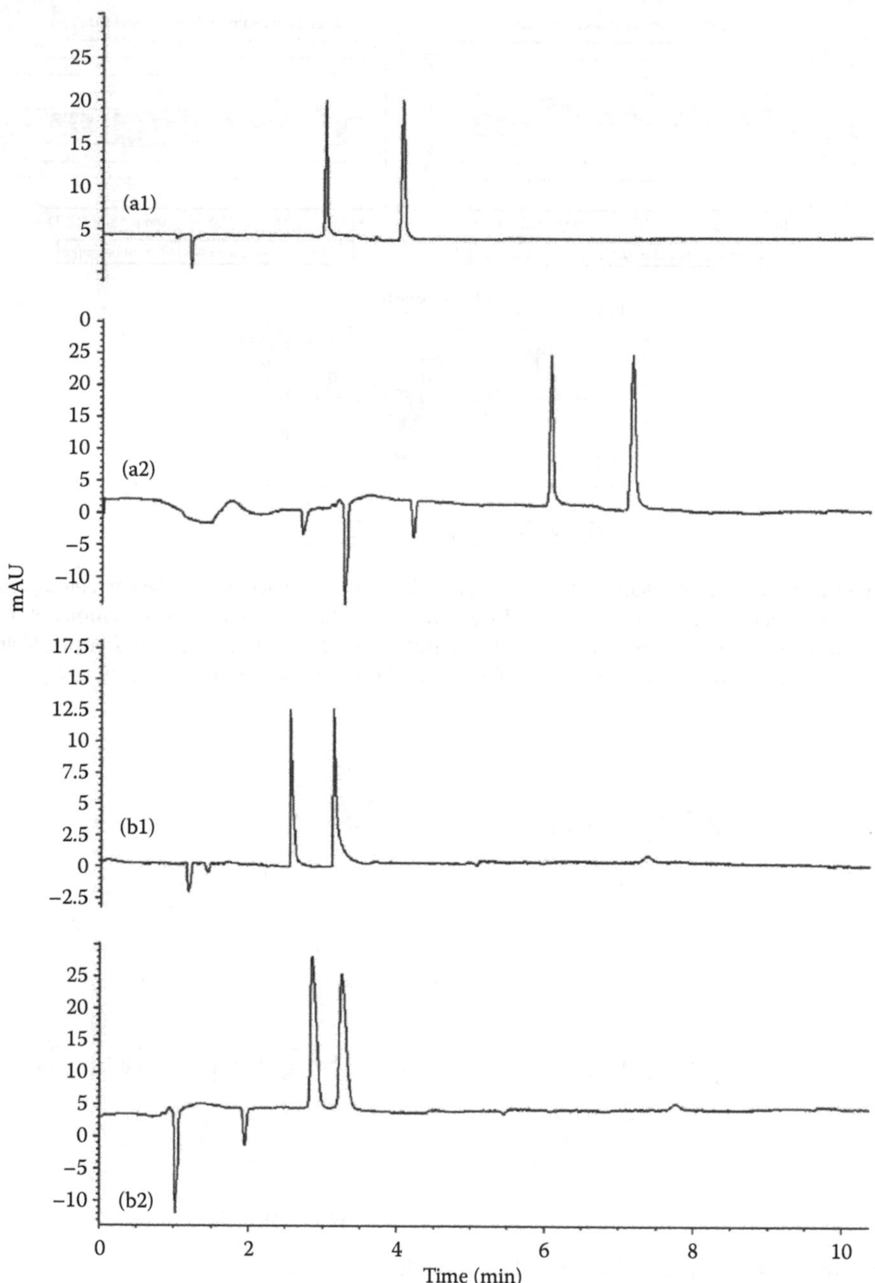

FIGURE 11.5 Chiral CD-modified MEEKC: Separation of (a) atropine and (b) scopolamine using the anionic β-CD 1) 5 mM HDMS-β-CD, 2) 5 mM Sulf-β-CD. Separation conditions: microemulsion BGE; 0.8% (w/w) octane, 6.6% (w/w) 1-butanol, 2.0% (w/w) SDS, and 90.6% (w/w) 10 mM sodium tetraborate buffer pH 9.2, applied voltage +15 kV, temperature 30°C, detection wavelength 195 nm, capillary fused-silica capillary (50 μm i.d. × 24 cm).

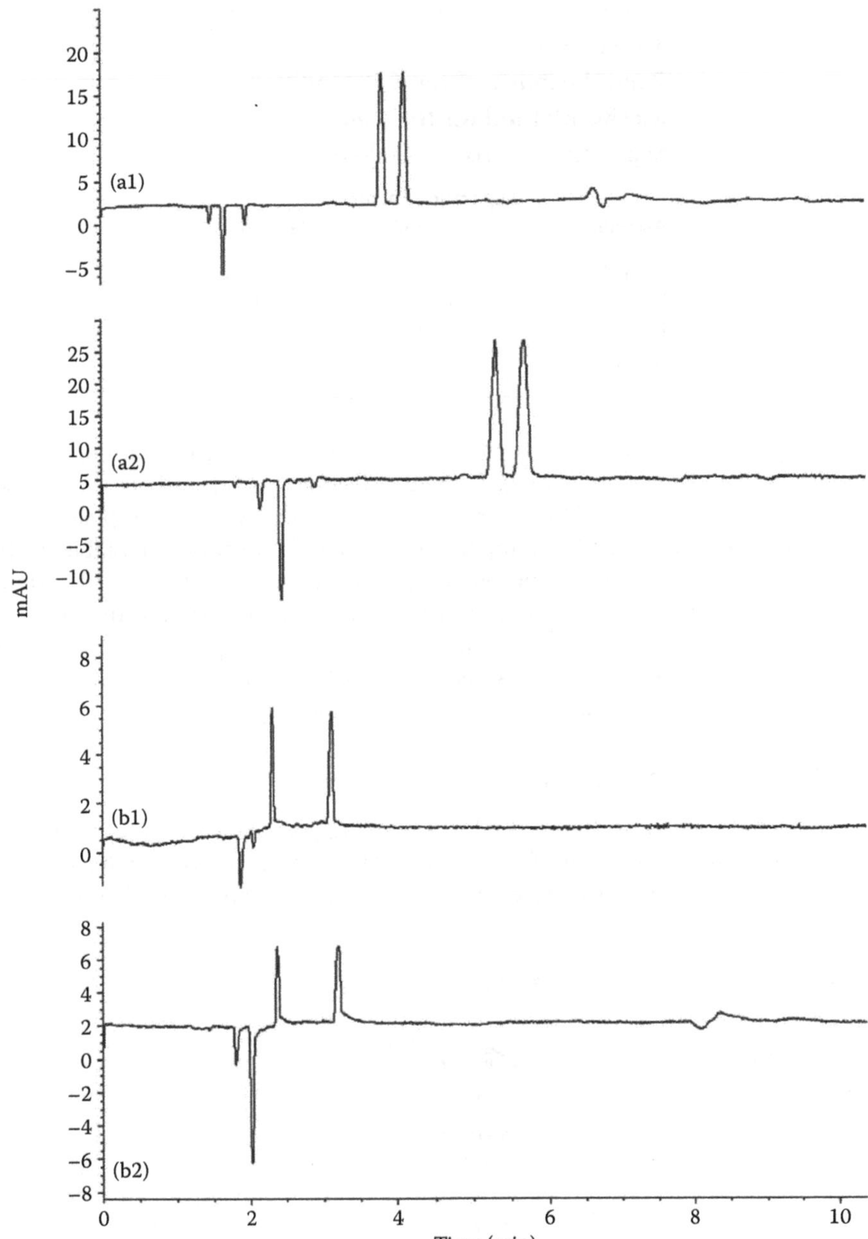

FIGURE 11.6 Chiral CD-modified MEEKC: Separation of (a) ipratropium and (b) homatropine using the anionic β-CD 1) 5 mM HDMS-β-CD, 2) 5 mM Sulf-β-CD. Other conditions: see Figure 11.5.

TABLE 11.4
Reproducibility of the CD-Modified-MEEKC Method for the Chiral Separation of Tropa Alkaloids

Alkaloid	Separation Factor α (n = 6)	RSD (%)
Atropine	1.36	0.87
Scopolamine	1.26	1.91
Ipratropium	1.09	1.37
Homatropine	1.36	1.86

The reproducibility of the CD-MEEKC method was investigated with respect to the RSD of the separation factor (α). Each test solution was injected six times. The RSDs with respect to the RMTs (separation factors) were 0.87% for atropine, 1.91% for scopolamine, 1.86% for homatropine, and 1.37% for ipratropium (Table 11.4). In the experiments, the microemulsion BGE solution was sonicated for 20–25 min before use, which could ensure that the reproducibility of RMTs was less than 2.0%.

Taken together, CD-MEEKC analysis was shown to be a powerful tool for the chiral analysis of hydrophobic compounds, for which CE methods are currently not used. Additionally, the validation of the method was not a problem.

11.5.3 Aziridine Derivatives

Aziridines 1–5, shown in Figure 11.7, are attracting interest as protease inhibitors which might be used, e.g., for treatment of parasitic diseases. A huge number of aziridines having two stereogenic centers were synthesized. Thus, a fast

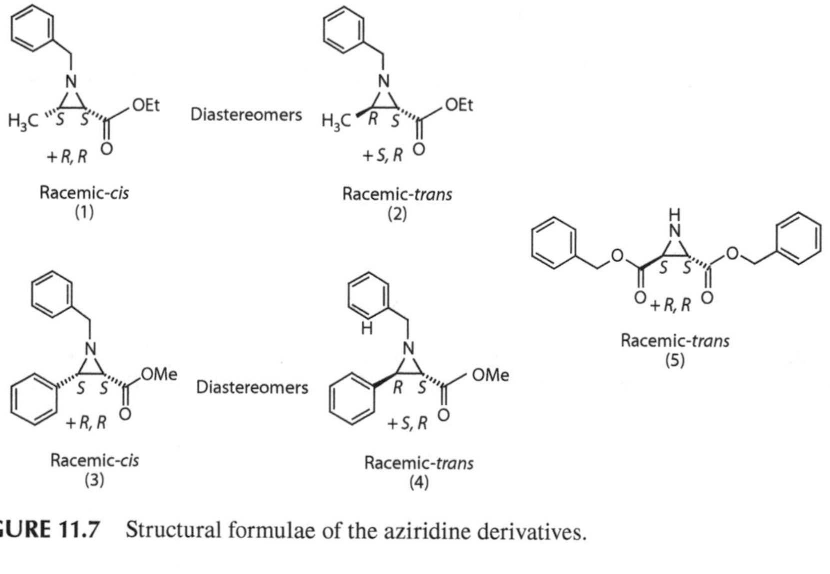

FIGURE 11.7 Structural formulae of the aziridine derivatives.

and reliable screening method for the evaluation of the isomeric composition was needed.

Applying CD-modified CE resulted in a baseline separation of the four isomers. The most robust separation was obtained by means of 2 mM Sulf-β-CD in 50 mM phosphate buffer of pH 2.5. Using this method, 0.25% of the *trans*-diastereomers aziridine could be precisely and accurately quantified in the presence of 99.75% of the *cis*-isomers (Figure 11.8b) [33]. Since the solubility of all aziridines is limited

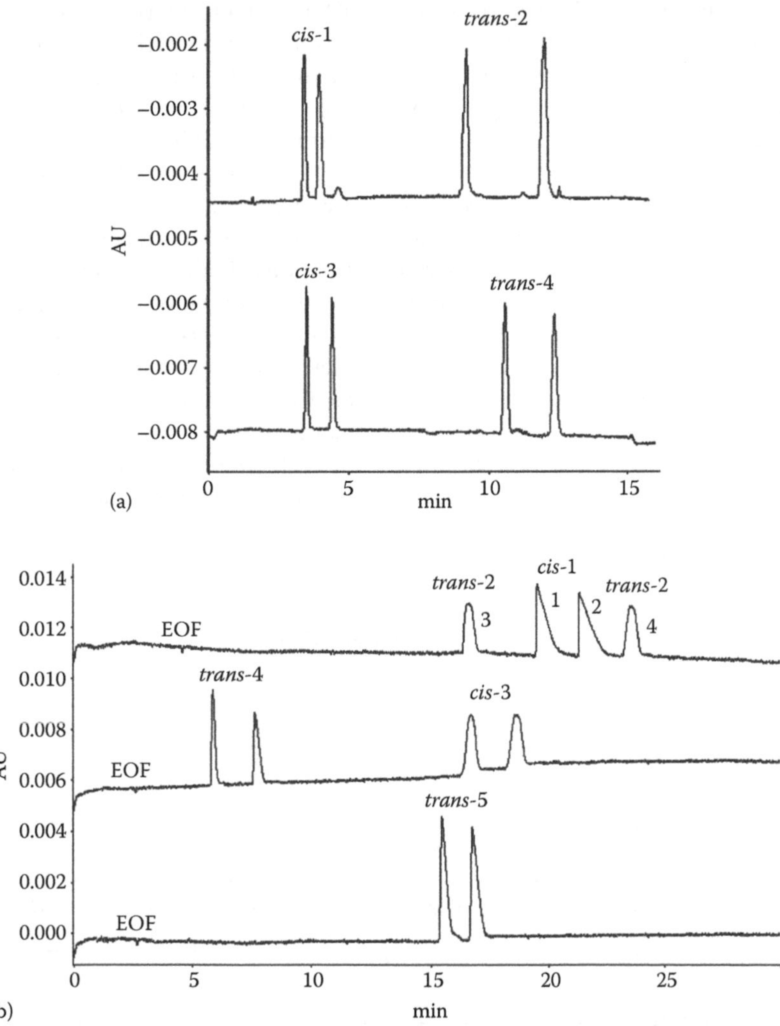

FIGURE 11.8 Electropherograms of enantioseparation of racemic *cis*-1, 3 and racemic *trans*-2, 4, and 5 using a 39/29 cm fused-silica capillary and UV detection at 214 nm (a) the nonaqueous system under optimized conditions: chiral agent (10 mM HDAS-ß-CD) dissolved in acidic methanol as BGE, 20°C capillary temperature, 10 kV applied voltage; (b) the aqueous system: chiral agent (2 mM Sulf-ß-CD) dissolved in the phosphate buffer (20 mM, pH 2.5) as BGE, 25°C capillary temperature, 15 kV applied voltage.

in aqueous phosphate buffer, the isomers were evaluated in nonaqueous systems. Preferably heptakis-(2,3-*O*-diacetyl-6-*O*-sulfo)-β-CD (HDAS-β-CD) was used, which resulted in very good resolution in a rather short analysis time. The mixtures of diastereomers and enantiomers of a given compound (1/2 and 3/4) can be performed in one run (Figure 11.8a) [34].

Nevertheless, the more robust aqueous method, which was employed for the determination of small amounts of racemic *trans* substance in racemic *cis* substance, was validated according to ICH guidelines [21] and *Ph. Eur.* [17] with respect to specificity, linearity, and range, as well as LOD and LOQ, precision, accuracy, and robustness. With respect to specificity, all four isomers were well separated (Figure 11.8b). By spiking the samples with the individual racemic substances the peaks of the electropherogams were assigned. In order to determine the *trans* isomer as the diastereomeric impurity in samples of the *cis* isomer, a calibration was performed using eight concentration levels in the range of 5.0–50.0 μg/mL which corresponds to a 0.25%–2.50% content of the *trans* substance in a *cis* sample of 2.0 mg/mL. Each standard was injected three times randomly and the mean corrected peak area of each enantiomer was evaluated. Calibration curve was obtained by plotting the percentage of diastereomeric impurity against the peak area ratios. The good calibration data are demonstrated by the linear equation $y = 0.0097x + 0.0002$ and showed a linearity confirmed by the coefficient of determination $r^2 = 0.9989$. Figure 11.9 shows the residuals plotted against the calculated *y*-values from the regression line. It can be seen that there is only random fluctuation in the residuals (i.e., no trends are observed) corroborating the linearity of the method. For the highest concentrations, a slightly larger elevation can be seen. This is most probably caused by random errors in the sample preparation.

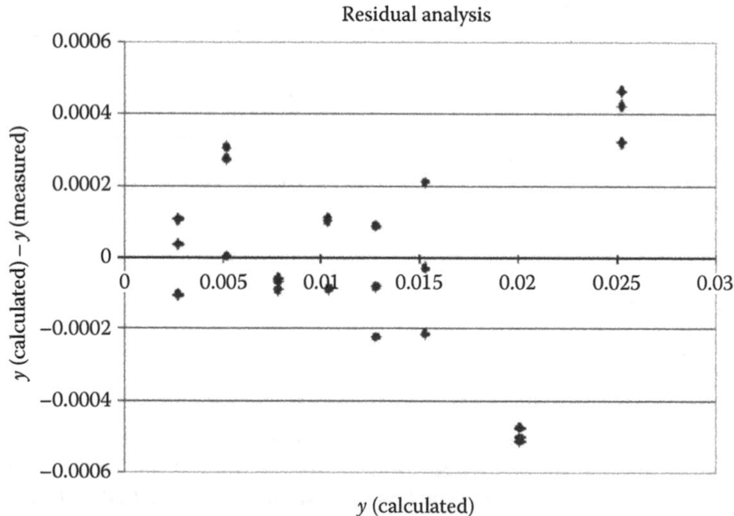

FIGURE 11.9 The plot of the residuals against the calculated *y*-values from the regression line of the determination of aziridine diastereomers.

The LOD was 2.0 μg/mL of the racemic *trans* substance corresponding to a relative concentration of 0.1% content of the *trans* substance in a *cis* sample. The LOQ amounted to 0.25% racemic *trans* substance.

The precision of the method was investigated with respect to repeatability and intermediate precision. Three different solutions of 10.0, 25.0, and 40.0 μg/mL of the racemic *trans* substance corresponding to the 0.5%, 1.25%, and 2.0%, respectively, in a *cis* sample solution of 2.0 mg/mL were prepared. Each solution was injected six times on one day (intraday precision) and on three different days, using two different CE instruments, two different buffer lots, and two different chiral selector lots, respectively (interday precision). The RMTs of the first peak of each enantiomer to the second peak as well as the corrected peak area ratio (*trans*/*cis*) are summarized in Table 11.5. The acceptable RSDs with respect to the RMTs were found to be between 0.16% and 0.40% for intraday precision and between 0.41% and 1.72% for interday precision as a mean of 3 days.

The accuracy of the method was investigated in a similar manner by injecting samples of racemic *trans* substance at a concentration of about 10.0, 25.0, and

TABLE 11.5
Determination of the *trans*-Diastereomers of Aziridine in the Presence of the *cis*-Isomers: Intra- and Interday Precision of the Corrected Peak Area Ratio and Relative Migration Times at Three Concentration Levels

Day	Conc. of Test Solution (%)	Peak Area Ratio	RSD (%)	RMT-*t*	RSD (%)	RMT-*c*	RSD (%)
Precision ($n = 6$)							
1	0.5	0.0989	0.76	0.699	1.06	0.903	0.99
	1.25	0.2609	1.43	0.700	0.69	0.914	0.96
	2.0	0.4183	1.11	0.696	0.99	0.898	1.20
2	0.5	0.1013	2.99	0.703	1.15	0.915	0.99
	1.25	0.2619	0.57	0.696	2.05	0.903	1.01
	2.0	0.4244	1.90	0.706	0.49	0.907	0.74
3	0.5	0.1028	2.01	0.710	1.37	0.903	1.57
	1.25	0.2622	1.36	0.697	0.68	0.908	0.91
	2.0	0.4123	1.35	0.702	0.56	0.919	0.66
Interday precision ($n = 18$)							
	0.5	0.1010	2.73	0.697	1.72	0.907	0.76
	1.25	0.2617	1.42	0.698	0.41	0.909	0.71
	2.0	0.4183	3.06	0.699	1.40	0.908	1.12
Intraday precision ($n = 6$)							
	0.5	0.1021	0.73	0.706	0.21	0.914	0.16
	1.25	0.2625	1.92	0.702	0.37	0.910	0.19
	2.0	0.4099	1.81	0.704	0.22	0.909	0.40

TABLE 11.6
Determination of the *trans*-Diastereomers of Aziridine in the Presence of the *cis*-Isomers: Accuracy of the Method Tested at Three Different Concentration Levels

		Accuracy (n = 6)					
Day	Test Solution (%)	Nominal Conc. Racemic-*trans* (μg/mL)	Accuracy (%)	RSD (%)	Nominal Conc. Racemic-*cis* (mg/mL)	Accuracy (%)	RSD (%)
1	0.5	10.5	99.06	0.79	2.01	101.00	1.58
	1.25	25.3	100.27	1.66	2.03	100.99	1.27
	2.0	41.4	102.64	0.70	2.01	101.60	0.55
2	0.5	10.2	101.61	0.47	2.00	100.47	0.46
	1.25	26.1	101.57	1.56	2.00	100.85	0.70
	2.0	41.3	102.51	1.32	2.02	100.88	1.03
3	0.5	10.4	99.61	3.15	2.01	101.91	1.46
	1.25	25.9	103.09	1.01	2.02	99.97	1.65
	2.0	41.3	102.55	2.65	2.00	101.33	2.05
Interday accuracy (n = 18)							
	0.5	10.4	100.09	1.29	2.01	101.13	1.81
	1.25	25.8	101.64	1.51	2.02	100.60	1.59
	2.0	41.3	102.57	3.03	2.01	101.27	1.26
Intraday accuracy (n = 6)							
	0.5	10.3	99.59	0.60	2.00	100.50	0.25
	1.25	25.9	102.06	2.02	2.02	100.69	0.32
	2.0	41.3	100.04	2.10	2.01	100.79	0.48

40.0 μg/mL corresponding to 0.5%, 1.25%, and 2.0%, respectively, in a racemic *cis* substance of 2.0 mg/mL. The data, which are summarized in Table 11.6, show acceptable accuracy of the method. The highest RSD values for the interday and intraday accuracy were found to be 3.03% and 2.02%, respectively, and, thus, within the required range of ±10% of the theoretical values of the concentration of minor component, the *trans* isomer, and within ±2% of the theoretical values of the concentration of the main component, the *cis* isomer.

The robustness of the method was checked with respect to the most critical parameters such as variation of the applied voltage between 14 and 16 kV; the capillary temperature ranging from 24°C–26°C, the pH of the BGE between 2.4 and 2.6, the buffer concentration in a range of 48–52 mM, the chiral selector concentration between 1.9 and 2.1 mM, considering the RMTs of each isomer to the first peak of the standard (Peak 1), the resolution factor (R_s) of the critical pair are the last two peaks in the electropherogram, and corrected peak area ratio. The solution containing 2.0 mg/mL of *cis* substance spiked with 1.5% of

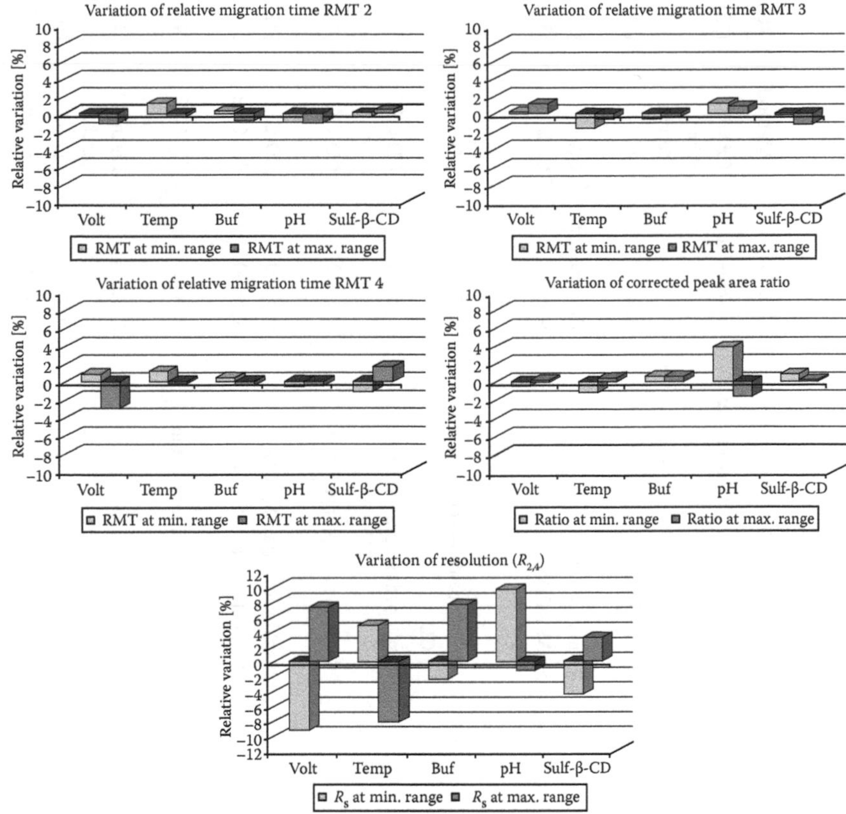

FIGURE 11.10 Determination of the *trans*-diastereomers of aziridine in the presence of the *cis*-isomers: relative variation of relative migration times (RMTs), the corrected peak area ratio, and the resolution factor ($R_{2,4}$) between the last two peaks at small variations of the standard conditions. The bars are marked with Volt, voltage; Temp, temperature; Buf, buffer concentration; pH, buffer pH; and Sulf-ß-CD, Sulf-ß-CD concentration.

the *trans* substance was analyzed six times under either condition. The data of robustness with respect to RMTs and resolution are displayed in Figure 11.10. Acceptable relative variations of the RMTs and resolution values (R_s) were found and the method can be considered to be very robust against small variations of the standard conditions.

11.6 CONCLUSION

CE is an appropriate method for chiral analysis, especially enantiomeric excess determination of enantiomeric drugs (further examples see Table 11.7). The methods can be easily validated with respect to linearity, accuracy, and precision. Moreover, the methods are often shown to be robust.

TABLE 11.7
Selected Examples of Chiral Separations by CE

Substance	Buffer	Chiral Selector	R_s/α	Method	Ref.
Glutamine acid, alanine (ee of 95%)	40 mM borate buffer, pH 6.0	OPA/NAC (ortho-phtalaldehyde/ *N*-acetyl-L-cysteine)	Baseline sep.	CE with chem. derivat.	[35]
14,15-Epoxyeicosatrienoic acids (ee of 97%)	50 mM sodium phosphate, pH 7.0 + 32% CH_3CN (v/v)	15 mM β-CD	R_s 1.3	CD-CZE	[36]
Histidine (ee of 82%)	20 mM borax buffer pH 10	2.5 mM β-CD	Baseline sep.	CAE	[37]
Glutamine acid, derivatization with fluorescein isothionate for detection (ee of 80%)	10 mM borax buffer, pH 10	5 mM γ-CD	n. d.	CAE	[38]
cis-2,3-Epoxy-1-hexanol (ee of 26.9%)	Tetradecane	2,3-di-*O*-benzyl-6-*O*-octanonyl-β-CD	R_s 0.78	CGC	[39]
Methyl-2,3-dihydroxy-3-phenylpropionate (ee > 99%)	200 mM borate buffer, pH 10.3 + 6% MeOH	50 mM hydroxypropyl-β-CD	$R_s > 2.0$	CD-CZE	[40]
1,2-Diphenyl-1,2-ethanediol, 2-Methyl-1-phenyl-1,2-ethanediol (ee > 99%)	200 mM borate buffer, pH 9.8 + 9% MeOH	20 mM hydroxypropyl-β-CD	$R_s > 2.2$	CD-CZE	[41]
Phenylglycidol	10 mM borate buffer, pH 10.0	20 mM succinylated-β-CD	$R_s > 2.0$	CD-CZE	[42]
Analogs of stavudine: *cis*- and *trans*-nucleosides	25 mM phosphate buffer, pH 2.5, capillary coated with PEO	4% highly sulfated β-CD	$R_s > 2.0$	CD-CZE	[43]
Asparagine (derivatization with fluorescein isothionate for detection)	80 mM borate buffer, pH 10.0	20 mM β-CD + 30 mM sodium taurocholate	Baseline sep.	CD-CZE	[44]
Synephedrine	80 mM L-tartaric acid + 40 mM Cu(II), pH 12.0		R_s 1.3	Ligand-exchange	[45]
Tryptophan Phenylalanine Tyrosine	5 mM ammonium acetate, 100 mM boric acid, 3 mM $ZnSO_4 {*} 7H_2O$, 6 mM L-lysine		R_s 7.09 R_s 3.69 R_s 2.47	Ligand-exchange	[46]

Catechin	0.1 M borate buffer, pH 8.5	1 M 6-*O*-α-maltosyl-β-CD	R_s 1.51	CD-CZE	[47]
Epicatechin		12 mM 2-hydroxypropyl-γ-CD	R_s 1.60		
		1 M 6-*O*-α-maltosyl-β-CD	R_s 1.30		
		12 mM 2-hydroxypropyl-γ-CD	R_s 1.62		
Linezolid (ee > 99%)	50 mM borate buffer, pH 9.0	27.5 mM HDAS-β-CD	R_s 3.7	CD-CZE	[48]
Propiconazol	25 mM phosphate buffer pH 7.0 + 50 mM SDS + MeOH-ACN (10:5, v/v)	30 mM hydroxypropyl-β-CD	$R_s > 1.5$	MEKC	[49]
Timolol (ee > 99%)	40 mM KOH in MeOH/EtOH (40:60, v/v)	100 mM 1*S*,4*R*-(+)-ketopinic acid	R_s 4.2	NACE	[50]
Alanine, methionine	20 mM borax buffer, pH 10	2.5 mM β-CD	Baseline sep.	CAE	[51]
Phenylalanine, tryptophan, kynurenine after derivatization with NDA (naphthalene-2,3-dicarboxaldehyde) (ee > 99%)	50 mM borate buffer, pH 9.5	25 mM DHADMP (3-[(3-Dehydroabietamidopropyl) dimethylammonio]-1-propane-sulfonate	$R_s > 1.5$	MEKC	[52]
Carprofen	60 mM sodium acetate buffer, pH 5.0 + 10% MeOH	30 mM trimethyl-β-CD + 10 mM EtChol ((*R*)-(−)-1-hydroxy-*N*,*N*,*N*-trimethylbutan-2-aminium bis(trifluoromethylsulfonyl)imide	R_s 2.90	CD-CZE	[53]
Suprofen			R_s 1.61		
Dansyl derivatives of α-amino acids:	20 mM NH_4Ac, pH 6.8	1.1 mM CDen (ethylenediamine-β-cyclodextrin)		CD-EKC	[54]
Aspartic acid			R_s 2.04		
Glutamine acid			R_s 2.35		
Warfarin	25 mM NH_4Ac, pH 5.5	25 mM poly-L,L-SULV (polysodium *N*-undecenoyl-L,L-leucyl-valinate)	R_s 1.9	MEKC	[55]

Note: CAE, capillary array electrophoresis; CD-CZE, cyclodextrin-modified capillary zone electrophoresis; CD-EKC, electrokinetic chromatography by cyclodextrins; ee, enantiomeric excess; NACE, nonaqueous capillary electrophoresis.

REFERENCES

1. Bitar, Y, Holzgrabe, U. 2007. Enantioseparation of chiral tropa alkaloids by means of cyclodextrin-modified microemulsion electrokinetic chromatography. *Electrophoresis* 28: 2693–2700.
2. Branch, SK. 2005. Guidelines from the International Conference on Harmonisation (ICH). *J Pharm Biomed Anal* 38: 798–805.
3. Schmitt, U, Branch, SK, Holzgrabe, U. 2002. Chiral separations by cyclodextrin-modified capillary electrophoresis—Determination of the enantiomeric excess. *J Sep Sci* 25: 959–974.
4. ICH Guidelines. Quality Topics. http://www.ich.org/cache/compo/276-254-1.html
5. Technical guide for the elaboration of monographs. http://www.edqm.eu/medias/fichiers/technical_guide_english.pdf
6. Ha, PTT, Hoogmartens, J, Van Schepdael, A. 2006. Recent advances in pharmaceutical applications of chiral capillary electrophoresis. *J Pharm Biomed Anal* 41: 1–11.
7. Malá, Z, Křivánková, L, Gebauer, P, Boček, P. 2007. Contemporary sample stacking in CE: A sophisticated tool based on simple principles. *Electrophoresis* 28: 243–253.
8. García-Ruiz, C, Marina, ML. 2006. Sensitive chiral analysis by capillary electrophoresis. *Electrophoresis* 27: 195–212.
9. Holzgrabe, U. 2008. The need for CE methods in pharmacopoeial monographs. In *Capillary Electrophoresis Methods for Pharmaceutical Analysis (Separation Science and Technology)*, Ahuja, S and Jimidar, MI (Eds.), pp. 299–315. London, U.K.: Elsevier Inc.
10. Clohs, L, McErlane, KM. 2003. Comparison between capillary electrophoresis and high-performance liquid chromatography for the stereoselective analysis of carvedilol in serum. *J Pharm Biomed Anal* 31: 407–412.
11. Holzgrabe, U, Brinz, D, Kopec, S, Weber, C, Bitar, Y. 2006. Why not using capillary electrophoresis in drug analysis? *Electrophoresis* 27: 2283–2292.
12. Song, S, Zhou, L, Thompsom, R, Yang, M, Ellison, D, Wyvratt, JM. 2002. Comparison of capillary electrophoresis and reversed-phase liquid chromatography for the determination of the enantiomeric purity of an M3 antagonist. *J Chromatogr A* 959: 299–308.
13. Rocheleau, MJ. 2005. Generic capillary electrophoresis for chiral assay in early pharmaceutical development. *Electrophoresis* 26: 2320–2329.
14. Jimidar, MI, Van Ael, W, Van Nyen, P, Peeters, M, Redlich, D, De Smet, M. 2004. A screening strategy for the development of enantiomeric separation methods in capillary electrophoresis. *Electrophoresis* 25: 2772–2785.
15. Borst, C, Holzgrabe, U. 2008. Enantioseparation of dopa and related compounds by cyclodextrin-modified microemulsion electrokinetic chromatography. *J Chromatogr A* 1204: 191–196.
16. Schmitt, U, Ertan, M, Holzgrabe, U. 2004. Chiral capillary electrophoresis: Facts and fiction on the reproducibility of resolution with randomly substituted cyclodextrins. *Electrophoresis* 25: 2801–2807.
17. *European Pharmacopoeia*. 6th edition. 2008. European Department for the Quality of Medicines. Strasbourg, France.
18. Wenzel, TJ, Wilcox, JD. 2003. Chiral reagents for the determination of enantiomeric excess and absolute configuration using NMR spectroscopy. *Chirality* 15: 256–270.
19. Shamsipur, M, Dastjerdi, LS, Haghgoo, S, Armsprach, D, Matt, D, Aboul-Enein, HY. 2007. Chiral selectors for enantioresolution and quantitation of the antidepressant drug fluoxetine in pharmaceutical formulations by ^{19}F NMR spectroscopic method. *Anal Chim Acta* 601: 130–138.
20. Schurig, V. 2002. Chiral separations using gas chromatography. *Trends Anal Chem* 21: 647–661.
21. ICH Guideline Quality Topic Q2(R1). http://www.ich.org/LOB/media/MEDIA417.pdf

22. Fabre, H, Altria, KD. 2001. Validating CE methods for pharmaceutical analysis. *LC-GC Europe* 14(5): 302, 304, 306, 308–310.
23. Harris, DC. 2003. *Quantitative Chemical Analysis 6th edition.* New York: W. H. Freeman.
24. Wätzig, H, Degenhardt, M, Kunkel, A. 1998. Strategies for capillary electrophoresis: method development and validation for pharmaceutical and biological applications. *Electrophoresis* 19: 2695–2752.
25. Altria, KD, Rudd, DR. 1995. An overview of method validation and system suitability aspects in capillary electrophoresis. *Chromatographia* 41: 325–331.
26. Chankvetadze, B, Lomsadze, K, Burjanadze, N, Breitkreutz, J, Pintore, G, Chessa, M, Bergander, K, Blaschke, G. 2003. Comparative enantioseparations with native β-cyclodextrin, randomly acetylated β-cyclodextrin and heptakis(2,3-di-O-acetyl)-β-cyclodextrin in capillary electrophoresis. *Electrophoresis* 24: 1083–1091.
27. *European Pharmacopoeia.* 6th edition. 2008. Monograph of Gluthatione (No. 1670). European Department for the Quality of Medicines. Strasbourg, France.
28. ICH Guideline Quality Topic Q3A(R2). http://www.ich.org/LOB/media/MEDIA422.pdf
29. Maruszak, W, Trojanowicz, M, Margasińska, M, Engelhardt, H. 2001. Application of carboxymethyl-β-cyclodextrin as a chiral selector in capillary electrophoresis for enantiomer separation of selected neurotransmitters. *J Chromatogr A* 926: 327–336.
30. Ha, PTT, Van Schepdael, A, Hauta-aho, T, Roets, E, Hoogmartens, J. 2002. Simultaneous determination of dopa and carbidopa enantiomers by capillary zone electrophoresis. *Electrophoresis* 23: 3404–3409.
31. *European Pharmacopoeia.* 6th edition. 2008. Monograph of Levodopa (No. 0038). European Department for the Quality of Medicines. Strasbourg, France.
32. Bitar, Y, Holzgrabe, U. 2006. Impurity profiling of atropine sulfate by microemulsion electrokinetic chromatography. *J Pharm Biomed Anal* 44: 623–633.
33. Bitar, Y, Degel, B, Schirmeister, T, Holzgrabe, U. 2005. Development and validation of a separation method for the diastereomers and enantiomers of aziridine-type protease inhibitors. *Electrophoresis* 26: 2313–2319.
34. Bitar, Y, Degel, B, Schirmeister, T, Holzgrabe, U. 2005. Comparison of the separation of aziridine isomers applying heptakis(2,3-di-O-methyl-6-sulfato)β-CD and heptakis (2,3-di-O-acetyl-6-sulfato)β-CD in aqueous and nonaqueous systems. *Electrophoresis* 26: 3897–3903.
35. Ptolemy, AS, Tran, L, Britz-McKibbin, P. 2006. Single-step enantioselective amino acid flux analysis by capillary electrophoresis using on-line sample preconcentration with chemical derivatization. *Anal Biochem* 354: 192–204.
36. Van der Noot, VA, Van Rollins, M. 2002. Capillary electrophoresis of cytochrome P-450 epoxygenase metabolites of arachidonic acid. 2. Resolution of Stereoisomers. *Anal Chem* 74: 5866–5870.
37. Wang, J, Liu, K, Sun, G, Bai, J, Wang, L. 2006. High-throughput screening for the asymmetric transformation reaction of L-histidine to D-histidine by capillary array electrophoresis. *Anal Chem* 78: 901–904.
38. Wang, J, Liu, KY, Wang, L, Bai, JL. 2006. Determination of enantiomeric excess of glutamic acids by lab-made capillary array electrophoresis. *Chin Chem Lett* 1: 49–52.
39. Shi, XY, Liang, P, Gao, XW. 2006. Determination enantiomer excess (e.e.%) of chiral sharpless epoxides with β-cyclodextrin derivatives as chiral stationary phases of capillary gas chromatography. *Chin Chem Lett* 17: 505–508.
40. Zhao, Y, Yang, X-B, Wang, Q-F, Nan, P-J, Jin, Y, Zhang, S-Y. 2007. Determination of enantiomeric excess for 2,3-dihydroxy-3-phenylpropionate compounds by capillary electrophoresis using hydroxypropyl-β-cyclodextrin as chiral selector. *Chirality* 19: 380–385.
41. Zhao, Y, Yang, X-B, Wang, Q-F, Sun, XL, Jiang, R, Zhang, S-Y. 2007. Determination of enantiomeric excess of aromatic 1,2-diols with HP-β-cyclodextrin as chiral selector by CE. *Chromatographia* 66: 601–605.

42. Morante-Zarcero, S, Crego, AL, Sierra, I, Fajardo, M, Marina, ML. 2004. Chiral capillary electrophoresis applied to the determination of phenylglycidol enantiomers obtained from cinnamyl alcohol by asymmetric epoxidation using new titanium(IV) alkoxide compounds as catalysts. *Electrophoresis* 25: 2745–2754.
43. Lipka, E, Len, C, Rabiller, C, Bonte, J-P, Vaccher, C. 2006. Enantioseparation of *cis* and *trans* nucleosides, aromatic analogues of stavudine, by capillary electrophoresis and high-performance liquid chromatography. *J Chromatogr A* 1132: 141–147.
44. Liu, M, Qiu, C, Guo, Z, Qi, L, Xie, M, Chen, Y. 2007. Degree of supersaturation-regulated chiral symmetry breaking in one crystal. *J Phys Chem B Lett* 111: 11346–11349.
45. Hödl, H, Krainer, A, Holzmüller, K, Koidl, J, Schmid, MG, Gübitz, G. 2007. Chiral separation of sympathomimetics and β-blockers by ligand-exchange CE using Cu(II) complexes of L-tartaric acid and L-threonine as chiral selectors. *Electrophoresis* 28: 2675–2682.
46. Qi, L, Han, Y, Zuo, M, Chen, Y. 2007. Chiral CE of aromatic amino acids by ligand-exchange with zinc(II)-L-lysine complex. *Electrophoresis* 28: 2629–2634.
47. Kofink, M, Papagiannopoulos, M, Galensa, R. 2007. Enantioseparation of catechin and epicatechin in plant food by chiral capillary electrophoresis. *Eur Food Res Technol* 225: 569–577.
48. Michalska, K, Pajchel, G, Tyski, S. 2008. Determination of enantiomeric impurity of linezolid by capillary electrophoresis using heptakis-(2,3-diacetyl-6-sulfato)-β-cyclodextrin. *J Chromatogr A* 1180: 179–186.
49. Ibrahim, WAW, Hermawan, D, Sanagi, MM. 2007. On-line preconcentration and chiral separation of propiconazole by cyclodextrin-modified micellar electrokinetic chromatography. *J Chromatogr A* 1170: 107–113.
50. Hedeland, Y, Lehtinen, J, Pettersson, C. 2007. Ketopinic acid and diisopropylideneketogulonic acid as chiral ion-pair selectors in capillary eletrophoresis enantiomeric impurity analysis of *S*-timolol and 1*R*,2*S*-ephedrine. *J Chromatogr A* 1141: 287–294.
51. Wang, J, Zhang, Y, Wang, L, Bai, J. 2007. Capillary array electrophoresis for the research of racemization reaction of L-amino acids. *J Chromatogr A* 1144: 279–282.
52. Zhao, S, Wang, H, Pan, Y, He, M, Zhao, Z. 2007. 3-[(3-Dehydroabietamidopropyl) dimethylammonio]-1-propane-sulfonate as a new type of chiral surfactant for enantiomer separation in micellar electrokinetic chromatography. *J Chromatogr A* 1145: 246–249.
53. François, Y, Varenne, A, Juillerat, E, Villemin, D, Gareil, P. 2007. Evaluation of chiral ionic liquids as additives to cyclodextrins for enantiomeric separations by capillary electrophoresis. *J Chromatogr A* 1155: 134–141.
54. Cucinotta, V, Giuffrida, A, Grasso, G, Maccarrone, G, Messina, M, Vecchio, G. 2007. High selectivity in new chiral separations of dansyl amino acids by cyclodextrin derivatives in electrokinetic chromatography. *J Chromatogr A* 1155: 172–179.
55. Hou, J, Zheng, J, Shamsi, SA. 2007. Separation and determination of warfarin enantiomers in human plasma using a novel polymeric surfactant for micellar electrokinetic chromatography-mass spectrometry. *J Chromatogr A* 1159: 208–216.

12 Analysis of Chiral Drugs in Body Fluids

Pierina Sueli Bonato, Valquiria Aparecida Polisel Jabor, and Anderson Rodrigo Moraes de Oliveira

CONTENTS

12.1 Introduction 341
12.2 Sample Preparation 342
12.2.1 Classical Techniques 343
12.2.1.1 Liquid–Liquid Extraction 343
12.2.1.2 Solid-Phase Extraction 343
12.2.1.3 Protein Precipitation 344
12.2.2 Microextraction Techniques 344
12.2.2.1 Solid-Phase Microextraction and Related Techniques 344
12.2.2.2 Liquid-Phase Microextraction 347
12.3 Enhanced Sensitivity 350
12.3.1 Sensitivity Improvements by Sample Concentration Methods 350
12.3.1.1 Sample Stacking 350
12.3.1.2 Isotachophoresis (ITP) 352
12.3.1.3 Sweeping 352
12.3.2 Improvements in Detection Systems 353
12.4 Method Validation 355
12.5 Concluding Remarks 358
Acknowledgments 358
References 358

12.1 INTRODUCTION

Understanding the contribution of stereochemistry to the action and disposition of drugs is essential to the successful development of chiral drugs as well as to establishing if a drug already developed or used as a racemic mixture might be replaced by a safer or more effective pure single isomer. Thus, the demand for analytical methods suitable for the accurate and precise determination of enantiomers of drugs and their metabolites in biological samples has increased significantly over the last years [1]. High-performance liquid chromatography (HPLC) has been frequently used in pharmacokinetic and metabolism studies as well as for other applications. In contrast to HPLC, there are relatively few reports regarding the application of electromigration techniques (capillary electrophoresis, CE and capillary electrochromatography, CEC) in the chiral separations of drugs in biological samples. However, the number of studies in this area has increased in the last years [2]. The increase in the interest in CE and CEC techniques is mainly supported by their high-efficiency separation power which allows baseline resolution of enantiomers even at low degree of enantioselectivity and prevents the interference of matrix compounds, metabolites and other coadministered drugs. Other features such as suitability for the analysis of polar and/or basic drugs and their metabolites, speed and reduced operational costs, simplicity of instrumentation, relatively easy method of development, and requirement of small amounts of samples make electromigration methods particularly attractive for this application [2].

The first step for the analysis of chiral drugs and metabolites in biological samples is to obtain the chiral resolution of the drug and metabolites. In CE, the chiral selector is added to the running buffer whereas in CEC it is present in the stationary phase. Among the several chiral selectors discussed in the previous chapters of this book, cyclodextrins (CD) and their derivatives are the most popular ones and they have been intensively used for the resolution of a large number of drugs in biological samples. In the specific case of chiral separations by CE, the optimization of the resolution can easily be obtained by changing the chiral selector, its concentration or other analytical conditions, including the combination of chiral selectors. In addition, the enantiomer migration order can be inverted by using simple procedures such as changing the chiral selector [2].

Considering that the chiral resolution has been extensively discussed in the previous chapters of this book, we will discuss here more specific topics regarding the analysis of drugs in biological samples, such as sample preparation techniques, enhanced sensitivity and method validation.

12.2 SAMPLE PREPARATION

The analysis of drugs and metabolites in biological matrix requires a well-designed sample preparation procedure in order to remove proteins and other endogenous compounds and to concentrate the analyte in the analyzed sample. Recently, the trends in sample preparation area have been directed toward the use of small sample sizes, great specificity or selectivity, increased potential for automation or for online methods and environmentally friendly approaches [3].

Sample preparation techniques for drug analysis in biological samples can be, didactically, divided into classical techniques (represented by liquid–liquid extraction (LLE), solid-phase extraction (SPE), and protein precipitation) and microextraction techniques (e.g., solid-phase microextraction, SPME and liquid-phase microextraction, LPME). Although there are other sample preparation techniques, only these techniques will be discussed in this chapter, since they are the most frequently used.

12.2.1 Classical Techniques

12.2.1.1 Liquid–Liquid Extraction

LLE is based on the relative solubility of an analyte in two immiscible phases: the aqueous phase (sample) and an organic solvent. The more frequently used organic solvents in LLE are hexane, dichloromethane, chloroform, ethyl acetate, or a mixture of them. The main drawbacks of LLE are the requirement of high amounts of organic solvents, difficulty of automatization, need for highly pure solvents, and difficulty in the extraction of highly hydrophilic compounds. Another problem is the occurrence of emulsion, which, sometimes, can be solved by the addition of salt or centrifugation [4].

Besides these drawbacks, LLE is the main extraction technique employed in the analyses of drugs in biological fluids. The reproducibility, easily handling, and relatively low cost make this technique the first choice. Approximately 60% of the developed methods for the chiral analysis of drugs and metabolites by CE published in the last 5 years were based on LLE.

12.2.1.2 Solid-Phase Extraction

SPE has been widely used for the analysis of drugs in biological fluids due to the following advantages over conventional LLE: higher recovery, less organic solvent used, no emulsion problems, shorter sample preparation time, easier operation, and capacity to be incorporated into an automated process [5].

In SPE, the analytes are extracted due to the partitioning between a liquid (sample) phase and a solid (extraction) phase whereby the intermolecular forces between the phases influence retention and elution. Retention may involve nonpolar, polar, or ionic interactions. The optimum conditions for the extraction of a particular drug from the matrix can be predicted from their physicochemical properties, such as the presence of polar functional groups. The wide range of SPE sorbents available provides a wide range of interactions [5].

SPE can be very selective by using new sorbents such as molecularly imprinted polymers (MIPs), restricted access media (RAM) or phases based on immunoaffinity (immunosorbents). The MIP is obtained through the preparation of polymers with synthetic recognition sites and it has a predetermined selectivity for one or more analytes. These recognition sites are obtained by the arrangement of the functional monomers around the molecules of the analyte. After that, the analyte is removed from the cavity, leaving gaps (sites of recognition) in the material that will show an affinity for the analyte. The restrict access media (RAM) has a biocompatible hydrophilic surface outside and a hydrophobic surface inside the sorbent pores. The separation mechanism is based on a combination of exclusion and partition. Large molecules, such as proteins, are unable to penetrate into pores and hence elute quickly. On the other hand, small molecules can enter into the pores and are retained for further elution and analysis. The phases based on immunoaffinity have a specific antibody immobilized on a solid support, such as agarose or silica. The basic principle of retention is based on the attraction of the antibody (SPE phase) and the antigen (drug). This makes this phase very specific [4,6]. Although these new materials are very attractive for sample preparation, their use is still restricted.

SPE can be coupled directly to the CE system, in the same way as in HPLC. In this case, the SPE device is placed directly within the capillary. Several approaches have been reported, including: (1) an open-tubular capillary coated with a sorbent, (2) a small (1–2 mm) packed section containing microsphere beads and retained by suitable frits at the inlet of the capillary, and (3) an immobilized particle loaded membrane [7]. A tee-split interface has also been used [8]. The biggest advantage in using this approach is the reduced risk of contamination and exposition to potentially hazardous biological materials.

12.2.1.3 Protein Precipitation

Protein precipitation is the simplest sample preparation technique. Employing this technique, the proteins present in the sample are precipitated by the addition of a precipitant agent, such as methanol, acetonitrile, or an acid solution. The disadvantages of protein precipitation are incomplete cleanup and sample dilution. Recently, Schappler et al. [9] developed a stereoselective method to analyze some amphetamines by CE–ESI/MS after protein precipitation. The precipitation procedure was performed by adding 400 μL of acetonitrile to 200 μL of plasma sample. Analysis time was relatively short (less than 7 min) and no matrix effect such as signal suppression occurred for the determination of the tested compounds.

12.2.2 Microextraction Techniques

Drug analysis in biological fluids employing microextraction techniques has increased in the last 10 years. Nowadays, there are several microextraction techniques that can be used, such as SPME, LPME, stir bar sorptive extraction (SBSE), single-drop microextraction (SDME) and so on. The main attractive aspect of these techniques is the low amount of organic solvent consumed during the extraction. Using these techniques, the preconcentration and extraction occur at the same time with no waste of time.

12.2.2.1 Solid-Phase Microextraction and Related Techniques

In SPME, a small amount of an extracting phase dispersed on a solid support is exposed to the sample for a well-defined period of time [10] (Figure 12.1). If the time is long enough, a concentration equilibrium is established between the sample matrix and the extracting phase. After that, exposing the fiber for further periods does not accumulate more analytes.

SPME is essentially a two-step process. The first step is the partitioning of the analytes between the sample matrix and the fiber. This step can be performed in two modes, i.e., direct extraction (Figure 12.1A) and headspace extraction (Figure 12.1B). In the direct mode, the fiber is directly immersed in the sample solution. This mode is useful for extracting nonvolatile and highly hydrophilic analytes. The headspace extraction is used to extract analytes that are present in the vapor phase, i.e., analytes that are volatile or became volatile by a derivatization reaction. The second step is the desorption of the analyte into the analytical instrument (HPLC, gas chromatography [GC]—online desorption) or in an appropriated solvent (off-line desorption) [10].

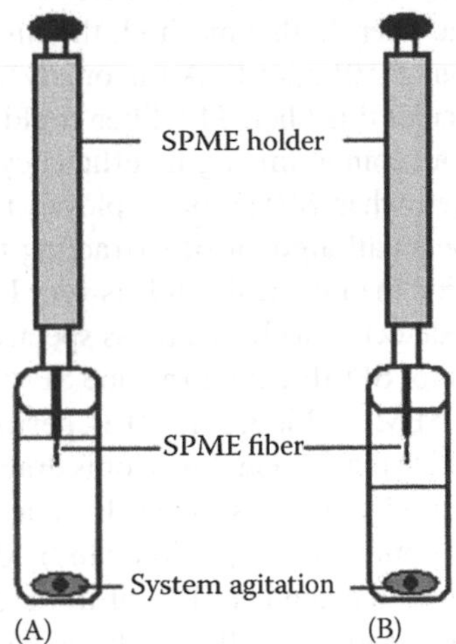

FIGURE 12.1 SPME device. (A) Direct extraction and (B) headspace extraction.

There are some steps to be optimized when SPME is employed, such as type of fiber, extraction time, salt addition, extraction temperature, sample pH, desorption solvent, desorption time, and evaluation of carryover. These steps are very critical for the development of the method and they should be performed before the validation step.

Table 12.1 shows some fibers employed in SPME that are marketed by Supelco Company [11] and are suitable for HPLC or CE analysis. Fibers based on other materials such as MIPs [12] and sol-gel technology [13] have also been developed and used for the analysis of drugs in biological samples. Fang et al. [14] developed a stereospecific method to analyze ephedrine stereoisomers in urine using a novel fiber obtained by the sol-gel and cross-linking technology [a copoly(butyl methacrylate/hydroxyl-terminated silicone oil)]. The method was linear over 20–5000 ng/mL with detection limits of 3, 5, and 5 ng/mL for (1*R*,2*S*)-ephedrine, (1*R*,2*R*)-pseudoephedrine, and (1*S*,2*S*)-pseudoephedrine, respectively. Another enantioselective method was

TABLE 12.1
Some Fibers Employed in SPME and Marketed by Supelco

Chemical Composition	Thickness (μm)	Description
Polydimetylsiloxane (PDMS)	100	Nonbonded
	30	Nonbonded
	7	Bonded
Polyacrylate (PA)	85	Bonded
Polydimetylsiloxane/divinylbenzene (PDMS/DVB)	60	Bonded
Carbowax/templated resin (CW/TPR)	50	Bonded

developed by Zhou et al. [15] to analyze propranolol enantiomers in urine using a sol-gel derived calix[4]arene fiber. In this method, the linear range was 0.05–10 μg/mL, with a limit of detection of 0.01 μg/mL. A major advantage of this method is the alkali- and solvent-resistance of this fiber. The fiber could be used in more than 150 extraction procedures without compromising its efficiency.

The main drawback faced when SPME is employed is the low recovery values obtained, mainly due to the small amount of extracting phase coated on the fiber. Usually, analyte concentration in biological fluids is very low. In this case, it is advisable to use a more sensitive detector, such as a mass spectrometer. Another approach is to use alternative designs for SPME called in-tube SPME or SBSE.

In-tube SPME (In-tube SPME) (Figure 12.2) is performed using an internally coated capillary, through which the sample flows, or is drawn repeatedly. The trapped analytes are then desorbed or eluted by a solvent. In-tube SPME is well suitable for automation; extraction, desorption, and injection can be done continuously using a standard autosampler. To show the applicability of in-tube SPME to the analysis of drugs in biological samples, Jinno et al. [16] analyzed four tricyclic antidepressant in urine using a Zylon® fiber-packed capillary [poly(*p*-phenylene-benzobisoxazole)] for the extraction. This method showed excellent enhanced sensitivity that was more than 100 times better than direct CE analysis without SPME process (Figure 12.3). More recently, Lin et al. [17] used in-tube SPME and CEC for the enantioselective analysis of propranolol in urine. Detection limits of 4 and 7 ng/mL were obtained for (*S*)- and (*R*)-propranolol, respectively.

The SBSE system consists of a glass stirrer bar coated with a thick bonded absorbent layer (polydimethylsiloxane or other material) to give a large surface area, which leads to a higher phase ratio and a better recovery value. The desorption can be performed thermally (in the GC injector) or in a solvent and further injected into a CE or HPLC equipment. The major advantage of stir bar technique is the high concentration factors that can be achieved [6]. Until now, there is no application of this extraction technique in CE or CEC analysis.

An aspect to be considered when using SPME is the price of the fibers and their durability, mainly when direct extraction is employed. The holder and the interface for HPLC are also quite expensive. Until now, no interface to couple CE and SPME is commercially available. However, there are papers showing that it is possible. One approach was designed that introduces the fiber with extracted compounds directly into the capillary (Figure 12.4). The most important advantage of this interface is its zero-dead volume and the minimum amount of desorption solvent, which leads to no band broadening peaks [18].

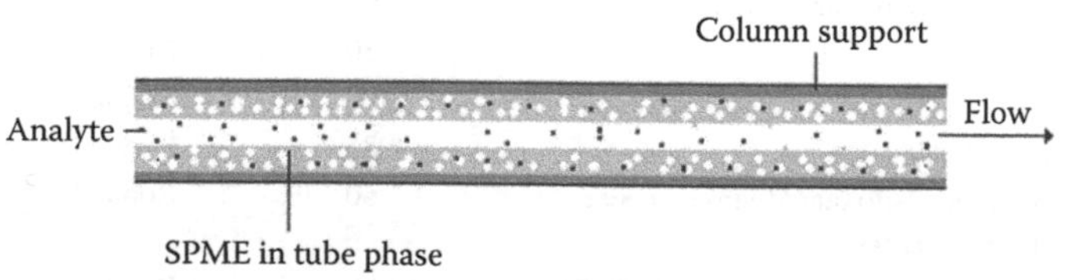

FIGURE 12.2 Schematic in-tube SPME.

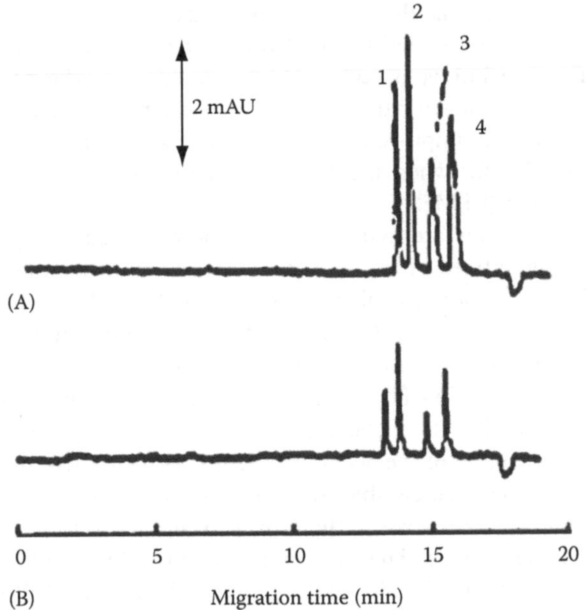

FIGURE 12.3 Electropherograms of a drug mixture obtained by fiber-in-tube SPME-CE (A) and direct CE analysis (B). 1, desipramine; 2, nortriptyline; 3, imipramine; 4, amitriptyline. (From Jinno, K. et al., *Electrophoresis*, 22, 3785, 2001. With permission.)

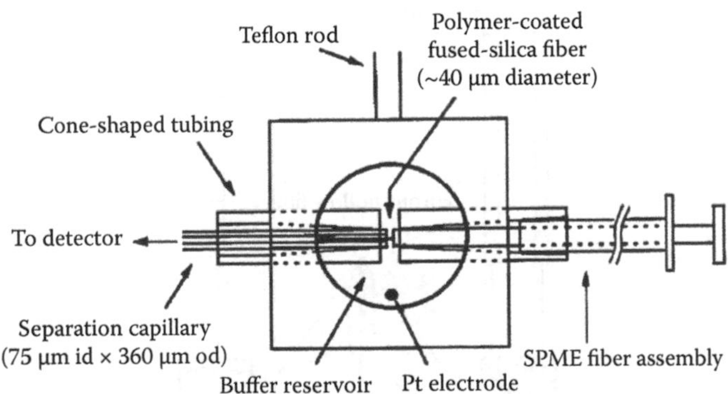

FIGURE 12.4 Schematic of the SPME-CE interface. (From Whang, C-W. and Pawliszyn, J., *Anal. Commun.*, 35, 353, 1998. With permission.)

12.2.2.2 Liquid-Phase Microextraction

LPME is a microextraction technique that was developed from another microextraction technique called SDME. In SDME the extraction phase is a drop of a water immiscible solvent suspended from the tip of a microsyringe needle or a teflon rod in an aqueous sample [19]. After suspending, the drop is immersed in the aqueous sample and the extraction is performed. After a predetermined time, the drop is withdrawn into the needle and the extract is injected directly into a GC system or the drop is evaporated to dryness and suspended in an appropriate solvent. As it can be seen, this extraction technique is not suitable for systems in which the extraction is performed with a magnetic stir bar, since the drops can be lost during the extraction, leading to bias. Trying to improve this system, Pedersen-Bjergaard and Rasmussen [20] developed a technique where the drop (acceptor phase) was protected by a polypropylene hollow fiber (LPME).

In LPME, the analytes are extracted from the biological samples through an organic solvent immobilized in the pores of a porous hollow fiber and into a microliter volume of an acceptor phase placed inside the hollow fiber [20]. There are two possible ways to perform LPME (Figure 12.5). In the conventional configuration, two microsyringe needles are inserted through a septum and the two ends are connected with a piece of a hollow fiber (Figure 12.5A). The other one, called "rodlike," the piece of a hollow fiber is connected to only one microsyringe which is used for collection and injection of the acceptor phase (Figure 12.5B). In addition, there are two sampling modes that can be used with LPME: two-phase and three-phase (Figure 12.6). If the acceptor phase is the same solvent used to impregnate the pores, it is called a two phase mode. The final extract of two-phase LPME is an organic phase, which is compatible with GC analysis. In the three-phase LPME, the analyte is extracted from an aqueous solution (sample solution) through the organic solvent immobilized in the pores of the hollow fiber (organic phase) into another aqueous

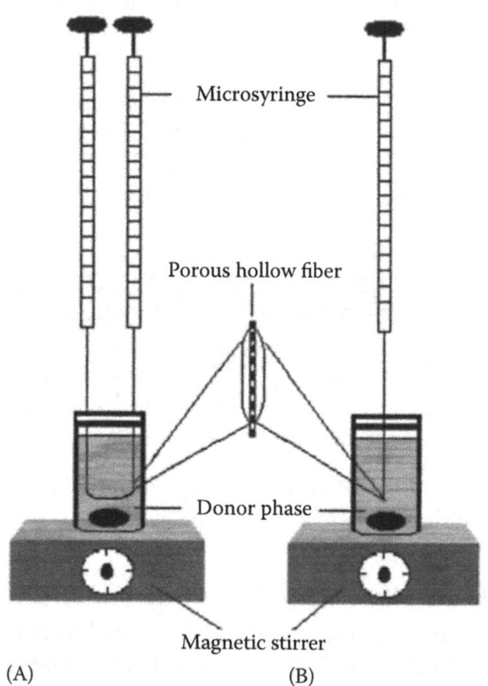

FIGURE 12.5 LPME device. Conventional device (A) and rodlike device (B).

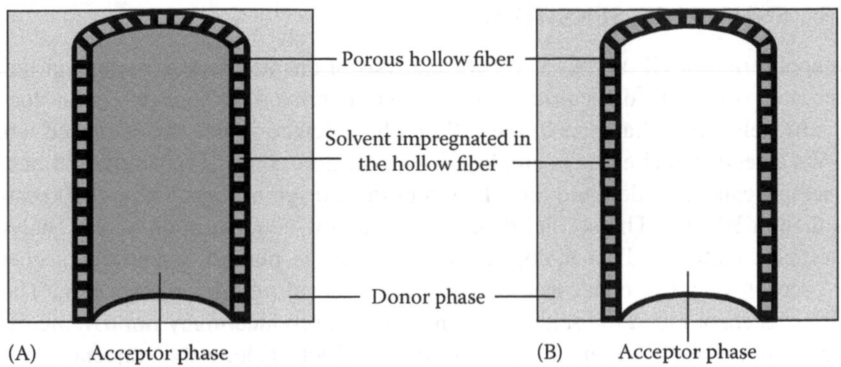

FIGURE 12.6 LPME modes. Two-phase mode (A) and three-phase mode (B).

phase (acceptor phase) present inside the lumen of the hollow fiber (Figure 12.6) [21]. Since the acceptor phase is an aqueous solution, it can be directly injected into the CE or HPLC system. In this mode, the pH of the acceptor phase is the opposite of the donor phase and the analytes in their neutral form migrate through the fiber into the acceptor phase. In the acceptor phase, the analytes are charged and they can not reach the donor phase. Therefore, they are concentrated.

When LPME is employed, some steps should be optimized to guarantee the maximum extraction efficiency (e.g., type of organic solvent, extraction time, salt addition, sample agitation, and compositions of donor and acceptor phase).

Despite the large number of publications employing LPME-CE (achiral), few papers have been published using LPME and chiral analysis by CE. Table 12.2 summarizes the stereoselective methods employing LPME and CE [22–25].

TABLE 12.2
Stereoselective Developed Methods Using LPME-CE

Analyte	Matrix	Acceptor Phase	LPME Mode	Solvent	Quantification Limit (ng/mL)	Reference
Mirtazapine (MTZ) and metabolites	Human urine	Acetic acid	Three-phase	Di-*n*-hexyl ether	62.5	[22]
Hydroxychloroquine (HCQ) and its metabolites	Human urine	HCl	Three-phase	1-Octanol	10.0 (HCQ) 21.0 (Metabolites)	[23]
Citalopram (CTP)/desmethylcitalopram (DCTP)	Human plasma	Phosphate buffer pH 7.7	Three-phase	Dodecyl acetate	4.8 (CTP) 10.0 (DCTP)	[24]
Mianserin	Human plasma	HCl	Three-phase	Di-*n*-hexyl ether	12.5	[25]

12.3 ENHANCED SENSITIVITY

The application of CE and CEC for the analysis of enantiomers in biological fluids encounters two main difficulties. The first one is related to poor detection due to the extremely small diameter of the cell in which detection must be achieved when UV-Vis detection technique is employed. Furthermore, only very small volumes of the sample can be loaded into the CE system in its original format (about 100 times less than in HPLC). The second difficulty is the low concentration of the analytes in complex matrices. In a biological matrix, such as plasma, enantiomers generally occur in very low concentrations, often at levels of ng/mL or lower [26]. These limitations encouraged the scientific community to produce many improvements in CE procedures aimed at enhancing sensitivity, which include the concentration of biological samples by using off-line or online concentration procedures, as well as the use of alternative detection systems.

12.3.1 Sensitivity Improvements by Sample Concentration Methods

The main techniques used for off-line sample concentration for chiral analysis were previously discussed in Section 12.2 of this chapter. Some of these procedures can be coupled directly to the CE system, in the same way as SPE is coupled to HPLC. Such a setup is promising as a way to automation but data are still scarce [27].

A different approach is to provide conditions for concentrating the analyte in the sample after it is loaded into the capillary, i.e., online sample concentration. In general these techniques are designed to compress the analytes in bands within the capillary, thereby increasing the volume of sample that can be injected without loss of efficiency [28]. The principle involved is to subject the analytes to the influence of two zones in the capillary, so that in one zone the analyte moves faster and in the other one it moves slower. Thus, the analytes accumulate on the limit of the two zones. There are many ways to obtain changes in the velocity of the analytes; generally they can be based on changes in field strength. In spite of many acronyms used by different authors, some techniques are very similar with only slight variation in either sample preparation methods or sample injection procedures.

12.3.1.1 Sample Stacking

Sample stacking is a rapid procedure that may be easily carried out with standard commercial instrumentation. With this technique it is possible not only to improve the separation efficiency, but also to increase the sensitivity between 10- and 1000-fold in comparison with the conventional sample injection [28–30].

Consider a sample dissolved in a low-conductivity solution, such as water, injected into the capillary and making contact with the background electrolyte (BGE). Under a high positive voltage, the electric field strength will be higher over the low-conductivity sample zone than over the BGE. Therefore, the velocity of the analyte will be high in the sample zone until it reaches the buffer interface. As the sample ions reach the high-concentration BGE they slow down and stack into a narrow zone (Figure 12.7). This online concentration procedure is termed field-amplified sample

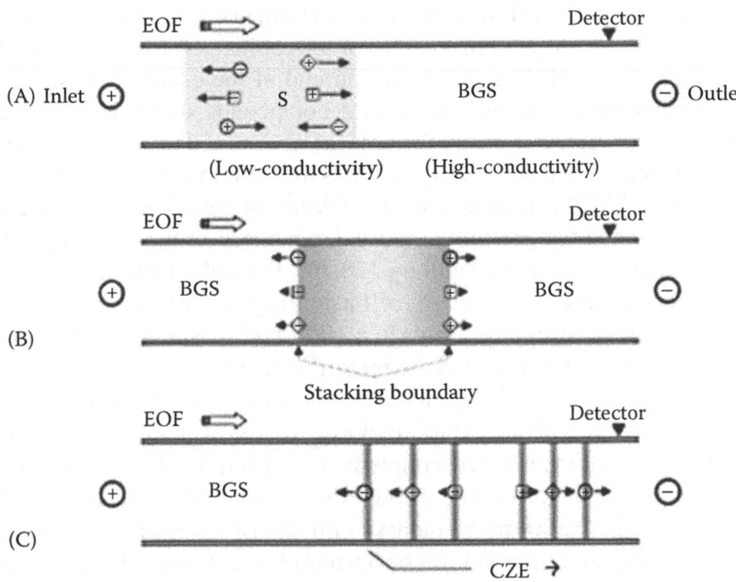

FIGURE 12.7 Schematic diagrams of the FASS model. (A) The capillary is conditioned with a BGS (a high-conductivity buffer), the sample, prepared in a low-conductivity matrix, is then injected to a certain length, and a high positive voltage is applied; (B) focusing of the analytes occurs near the boundaries between the sample zone and the BGS because of their mobility changes; (C) stacked analytes migrate and are separated by the CZE mode. (From Lin, C.-H. and Kaneta, T., *Electrophoresis*, 25, 4058, 2004. With permission.)

stacking (FASS). Very often a tenfold difference in ionic strength between the BGE and sample provides satisfactory stacking efficiency [30]. The need for a large difference in ionic strength between sample and BGE is a limitation when samples of high salt concentration (near 100 mM) are considered.

Enantioselective analysis of drugs and metabolites in biological samples carried out by our group employing this stacking condition, combined with other techniques of sample preconcentration, provided limits of quantification (LOQs) similar to those obtained by HPLC-UV. Therefore, it was possible to employ the methods for kinetic disposition studies [23,31,32].

A variant of FASS is the stacking performed in micellar electrokinetic chromatography (MEKC) for enhancing the detection of neutral analytes. In MEKC-FASS the sample is injected hydrodynamically in a low-conductivity micellar solution into the capillary containing a high-conductivity micellar BGE. The analytes may be stacked in either normal or reverse polarity mode [28,30].

Another simple stacking method consists in the addition of two sample volumes of organic modifiers, such as acetonitrile [29]. In this case, the addition of acetonitrile could also promote protein precipitation, acting as an off-line sample preparation procedure. However, this method of sample preparation does not allow the concentration of the analyte so it is not suitable for the quantification of analytes at low plasmatic concentration levels.

Large-volume sample stacking (LVSS) is performed by dissolving the sample in water and hydrodynamically filling 1/3–1/2 of the capillary with the sample. Reverse polarity is applied with BGE at the detection end of the capillary. Under reversed polarity, sample solvent, cations, and neutral compounds should exit the capillary into the waste buffer reservoir before the polarity is returned to normal. Applications of this method include analysis of drugs with 2- to 100-fold enhancement in sensitivity. A variation on LVSS is termed double stacking, where all potentials are applied in normal polarity and the pressure at the detection side of the capillary is used to back the sample plug out of the capillary [28,30]. Recently, Denola et al. [33] used LVSS for the enantioselective analysis of basic drugs in serum. In comparison to conventional injection, several hundreds to a 1000-fold sensitivity enhancement was achieved, depending on the drug and the injection mode.

One method for dealing with high ionic strength samples is to neutralize the high conductivity matrix with pH-mediated stacking. This is a technique in which FASS is triggered by titrating the injected sample zone to neutral pH, thus creating a low-conductivity region. This procedure allows the sample cations to migrate quickly through the titrated zone to the boundary with the BGE, where they stack into a narrow band [28,29]. A detection limit of 0.6 ng/mL was achieved in the analysis of isoproterenol enantiomers in dialysate using pH-mediated stacking and amperometric detection [34].

12.3.1.2 Isotachophoresis (ITP)

ITP is a particularly useful method of preconcentration as it can also function as a purification method, making it suitable for the analysis of high-ionic-strength samples including blood, plasma, and urine [28]. In this method, the sample is injected between a leading and a terminating electrolyte, as illustrated in Figure 12.8. The leading electrolyte contains a high concentration of a high-mobility ion with the same charge as that of the analytes of interest and the terminating electrolyte contains an ion of lower mobility than that of the analytes. When high voltage is applied, a potential gradient is established over the electrolyte and sample zones, with the field strength over each zone being inversely proportional to the mobility of the ion in that zone. Thus, high field strength will fall across a zone with a low-mobility ion and low field strength over a zone with a high-mobility ion. If an analyte moves too slowly and enters a slower zone, then the ion will accelerate until it reaches its own zone due to the higher field strength. At equilibrium, each analyte moves at a discrete band according to its own mobility, with high-mobility ions migrating earlier than low-mobility ions. Isotachophoretic preconcentration can be performed in a coupled-capillary or single-capillary approach [28].

Coupled-capillary ITP allows sample volumes greater than the total volume of the CE capillary. Limits of detection at ng/mL concentration levels have been reported with this technique for enantioselective analysis of tryptophan in urine sample [35].

12.3.1.3 Sweeping

A technique for on-column preconcentration of samples of nonpolar molecules is termed "sweeping" and it is based on the ability of analytes to partition with the

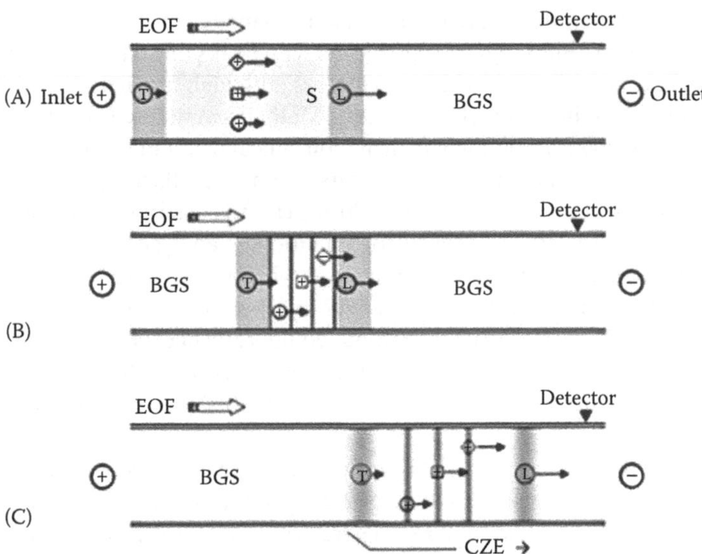

FIGURE 12.8 Schematic diagrams of a transient-ITP model for cations. (A) The capillary is conditioned with a BGS, the leading electrolyte, sample solution, and terminating electrolyte are then injected in turn, and a high positive voltage is applied; (B) concentration of the analytes occurs between the leading and the terminating ions during transient-ITP migration; (C) the concentrated analyte zones are separated by the CZE mode. (From Lin, C.-H. and Kaneta, T., *Electrophoresis*, 25, 4058, 2004. With permission.)

pseudostationary phase in MEKC and to produce a focusing effect. Samples are injected onto the capillary in a buffer solution with a conductivity similar to the one used in the BGE, but in the absence of a pseudostationary phase (micelles). The capillary inlet is then placed into an anionic micellar BGE solution and the separation is performed in the reverse polarity mode. The BGE is kept at low pH to suppress the electroosmotic flow (EOF), allowing the anionic micelles to migrate toward the detector. As the micelles migrate they "sweep" the neutral analytes along. The effect is dependent on a uniform electric field and the absence of micelles in the sample solution. The effectiveness of this sample preconcentration has been shown to be dependent on the affinity of the analytes for the pseudostationary phase. This technique allows to obtain up to a 1000-fold lower limits of detection (LODs) [27,28].

12.3.2 Improvements in Detection Systems

The sensitivity of CE when on-column UV-Vis detection is used, is limited by the short optical path, which is the capillary inner diameter itself. One way to increase UV-Vis detection sensitivity is to use capillaries with extended light path, which should increase the signal-to-noise ratio simply as a result of Beer's law. The capillary can be angled into a "z" shape (z-cell) or extended (bubble-cell), and both designs are commercially available. Sensitivity improvement is in the range of —two to threefold. However, the localized increase in diameter of the capillary results in decreased resolution; this may cause problems with peaks migrating closely together, as it is almost always the case with enantiomers, although it is not difficult to adjust the separation factor in chiral CE [26,27,36]. Methods were described for chiral and achiral drugs and their metabolites in urine, plasma and serum samples, but they are much less numerous than those utilizing conventional capillaries [2].

Alternative detection systems for on column UV-Vis absorption can provide very low limits of quantifications (LOQs) for enantiomer analysis in biological fluids. Further improvement in sensitivity is obtained when using these systems in combination with online sample concentration. The most sensitive and selective detectors introduced in the CE field include laser-induced fluorescence, electrochemical detection and mass spectrometry (MS). In each of these cases, modification to the basic CE instrument is required; they are still research tools rather than routine detection techniques.

Laser induced fluorescence (LIF) is a very attractive detection system for biological analysis, but it has some limitations for drug analysis since it is only applicable to molecules containing fluorophores. There are few chiral drugs showing native fluorescence and allowing direct LIF detection [37,38]. The sensitivity of fluorescence detection is proportional to the intensity of the induced light emitted under the exciting laser. In a LIF detector, the high intensity of the incident light from a laser is accurately focused by a lens onto the inner diameter of the capillary. Thus, for a particular analyte, the sensitivity and selectivity is often higher than the one obtained with UV-Vis absorption detector. A variety of lasers are available, but the most useful lasers for LIF detection are the argon–ion, helium–cadmium, helium–neon, and diode-lasers [39]. An alternative to direct detection is derivatization of the enantiomers with fluorescent tags even if derivatization implies loss in selectivity. Several methods have been described with this procedure for the enantioselective analysis of drugs and their metabolites in plasma and urine samples with acceptable limits of quantification [26,27].

Electrochemical detection, in the amperometric mode, has been shown to allow a highly selective and sensitive detection of many biologically important molecules, including catecholamines and indoleamines. One of the main advantages of electrochemical detection is its applicability to many electroactive species without prior derivatization. The use of amperometric detection (with carbon or metallic electrode of special geometries) yields in some cases exceptionally low detection limits up to 10^{-9} M. An example in chiral analysis is the enantioseparation of neurotransmitters, such as norepinephrine, epinephrine, and isoproterenol employing end-column micro-fabricated interdigitated electrodes [40]. This detector system is very promising in microchip CE due to its compatibility with micro-fabricated techniques [41].

Mass spectrometry is a universal and selective detection system that also provides mass and structural information. The coupling of CE to MS combines the advantages of both techniques and it is considered a very promising technique in the determination and characterization of chiral drugs in complex biological samples. However, CE–MS still represents an area where new developments in instrumentation (interfaces, ionization sources, mass analyzers) are needed and are continually being produced [42]. The difficulty involved in coupling CE to MS is that CE operates with aqueous buffers whereas the MS instrument operates at high vacuum. Thus, CE–MS requires reliable interfaces that effectively transfer the analytes from solution to the vapor phase without thermal degradation and without contamination. Currently, electrospray ionization (ESI) serves as the most common interface between CE and MS, as it can produce ions with high sensitivity and selectivity [27,42]. CE–MS for chiral analysis faces additional difficulties due to the presence of chiral selectors in the ion-source of the MS [26,42]. In order to avoid the introduction of chiral selectors into the MS, partial-filling and countercurrent migration techniques have been used. In the partial-filling technique, a zone of the capillary where the enantiomeric separation takes place is filled with buffer containing the chiral selector, while the zone close to the MS detector, which the separated enantiomers cross before being introduced in the detector, is filled with buffer without chiral selector. The countercurrent migration principle is based on the control of the EOF to avoid the chiral selector from reaching the MS detector [26,27,41]. Intense effort in developing new chiral selectors compatible with MS is also taking place, promising a bright future for chiral CE–MS. Another approach to solve the problem is the use of CEC with chiral stationary phase. A CEC capillary packed with (3*R*,4*S*)-Whelk-01 chiral stationary phase was used by Zheng and Shamsi [43] for the enantioselective assay of warfarin in human plasma. Further information concerning the coupling of CE to MS detection can be found in Chapter 13.

LODs in the range of approximately 10^{-6} to 10^{-9} M were obtained by CE–ESI–MS for the analyses of chiral compounds in biological samples [27]. In fact, CE coupled to ESI–MS provides enough sensitivity and selectivity to control chiral impurity, to check *in vivo* racemization processes as well as to perform pharmacokinetic studies [42]. Due to its ability of peak identification, its ability of performing peak purity tests and overcoming matrix effects, CE–MS coupling has become a potential powerful tool for the elucidation of chiral metabolic studies. Difficulties with micro-preparative fraction collection in CE for their further offline identification make direct chiral CE–MS even more significant for metabolic studies [26].

12.4 METHOD VALIDATION

The validation of a bioanalytical method includes all procedures required to demonstrate that it does reliably accomplish its goal. We deal here with quantifying, with adequate precision and accuracy, the concentration of one analyte or a series of analytes (drugs and their metabolites) in a biological matrix. The validation procedures are carried out according to generally accepted guidelines from institutional bodies such as the International Conference on Harmonization (ICH) [44] or the Food and Drug Administration (FDA) [45]. Classical validation parameters include linearity, precision, accuracy, selectivity, LODs and LOQs, recovery, and stability. Stability studies are especially important in enantioselective analysis because of the possibility of stereoconversion and racemization during sample storage and/or processing [46].

A detailed step-by-step guide to analytical method validation is not intended here since several reviews have covered this subject [47–51]. A discussion of characteristics inherent to CE techniques is mainly focused instead, and some suggestions are given for better results of precision and accuracy in CE routine analysis in biological fluids.

When CE was introduced, CE equipments had poor precision and sensitivity [52–54]. Improvements in the CE equipment and in the optical detection systems introduced along the last decade made it possible to obtain acceptable quantitative results. However, the deficiency in CE precision is not only due to limitations in equipment but also to operational variables [53]. Generally, precision depends on four main factors: buffer composition, injection mode, capillary properties, and temperature. These factors interact with each other and affect migration times and size of the peaks. Stringent attention should be given to all these sources of variation so that a set of measures is adopted to be consistently used in the validation procedure.

Buffers must be rigorously prepared daily with high-purity water and HPLC-grade reagents and solvents, and the choice of the BGE should favor low conductivity [53]. The pH adjustment and addition of chiral selectors in the running buffer must be done with analytical care as small changes of buffer composition or pH may result in large changes of migration times. The running buffer containing the chiral selector must be always freshly prepared, filtered, and degassed. Air bubbles in the running buffer cause short-term EOF and current changes, leading to imprecise results of peak areas. In the case of organic solvent addition to the running buffer, the possibility of losses by evaporation should be kept in mind [52]. On the other hand, buffer additives (e.g., hexadodecyltrimethylammonium bromide) may contribute to precision not only by reducing the adsorption of particulate material and proteins on the capillary surface, but also by diminishing analyte-wall interactions. Furthermore, they stabilize the EOF and lead to more repeatable migration times [52].

Buffer solutions change during CE analysis due to electrolysis, resulting in a pH gradient across the capillary that affects separation efficiency and migration times. To minimize this effect, previous attention is necessary with regard to buffering capacity, buffer concentration, ionic strength, pH, volume of the reservoirs, temperature, generated current, and total run time. The maximum number of analyses that can be performed with a unique vial of running buffer without significant depletion effects should be determined beforehand [52].

It is generally recognized that capillary preconditioning is important for high separation efficiency as well as high run-to-run repeatability [52]. In the analysis of biological samples, the cleanup step by itself is not enough to eliminate proteins and fats that could cause chemical instability of the capillary wall. It is therefore common practice, when analyzing body fluids, to precondition the capillary before each run in order to achieve reproducible migration time. Different washing steps with 0.5–0.1 M solution of sodium hydroxide, water, and buffer should be performed to find the optimum conditions for rinsing the capillary. The duration of each washing should also be noted. The vial of buffer solution for rinsing and the vial for the running buffer must be used separately so as to avoid changes in the latter [53]. When lack of reproducibility is observed, the running buffer reservoir should be replaced by a freshly prepared one.

The number of runs supported by the same capillary also affects migration time reproducibility, because the changes in the capillary's inner surface produced by the repeated washings result in variation in the EOF and in tailing peaks. It is possible to improve the capillary tube's chemical and mechanical stability by coating its inner surface with polyvinyl alcohol, methylcellulose, and polyacrylamide [2]. It is even possible to employ capillaries made of glass or teflon; these capillaries release Joule heat more efficiently than the fused silica ones [53].

The variability in the injection procedure in CE has a strong effect on the final precision as the volumes involved are in the nanoliter range. The robustness of the injection system is still not satisfactory, but temperature itself largely affects the viscosity of the liquids and, therefore, the amount of sample that is being handled [52]. The hydrodynamic injection mode (short injection time at high pressure) is more precise and robust than the electrokinetic mode and it is the preferred one by the analysts. The hydrodynamic injection mode is especially used when biological matrices such as plasma or urine are considered, because they vary in composition and conductivity.

Temperature crucially affects numerous variables pertaining to the analytes (net charge, stability, protein denaturizing, and configuration changes), to the running buffer (viscosity, conductivity, pH, air, and vapor bubble formation), and to the chemical equilibrium (ionization of the capillary surface, EOF, micelle partitioning, kinetics between the transient complex of analyte-chiral selector). If temperature is not adequately controlled, the migration times are less reproducible, analyte identification becomes more difficult and quantification becomes less accurate because of peak size variation. Room temperature must be kept stable and should not differ by more than 2°C from the temperature of the CE equipment itself. This is particularly important when CE is hyphenated with detection methods, such as a MS detector, where the capillary leads out of the CE instrument. Temperature control is also provided for the autosampler tray so that the samples are protected from the influence of heat sources inherent to the equipment such as the capillary itself and the detector lamp [52].

In CE, calibration is made in the same manner used for chromatography, by means of internal or external standardization. It is highly advisable to use internal standards (IS) whenever possible. A good choice of an IS favors the correction of errors due to variation of injection volume, voltage, or EOF; furthermore, it helps to improve injection repeatability [52].

Either the area or the height of a peak can be used for quantitation. When peak areas are used, further improvement of precision is possible by means of area normalization, which corrects the influence of drift of migration times; the corrected area is the ratio of the peak area and the corresponding migration time. Peak height sometimes gives better precision for sharp peaks at low concentrations because the determination of peak height is less subjected to integration errors than the determination of areas [52,55]. For each new procedure the analyst should evaluate which of the two values is more adequate.

12.5 CONCLUDING REMARKS

Due to the high resolution power, electromigration techniques have been increasingly used for the stereoselective analysis of drugs and metabolites in biological samples. Reliable methods have been obtained after elaborated validation procedures. Precision requirements have been attained by using well-designed capillary washing procedures before each run, by using an internal standard and/or by taking careful control of analytical conditions, including the prevention of buffer depletion and temperature control. The sensitivity required for the analysis of trace amounts of chiral drugs and their metabolites in biological samples has been enhanced by using more sensitive detection systems, by stacking procedures and/or by sample preconcentration. Finally, we can conclude that electromigration techniques might be powerful alternatives to the well-established HPLC-based methods for the enantioselective analysis of chiral drugs and their metabolites in biological samples, with the advantages of low costs, low-waste production, short analysis time, and easy method development.

ACKNOWLEDGMENTS

The authors wish to express special gratitude to Professor Dr. Heni Sauaia for his important encouragement and valuable suggestions and also to Dr. Franciane M. de Oliveira for her assistance in the revision which provided essential help in the preparation of this chapter. The authors are grateful to FAPESP and CNPq.

REFERENCES

1. Brocks, DR. 2006. Drug disposition in three dimensions: An update on stereoselectivity in pharmacokinetics. *Biopharm Drug Dispos* 27: 387–406.
2. Bonato, PS. 2003. Recent advance in the determination of enantiomeric drugs and their metabolites in biological fluids by capillary electrophoresis-mediated microanalysis. *Electrophoresis* 24: 4078–4094.
3. Smith, RM. 2003. Before the injection—modern methods of sample preparation for separation techniques. *J Chromatogr A* 1000: 3–27.
4. Queiroz, SCN, Collins, CH, Jardim, ICSF. 2001. Métodos de extração e/ou concentração de compostos encontrados em fluidos biológicos para posterior determinação cromatográfica. *Química Nova* 24: 68–76.
5. Kataoka, H. 2003. New trends in sample preparation for clinical and pharmaceutical analysis. *Trends Anal Chem* 22: 232–244.
6. Ridgway, K, Lalljie, SPD, Smith, RM. 2007. Sample preparation techniques for the determination of trace residues and contaminants in foods. *J Chromatogr A* 1153: 36–53.
7. Schaller, D, Hilder, EF, Haddad, PR. 2006. Separation of antidepressants by capillary electrophoresis with in-line solid-phase extraction using a novel monolithic adsorbent. *Anal Chim Acta* 556: 104–111.
8. Puig, P, Tempels, FWA, Borrull, F, Calull, M, Aguilar, C, Somsen, GW, De Jong, GJ. 2007. On-line coupling of solid-phase extraction and capillary electrophoresis for the determination of cefoperazone and ceftiofur in plasma. *J Chromatogr B* 856: 365–370.

9. Schappler, J, Guillarme, D, Prat, J, Veuthey, J-L, Rudaz, S. 2006. Enhanced method performances for conventional and chiral CE-ESI/MS analyses in plasma. *Electrophoresis* 27: 1537–1546.
10. Lord, H, Pawliszyn, J. 2000. Evolution of solid-phase microextraction technology. *J Chromatogr A* 885: 153–193.
11. Mani, V. 1999. Properties of commercial SPME coatings. In *Applications of Solid-Phase Microextraction*. ed. J. Pawliszyn, Cambridge: The Royal Society of Chemistry, pp. 57–72.
12. Tamayo, FG, Turiel, E, Martín-Esteban, A. 2007. Molecularly imprinted polymers for solid-phase extraction and solid-phase microextraction: Recent developments and future trends. *J Chromatogr A* 1152: 32–40.
13. Dietz, C, Sanz, J, Cámara, C. 2006. Recent developments in solid-phase microextraction coatings and related techniques. *J Chromatogr A* 1103: 183–192.
14. Fang, H, Liu, M, Zeng, Z. 2006. Solid-phase microextraction coupled with capillary electrophoresis to determine ephedrine derivatives in water and urine using a sol-gel derived butyl methacrylate/silicone fiber. *Talanta* 68: 979–986.
15. Zhou, X, Li, X, Zeng, Z. 2006. Solid-phase microextraction coupled with capillary electrophoresis for the determination of propranolol enantiomers in urine using a sol-gel derived calix[4]arene fiber. *J Chromatogr A* 1104: 359–365.
16. Jinno, K, Kawazoe, M, Saito, Y, Takeichi, T, Hayashida, M. 2001. Sample preparation with fiber-in-tube solid-phase microextraction for capillary electrophoretic separation of tricyclic antidepressant drugs in human urine. *Electrophoresis* 22: 3785–3790.
17. Lin, B, Zheng, M-M, Ng, S-C, Feng, Y-Q. 2007. Development of in-tube solid-phase microextraction coupled to pressure-assisted CEC and its application to the analysis of propanolol enantiomers in human urine. *Electrophoresis* 28: 2771–2780.
18. Whang, C-W, Pawliszyn, J. 1998. Solid phase microextraction coupled to capillary electrophoresis. *Anal Commun* 35: 353–356.
19. Psillakis, E, Kalogerakis, N. 2002. Developments in single-drop microextraction. *Trends Anal Chem* 21: 53–63.
20. Pedersen-Bjergaard, S, Rasmussen, KE. 1999. Liquid-liquid-liquid microextraction for sample preparation of biological fluids prior to capillary electrophoresis. *Anal Chem* 71: 2650–2656.
21. Psillakis, E, Kalogerakis, N. 2003. Developments in liquid-phase microextraction. *Trends Anal Chem* 22: 565–574.
22. De Santana, FJM, Lanchote, VL, Bonato, PS. 2008. Capillary electrophoretic chiral determination of mirtazapine and its main metabolites in human urine after enzymatic hydrolysis. *Electrophoresis* 29: 3924–3932.
23. De Oliveira, ARM, Cardoso, CD, Bonato, PS. 2007. Stereoselective determination of hydroxychloroquine and its metabolites in human urine by liquid-phase microextraction and CE. *Electrophoresis* 28: 1081–1091.
24. Andersen, S, Halvorsen, TG, Pedersen-Bjergaard, S, Rasmussen, KE, Tanum, L, Refsum, H. 2003. Stereospecific determination of citalopram and desmethylcitalopram by capillary electrophoresis and liquid-phase microextraction. *J Pharm Biomed Anal* 33: 263–273.
25. Andersen, S, Halvorsen, TG, Pedersen-Bjergaard, S, Rasmussen, KE. 2002. Liquid-phase microextraction combined with capillary electrophoresis, a promising tool for the determination of chiral drugs in biological matrices. *J Chromatogr A* 963: 303–312.
26. Blaschke, G, Chankvetadze, B. 2000. Enantiomer separation of drugs by capillary electromigration techniques. *J Chromatogr A* 875: 3–25.
27. García-Ruiz, C, Marina, ML. 2006. Sensitive chiral analysis by capillary electrophoresis. *Electrophoresis* 27: 195–212.
28. Osbourn, DM, Weiss, DJ, Lunte, CE. 2000. On-line preconcentration methods for capillary electrophoresis. *Electrophoresis* 21: 2768–2779.

29. Sentellas, S, Puignou, L, Galceran, MT. 2002. Capillary electrophoresis with on-line enrichment for the analysis of biological samples. *J Sep Sci* 25: 975–987.
30. Lin, C-H, Kaneta, T. 2004. On-line sample concentration techniques in capillary electrophoresis: Velocity gradient techniques and sample concentration techniques for biomolecules. *Electrophoresis* 25: 4058–4073.
31. Jabor, VAP, Lanchote, VL, Bonato, PS. 2001. Simultaneous determination of disopyramide and mono-*N*-dealkyldisopyramide enantiomers in human plasma by capillary electrophoresis. *Electrophoresis* 22: 1406–1412.
32. Jabor, VAP, Lanchote, VL, Bonato, PS. 2002. Enantioselective analysis of ibuprofen in human plasma by anionic cyclodextrin-modified electrokinetic chromatography. *Electrophoresis* 23: 3041–3047.
33. Denola, NL, Quiming, NS, Saito, Y, Jinno, K. 2007. Simultaneous enantioseparation and sensitivity enhancement of basic drugs using large-volume sample stacking. *Electrophoresis* 28: 3542–3552.
34. Hadwiger, ME, Torchia, SR, Park, S, Biggin, ME, Lunte, CE. 1996. Optimization of the separation and detection of the enantiomers of isoproterenol in microdialysis samples by cyclodextrin-modified capillary electrophoresis using electrochemical detection. *J Chromatogr B* 681: 241–249.
35. Danková, M, Kaniansky, D, Fanali, S, Iványi, F. 1999. Capillary zone electrophoresis separations of enantiomers present in complex ionic matrices with on-line isotachophoretic sample pretreatment. *J Chromatogr A* 838: 31–43.
36. Hempel, G. 2000. Strategies to improve the sensitivity in capillary electrophoresis for the analysis of drugs in biological fluids. *Electrophoresis* 21: 691–698.
37. Horstkotter, C, Blaschke, G. 2001. Stereoselective determination of ofloxacin and its metabolites in human urine by capillary electrophoresis using laser-induced fluorescence detection. *J Chromatogr B* 754: 169–178.
38. Soetebeer, UB, Schierenberg, M-O, Schulz, H, Andresen, P, Blaschke, G. 2001. Direct chiral assay of tramadol and detection of the phase II metabolite *O*-demethyl tramadol glucuronide in human urine using capillary electrophoresis with laser-induced native fluorescence detection. *J Chromatogr B* 765: 3–13.
39. Weinberg, R. 2000. *Practical Capillary Electrophoresis*. San Diego: Academic Press.
40. Male, KB, Luong, JHT. 2003. Chiral analysis of neurotransmitters using cyclodextrin-modified capillary electrophoresis equipped with microfabricated interdigitated electrodes. *J Chromatogr A* 1003: 167–178.
41. Altria, KD, Elder, D. 2004. Overview of the status and applications of capillary electrophoresis to the analysis of small molecules. *J Chromatogr A* 1023: 1–14.
42. Shamsi, SA, Miller, BE. 2004. Capillary electrophoresis-mass spectrometry: Recent advances to the analysis of small achiral and chiral solutes. *Electrophoresis* 25: 3927–3961.
43. Zheng, J, Shamsi, SA. 2003. Combination of chiral capillary electrochromatography with electrospray ionization mass spectrometry: Method development and assay of warfarin enantiomers in human plasma. *Anal Chem* 75: 6295–6350.
44. ICH Guidelines. 1994. Validation of analytical procedures, ICH Steering Committee. http://www.ich.org/cache/compo/363-272-1.html (accessed March 24, 2008).
45. Guidance for industry. Bioanalytical method validation. 2001. http://www.fda.gov/cder/guidance/index.htm (accessed March 24, 2008).
46. Ducharme, J, Fernandez, C, Gimenez, F, Farinotti, R. 1996. Critical issues in chiral drug analysis in biological fluids by high-performance liquid chromatography. *J Chromatogr B* 686: 65–75.
47. Causon, R. 1997. Validation of chromatographic methods in biomedical analysis. Viewpoint and discussion. *J Chromatogr B* 689: 175–180.

48. Hartmann, C, Smeyers-Verbeke, J, Massart, DL, MacDowall, RD. 1998. Validation of bioanalytical chromatographic methods. *J Pharm Biomed Anal* 17: 193–218.
49. Hubert, Ph, Chiap, P, Crommen, J, Boulanger, B, Chapuzet, E, Mercier, N, Bervoas-Martin, S, Chevalier, P, Grandjean, D, Lagorce, P, Lallier, M, Laparra, MC, Laurentie, M, Nivet, JC. 1999. The SFSTP guide on the validation of chromatographic methods for drug bioanalysis: From the Washington Conference to the laboratory. *Anal Chim Acta* 391: 135–148.
50. Bressolle, F, Bromet-Petit, M, Audran, M. 1996. Validation of liquid chromatographic and gas chromatographic methods. Applications to pharmacokinetics. *J Chromatogr B* 686: 3–10.
51. González, AG, Herrador, MÁ. 2007. A practical guide to analytical method validation, including measurement uncertainly and accuracy profiles. *Trends Anal Chem* 26: 227–238.
52. Mayer, BX. 2001. How to increase precision in capillary electrophoresis. *J Chromatogr A* 907: 21–37.
53. Ali, I, Aboul-Enein, H, Gupta, VK. 2006. Precision in capillary electrophoresis. *Anal Letters* 39: 2345–2357.
54. Thi, TD, Pomponio, R, Gotti, R, Saevels, J, Hove, BV, Ael, WV, Matthijs, N, Vander Heyden, Y, Djan'geing'a, RM, Chiap, P, Hubert, P, Crommen, J, Fabre, H, Dehouck, P, Hoogmartens, J, Van Schepdael, A. 2006. Precision study on capillary electrophoresis methods for metacycline. *Electrophoresis* 27: 2317–2329.
55. Wätzig, H. 1995. Appropriate calibration functions for capillary electrophoresis. I. Precision and sensitivity using peak areas and heights. *J Chromatogr A* 700: 1–7.

13 Chiral CE–MS

Serge Rudaz, Jean-Luc Veuthey, and Julie Schappler

CONTENTS

13.1 Introduction 363
13.2 CE–MS Coupling and Instrumentation 364
13.2.1 Interfaces 365
13.2.1.1 Sheath–Flow Interfaces 365
13.2.1.2 Sheathless Interfaces 366
13.2.2 Ionization Sources 367
13.2.2.1 Atmospheric Pressure Ionization Sources 367
13.2.2.2 Other Ionization Sources 369
13.2.3 Analyzers 369
13.3 Chiral CE 370
13.4 Chiral CE–MS 371
13.4.1 Partial Filling Technique 373
13.4.2 Partial Filling Countercurrent Technique 374
13.4.3 Crown Ethers, Chiral Micellar Electrokinetic Chromatography, and Chiral Nonaqueous Electrophoresis with MS Detection 377
13.5 Requirements for Biological Sample Quantitation with Chiral CE–MS 378
13.6 Discussion 382
List of Abbreviations 382
References 383

13.1 INTRODUCTION

Due to the simplicity of the online configuration, UV-Vis spectrophotometry is the most widely used detection technique in capillary electrophoresis (CE). However, the sensitivity of the CE–UV coupling is relatively low (i.e., in the μg/mL range) due to the optical pathlength afforded by the small internal diameter of the capillaries. Relatively high analyte concentrations are thus required for UV detection, which is often unsuitable for numerous applications. Consequently, the major evolutions in recent CE-chiral methodologies consist of improving the detection limits, and, in particular, extending the detection range of analytes.

Moreover, many pharmaceutical and biological substances do not present sufficient chromophore moieties, and a derivatization procedure can be mandatory for their optical detection. This also occurs when fluorescence, and especially laser-induced fluorescence (LIF), is employed. Concerning LIF, only a limited number of substances are fluorescent at the available laser wavelength. The required derivatization procedure often compromises both the fast analysis time and the small sample volume capabilities naturally afforded by CE [1,2]. Other detection techniques for CE include electrochemical detection, but the latter is limited to electroactive substances [3]. Furthermore, the interfacing with CE is not trivial since it requires a decoupling of the electrical fields and a tedious alignment of the electrode with the capillary. It has to be noted that with all these detection methods, the peak identity is generally assessed using migration time, which is often insufficient to unequivocally identify compounds of interest. This point is particularly important with complex samples (e.g., biological) that may contain unknown analytes, and/or in the case of nonrobust analytical methods [4,5].

In this context, the use of mass spectrometry (MS), which possesses the ability to identify analytes, is of utmost interest. This selective and highly sensitive detection mode, routinely used in gas chromatography (GC) and liquid chromatography (LC), is also compatible with CE and, therefore, provides a powerful combination for performing rapid, efficient, and sensitive analysis. In this chapter, particular attention is paid to the instrumental aspects necessary for the successful hyphenation of CE with MS, regarding interfaces and ionization sources, particularly with the constraints afforded by the presence of chiral selectors in the background electrolyte (BGE).

13.2 CE–MS COUPLING AND INSTRUMENTATION

Due to its high sensitivity and specificity, MS is the detector of choice in microseparation techniques such as CE. This selective detector provides additional capacities by allowing high speed analysis, giving information about the mass, and potentially, the structure of the separated compounds. This information is highly desirable for an unambiguous identification and confirmation of the components in complex mixtures [6]. Even if the hyphenation of MS with chromatography is considered as the method of choice for numerous analytical purposes, CE coupled to MS has recently gained importance with the emergence of new approaches such as the "omics" [7,8]. Each methodology possesses distinct advantages, and the resulting information can help in obtaining a better characterization of the tested compounds (e.g., lipidomics, proteomics, glycomics, metabolomics, and so on). Hence, complementary approaches to the well established GC–MS and LC–MS techniques, such as CE–MS, have their place in numerous research fields.

This chapter does not intend to cover all of the recent technical developments in achiral CE–MS. Such dedicated applications are summarized in recent reviews and can be consulted for a more systematic coverage of the field [9]. Recently, CE–MS applications were successfully reported in numerous fields with or without commercially available setups [9–11]. Improvements in homebuilt instruments continue to occur, driving new strategies for technological developments in instrumentation. Homebuilt approaches still represent interesting features for dedicated problems and offer numerous advantages in terms of interface modifications, applied voltage, capillary dimension, and reduced cost. However, for routine applications, commercially available setups are preferred, and several manufacturers propose user-ready CE–MS to be employed in the field of analytical chemistry.

Concerning the coupling of CE with MS, it should be noted that the majority of the developed interfaces were initially developed for LC hyphenation but adapted to the constraints of CE analysis [12,13]. In fact, additional difficulties in the coupling of both techniques should be considered.

1. The typical flow rates in the CE capillary are below 100 nL/min, even with an important electroosmotic flow (EOF). This is not compatible with conventional LC–MS interfaces. Therefore, either a miniaturized electrospray system should be implemented or a makeup liquid should be used.
2. The electrical connection at the interface side of the separation capillary has to be achieved with the end of the capillary directly inserted into the MS interface, without direct connection with the BGE.
3. The analytes must be released from the solvent molecules and brought into the gaseous phase. However, selectivity modifier additives, often present in the BGE as nonvolatile constituents, may be detrimental to the MS performance, and particularly to the ionization process, due to ion source or analyzer contamination. This point will be discussed in detail as it specifically concerns the dynamic mode in chiral separations.

13.2.1 Interfaces

Unlike on-capillary detection with conventional spectroscopic detectors, MS hyphenation is achieved in an open mode. The end of the separation capillary is removed from the outlet vial and positioned in front of the MS inlet. This interfacing must be performed without peak broadening or efficiency alteration and by ensuring electrical continuity in order to maintain the electrophoretic and ionization electrical fields. Therefore, the interface between CE and MS is one of the keys to the success of this coupling. Two main configurations are distinguished for adequate CE–MS interfacing. The first one is based on the addition of a makeup liquid at the interface (sheath-flow approach), while the second one consists of a miniaturized interface without the need of an additional liquid (sheathless approach). The sheath-flow interface still represents the most common approach when hyphenating CE with MS due to its instrumental simplicity, versatility, and robustness. According to its improved sensitivity, a growing interest emerges in the development and use of sheathless interfaces. With both interfaces, the electrophoretic flow rate, as well as the electrical connection issues are resolved.

13.2.1.1 Sheath-Flow Interfaces

Because liquid flow rates in CE are in the nL/min range, the evaporation process remains quite easy. However, at the tip of the CE capillary, droplets can be formed leading to an unstable flow to the MS. Makeup flows are thus used to overcome this drawback. While it is relatively easy to implement a sheath liquid system, optimizing the operational parameters (e.g., sheath liquid flow rate and composition, capillary tip position, etc.) in order to obtain a stable and reproducible spray can be quite complex [14].

Two types of sheath-flow interfaces are used: the coaxial sheath-liquid interface and the liquid-junction interface. The difference is based on the makeup liquid integration: liquid addition takes place proximal to the MS orifice in the coaxial configuration, while the liquid junction geometry places the makeup liquid distal to the sprayer tip.

With the coaxial configuration, the sheath-flow is mixed with the BGE at the end of the capillary [15,16]. In this arrangement, the CE capillary is located in the center of the sprayer. The middle capillary (usually in stainless steel) provides the makeup flow, ensuring a stable electrospray, as well as electrical contact at the capillary outlet. The sheath liquid is generally introduced in the μL/min flow rate range, which is significantly higher than the typically CE flow rates. Because the sheath-flow surrounds the CE capillary, the problem of irregular droplet formation is circumvented. Finally, the outer capillary (also in stainless steel) supplies the nebulizing gas that assists in spray formation. The coaxial sheath-liquid interface is most commonly used due to numerous advantages. It is easily implemented and gives appropriate flow and solvent conditions for ionization and evaporation. This results in excellent spray stability, independent of the nature of the CE buffer solution. High buffer-salt concentrations could be employed since a dilution with the sheath liquid occurs. As indicated in the literature [17], this dilution does not significantly affect the detection sensitivity since the sheath liquid is simultaneously evaporated with the BGE during the spray process. A number of papers have described the practical aspects of optimizing and setting up the coaxial sheath-liquid interface [18].

Initially proposed by Henion and coworkers [19,20] and further modified [21–23], the liquid junction interface establishes an electrical connection at the junction between the separation capillary and the capillary emitter, in front of the mass spectrometer. The junction between both capillaries is placed in a liquid reservoir, which acts as makeup flow, together with the electrode. The analytes move towards the detector, both with the action of the electric field and the siphoning effect of the nebulizing gas. Because the liquid junction interface is physically and electrically disconnected from the emitter, it can be easily replaced. However, the gap between both capillaries is a critical zone since it can introduce significant external peak broadening and decrease separation efficiency. Hence, extreme care must be taken for proper alignment.

13.2.1.2 Sheathless Interfaces

Two types of sheathless interfaces can be distinguished. The first one consists of a direct connection from the separation capillary to a nanospray needle. In this configuration, the spray needle can be replaced independently of the separation capillary [24–27]. With the second interface, the separation capillary tip is directly used as an emitter via a conductive coating [28–30] or by inserting a conductive wire into the capillary outlet [31,32]. Different procedures for modifying the capillary tip and applying the conductive layer were reported [33,34]. As previously indicated, sheathless interfaces are particularly efficient in terms of sensitivity. The reduced flow rate generally improves analyte ionization and generates smaller droplets. The proportion of ions reaching the MS analyzer is important because the nanospray tip is more closely positioned to the MS orifice compared to the sheath-flow interfaces.

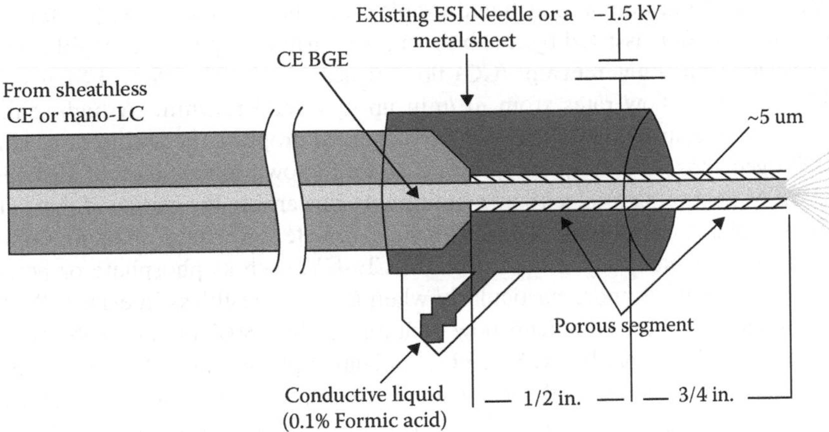

FIGURE 13.1 Schematic representation of the sheathless interface using a porous tip. (From Moini, M., *Anal. Chem.*, 79, 4241, 2007. With permission.)

In contrast, sheathless interfaces appear less stable due to coating deposits or flaking, and electrophoretic and ionization currents are not easily tuned due to low liquid flow rates eluting from the separation capillary. Among the last improvements to connect low-flow separation techniques such as CE or nanoliquid chromatography, the porous junction design has recently emerged [35]. As presented in Figure 13.1, ion transport through the porous section of the capillary, which is in contact with the conductive solution in the electrospray ionization (ESI) needle, provides voltage to the solution inside the capillary for ESI and closes the CE electrical circuit. The porous section is obtained by removing a small part of the polyimide coating of the capillary outlet and etching by immersion into a fresh acid solution until it becomes porous. The latter is inserted into an existing ESI needle, filled with a conductive solution, and positioned in front of the MS inlet. This interface was successfully employed for the detection of amino acids, or intact and digested protein mixtures, and it did not induce deterioration of the CE peak shapes due to the porous segment [36].

13.2.2 Ionization Sources

In order to allow mass separation and detection, the analytes must be charged (ionization) and liberated from their associated solvent molecules (desolvation). Both processes take place thanks to several ionization methods such as atmospheric pressure ionization (API), matrix-assisted laser desorption/ionization (MALDI), and inductively coupled plasma (ICP) ionization. Of these techniques, API is the most widespread ionization source as it covers a wide range of analyte polarities.

13.2.2.1 Atmospheric Pressure Ionization Sources

Undoubtedly, electrospray (ESI) is the predominant ionization method for online CE–MS. ESI is perfectly suited for the analysis of ionizable or polar compounds with a mass range from 10^2 to 10^5 Dalton. ESI is a soft ionization method that produces ions from evaporating liquid droplets in a high electric field [37,38]. The nebulization is often assisted by a nebulizing gas surrounding the CE capillary and the presence of a countercurrent (CC) flow of heated gas. Therefore, ESI is compatible with large flow rates from nL/min up to several mL/min. A high ionization efficiency regarding ions released from charged droplets, the ability to produce multicharged ions, and simplicity are other well-known advantages of ESI [39]. CE–ESI–MS coupling induces some limitations concerning the choice of the BGE. As in LC, volatile electrolytes, such as formate, acetate, carbonate, and ammonium, are mandatory. Commonly used buffers in CE–UV, such as phosphate or borate, are generally avoided, more particularly when using a sheathless interface. With a sheath liquid interface, a makeup flow containing 50%–80% organic solvent may encompass this incompatibility. Since the ionization process in ESI is impaired by highly conductive solutions, the BGE ionic strength must be low. Therefore, the electrophoretic performance could be modified when transferring analytical methods from CE–UV to CE–ESI–MS. The optimization of a CE–ESI–MS procedure often results from a compromise in adjustments. Selectivity modifiers, including chiral selectors, micelles, microemulsions, and ion-pairing agents may contaminate the MS ionization source and cause significant ion suppression [40–42]. In order to address these issues, different strategies such as the partial filling technique (PFT), capillary electrochromatography (CEC) (see Chapter 14) or the use of less sensitive ionization sources have been reported.

In this context, other API sources such as atmospheric pressure chemical ionization (APCI) or atmospheric pressure photoionization (APPI) can be envisioned. In both techniques, the effluent is vaporized in a heated nebulizer before ionization, allowing complete evaporation of the solvent. Nonvolatile salts or additives can be easily removed during this step. In APCI, the chemical ionization process is achieved via a "corona discharge" electrode placed in the source next to the vaporizer which produces electrons and initiates the ionization of the gas formed from the evaporated effluent [43]. In APPI, the ionization is afforded by the analyte absorption of a photon generally emitted from a krypton lamp, which can occur if the lamp energy is higher than the analytes ionization potential (IP). Both sources are, therefore, very similar as far as geometry is concerned, but numerous differences were recently evidenced in their respective use. Some assays demonstrated poor sensitivities for CE–APCI–MS, since most of the APCI interfaces were initially designed for LC–MS [44,45]. Because the APCI source depends on the mass-flow, its sensitivity is impaired with the low flow rates conventionally encountered in CE. However, successful work was achieved by implementing APPI, which can lead to an improvement in sensitivity in several cases. APPI was first introduced by Robb et al. [46,47] in 2000 as a complementary ionization technique for the analysis of nonpolar compounds and commercialized under the trademark PhotoSpray®. Another source was developed in an orthogonal geometry under the name PhotoMate®. The main difference between both sources is related to the addition of a doping agent to significantly improve the ionization process via a charge or proton transfer mechanism to the compound of interest [48,49]. APPI presents excellent results in the coupling with CE. First, common solvents (water, ACN, or MeOH) are not affected as their IPs are above the conventional lamp energy of 10 eV, resulting in a low background noise. Second, compared to APCI and ESI, the APPI source appears to be less sensitive to ion suppression [50]. Various electrolytes, buffers, and additives can be used for the analysis of both polar and nonpolar compounds without background noise and source contamination. As reported by Mol and coworkers [51] and Schappler et al. [52], APPI is a good solution for MEKC-MS and MEEKC-MS, respectively, because SDS has a lower effect on the photoionization efficiency than on electrospray ionization (ESI). Finally, APPI achieves significantly better sensitivity than APCI at very low flow rates generated by CE and can also be considered in the sheathless configuration [53]. To date, very few publications on CE–APPI–MS have been published specifically on a conventional CE–MS system equipped with the sheath flow configuration, where the dopant is added within the sheath liquid. To adapt the APPI to this interface, a spacer should be positioned between the nebulizer and the vaporizer.

13.2.2.2 Other Ionization Sources

Two other ionization sources should be cited in the context of CE–MS: MALDI and ICP ionization. In MALDI, laser energy is used to ionize the analyte and vaporize the CE effluent. A light-absorbing matrix is needed for energy absorption, and, ideally, it should act as the CE electrolyte and absorber. Because simultaneous tasks are difficult to fulfil, the offline mode is often preferred. The latter requires either direct sample deposition onto a MALDI target or CE fraction collection for subsequent MALDI–MS analysis. This approach has been recently reviewed [54]. Some successful applications were reported for the characterization of large molecules with molecular masses up to several hundred kDa [55], but the complexity as well as the equipment cost has prevented its widespread use. In ICP, the analyte of interest is fully fragmented to determine its elementary composition. This is particularly useful for the selective detection of specific elements, such as metal analysis, where CE–ICP–MS attains very low detection limits. Therefore, applications in the field of trace-element speciation analysis are currently increasing with a strong interest in the analysis of bioinorganic (selenium and arsenic speciation), inorganic (metal ions determination), or organometallic (mercury speciation) compounds [56,57].

13.2.3 Analyzers

The hyphenation of CE with MS can be potentially achieved with various mass analyzers, but the high efficiency achieved in CE leads to a very short analysis time window. The signal given by the analytes should be recorded in only a few seconds. Therefore, the MS sampling rates must be compatible in order to achieve a correct peak acquisition. For example, the full scan acquisition over a large mass range with a single quadrupole should be avoided, taking into account both peak definition and sensitivity due to the relatively slow cycle time afforded by this analyzer (>1s). It should be noted that important progress has been made in terms of the acquisition rates, an issue that will be resolved in the next few years. The majority of CE–MS applications have been performed with single quadrupole analyzers [12,13], but hyphenation with ion trap [58,59], time-of-flight (TOF) [60], Fourier transform-ion cyclotron resonance (FT-ICR) [61,62] or tandem mass analyzers such as triple-quadrupole [63,64], Q-TOF and so on, has also been achieved.

Quadrupole MS is mostly encountered in CE–MS because instruments are available at a relatively low cost, possess small dimensions, and are easy to operate. They are generally used for quantitative determination or as low resolution analyzers. In scan mode, the cycle time is relatively slow, making this mode not suitable with very narrow CE peaks. Alternatively, the use of selected ion monitoring (SIM) mode is perfectly adapted to CE. This mode improves the sensitivity, although it could be problematic for the detection of complex mixtures [65].

Ion traps or "quistors" (quadrupole ion storage traps), are three-dimensional analogs of the quadrupole systems. They present very fast scanning rates and are able to record a large number of spectra per second. Compared to the single quadrupole, this analyzer can accumulate ions of pre-selected mass to charge ratio (*m/z*) values in a trap, with a resulting gain in sensitivity. In fact, ions of different *m/z* could be stored or released one at a time by applying appropriate voltages. Therefore, they offer the possibility of MS^n experiments and multiple-stage fragmentation of the analytes, which provide additional structural information on the analyte of interest [66,67].

FT-ICR analyzers are the most powerful analyzers concerning the online coupling with CE when considering MS^n capabilities, sensitivity (one molecule detection), mass resolution (in excess of 10^5), mass accuracy (<2 ppm), and scan speed. However, several issues, such as its technical demands (vacuum, etc.) and high price, limit their use by conventional laboratories.

Finally, the new generation of TOF instruments that allow high resolution, high mass accuracy, and fast acquisition rates, are one of the most promising analyzers for hyphenation with CE. The basic principle of a TOF mass spectrometer, which involves the TOF determination of an ion through the MS yielding its *m/z* value, is now recognized to be perfectly compatible with electrodriven-based separation techniques, such as CE. The characterization of peaks with the additional advantages of high mass resolution, high mass accuracy (<10 ppm), extended mass range, and high sensitivity (sub-fmol detection limits) is easily achieved with the fast acquisition rates. The latter technique provides extremely short times to generate a mass spectrum, which is extremely useful as demonstrated in numerous applications in proteomics and metabolomics. It has to be noted that numerous CE manufacturers have recently proposed commercial ready-to-use solutions for the CE–TOF/MS coupling. For a more complete description of the MS analyzer, the reader is referred to dedicated literature [68,69].

13.3 CHIRAL CE

Since its introduction, CE has rapidly become a powerful separation technique with high efficiency, short analysis time, rapid method development, simple instrumentation, low cost per analysis, and low sample requirement [70,71]. For chiral separations, the most applied electrophoretic-based techniques are capillary zone electrophoresis (CZE), micellar electrokinetic chromatography (MEKC), and chiral capillary electrochromatography (CEC). In this context, the following strategies could be discerned for the enantiomer resolution:

- Enantiomers are directly separated on a chiral stationary phase (CSP) containing an immobilized stereoselective selector. In this case, labile diastereomers are formed on the support (CEC). This technique, as well as the coupling with

MS, will be extensively discussed in Chapters 14 and 15 where the enantioseparations and the recent applications with CEC are presented, respectively.
- Separation of the enantiomers is carried out by directly adding a chiral selector in the BGE for in-situ formation of diastereomeric derivatives (CZE, MEKC). This technique is known as the dynamic mode.

In all cases, a stereoselective environment is created by the chiral selector, enabling enantiomer separation. It should be noted that the indirect method that separates enantiomers via diastereomeric derivatives, after a prederivatization step with an optically pure reagent is out of the scope of the present chapter as it does not differ from conventional CE–MS analysis.

Most of the published chiral separations in CE–MS concern the dynamic mode. One major advantage during method development of CZE and MEKC for chiral analysis is that the separation media can be easily tuned to screen the most appropriate stereoselective environment by changing the chiral selector type and concentration. Secondary optimum operating conditions can be rapidly found at low cost according to the small amount of additives and solvents required. The possibility of a reversal of the enantiomeric migration order by simple procedures such as changing the chiral selector nature or system polarity is another decisive advantage of CE towards CEC. This important feature can induce a migration of the minor enantiomer in front of the major one, which allows for better quantitation in the case of resolving nonracemic mixtures.

Various chiral selectors, such as crown ethers, ligand exchangers, proteins and peptides, macrocyclic antibiotics, bile salts, oligo- and polysaccharides, chiral micelles, and ergot alkaloids, have been reported in the literature. An extensive review of the numerous selectors of CE is presented throughout various chapters of the present book. Nevertheless, the most widely used selectors in chiral CE–MS are cyclodextrins (CD) [1,72–74]. CDs are nonionic cyclic oligosaccharides consisting of six, seven, or eight glucose units and are called α-, β-, and γ-CD, respectively. As presented in Chapter 3, neutral and charged CD derivatives, with various functional groups, have been developed. Their use in CE was first reported in 1985 by Terabe for the separation of positional isomers [75]. In 1989, Fanali obtained a noteworthy chiral separation in CZE by simply adding CD to the BGE [76]. In 1995, Henion and coworkers [77] reported the first chiral analysis with selectivity and sensitivity advantages of using MS as the CE detector.

13.4 CHIRAL CE–MS

In the work of Henion et al., terbutaline and ephedrine enantiomers were separated using a methylated CD (heptakis(2,6-di-*O*-methyl)-β-CD (DM-β-CD)) as the chiral selector. In this initial application, an important increase in sensitivity was observed towards UV detection for the analysis of urine samples. With the simple addition of DM-β-CD in the BGE, the analyte-CD complex and the free drug were detected by MS. The simultaneous presence of analytes and chiral selectors on the ionization side was observed. The drug free signal was determined to be eight times higher than the enantiomer-CD inclusion complexes but an important signal alteration in the presence of the chiral selector in the ionization chamber was revealed. In another pioneering study for the analysis of basic compounds including terbutaline, a significant decrease of ionic current abundance was observed when increasing the concentration

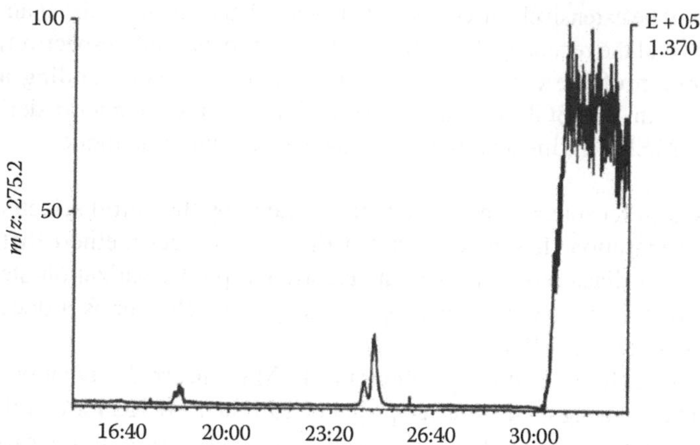

FIGURE 13.2 Chiral CE–ESI–MS analysis of ropivacaine with DM-β-CD as the chiral selector. The analysis was not stopped after enantiomeric detection, which resulted in an increased baseline presumably due to the introduction of DM-β-CD into the mass spectrometer. (From Lamoree, M.H. et al., *J. Chromatogr. A.*, 742, 235, 1996. With permission.)

of DM-β-CD and 2-hydoxypropyl-β-CD (HP-β-CD) in the BGE. Minor quantities of organic modifier were added to improve ESI stability and to enhance relative abundance [78]. In a previous study on ropivacaine analysis, Lamoree et al. clearly demonstrated numerous problems encountered with the introduction of DM-β-CD into the MS [79]. As presented in Figure 13.2, the analysis was not stopped after the detection of (*R*)- and (*S*)-ropivacaine, and introduction of the chiral selector at the ESI interface seriously hampered MS detection. After this experiment, the ESI interface had to be cleaned thoroughly; the electrospray needle and the tubelens, which were covered with a layer of DM-β-CD, also had to be cleaned thoroughly. The authors concluded that the presence of a constant flow of chiral selector was not advisable for the long-term stable operation of the MS.

Interesting results were obtained by Otsuka et al. [80] for the analysis of phenoxy-acid enantiomers with heptakis(2,3,6-tri-*O*-methyl)-β-cyclodextrin (TM-β-CD) as the chiral selector with nonaqueous solvent as the BGE. Later, Iwata et al. [81] employed a negatively charged highly sulfated-γ-CD (HS-γ-CD) for the simultaneous stereoselective separation of amphetamine derivatives with MS/MS detection. In this work, high HS-γ-CD concentrations were used in the BGE at acidic pH in the reversed polarity mode. Under these conditions of low or negligible EOF, the cationic analytes migrated toward the anodic MS detection end as a negatively charged complex with HS-γ-CD derivatives. All 18 individual enantiomers were identified in positive detection mode by their mass spectra. Their specific fragmentation pattern indicated that the migrated amphetamines–CD complex was dissociated at the ESI interface. An improved sensitivity was obtained, but the presence of relatively high concentrations of charged chiral selector impaired the use of buffer in the BGE to decrease the generated current. In fact, the coupling of chiral CE with MS is severely hindered by the presence of nonvolatile additives, leading to the loss in ionization efficiency and/or ionization chamber contamination. Much progress has been recently achieved to circumvent these limitations, increasing the range of applications for chiral CE–MS in the dynamic mode. For this purpose, the PFTs have emerged as the most simple, straightforward, and efficient alternatives to avoid potential interferences of the chiral selector with MS detection.

13.4.1 Partial Filling Technique

The PFT was originally developed by Valtcheva et al. [82] and modified later by Tanaka and Terabe [83], to improve the sensitivity when proteins [84,85] or macrocyclic antibiotics [86] were used as chiral selectors. The latter strongly reduced the detector response in UV according to the numerous chromophoric moieties. In PFT, a discrete zone of the capillary (i.e., partial filling) is filled with the BGE containing an appropriate amount of chiral selector (where the stereoselective separation takes place), while the zone closest to the detector contains only BGE. PFT was particularly well suited for chiral CE–MS determination, avoiding the presence of chiral selectors at the separation capillary end, where the MS detection takes place (Figure 13.3).

With a neutral selector, the main issue encountered with PFT consists of preventing the chiral selector from entering the ionization source when a flow into the separation capillary is present. First, a hydrodynamic flow induced by the Venturi effect of the coaxial sheath gas can aspirate the selector towards the ionization chamber when a sheath liquid interface is used [87]. Second, an important EOF is present at high pH conditions. This can be partially corrected by the use of an acidic buffer, a coated capillary, or the combination of both.

Javerfalk et al. [88] used a capillary coated with polyacrylamide for the separation of racemic bupivacaine and ropivacaine. The PFT was achieved with an acidic BGE (50 mM acetate buffer at pH 3) containing methyl-β-CD. After migration of the

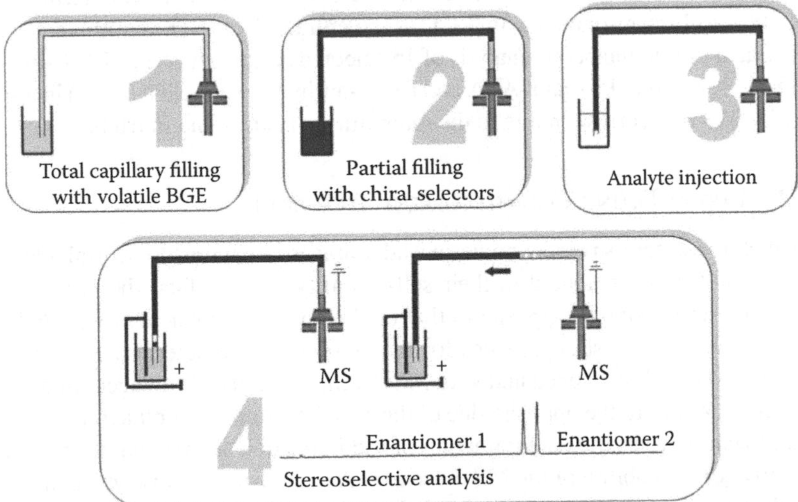

FIGURE 13.3 Schematic representation of the partial filling technique for coupling chiral CE with ESI–MS.

analytes through the zone containing the neutral selector, acting as a pseudostationary phase, very low relative concentrations of *R*-ropivacaine (0.25%) were detected in the presence of *S*-ropivacaine. Cherkaoui et al. employed a polyvinyl alcohol (PVA)-coated capillary for the separation of five amphetamine derivatives [89]. The PFT was achieved with an acidic BGE (40 mM formate buffer at pH 3) containing HP-β-CD. In order to prevent a detrimental effect on the chiral separation, the separation was achieved without nebulization pressure at the MS interface. Other studies demonstrated the possibility of reaching very low detection limits (LOD) with tandem mass spectrometry (MS/MS) detection or using a dedicated sample preparation technique before chiral analysis with PFT. With HP-β-CD dissolved in a volatile 50 mM formate buffer at pH 4, Grard et al. [90] obtained a LOD of 5 ng/mL for the stereoselective separation of adrenoreceptor antagonist drugs after optimization of the capillary zone length containing the CD. The advantage of a high selective and sensitive detection mode such as MS/MS, hyphenated to CE was shown towards single MS or UV.

After solid-phase extraction (SPE), the enantiomers of clenbuterol in plasma were separated by the PFT with an acidic BGE containing DM-β-CD as the chiral selector by Toussaint et al. [91]. Urine analysis of four amphetamine derivatives after SPE extraction was achieved by Lio et al. [92] using a CD dual system made of a mixture of β-CD and DM-β-CD. By using the selective sample preparation, a detection level of 30 ng/mL was reached for some of the selected analytes with a migration time repeatability of less than 0.05% after correction with an internal standard (IS). The development of a CE–MS method for the separation of D- and L-amino acids in food was shown by Rizzi and coworkers [93]. In this interesting work, the focus was placed on the critical experimental parameters that decreased chiral resolution. A complete optimization of the CE parameters (pH, type and concentration of chiral selector, and capillary inner diameter) and MS parameters (sheath liquid nature and flow rate, and nebulizer pressure) was carried out. Two different derivatization protocols of amino acids using dansyl chloride and fluorescein isothiocyanate (FITC) were compared in terms of MS sensitivity and chiral resolution with β-CD as the chiral selector. The optimized method demonstrated the simultaneous analysis of 15 selected amino acids (e.g., FITC-d/l-Asp, -Glu, -Ser, -Asn, -Ala, -Pro, and -Arg), and FITC-aminobutyric acid (GABA) in a single chiral CE–MS run, corresponding to the main amino acids found in fruits.

13.4.2 Partial Filling Countercurrent Technique

Charged chiral selectors present numerous advantages in contrast to neutral additives, qualities that can be attributed to their self-mobility [94]. In fact, the use of a chiral selector with a mobility opposite to that of the analyte further improves PFT, and represents one of the best approaches for the stereoselective determination of chiral compounds when CE is hyphenated with MS. In this case, the chiral agent migrates in its ionized state toward the opposite side of the MS detection (i.e., countercurrent, CC). This approach prevents the contamination of the ionization source and improves both the sensitivity and stability of the MS detection. As demonstrated elsewhere, the additional electrostatic interactions afforded by the charged selectors often demonstrate a higher resolution than that of neutral chiral selectors. Enhanced chiral resolution can be obtained via this CC process, as the mobility difference between the free and complexed analytes is increased [95,96]. A low concentration of selector is employed, and therefore low electrophoretic currents are generated [97]. Successful applications for anionic and particularly cationic analytes were published with the partial filling countercurrent technique (PFT-CC), using both positively and negatively charged chiral selectors. Because more than 85% of the ionizable analytes possess a basic function, the PFT-CC with a negatively charged chiral selector was by far the most studied approach. Carboxymethyl-β-CD (CM-β-CD), a negatively charged chiral selector at pH>3, and sulfobutylether-β-CD (SBE-β-CD) were the most used negatively charged selectors during the early times of chiral CE–MS. Later, highly sulfated cyclodextrins (HS-CD) introduced by Vigh and coworkers [98–100] as single isomers, and from commercial sources as mixtures of randomly sulfated CD [67,68,101,102], emerged as the first choice for the separation of cationic compounds.

CM-β-CD and SBE-β-CD were used for the first time in 1998 for the separation of etilefrine, mianserine, dimethindene, and chloropheniramine by Blaschke and coworkers [103]. The chiral resolution of venlafaxine, a second generation antidepressant drug and its major active metabolite, *O*-desmethyl venlafaxine was successfully achieved with CM-β-CD by Cherkaoui et al. [89]. CM-β-CD was also employed for the analysis of methadone (MTD), for which the use of a PVA-coated capillary was mandatory to avoid the presence of the CD in the MS source. The sensitivity was increased by a factor of 10 compared to UV detection, allowing for the therapeutic drug monitoring of MTD. Some time later, MTD was selected by Rudaz et al. [104] as a model compound for a chemometric study of the parameters involved in enantioseparation by CE–ESI–MS using three different CD, including CM-β-CD and SBE-β-CD. The evaluation of three important experimental parameters regarding PFT, namely, nebulization gas pressure, CD concentration, and separation zone length, was carried out. A sequential experimental design strategy was achieved first using a full factorial design. In a second step, additional experiments were completed to fully model the chiral resolution and to obtain response surfaces thanks to a second degree design. The results obtained confirmed that the nebulization pressure dramatically affected the quality of the enantiomeric separation, while the separation zone length was of minor importance. The hydrodynamic flow induced by the sheath-liquid interface was less detrimental when a negatively charged CD was used, demonstrating another benefit of the CC migration process towards the regular PFT. It should be noted that CM-β-CD can be only partially negatively charged and used as a neutral or charged selector, while the SBE-β-CD is ionized regardless of the pH in the BGE. SBE-β-CD was chosen for the stereoselective PFT-CC analysis of tramadol and its phase I metabolites in biological matrices [105]. The SIM acquisition allowed the unambiguous determination of each analyte in spite of the peak overlapping, and it demonstrated the high selectivity of MS compared to conventional detectors. Among other investigated CDs, Cherkaoui and Veuthey [106] demonstrated that SBE-β-CD was well suited for the simultaneous enantioseparation of bupivacaine, mepivacaine, prilocaine, and ketamine. The enantiomeric determination of atropine in plant extracts without any additional preconcentration step was achieved with SBE-β-CD [89]. By using MS, a 10^3 factor sensitivity was achieved compared to UV, with a detection concentration at the ppb level. After correction with homatropine as IS, the migration time and peak area repeatabilities were found to be 0.2% and 5.5%, respectively.

As previously mentioned, the powerful resolving power of HS-CD has largely contributed to their use in PFT-CC. These selectors were rapidly considered as a

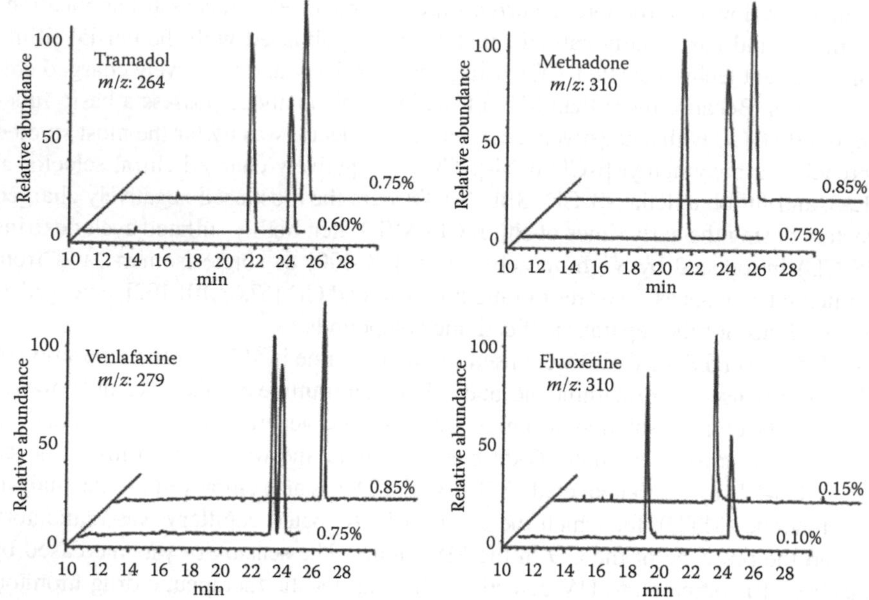

FIGURE 13.4 Chiral CE–ESI–MS analysis of tramadol, methadone, venlafaxine, and fluoxetine with HS-γ-CD as the chiral selector. Lower traces: enantioseparations under optimal conditions. Upper traces: infinite enantioseparations. The infinite enantioselectivity was achieved with different chiral selector concentrations for each enantiomer, but the preferentially bonded enantiomer never reached the MS source. (From Rudaz, S. et al., *Electrophoresis*, 24, 2633, 2003. With permission.)

first choice in CE–UV when a rapid screening of various chiral compounds (acidic, neutral, and basic) is needed [107,108]. Due to their high affinity toward basic compounds, excellent stereoselective resolutions were obtained in CE–MS with HS-γ-CD for several cationic pharmaceutical compounds including tramadol and its phase I metabolites, MTD, venlafaxine, and fluoxetine [109]. An increase in the HS-γ-CD concentration counter-balanced the negative influence of the nebulization gas pressure, allowing a high signal-to-noise ratio as well as a stable electrospray current [110,111]. When studying the impact of the selector concentration on the chiral resolution, the authors were able to determine a CD concentration, which allowed for the detection of only one enantiomer. In fact, a fine-tuning of the HS-γ-CD concentration led to a situation where one enantiomer migrated toward the cathode while the other one migrated toward the anode, as presented in Figure 13.4. This phenomenon could be simply explained by differences in their respective charge, i.e., without the need of any other counter-effect, such as counter-pressure or counter-electroosmotic flow. This infinite enantiomeric resolution was anticipated by Williams and Vigh [112] within the "charged resolving agent migration" (CHARM) model. It corresponds to the chiral selector concentration where the apparent enantiomeric mobilities are opposite for the two enantiomers. One enantiomer migrates in the MS side while the other counter-migrates with the negatively charged CD. In the PFT, the separation process occurs inside the plug of the BGE containing the chiral selector.

It is, therefore, interesting to simultaneously optimize the chiral selector concentration and the selector plug length to achieve an infinite enantioresolution with the shortest plug length. Rudaz et al. [113] demonstrated the resolution of tramadol enantiomers with a proper amount of HS-γ-CD and a separation zone of 3 mm, corresponding to only 0.56% of the capillary length. The infinite value of separation selectivity was thus obtained with an ultrashort plug length and was reached when the lower-binding enantiomers still migrated toward the cathode. Therefore, the principle of the infinite resolution can be envisaged for chiral discrimination in miniaturized devices.

As indicated above, the PFT-CC is also used for the separation of negatively charged analytes with positively charged chiral selectors. In this case, vancomycin is one of the most employed chiral selectors in CE–UV [114] and was evaluated for CE–MS as well. The stereoselective analysis of anionic pharmaceutical drugs (chiral anionic arylpropionic acids), such as ibuprofen, as well as etodolac, and its phase I metabolites, was achieved by Fanali et al. [115]. The MS detection allowed for an unambiguous identification of unresolved peaks of these compounds in biological matrices. The same selector was employed by Tanaka et al. [116] for the separation of a racemic isocitric acid lactone. Other positively charged antibiotics, such as avidin or quaternary ammonium β-CD, were also investigated for the separation of ibuprofen, ketoprofen, warfarin, and camphor sulfonic acid [117].

13.4.3 Crown Ethers, Chiral Micellar Electrokinetic Chromatography, and Chiral Nonaqueous Electrophoresis with MS Detection

Initially synthesized in 1967, crown ethers are macrocyclic polyethers widely used for the separation of cations according to the formation of stable inclusion complexes. Similarly to CDs, numerous modifications were achieved on the native structures, and, among them, the crown ethers derivatized with carboxylic groups (18-crown-6-tetracarboxylic acid) introduced by Cram and coworkers [118] were used in CE–MS. The separation of racemic 3-aminopyrrolidine and racemic α-amino-ε-caprolactam was achieved by Tanaka et al. with high sensitivity [119]. The chiral selector concentration and the separation zone length were carefully investigated, and they demonstrated an increase in chiral resolution with a high concentration of crown ether and a long separation zone. The same selector was successfully used for the separation of real pharmaceutical analytes by Zhou et al. The possible mechanisms for the interaction between the studied compounds and the optically pure chiral crown ethers were investigated using CE-ESI-MS, and ^{1}H-NMR spectroscopy [120].

Because MEKC extends CZE applications to electrically neutral substances, many surfactants including sodium dodecyl sulfate (SDS), amino acid-derived synthetic surfactants as well as bile salts, an abundant source of chiral surfactants, have been used. For enantiomeric separation, Shamsi was the first to report on chiral surfactants in MEKC-MS [121–123]. The baseline chiral resolution of (±)-1,1′-bi-2-naphtol was reported later by the same author [41]. The use of molecular micelles offers important advantages such as a greater structural stability due to covalent bonds formed between the monomers. Furthermore, even if the molecular micelles are ionized, they do not interfere with the low molecular weight analyte signal. Another possibility was presented by Terabe et al., who used a combination of CD with ionic micelles [124]. The addition of CD to the BGE containing micelles modified the apparent distribution coefficients between micellar and nonmicellar phases, due to the analytes' hydrophobic inclusion into the CD cavity. The hyphenation of MEKC with ESI-MS remains problematic due to nonvolatile micelles that contaminate the ionization source and the MS detector, resulting in increased baseline noise and reduced sensitivity. Because APCI and APPI processes are less affected by nonvolatile salts, they also provide several possibilities for obtaining satisfactory results using MEKC-MS. However, to our knowledge no chiral separation was published using a CD-MEKC BGE solution with an APPI or APCI interface.

Different selectivity, enhanced efficiency, reduced analysis time, lower Joule heating, and better solubility, or stability of some compounds in organic solvent than in water are the main reasons for the success of nonaqueous capillary electrophoresis (NACE), as presented in Chapter 10. The online coupling of NACE with MS is particularly interesting, due to the low evaporation temperature and surface tension of organic solvents that are favorable for both the sensitivity and spray stability [110,125,126]. Furthermore, the generated electric currents are lower than in water-based BGE, allowing stable CE–MS conditions [127,128]. The potential of chiral NACE hyphenated with ESI–MS was studied by Servais et al. [129]. CD-NACE was presented as a valuable tool for detecting drug enantiomers without MS signal suppression, although the BGE contained nonvolatile components. Somsen et al. recently investigated the coupling of nonaqueous electrokinetic chromatography (NAEKC) using CD with ESI–MS [130,131]. Due to ionization suppression, analyte concentrations were detected at the sub-µg/mL level.

13.5 REQUIREMENTS FOR BIOLOGICAL SAMPLE QUANTITATION WITH CHIRAL CE–MS

Several analytical problems, such as the determination of enantiomeric purity of food components, or the determination of chiral compounds in biological samples, need an important level of sensitivity with acceptable quantitative performance [132,133]. However, the validation of chiral CE–MS methods according to recognized guidelines (ICH, ISO, FDA, etc.) has scarcely been achieved, and only a few complete validation methods have been reported in the literature [134]. In fact, some technical aspects related to CE–MS interfacing have to be considered when a quantitative determination is envisaged. Furthermore, specific requirements should be fulfilled when complex matrices, such as biological samples, are analyzed [135]. A sample preparation has to be performed to remove interferences coming from the matrix and to preconcentrate the compounds of interest before analysis. Regarding MS detection, the SIM mode should be preferred for the quantitative determination of low molecular weight analytes, according to its high selectivity and sensitivity. Tandem mass spectrometry (MS^2) is also advantageous for quantitative purposes [81,90] due to its gain in selectivity.

As indicated above, stable conditions are mandatory to achieve good quantitative results. Problems encountered with stability are generally overcome with an optimized interface, which is the critical part in connecting both instruments and the main reason why CE–MS is not considered a routine technique. Parameters such as sheath liquid flow rate and composition, capillary outlet position, ionization voltage, nebulizing gas pressure, drying gas flow rate, and temperature should be optimized. According to the numerous experimental factors and the potential for parameter interaction, an experimental design methodology with multivariate optimization should be preferentially selected [18,52]. The choice of an appropriate sheath liquid and its flow rate is a compromise to maintain an efficient electrophoretic separation and to assist both droplet formation and spray stability. The sheath liquid composition is generally composed of a mixture of water-organic solvent containing an electrolyte. The optimization can be done with a conventional CE–UV apparatus by placing the sheath liquid into the outlet vial and evaluating its consequence on the chiral separation. The most important point concerns the solubility of the chiral selector to avoid its precipitation at the spray needle. The capillary outlet position has to be slightly adjusted to achieve good quantitative results according to the instrument and the interface geometry. The optimal sprayer positioning inside the MS source avoids unwanted corona discharges and stabilizes the electric fields inside the ionization source. Their intensity can also be adjusted by modifying the applied ionization voltage. The monitoring of the electrophoretic current is another interesting method used to control the chiral selector entrance into the ionization chamber [136]. As indicated throughout this chapter, the nebulizing gas pressure is also of utmost importance to obtain a stable electrospray process without producing a suction effect inside the capillary, which is particularly detrimental to chiral efficiency and resolution. The nebulizing gas pressure should be turned off during the injection process in order to avoid air entering into the capillary, as the suction effect that is potentially induced can be another variability issue. It should be noted that the use of a capillary with a relatively long length seems to be preferential in maintaining stability.

When biological fluids are analyzed by CE–ESI–MS, MS suppression or enhancement effects, termed matrix effects, must also be considered [137]. Inspired by the work of Bonfiglio et al. [138] on LC–ESI–MS experiments with the a post-column infusion configuration, Schappler et al. [139] used a similar set-up with the sheath liquid interface as the analyte delivering device, as presented in Figure 13.5. The latter allowed for the characterization of the matrix effect after various sample preparation procedures prior to chiral separation of pharmaceutical drugs by CE–ESI–MS. Protein precipitation (PP) and liquid–liquid extraction (LLE) were compared with the postcapillary infusion system and interfering substances that presented a strong influence on ESI were highlighted. With this setup, the exact time window, in which the matrix effect occurred was revealed in regard to sample preparation. After PP, many remaining substances were found to be polar. Therefore, a special attention had to be paid in CE–ESI–MS to determine the time window where potential interference could occur and to exclude comigration with the analytes of interest. After LLE, identical and stable profiles of BGE, extracted water, and extracted plasma were observed, leading the authors to conclude that a minimal matrix effect due to endogenous compounds occurred. To obtain the lowest LOD, on-capillary preconcentration based on electrokinetic injection was necessary. In case of PFT-CC, the electrokinetic injection process had to be carefully adjusted due to the presence of the charged chiral selector in the BGE. In fact, the mobility of the chiral selector opposite to the analyte, led to numerous drawbacks during the electric field application. The applied voltage enabled the introduction of cationic analytes into the capillary,

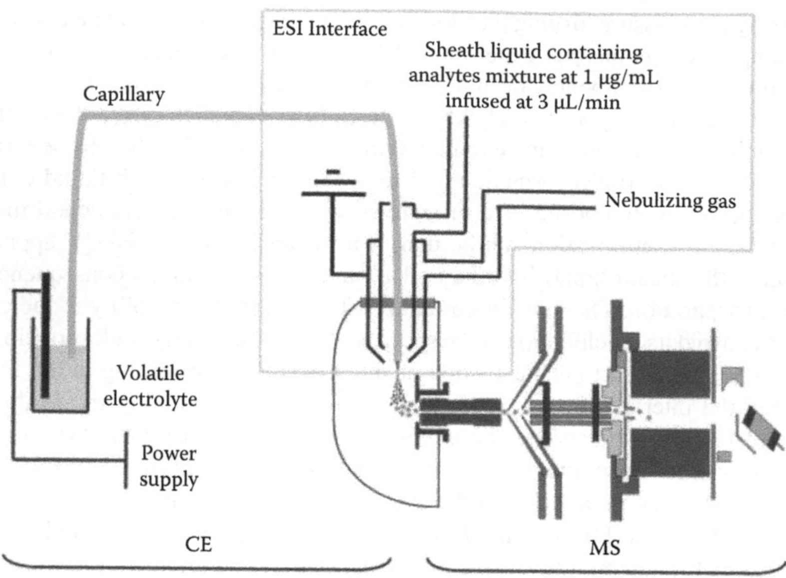

FIGURE 13.5 Schematic representation of the coaxial sheath-flow interface used as a postcapillary infusion device for matrix effect evaluation. (From Schappler, J. et al., *Electrophoresis*, 27, 1537, 2006. With permission.)

but simultaneously started the displacement of negatively charged CD in the injection vial. Besides sample contamination, the chiral selector quantity in the capillary decreased in an uncontrolled manner, and the analysis time variability increased. In order to address this issue, the injection of an appropriate buffer plug length between the zone, containing the chiral selector and the injected analyte, was mandatory. A concentration of 1 ppb, corresponding to an enantiomeric concentration of 0.5 ppb, of several basic analytes was detected in plasma samples after LLE.

Considering the quantitative performance of chiral CE–ESI–MS analyses, the method accuracy can be greatly improved by the choice of an appropriate IS to reduce errors due to sample preparation and injection. An IS also decreases the peak area variability and the migration time drift, particularly considering the low concentration level of drugs in bioanalysis and the importance of matrix effects when dealing with biological fluids. In order to eliminate the problems associated with quantitation in CE–MS and to reduce the impact of system variability on method performance, an isotopically labeled (deuterated) IS should be used [140,141]. Consequently, the analytes and IS reach the ionization source at identical times and the short-term variation of the ionization process, particularly present when ESI is used, can be corrected. Thanks to the isotopically labeled IS, which possesses identical ionization response and fragmentation patterns, the correction of both the overall method variability (i.e., sample preparation, injection, electrophoretic process, etc.) as well as matrix effects can be achieved. For instance, a complete validation was performed on ecstasy (MDMA) and MTD as model compounds and their deuterated analogues were used as IS. The quantitative performance was estimated according to SFSTP validation guidelines based on the total error concept and accuracy profiles [142–144]. The ability of the method to quantify MDMA and MTD in plasma was demonstrated over a broad concentration range. Different validation criteria were evaluated by applying an appropriate regression model. The limit of quantification, as well as the selectivity, trueness, and precision were determined, and the methodology demonstrated good accuracy performance included within the 30% acceptance range. The stereoselective CE–ESI–MS method finally proved to be valid for quantifying MDMA and MTD in biological samples, and real case analyses were successfully achieved, as presented in Figure 13.6.

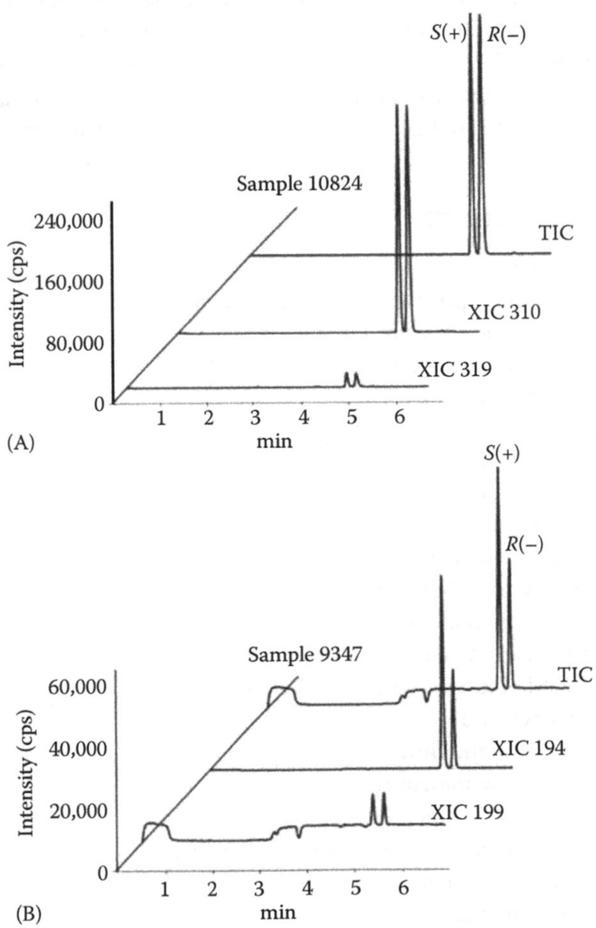

FIGURE 13.6 Chiral CE–ESI–MS analysis of real blood samples containing methadone (A) sample 10824 and ecstasy (B) sample 9347 with HS-γ-CD as the chiral selector. Blood samples were treated by a liquid–liquid extraction procedure and electrokinetically injected. Limits of quantitation, as low as 0.50 ng/mL per enantiomer were achieved, and the enantiomeric ratio was determined. (From Schappler, J. et al., *Electrophoresis*, 29, 2193, 2008. With permission.)

13.6 DISCUSSION

Chiral discrimination by CE, achieved with the dynamic mode where the chiral selector is simply added to the BGE, is a powerful analytical technique for resolving enantiomers. The hyphenation of CE with MS appears appropriate for the stereoselective analysis of numerous compounds in various matrices, and it overcomes the lack of sensitivity of CE. Because it is the most adapted ionization method for ionizable and polar compounds, ESI is generally used. However, its limitation concerns the strict use of volatile electrolytes and selectivity modifiers, such as chiral selector. Among the various chiral selectors that can be applied, neutral and charged CDs are the most employed. To decrease the negative effect of BGE composition, different strategies have been reported. The PFT, with or without CC, has proved to be a suitable and efficient approach to avoid MS source contamination, as well as signal suppression due to nonvolatile additives. Therefore, the PFT is particularly adapted for use with chiral selectors added to the electrolyte solution. Because of the CC contribution, charged chiral selectors are found more suitable for the online MS detection of separated enantiomers.

LIST OF ABBREVIATIONS

ACN	acetonitrile
APCI	atmospheric pressure chemical ionization
API	atmospheric pressure ionization
APPI	atmospheric pressure photoionization
BGE	background electrolyte
CC	countercurrent
CD	cyclodextrin
CE	capillary electrophoresis
CEC	capillary electrochromatography
CHARM	charged resolving agent migration
CM-β-CD	carboxymethyl-β-cyclodextrin
CSP	chiral stationary phase
CZE	capillary zone electrophoresis
DM-β-CD	heptakis(2,6-di-*O*-methyl)-β-cyclodextrin
EK	electrokinetic
EOF	electroosmotic flow
ESI	electrospray ionization
FITC	fluorescein isothiocyanate
FT-ICR	fourier transform-ion cyclotron resonance
GABA	aminobutyric acid
GC	gas chromatography
H/N	height to noise ratio
HP-β-CD	hydroxypropyl-β-cyclodextrin
HS-CD	highly sulfated cyclodextrin
HS-γ-CD	highly sulfated-γ-cyclodextrin
ICP	inductively coupled plasma ionization
IP	ionization potential

IS	internal standard
LC	liquid chromatography
LIF	laser-induced fluorescence
LLE	liquid–liquid extraction
LOD	limit of detection
m/z	mass to charge ratio
MALDI	matrix-assisted laser desorption/ionization
MDMA	ecstasy
MEEKC	microemulsion electrokinetic chromatography
MEKC	micellar electrokinetic chromatography
MeOH	methanol
MS	mass spectrometry
MTD	methadone
NACE	nonaqueous capillary electrophoresis
NAEKC	nonaqueous electrokinetic chromatography
PFT	partial filling technique
PFT-CC	partial filling countercurrent technique
PP	protein precipitation
PVA	polyvinyl alcohol
SBE-β-CD	sulfobutylether-β-cyclodextrin
SDS	sodium dodecyl sulfate
SFSTP	societe française des sciences et techniques pharmaceutiques
SIM	selected ion monitoring
SPE	solid phase extraction
TOF	time-of-flight
TM-β-CD	heptakis(2,3,6-tri-*O*-methyl)-β-cyclodextrin
TMD	tramadol
UV/Vis	ultraviolet/Visible

REFERENCES

1. Chankvetadze, B, Blaschke, G. 2001. Enantioseparations in capillary electromigration techniques: Recent developments and future trends. *J Chromatogr A* 906: 309–363.
2. Kavran-Belin, G, Rudaz, S, Veuthey, JL. 2005. Enantioseparation of baclofen with highly sulfated beta-cyclodextrin by capillary electrophoresis with laser-induced fluorescence detection. *J Sep Sci* 28: 2187–2192.
3. Xu, X, Li, L, Weber, SG. 2007. Electrochemical and optical detectors for capillary and chip separations. *Trends Anal Chem* 26: 68–79.
4. Servais, AC, Crommen, J, Fillet, M. 2006. Capillary electrophoresis-mass spectrometry, an attractive tool for drug bioanalysis and biomarker discovery. *Electrophoresis* 27: 2616–2629.
5. Erny, GL, Cifuentes, A. 2006. Liquid separation techniques coupled with mass spectrometry for chiral analysis of pharmaceuticals compounds and their metabolites in biological fluids. *J Pharm Biomed Anal* 40: 509–515.
6. Baldacci, A, Theurillat, R, Caslavska, J, Pardubska, H, Brenneisen, R, Thormann, W. 2003. Determination of γ-hydroxybutyric acid in human urine by capillary electrophoresis with indirect UV detection and confirmation with electrospray ionization ion-trap mass spectrometry. *J Chromatogr A* 990: 99–110.

7. Ramautar, R, Demirci, A, de Jong, GJ. 2006. Capillary electrophoresis in metabolomics. *Trends Anal Chem* 25: 455–466.
8. Simpson, DC, Smith, RD. 2005. Combining capillary electrophoresis with mass spectrometry for applications in proteomics. *Electrophoresis* 26: 1291–1305.
9. Klampfl, CW, Buchberger, W. 2007. Coupling of capillary electroseparation techniques with mass spectrometric detection. *Anal Bioanal Chem* 388: 533–536.
10. Klampfl, CW. 2006. Recent advances in the application of capillary electrophoresis with mass spectrometric detection. *Electrophoresis* 27: 3–34.
11. Campa, C, Coslovi, A, Flamigni, A, Rossi, M. 2006. Overview on advances in capillary electrophoresis-mass spectrometry of carbohydrates: A tabulated review. *Electrophoresis* 27: 2027–2050.
12. Schmitt-kopplin, P, Englmann, M. 2005. Capillary electrophoresis—mass spectrometry: Survey on developments and applications 2003–2004. *Electrophoresis* 26: 1209–1220.
13. Schmitt-kopplin, P, Frommberger, M. 2003. Capillary electrophoresis—mass spectrometry: 15 years of developments and applications. *Electrophoresis* 24: 3837–3867.
14. Olivares, JA, Nguyen, NT, Yonker, CR, Smith, RD. 1987. On-line mass spectrometric detection for capillary zone electrophoresis. *Anal Chem* 59: 1230–1232.
15. Smith, RD, Barinaga, CJ, Udseth, HR. 1988. Improved electrospray ionization interface for capillary zone electrophoresis-mass spectrometry. *Anal Chem* 60: 1948–1952.
16. Smith, RD, Olivares, JA, Nguyen, NT, Udseth, HR. 1988. Capillary zone electrophoresis-mass spectrometry using an electrospray ionization interface. *Anal Chem* 60: 436–441.
17. Schappler, J, Guillarme, D, Prat, J, Veuthey, JL, Rudaz, S. 2007. Coupling CE with atmospheric pressure photoionization MS for pharmaceutical basic compounds: Optimization of operating parameters. *Electrophoresis* 28: 3078–3087.
18. Nilsson, SL, Bylund, D, Joernten-Karlsson, M, Petersson, P, Markides, KE. 2004. A chemometric study of active parameters and their interaction effects in a nebulized sheath-liquid electrospray interface for capillary electrophoresis-mass spectrometry. *Electrophoresis* 25: 2100–2107.
19. Lee, ED, Mueck, W, Henion, JD, Covey, TR. 1989. Liquid junction coupling for capillary zone electrophoresis/ion spray mass spectrometry. *Biomed Environ Mass Spectrom* 18: 844–850.
20. Lee, ED, Mueck, W, Henion, JD, Covey, TR. 1988. On-line capillary zone electrophoresis-ion spray tandem mass spectrometry for the determination of dynorphins. *J Chromatogr* 458: 313–321.
21. Wachs, T, Sheppard, RL, Henion, J. 1996. Design and applications of a self-aligning liquid junction-electrospray interface for capillary electrophoresis-mass spectrometry. *J Chromatogr B* 685: 335–342.
22. Jussila, M, Sinervo, K, Porras, SP, Riekkola, ML. 2000. Modified liquid-junction interface for nonaqueous capillary electrophoresis-mass spectrometry. *Electrophoresis* 21: 3311–3317.
23. Varesio, E, Rudaz, S, Krause, KH, Veuthey, JL. 2002. Nanoscale liquid chromatography and capillary electrophoresis coupled to electrospray mass spectrometry for the detection of amyloid-b peptide related to Alzheimer's disease. *J Chromatogr A* 974: 135–142.
24. Alexander, JN, Schultz, GA, Polil, JB. 1998. Development of a nano-electrospray mass spectrometry source for nanoscale liquid chromatography and sheathless capillary electrophoresis. *Rapid Commun Mass Spectrom* 12: 1187–1191.
25. Bateman, KP, White, RL, Thibault, P. 1997. Disposable emitters for online capillary zone electrophoresis/nanoelectrospray mass spectrometry. *Rapid Commun Mass Spectrom* 11: 307–315.
26. Gucek, M, Vreeken, RJ, Verheij, ER. 1999. Coupling of capillary zone electrophoresis to mass spectrometry (MS and MS/MS) via a nanoelectrospray interface for the characterization of some b-agonists. *Rapid Commun Mass Spectrom* 13: 612–619.

27. Kele, Z, Ferenc, G, Klement, E, Toth, GK, Janaky, T. 2005. Design and performance of a sheathless capillary electrophoresis/mass spectrometry interface by combining fused-silica capillaries with gold-coated nanoelectrospray tips. *Rapid Commun Mass Spectrom* 19: 881–885.
28. Barnidge, DR, Nilsson, S, Markides, KE, Rapp, H, Hjort, K. 1999. Metallized sheathless electrospray emitters for use in capillary electrophoresis orthogonal time-of-flight mass spectrometry. *Rapid Commun Mass Spectrom* 13: 994–1002.
29. Chang, YZ, Chen, YR, Her, GR. 2001. Sheathless capillary electrophoresis/electrospray mass spectrometry using a carbon-coated tapered fused-silica capillary with a beveled edge. *Anal Chem* 73: 5083–5087.
30. Mazereeuw, M, Hofte, AJP, Tjaden, UR, Vandergreef, J. 1997. A novel sheathless and electrodeless microelectrospray interface for the on-line coupling of capillary zone electrophoresis to mass spectrometry. *Rapid Commun Mass Spectrom* 11: 981–986.
31. Herring, CJ, Qin, J. 1999. An online preconcentrator and the evaluation of electrospray interfaces for the capillary electrophoresis-mass spectrometry of peptides. *Rapid Commun Mass Spectrom* 13: 1–7.
32. McComb, ME, Perreault, H. 2000. Design of a sheathless capillary electrophoresis-mass spectrometry probe for operation with a Z-spray ionization source. *Electrophoresis* 21: 1354–1362.
33. Samskog, J, Wetterhall, M, Jacobsson, S, Markides, K. 2000. Optimization of capillary electrophoresis conditions for coupling to a mass spectrometer via a sheathless interface. *J Mass Spectrom* 35: 919–924.
34. Janini, GM, Conrads, TP, Wilkens, KL, Issaq, HJ, Veenstra, TD. 2003. A sheathless nanoflow electrospray interface for on-line capillary electrophoresis mass spectrometry. *Anal Chem* 75: 1615–1619.
35. Whitt, JT, Moini, M. 2003. Capillary electrophoresis to mass spectrometry interface using a porous junction. *Anal Chem* 75: 2188–2191.
36. Moini, M. 2007. Simplifying CE–MS operation. 2. Interfacing low-flow separation techniques to mass spectrometry using a porous tip. *Anal Chem* 79: 4241–4246.
37. Kebarle, P, Tang, L. 1993. From ions in solution to ions in the gas phase—the mechanism of electrospray mass spectrometry. *Anal Chem* 65: 972A-986A.
38. Bruins, AP. 1998. Mechanistic aspects of electrospray-ionization. *J Chromatogr A* 794: 345–357.
39. Lazar, IM, Lee, ED, Rockwood, AL, Lee, ML. 1998. General considerations for optimizing a capillary electrophoresis-electrospray ionization time-of-flight mass-spectrometry system. *J Chromatogr A* 829: 279–288.
40. Somsen, GW, Mol, R, de Jong, GJ. 2003. On-line micellar electrokinetic chromatography-mass spectrometry: Feasibility of direct introduction of non-volatile buffer and surfactant into the electrospray interface. *J Chromatogr A* 1000: 953–961.
41. Shamsi, SA. 2002. Chiral capillary electrophoresis-mass spectrometry: Modes and applications. *Electrophoresis* 23: 4036–4051.
42. Lu, W, Shamsi, SA, McCarley, TD, Warner, IM. 1998. Online capillary electrophoresis-electrospray ionization mass spectrometry using a polymerized anionic surfactant. *Electrophoresis* 19: 2193–2199.
43. Carroll, DI, Dzidic, I, Stillwell, RN, Haegele, KD, Horning, EC. 1975. Atmospheric pressure ionization mass spectrometry. Corona discharge ion source for use in a liquid chromatograph-mass spectrometer-computer analytical system. *Anal Chem* 47: 2369–2372.
44. Tanaka, Y, Otsuka, K, Terabe, S. 2003. Evaluation of an atmospheric pressure chemical ionization interface for capillary electrophoresis-mass spectrometry. *J Pharm Biomed Anal* 30: 1889–1895.

45. Isoo, K, Otsuka, K, Terabe, S. 2001. Application of sweeping to micellar electrokinetic chromatography-atmospheric pressure chemical ionization-mass spectrometric analysis of environmental pollutants. *Electrophoresis* 22: 3426–3432.
46. Robb, DB, Covey, TR, Bruins, AP. 2000. Atmospheric pressure photoionization: An ionization method for liquid chromatography-mass spectrometry. *Anal Chem* 72: 3653–3659.
47. Robb, DB, Covey, TR, Bruins, AP. 2001. Atmospheric pressure photoionization (APPI): A new ionization technique for LC/MS. *Adv Mass Spectrom* 15: 391–392.
48. Raffaelli, A, Saba, A. 2003. Atmospheric pressure photoionization mass spectrometry. *Mass Spectrom Rev* 22: 318–331.
49. Syage, JA. 2004. Mechanism of $[M + H]^+$ formation in photoionization mass spectrometry. *J Am Soc Mass Spectrom* 15: 1521–1533.
50. Marchi, I, Rudaz, S, Selman, M, Veuthey, JL. 2007. Evaluation of the influence of protein precipitation prior to on-line SPE-LC-API/MS procedures using multivariate data analysis. *J Chromatogr B* 845: 244–252.
51. de Jong, GJ, Mol, R, Somsen, GW. 2004. Coupling of micellar electrokinetic chromatography and mass spectrometry for the impurity profiling of drugs. *Chemie Magazine* 4: 18–23.
52. Schappler, J, Guillarme, D, Rudaz, S, Veuthey, JL. 2008. Microemulsion electrokinetic chromatography hyphenated to atmospheric pressure photoionization mass spectrometry. *Electrophoresis* 29: 11–19.
53. Syage, JA, Hanold, KA, Lynn, TC, Horner, JA, Thakur, RA. 2004. Atmospheric pressure photoionization. II. Dual source ionization. *J Chromatogr A* 1050: 137–149.
54. Huck, CW, Bakry, R, Huber, LA, Bonn, GK. 2006. Progress in capillary electrophoresis coupled to matrix-assisted laser desorption/ionization—time of flight mass spectrometry. *Electrophoresis* 27: 2063–2074.
55. Stutz, H. 2005. Advances in the analysis of proteins and peptides by capillary electrophoresis with matrix-assisted laser desorption/ionization and electrospray-mass spectrometry detection. *Electrophoresis* 26: 1254–1290.
56. Michalke, B. 2005. Capillary electrophoresis-inductively coupled plasma-mass spectrometry: A report on technical principles and problem solutions, potential, and limitations of this technology as well as on examples of application. *Electrophoresis* 26: 1584–1597.
57. Alvarez-Llamas, G, Fernandez de laCampa, MdR, Sanz-Medel, A. 2005. ICP-MS for specific detection in capillary electrophoresis. *Trends Anal Chem* 24: 28–36.
58. Bach, GA, Henion, J. 1998. Quantitative capillary electrophoresis-ion-trap mass spectrometry determination of methylphenidate in human urine. *J Chromatogr B* 707: 275–285.
59. Chen, YR, Wen, KC, Her, GR. 2000. Analysis of coptisine, berberine, and palmatine in adulterated Chinese medicine by capillary electrophoresis-electrospray ion trap mass spectrometry. *J Chromatogr A* 866: 273–280.
60. Wu, JT, Qian, MG, Li, MX, Zheng, K, Huang, P, Lubman, DM. 1998. On-line analysis by capillary separations interfaced to an ion trap storage/reflectron time-of-flight mass spectrometer. *J Chromatogr A* 794: 377–389.
61. Hofstadler, SA, Wahl, JH, Bruce, JE, Smith, RD. 1993. On-line capillary electrophoresis with Fourier transform ion cyclotron resonance mass spectrometry. *J Am Chem Soc* 115: 6983–6984.
62. Hofstadler, SA, Severs, JC, Smith, RD, Swanek, FD, Ewing, AG. 1996. High performance Fourier transform ion cyclotron resonance mass spectrometric detection for capillary electrophoresis. *J High Res Chromatogr* 19: 617–621.
63. Figeys, D, Ducret, A, Yates, JR III, Aebersold, R. 1996. Protein identification by solid phase microextraction-capillary zone electrophoresis-microelectrospray-tandem mass spectrometry. *Nature Biotechnology* 14: 1579–1583.

64. Figeys, D, van Oostveen, I, Ducret, A, Aebersold, R. 1996. Protein identification by capillary zone electrophoresis/microelectrospray ionization-tandem mass spectrometry at the subfemtomole level. *Anal Chem* 68: 1822–1828.
65. Smyth, WF, Brooks, P. 2004. A critical evaluation of high performance liquid chromatography-electrospray ionization-mass spectrometry and capillary electrophoresis-electrospray-mass spectrometry for the detection and determination of small molecules of significance in clinical and forensic science. *Electrophoresis* 25: 1413–1446.
66. Klampfl, CW. 2004. Review coupling of capillary electrochromatography to mass spectrometry. *J Chromatogr A* 1044: 131–144.
67. Baldacci, A, Prost, F, Thormann, W. 2004. Identification of diphenhydramine metabolites in human urine by capillary electrophoresis-ion trap-mass spectrometry. *Electrophoresis* 25: 1607–1614.
68. Hernandez-Borges, J, Neusuess, C, Cifuentes, A, Pelzing, M. 2004. On-line capillary electrophoresis-mass spectrometry for the analysis of biomolecules. *Electrophoresis* 25: 2257–2281.
69. Weissinger, EM, Hertenstein, B, Mischak, H, Ganser, A. 2005. Online coupling of capillary electrophoresis with mass spectrometry for the identification of biomarkers for clinical diagnosis. *Exp Rev Proteomics* 2: 639–647.
70. Fanali, S, Desiderio, C. 1997. Enantiomeric resolution study by capillary electrophoresis—selection of the appropriate chiral selector. *J Chromatogr A* 772: 185–194.
71. Fanali, S, Cristalli, M, Vespalec, R, Bocek, P. 1994. Chiral separations in capillary electrophoresis. In *Advances in Electrophoresis*, eds. A Chrambach, MJ Dunn and BJ Radola, Weinheim: VCH, pp. 1–86.
72. Fanali, S. 2000. Enantioselective determination by capillary electrophoresis with cyclodextrins as chiral selectors. *J Chromatogr A* 875: 89–122.
73. Fanali, S. 1996. Identification of chiral drug isomers by CE. *J Chromatogr A* 735: 77–121.
74. Chankvetadze, B, Kartozia, I, Burjanadze, N, Bergenthal, D, Luftmann, H, Blaschke, G. 2001. Enantioseparation of chiral phenothiazine derivatives in capillary electrophoresis using cyclodextrin type chiral selectors. *Chromatographia* 53: S290-S295.
75. Terabe, S, Ozaki, H, Otsuka, K, Ando, T. 1985. Electrokinetic chromatography with 2-*O*-carboxymethyl-β-cyclodextrin as moving "stationary" phase. *J Chromatogr* 332: 211–217.
76. Fanali, S. 1989. Separation of optical isomers by capillary zone electrophoresis based on host-guest complexation. *J Chromatogr* 474: 441–446.
77. Sheppard, RL, Tong, X, Cai, J, Henion, JD. 1995. Chiral separation and detection of terbutaline and ephedrine by capillary electrophoresis coupled with ion spray mass spectrometry. *Anal Chem* 67: 2054–2058.
78. Lu, W, Cole, RB. 1998. Determination of chiral pharmaceutical compounds, terbutaline, ketamine and propranolol, by on-line capillary electrophoresis-electrospray ionization mass spectrometry. *J Chromatogr B* 714: 69–75.
79. Lamoree, MH, Sprang, AFH, Tjaden, UR, Van der Greef, J. 1996. Use of heptakis(2,6-di-*O*-methyl)-β-cyclodextrin in online capillary zone electrophoresis-mass spectrometry for the chiral separation of ropivacaine. *J Chromatogr A* 742: 235–242.
80. Otsuka, K, Smith, CJ, Grainger, J, Barr, JR, Patterson, J, Tanaka, N, Terabe, S. 1998. Stereoselective separation and detection of phenoxy acid herbicide enantiomers by cyclodextrin-modified capillary zone electrophoresis-electrospray ionization mass spectrometry. *J Chromatogr A* 817: 75–81.
81. Iwata, YT, Kanamori, T, Ohmae, Y, Tsujikawa, K, Inoue, H, Kishi, T. 2003. Chiral analysis of amphetamine-type stimulants using reversed-polarity capillary electrophoresis/positive ion electrospray ionization tandem mass spectrometry. *Electrophoresis* 24: 1770–1776.

82. Valtcheva, L, Mohammed, J, Pettersson, G, Hjerten, S. 1993. Chiral separation of β-blockers by high performance capillary electrophoresis based on nonimmobilized cellulase as enantioselective protein. *J Chromatogr* 638: 263–267.
83. Tanaka, Y, Terabe, S. 1997. Separation of the enantiomers of basic drugs by affinity capillary electrophoresis using a partial filling technique and α1-acid glycoprotein as chiral selector. *Chromatographia* 44: 119–128.
84. Amini, A, Paulsen-Sorman, U. 1997. Enantioseparation of local anaesthetic drugs by capillary zone electrophoresis with cyclodextrins as chiral selectors using a partial filling technique. *Electrophoresis* 18: 1019–1025.
85. Amini, A, Paulsensorman, U, Westerlund, D. 1999. Principle and applications of the partial filling technique in capillary electrophoresis. *Chromatographia* 50: 497–506.
86. Armstrong, DW, Nair, UB. 1997. Capillary electrophoretic enantioseparations using macrocyclic antibiotics as chiral selectors. *Electrophoresis* 18: 2331–2342.
87. Geiser, L, Cherkaoui, S, Veuthey, JL. 2000. Simultaneous analysis of some amphetamine derivatives in urine by nonaqueous capillary electrophoresis coupled to electrospray ionization mass spectrometry. *J Chromatogr A* 895: 111–121.
88. Javerfalk, EM, Amini, A, Westerlund, D, Andren, PE. 1998. Chiral separation of local anesthetics by a capillary electrophoresis/partial filling technique coupled online to micro-electrospray mass spectrometry. *J Mass Spectrom* 33: 183–186.
89. Cherkaoui, S, Rudaz, S, Varesio, E, Veuthey, JL. 2001. On-line capillary electrophoresis-electrospray mass spectrometry for the stereoselective analysis of drugs and metabolites. *Electrophoresis* 22: 3308–3315.
90. Grard, S, Morin, P, Dreux, M, Ribet, JP. 2001. Efficient applications of capillary electrophoresis-tandem mass spectrometry to the analysis of adrenoreceptor antagonist enantiomers using a partial filling technique. *J Chromatogr A* 926: 3–10.
91. Toussaint, B, Palmer, M, Chiap, P, Hubert, P, Crommen, J. 2001. On-line coupling of partial filling-capillary zone electrophoresis with mass spectrometry for the separation of clenbuterol enantiomers. *Electrophoresis* 22: 1363–1372.
92. Lio, R, Chinaka, S, Tanaka, S, Takayama, N, Hayakawa, K. 2003. Simultaneous chiral determination of methamphetamine and its metabolites in urine by capillary electrophoresis-mass spectrometry. *Analyst* 128: 646–650.
93. Simo, C, Rizzi, A, Barbas, C, Cifuentes, A. 2005. Chiral capillary electrophoresis-mass spectrometry of amino acids in foods. *Electrophoresis* 26: 1432–1441.
94. Chankvetadze, B, Endresz, G, Blaschke, G. 1994. About some aspects of the use of charged cyclodextrins for capillary electrophoresis enantioseparation. *Electrophoresis* 15: 804–807.
95. Wren, SAC, Rowe, RC. 1992. Theoretical aspects of chiral separation in CE. I. Initial evaluation of a model. *J Chromatogr* 603: 235–241.
96. Wren, SAC, Rowe, RC. 1992. Theoretical aspects of chiral separation in CE II. The role of organic solvent. *J Chromatogr* 609: 363–367.
97. Chankvetadze, B. 1997. Separation selectivity in chiral capillary electrophoresis with charged selectors. *J Chromatogr A* 792: 269–295.
98. Vincent, JB, Kirby, DM, Nguyen, TV, Vigh, G. 1997. A family of single-isomer chiral resolving agents for capillary electrophoresis. 2. Hepta-6-sulfato-beta-cyclodextrin. *Anal Chem* 69: 4419–4428.
99. Cai, H, Nguyen, TV, Vigh, G. 1998. A family of single-isomer chiral resolving agents for capillary electrophoresis. 3. Heptakis(2,3-dimethyl-6-sulfato)-β-cyclodextrin. *Anal Chem* 70: 580–589.
100. Zhu, W, Vigh, G. 2001. A family of single-isomer, sulfated gamma-cyclodextrin chiral resolving agents for capillary electrophoresis. 1. Octakis(2,3-diacetyl-6-sulfato)-gamma-cyclodextrin. *Anal Chem* 72: 310–317.
101. Verleysen, K, Sandra, P. 1998. Separation of chiral compounds by capillary electrophoresis. *Electrophoresis* 19: 2798–2833.

102. Chen, F-TA, Shen, G, Evangelista, RA. 2001. Characterization of highly sulfated cyclodextrins. *J Chromatogr A* 924: 523–532.
103. Schulte, G, Heitmeier, S, Chankvetadze, B, Blaschke, G. 1998. Chiral capillary electrophoresis-electrospray mass spectrometry coupling with charged cyclodextrin derivatives as chiral selectors. *J Chromatogr A* 800: 77–82.
104. Rudaz, S, Cherkaoui, S, Gauvrit, JY, Lanteri, P, Veuthey, JL. 2001. Experimental designs to investigate capillary electrophoresis-electrospray ionization-mass spectrometry enantioseparation with the partial-filling technique. *Electrophoresis* 22: 3316–3326.
105. Rudaz, S, Cherkaoui, S, Dayer, P, Fanali, S, Veuthey, JL. 2000. Simultaneous stereoselective analysis of tramadol and its main phase I metabolites by on-line capillary zone electrophoresis-electrospray ionization mass spectrometry. *J Chromatogr A* 868: 295–303.
106. Cherkaoui, S, Veuthey, JL. 2001. Use of negatively charged cyclodextrins for the simultaneous enantioseparation of selected anesthetic drugs by capillary electrophoresis-mass spectrometry. *J Pharm Biomed Anal* 27: 615–626.
107. Matthijs, N, Perrin, C, Maftouh, M, Massart, DL, Vander Heyden, Y. 2001. Knowledge-based system for method development of chiral separations with capillary electrophoresis using highly-sulphated cyclodextrins. *J Pharm Biomed Anal* 27: 515–529.
108. Perrin, C, Vander Heyden, Y, Maftouh, M, Massart, DL. 2001. Rapid screening for chiral separations by short-end injection capillary electrophoresis using highly sulfated cyclodextrins as chiral selectors. *Electrophoresis* 22: 3203–3215.
109. Rudaz, S, Calleri, E, Geiser, L, Cherkaoui, S, Prat, J, Veuthey, JL. 2003. Infinite enantiomeric resolution of basic compounds using highly sulfated cyclodextrin as chiral selector in capillary electrophoresis. *Electrophoresis* 24: 2633–2641.
110. Cherkaoui, S, Rudaz, S, Varesio, E, Veuthey, JL. 1999. Online capillary electrophoresis-electrospray mass spectrometry for the analysis of pharmaceuticals. *Chimia* 53: 501–505.
111. Cai, JY, Henion, J. 1995. Capillary electrophoresis mass spectrometry [review]. *J Chromatogr A* 703: 667–692.
112. Williams, BA, Vigh, G. 1997. Dry look at the charm (charged resolving agent migration) model of enantiomer separations by capillary electrophoresis. *J Chromatogr A* 777: 295–309.
113. Rudaz, S, le Saux, T, Prat, J, Gareil, P, Veuthey, JL. 2004. Ultrashort partial-filling technique in capillary electrophoresis for infinite resolution of tramadol enantiomers and its metabolites with highly sulfated cyclodextrins. *Electrophoresis* 25: 2761–2771.
114. Armstrong, DW, Rundlett, KL, Chen, JR. 1994. Evaluation of the macrocyclic antibiotic vancomycin as a chiral selector for capillary electrophoresis. *Chirality* 6: 496–509.
115. Fanali, S, Desiderio, C, Schulte, G, Heitmeier, S, Strickmann, D, Chankvetadze, B, Blaschke, G. 1998. Chiral capillary electrophoresis electrospray mass-spectrometry coupling using vancomycin as chiral selector. *J Chromatogr A* 800: 69–76.
116. Tanaka, Y, Kishimoto, Y, Otsuka, K, Terabe, S. 1998. Separation of enantiomers by capillary electrophoresis-mass spectrometry. *Chromatography* 19: 76–77.
117. Tanaka, Y, Kishimoto, Y, Terabe, S. 1998. Separation of acidic enantiomers by capillary electrophoresis-mass spectrometry employing a partial filling technique. *J Chromatogr A* 802: 83–88.
118. Kyba, EP, Timko, JM, Kaplan, LJ, De Jong, F, Gokel, GW, Cram, DJ. 1978. Host-guest complexation. 11. Survey of chiral recognition of amine and amino ester salts by dilocular bisdinaphtyl hosts. *J Am Chem Soc* 1000: 4555–4568.
119. Tanaka, Y, Otsuka, K, Terabe, S. 2000. Separation of enantiomers by capillary electrophoresis-mass spectrometry employing a partial filling technique with a chiral crown ether. *J Chromatogr A* 875: 323–330.
120. Zhou, L, Lin, Z, Reamer, RA, Mao, B, Ge, Z. 2007. Stereoisomeric separation of pharmaceutical compounds using CE with a chiral crown ether. *Electrophoresis* 28: 2658–2666.

121. Shamsi, SA, Macossay, J, Warner, IM. 1997. Improved chiral separations using a polymerized dipeptide anionic chiral surfactant in electrokinetic chromatography: Separations of basic, acidic, and neutral racemates. *Anal Chem* 69: 2980–2987.
122. Shamsi, SA, Warner, IM. 1997. Monomeric and polymeric chiral surfactants as pseudostationary phases for chiral separations. *Electrophoresis* 18: 853–872.
123. Shamsi, SA. 2001. Micellar electrokinetic chromatography-mass spectrometry using a polymerized chiral surfactant. *Anal Chem* 73: 5103–5108.
124. Terabe, S, Miyashita, Y, Shibata, O. 1990. Separation of highly hydrophobic compounds by cyclodextrin- modified MEKC. *J Chromatogr* 516: 23–31.
125. Anderson, MS, Wan, H, Abdel-Rehim, M, Blomberg, LG. 1999. Characterization of lidocaine and its metabolites in human plasma using capillary electrophoresis. *J Microcol Sep* 11: 620–626.
126. Cherkaoui, S, Rudaz, S, Veuthey, JL. 2001. Nonaqueous capillary electrophoresis-mass spectrometry for separation of venlafaxine and its phase I metabolites. *Electrophoresis* 22: 491–496.
127. Geiser, L, Veuthey, JL. 2007. Nonaqueous capillary electrophoresis in pharmaceutical analysis. *Electrophoresis* 28: 45–57.
128. Tomlinson, AJ, Benson, LM, Naylor, S. 1995. Advantages of nonaqueous solvents in the analysis of drug metabolites using CE and on-line CE-MS. *LC-GC Int* 8: 210–216.
129. Servais, AC, Fillet, M, Mol, R, Somsen, GW, Chiap, P, de Jong, GJ, Crommen, J. 2006. On-line coupling of cyclodextrin mediated nonaqueous capillary electrophoresis to mass spectrometry for the determination of salbutamol enantiomers in urine. *J Pharm Biomed Anal* 40: 752–757.
130. Mol, R, de Jong, GJ, Somsen, GW. 2008. Coupling of non-aqueous electrokinetic chromatography using cationic cyclodextrins with electrospray ionization mass spectrometry. *Rapid Commun Mass Spectrom* 22: 790–796.
131. Mol, R, Servais, AC, Fillet, M, Crommen, J, de Jong, GJ, Somsen, GW. 2007. Nonaqueous electrokinetic chromatography-electrospray ionization mass spectrometry using anionic cyclodextrins. *J Chromatogr A* 1159: 51–57.
132. Sanchez-Hernandez, L, Crego, AL, Marina, ML, Garcia-Ruiz, C. 2008. Sensitive chiral analysis by CE: An update. *Electrophoresis* 29: 237–251.
133. Garcia-Ruiz, C, Marina, ML. 2006. Sensitive chiral analysis by capillary electrophoresis. *Electrophoresis* 27: 195–212.
134. Kindt, EK, Kurzyniec, S, Wang, SC, Kilby, G, Rossi, DT. 2003. Quantitative bioanalysis of enantiomeric drugs using capillary electrophoresis and electrospray mass spectrometry. *J Pharm Biomed Anal* 31: 893–904.
135. Watzig, H, Degenhardt, M, Kunkel, A. 1998. Strategies for capillary electrophoresis: Method development and validation for pharmaceutical and biological applications. *Electrophoresis* 19: 2695–2752.
136. Geiser, L, Rudaz, S, Veuthey, JL. 2003. Validation of capillary electrophoresis—mass spectrometry methods for the analysis of a pharmaceutical formulation. *Electrophoresis* 24: 3049–3056.
137. Souverain, S, Rudaz, S, Veuthey, JL. 2004. Matrix effect in LC-ESI-MS and LC-APCI-MS with off-line and on-line extraction procedures. *J Chromatogr A* 1058: 61–66.
138. Bonfiglio, R, King, RC, Olah, TV, Merkle, K. 1999. The effects of sample preparation methods on the variability of the electrospray ionization response for model drug compounds. *Rapid Commun Mass Spectrom* 13: 1175–1185.
139. Schappler, J, Guillarme, D, Prat, J, Veuthey, JL, Rudaz, S. 2006. Enhanced method performances for conventional and chiral CE-ESI/MS analyses in plasma. *Electrophoresis* 27: 1537–1546.
140. Ohnesorge, J, Saenger-van de Griend, C, Waetzig, H. 2005. Quantification in capillary electrophoresis-mass spectrometry: Long- and short-term variance components and their compensation using internal standards. *Electrophoresis* 26: 2360–2375.

141. Liu, RH, Lin, DL, Chang, WT, Liu, C, Tsay, WI, Li, JH, Kuo, TL. 2002. Issues to address when isotopically labeled analogues of analytes are used as internal standards. *Anal Chem* 74: 618A–626A.
142. Boulanger, B, Chiap, P, Dewe, W, Crommen, J, Hubert, P. 2003. An analysis of the SFSTP guide on validation of chromatographic bioanalytical methods: Progresses and limitations. *J Pharm Biomed Anal* 32: 753–765.
143. Hubert, P, Chiap, P, Crommen, J, Boulanger, B, Chapuzet, E, Mercier, N, Bervoas-Martin, S, Chevalier, P, Grandjean, D, Lagorce, P, Lallier, M, Laparra, MC, Laurentie, M, Nivet, JC. 1999. The SFSTP guide on the validation of chromatographic methods for drug bioanalysis: From the Washington Conference to the laboratory. *Anal Chim Acta* 391: 135–148.
144. Hubert, P, Nguyen-Huu, JJ, Boulanger, B, Chapuzet, E, Chiap, P, Cohen, N, Compagnon, PA, Dewe, W, Feinberg, M, Lallier, M, Laurentie, M, Mercier, N, Muzard, G, Nivet, C, Valat, L. 2004. Harmonization of strategies for the validation of quantitative analytical procedures. *J Pharm Biomed Anal* 36: 579–586.
145. Schappler, J, Guillarme, D, Prat, J, Veuthey, JL., Rudaz, S. 2008. Validation of chiral capillary electrophoresis electrospray ionization-mass spectrometry methods for ecstasy and methadone in plasma. *Electrophoresis* 29: 2193–2202.

14 Enantioseparations by Capillary Electrochromatography

Michael Lämmerhofer

CONTENTS

14.1 Introduction ... 393
14.2 Principal Approaches ... 394
14.3 Electroosmotic Flow ... 395
14.4 Migration and Separation Mechanism ... 399
14.5 Pressurized CEC ... 403
14.6 Zone Dispersion ... 405
14.7 Mobile Phase Modes ... 407
14.8 Column Technologies ... 409
- 14.8.1 Open-Tubular Columns ... 409
- 14.8.2 Packed Capillary Columns ... 415
- 14.8.3 Monolithic Columns ... 426
  - 14.8.3.1 Silica Monoliths ... 428
  - 14.8.3.2 Organic Polymer Monoliths ... 431
  - 14.8.3.3 Hybrid Monoliths from Particles ... 440

14.9 CEC–MS ... 441
14.10 Concluding Remarks ... 443
List of Abbreviations ... 444
References ... 444

14.1 INTRODUCTION

Capillary electrochromatography (CEC) [1] is, according to the definition by IUPAC, "a special case of capillary liquid chromatography (CLC), where the movement of the mobile phase through a capillary that is filled, packed or coated with a stationary phase is achieved by electroosmotic flow (EOF) (which may be assisted by pressure)." In this electric field driven system, "the retention time is determined by a combination of electrophoretic migration and chromatographic retention [2]." CEC experiments are typically performed in capillary electrophoresis (CE) equipment that allow the application of voltages up to ±30 kV and external pressures up to 12 bar at each end of the capillary column. If equal pressures are applied at both ends, no pressurized flow is generated, yet bubble formation in the capillary can be successfully prevented. Despite the availability of an external pressure system, sample injections are usually carried out electrokinetically due to backpressure restrictions arising from the packing material in the column. In the normal arrangement on-column UV detection is performed near the outlet end leading to longer separation zones (typically 25 cm), while in the so-called short-end injection the separation is performed by injecting the sample at the outlet end yielding a short migration distance to the detector (commonly ~8.5 cm) and hence faster separation (note, both injection and separation are carried out with reversed polarity in this case).

CEC became popular in the late 1990s due to its capability to outperform high-performance liquid chromatography (HPLC) in terms of separation efficiencies and reportedly offers more flexibility for selectivity adjustment via the stationary phase than CE where mobility differences are the key to separation. It turned out, however, that the high efficiencies are not that easy to realize, especially in real samples in which matrix components may adsorb and lead to a modification of the surface charge and the flow properties. Moreover, the complexity of the separation mechanisms, difficulties to control retentions and EOF as well as the lack of robustness, compared to HPLC, seriously hampered the implementation in routine analysis laboratories. Nowadays, CEC still attracts considerable interest in academic environment. Some of the basic principles and practical aspects will be discussed in the following chapter.

14.2 PRINCIPAL APPROACHES

Separation of enantiomers can be accomplished by three basically different ways:

1. The indirect approach
2. The chiral mobile phase additive (CMPA) approach
3. The chiral stationary phase (CSP) approach

In the "indirect approach," analyte enantiomers are chemically modified with a highly enantiomerically pure chiral derivatizing agent yielding diastereomers which owing to their different physicochemical properties may be separated by CEC methods employing achiral stationary phases [3]. For example, β-blockers like propranolol, metoprolol, and atenolol have been separated by CEC on 3 µm octadecyl-modified silica (ODS) stationary phases after precolumn derivatization with (+)-1-(9-fluorenyl) ethyl chloroformate (FLEC) [3]. The indirect approach is associated with a number of drawbacks, limitations, problems, and caveats [4] which make it, like in HPLC, a second choice methodology in CEC enantiomer separation. This is reflected in the negligible number of studies found in the literature.

In one mode of the direct approach, the chiral selector is added in enantiomeric form to the mobile phase and background electrolyte (BGE), respectively ("CMPA approach"). Enantiomeric separation takes place on an achiral column such as ODS and mobility differences arise due to enantioselective complexation of the analyte enantiomers with the chiral selector. Migration differences of the enantiomers are created as a result of distinct complexation, mobility and retention differences of free and complexed solute species and/or of the formed diastereomeric associates [5–9]. Mechanistic studies have been published by Deng et al. [5]. In most cases, cyclodextrins (CD) or its derivatives were employed as chiral additives and the capillary columns were packed with 3 μm ODS [5,7–9]. In one example, a chiral ion-pair selector, namely *O*-9-*tert*-butylcarbamoylquinine, was utilized as CMPA for *N*-derivatized amino acids [6]. Due to the capillary format, the CMPA approach is a viable technique. However, detection issues such as interferences of the selector with solute detection, distinct detector response for individual enantiomers (which are detected in the "detection cell" as diastereomeric associates), and limitations in terms of robustness make it also a second choice methodology in CEC enantioseparation.

In the preferred direct CEC enantiomer separation mode, capillary columns containing a CSP are used and the solute enantiomers differently interact with the chiral selector of the CSP leading to diastereomeric associates on the surface ("CSP approach"). The enantiomer forming the more stable complex elutes later leading to the separation of the individual enantiomers of the mixture.

The first two approaches ("indirect method" and "CMPA approach") which have little practical importance despite their operational simplicity and straightforward implementation are not further treated herein. The following sections solely focus on the "CSP approach" of direct CEC enantioseparation with capillary columns that accommodate a CSP.

14.3 ELECTROOSMOTIC FLOW

CEC utilizes electroosmosis for the generation of bulk mobile phase flow instead of a pressure gradient and this is the most significantly distinctive characteristic in comparison to nano-HPLC. Electroosmosis is an electrokinetic phenomenon in which the bulk liquid moves tangentially relative to an adjacent charged surface upon application of an electric field. It is the result of an electrical double layer between a charged surface and an electrolyte solution being in contact with each other. The double layer is characterized by the ζ-potential, which is the potential at the shear distance from the surface [10]. More details on the theory of the electrical double layer (EDL) can be found in Ref. [10].

EOF has certain advantages over hydrodynamic flow, such as absence of backpressure and independence of its velocity from the diameter of the flow channels which allows, e.g., the use of small-particle diameter columns (even in the submicrometer range). Moreover, it is characterized by a flat flow profile as compared to the parabolic profile of pressurized flow being beneficial in terms of zone dispersion [11].

Like in CE, the von Smoluchowski equation (Equation 14.1) is commonly invoked to describe EOF in CEC.

$$\mu_{eo} = -\frac{\varepsilon_0 \cdot \varepsilon_r \cdot \zeta}{\eta} \tag{14.1}$$

It is evident that the electroosmotic mobility μ_{eo} is proportional to the ζ-potential ($\zeta = \sigma_0\kappa^{-1}/\varepsilon_0\varepsilon_r$), and depends—besides the surface charge density σ_0—on a number of other factors such as electrolyte concentration (ionic strength I), dielectric permittivity of the medium ε, viscosity η, and temperature T as given by Equation 14.2

$$\mu_{eo} = \frac{\sigma_0}{\eta} \cdot \sqrt{\frac{\varepsilon_0 \cdot \varepsilon_r \cdot R \cdot T}{2F^2}} \cdot \frac{1}{\sqrt{I}} \tag{14.2}$$

wherein

κ is the Debye–Hückel parameter
R is the gas constant (1.38×10^{-23} J/K)
T is the temperature in K
F is the Faraday constant
ε_r is the relative permittivity (78.54 for water)

the ionic strength $I = \frac{1}{2} \sum c_i z_i^2$, where c_i is the molar ionic concentration and z_i the charge number of the particular ion.

Since the charge of surfaces with ionizable groups such as siliceous materials or modified silica materials depends on the pH, mobile phase pH is a major determinant of the strength of the EOF. The von Smoluchowski equation is, however, an approximation and other factors may influence the electroosmotic mobility in CEC as well. Besides the above factors, double-layer overlap in narrow pores (if $\kappa r < 10$, wherein r denotes the pore radius), concentration polarization (CP) in mesoporous media, packing factors, and wall effects make the situation in CEC more complicated [12–15].

It is evident from Equation 14.1 that the EOF velocity ($\nu_{eo} = \mu_{eo} E$, with E being the electric field strength) is dependent on both mobile phase conditions and stationary phase effects. Strong and stable EOF requirement dictates that the mobile phase is composed of a medium with high dielectric-viscosity ratio and this is best realized by acetonitrile-rich aqueous or methanolic solutions [16]. In contrast, typical normal-phase (NP) conditions such as hexane-2-propanol mixtures do not fulfill this requirement because of the low dielectric constant of such media. The insufficient solubility of ionic species for ionic conduction is a further discouraging factor to implement NP conditions in CEC. Unfortunately, optimal conditions with regard to EOF strength are not always ideally compatible with the enantiomer separation capability of the chiral selector system so that compromises in one or the other property have often to be accepted. In order to minimize currents and Joule heating, ionic strength of the eluents is kept low in CEC and good starting conditions in this respect are ionic strengths of around 10 mmol/L. If ionic solute–adsorbent interactions are involved in adsorbate-formation, such improperly counterbalanced ionic interactions may lead to strong adsorption and long retention times as well as poor peak shapes. Increasing the counterion concentration may be necessary leading to faster elution at the expense of higher currents (which should be kept below 20 μA). Nonaqueous polar organic conditions with polar solvents like acetonitrile and methanol in combination with relatively high concentrations of organic additives (i.e., organic acids and bases serving as co and counterions) turned out to be advantageous in such cases, e.g., with chiral ion-exchanger CSPs [17,18].

Regarding enantioselective separation systems, both the chiral selector and the supporting carrier material on which the chiral selector is bonded may principally contribute to surface charge and ζ-potential, and thus to EOF generation. Hence, we may classify the CSPs into those where:

1. The chiral selector is neutral and only the support (usually silica or aminopropylated silica) provides charged groups as source for EOF generation. Polysaccharide-based CSPs, cyclodextrin-based CSPs, and Pirkle-type CSPs do belong to this group of CSPs.
2. The chiral selector is ionizable, but due to its limited molar surface concentration does not significantly contribute as a source for EOF generation. EOF characteristics of such stationary phases is thus dominated by the support, i.e., usually residual silanols. Protein-based CSPs and macrocyclic antibiotic CSPs may be classified under such a system.
3. The chiral selector is ionizable and significantly contributes or dominates the EOF behavior of the stationary phase which is, for example, the case for chiral ion-exchangers.

Although it has been shown that even neutral polymeric materials that were devoid of any charged functional groups on the surface can produce reasonable EOF when they are in contact with an electrolyte solution, such a system has not been reported for enantioselective CEC yet.

CEC columns of class (1) and (2) do not significantly differ from each other in terms of EOF characteristics. The trends are the same as in both systems the dissociation of the residual silanols determines the overall EOF behavior. A typical example of pH dependency for such CSPs is depicted in Figure 14.1.

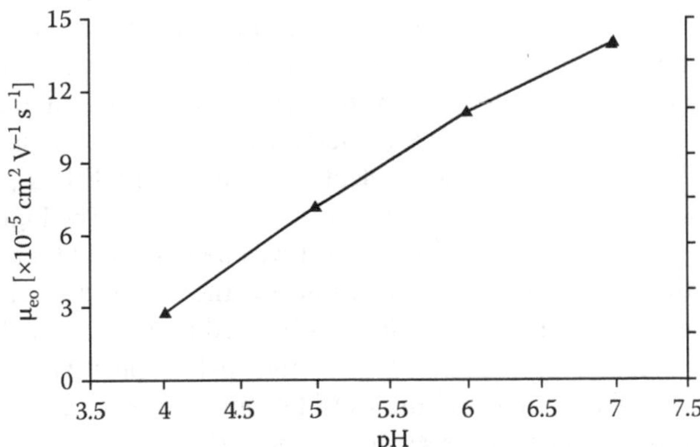

FIGURE 14.1 Dependence of the EOF on the pH for vancomycin-based particulate packed capillary column. Experimental conditions: Mobile phase, acetonitrile-0.1% triethylammonium acetate (30:70; v/v); capillary dimension, 335 mm × 75 μm i.d. (L_{eff}, 250 mm); applied voltage, 25 kV; 2s injection at 10 kV; 10 bar at both inlet and outlet vials; detection at 214 nm; EOF marker, acetone. (Reprinted from Karlsson, C. et al., *Anal. Chem.*, 72, 4394, 2000. With permission.)

A silica-supported vancomycin-based CSP (Chirobiotic V, 5 μm particles) has been packed into 75 μm i.d. capillary and was tested in CEC [19]. The vancomycin selector prior to immobilization has a number of ionizable groups (six with different p*K* in the range of 2.9–11.7 yielding p*I* of 7.2). However, the plot of the electroosmotic mobility vs. pH shows the typical pH-curve known from siliceous surfaces with maximal EOF above pH 6 because the EOF is dominated by the residual silanols of the modified silica particles (see Figure 14.1). Hence, CSPs of type (1) and (2) which behave quite similar should be operated at pH values above 4, ideally above 6, to benefit from a maximal EOF strength. Sometimes, such selectors, e.g., polysaccharides [20] are immobilized onto 3-aminopropylated silica (APS) which leads to altered EOF characteristics, namely, stronger EOF in the low pH region. In some instances, an ionic spacer has been introduced to accelerate the EOF [21].

Ion-exchangers, on the other hand, may be the preferred stationary phases in CEC owing to the capability of the bonded ionic groups to support EOF. Such enantioselective ion-exchangers with ionizable low molecular chiral selectors have been successfully adapted for the CEC enantiomer separation [17,22–24]. These type (3) CSPs show considerably distinct EOF behavior because the EOF does not rely on the residual silanols (Figure 14.2). Figure 14.2a and b shows the structures of a quinine-carbamate based weak chiral anion-exchanger and of a β-aminosulfonic acid derived strong chiral cation-exchanger, respectively, which were both immobilized onto thiol-modified silica particles (3.5 μm, 100 Å pore size). The pH-dependency of the resulting ζ-potential of these chiral ion-exchangers is shown in Figure 14.2c along with the corresponding plain silica and thiol-modified silica particles (note that the EOF mobility is, according to Equation 14.1, directly proportional to the ζ-potential) [12]. The ζ-potential has been measured in an aqueous 10 mM KCl solution in 1 mM buffer over the pH range between 3 and 10 (thus, $\mu_{eo} = 0.78\zeta$; with μ_{eo} given in 10^{-9} m^2/V s and ζ in mV). As can be seen, at high pH the plain silica particles have the largest ζ-potential which drops significantly as the pH is lowered. The thiol-modified particles have throughout the investigated pH range a somewhat lower ζ-potential because a significant portion of all silanols have been derivatized with 3-mercaptopropylsilane. The introduction of the sulfonic acid selector makes the EOF less pH-dependent, and a high and stable EOF can be maintained also at low pH (see Figure 14.2c). For all of these stationary phases, the ζ-potential was negative and therefore the EOF is directed to the cathode which is favorable for the analysis of cations. On the other hand, when the quinine carbamate was bonded to the surface the ζ-potential of the particles changed from negative at high pH to positive at low pH (<8) (see Figure 14.2c). Thus, at low pH the EOF becomes anionic, i.e., it is directed to the anode which is favorable for the separation of acids and anions, respectively. The useful pH range of such silica-supported weak chiral anion-exchanger particles is in this lower acidic pH range, while the CEC system is highly instable in particular around the isoelectric point of the particles. Anyhow, it can be clearly seen that for both chiral cation and anion-exchangers the EOF was dominated by the charged chiral selector moieties.

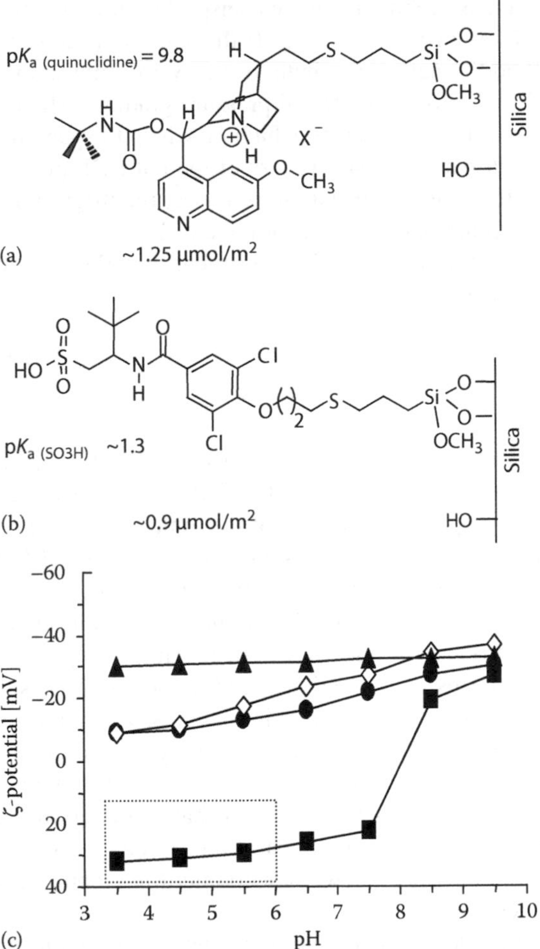

FIGURE 14.2 Chiral anion-exchanger (a), chiral cation-exchanger (b), and their ζ-potential (c) as measured by microelectrophoresis, of native and modified spherical silica particles. Experimental conditions: 10 mM KCl in 1 mM buffer; supporting particles, Kromasil 100 Å, 3.5 µm (320 m²/g). Rhombi, silica; circles, 3-mercaptopropyl-modified silica; squares, chiral anion-exchanger (a); triangles, chiral cation-exchanger (b). Dotted line indicates recommended pH range for chiral anion-exchanger CSP to obtain robust conditions. (Modified from Munoz, O.L.S. et al., *Electrophoresis*, 24, 390, 2003. With permission.)

14.4 MIGRATION AND SEPARATION MECHANISM

In the absence of electrokinetic phenomena of the second kind [14,15,25] (which is typically realized at common ionic strengths of 10 mmol/L or higher) the observed linear migration velocity ν_{CEC} (usually given in mm/s) in CEC can be simply approximated by Equation 14.3 [5,25].

$$\nu_{CEC} = \left(\nu_{eo} + \nu_{ep} + \nu_{press}\right) \cdot \frac{1}{\left(1 + k_{CLC}\right)} \qquad (14.3)$$

Both analytes as well as mobile phase are transported through the capillary column containing the CSP by virtue of EOF ν_{eo}. If the solutes are ionized, electrophoretic migration ν_{ep} will additionally contribute to the overall migration velocity. In pressure-assisted CEC (pCEC) an additional pressure gradient which generates a pressurized flow ν_{press} is externally applied which may help to speed up the analysis and stabilize the flow, but partially leads to loss of the flat EOF profile [1]. If the solute enantiomers can interact with the stationary phase, the migration speed is reduced depending on the chromatographic retention factor k_{CLC}.

Under conditions of a typical CEC experiment, i.e., the absence of pressurized flow Equation 14.3 becomes

$$\nu_{CEC} = (\nu_{eo} + \nu_{ep}) \cdot \frac{1}{(1 + k_{CLC})} = \frac{(\mu_{eo} + \mu_{ep,eff}) \cdot E}{1 + k_{CLC}} \tag{14.4}$$

wherein E is the electric field strength and μ_{CEC} (=ν_{CEC}/E), μ_{eo}, and $\mu_{ep,eff}$ are the observed mobilities of the ionic solute in CEC, the electroosmotic mobility in CEC and the effective electrophoretic mobility of the ionic solute, respectively.

Since the individual processes for solute migration in CEC (without applied pressure gradient), ν_{eo} and ν_{ep}, may have distinct directions, it may be distinguished between a codirectional and a counterdirectional separation process in which electroosmotic and electrophoretic mobilities are in same or opposite directions [26]. It is evident that in terms of separation speed a codirectional separation mode is highly desirable. This is documented by Figure 14.3a, where the separation by CLC on the same column with bulk mobile phase flow adjusted to the same velocity (note the equal t_0 is CEC and CLC) served as reference for a system with absence of electrophoretic migration contributions. It is seen that in the codirectional mode with electrophoretic migration support the separation can be greatly accelerated (Figure 14.3a), while it would be slowed down in a counterdirectional separation leading to longer analysis times.

Preinerstorfer et al. deconvoluted the individual contributions of the CEC retention process in accordance with Equation 14.3, by separate measurement of chromatographic retention factors k_{CLC} and electrophoretic mobilities $\mu_{ep,eff}$ under identical mobile phase conditions [25]. Reasonable agreements for CEC velocities determined experimentally and calculated from individual migration contributions by Equation 14.3 were found in the absence of nonlinear electrokinetic effects. For example, the electrophoretic mobility of pronethalol under conditions as given in Figure 14.3 was determined to be 22.4×10^{-5} cm^2/V s and thus it contributed almost equally as the EOF ($\mu_{eo} = 26.7 \times 10^{-5}$ cm^2/V s) to solute transport through the column. Liquid chromatographic retention factors for the two enantiomers were 1.96 ± 0.02 and 2.28 ± 0.02 for the first and second eluted enantiomers yielding electrochromatographic velocities that were reduced to about 36% and 32% of the combined migration velocities (sum of ν_{eo} and ν_{ep}) with absence of retention for first and second migrating enantiomers. It is obvious that valuable insights into complex CEC separation mechanisms may be gained by such a deconvolution approach.

The typical capillary column in CEC contains a porous CSP (see below) that is packed into two distinct zones yielding a dual character with an open and a packed segment.

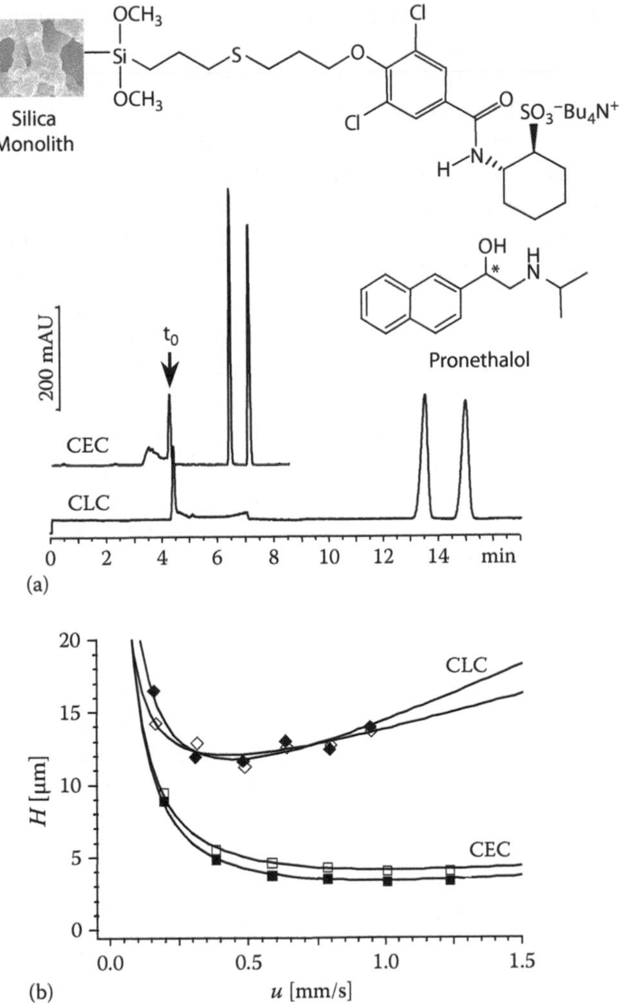

FIGURE 14.3 CEC and CLC enantiomeric separations of pronethalol (a) and *H–u* curves of mefloquine (b) on a monolithic strong chiral cation-exchanger CSP. Deconvoluted migration contributions for pronethalol in CEC: $\mu_{eo} = 26.7 \pm 0.9 \times 10^{-5}$ cm^2/V s; $\mu_{ep,eff} = 22.4 \pm 0.6 \times 10^{-5}$ cm^2/V s; $k_{CLC,1} = 1.96 \pm 0.02$; $k_{CLC,2} = 2.28 \pm 0.02$; $\mu_{CEC,1} = 17.6 \pm 0.4 \times 10^{-5}$ cm^2/V s; $\mu_{CEC,2} = 15.8 \pm 0.30 \times 10^{-5}$ cm^2/V s. Experimental conditions: Mobile phase, ACN-MeOH (80:20, v/v) containing 25 mM formic acid and 12.5 mM 2-aminobutanol; applied voltage, 20 kV (a) and 4–24 kV (b), respectively; electrokinetic injection at 10 kV/10 s; capillary temperature, 20°C; UV detection at 216 nm; capillary dimension, L_{tot} = 33.5 cm, L_{eff} = 25 cm. (Modified from Preinerstorfer, B. et al., *J. Sep. Sci.*, 31, 3065, 2008. With permission.) (b) CLC: applied pressures, 2–12 bar. (Modified from Preinerstorfer, B. et al., *Electrophoresis*, 29, 1626, 2008. With permission.)

It must be borne in mind that in such a dual column with distinct conductivities in the open and packed segment the voltage does not drop linearly and thus the actual electric field strength is different in both of the segments that may lead to inappropriateness of Equation 14.1 [13,25]. Another complication has been described by Tallarek and coworkers [14,15,25]. The above model may be seriously impaired by nonlinear electrokinetic effects (electrokinetic phenomena of the second kind) such as arising from CP. Since EOF is originating from the EDL on the CSP surface, double layer overlap in mesopores of stationary phases such as particulate and monolithic materials (in particular ion-exchangers) may occur at low ionic strength conditions (e.g., ionic strengths <1 mM). As a consequence, they become charge-selective and thus develop ion-permselectivity which means that they exclude coions and enrich counterions in the intraparticulate pore spaces [14,15]. Upon application of high voltages typically used in CEC, CP may arise leading to nonlinear electrokinetic effects. In such cases when CP is significant, Equations 14.3 and 14.4, respectively, do not adequately describe the CEC process. Also other factors such as Joule heating may disturb the validity of Equation 14.3 [28].

The goal of enantioselective CEC is the separation of corresponding enantiomers and this can be achieved if the observed migration velocities v_{CEC} differ from each other. It is evident from Equation 14.1 that different retention factors ($k_{CLC,1} \neq k_{CLC,2}$) are the only source for enantioselectivity. Like in HPLC, differences in k_{CLC} of the two enantiomers may be explained based on thermodynamic grounds and nonequal Gibbs free energies of the enantiomers for the transfer of the solute from the mobile phase to the CSP [4]. In contrast, electrophoretic mobilities which are identical for enantiomers are nonselective contributions in view of generating chiral separation [26,29]. However, mobility differences between chemically distinct compounds may provide valuable chemoselectivity contributions in more complex separations, e.g., if (phase 1) metabolites of chiral drugs need to be separated besides parent drug enantiomers.

The terminology that is commonly used to describe enantiomer separations in CEC is more or less the same as in enantioselective HPLC, although it is not always adequate [18]. Retention factors are commonly calculated as in HPLC [30], namely

$$k_{CEC} = \frac{t_e - t_0}{t_0} = k_{CLC} - \mu_r \cdot k_{CLC} - \mu_r \tag{14.5}$$

wherein t_e and t_0 are the elution times of the solute and of a neutral nonretained flow marker, respectively. While for neutral compounds, the meaning of k_{CEC} is the same as in HPLC, for ionic solutes it has the meaning of a peak locator only because it depends on the electrophoretic mobility contribution as described by Equation 14.5. In this equation the reduced mobility μ_r describes the relative contribution of electrophoretic mobility to the sum of both driving forces $\mu_{ep,eff}$ and μ_{eo} as defined by Equation 14.6

$$\mu_r = \frac{\mu_{ep,eff}}{\mu_{ep,eff} + \mu_{eo}} \tag{14.6}$$

The retention window for the separation in CEC involving electrophoretic migration is thus spanned by

$$-1 < k_{\mathrm{CEC}} < \infty \tag{14.7}$$

For the separation factors α_{CEC} it then holds

$$\alpha_{\mathrm{CEC}} = \frac{k_{\mathrm{CLC},2} - \mu_r \cdot k_{\mathrm{CLC},2} - \mu_r}{k_{\mathrm{CLC},1} - \mu_r \cdot k_{\mathrm{CLC},1} - \mu_r} \tag{14.8}$$

They contain the influence of the electrophoretic mobility contribution, which depending on its relative magnitude exerts an alteration on the calculated separation factor α_{CEC}, as compared to HPLC separation factors. Obscure numbers may be attained and α_{CEC} may adopt even negative values if one enantiomer is eluted before and the second after the neutral marker.

Thus, if ionic solutes elute very close to or even before the EOF marker the chromatographic terminology to calculate separation factors is inappropriate, and the electrophoretic formalism, where α_{CE} is calculated as the ratio of the mobilities or linear velocities or simply elution times should be adopted.

$$\alpha_{\mathrm{CE}} = \frac{v_{\mathrm{CEC},1}}{v_{\mathrm{CEC},2}} = \frac{(1 + k_{\mathrm{CLC},2})}{(1 + k_{\mathrm{CLC},1})} \tag{14.9}$$

Such calculated separation factors are void of strong electrophoretic bias and depend only on the chromatographic retention factors. Yet, they are lower than the liquid chromatographic separation factors unless k_{CLC} values become large compared to 1.

Both of the above definitions of α-values have the serious disadvantage that they are not directly related to the thermodynamic selectivity of the solute–selector interaction, i.e., the intrinsic selectivity of the chiral selector system. Preinerstorfer et al. showed that, if CP, Joule heating, and other disturbing effects are absent, meaningful chromatographic α-values may be easily derived from CEC runs and CE experiments under identical conditions using Equation 14.2 that agree very well with those of a chromatographic experiment and are related to the Gibbs free energy of the solute–sorbent interaction [25].

14.5 PRESSURIZED CEC

A number of papers dealt with the application of pressure-supported CEC for enantiomer separation [31–44]. The separation process may become even more complicated in pCEC enantiomer separation because pressurized flow may be co- or counterdirectional with respect to EOF and/or electrophoretic migration of ionic solutes. Thus, pressure support introduces an additional degree of freedom for retention tuning of solutes and offers even greater flexibility for method development as outlined by Wistuba et al. [38], amongst others. Equation 14.1 may provide a helpful framework for method development in this case and for rationalization of the individual contributions. pCEC retention and selectivity has also been modeled by Yao et al. who utilized a commercial vancomycin phase (Chirobiotic V, 5 μm), electrokinetically packed, for the separation of several basic drugs on a modified version of the commercial TriSep™ 2000 pressurized CEC system (from Unimicro, Pleasanton, CA) [37]. In this paper, the authors investigated the influence of varying pressure as well as electric field strength on separation performance.

Pressure-assistance was found to be an appropriate means to make elution less dependent on EOF and to accelerate the separation [33,35]. For example, the EOF in the NP CEC enantiomer separation shown in Figure 14.4 is weak (due to low polarity of the NP eluent). Therefore, the separation (of benzoin on Whelk O1) was slow. Compared to both CEC and capillary HPLC separations (0.2 mL/min), under same conditions the run time of the pCEC separation was greatly reduced and the resolution improved (see Figure 14.4). Moreover, pCEC is an effective way to stabilize the flow and make the CEC system more robust, which seems to be of particular importance when the EOF is weak (like in effectively bonded or coated CSPs without ionic groups in the bonding and coating, respectively).

Compared to a pure CEC mode, chromatographic performance is usually compromised only to a minor degree. The *H*/*u*-curves (Van Deemter curves) are shifted so that the *H*-minima (mimima of theoretical plate heights H_{min}) are located at slightly

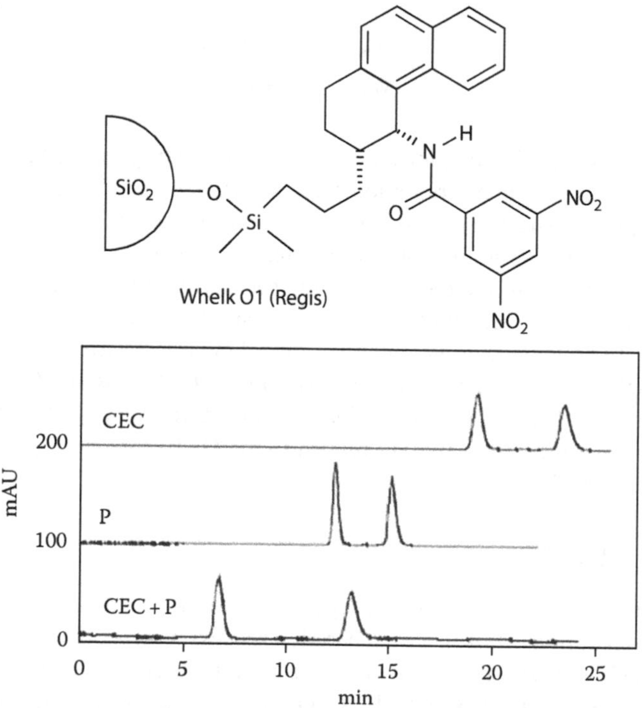

FIGURE 14.4 Chromatograms showing the separation of the chiral analyte benzoin by CEC, pressure (P), and pressure-assisted CEC (CEC + P). Mobile phase: Hexane/2-propanol (1:1, v/v) + 5% (v/v) water. CSP: Whelk O-1. (Reprinted from Vickers, P.J. and Smith, N.W., *J. Sep. Sci.*, 25, 1284, 2002. With permission.)

higher linear flow velocities [34]. *A*-terms are slightly larger and H_{min} increases in the following order $H_{min,CEC} < H_{min,pCEC} < H_{min,HPLC}$ [34]. Such shifts of the *H*-minima to higher flow velocities are certainly favorable for fast separations.

Last but not least, it should be emphasized that hyphenation of CEC with mass-spectrometric detection might benefit from the additional application of a pressure at the inlet end. Capillary columns for MS have to be quite long to allow the introduction of the effluent to the ESI-sprayer. This leads to low field strengths and poor flow stability, and pressure-assistance may help to partly overcome this problem [45].

14.6 ZONE DISPERSION

The major driving force to use CEC is to achieve more efficient separations than with HPLC. Various contributions can be made responsible for a reduced dispersion of analyte zones in CEC (see below). The Van Deemter equation (Equation 14.10), which describes the dependency of the theoretical plate height *H* of a test compound on the linear flow velocity *u* in the HPLC mode as the sum of different contributions (*A*-term, Eddy diffusion; *B*-term, longitudinal diffusion; *C*-term, resistance to mobile phase and stationary phase mass-transfer) is usually employed also for CEC to characterize the performance of packed columns [46,47].

$$H = A \cdot u^{1/3} + B/u + C \cdot u \tag{14.10}$$

wherein $A = 2\lambda d_p$, $B = 2\gamma D_m$, and $C = [k/(1 + k)^2]d_p^2/D_s$ with d_p being the particle diameter, D_m and D_s the diffusion coefficients in mobile and stationary phases, *k* the retention factor, λ the packing factor, and γ an obstruction factor for diffusion. Kinetic plot methods (Poppe plot etc.) have later been proposed to compare separation columns filled with materials of different morphologies in HPLC as well as CEC [46,48,49].

In CE enantiomer separation, there is no packing in the capillary and hence Eddy diffusion as well as mass transfer resistances are absent. Enantioselectivity is created by differential selector-analyte complexation of individual enantiomers. If the selector-analyte association–dissociation process is reasonably fast at the time scale of the CE experiment which is usually the case and external factors such as electrokinetic dispersion, Joule heating, wall adsorption, etc. are absent, Equation 14.10 reduces to $H = B/u$ which is more or less tantamount to the Einstein equation and theoretical plate heights are proportional to solute diffusion coefficients in the BGE. Hence, it is obvious that higher peak efficiencies can be achieved in CE enantiomer separation than with both HPLC and CEC.

On the other hand, zone dispersion effects in CEC enantiomer separation are largely the same as in achiral CEC. One important factor for improved efficiencies in CEC, as compared to HPLC, is the flat flow profile of EOF as opposed to the parabolic velocity profile of pressure driven flow, as already mentioned above. According to Knox and Grant, this favorable flow pattern is, however, partly lost upon significant double layer overlap in the interstitial flow channels which may occur if the diameter of the flow channels d_c is small relative to the thickness of the

EDL δ, e.g., $d_c/\delta < 20$ or $\kappa r < 10$ [50]. The interstitial channel diameter in packed beds may be approximated by $d_c = 0.28d_p$ [51]. Consequently, with typical electrolyte systems of an ionic strength of around 10 mmol/L (δ ~ 30 Å in aqueous solution) no double layer overlap will occur in the interstitial flow channels and a flat flow profile with maximal velocities across the entire flow channel will be maintained [50]. On the other hand, in the intraparticulate pores double layers are partly overlapping under such conditions if a 100 Å material is considered ($d_c/\delta \sim 3$). This reduces the EOF in the pores and thus is supposed to increase *C*-term contributions. Tallarek and coworkers showed, however, that double layer overlap in the mesoporous spaces (i.e., intraparticulate pores) of ion-exchanger particles is associated with phenomena like ion-permselectivity and CP yielding nonlinear electrokinetic effects [15]. As a matter of consequence, even enhanced peak efficiencies could be measured in comparison to conditions under which CP was absent [14]. However, the favorable effect of intraparticulate double layer overlap, CP, and nonlinear kinetics on plate numbers has not been demonstrated yet for enantioselective separation systems.

As can be seen from Equation 14.10, a decrease in *H* and a gain in efficiencies can be simply furnished by reduction in particle diameter. If the traveling distances to reach the adsorption sites inside the pores are reduced, the *C*-term (i.e., resistance to mass transfer) is greatly reduced as well. Since CEC is pressure independent, longer columns than in HPLC even (typically 25 cm) packed with small particles such as 3 μm beads can be used with standard equipment. Since the introduction of UPLC and columns with sub 2 μm particles, this issue is no longer a strong argument for CEC. Yet, the favorable mass transfer in CEC may be partly also caused by intraparticulate pore flow. Since EOF is generated on the surface of the packing, mass transfer in the pores (which in HPLC is by diffusion only) may be enhanced by convection (convective pore flow) if double layer overlaps in the mesopores are absent such as in wide-pore particles [52,53]. This beneficial effect has been proven by Chankvetadze et al. for enantioselective CEC with polysaccharide CSPs (see below) ([54,55]. The result of the better mass-transfer (smaller *C*-term) in CEC in comparison to HPLC are flatter *H–u* curves (which is illustrated in Figure 14.3b). In this example, the *A*-term was by a factor of about 2–3 lower in CEC, while the *C*-term was improved by a factor of about 6 and was insignificant in CEC (Figure 14.3b). As a rule of thumb, in CEC H_{min} values of 1–2 d_p can be reached while for HPLC H_{min} between 2 and 5 d_p is typical [47]. Thus, with particle packed columns, theoretical plate numbers of 100,000–300,000/m can readily be achieved on 3.5 μm particulate packed columns (corresponding to 25,000–75,000 plates per 25 cm column that are commonly employed). Yet, plate numbers up to 400,000/m (corresponding to reduced plate heights h_{red} of less than unity) have also been reported, e.g., for enantioselective CEC with polysaccharide-based CSPs [56] and ion-exchanger CSPs [24]. Moreover, Enlund et al. have shown that by peak compression effects (i.e., focusing effects) plate numbers can be further increased, e.g., from approximately 100,000 to 1.4–1.6 million plates/m on a vancomycin CSP [57]. Such a reproducible focusing effect was accomplished, when the composition of the mobile phase was adjusted so that the target analyte eluted within either one of two system zones originating from the sample solution. This focusing effect could be exploited for impurity profiling. Thus, it could be demonstrated that a tenfold improvement in the quantification limit for the minor enantiomer can be furnished in comparison to elution under nonfocused conditions.

For open-tubular electrochromatography (OT-EC), the *A*-term contribution is absent since there is no packing in the capillary tube and the (extended) Golay equation is usually invoked to describe peak dispersion (Equation 14.11) [46,58]

$$H = \frac{2D_{\text{m}}}{u} + C_{\text{m}} \cdot \frac{r_{\text{c}}^2}{D_{\text{m}}} \cdot u + \frac{2k}{3(1+k)^2} \cdot \frac{d_{\text{f}}^2}{D_{\text{s}}} \cdot u \tag{14.11}$$

wherein

C_{m} is the mobile phase mass transfer resistance factor
r_{c} the capillary radius
d_{f} the film thickness of the stationary phase (coating) and the other parameters are as defined above

The most significant difference for electro-driven and pressure-driven OT separations are then the mobile phase mass transfer terms C_{m} which are different for the flat flow profile in OT-EC and the parabolic profile of OT-LC

$$C_{\text{m,flat}} = \frac{k^2}{4(1+k)^2} \tag{14.12}$$

$$C_{\text{m,parabolic}} = \frac{11k^2 + 6k + 1}{24(1+k)^2} \tag{14.13}$$

A detailed interpretation of the individual parameters of these equations in terms of their contribution to plate height can be found in Ref. [58]. Generally, the difference in $C_{\text{m,flat}}$ and $C_{\text{m,parabolic}}$ represents the essence of the difference in electro-driven and pressure-driven OT chromatography modes. Roughly speaking, for $k < 1$ $C_{\text{m,flat}}$ is about a factor 2 smaller than $C_{\text{m,parabolic}}$ [58]. Requirements for efficient separations in OT formats are the use of very narrow bore capillaries (20 μm i.d. or smaller), thin stationary phase films or coatings (e.g., <0.2 μm thick films), small retention factors that are frequently contradictive with effective enantiomer separations, and slow mobile phase velocities. More details and limitations in the context of enantiomer separations will be discussed later.

14.7 MOBILE PHASE MODES

The necessity to maintain a stable current by ionic conduction through the bulk mobile phase requires reasonable solubility of electrolytes in the eluent. Hence, it is evident that reversed-phase conditions, i.e., aqueous or hydroorganic eluents, have been the first choice in CEC enantiomer separation in terms of mobile phase mode especially if chiral recognition involves hydrophobic interactions which are strengthened in aqueous media. In order to achieve a high EOF velocity and a stable flow, acetonitrile-rich eluents with 80%–95% organic modifier are most favorable due to a high dielectric/viscosity ratio, as mentioned earlier. This of course may compromise enantioselectivity and therefore it may be necessary to switch to alternative modifiers such as methanol or ethanol with less strong EOF. CSPs that appear to be well compatible with such RP conditions include macrocyclic antibiotics CSPs, cyclodextrin-based CSPs, protein-type CSPs, ligand-exchange type CSPs, chiral ion-exchangers, and crown-ether-based CSPs. Also polysaccharide-based CSPs can be employed under RP conditions. Yet, enantioseparation factors are often lower than under NP conditions and the application spectrum is much narrower. The same applies to Pirkle-type CSPs (such as Whelk O1). RP conditions seem to be also prone to outgassing and interruption of current due to bubble formation (vide infra).

An alternative to RP conditions represents the polar organic mode, which makes use of nonaqueous mobile phases that are composed of polar organic solvents such as methanol, acetonitrile, or mixtures thereof. Organic acids (acetic acid, formic acid) and bases (ammonia, triethylamine) can be readily dissolved in these media. These weak electrolytes are still sufficiently dissociated in these polar solvents and therefore capable of supporting ionic conduction and maintaining a stable current. In this context, it is worth noting that ionization and acid–base equilibria, respectively, are strongly shifted in nonaqueous media [59]. Hydrophobic interactions are diminished in this mode while hydrophilic interactions like H-bonding and dipole–dipole interactions are strengthened, especially if the acetonitrile content is high. Chiral ion-exchanger type CSPs, macrocyclic antibiotics CSPs, and also polysaccharide CSPs all turned out to be readily suitable for this mobile phase mode. In terms of maximal EOF velocity and peak performances, an acetonitrile-methanol mixture (60:40; v/v) with ammonium acetate as electrolyte turned out to be optimal for a quinine-carbamate based chiral anion-exchanger CSP [17]. Similar ACN–MeOH mixtures were successfully employed as mobile phases for macrocyclic antibiotics CSPs [19,60–62]. In sharp contrast, for polysaccharide CSPs methanol with ammonium acetate as electrolyte had to be utilized because of the deleterious effect of ACN on enantioselectivity [20,54,63–65]. A number of advantages have been claimed to exist for nonaqueous CEC enantiomer separations in the polar organic mode [18,56]. On the one hand, bubble formation seems to be less of an issue, currents appear to be more stable leading to more robust CEC separation systems and less noisy baseline [17]. For specific solutes, significantly enhanced enantioselectivities resulted and in general complementary selectivity profiles as compared to RP-elution mode were obtained for macrocyclic antibiotics CSPs [19,60]. In general, this elution mode seemingly has become the most frequently employed one in CEC enantiomer separation. More details have been reviewed in Refs. [18,56].

The majority of HPLC separations on polysaccharide- and Pirkle-type CSPs (like Whelk O1) is performed under NP conditions. Such conditions promote H-bonding interactions, π–π interactions, and dipole–dipole interactions. The typical mobile phases, e.g., hexane-based eluents with 2-propanol do not allow the generation of a strong and stable EOF in CEC. Instable current and strong susceptibility for bubble formation are further disadvantages for this elution mode. Hence, NP-conditions are rarely used. However, Vickers and Smith showed that by the addition of low amounts of water in the range of 1%–5% to, e.g., hexane–2-propanol (1:1; v/v) a noteworthy EOF can be obtained [35]. Yet, pressure-support was necessary because the generated EOF was still not strong enough (see Figure 14.4). In a number of papers, Zou and coworkers tested the NP mode for CEC enantioseparations on immobilized polysaccharide CSPs [21,66,67]. Typical mobile phases consisted of hexane–methanol–2-propanol–tetrahydrofuran (THF) (50:35:10:5; v/v/v/v) with 5 mM acetic acid-triethylamine. Such eluents apparently yielded reasonably stable EOF and decent enantiomer separations for a number of drug racemates [21,66,67].

14.8 COLUMN TECHNOLOGIES

CEC enantiomer separations have been performed with different column types:

1. *Open-tubular columns* (with dynamically adsorbed selectors, polyelectrolyte multilayer (PEM) coatings, covalently linked selectors, wall-coated polymeric selectors, chiral porous-layers).
2. *Packed columns* (with usually silica-based microparticulate CSPs).
3. Organic polymer or inorganic silica *monolith columns* (with selectors either incorporated by in situ copolymerization of a chiral monomer or attached by derivatization of the monolithic support) or hybrid monolithic-particulate columns.

Each of these columns has its strengths and limitations which have been described in detail in various reviews and books [1,26,53,68–79] and will be briefly discussed in the following section.

14.8.1 OPEN-TUBULAR COLUMNS

The history of enantioselective OT-EC dates back to the early 1990s when Mayer and Schurig tested the GC column Chirasil-Dex in the electrochromatographic mode [80]. Thus, this approach, i.e., OT-EC, was actually historically employed first before CEC with packed and monolithic columns came into the game. The interest in OT columns was then lost somehow. Nowadays, though, open-tubular capillary columns experience a kind of renaissance and are sometimes suggested or employed as alternatives to the troublesome packed columns (see below) because they feature some advantageous properties including no need for retaining frits, ease of preparation, and the absence of bubble formation during CEC runs. The use of long capillaries (e.g., in pCEC) and the ability of quickly exchanging the mobile phases are some further benefits. The OT column technology is also most straightforward to transfer to miniaturized microfluidic channels of microchips.

Main problems of this column technology are a small surface area and a low phase ratio of the OT columns with thin stationary phase films which are associated with a resultant low sample capacity. As long as working under idealized conditions with clean standard solutions, this problem can be kept under control and efficient separations may be achieved. Yet, with real samples and under conditions of practical applications the capacity is easily overloaded. Pollution of the limited surface associated with uncontrolled EOF changes may also become a serious problem. Hence, one has to be critical with regard to the practical usability of this OT-EC approach. For example, the analysis of enantiomeric impurities of less than 0.1% which requires typically overloading of the main enantiomer seems to be hardly achievable.

To overcome the drawback of the low phase ratio, narrow inner diameter columns (<50 μm i.d.) must be employed and favorably 20 or even 10 μm i.d. capillaries are recommended [58]. Another benefit that can be gained by the use of narrower capillaries is the reduction of the diffusion distances for the radial mass transfer from the center of the bore to the stationary phase at the capillary surface which according to the Golay Equation 14.11 should yield an improved mobile phase mass-transfer and thus improved efficiencies, i.e., smaller plate heights. As a trade-off of narrower capillaries, sensitivity of on-column UV detection is, according to Lambert–Beer law, compromised owing to a shorter detection path length in comparison to the typical 100 μm i.d. packed and monolithic capillary columns. This issue might be of less relevance with highly sensitive mass spectrometric detection, although in a recent study this hypothesis could not be confirmed for ESI–MS [81]. In order to increase the detection path length and sensitivity, respectively, for UV detection, Liu et al. used 20 μm i.d. capillaries with extended light path (150 μm bubble cell) for the preparation of OT capillary columns with adsorptively coated avidin as chiral selector yielding a 12-fold increase in peak areas [82]. It is also evident that clogging may represent a more serious problem in narrower capillaries (10 or 20 μm i.d.) which was found to be problematic during column preparation of wall-coated polysaccharide capillaries [83]. Pesek et al. suggested etching of the fused silica surface with ammonium hydrogendifluoride before functionalization with chiral selector as a strategy to increase the phase ratio [84]. The etching process creates a rough surface that is associated with an increase in the surface area resulting in higher loadings with ligands and higher sample capacity. For the same purpose, porous-layer open-tubular (PLOT) column technology, known from GC, has been introduced [85].

Besides avoiding overloading and using narrow i.d.-columns, a number of other requirements must be fulfilled to achieve efficient separations in OT-EC as can be derived from the Golay equation (Equation 14.11) [26]. Retention factors should be kept as low as possible for complete separation of enantiomers because (both mobile and stationary phase) mass transfer resistances, and thus theoretical plate heights as well, rapidly increase with retention factors [83]. The Golay equation further illustrates that the stationary phase film thickness contributes with the square to the stationary phase mass transfer resistance and thus to band broadening, and this has been confirmed experimentally [80,83].

A variety of immobilization and surface modification chemistries for the preparation of enantioselective OT-EC capillary columns have been proposed [86]. They are summarized in Table 14.1. The technically simplest one is the adsorptive coating of chiral selectors onto the surface of fused silica capillaries [82,87–90]. The strong adhesion of basic proteins such as lysozyme [87,88] and avidin ($pI \sim 10$) [82,89,90] to the silanols by electrostatic interactions has been exploited to prepare protein-bonded OT columns. The immobilization of the protein occurs on simple rinsing of the capillaries for several minutes (up to 90 min or so) in an aqueous buffered

TABLE 14.1
Open-Tubular Columns Used in CEC Enantiomer Separations

Type of Immobilization	Selector-Type	Ref.
Adsorptive noncovalent bonding	Proteins (lysozyme, avidin)	[82,87–90]
	Biospecific binding of avidin to biotinylated phospholipid coatings	[91]
Polyelectrolyte multilayers	Poly(*N*-undecanoyl-dipeptides)	[92,93]
Covalent linkage	Proteins (BSA, avidin)	[81,94,95]
Polymer coating (WCOT)	Cyclodextrin-bonded polysiloxane	[80]
	Terguride-bonded polysiloxane	[96]
	Polysaccharide	[83]
Porous-layer (PLOT type)	MIP	[85]

solution. The straightforward preparation procedure facilitates the use of 20 or even 10 μm i.d. capillaries that lead to improved column efficiencies. Liu et al. reported on avidin-adsorbed columns (28 μm i.d.) and the amount of avidin adsorbed on the capillary wall was estimated by frontal analysis to be of the order of magnitude of 10^{-12} mol for a capillary of 50 cm effective length, corresponding to 10^{-8} mol/m^2 [90]. A number of pharmaceuticals could be baseline resolved into enantiomers with 10 mM phosphate buffer (pH 6) containing up to 5% (v/v) methanol with reasonable plate numbers for the first migrating enantiomers (10–70,000 plates/m). The second migrating enantiomer exhibited much lower plate counts and a significant tailing which is not uncommon for separations on protein CSPs. The drifting baseline in these separations might be an indication for bleeding of the selector.

In a similar adsorptive column fabrication approach phospholipid-coated fused silica capillaries (50 μm i.d.) with immobilized avidin were recently prepared by Han et al. [91]. These OT columns were obtained in a two step procedure in which first a phospholipids coating was generated by rinsing with 1 mM liposome solution synthesized from phosphatidylserine and biotinylated phospholipids for 10–30 min. Avidin was then adsorbed onto the biotin-residues by rinsing for about 30 min. These OT-columns showed highly efficient separations for PTH-amino acids (Figure 14.5).

A related procedure with great potential for OT-EC has been proposed by Warner's group [97]. They make use of a PEM coating procedure for the fabrication of enantioselective capillary columns [92,93]. The technique is based on an electrostatic layer-by-layer deposition of positively and negatively charged polymers. The coatings are fabricated in situ by rinsing a fused silica capillary alternating with cationic and anionic polyelectrolytes starting with the cationic polymeric surfactant. One cationic and one anionic layer constitute a so-called bilayer. Since the last layer was usually prepared by the negatively charged polymer, the authors always observed cathodic EOF. Several bi-layers can be applied which was shown to increase the separation factors as illustrated for 1,1′-binaphthyl-2,2′-dihydrogenphosphate in Figure 14.6. It constitutes a straightforward approach to adjust the enantioselectivity by the demand of the separation problem. In their studies, the authors utilized

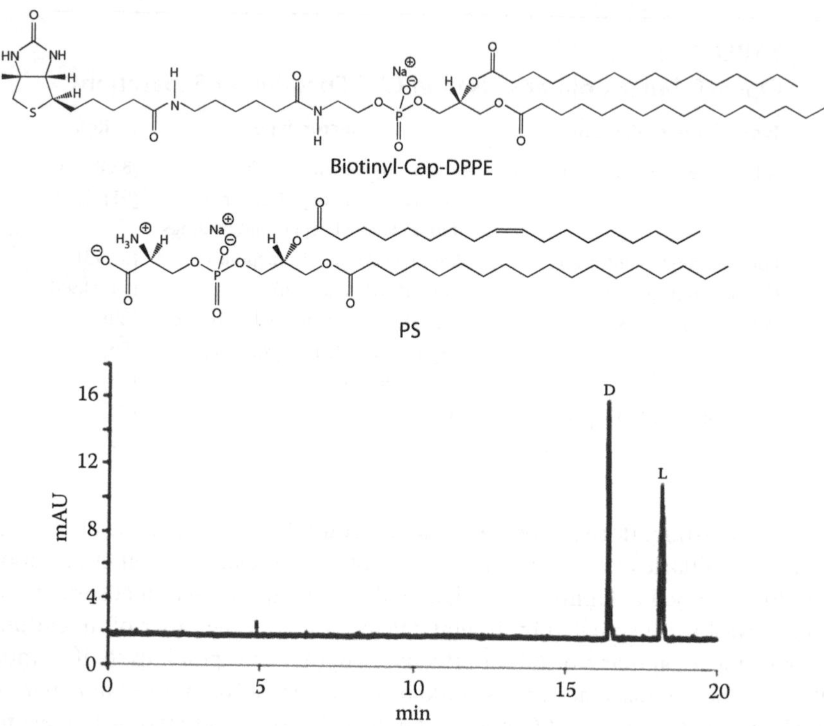

FIGURE 14.5 Chiral separation of D- and L-PTH-Thr with a phospholipid/avidin-coated OT capillary column. Biotinyl-Cap-DPPE/PS (80:20 mol%). Experimental conditions: 50 μm i.d. fused silica capillary; L_{tot}, 48.5 cm; L_{eff}, 40 cm; capillary temperature 25°C; applied voltage, 20 kV; inj. 10 s at 50 mbar; UV 214 nm; BGE solution Tris (ionic strength 20 mM) at pH 7.4. (Modified from Han, N.-Y. et al., *Electrophoresis*, 27, 1502, 2006. With permission.)

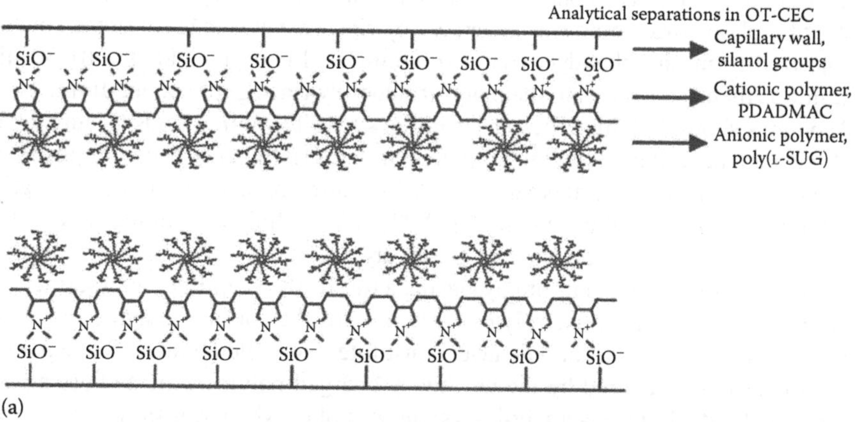

FIGURE 14.6 Scheme for PEM-coated capillary (a) and effect of bilayer number on the chiral separation of 1,1′-binaphthyl-2,2′-dihydrogenphosphate (b) (Modified from Kapnissi-Christodoulou, C.P. et al., *Electrophoresis*, 24, 3917, 2003. With permission.)

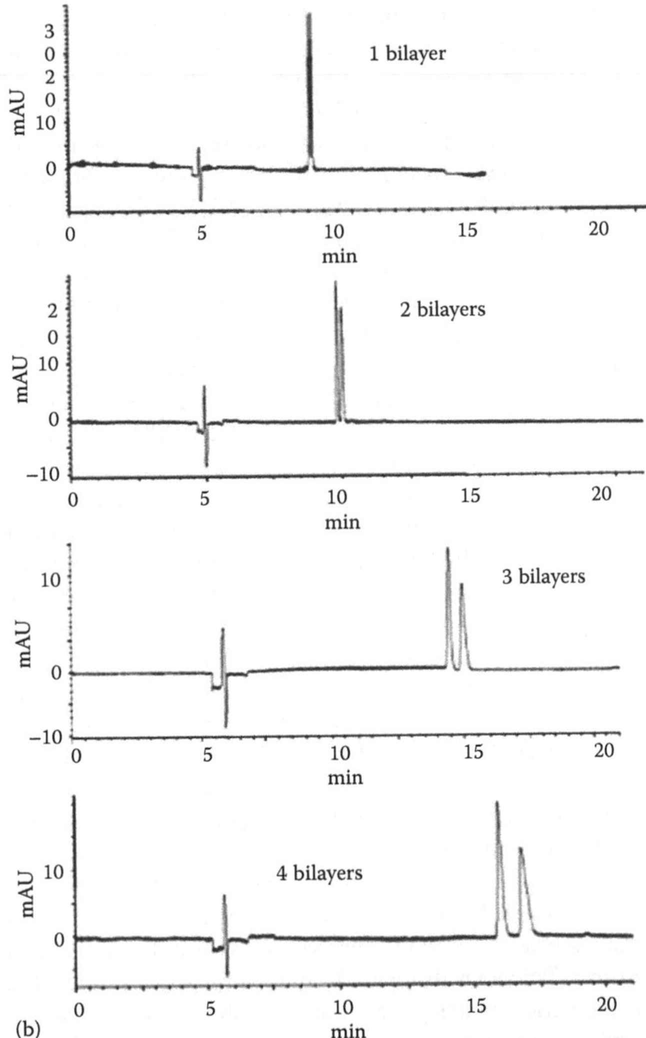

FIGURE 14.6 (continued) (b) Conditions: 0.02% (w/v) poly(L-lysine) with 0.5 M NaCl and 0.25% (w/v) poly(L-SULA); BGE, 100 mM Tris and 10 mM $Na_2B_4O_7$ (pH 10.2); pressure injection, 30 mbar for 5 s; applied voltage, 30 kV; current, 27.4 μA; temperature, 15°C; capillary, 57 cm (50 cm effective length) × 50 μm i.d.; detection, 220 nm. Poly(L-SULA): Poly(sodium *N*-undecanoyl-L-leucyl-L-alaninate). (Modified from Kamande, M.W. et al., *Anal. Chem.*, 76, 6681, 2004. With permission.)

poly(diallyldimethylammonium chloride) or poly(L-lysine) as cationic polymers and poly(undecanoyl-dipeptides) such as poly(sodium *N*-undecanoyl-L-leucyl-L-valinate), poly(sodium *N*-undecanoyl-L-leucine-L-alaninate) or poly(sodium *N*-undecanoyl-L-alanine-L-leucinate) as anionic polymer. The PEM coatings seem to be particularly stable, even against basic pH, e.g., 0.1 M NaOH [98].

Covalent binding of proteins onto the fused silica surface has been another utilized technology for the preparation of enantiomer-selective OT capillary columns [81,94,95]. Hofstetter's linking chemistry involved activation of the fused silica capillaries with glycidoxypropylsilane, hydrolysis of epoxides to hydroxyl-surface and tresyl-activation before immobilization of bovine serum albumin (BSA) [94]. The column performed reasonably well allowing the enantiomeric separation of 3-hydroxy-1,4-benzodiazepines as well as *N*-2,4-dinitrophenyl(DNP)-amino acids (amino acids derivatized with Sanger's reagent). Schiff-base type chemistry with 3-aminopropylsilanization of the fused silica wall, glutardialdehyde activation followed by the bonding of avidin has been another common multistep immobilization procedure [81,95] and allowed the enantiomeric separation of abscisic acid and arylpropionic acids (profens) [81] as well as of omeprazole and its metabolite 5-hydroxyomeprazole [95], respectively. The more stable bonding and less bleeding as compared to the analogous columns prepared by the adsorption of the proteins are paid off by the more complicated multistep preparation procedures.

In the early works on OT-EC, Schurig and coworkers made use of a wall-coated open-tubular (WCOT) column approach, a technology commonly employed for GC enantiomeric separation, in which a chirally modified-polysiloxane film was coated onto the fused silica surface [80]. Such polymer coatings must have a thin submicrometer polymer film, otherwise efficiencies would be seriously deteriorated. The narrower the capillary used, the more prone it is to clogging in the course of column fabrication and the more difficult it is to get a well-performing column with homogenous CSP layer. For example, Schurig used 80–100 cm × 50 μm i.d. fused silica capillaries coated with a 0.2 μm film of Chirasil-Dex, a poly(dimethylsiloxane) with covalently bonded permethylated β-cyclodextrin (ca. 16% (w/w)). Although the silanols of the silica surface were shielded by the polysiloxane coating, a sufficient amount of these charged functionalities remained accessible to achieve satisfactory EOF velocities. Due to the thin CSP film, separation factors for aromatic compounds were low (between 1.1 and 1.2), yet efficiencies of up to 300,000 theoretical plates/m enabled baseline separations. When the film thickness was increased from 0.2 to 0.8 μm, efficiencies dropped dramatically as theory predicts, e.g., from 30,500 to 9,500 theoretical plates per column. The above OT-capillary column with 0.2 μm film permitted in a "unified enantioselective chromatography approach" the separation of enantiomers by gas chromatography, supercritical fluid chromatography, open-tubular liquid chromatography, and electrochromatography in the same column [99,100]. Later and newer developments in this field have recently been reviewed [79].

Similarly, Sinibaldi et al. prepared coated 100 μm i.d. capillaries by a two-step procedure [96]. In the first step, a poly(vinyl siloxane) film was covalently bonded to the capillary wall. In the second step, 1-allyl-terguride was chemically bonded by a free-radical addition reaction. The positively charged ergoline moieties produced an anodic EOF at low pH which was found to be favorable for the analysis of acidic compounds such as dansylated amino acids and flobufen.

The WCOT technology can also be applied to other type of polymers. Francotte and Jung, for example, prepared enantioselective OT columns by coating cellulose tris(3,5-dimethylphenylcarbamate) and cellulose tris(*para*-methylbenzoate) onto the surface of 50 μm i.d. capillaries. Its feasibility could be demonstrated by the OT-EC enantiomer separation of a variety of chiral pharmaceuticals [83]. Largely in agreement with the trends found by Mayer and Schurig, a film thickness of only 0.025 μm furnished a high efficiency at low retention and separation factors while thicker films of 0.1 μm afforded much better enantioselectivities at unacceptable peak performance unfortunately. As another drawback, they reported on strongly reduced EOF by the cellulose polymer coating.

In another approach, thin porous films of highly cross-linked molecularly imprinted polymers (MIP) anchored to the inner walls of 25 μm i.d. fused silica capillaries were prepared by an in situ polymerization process [85]. The MIPs were prepared from methacrylic acid or 2-vinyl pyridine as functional monomers and ethylene dimethacrylate or trimethylolpropane trimethacrylate as cross-linkers in the presence of toluene and acetonitrile as porogenic solvents. *N*-dansyl (*S*)-phenylalanine served as the template molecule. This column type resembles a porous-layer open tubular (PLOT) column in which through the creation of a porous polymer layer the surface area can be significantly enhanced. Although some success could be achieved, the column suffered from the limited kinetic performance for the peak corresponding to the templated enantiomer, which is a common problem for separation materials whose selectivity originates from MIP technology.

Although the feasibility of enantiomer separation by electrochromatography with enantioselective open-tubular columns has been convincingly demonstrated, the window of experimental conditions that give highly efficient separations seems to be narrow. Especially the problem of easy overloading is hard to overcome and very seldom separations other than that of standard solutions have been shown. More applications of the OT-EC approach can be found in Chapter 15.

14.8.2 Packed Capillary Columns

Packed capillary columns are derived from the typical HPLC column packing technologies and often the same (commercial) porous particles (3–5 μm diameter) as employed for HLPC are slurry packed by pressure, supercritical-fluid chromatography, or electrokinetically into fused silica tubes of 75–100 μm i.d. The packed bed is stabilized with retaining frits on both ends to keep the packing in the capillary during the CEC runs. Packed chiral CEC capillary columns are nowadays not commercially available. Thus, research groups being active in the field have established their own dedicated packing protocols. Most follow a procedure similar to the one outlined in Figure 14.7. For example, the chirally modified particles are slurried in suitable media such as acetone or ethanol and then pumped with high pressures into the empty fused silica capillary tube which has a temporary retaining frit attached at one end [101]. After exchange of the organic solvent by an aqueous solution containing sodium ions, retaining frits are sintered on the silica-supported column packing at about 500°C thereby achieving partial fusion with grain boundary formation between the particles. The sodium ions, without which no sufficiently stable frits can be obtained, support this process of joining the particles to one another by grain boundaries because they help to overcome the repulsive electrostatic interaction between the silica beads and promote the sodium silicate matrix formation at the

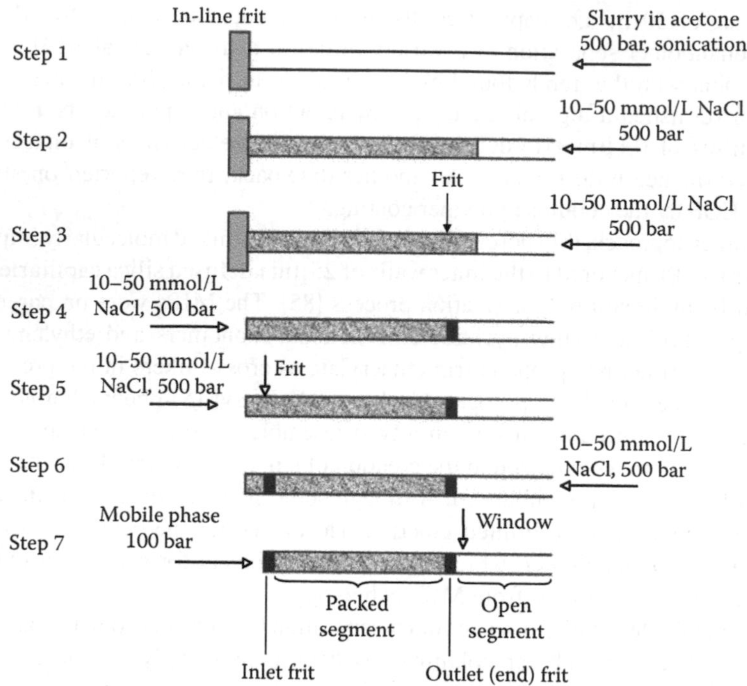

FIGURE 14.7 Typical multistep procedure for capillary column packing including frit fabrication. (Reprinted from Tobler, E., Development of enantioselective separation systems for analysis of ionic compounds by capillary electrochromatography. PhD thesis, Austria, University of Vienna, 2001. With permission.)

contact points of the particles. The frits are usually directly sintered on the chiral packing material [23,64,102]. If the modified silica does not allow to produce stable frits, a short segment is packed on each side with a different material such as silica, diol-silica, or a mixture of both on which stable frits can be sintered [103].

Such packed columns possess usually a dual column architecture as outlined in Figure 14.7 in which the separation takes place in the packed segment and UV detection is carried out in the open segment immediately after the end-frit. It must be emphasized that in such a column, the applied voltage does not drop linearly and thus distinct actual field strengths are imposed on the solutes in the two zones which makes CEC theories even more complicated.

The most serious problem of CEC, being mainly attributed to packed columns, is bubble formation inside of the capillary column. This bubble formation occurs usually at the end-frit on the transition from the packed to the open segment due to a slight depressurization effect (as explained by a flow equalizing intersegmental pressure difference [13]). As a consequence of outgassing in this region and bubble formation, current instabilities and finally even current breakdown may result. In order to avoid or minimize the risk for such a failure of the CEC experiment, a low pressure typically between 6 and 12 bar is applied at both ends of the capillaries which is not needed with open-tubular columns and may not be necessary for monolithic packings (see below).

Overall, the fabrication of packed capillary columns is tedious and may lack reproducibility. Performance and stability of particle-packed columns strongly depend on the integrity and homogeneity of the packing and in particular of the quality of the frits at both ends of the packed bed. During the heat sintering process, a compromise between the stability and the permeability of the frits has to be found. Thus, the frits are the critical parts in packed columns. Furthermore, the heat treatment during the sintering process may destroy both the organic chiral surface modification and the external protective polyimide layer which makes the column fragile. The frit region represents therefore a bed inhomogeneity in which ζ-potential, conductivity, and (electrokinetic) permeability may be altered as compared to the chromatographic bed. This heterogeneity in the chromatographic bed may seriously deteriorate the chromatographic performance of the column. Moreover, the frit itself tends to disintegrate during use leading to limited column longevity of packed columns which is actually the most painful problem of this column technology. In order to overcome the problems associated with frits tapered columns (see Figure 15.2) are nowadays sometimes used, in particular for CEC–MS coupling [104–106].

In spite of limitations of this column technology, a substantial number of studies have been published that utilized particulate CSPs in packed beds for CEC applications. Pioneering work dates back to the early 1990s when Lloyd and coworkers investigated the potential of protein-based CSPs (Chiral AGP) [107,108] and cyclodextrin-based CSPs [109,110] in CEC. Besides the above shortcomings, a major benefit of the packed column technology emerges from the possibility to employ chromatographic particles that are utilized in standard HPLC and which are usually commercially available. Moreover, this technology may take profit from the huge knowledge on the multitude of well-proven separation selectivities. In fact, most of the particulate CSPs commonly used for HPLC enantiomer separations have been tested in CEC (see Table 14.2). Applications of these packed-bed capillary columns for CEC enantiomer separations are described in Chapter 15. A few specifics of these particulate packed CSPs are outlined hereafter.

If any, the probably most significant difference of the CSP particles employed in CEC as compared to HPLC is related to the particle diameter. If chemistry and selector coverage can be kept constant, the EOF velocities will be similar in packed columns with different particle diameters [144]. This is supported by CEC theory, which states that flow velocities in CEC are independent of interstitial flow channel diameter d_c ($d_c \sim 0.28\ d_p$) as long as double layer overlap is avoided. This has been substantiated experimentally in our group for enantioselective CEC. Regression lines of plots of EOF velocities versus field strength for columns packed with 3.5 and 5 µm CSP particles (*N*-4-allyloxy-3,5-dichlorobenzoylcysteic acid-modified silica) [139] could be more or less superimposed upon each other (unpublished data). This finding has been confirmed by a study of Fanali's group for columns packed with 3.5 and 5 µm silica-supported teicoplanin CSPs in which they found even slightly higher EOF velocities for the 3.5 µm packed column [126]. Since there are no backpressure limitations for the chromatographic bed length due to pressure independence of CEC,

TABLE 14.2
Particulate CSPs Evaluated in CEC Enantiomer Separations

CSP	Ref.
Coated polysaccharide CSPs (Figure 14.8a) Including cellulose tris(3,5-dimethylphenylcarbamate) (Chiralcel OD), amylose tris(3,5-dimethylphenylcarbamate) (Chiralpak AD), cellulose tris(4-methylbenzoate) (Chiralcel OJ), cellulose tris(3,5-dichlorophenylcarbamate), cellulose tris(3-chloro-4-methylphenylcarbamate)	[20,54,55,63–65,111–119]
Immobilized polysaccharide CSPs (Figure 14.8b)	[21,66,67]
Protein CSPs	[107,108]
Macrocyclic antibiotics CSPs	
including vancomycin (Figure 14.9a)	[19,61,120–124]
teicoplanin (Figure 14.9b)	[60,62,125,126]
teicoplanin aglycone (Figure 14.9b)	[36,127]
Cyclodextrin CSPs (e.g., Figure 14.10)	[31,32,39,110,128–132]
Donor-acceptor (Pirkle-type) CSPs (e.g., Whelk O 1; see Figure 14.4)	[35,133–137]
Chiral ion-exchangers	
Chiral anion-exchangers (e.g., Figure 14.2a)	[17,22,102,138]
Chiral cation-exchangers (e.g., Figure 14.2b)	[23,24,139–141]

longer packed columns can be used (typical 25 cm) than in standard HPLC with 3 μm particles. (Note, UPLC was not available when CEC technology was developed; hence this was a striking argument.) Together with less flow maldistribution, i.e., a reduced *A*-term contribution to band broadening, as well as a relaxed-mass transfer resistance (smaller *C*-term leading to flatter *H–u* curves) 3.5 μm particles lead to greatly improved separation efficiencies in CEC. 30–75,000 theoretical plates per column of 25 cm length can readily be obtained as already mentioned before [24,126,140]. Unfortunately, common CSPs like polysaccharide-based CSPs, macrocyclic antibiotics CSPs were formerly commercially available only as 5 μm beads and only since recently as 3 μm beads as well.

Enantiomeric separation deals with low molecular mass compounds and hence virtually all CSP particles but the polysaccharide CSPs are based on silica beads with about 100 Å pore diameter. This pore diameter is appropriate for most separations and has the advantage of a large surface area and high adsorption capacity which may be required in analytical separations as well, e.g., when overloading is necessary to detect trace enantiomeric impurities [140].

However, for the coating of polysaccharide-type biopolymers onto the support surface wide pore silica (1000–4000 Å) appears to be advantageous because the risk for clogging of intraparticulate pores in the course of the coating step is reduced. In such wide pore materials, convective flow in the mesopores may favorably contribute to the flow characteristics and dynamics in CEC and lead to an enhancement of relative EOF velocities and mass transfer through convection [54]. In fact, electroosmotic

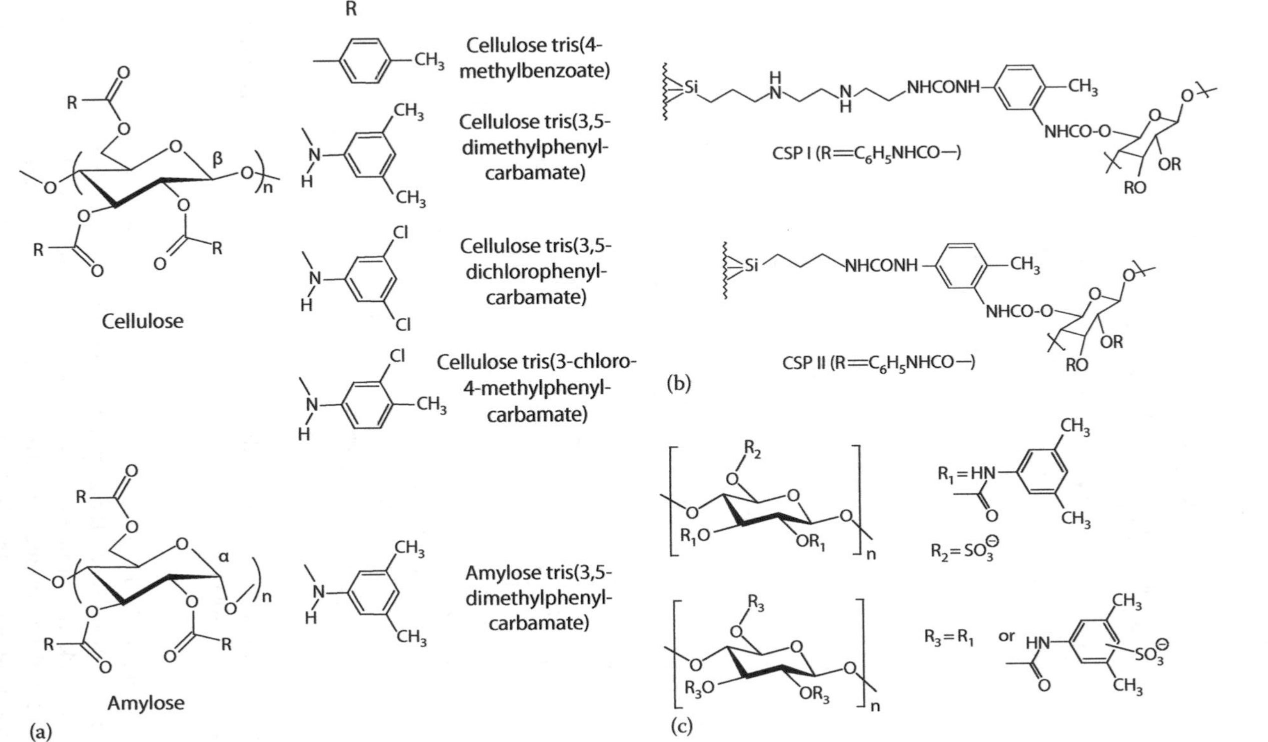

FIGURE 14.8 Structures of polysaccharide-based CSPs. (a) Coated-polysaccharide CSPs. (b) Regioselectively immobilized polysaccharide CSPs. (Modified from Chen, X. et al., *Electrophoresis*, 24, 2559, 2003. With permission.) (c) Sulfated and sulfonated coated-polysaccharide CSPs. (Modified from Zheng, J. et al., *J. Chromatogr. A*, 1216, 857, 2009. With permission.)

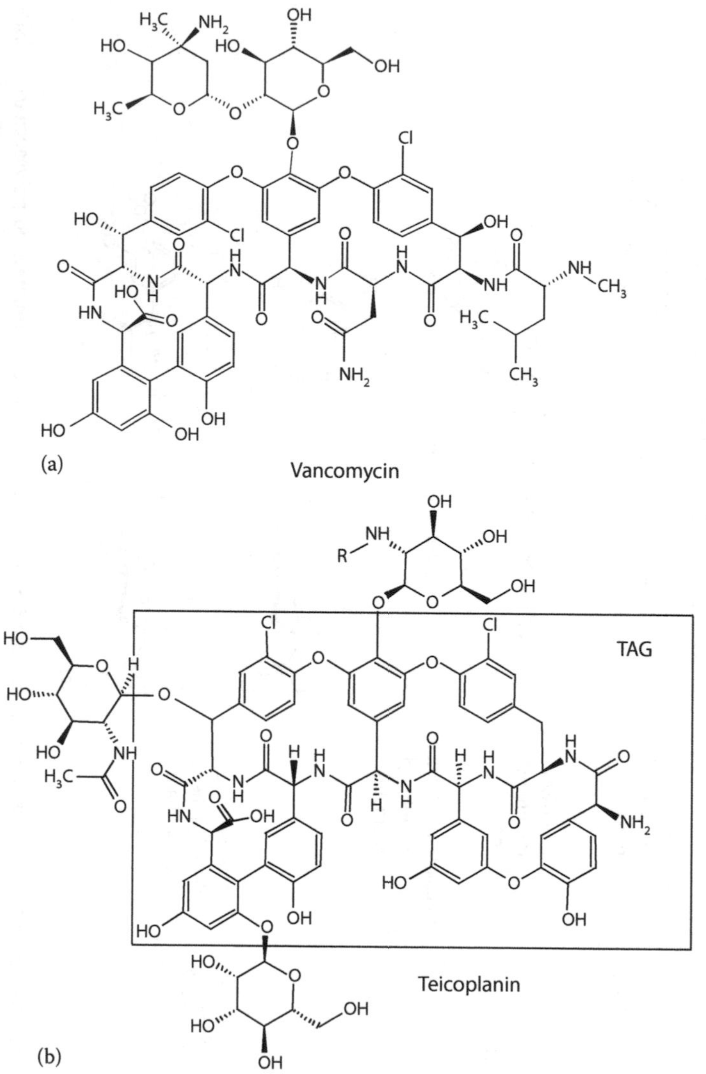

(a) Vancomycin

(b) Teicoplanin

TAG

FIGURE 14.9 Macrocyclic antibiotics frequently employed as chiral selectors: (a) vancomycin, and (b) teicoplanin as well as teicoplanin aglycone TAG.

mobilities were shown to be by a factor of about 3 higher with 1000 Å wide-pore particles as compared to a 60 Å material, presumably due to the double layer overlap in the latter (Figure 14.11a). Simultaneously, very flat *H*/*u*-curves of the 1000 and 2000 Å particles could be obtained (Figure 14.11b) indicating better mass transfer characteristics for wide pore materials that have been attributed to convective transport in the mesopores of these wide-pore materials [54]. The higher efficiencies of the wide pore materials could overcome their diminished retentivity (due to a lower adsorption surface) yielding optimal resolution values despite significantly dropped

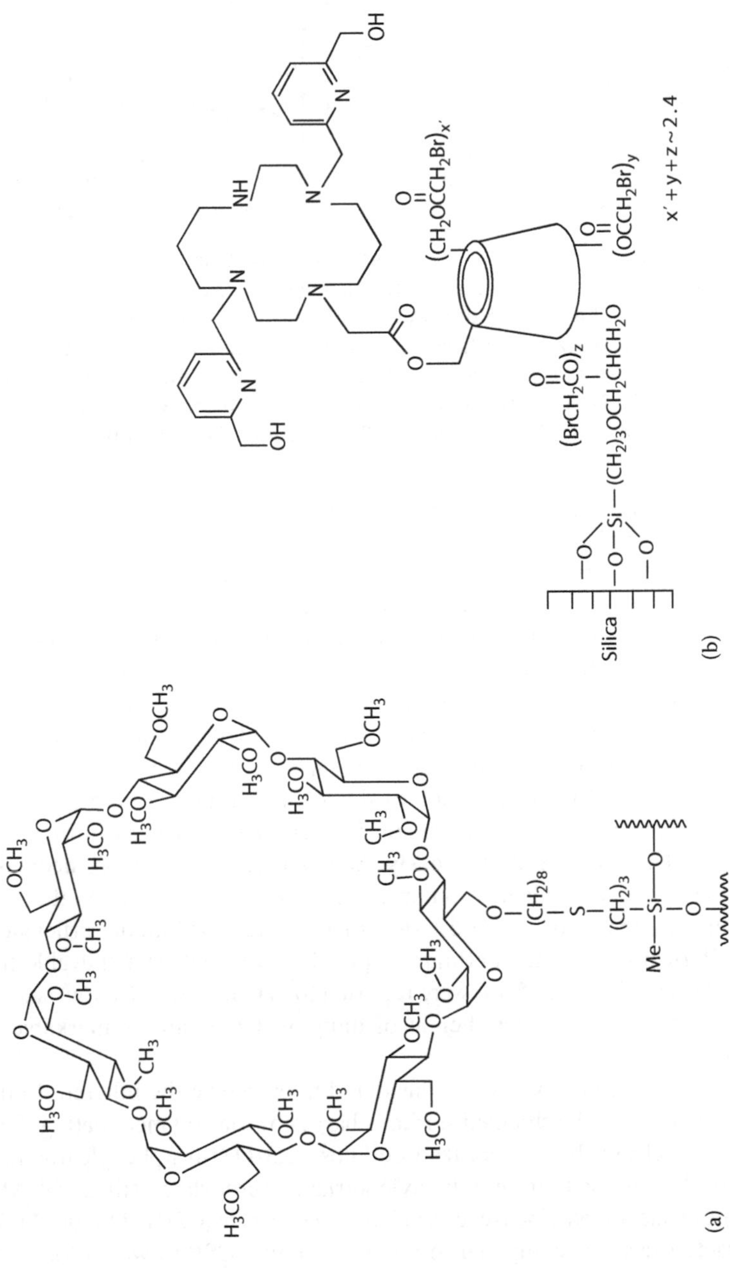

FIGURE 14.10 Cyclodextrin-based CSPs: (a) permethyl-β-cyclodextrin-modified silica (Reprinted from Wistuba, D. et al., *J. Chromatogr. A*, 815, 183, 1998. With permission.) (b) cyclam-capped β-cyclodextrin-bonded silica particles as a CSP. (Reprinted from Gong, Y. and Lee, H.K., *Anal. Chem.*, 75, 1348, 2003. With permission.)

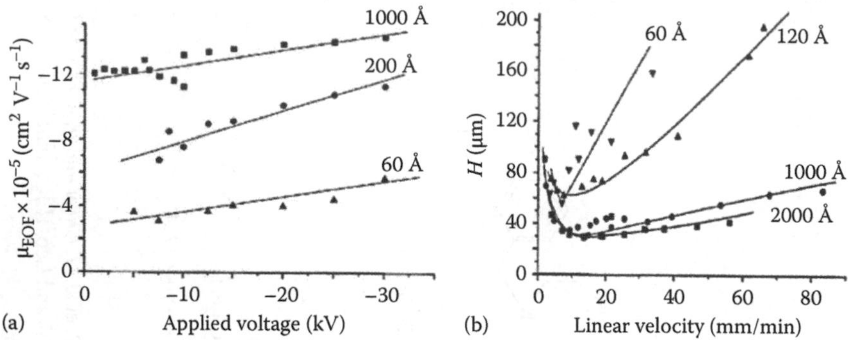

FIGURE 14.11 Effect of pore diameter of supporting silica particles (a) on EOF velocities, and (b) on Van Deemter plots for the second eluted enantiomer of *trans*-stilbene oxide. CSPs prepared by coating of cellulose tris(3,5-dimethylphenylcarbamate) (20%; w/w) on aminopropylated silica of various pore sizes. Capillary, fused silica (24/32.5 cm × 100 µm i.d.); particle size of silica, 5 µm; applied voltage, −1 kV to −30 kV; separation medium, 2.5 mM ammonium acetate in methanol (pH 7.7). A pressure of 8 bar was applied on the inlet and the outlet vial. (Modified from Girod, M. et al., *Electrophoresis*, 22, 1282, 2001. With permission.)

separation factors. One problem, though, constituted the availability of wide-pore materials with particle diameter smaller than 5 µm.

Polysaccharide selectors are commonly coated onto wide-pore silica particles in a relatively thick film at about 20% (w/w). Two problems emerge from such efficient coatings. First, the polysaccharide selectors are neutral and do not support EOF generation. Moreover, the polymer coating shields the silanols which yields weak EOF. No clear preference for tested silica and aminopropyl-silica supports was claimed [63]. However, aminopropyl-silica may have some advantages under acidic conditions [20]. When less selector was coated onto the surface, the EOF was stronger, e.g., by more than a factor of 2 upon reduction of the loading from 20% to 5% (w/w) of cellulose tris(3,5-dimethylphenylcarbamate) on 5 µm 2000 Å silica and 10 mM ammonium acetate as eluent [64]. Simultaneously, however, retention factors decreased and enantioselectivities were clearly compromised (see Figure 14.12). On the other hand, CSPs with a thinner coating of 5% or 10% (w/w) revealed higher efficiencies and very flat *H*/*u*-curves with the optimum typically at the highest achievable flow rate. Plate heights *H* of less than 5 µm were reported for a CSP with 5% (w/w) selector being equivalent to a reduced plate height of unity, which is quite remarkable for enantiomer separations.

Further, two other approaches were considered to overcome the inherent limitation of a weak EOF due to the shielded surface charges by the polymer coating. First, Zou's group regioselectively immobilized cellulose 2,3-*O*,*O*-bis(phenylcarbamate) at position 6 of the glucose unit onto diethylenetriaminopropylated silica (DEAPS) using 2,4-diisocyanato-1-methylbenzene as bifunctional reagent (CSP I of Figure 14.8b). Its EOF characteristics was compared to a corresponding CSP in which the selector was immobilized onto 3-aminopropyl-silica (CSP II in Figure 14.8b) [21]. The DEAPS-based CSP with additional amino groups in the spacer that act as additional EOF sources showed indeed up to 60% higher EOF. Second, Shamsi and coworkers

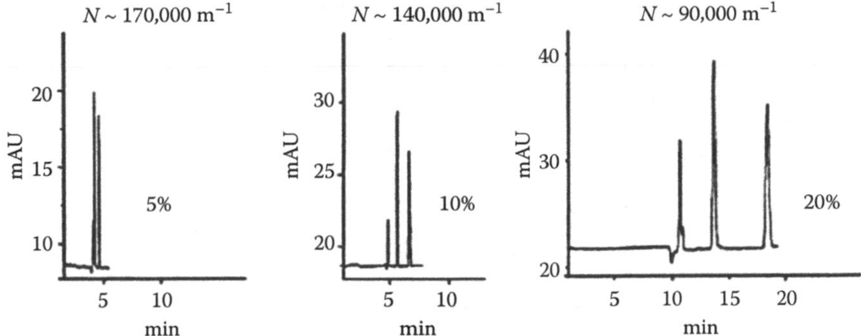

FIGURE 14.12 Enantioseparation of Tröger's base by CEC using CSPs with various loading of cellulose tris(3,5-dimethylphenylcarbamate) on silica gel. Separation capillary was 100 μm i.d. fused silica of 32.5 cm total length and with 24 cm packed bed. The CSPs were silica gel (Daisogel, 5 μm, 200 nm pore size) coated with 5%, 10%, and 20% of cellulose tris(3,5-dimethylphenylcarbamate). Applied voltage in CEC, 25 kV; mobile phase, 10 mM ammonium acetate dissolved in methanol. (Modified from Chankvetadze, L. et al., *Electrophoresis*, 23, 486, 2002. With permission.)

attempted to alleviate the problem of limited EOF by introducing negatively charged sulfate and sulfonate groups, respectively, as residues on the polysaccharide selectors (Figure 14.8c) [142]. Especially, the sulfonated CSP (sulfur content 1.76%, w/w) provided both a fast analysis and a maintained resolving power.

In CEC, polysaccharide CSPs are most favorably run in the polar organic mode [20,54,56,64,65], whilst both RP [115] and NP with immobilized CSPs [21,66,67] have been shown to be also useful elution modes for specific applications.

Macrocyclic antibiotics CSPs including particulate CSPs with immobilized vancomycin, teicoplanin, and teicoplanin aglycone (TAG) (Figure 14.9) are another class of CSPs that have been very popular in packed bed CEC owing to their broad applicability spectra ranging from acids and bases to amino acids and peptides as well as their optimal compatibility with common CEC elution modes [19,60,61,103,125,127,145–148] (see Chapter 15). They have been applied in the reversed-phase and polar organic mode [19,60]. Both exhibit distinct selectivity profiles, yet the polar organic mode seems to be preferred. It provides often higher enantioselectivities and/or faster separations (see Figure 14.13).

In a number of papers, Schurig's group investigated the potential of particulate cyclodextrin-based CSPs (amongst others permethyl-β-cyclodextrin-bonded silica; Figure 14.10a) for CEC [31,129]. They also elucidated the packed columns by nano-HPLC with typical efficiency losses by a factor of 2–3 as compared to CEC [31]. Low EOF which depends on residual silanols because of the nonionic selectors and bubble formation were both circumvented by the application of pressures of about 10 bar to the inlet vial (i.e., by use of pCEC). Cyclodextrin-based CSPs exhibit broad enantiorecognition capabilities for chiral aromatic compounds under aqueous conditions because of molecular recognition due to inclusion complexation into the CD cavity which is driven by hydrophobic interactions. Unfortunately, loss of enantioselectivities may be observed with the addition of higher percentages of acetonitrile

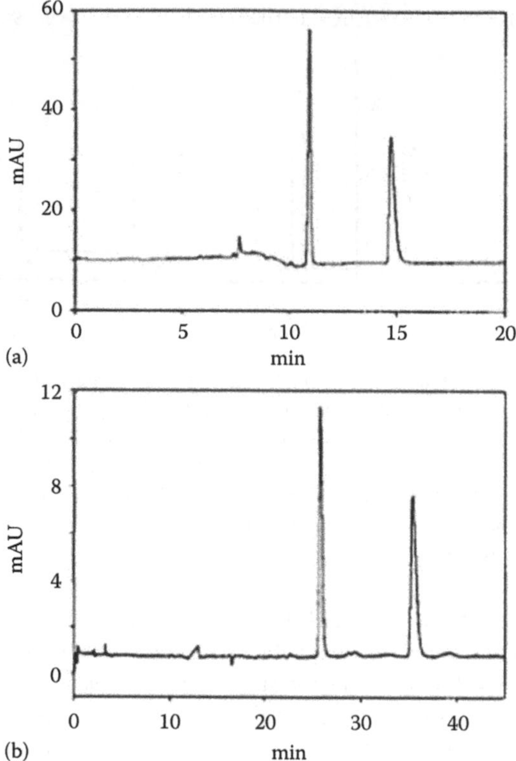

FIGURE 14.13 Comparison of CEC enantiomer separations of thalidomide on vancomycin CSP in nonaqueous polar organic mode (a) as well as hydroorganic reversed-phase mode (b). Conditions: (a) Methanol–acetonitrile–triethylamine–acetic acid (80:20:0.1/0.1, v/v/v/v); voltage, 20kV; T, 15°C. (b) Eluent, acetonitrile–0.1% triethylammonium acetate buffer (pH 4) (30:70, v/v); voltage, 25kV; T, 15°C. (Reprinted from Karlsson, C. et al., *Anal. Chem.*, 72, 4394, 2000. With permission.)

which would be favorable in view of EOF strength. A similar trend, namely gradually decreasing separation factors, was found when increasing methanol concentrations were added to the eluent (see Figure 14.14). El Rassi's group proposed as a solution for weak EOF of such type of CSPs the introduction of a hydrophilic sulfonated sublayer to the support onto which a chiral top-layer with surface-bound hydroxypropyl-β-cyclodextrin moieties was immobilized [128]. Other researchers later employed carbamoylated cyclodextrins [131], azido-functionalized cyclodextrin derivative immobilized by click chemistry [132], or specifically decorated CDs, such as crown-ether capped and cyclam-capped cyclodextrins, as selectors [130,143] (Figure 14.10b). They showed modified applicability profiles from cooperative molecular recognition effects originating from the introduced new functionalities and the inclusion capability of the cyclodextrin cavity.

Only a limited number of studies addressed the use of donor-acceptor phases (Pirkle-type) CSPs [35,133–137]. Residual silanols are responsible for EOF in columns packed with such CSPs because the selectors are neutral. Pirkle-type CSPs

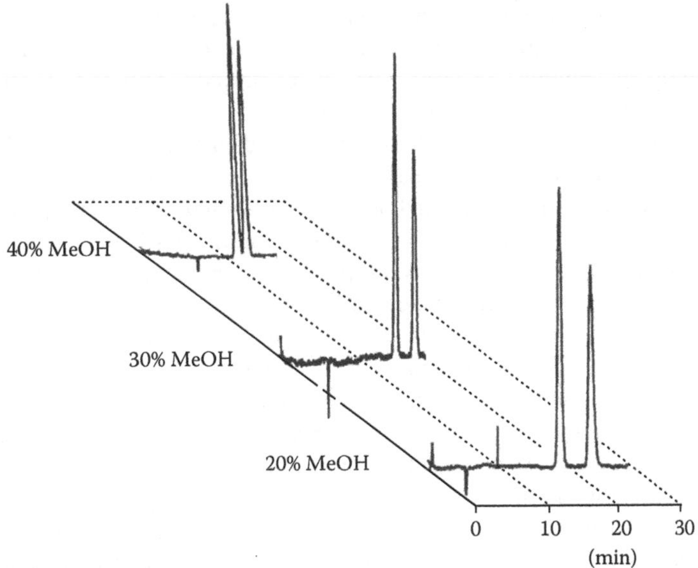

FIGURE 14.14 Enantiomer separation of mephobarbital by CEC. Conditions: 23.5 cm (overall length, 40 cm) × 100 μm i.d. capillary packed with permethyl-β-CD-bonded silica; buffer: phosphate (5 mM, pH 7.0), 15 kV, 10–15 bar (inlet vial); UV detection, 230 nm; modifier, 20%, 30%, and 40% methanol. (Reprinted from Karlsson, C. et al., *Anal. Chem.*, 72, 4394, 2000. With permission.)

are normally employed in the NP mode in which the driving interaction forces such as hydrogen-bonding, dipole–dipole interactions, and π–π interactions are strong. Typically, they show a decrease in enantioselectivity upon switching to the RP mode and are usually virtually ineffective in the polar organic mode. Hence, their applicability spectra in CEC are seriously restricted.

In contrast, the ionic counterparts of the Pirkle-selectors, the chiral ion-exchangers, turned out to be highly suitable for CEC enantiomer separation. Chiral anion-exchangers based on quinine and quinidine carbamate derivatives [17,22,102,138] (see for example Figure 14.2a) can be employed for the separation of chiral acids while chiral cation exchangers derived from amino sulfonic, phosphonic, carboxylic acids, or sulfopeptides [23,24,139–141] (see for example Figure 14.2b) enabled the separation of chiral bases. They are particularly suitable for CEC because the ion-exchange site is not only the primary interaction site of the active molecular recognition motif, but also nucleation site for EOF generation, as discussed above. Through its dominating nature for EOF generation, a favorable faster codirectional separation process can be accomplished. Under specific conditions, the solute enantiomers may even migrate before the EOF marker, yet being still resolved from each other [17]. For these chiral ion-exchangers, the polar organic mode turned out to be highly preferable over the reversed-phase mode. The nonaqueous eluent composed of methanol–acetonitrile mixtures, though, need to contain reasonably high electrolyte concentrations such as acetic or formic acid and triethylamine or ammonia which represent the counter- and coions of the system, to allow proper balance of the otherwise strong ionic interactions and thus short run times. Amongst the chiral cation-exchangers, sulfonic acid-based analogs turned out to be advantageous over corresponding phosphonic and carboxylic acid analogs due to both broader selectivity profiles and better peak performances [24].

A large number of other CSPs have been tested in CEC as well, including for example poly-*N*-acryloyl-L-phenylalanineethylester (Chiraspher) [111], helically chiral poly(diphenyl-2-pyridylmethylmethacrylate)-based CSP [149], particulate MIP [150–152], and an aptamer-based CSP [153]. In this particular case, an L-RNA aptamer with an affinity for 2-phenoxypropionic acid obtained by the in vitro evolutionary selection strategy systematic evolution of ligands by exponential enrichment (SELEX) was employed as chiral selector. This L-RNA aptamer, which represents the mirror image of natural D-RNA aptamers is biostable against RNase cleavage and was biotinylated and immobilized onto 3.5 µm streptavidin-modified Kromasil silica gel via strong biotin–streptavidin interaction. Such aptamer-based CSPs have evinced some interest because artificial receptors for a variety of solutes may be developed by the SELEX strategy and the microscale separation format is particularly suitable for this kind of biomolecular selectors that are accessible in limited amounts only.

Many more studies reported interesting results with packed bed capillary columns and their applicability profiles are described in the next chapter. Overall, it can be concluded that highly efficient separations may be obtained with packed columns by CEC. Yet, the most problematic issue, namely the limited column longevities and problems with reproducibility of column fabrication have decreased the interest in this column technology and led to an increasing trend toward monolithic columns.

14.8.3 Monolithic Columns

In order to depart from retaining frits and alleviate the connected problems, such as poor reproducibility of column fabrication, susceptibility for bubble formation, fragility, and limited column longevity, researchers are more and more switching to monolithic capillary column technologies in CEC. Monolithic stationary phases consist of a single piece of a bicontinuous porous material in which highly interconnected macropores allow the flow-through of the mobile phase [68]. The polymer skeleton may be mesoporous which leads to enhanced adsorption capacities. In the literature such monolithic stationary phases are also sometimes termed as continuous beds or gels [33,154–156]. In general, monolithic stationary phases can be divided into three types:

1. Inorganic silica-based monoliths prepared by sol–gel process.
2. Organic polymer monoliths prepared by polymerization of organic monomers in the presence of porogens.
3. Hybrid monoliths obtained from particles that are entrapped or fixed in a monolithic matrix.

Table 14.3 gives an overview of the various types of enantioselective monoliths which have been described in the literature. Thereby, in the first two types of monoliths the chiral functionalities may be introduced by two distinct concepts, namely (1) in situ preparation of chirally-functionalized monoliths, e.g., through

TABLE 14.3
Summary of Various Approaches of Enantioselective Monoliths Used for CEC

Matrix	Selector[a]	Ref.
Monolithic silica		
Sol–gel matrix (postmodification)	Ligand exchange	[157–160]
	Cyclodextrins	[161,162]
	Physically adsorbed avidin	[163]
	Cation-exchange type	[25,27,164]
	Anion-exchange type	[165]
	Coated polysaccharide	[166]
	Immobilized polysaccharide	[167]
	Vancomycin	[168]
	MIP	[169]
Sol–gel matrix (in situ)	Encapsulated proteins	[170–172]
Sol–gel/organic hybrid composite material (in situ)	Encapsulated proteins	[173]
	MIP	[174]
Organic polymer (in situ)		
Poly(meth)acrylamide	Entrapped polymeric cyclodextrins	[175]
	Cyclodextrins	[155,176–179]
	Crown ether	[180]
	Ligand-exchange type	[33,181,182]
	Human serum albumin	[183]
Poly(meth)acrylate	Pirkle-type	[184]
	Anion-exchange type	[72,185–188]
	Cyclodextrin	[189]
	MIP	[190–197]
Organic/inorganic composite material (silica/vinylpyridine/methacrylate)	MIP	[198]
Organic polymer (postmodification)		
Polyacrylamide	Vancomycin	[156,199]
	Coated polysaccharide	[200]
Polymethacrylate	Anion-exchange type	[201]
	Cation-exchange type	[202]
	Pirkle-type	[203,204]
Hybrid monoliths made from particles		
Consolidated packed-bed of silica particles	Cyclodextrin	[205]
Silicate matrix	MIP	[206]
Siliceous sol-gel matrix	Pirkle-type particles (5 μm)	[207]
	Cyclodextrin-modified silica particles	[38]
Poly(meth)acrylamide	MIP	[208]
	Macrocyclic antibiotic CSPs (3 μm)	[209]
	Ligand-exchange type particles (3 μm)	[210]
ROMP	Macrocyclic antibiotic CSPs (3 μm)	[211]

[a] Unless otherwise stated covalently bonded.

copolymerization of a chiral monomer (mainly pursued with organic polymer monoliths), and (2) postmodification of a monolithic support by dedicated derivatization with chiral selector moieties (which is typically the method of choice for inorganic silica monolithic materials). In the hybrid type monoliths, particles with the desired chromatographic properties, i.e., particulate CSPs are embedded or entrapped in a monolithic matrix.

14.8.3.1 Silica Monoliths

Silica monoliths are prepared by a sol–gel process following usually a protocol published by Nakanishi and Soga in the early 1990s [212]. This process involves the acidic hydrolysis and subsequent polycondensation of silica alkoxide precursors such as tetramethoxysilane. Conditions must be carefully adjusted [i.e., content of solvent, precursor and additives such as poly(ethylene glycol) (PEG) which induces phase separation in the reaction mixture and controls macropore formation] so that a co-continuous structure of silica skeleton and macropores is formed [213]. Mesopore formation in the silica skeleton (induced by the decomposition of urea that was admixed to the sol solution) as well as aging and drying steps follow furnishing the typical silica monolith structure. For capillaries, the protocol has been slightly modified to account for the co-condensation and cross-linking of the silica matrix with the fused silica wall [214]. In most cases, 100 μm i.d. capillaries were employed, however, narrower capillaries might be of advantage in terms of attachment of the silica to the fused silica wall and have, for example, been used by Zou's group [166–168]. (Note: The risk for the formation of gaps between silica monolith and fused silica wall is minimized by narrower capillaries; for wider capillaries a trialkoxysilane, such as trimethoxymethylsilane, is admixed to the reaction mixture which may prevent the formation of voids [213].)

Silica monoliths typically possess a bimodal pore structure consisting of a large number of macropores and mesoporous silica skeleton (see Figure 14.15 for

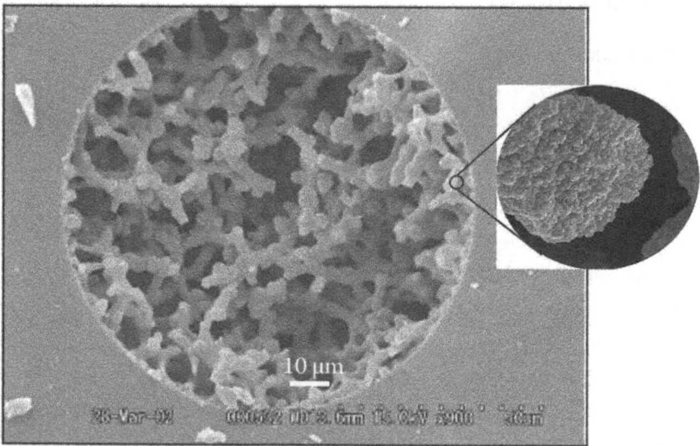

FIGURE 14.15 Scanning electron microscopy image of a typical silica monolith in a capillary column. Insert illustrates the mesoporous nature of the silica skeleton. (Modified from Preinerstorfer, B. et al., *J. Chromatogr. A*, 1106, 94, 2006. With permission.)

morphology of a typical silica monolith as prepared by the Nakanishi method [214]) [164]. The macropores enable the convective bulk mobile phase flow. The mesopores, on the other hand, provide a large active surface for the chromatographic interaction of the solutes with the stationary phase, similar to mesoporous silica beads. The thin silica skeleton provides short diffusion distances and hence guarantees effective mass transfer. Efficiencies similar to 3 μm silica particles may readily be achieved [164]. One advantage, as opposed to some compressible organic polymer gel monoliths, is the pressure stability so that monolithic silica columns can be employed for both CEC and HPLC applications [157]. The high permeability of these columns originates from macropore diameters of ~2 μm (or even wider) and total porosities >90%. It permits the use of the external pressurization system of common CE instruments (like that of Agilent which provides pressures up to 12 bar) to perform nano-HPLC separations with CE equipment [25]. With on-column detection any extra-column peak broadening effects can be precluded, which attenuates bias in CEC and HPLC comparisons (see Figure 14.3). Column stability and system robustness are greatly improved compared to the packed capillary technology resulting in enhanced column longevity. Owing to these advantages, stationary phases based on monolithic silica materials received a lot of interest over the last years and seem to become the support of choice for small molecule enantiomer separations by CEC.

One of the drawbacks is that the preparation of enantioselective monolithic silica capillaries is a multi-step procedure. Besides sol–gel process, aging and time-consuming drying steps, chiral selector functionalities have to be introduced which typically involves activation of the silica matrix with reactive linkers that enables the bonding of the chiral selector. Pioneering works in this field have been performed by Chen and coworkers [157–160], Schurig's group [161] as well as Otsuka and Terabe [163].

The in situ modification of the silica support with silane, i.e., the silanization in the capillary is commonly employed in a flow-through mode [164]. Various activation chemistries known from silica beads have been employed including, for example, silanization with 3-glycidoxypropyl silane [158,162], 3-mercaptopropyl silane [164], or γ-methacryloyloxypropyl silane [169]. Chiral selectors were then attached in the following step, e.g., amino acid amides (ligand exchange selectors) [158] and cyclodextrins [162] onto glycidoxypropyl-modified silica monoliths, vinyl-group bearing cation-exchange selectors (derived from aminosulfonic and phosphonic acids) (see Figure 14.3) [25,27,164] and anion-exchange selectors (*tert*-butylcarbamoylquinidine) [165] onto thiol-modified silica monoliths. MIPs were attached to the surface in the course of the imprinting process by copolymerization with the γ-methacryloyloxypropyl-modified silica monolith [169]. Although this surface-molecular imprinting process on a silica monolith column was successful, the results were not groundbreaking as compared to later described MIP-type polymethacrylate monoliths. For the immobilization of the vancomycin selector to monolithic silica support, Zou's group followed a three-step approach published earlier by Wikström for packed columns [61]. It involved acidic hydrolysis of the epoxy functionalities of the 3-glycidoxypropyl-modified silica monolith surface, followed by the oxidation of secondary hydroxyl groups of the diol-surface to aldehydes with sodium periodate, and finally the bonding of vancomycin with its amino group to the surface-aldehydes by reductive amination with sodium cyanoborohydride [168].

Such complicated multistep immobilization procedures were circumvented by bonding an avidin selector through physical adsorption, although at the expense of risk for bleeding and worse column stability [163]. The coating of polymeric selectors onto the surface was shown to be also a viable route to chiral monolith capillary columns [161,166]. Thus, the sol–gel derived silica monolith was coated by Schurig's group with Chirasil-β-Dex which represents a chiral polymer prepared by grafting permethyl-β-cyclodextrin to polymethylsiloxane via an octamethylene spacer [161]. Immobilization of Chirasil-β-Dex was performed by heat treatment at 120°C for 48 h to give a nonextractable coating. Zou's group followed a coating procedure for the noncovalent immobilization of cellulose tris(3,5-dimethylphenylcarbamate) published earlier by Chankvetadze et al. using a solvent evaporation protocol from acetone solution [215]. The obtained coated polysaccharide capillary (50 μm i.d.) showed highly efficient separations for a variety of test compounds. Up to 240,000 theoretical plates/m have been obtained on a 20 cm long monolith (Figure 14.16a). Short-end injection with 10 cm bed length and high voltage (30 kV) gave rise to fast separations (Figure 14.16b). Just recently, the same group described the fabrication and evaluation of a covalently bonded cellulose tris(3,5-dimethylphenylcarbamate) monolithic silica capillary column [167] following again a protocol published earlier by Chankvetadze et al. [216].

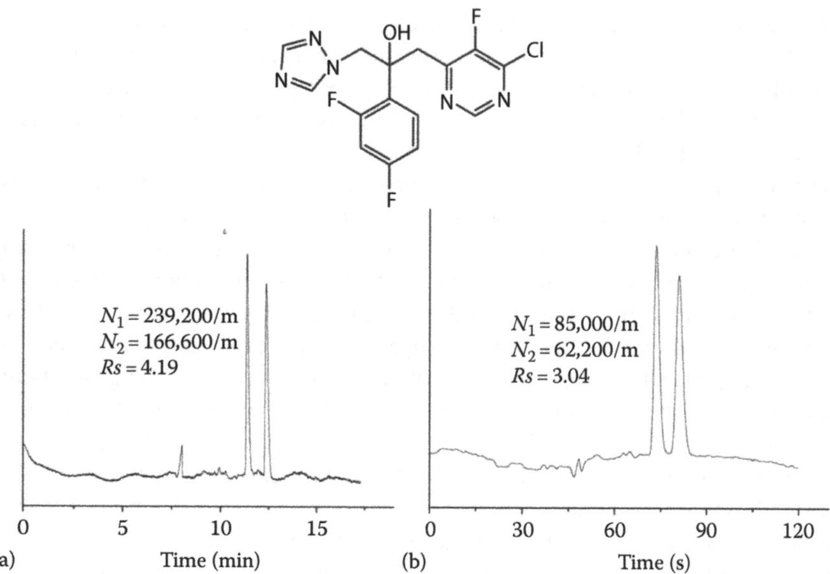

FIGURE 14.16 Electrochromatograms showing the enantioresolution of a drug candidate on a monolithic silica capillary with coated cellulose tris(3,5-dimethylphenylcarbamate) as chiral selector. Mobile phase: acetonitrile/phosphate buffer (2 mM, pH 6.80) (60: 40, v/v). Capillary: 30.2 cm (total length) × 50 μm i.d.; (a) Effective length, 20 cm; voltage, 10 kV. (b) effective length, 10.2 cm; voltage, 30 kV. (Reprinted from Qin, F. et al., *J. Sep. Sci.*, 29, 1332, 2006. With permission.)

Such immobilized polysaccharide columns allow the use of nonstandard solvents such as THF which appears, however, to be of less importance for CEC since NP conditions are less favorable for this electrokinetic technique, as outlined above.

In situ protein-encapsulation into a sol–gel matrix represents a conceptionally different approach toward enantioselective silica-based monoliths which was first presented for the CEC enantiomer separation by Kato et al. [170]. The protein (BSA or ovomucoid) was admixed to a fully or partially hydrolyzed silica precursor, injected into the capillary and the protein was finally trapped in the silica network. These capillaries were useful for CEC enantiomer separations. Later, a similar methodology was employed by encapsulation of a protein selector (BSA) in a hybrid sol–gel/organic polymer matrix (in which gelatin or chitosan constituted the organic polymer material) without any evident striking advantages in terms of chromatographic performances [173].

14.8.3.2 Organic Polymer Monoliths

Organic polymer monoliths are typically synthesized from acrylamide, methacrylate, or styrene monomers or are prepared by ring-opening metathesis polymerization (ROMP) [68]. Styrenic matrices seem to be less suitable for enantiomer separation due to nonspecific interactions with the aromatic residues of the matrix and no reports on the CEC use of such enantioselective materials can be found. ROMP-based organic polymers have been prepared for HPLC, but not for CEC enantioseparation [218]. On the other hand, a number of studies addressed the synthesis and CEC evaluation of enantioselective organic polymer monoliths with poly(meth)acrylamide and polymethacrylate backbone (see Table 14.3).

They are prepared from a homogenous polymerization mixture containing besides the (functional) monomer(s) a suitable cross-linker, a polymerization initiator, and porogens that allow the tailoring of the macropore diameter. The solution of the reaction mixture is filled into the fused silica capillary with a syringe, which is easy to implement for any other mold such as the channels of a microfluidic device. The copolymerization of the vinyl monomers is started by some kind of initiation process (addition of redox catalyst, thermal, or UV initiation) and finally leads to a macroporous microglobular monolithic polymer matrix with typically unimodal pore distribution. Cross-linking to the capillary wall may be accomplished by copolymerization with vinyl groups that have been previously attached to the fused silica capillary wall. The introduction of the active chiral recognition sites can be achieved by two basically different approaches:

1. The in situ approach.
2. The postmodification strategy.

The former case represents a single-step capillary column fabrication in which the chiral recognition sites are introduced by the copolymerization of a functional monomer bearing chiral moieties. In situ incorporation of active chiral recognition sites may, however, also be accomplished by creation of chiral cavities from achiral functional monomers that are noncovalently assembled around an enantiomeric template molecule during polymerization yielding molecularly imprinted polymers with enantioselective rebinding capabilities for the template and/or structurally closely related molecules.

The postmodification strategy makes use of a monolithic support with reactive moieties that are derivatized and chirally functionalized, respectively, in a second step.

Pioneering works for chiral polyacrylamide gels were performed by Koide and Ueno [175,176], for enantioselective polymethacrylate monoliths by Svec's group [184] and for monolithic MIPs by Lin et al. [190] and Nilsson and coworkers [191,192]. A few details are outlined hereafter.

In 1998, Peters et al. prepared Pirkle-type polymethacrylate monoliths by in situ copolymerization of *N*-[(2-methacryloyloxyethyl)oxycarbonyl]-(*S*)-valine-3,5-dimethylanilide as chiral monomer with ethylenedimethacrylate (EDMA), butylmethacrylate or glycidylmethacrylate (GMA) as comonomer and 2-acrylamido-2-methyl-1-propanesulfonic acid (AMPS) as an ionizable comonomer for EOF generation in the presence of a porogenic solvent consisting of 1-propanol, 1,4-butanediol, and water. The polymerization was carried out by thermal initiation with α,α'-azobis(isobutyronitrile) (AIBN) as radical initiator. These experiments toward enantioselective monolithic capillary columns were successful in that they enabled the baseline separation of a model test solute. However, only the monolith synthesized from the GMA-comonomer yielded reasonable efficiencies when its epoxide groups were hydrolyzed into a diol surface.

The above in situ molding approach is very flexible and allows the tuning of the polymer in terms of porosity, macropore diameter, and surface chemistry in order to meet the specific requirements of CEC and enantiomer separation simultaneously. This was shown by detailed studies on chiral anion-exchange type polymer monoliths [185–188]. Copolymerization of a hydrophilic comonomer such as 2-hydroxyethylmethacrylate (HEMA) reducing nonspecific interactions with the polymer matrix as well as the preclusion of the ionizable comonomer from the polymerization mixture and use of an ionizable chiral monomer (**1**) or (**2**) (see Figure 14.17) constituted major advancements toward efficient in situ prepared polymethacrylate type chiral monoliths for CEC application. Highly suitable monolithic capillary columns for the separation of chiral acidic compounds could be obtained by a single-step thermally initiated or photo-polymerization reaction of chiral monomer, HEMA, and EDMA. Characteristically, every monomer system requires specific optimization of the porogen composition and detailed studies in this regard have been performed with monomers **1** [185] and **2** [188]. Changes in the porogen composition, e.g., variations of the 1-dodecanol/cyclohexanol ratio which served as pore-forming agents during the preparation of these monolithic columns, allow for adjusting macropore diameter of the materials (Figure 14.17c). Characteristic morphologies for such organic polymer monoliths [68] are shown in Figure 14.17b. Individual polymer microglobules are irregularly clustered together and cross-linked. By decreasing the macropore diameter (e.g., in the given example of Figure 14.17b by decreasing the dodecanol content in the polymerization mix from 60% (w/w) to 47% (w/w) yielding materials with 3.2 μm and 0.8 μm pore size), the microglobules get smaller and the polymer morphology appears more homogenous. Associated with such a decrease in the pore diameter is thus an improvement in the chromatographic efficiencies (mainly due to a reduced *A*-term) (see below). These morphological aspects are therefore of key importance because they determine the kinetic behavior and flow properties of the monolithic column.

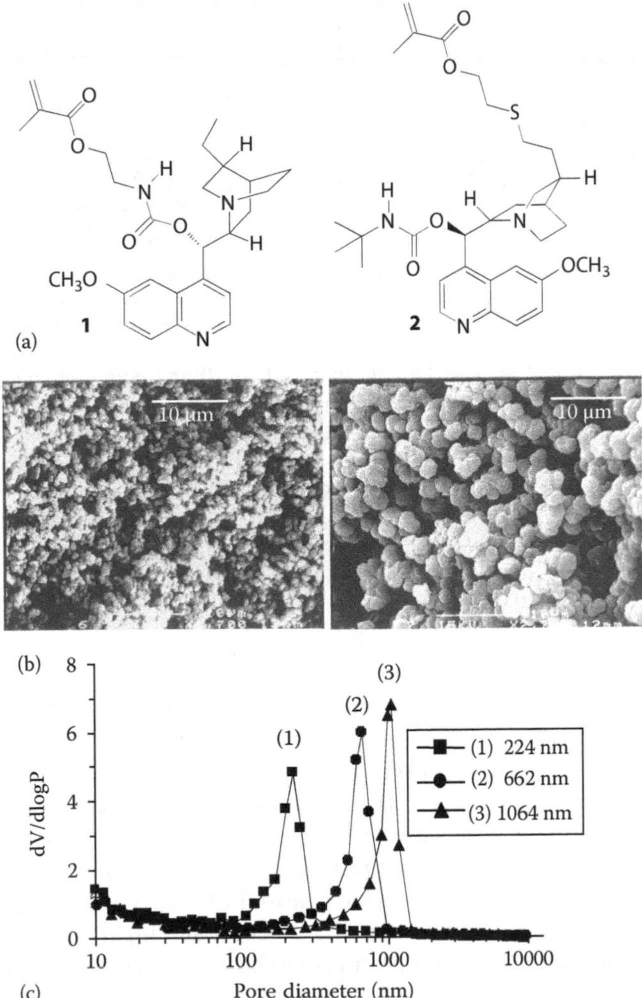

FIGURE 14.17 Polymethacrylate-type organic polymer monolith: (a) Quinidine (**1**) and quinine (**2**) derived chiral monomers. (Modified from Laemmerhofer, M. et al., *Electrophoresis*, 24, 2986, 2003. With permission.) (b) Morphologies of poly(**1**-*co*-HEMA-*co*-EDMA) monoliths with 0.8 and 3.2 µm macropore diameter. (Modified from Lammerhofer, M. et al., *J. Microcol. Sep.* 12, 597, 2000. With permission.) (c) Pore size distribution profiles of poly(**1**-*co*-HEMA-*co*-EDMA) monoliths with 0.2, 0.7, 1.1 µm macropore diameter obtained by increasing the dodecanol content from 39% over 47% to 59% in the binary dodecanol-cyclohexanol porogen mixture. (Modified from Laemmerhofer, M. et al., *Anal. Chem.*, 72, 4614, 2000. With permission.)

Besides the porogen composition, the degree of cross-linking also affects the performance of the monoliths (see Figure 14.18). When the macropore diameter was held constant at about 1.8 µm, a reduction of the percentage of cross-linker in the reaction mixture (e.g., from 40 over 20% to 10% related to the total monomer

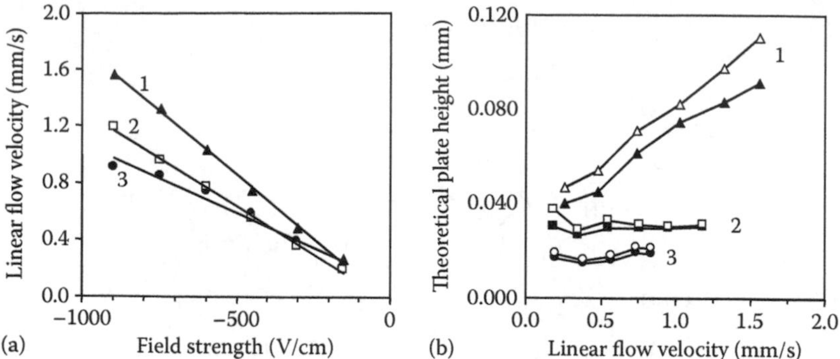

FIGURE 14.18 Effect of cross-linking (curves 1 vs. 2) and pore size (curves 2 vs. 3) for poly(**2**-*co*-HEMA-*co*-EDMA) monoliths: (a) Dependence of linear flow velocities on electric field strengths and (b) dependence of theoretical plate heights on linear flow velocities. Curve 1, 20% cross-linked with macropore diameter of 1.7 μm; curve 2, 10% cross-linked with macropore size of 1.8 μm; curve 3, 10% cross-linked with macropore size of 1.0 μm. (b) Closed symbols, (*R*)-enantiomer; open symbols, (*S*)-enantiomer. Capillary dimension, L_{total} = 33.5 cm, $L_{monolith}$ = 25 cm, $L_{effective}$ = 25 cm, 100 μm i.d., EOF marker, acetone. Sample, *N*-3,5-dinitrobenzyloxycarbonyl-leucine. (Modified from Laemmerhofer, M. et al., *Electrophoresis*, 24, 2986, 2003. With permission.)

content) dramatically improved the mass transfer properties in the monolithic bed as indicated by much flatter *H*/*u*-curves (Figure 14.18b). When the macropore diameter of the 10%-cross-linked materials was decreased from 1.8 μm to about 1 μm, the *A*-term contribution to zone dispersion could be reduced by a factor of 2 or so (Figure 14.18b). Unfortunately, less cross-linking in combination with smaller macropore diameters negatively impacts EOF velocities (Figure 14.18a) and such capillary columns are less suitable for nano-HPLC due to limited pressure resistance. For CEC application, it is less of a problem and highly efficient enantiomer separations with up to 240,000 plates/m could be achieved with an optimized in situ prepared polymethacrylate monolith (Figure 14.19a).

Overall, the effects of porogen and monomer compositions on chromatographic performances are remarkable for organic polymer monoliths and thus it seems worthwhile to carefully optimize both if the optimum in performance should be reached. This also holds for the other organic polymer monoliths.

Instead of incorporating a chiral monomer into the polymerization mixture, enantioselective binding sites with predetermined selectivity for a given solute, or structurally closely related analogs, may be obtained by the molecular imprinting process. If the imprinting process is performed directly in the capillary in the presence of an appropriate porogen, macroporous monolithic MIP columns can be obtained as first shown by Lin et al. [190] and Nilsson and coworkers [191,192]. A substantial number of papers on this topic have been published (see Table 14.3). Although electrically driven flow significantly enhanced column efficiencies compared to HPLC separations with similar stationary phases, the deleterious effect of both polydispersity of the binding sites created by the imprinting process and

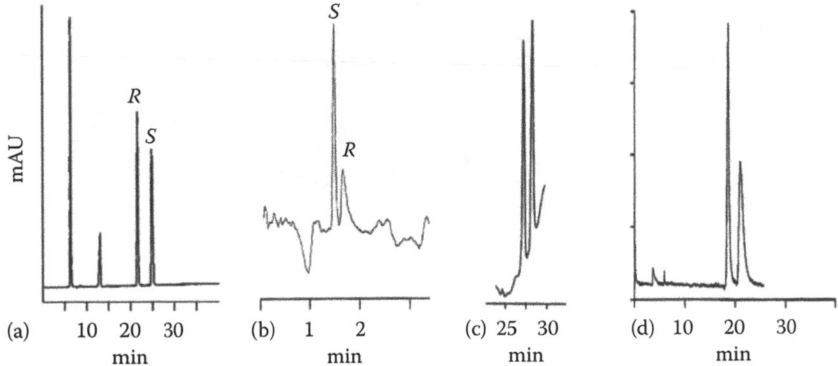

FIGURE 14.19 CEC enantiomer separations on various organic polymer monolith capillary columns. (a) *N*-2,4-Dinitrophenyl-valine on poly(**1**-*co*-HEMA-*co*-EDMA) monolith (see Figure 14.17a). (Reprinted from Laemmerhofer, M. et al., *Anal. Chem.*, 72, 4623, 2000. With permission.) (b) Propranolol on MIP-type polymethacrylate monolith. (Reprinted from Schweitz, L. et al., *Anal. Chem.*, 69, 1179, 1997. With permission.) (c) Warfarin on β-cyclodextrin-bonded polyacrylamide monolith. (Reprinted from Vegvari, A. et al., *Electrophoresis*, 21, 3116, 2000. With permission.) (d) Phenylalanine on hydroxyproline-bonded polymethacrylamide monolith in ligand-exchange mode. (Reprinted from Schmid, M.G. et al., *Electrophoresis*, 21, 3141, 2000. With permission.)

mass transfer limitations on peak performance, especially for the high affinity enantiomer, still persists (Figure 14.19b).

Soft polyacrylamide monoliths with covalently bonded β-cyclodextrin [176–179] and crown ether [180] moieties, respectively, were developed by Koide and Ueno (Figure 14.20), and at about the same time also by Hjerten's group [155], using acrylamide as monomer and *N*,*N*′-bis-methylenebisacrylamide as cross-linker. Such polyacrylamide matrices are typically synthesized from aqueous solutions at room temperature with a redox initiator (ammonium persulfate/*N*,*N*,*N*′,*N*′-tetramethylethylenediamine, TEMED). In this approach, pore sizes can be altered, for example, via the concentration of ammonium sulfate in the reaction mixture. Due to low amounts of total monomers (5%–10%) in the polymerization mixture and low cross-linking degree the hydrophilic polyacrylamide matrix is highly swollen in the aqueous media and can be easier operated by EOF than pressurized flow. Both negatively and positively charged gels were prepared that afforded either cathodic or anodic EOF dependent on the requirements dictated by the analytes (i.e., negatively charged gels with cathodic EOF for cationic solutes and positively charged gels with anodic EOF for analysis of anionic solutes). A hydrophilic backbone such as a polyacrylamide matrix is paramount for separations with cyclodextrin-containing monoliths, because it minimizes nonstereoselective hydrophobic interactions between lipophilic residues of solutes and polymer chains. They would interfere with the inclusion complexation of aromatic groups in the hydrophobic cyclodextrin cavity leading to decreased enantioselectivities. A number of chiral solutes could be resolved into enantiomers with plate numbers in excess of 100,000/m (Figure 14.19c), yet sometimes with excessively long run times due to slow flow in such gel columns.

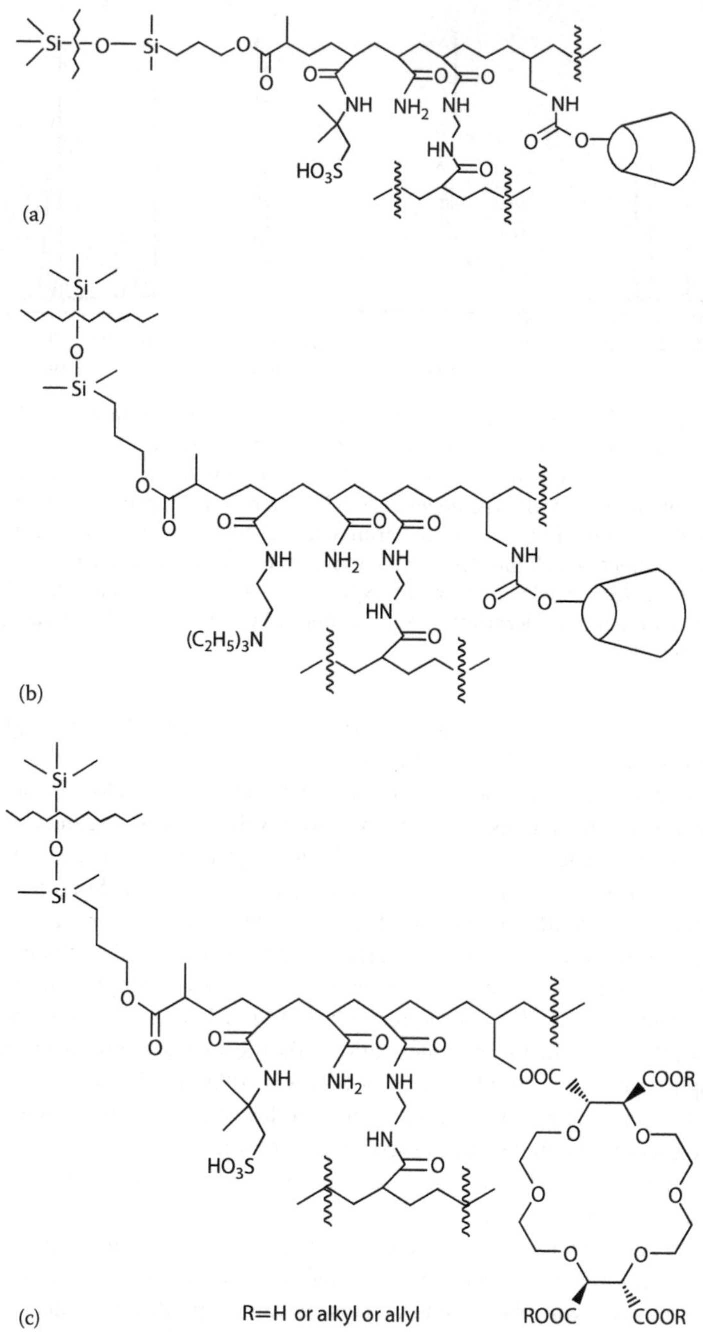

(a)

(b)

(c)

R=H or alkyl or allyl

FIGURE 14.20 β-Cyclodextrin-bonded (a) negatively charged and (b) positively charged polyacrylamide gels and (c) crown ether-bonded gel. (Reprinted from Fujimoto, C. et al., *Anal. Sci.*, 18, 19, 2002. With permission.)

Similarly, polyacrylamide monoliths made from methacrylamide and piperazine diacrylamide with copolymerized allyl-modified 4-hydroxyproline as chelating selector were used for enantioselective CEC separations of underivatized amino acids (Figure 14.19d) [33] and hydroxy carboxylic acids [181] in ligand-exchange mode using copper-ion containing eluents. While vinylsulfonic acid provided the appropriate charge for codirectional separation of amino acids, it needed to be eliminated for hydroxy acids [181] or better replaced by a cationic ionizable monomer such as diallyldimethylammonium chloride [182] to achieve faster separations.

All these as well as further studies with other selectors (see Table 14.3) demonstrate the great flexibility in terms of surface chemistries that would allow the adaption of the organic polymer monolith approach to virtually every separation problem. However, it was found to be tedious and time consuming to reoptimize the polymerization mixture and especially the porogenic solvent with every exchange of selector or other constituents of the polymerization mix. Hence, more generic on-column postmodification strategies were developed later by several groups [156,199–204] that allow the immobilization of various chromatographic ligands onto preoptimized polyacrylamide- [156,199,200] or polymethacrylate-type [201–204] monolithic supports. Such a strategy usually entails the preparation of an organic monolith with reactive groups that can be efficiently derivatized under mild conditions in the capillary, either in the flow-through mode or statically. Few concepts in this direction have been reported which includes the synthesis of epoxy-containing polymer monoliths [201–204]. They have been activated to thiol-monoliths by rinsing with sodium hydrogen sulfide solution followed by the immobilization of anion-exchange type (i.e., quinine carbamate) [201] or cation-exchange type (i.e., *N*-derivatized aminophosphonic acid) [202] chiral selector moieties by radical addition reaction. Messina et al. [203] directly derivatized the epoxy functionalities of the monolith with an amino group containing chiral selector, namely 1-(4-aminobutyl)-terguride (see Figure 14.21). Kornysova et al. [156] suggested a postmodification of polyacrylamide monoliths with vancomycin by the transformation of incorporated hydroxymethyl-groups stemming from the functional monomer to aldehydes followed by immobilization of the chiral selector through reductive amination. Even plain monoliths such as polyacrylamide monoliths without reactive groups may be useful, e.g., for coating of polymeric selectors such as cellulose tris(3,5-dimethylphenylcarbamate) [200]. All of these concepts have led to useful enantioselective capillary columns. For example, the terguride column by Messina et al. produced plate numbers of about 100,000/m for the herbicide dichlorprop and was utilized to study its stereoselective degradation in soil [204].

The postmodification strategy has the advantage that only minute amounts of chiral selector are needed, while in sharp contrast substantial amounts of chiral monomer may be consumed in the course of the full optimization of an in situ polymerization protocol. However, the yields for the derivatization reaction in the capillary under unstirred conditions are usually worse than, e.g., for particles that can be synthesized in stirred reaction vessels. Lower surface coverages are usually the result, although they have seldom and only indirectly been determined [201].

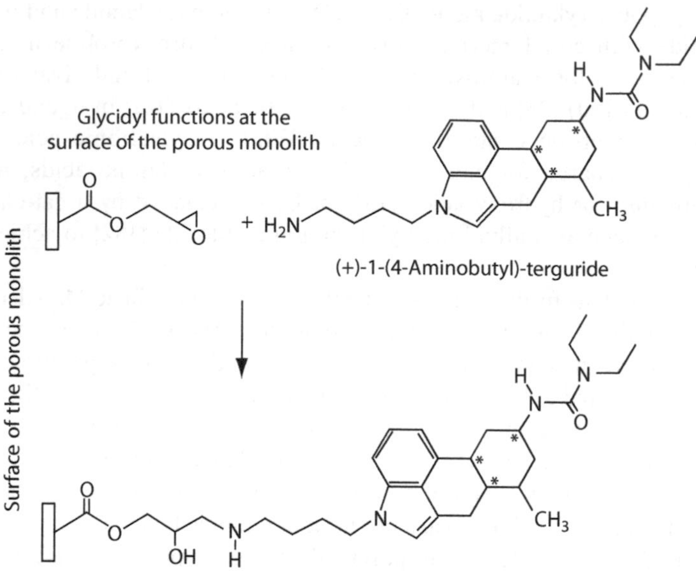

FIGURE 14.21 Postmodification strategy for the synthesis of terguride-modified polymethacrylate capillary columns. (Reprinted from Messina, A. et al., *J. Chromatogr. A*, 1120, 69, 2006. With permission.)

Moreover, reproducibility of multistep surface functionalizations may be a critical factor of such column types, yet they were found to be satisfactory [219].

To get an idea about the relative performance, Preinerstorfer et al. [164] compared the chromatographic characteristics of capillary columns based on different supports, namely, monolithic and particulate silica as well as monolithic polymethacrylate, always derivatized with the same chiral selector, i.e., a *N*-4-alkoxy-bonded-3,5-dichlorobenzoyl-2-amino-3,3-dimethylbutanephosphonic acid (Figure 14.22). With all three capillaries, baseline separations could be achieved. However, the capillary packed with silica beads showed slightly higher enantioselectivities and stronger EOF. The 100 Å particles have a large surface area and the chemical modification of the beads in the stirred reaction vessel yielded a relatively high surface coverage with selector moieties (about 0.21 mmol selector/g CSP) [24]. Hence retention factors are much larger and the ionic strength had to be increased by a factor of 2, in comparison to the other separations, to allow the elution of the solute from the column within reasonable run time. The monolithic silica has also a reasonable surface area due to its mesoporous nature of the silica skeleton (~16 nm [164]). However, owing to the large total porosity (>90%) the total amount of silica and therefore the absolutely available adsorption surface in the monolithic silica capillary is much lower than in the packed capillary. Moreover, it seems that the two step modification for the introduction of selector moieties had yielded lower surface coverages for the monolithic silica capillary as can be derived from the weaker EOF in spite of a lower ionic strength of the mobile phase. Hence, further improvements of the surface chemistry seem possible which is expected to yield better results for monolithic silica capillaries in terms of

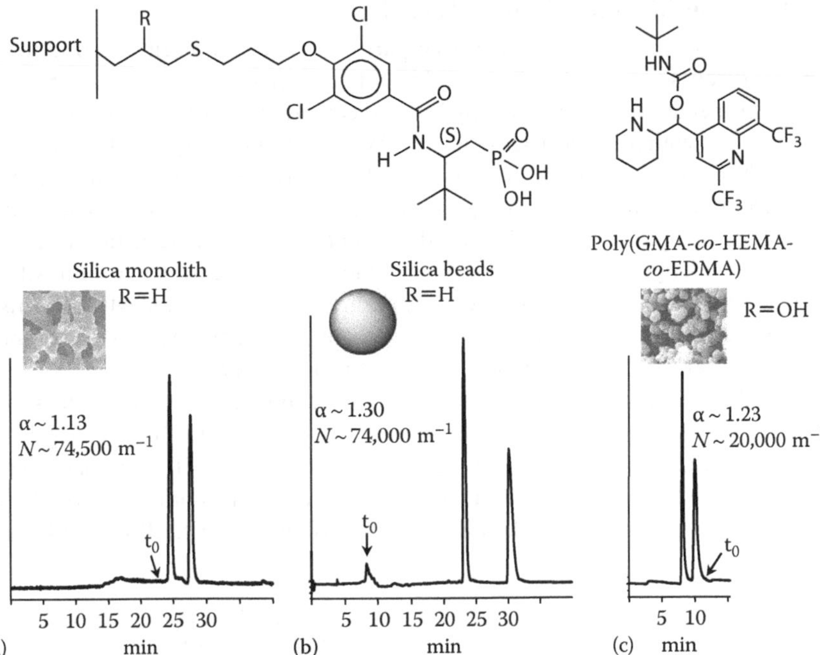

FIGURE 14.22 Effect of the support material and morphology on the CEC separation of O-(*tert*-butylcarbamoyl)mefloquine: (a) monolithic silica, (b) silica particles (100 Å, 3.5 μm), (c) monolithic poly(GMA-*co*-HEMA-*co*-EDMA) (40:40:20; w/w/w), all functionalized with β-aminophosphonic acid-derived strong cation-exchange moieties. Experimental conditions: Mobile phase: ACN-MeOH (80:20; v/v) containing 25 mM (a,c) and 50 mM (b) formic acid, and 12.5 mM (a,c) and 25 mM (b) 2-amino-1-butanol; applied voltage, +7 kV (9 μA) (a), +15 kV (15 μA) (b), +12 kV (10 μA) (c).

enantioselectivities and EOF strength. Later experiments with a different selector have confirmed this conclusion [25]. At least what chiral ion-exchange type CSPs is concerned, the monolithic silica column appears to be advantageous with regard to speed of the separations.

On the other hand, typical organic polymer monoliths do not possess a significant and tailored mesoporous structure. Hence, their surfaces are low (usually $<10\,m^2/g$ [185]). It is therefore not surprising that the organic polymer monolith column obtained by postmodification of a poly(GMA-*co*-HEMA-*co*-EDMA) monolith provides the fastest separation. Unfortunately, a compromise between higher efficiencies (obtained with monoliths devoid of HEMA) and higher separation factors (with monoliths made from HEMA as comonomer) had to be accepted and thus the plate numbers achieved with the organic polymer column are significantly lower in this example. Here it must be noted that the phosphonic acid moiety itself imparts some limitations with respect to maximally achievable plate numbers because it seems that the desorption kinetics of cationic compounds from the bis-acidic ion-exchange site is substantially impaired. Overall, considering all factors the silica monolith technology is expected to be the most promising one, and the increasing number of papers seems to confirm this assessment.

14.8.3.3 Hybrid Monoliths from Particles

A third kind of monoliths are made from particles. They represent a hybrid form of the above discussed packed and monolithic columns. Such monoliths have been first described for enantioselective CEC by Lin and coworkers in 1996 in the form of capillaries packed with MIP particles that were fixed with a polyacrylamide matrix [208] and later by Chirica and Remcho who embedded MIP particles in a silicate matrix [206]. In another early paper, Schurig's group proposed the consolidation of the chromatographic bed packed with silica beads by sintering the particles in the capillary over the entire length [205]; the subsequent coating with selectors like Chirasil-Dex, a dimethylpolysiloxane derivative bearing permethylated β-cyclodextrin, furnished an enantioselective monolithic column in a second step. The complexity of fabrication of the latter type of column prevented its further consideration.

Two other concepts, however, attracted more interests. In one, the stationary phase particles are slurry packed into the capillary and the packing is then stabilized in the capillary by a sol–gel or polymer matrix. If silica particles are used, the particles may be fixed (glued together at their contact points) through co-condensation of residual silanols with the sol–gel matrix (Figure 14.23a) or

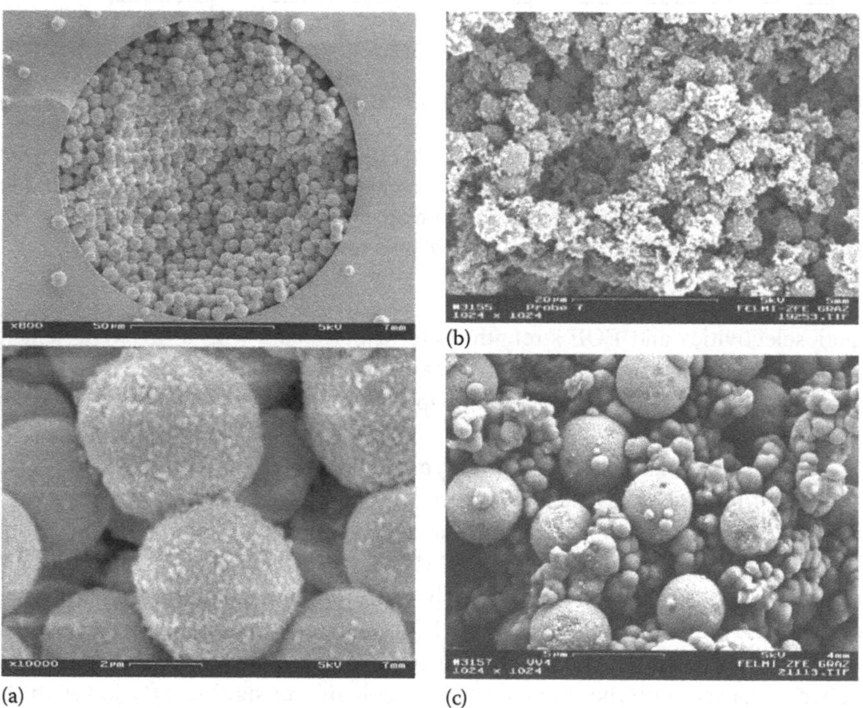

FIGURE 14.23 SEM images of hybrid monoliths: (a) Chira-Dex-silica particle-glued monolith. Magnification, 800 × (top) and 10,000 × (bottom). (Reprinted from Wistuba, D. et al., *Electrophoresis*, 26, 2019, 2005. With permission.) (b) Particle-loaded polyacrylamide-based monolith. (Reprinted from Schmid, M.G. et al., *Electrophoresis*, 25, 3195, 2004. With permission.) (c) ROMP-based monolithic stationary phase. (Reprinted from Gatschelhofer, C. et al., *J. Biochem. Biophys. Methods*, 69, 67, 2006. With permission.)

they are simply entrapped in a polymer matrix. Schurig's group used such prepared Chira-Dex-silica (3 or 5 μm) particle-glued capillary columns successfully for pCEC enantiomer separations of various aromatic solutes [38]. The tedious column packing procedure that is still involved in this approach, however, makes it less attractive compared to other technologies. In the other concept, particle-loaded monoliths are prepared from beads which are suspended in a sol–gel precursor solution [207] or organic monomer mixture [209–211] at typically 10%–25% (w/w). The suspension is introduced into the capillary with a syringe and then subjected to sol–gel matrix formation or polymerization. The particles which are entrapped or encapsulated in the matrix are less densely and regularly packed than when the particles are packed prior to the matrix formation (as can be seen from comparison of the SEM images in Figure 14.23). Schmid et al. employed this methodology for embedding various CSPs (3 μm) including macrocyclic antibiotics and ligand exchange type CSPs into polyacrylamide- [209,210] (Figure 14.23b) or ROMP-type polymer matrix [211] (Figure 14.23c). The approach is straightforward and broadly applicable, yet efficiencies in CEC were usually somewhat lower than with other column technologies.

It is common to both approaches that the particles provide the desired chromatographic properties, i.e., a high surface area for adsorption and the chromatographic selectivity. The matrix entraps the particles and may contain functionalities which support EOF generation. Unlike with other particle-based capillary columns (i.e. the packed counterparts), cumbersome frits are not required and this is the main advantage.

14.9 CEC–MS

While enantioselective CE–MS has developed into a fairly well accepted technology despite the complications caused by the chiral additive in the BGE (see also Chapter 13), only a few reports exist in the literature on the hyphenation of enantioselective CEC with mass spectrometry [79,81,104–106,136,220] (for general reviews on CEC-MS see Ref. [45,221,222]). This may come as a surprise because CEC has the advantage that nonvolatile chiral additives can be eliminated from the BGE since the selector is immobilized in the capillary column. Thus, ion-source contamination and sensitivity loss due to nonvolatile chiral selectors in the BGE are avoided in CEC–MS. On the other hand, one of the major obstacles may be that long capillaries are required to connect the column with the MS interface. Commercial instrumentation typically needs about 60 cm long capillaries to achieve a connection. In CEC–MS, this capillary is usually completely filled with the stationary phase. As a consequence, low resultant electric field strengths produce slow separations, which is a considerable limitation. Moreover, homogeneously packing such long capillary columns with small μm-sized beads is not a trivial task and the chromatographic performance may be partly compromised. Hyphenation of microchips containing a CSP with MS appears to be more promising in this regard.

In principle, a variety of different instrumentations are available for CEC–MS. However, all of the few reported studies used electrospray ionization mass spectrometry (ESI–MS) with a (coaxial) sheath–flow interface [221,222]. This instrumental arrangement allows the possibility of providing electrical contact without current breakdown at the column outlet, relative robustness, effective ion production, and the ability to be used at the low flow rates of CEC (<1 μL/min, which is too low to support a stable electrospray without sheath-flow [221]). Nevertheless, further stabilization of the system and improvement of the robustness can be accomplished by applying low pressures at the inlet end (e.g., 12 bar). In fact, using such a pCEC methodology has become the method of choice with respect to the separation part [79,104–106,136,220]. Notwithstanding, pCEC with nanospray-MS should have also promising potential [222].

Amongst the different above described column types, packed CEC columns have typically been utilized. Shamsi and coworkers tested three different column designs: (1) untapered, (2) externally tapered, and (3) internally tapered packed columns (see Figure 15.2). In the untapered column, a short unpacked section (of about 1 cm) served as the electrospray tip. The externally and internally tapered columns were produced from capillaries with either reduced outer or inner diameter [104,136]. In such tapered columns, the packing material can be retained without outlet frit. Moreover, such capillary columns are fully packed and the tapered end is directly exposed to the ESI source. The tip opening is also an important parameter and inner diameters of the tip of about 7–10 μm performed well [106]. A typical CEC–MS column used by Shamsi's group consists of 60 cm packed and 3 cm unpacked segment [106].

Untapered columns were shown to have problems with run time reproducibility in pCEC–MS, presumably due to bubble formation at the end-frit. In contrast, with tapered columns the run-to-run repeatability could be dramatically improved from about 35% RSD with the untapered column to about 5% RSD with the tapered column, both packed with Whelk O1 for warfarin enantiomer separation by pCEC–ESI–MS [136]. Apparently, the tapered columns suppress problems with bubble formation and current instability at the spray tip. Moreover, the internally tapered column turned out to be more robust compared to the externally tapered one [106,222]. Hence, such internally tapered packed columns may become the standard for CEC–MS if not open-tubular columns or monolithic columns are employed which seem more promising and easier to implement, especially at a length of 60 cm or longer. While the usability for the former has already been demonstrated for enantioselective CEC–MS [79,81], no reports on the use of enantioselective monolithic columns could be found. Yet, there exists positive experience for hyphenation of monolithic columns with CEC–MS from nonenantioselective applications [222].

For the pCEC separation, mobile phase conditions have to be adopted which are compatible with the ESI interface, i.e., nonvolatile components such as phosphate buffers should be avoided, although the sheath-liquid that is admixed at higher flow rates than the flow from the column relaxes constraints in this respect a little bit. Aqueous eluents containing organic modifiers such as acetonitrile or methanol and volatile buffers such as ammonium acetate or ammonium formate as well as nonaqueous polar organic eluents with organic acids and bases as electrolytes are well suitable for CEC-MS. For a maximal ESI signal intensity and stability, the composition of the sheath-liquid (water/alcohol ratio and pH) as well as its flow rate need to be carefully optimized as demonstrated by Shamsi and coworkers [106,136]. It was claimed that the sheath-liquid has no significant effect on the pCEC separation performance (such as resolution and analysis time) [221].

Hyphenation of pCEC with coordination ion spray-mass spectrometry (CIS–MS) was found to be a way to perform CEC–MS of nonpolar or weakly polar solutes which exhibit low ionization yields by common ESI [220]. A modified sheath–liquid interface was used in this study which allowed the introduction of complexing ions such as silver(I), cobalt(II), copper(II), and lithium(I) via sheath flow. pCEC–CIS–MS was applied for enantiomeric separation of barbiturates and chlorinated alkyl phenoxypropanoates on a permethylated β-cyclodextrin stationary phase showing increased sensitivity and high selectivity. The superiority of pressure-supported CEC compared to HPLC in the hyphenation with CIS–MS has also been demonstrated [220].

14.10 CONCLUDING REMARKS

Although an increasing number of studies continuously demonstrate the better chromatographic performance of CEC as compared to HPLC, the technology remained mainly of academic interest. It is hard to find applications that cannot be performed by either CE or HPLC. In the context of enantiomeric separation, the elimination of the selector additive from the BGE thereby facilitating the hyphenation to MS is probably one of the more striking advantages over CE. Yet, only a few studies in this direction have been performed. The method of choice in this field is currently pCEC–ESI–MS with coaxial sheath–liquid interface using internally tapered packed columns, while monolithic as well as open-tubular columns should be well suitable as well. On the other hand, the standard in HPLC enantiomer separation in terms of column technology is extremely advanced and currently not equally existing with respect to commercially available capillary columns. Moreover, with UPLC and fused-core particle technology new competing technologies are visible on the horizon which, although not yet available for enantiomer separation, may be available in the near future. They may outperform CEC easily, especially what concerns system robustness.

As preferred column technology for CEC, monolithic silica capillaries with various chiral ligands or coatings seem to become more popular, which is substantiated by an increasing number of reports in the literature. The surface chemistries known from particles employed in HPLC may be readily transferred, but need to be carefully reoptimized for in-capillary functionalization. A lot needs to be done in this respect. Such enantioselective monolithic silica columns, on the other hand, are also very useful for nano-HPLC and hence research in this field is worthwhile. Such miniaturized separation formats are the demand in the future to avoid organic waste on the one hand and provide appropriate column formats for hyphenation with highly sensitive MS technologies such as nanospray MS. Moreover, the electrically driven liquid chromatography mode may have some potential for enantiomeric separations in microchip formats. Open-tubular and organic polymer monolith column technologies would be easier to implement in the channels of microfluidic devices although the packing of particles into chip channels has brought some major advancements and even commercial utilization. Overall, CEC and pCEC will remain methodologies of significant academic interest in the future, at least in the one or the other facet.

LIST OF ABBREVIATIONS

ACN	acetonitrile
AIBN	α,α′-azobis(isobutyronitrile)
AMPS	2-acrylamido-2-methyl-1-propanesulfonic acid
APS	3-aminopropylated silica
BGE	background electrolyte
BSA	bovine serum albumin
CD	cyclodextrin
CE	capillary electrophoresis
CEC	capillary electrochromatography
CLC	capillary liquid chromatography
CMPA	chiral mobile phase additive
CP	concentration polarization
CSP	chiral stationary phase
DEAPS	diethylenetriaminopropylated silica
DNP	2,4-dinitrophenyl
EDL	electrical double layer
EDMA	ethylenedimethacrylate
EOF	electroosmotic flow
GMA	glycidylmethacrylate
HEMA	2-hydroxyethylmethacrylate
i.d.	inner diameter
MeOH	methanol
MIP	molecularly imprinted polymer
ODS	octadecyl-modified silica
OT	open-tubular
OT-EC	open-tubular electrochromatography
OT-LC	open-tubular liquid chromatography
pCEC	pressure-assisted CEC
PEG	poly(ethylene glycol)
PEM	polyelectrolyte multilayer (coating)
PLOT	porous-layer open-tubular (column)
PTH	phenylthiohydantoin
SELEX	systematic evolution of ligands by exponential enrichment
SEM	scanning electron microscopy
TAG	teicoplanin aglycone
UPLC	ultra-performance liquid chromatography
WCOT	wall-coated open-tubular (column)

REFERENCES

1. Deyl, Z, Svec, F. 2001. *Capillary Electrochromatography. Journal of Chromatography Library*, volume 62, Amsterdam: Elsevier Science Ltd.
2. Riekkola, M-L, Jönsson, JÅ, Smith, RM. 2004. Terminology for analytical capillary electromigration techniques. *Pure Appl Chem* 76: 443–451.
3. Aturki, Z, Vichi, F, Messina, A, Sinibaldi, M. 2004. Indirect resolution of β-blocker agents by reversed-phase capillary electrochromatography. *Electrophoresis* 25: 607–614.

4. Lämmerhofer, M, Maier, NM, Lindner, W. 2009. Enantiomer separation. In *Introduction to Modern Liquid Chromatography, 3rd edition*, eds. LR Snyder, JJ Kirkland and JW Dolan, Chapter 14. Hoboken, NJ: John Wiley.
5. Deng, Y, Zhang, J, Tsuda, T, Yu, PH, Boulton, AA, Cassidy, RM. 1998. Modeling and optimization of enantioseparation by capillary electrochromatography. *Anal Chem* 70: 4586–4593.
6. Lammerhofer, M, Lindner, W. 1999. High-efficiency enantioseparations of N-derivatized amino acids by packed capillary electrochromatography using ODS silica and a quinine-derived chiral selector as ion-pair agent. *J Chromatogr A* 839: 167–182.
7. Wei, W, Luo, G, Xiang, R, Yan, C. 1999. Enantiomer separations of phenylephrine and synephrine by capillary electrochromatography on are silica stationary phase using hydroxypropyl-β-cyclodextrin as a mobile phase additive. *J Microcol Sep* 11: 263–269.
8. Zhang, M, El Rassi, Z. 2000. Enantiomeric separation by capillary electrochromatography I. Chiral separation of dansyl amino acids and organochlorine pesticides on a diol-silica dynamically coated with hydroxypropyl-β-cyclodextrin. *Electrophoresis* 21: 3126–3134.
9. O'Mahony, T, Moore, S, Brosnan, B, Glennon, JD. 2003. Monitoring the supercritical fluid extraction of pyrethroid pesticides using capillary electrochromatography. *Int J Environ Anal Chem* 83: 681–691.
10. Delgado, AV, Gonzalez-Caballero, F, Hunter, RJ, Koopal, LK, Lyklema, J. 2005. Measurement and interpretation of electrokinetic phenomena. *Pure Appl Chem* 77: 1753–1805.
11. Paul, PH, Garguilo, MG, Rakestraw, DJ. 1998. Imaging of pressure- and electrokinetically driven flows through open capillaries. *Anal Chem* 70: 2459–2467.
12. Munoz, OLS, Hernandez, EP, Lammerhofer, M, Lindner, W, Kenndler, E. 2003. Estimation and comparison of ζ-potentials of silica-based anion-exchange type porous particles for capillary electrochromatography from electrophoretic and electroosmotic mobility. *Electrophoresis* 24: 390–398.
13. Rathore, AS, Horvath, C. 2001. Migration of charged sample components and electroosmotic flow in packed capillary columns. In *Capillary Electrochromatography. Journal of Chromatography Library*, Volume 62, ed. Z Deyl, F Svec, pp. 1–38. Amsterdam: Elsevier Science Ltd.
14. Nischang, I, Tallarek, U. 2004. Nonlinear electroosmosis in hierarchical monolithic structures. *Electrophoresis* 25: 2935–2945.
15. Tallarek, U, Leinweber, FC, Nischang, I. 2005. Perspective on concentration polarization effects in electrochromatographic separations. *Electrophoresis* 26: 391–404.
16. Wright, PB, Lister, AS, Dorsey, JG. 1997. Behavior and use of nonaqueous media without supporting electrolyte in capillary electrophoresis and capillary electrochromatography. *Anal Chem* 69: 3251–3259.
17. Tobler, E, Lammerhofer, M, Lindner, W. 2000. Investigation of an enantioselective non-aqueous capillary electrochromatography system applied to the separation of chiral acids. *J Chromatogr A* 875: 341–352.
18. Laemmerhofer, M. 2005. Chiral separations by capillary electromigration techniques in nonaqueous media. II. Enantioselective nonaqueous capillary electrochromatography. *J Chromatogr A* 1068: 31–57.
19. Karlsson, C, Karlsson, L, Armstrong, DW, Owens, PK. 2000. Evaluation of a vancomycin chiral stationary phase in capillary electrochromatography using polar organic and reversed-phase modes. *Anal Chem* 72: 4394–4401.
20. Girod, M, Chankvetadze, B, Blaschke, G. 2000. Enantioseparations in non-aqueous capillary electrochromatography using polysaccharide type chiral stationary phases. *J Chromatogr A* 887: 439–455.
21. Chen, X, Jin, W, Qin, F, Liu, Y, Zou, H, Guo, B. 2003. Capillary electrochromatographic separation of enantiomers on chemically bonded type of cellulose derivative chiral stationary phases with a positively charged spacer. *Electrophoresis* 24: 2559–2566.

22. Lammerhofer, M, Lindner, W. 1998. High-efficiency chiral separations of N-derivatized amino acids by packed-capillary electrochromatography with a quinine-based chiral anion-exchange type stationary phase. *J Chromatogr A* 829: 115–125.
23. Tobler, E, Lammerhofer, M, Wuggenig, F, Hammerschmidt, F, Lindner, W. 2002. Low-molecular-weight chiral cation exchangers: novel chiral stationary phases and their application for enantioseparation of chiral bases by nonaqueous capillary electrochromatography. *Electrophoresis* 23: 462–476.
24. Zarbl, E, Lammerhofer, M, Woschek, A, Hammerschmidt, F, Parenti, C, Cannazza, G, Lindner, W. 2002. Strong versus weak chiral cation exchangers: Comparative evaluation for enantiomer separation of chiral bases by non-aqueous CEC. *J Sep Sci* 25: 1269–1283.
25. Preinerstorfer, B, Laemmerhofer, M, Hoffmann, CV, Lubda, D, Lindner, W. 2008. Deconvolution of electrokinetic and chromatographic contributions to solute migration in stereoselective ion-exchange capillary electrochromatography on monolithic silica capillary columns. *J Sep Sci* 31: 3065–3078.
26. Lammerhofer, M, Svec, F, Frechet, JMJ, Lindner, W. 2000. Separation of enantiomers by capillary electrochromatography. *Trends Anal Chem* 19: 676–698.
27. Preinerstorfer, B, Hoffmann, C, Lubda, D, Laemmerhofer, M, Lindner, W. 2008. Enantioselective silica-based monoliths modified with a novel aminosulfonic acid-derived strong cation exchanger for electrically driven and pressure-driven capillary chromatography. *Electrophoresis* 29: 1626–1637.
28. Poole, CF, Poole, SK. 2009. Foundations of retention in partition chromatography. *J Chromatogr A* 1216: 1530–1550.
29. Chankvetadze, B. 2001. Enantioseparation of chiral drugs and current status of electromigration techniques in this field. *J Sep Sci* 24: 691–705.
30. Dittmann, MM, Masuch, K, Rozing, GP. 2000. Separation of basic solutes by reversed-phase capillary electrochromatography. *J Chromatogr A* 887: 209–221.
31. Wistuba, D, Czesla, H, Roeder, M, Schurig, V. 1998. Enantiomer separation by pressure-supported electrochromatography using capillaries packed with a permethyl-β-cyclodextrin stationary phase. *J Chromatogr A* 815: 183–188.
32. Wistuba, D, Schurig, V. 1999. Enantiomer separation by pressure-supported electrochromatography using capillaries packed with Chirasil-Dex polymer-coated silica. *Electrophoresis* 20: 2779–2785.
33. Schmid, MG, Grobuschek, N, Tuscher, C, Gubitz, G, Vegvari, A, Machtejevas, E, Maruska, A, Hjerten, S. 2000. Chiral separation of amino acids by ligand-exchange capillary electrochromatography using continuous beds. *Electrophoresis* 21: 3141–3144.
34. Wistuba, D, Schurig, V. 2001. Pressure supported CEC. High-efficiency technique for enantiomer separation. In *Capillary Electrochromatography. Journal of Chromatography Library 62*, ed. Z Deyl, FA Svec, pp. 317–339, Amsterdam: Elsevier.
35. Vickers, PJ, Smith, NW. 2002. Normal-phase chiral separations by pressure assisted capillary electrochromatography using the Pirkle type stationary phase Whelk-O 1. *J Sep Sci* 25: 1284–1290.
36. Grobuschek, N, Schmid, MG, Koidl, J, Gubitz, G. 2002. Enantioseparation of amino acids and drugs by CEC, pressure supported CEC, and micro-HPLC using a teicoplanin aglycone stationary phase. *J Sep Sci* 25: 1297–1302.
37. Yao, C, Tang, S, Gao, R, Jiang, C, Yan, C. 2004. Enantiomer separations on a vancomycin stationary phase and retention mechanism of pressurized capillary electrochromatography. *J Sep Sci* 27: 1109–1114.
38. Wistuba, D, Banspach, L, Schurig, V. 2005. Enantiomeric separation by capillary electrochromatography using monolithic capillaries with sol-gel-glued cyclodextrin-modified silica particles. *Electrophoresis* 26: 2019–2026.

39. Zhou, A, Lu, X, Xie, Y, Yan, C, Gao, R. 2005. Chromatographic evaluation of perphenylcarbamoylated β-cyclodextrin bonded stationary phase for micro-high performance liquid chromatography and pressurized capillary electrochromatography. *Anal Chim Acta* 547: 158–164.
40. Lin, B, Shi, Z-G, Zhang, H-J, Ng, S-C, Feng, Y-Q. 2006. Perphenylcarbamoylated β-cyclodextrin bonded-silica particles as chiral stationary phase for enantioseparation by pressure-assisted capillary electrochromatography. *Electrophoresis* 27: 3057–3065.
41. Lin, B, Zheng, M-M, Ng, S-C, Feng, Y-Q. 2007. Development of in-tube solid-phase microextraction coupled to pressure-assisted CEC and its application to the analysis of propranolol enantiomers in human urine. *Electrophoresis* 28: 2771–2780.
42. Chen, Z, Zeng, S, Yao, T. 2007. Separation of β-receptor blockers and analogs by capillary liquid chromatography (CLC) and pressurized capillary electrochromatography (pCEC) using a vancomycin chiral stationary phase column. *Pharmazie* 62: 585–592.
43. Dai, R, Tang, L, Li, H, Deng, Y, Fu, R, Parveen, Z. 2007. Synthesis and characterization of β-CD derivatized bovine serum albumin protein as chiral selector in pressurized capillary electrochromatography. *J Appl Polym Sci* 106: 2041–2046.
44. Lin, B, Ng, S-C, Feng, Y-Q. 2008. Chromatographic evaluation and comparison of three beta-cyclodextrin-based stationary phases by capillary liquid chromatography and pressure-assisted capillary electrochromatography. *Electrophoresis* 29: 4045–4054.
45. Klampfl, CW. 2004. Review: Coupling of capillary electrochromatography to mass spectrometry. *J Chromatogr A* 1044: 131–144.
46. Poppe, H. 1997. Some reflections on speed and efficiency of modem chromatographic methods. *J Chromatogr A* 778: 3–21.
47. Luedtke, S, Adam, T, Doehren, N, Unger, KK. 2000. Towards the ultimate minimum particle diameter of silica packings in capillary electrochromatography. *J Chromatogr A* 887: 339–346.
48. Eeltink, S, Gzil, P, Kok, WT, Schoenmakers, PJ, Desmet, G. 2006. Selection of comparison criteria and experimental conditions to evaluate the kinetic performance of monolithic and packed-bed columns. *J Chromatogr A* 1130: 108–114.
49. Eeltink, S, Desmet, G, Vivó-Truyols, G, Rozing, GP, Schoenmakers, PJ, Kok, WT. 2006. Performance limits of monolithic and packed capillary columns in high-performance liquid chromatography and capillary electrochromatography. *J Chromatogr A* 1104: 256–262.
50. Knox, JH, Grant, IH. 1991. Electrochromatography in packed tubes using 1.5 to 50 μm silica gels and ODS bonded silica gels. *Chromatographia* 32: 317–328.
51. Giddings, JC. 1965. *Dynamics of Chromatography*. New York: Marcel Dekker.
52. Stol, R, Poppe, H, Kok, WT. 2001. Effects of pore flow on the separation efficiency in capillary electrochromatography with porous particles. *Anal Chem* 73: 3332–3339.
53. Fanali, S, Catarcini, P, Blaschke, G, Chankvetadze, B. 2001. Enantioseparations by capillary electrochromatography. *Electrophoresis* 22: 3131–3151.
54. Girod, M, Chankvetadze, B, Blaschke, G. 2001. Enantioseparations using nonaqueous capillary electrochromatography on cellulose and amylose tris(3,5-dimethylphenylcarbamates) coated on silica gels of various pore and particle size. *Electrophoresis* 22: 1282–1291.
55. Chankvetadze, B, Kartozia, I, Okamoto, Y, Blaschke, G. 2001. The effect of pore size of silica gel and concentration of buffer on capillary chromatographic and capillary electrochromatographic enantioseparations using cellulose tris(3,5-dichlorophenylcarbamate). *J Sep Sci* 24: 635–642.
56. Chankvetadze, B, Blaschke, G. 2000. Enantioseparations using capillary electromigration techniques in nonaqueous buffers. *Electrophoresis* 21: 4159–4178.
57. Enlund, AM, Andersson, ME, Hagman, G. 2004. Improved quantification limits in chiral capillary electrochromatography by peak compression effects. *J Chromatogr A* 1028: 333–338.

58. Vindevogel, J, Sandra, P. 1994. On the possibility of performing chiral wall-coated open-tubular electrochromatography in 50 mm internal diameter capillaries. *Electrophoresis* 15: 842–847.
59. Sarmini, K, Kenndler, E. 1999. Ionization constants of weak acids and bases in organic solvents. *J Biochem Biophys Methods* 38: 123–137.
60. Karlsson, C, Wikstrom, H, Armstrong, DW, Owens, PK. 2000. Enantioselective reversed-phase and non-aqueous capillary electrochromatography using a teicoplanin chiral stationary phase. *J Chromatogr A* 897: 349–363.
61. Wikstrom, H, Svensson, LA, Torstensson, A, Owens, PK. 2000. Immobilization and evaluation of a vancomycin chiral stationary phase for capillary electrochromatography. *J Chromatogr A* 869: 395–409.
62. Carlsson, E, Wikstrom, H, Owens, PK. 2001. Validation of a chiral capillary electrochromatographic method for metoprolol on a teicoplanin stationary phase. *Chromatographia* 53: 419–424.
63. Chankvetadze, B, Kartozia, I, Breitkreutz, J, Girod, M, Knobloch, M, Okamoto, Y, Blaschke, G. 2001. Comparative capillary chromatographic and capillary electrochromatographic enantioseparations using cellulose tris(3,5-dichlorophenylcarbamate) as chiral stationary phase. *J Sep Sci* 24: 251–257.
64. Chankvetadze, L, Kartozia, I, Yamamoto, C, Chankvetadze, B, Blaschke, G, Okamoto, Y. 2002. Enantioseparations in nonaqueous capillary liquid chromatography and capillary electrochromatography using cellulose tris(3,5-dimethylphenylcarbamate) as chiral stationary phase. *Electrophoresis* 23: 486–493.
65. Girod, M, Chankvetadze, B, Okamoto, Y, Blaschke, G. 2001. Highly efficient enantioseparations in non-aqueous capillary electrochromatography using cellulose tris(3,5-dichlorophenylcarbamate) as chiral stationary phase. *J Sep Sci* 24: 27–34.
66. Chen, X, Zou, H, Ye, M, Zhang, Z. 2002. Separation of enantiomers by nanoliquid chromatography and capillary electrochromatography using a bonded cellulose trisphenylcarbamate stationary phase. *Electrophoresis* 23: 1246–1254.
67. Chen, X, Qin, F, Liu, Y, Liang, K, Zou, H. 2004. Preparation of a positively charged cellulose derivative chiral stationary phase with copolymerization reaction for capillary electrochromatographic separation of enantiomers. *Electrophoresis* 25: 2817–2824.
68. Svec, F, Tennikova, TB, Deyl, Z. 2003. *Monolithic Materials: Preparation, Properties and Applications. Journal of Chromatography Library*, volume 67. Amsterdam: Elsevier.
69. Gubitz, G, Schmid, MG. 2000. Chiral separation by capillary electrochromatography. *Enantiomer* 5: 5–11.
70. Kato, M, Toyo'oka, T. 2001. Enantiomeric separation by CEC using chiral stationary phases. *Chromatography* 22: 159–170.
71. Fujimoto, C. 2002. Enantiomer separation by capillary electrochromatography using fritless packed columns. *Anal Sci* 18: 19–25.
72. Laemmerhofer, M. 2005. Chirally-functionalized monolithic materials for stereoselective capillary electrochromatography. *Anal Bioanal Chem* 382: 873–877.
73. Preinerstorfer, B, Laemmerhofer, M. 2007. Recent accomplishments in the field of enantiomer separation by CEC. *Electrophoresis* 28: 2527–2565.
74. Wistuba, D, Schurig, V. 2000. Enantiomer separation of chiral pharmaceuticals by capillary electrochromatography. *J Chromatogr A* 875: 255–276.
75. Wistuba, D, Schurig, V. 2000. Recent progress in enantiomer separation by capillary electrochromatography. *Electrophoresis* 21: 4136–4158.
76. Kang, J, Wistuba, D, Schurig, V. 2002. Recent progress in enantiomeric separation by capillary electrochromatography. *Electrophoresis* 23: 4005–4021.
77. Wistuba, D, Schurig, V. 2006. Comparison of monolithic approaches for enantioselective capillary electrochromatography involving cyclodextrins. *J Sep Sci* 29: 1344–1352.

78. Zou, H, Ye, M. 2000. Capillary electrochromatography with physically and dynamically absorbed stationary phases. *Electrophoresis* 21: 4073–4095.
79. Schurig, V, Mayer, S. 2001. Separation of enantiomers by open capillary electrochromatography on polysiloxane-bonded permethyl-β-cyclodextrin. *J Biochem Biophys Methods* 48: 117–141.
80. Mayer, S, Schurig, V. 1992. Enantiomer separation by electrochromatography on capillaries coated with Chirasil-Dex. *J High Resol Chromatogr* 15: 129–131.
81. Kitagawa, F, Inoue, K, Hasegawa, T, Kamiya, M, Okamoto, Y, Kawase, M, Otsuka, K. 2006. Chiral separation of acidic drug components by open tubular electrochromatography using avidin immobilized capillaries. *J Chromatogr A* 1130: 219–226.
82. Liu, Z, Otsuka, K, Terabe, S. 2002. Evaluation of extended light path capillary and etched capillary for use in open tubular capillary electrochromatography. *J Chromatogr A* 961: 285–291.
83. Francotte, E, Jung, M. 1996. Enantiomer separation by open-tubular liquid chromatography and electrochromatography in cellulose-coated capillaries. *Chromatographia* 42: 521–527.
84. Pesek, JJ, Matyska, MT. 2000. Open tubular capillary electrokinetic chromatography in etched fused-silica tubes. *J Chromatogr A* 887: 31–41.
85. Tan, ZJ, Remcho, VT. 1998. Molecular imprint polymers as highly selective stationary phases for open tubular liquid chromatography and capillary electrochromatography. *Electrophoresis* 19: 2055–2060.
86. Kapnissi-Christodoulou, CP, Zhu, X, Warner, IM. 2003. Analytical separations in open-tubular capillary electrochromatography. *Electrophoresis* 24: 3917–3934.
87. Liu, Z, Zou, H, Ni, JY, Zhang, Y. 1999. Open tubular capillary electrochromatography with adsorbed stationary phase. *Anal Chim Acta* 378: 73–76.
88. Liu, Z, Zou, H, Ye, M, Ni, J, Zhang, Y. 1999. Study of physically adsorbed stationary phases for open tubular capillary electrochromatography. *Electrophoresis* 20: 2891–2897.
89. Liu, Z, Otsuka, K, Terabe, S. 2000. Study on chiral separation by open tubular capillary electrochromatography with adsorbed avidin stationary phase. *Chromatography* 21: 302–305.
90. Liu, Z, Otsuka, K, Terabe, S. 2001. Chiral separation by open tubular capillary electrochromatography with adsorbed avidin as a stationary phase. *J Sep Sci* 24: 17–26.
91. Han, N-Y, Hautala, JT, Bo, T, Wiedmer, SK, Riekkola, M-L. 2006. Immobilization of phospholipid-avidin on fused-silica capillaries for chiral separation in open-tubular capillary electrochromatography. *Electrophoresis* 27: 1502–1509.
92. Kamande, MW, Zhu, X, Kapnissi-Christodoulou, C, Warner, IM. 2004. Chiral separations using a polypeptide and polymeric dipeptide surfactant polyelectrolyte multilayer coating in open-tubular capillary electrochromatography. *Anal Chem* 76: 6681–6692.
93. Kapnissi, CP, Valle, BC, Warner, IM. 2003. Chiral separations using polymeric surfactants and polyelectrolyte multilayers in open-tubular capillary electrochromatography. *Anal Chem* 75: 6097–6104.
94. Hofstetter, H, Hofstetter, O, Schurig, V. 1998. Enantiomer separation using BSA as chiral stationary phase in affinity OTEC and OTLC. *J Microcol Sep* 10: 287–291.
95. Olsson, J, Blomberg, LG. 2008. Enantioseparation of omeprazole and its metabolite 5-hydroxyomeprazole using open tubular capillary electrochromatography with immobilized avidin as chiral selector. *J Chromatogr B* 875: 329–332.
96. Sinibaldi, M, Vinci, M, Federici, F, Flieger, M. 1997. Electrokinetic separation of enantiomers using a capillary coated with poly-terguride. *Biomed Chromatogr* 11: 307–310.
97. Kamande, MW, Fletcher, KA, Lowry, M, Warner, IM. 2005. Capillary electrochromatography using polyelectrolyte multilayer coatings. *J Sep Sci* 28: 710–718.

98. Kapnissi-Christodoulou, CP, Lowry, M, Agbaria, RA, Geng, L, Warner, IM. 2005. Investigation of the stability of polyelectrolyte multilayer coatings in open-tubular capillary electrochromatography using laser scanning confocal microscopy. *Electrophoresis* 26: 783–789.
99. Schurig, V. 1996. Unified enantioselective chromatography involving a permethylated β-cyclodextrin bonded to a dimethylpolysiloxane (CHIRASIL-DEX). *Proceedings of the Eighth International Symposium on Cyclodextrins*, Budapest, Mar. 31–Apr. 2, 1996: 641–647.
100. Schurig, V, Jung, M, Mayer, S, Fluck, M, Negura, S, Jakubetz, H. 1995. Unified enantioselective capillary chromatography on a Chirasil-DEX stationary phase. Advantages of column miniaturization. *J Chromatogr A* 694: 119–128.
101. Tobler, E. 2001. Development of enantioselective separation systems for analysis of ionic compounds by capillary electrochromatography. PhD thesis, Austria: University of Vienna.
102. Tobler, E, Lammerhofer, M, Mancini, G, Lindner, W. 2001. On-column deracemization of an atropisomeric biphenyl by quinine-based stationary phase and determination of rotational energy barrier by enantioselective stopped-flow HPLC and CEC. *Chirality* 13: 641–647.
103. Fanali, S, Catarcini, P, Presutti, C, Quaglia, MG, Righetti, P-G. 2003. A glycopeptide antibiotic chiral stationary phase for the enantiomer resolution of hydroxy acid derivatives by capillary electrochromatography. *Electrophoresis* 24: 904–912.
104. Zheng, J, Norton, D, Shamsi, SA. 2006. Fabrication of internally tapered capillaries for capillary electrochromatography electrospray ionization mass spectrometry. *Anal Chem* 78: 1323–1330.
105. Bragg, W, Norton, D, Shamsi, SA. 2008. Optimized separation of β-blockers with multiple chiral centers using capillary electrochromatography-mass spectrometry. *J Chromatogr B* 875: 304–316.
106. Zheng, J, Shamsi, SA. 2006. Simultaneous enantioseparation and sensitive detection of eight β-blockers using capillary electrochromatography-electrospray ionization-mass spectrometry. *Electrophoresis* 27: 2139–2151.
107. Li, S, Lloyd, DK. 1993. Direct chiral separations by capillary electrophoresis using capillaries packed with an α1-acid glycoprotein chiral stationary phase. *Anal Chem* 65: 3684–3690.
108. Lloyd, DK, Li, S, Ryan, P. 1995. Protein chiral selectors in free-solution capillary electrophoresis and packed-capillary electrochromatography. *J Chromatogr A* 694: 285–296.
109. Li, S, Lloyd, DK. 1994. Packed-capillary electrochromatographic separation of the enantiomers of neutral and anionic compounds using β-cyclodextrin as a chiral selector. Effect of operating parameters and comparison with free-solution capillary electrophoresis. *J Chromatogr A* 666: 321–335.
110. Lelievre, F, Yan, C, Zare, RN. 1996. Capillary electrochromatography: Operating characteristics and enantiomeric separations. *J Chromatogr A* 723: 145–156.
111. Krause, K, Girod, M, Chankvetadze, B, Blaschke, G. 1999. Enantioseparations in normal- and reversed-phase nano-high-performance liquid chromatography and capillary electrochromatography using polyacrylamide and polysaccharide derivatives as chiral stationary phases. *J Chromatogr A* 837: 51–63.
112. Otsuka, K, Mikami, C, Terabe, S. 2000. Enantiomer separations by capillary electrochromatography using chiral stationary phases. *J Chromatogr A* 887: 457–463.
113. Kawamura, K, Otsuka, K, Terabe, S. 2001. Capillary electrochromatographic enantioseparations using a packed capillary with a 3 μm OD-type chiral packing. *J Chromatogr A* 924: 251–257.
114. Chankvetadze, B, Kartozia, I, Breitkreutz, J, Okamoto, Y, Blaschke, G. 2001. Effect of organic solvent, electrolyte salt and a loading of cellulose tris(3,5-dichlorophenylcarbamate) on silica gel on enantioseparation characteristics in capillary electrochromatography. *Electrophoresis* 22: 3327–3334.

115. Chankvetadze, L, Kartozia, I, Yamamoto, C, Chankvetadze, B, Blaschke, G, Okamoto, Y. 2002. Enantioseparations in capillary liquid chromatography and capillary electrochromatography using amylose tris(3,5-dimethylphenylcarbamate) in combination with aqueous organic mobile phase. *J Sep Sci* 25: 653–660.
116. Chankvetadze, B, Kartozia, I, Yamamoto, C, Okamoto, Y, Blaschke, G. 2003. Comparative study on the application of capillary liquid chromatography and capillary electrochromatography for investigation of enantiomeric purity of the contraceptive drug levonorgestrel. *J Pharm Biomed Anal* 30: 1897–1906.
117. Chankvetadze, B. 2004. Enantioseparation in capillary chromatography and capillary electrochromatography using polysaccharide-type chiral stationary phases. In *Chiral Separations: Methods and Protocols (Methods in Molecular Biology, Vol. 243)*, eds. G Gubitz and MG Schmid, Totowa: Humana Press, pp. 387–399.
118. Fanali, S, D'Orazio, G, Lomsadze, K, Chankvetadze, B. 2008. Enantioseparations with cellulose tris(3-chloro-4-methylphenylcarbamate) in nano-liquid chromatography and capillary electrochromatography. *J Chromatogr B* 875: 296–303.
119. Mangelings, D, Hardies, N, Maftouh, M, Suteu, C, Massart, DL, Vander Heyden, Y. 2003. Enantioseparations of basic and bifunctional pharmaceuticals by capillary electrochromatography using polysaccharide stationary phases. *Electrophoresis* 24: 2567–2576.
120. Dermaux, A, Lynen, F, Sandra, P. 1998. Chiral capillary electrochromatography on a vancomycin stationary phase. *J High Resol Chromatogr* 21: 575–576.
121. Desiderio, C, Aturki, Z, Fanali, S. 2001. Use of vancomycin silica stationary phase in packed capillary electrochromatography. I. Enantiomer separation of basic compounds. *Electrophoresis* 22: 535–543.
122. Desiderio, C, Rudaz, S, Veuthey, J-L, Raggi, MA, Fanali, S. 2002. Use of vancomycin silica stationary phase in packed capillary electrochromatography. Part IV: Enantiomer separation of fluoxetine and norfluoxetine employing UV high sensitivity detection cell. *J Sep Sci* 25: 1291–1296.
123. Fanali, S, Catarcini, P, Quaglia, MG. 2002. Use of vancomycin silica stationary phase in packed capillary electrochromatography: III. Enantiomeric separation of basic compounds with the polar organic mobile phase. *Electrophoresis* 23: 477–485.
124. Fanali, S, Rudaz, S, Veuthey, JL, Desiderio, C. 2001. Use of vancomycin silica stationary phase in packed capillary electrochromatography. II. Enantiomer separation of venlafaxine and O-desmethylvenlafaxine in human plasma. *J Chromatogr A* 919: 195–203.
125. Carter-Finch, AS, Smith, NW. 1999. Enantiomeric separations by capillary electrochromatography using a macrocyclic antibiotic chiral stationary phase. *J Chromatogr A* 848: 375–385.
126. Catarcini, P, Fanali, S, Presutti, C, D'Acquarica, I, Gasparrini, F. 2003. Evaluation of teicoplanin chiral stationary phases of 3.5 and 5 µm inside diameter silica microparticles by polar-organic mode capillary electrochromatography. *Electrophoresis* 24: 3000–3005.
127. Schmid, MG, Grobuschek, N, Pessenhofer, V, Klostius, A, Gubitz, G. 2003. Enantioseparation of dipeptides by capillary electrochromatography on a teicoplanin aglycone chiral stationary phase. *J Chromatogr A* 990: 83–90.
128. Zhang, M, El Rassi, Z. 2000. Enantiomeric separation by capillary electrochromatography II. Chiral separation of dansyl amino acids and phenoxy acid herbicides on sulfonated silica having surface-bound hydroxypropyl-β-cyclodextrin. *Electrophoresis* 21: 3135–3140.
129. Wistuba, D, Cabrera, K, Schurig, V. 2001. Enantiomer separation by nonaqueous and aqueous capillary electrochromatography on cyclodextrin stationary phases. *Electrophoresis* 22: 2600–2605.
130. Gong, Y, Lee, HK. 2002. Enantiomeric separations in capillary electrochromatography with crown ether-capped β-cyclodextrin-bonded silica particles as chiral stationary phase. *Helv Chim Acta* 85: 3283–3293.

131. Kartozia, I, D'Orazio, G, Chankvetadze, B, Fanali, S. 2005. Evaluation of cyclodextrins modified with dichloro-, dimethyl-, and chloromethylphenylcarbamate groups as chiral stationary phases for capillary electrochromatography. *J Capill Electrophor Microchip Technol* 9: 31–38.
132. Wang, Y, Xiao, Y, Yang Tan, TT, Ng, S-C. 2008. Click chemistry for facile immobilization of cyclodextrin derivatives onto silica as chiral stationary phases. *Tetrahedron Lett* 49: 5190–5191.
133. Wolf, C, Spence, PL, Pirkle, WH, Derrico, EM, Cavender, DM, Rozing, GP. 1997. Enantioseparations by electrochromatography with packed capillaries. *J Chromatogr A* 782: 175–179.
134. Wolf, C, Spence, PL, Pirkle, WH, Cavender, DM, Derrico, EM. 2000. Investigation of capillary electrochromatography with brush-type chiral stationary phases. *Electrophoresis* 21: 917–924.
135. Honzatko, A, Aturki, Z, Flieger, M, Messina, A, Sinibaldi, M. 2003. Chiral separations by capillary electrochromatography on multiple-interaction based stationary phases. *Chromatographia* 58: 271–275.
136. Zheng, J, Shamsi, SA. 2003. Combination of chiral capillary electrochromatography with electrospray ionization mass spectrometry: Method development and assay of warfarin enantiomers in human plasma. *Anal Chem* 75: 6295–6305.
137. Zheng, J, Shamsi, SA. 2003. Brush-type chiral stationary phase for enantioseparation of acidic compounds. Optimization of chiral capillary electrochromatographic parameters. *J Chromatogr A* 1005: 177–187.
138. Lammerhofer, M, Tobler, E, Lindner, W. 2000. Chiral anion exchangers applied to capillary electrochromatography enantioseparation of oppositely charged chiral analytes: investigation of stationary and mobile phase parameters. *J Chromatogr A* 887: 421–437.
139. Constantin, S, Bicker, W, Zarbl, E, Lammerhofer, M, Lindner, W. 2003. Enantioselective strong cation-exchange molecular recognition materials: Design of novel chiral stationary phases and their application for enantioseparation of chiral bases by nonaqueous capillary electrochromatography. *Electrophoresis* 24: 1668–1679.
140. Bicker, W, Hebenstreit, D, Laemmerhofer, M, Lindner, W. 2003. Enantiomeric impurity profiling in ephedrine samples by enantioselective capillary electrochromatography. *Electrophoresis* 24: 2532–2542.
141. Hebenstreit, D, Bicker, W, Laemmerhofer, M, Lindner, W. 2004. Novel enantioselective strong cation exchangers based on sulfodipeptide selectors: Evaluation for enantiomer separation of chiral bases by nonaqueous capillary electrochromatography. *Electrophoresis* 25: 277–289.
142. Zheng, J, Bragg, W, Hou, J, Lin, N, Chandrasekaran, S, Shamsi, SA. 2009. Sulfated and sulfonated polysaccharide as chiral stationary phases for capillary electrochromatography and capillary electrochromatography-mass spectrometry. *J Chromatogr A* 1216: 857–872.
143. Gong, Y, Lee, HK. 2003. Application of cyclam-capped β-cyclodextrin-bonded silica particles as a chiral stationary phase in capillary electrochromatography for enantiomeric separations. *Anal Chem* 75: 1348–1354.
144. Adam, T, Ludtke, S, Unger, KK. 1999. Packing and stationary phase design for capillary electroendosmotic chromatography. *Chromatographia* 49(Suppl. 1): S49–S55.
145. Fanali, S, Catarcini, P, Presutti, C. 2003. Enantiomeric separation of acidic compounds of pharmaceutical interest by capillary electrochromatography employing glycopeptide antibiotic stationary phases. *J Chromatogr A* 994: 227–232.
146. Fanali, S, Catarcini, P, Presutti, C, Stancanelli, R, Quaglia, MG. 2003. Use of short-end injection capillary packed with a glycopeptide antibiotic stationary phase in electrochromatography and capillary liquid chromatography for the enantiomeric separation of hydroxy acids. *J Chromatogr A* 990: 143–151.

147. Fanali, S, D'Orazio, G, Quaglia, MG, Rocco, A. 2004. Use of a Hepta-Tyr antibiotic modified silica stationary phase for the enantiomeric resolution of D,L-loxiglumide by electrochromatography and nano-liquid chromatography. *J Chromatogr A* 1051: 247–252.
148. Schmid, MG, Grobuschek, N, Pessenhofer, V, Klostius, A, Guebitz, G. 2003. Chiral resolution of diastereomeric di- and tripeptides on a teicoplanin aglycone phase by capillary electrochromatography. *Electrophoresis* 24: 2543–2549.
149. Krause, K, Chankvetadze, B, Okamoto, Y, Blaschke, G. 2000. Enantioseparations in nonaqueous and aqueous capillary electrochromatography using helically chiral poly(diphenyl-2-pyridylmethylmethacrylate) as chiral stationary phase. *J Microcol Sep* 12: 398–406.
150. Lin, J-M, Nakagama, T, Uchiyama, K, Hobo, T. 1997. Temperature effect on chiral recognition of some amino acids with molecularly imprinted polymer filled capillary electrochromatography. *Biomed Chromatogr* 11: 298–302.
151. Quaglia, M, De Lorenzi, E, Sulitzky, C, Massolini, G, Sellergren, B. 2001. Surface initiated molecularly imprinted polymer films: A new approach in chiral capillary electrochromatography. *Analyst* 126: 1495–1498.
152. Quaglia, M, De Lorenzi, E, Sulitzky, C, Caccialanza, G, Sellergren, B. 2003. Molecularly imprinted polymer films grafted from porous or nonporous silica: Novel affinity stationary phases in capillary electrochromatography. *Electrophoresis* 24: 952–957.
153. Andre, C, Berthelot, A, Thomassin, M, Guillaume, Y-C. 2006. Enantioselective aptameric molecular recognition material: design of a novel chiral stationary phase for enantioseparation of a series of chiral herbicides by capillary electrochromatography. *Electrophoresis* 27: 3254–3262.
154. Hjerten, S, Vegvari, A, Srichaiyo, T, Zhang, H-X, Ericson, C, Eaker, D. 1998. An approach to ideal separation media for (electro)chromatography. *J Capill Electrophor* 5: 13–26.
155. Vegvari, A, Foldesi, A, Hetenyi, C, Kocnegarova, O, Schmid, MG, Kudirkaite, V, Hjerten, S. 2000. A new easy-to-prepare homogeneous continuous electrochromatographic bed for enantiomer recognition. *Electrophoresis* 21: 3116–3125.
156. Kornysova, O, Owens, PK, Maruska, A. 2001. Continuous beds with vancomycin as chiral stationary phase for capillary electrochromatography. *Electrophoresis* 22: 3335–3338.
157. Chen, Z, Hobo, T. 2001. Chemically L-prolinamide-modified monolithic silica column for enantiomeric separation of dansyl amino acids and hydroxy acids by capillary electrochromatography and μ-high performance liquid chromatography. *Electrophoresis* 22: 3339–3346.
158. Chen, Z, Hobo, T. 2001. Chemically L-phenylalaninamide-modified monolithic silica column prepared by a sol-gel process for enantioseparation of dansyl amino acids by ligand exchange-capillary electrochromatography. *Anal Chem* 73: 3348–3357.
159. Chen, Z, Niitsuma, M, Uchiyama, K, Hobo, T. 2003. Comparison of enantioseparations using Cu(II) complexes with L-amino acid amides as chiral selectors or chiral stationary phases by capillary electrophoresis, capillary electrochromatography and micro liquid chromatography. *J Chromatogr A* 990: 75–82.
160. Chen, Z, Nishiyama, T, Uchiyama, K, Hobo, T. 2004. Electrochromatographic enantioseparation using chiral ligand exchange monolithic sol-gel column. *Anal Chim Acta* 501: 17–23.
161. Kang, J, Wistuba, D, Schurig, V. 2002. A silica monolithic column prepared by the sol-gel process for enantiomeric separation by capillary electrochromatography. *Electrophoresis* 23: 1116–1120.
162. Chen, Z, Ozawa, H, Uchiyama, K, Hobo, T. 2003. Cyclodextrin-modified monolithic columns for resolving dansyl amino acid enantiomers and positional isomers by capillary electrochromatography. *Electrophoresis* 24: 2550–2558.

163. Liu, Z, Otsuka, K, Terabe, S, Motokawa, M, Tanaka, N. 2002. Physically adsorbed chiral stationary phase of avidin on monolithic silica column for capillary electrochromatography and capillary liquid chromatography. *Electrophoresis* 23: 2973–2981.
164. Preinerstorfer, B, Lubda, D, Lindner, W, Laemmerhofer, M. 2006. Monolithic silica-based capillary column with strong chiral cation-exchange type surface modification for enantioselective non-aqueous capillary electrochromatography. *J Chromatogr A* 1106: 94–105.
165. Preinerstorfer, B, Lubda, D, Mucha, A, Kafarski, P, Lindner, W, Lammerhofer, M. 2006. Stereoselective separations of chiral phosphinic acid pseudodipeptides by CEC using silica monoliths modified with an anion-exchange-type chiral selector. *Electrophoresis* 27: 4312–4320.
166. Qin, F, Xie, C, Feng, S, Ou, J, Kong, L, Ye, M, Zou, H. 2006. Monolithic silica capillary column with coated cellulose tris(3,5-dimethylphenylcarbamate) for capillary electrochromatographic separation of enantiomers. *Electrophoresis* 27: 1050–1059.
167. Dong, X, Wu, R, Dong, J, Wu, M, Zhu, Y, Zou, H. 2008. The covalently bonded cellulose tris(3,5-dimethylphenylcarbamate) on a silica monolithic capillary column for enantioseparation in capillary electrochromatography. *J Chromatogr B* 875: 317–322.
168. Dong, X, Dong, J, Ou, J, Zhu, Y, Zou, H. 2007. Preparation and evaluation of a vancomycin-immobilized silica monolith as chiral stationary phase for CEC. *Electrophoresis* 28: 2606–2612.
169. Ou, J, Li, X, Feng, S, Dong, J, Dong, X, Kong, L, Ye, M, Zou, H. 2007. Preparation and evaluation of a molecularly imprinted polymer derivatized silica monolithic column for capillary electrochromatography and capillary liquid chromatography. *Anal Chem* 79: 639–646.
170. Kato, M, Sakai-Kato, K, Matsumoto, N, Toyo'oka, T. 2002. A protein-encapsulation technique by the sol-gel method for the preparation of monolithic columns for capillary electrochromatography. *Anal Chem* 74: 1915–1921.
171. Kato, M, Matsumoto, N, Sakai-Kato, K, Toyo'oka, T. 2003. Investigation of chromatographic performances and binding characteristics of BSA-encapsulated capillary column prepared by the sol-gel method. *J Pharm Biomed Anal* 30: 1845–1850.
172. Sakai-Kato, K, Kato, M, Nakakuki, H, Toyo'oka, T. 2003. Investigation of structure and enantioselectivity of BSA-encapsulated sol-gel columns prepared for capillary electrochromatography. *J Pharm Biomed Anal* 31: 299–309.
173. Kato, M, Saruwatari, H, Sakai-Kato, K, Toyo'oka, T. 2004. Silica sol-gel/organic hybrid material for protein encapsulated column of capillary electrochromatography. *J Chromatogr A* 1044: 267–270.
174. Wang, H-F, Zhu, Y-Z, Lin, J-P, Yan, X-P. 2008. Fabrication of molecularly imprinted hybrid monoliths via a room temperature ionic liquid-mediated nonhydrolytic sol-gel route for chiral separation of zolmitriptan by capillary electrochromatography. *Electrophoresis* 29: 952–959.
175. Koide, T, Ueno, K. 1998. Enantiomeric separations of cationic and neutral compounds by capillary electrochromatography with charged polyacrylamide gels incorporating chiral selectors. *Anal Sci* 14: 1021–1023.
176. Koide, T, Ueno, K. 1999. Enantiomeric separations of cationic and neutral compounds by capillary electrochromatography with β-cyclodextrin-bonded charged polyacrylamide gels. *Anal Sci* 15: 791–794.
177. Koide, T, Ueno, K. 2000. Enantiomeric separations by capillary electrochromatography with charged polyacrylamide gels incorporating chiral selectors. *Anal Sci* 16: 1065–1070.
178. Koide, T, Ueno, K. 2000. Enantiomeric separations of cationic and neutral compounds by capillary electrochromatography with monolithic chiral stationary phases of β-cyclodextrin-bonded negatively charged polyacrylamide gels. *J Chromatogr A* 893: 177–187.

179. Koide, T, Ueno, K. 2000. Enantiomeric separations of acidic and neutral compounds by capillary electrochromatography with β-cyclodextrin-bonded positively charged polyacrylamide gels. *J High Resol Chromatogr* 23: 59–66.
180. Koide, T, Ueno, K. 2001. Enantiomeric separations of primary amino compounds by capillary electrochromatography with monolithic chiral stationary phases of chiral crown ether-bonded negatively charged polyacrylamide gels. *J Chromatogr A* 909: 305–315.
181. Schmid, MG, Grobuschek, N, Lecnik, O, Gubitz, G, Vegvari, A, Hjerten, S. 2001. Enantioseparation of hydroxy acids on easy-to-prepare continuous beds for capillary electrochromatography. *Electrophoresis* 22: 2616–2619.
182. Lecnik, O, Guebitz, G, Schmid, MG. 2003. Role of the charge in continuous beds in the chiral separation of hydroxy acids by ligand-exchange capillary electrochromatography. *Electrophoresis* 24: 2983–2985.
183. Machtejevas, E, Maruska, A. 2002. A new approach to human serum albumin chiral stationary phase synthesis and its use in capillary liquid chromatography and capillary electrochromatography. *J Sep Sci* 25: 1303–1309.
184. Peters, EC, Lewandowski, K, Petro, M, Svec, F, Frechet, JMJ. 1998. Chiral electrochromatography with a "molded" rigid monolithic capillary column. *Anal Commun* 35: 83–86.
185. Laemmerhofer, M, Peters, EC, Yu, C, Svec, F, Frechet, JMJ, Lindner, W. 2000. Chiral monolithic columns for enantioselective capillary electrochromatography prepared by copolymerization of a monomer with quinidine functionality. 1. Optimization of polymerization conditions, porous properties, and chemistry of the stationary phase. *Anal Chem* 72: 4614–4622.
186. Laemmerhofer, M, Svec, F, Frechet, JMJ, Lindner, W. 2000. Chiral monolithic columns for enantioselective capillary electrochromatography prepared by copolymerization of a monomer with quinidine functionality. 2. Effect of chromatographic conditions on the chiral separations. *Anal Chem* 72: 4623–4628.
187. Lammerhofer, M, Svec, F, Frechet, JMJ, Lindner, W. 2000. Monolithic stationary phases for enantioselective capillary electrochromatography. *J Microcol Sep* 12: 597–602.
188. Laemmerhofer, M, Tobler, E, Zarbl, E, Lindner, W, Svec, F, Frechet, JMJ. 2003. Macroporous monolithic chiral stationary phases for capillary electrochromatography: New chiral monomer derived from cinchona alkaloid with enhanced enantioselectivity. *Electrophoresis* 24: 2986–2999.
189. Pumera, M, Jelinek, I, Jindrich, J, Benada, O. 2002. β-Cyclodextrin-modified monolithic stationary phases for capillary electrochromatography and nano-HPLC chiral analysis of ephedrine and ibuprofen. *J Liq Chromatogr Rel Technol* 25: 2473–2484.
190. Lin, J-M, Nakagama, T, Uchiyama, K, Hobo, T. 1997. Capillary electrochromatographic separation of amino acid enantiomers using on-column prepared molecularly imprinted polymer. *J Pharm Biomed Anal* 15: 1351–1358.
191. Schweitz, L, Andersson, LI, Nilsson, S. 1997. Capillary electrochromatography with molecular imprint-based selectivity for enantiomer separation of local anesthetics. *J Chromatogr A* 792: 401–409.
192. Schweitz, L, Andersson, LI, Nilsson, S. 1997. Capillary electrochromatography with predetermined selectivity obtained through molecular imprinting. *Anal Chem* 69: 1179–1183.
193. Schweitz, L, Andersson, LI, Nilsson, S. 2001. Rapid electrochromatographic enantiomer separations on short molecularly imprinted polymer monoliths. *Anal Chim Acta* 435: 43–47.
194. Schweitz, L, Andersson, LI, Nilsson, S. 2002. Molecularly imprinted CEC sorbents: investigations into polymer preparation and electrolyte composition. *Analyst* 127: 22–28.

195. Liu, Z-S, Xu, Y-L, Wang, H, Yan, C, Gao, R-Y. 2004. Chiral separation of binaphthol enantiomers on molecularly imprinted polymer monolith by capillary electrochromatography. *Anal Sci* 20: 673–678.
196. Xu, Y-L, Liu, Z-S, Wang, H-F, Yan, C, Gao, R-Y. 2005. Chiral recognition ability of an (S)-naproxen-imprinted monolith by capillary electrochromatography. *Electrophoresis* 26: 804–811.
197. Ou, J, Dong, J, Tian, T, Hu, J, Ye, M, Zou, H. 2007. Enantioseparation of tetrahydropalmatine and Troeger's base by molecularly imprinted monolith in capillary electrochromatography. *J Biochem Biophys Methods* 70: 71–76.
198. Deng, Q-L, Lun, Z-H, Gao, R-Y, Zhang, L-H, Zhang, W-B, Zhang, Y-K. 2006. (S)-Ibuprofen-imprinted polymers incorporating gamma-methacryloxypropyl-trimethoxysilane for CEC separation of ibuprofen enantiomers. *Electrophoresis* 27: 4351–4358.
199. Kornysova, O, Jarmalaviciene, R, Maruska, A. 2004. A simplified synthesis of polymeric nonparticulate stationary phases with macrocyclic antibiotic as chiral selector for capillary electrochromatography. *Electrophoresis* 25: 2825–2829.
200. Dong, X, Wu, R, Dong, J, Wu, M, Zhu, Y, Zou, H. 2008. Polyacrylamide-based monolithic capillary column with coating of cellulose tris(3,5-dimethylphenyl-carbamate) for enantiomer separation in capillary electrochromatography. *Electrophoresis* 29: 919–927.
201. Preinerstorfer, B, Bicker, W, Lindner, W, Lammerhofer, M. 2004. Development of reactive thiol-modified monolithic capillaries and in-column surface functionalization by radical addition of a chromatographic ligand for capillary electrochromatography. *J Chromatogr A* 1044: 187–199.
202. Preinerstorfer, B, Lindner, W, Laemmerhofer, M. 2005. Polymethacrylate-type monoliths functionalized with chiral amino phosphonic acid-derived strong cation exchange moieties for enantioselective nonaqueous capillary electrochromatography and investigation of the chemical composition of the monolithic polymer. *Electrophoresis* 26: 2005–2018.
203. Messina, A, Flieger, M, Bachechi, F, Sinibaldi, M. 2006. Enantioseparation of 2-aryloxypropionic acids on chiral porous monolithic columns by capillary electrochromatography. *J Chromatogr A* 1120: 69–74.
204. Messina, A, Sinibaldi, M. 2007. CEC enantioseparations on chiral monolithic columns: a study of the stereoselective degradation of (R/S)-dichlorprop [2-(2,4-dichlorophenoxy)propionic acid] in soil. *Electrophoresis* 28: 2613–2618.
205. Wistuba, D, Schurig, V. 2000. Enantiomer separation by capillary electrochromatography on a cyclodextrin-modified monolith. *Electrophoresis* 21: 3152–3159.
206. Chirica, G, Remcho, VT. 1999. Silicate entrapped columns. New columns designed for capillary electrochromatography. *Electrophoresis* 20: 50–56.
207. Kato, M, Dulay, MT, Bennett, B, Chen, JR, Zare, RN. 2000. Enantiomeric separation of amino acids and nonprotein amino acids using a particle-loaded monolithic column. *Electrophoresis* 21: 3145–3151.
208. Lin, J-M, Uchiyama, K, Hobo, T. 1996. Enantiomeric resolution of dansyl amino acids by capillary electrochromatography based on molecular imprinting method. *Chromatographia* 47: 625–629.
209. Schmid, MG, Koidl, J, Freigassner, C, Tahed, S, Wojcik, L, Beesley, T, Armstrong, DW, Guebitz, G. 2004. New particle-loaded monoliths for chiral capillary electrochromatographic separation. *Electrophoresis* 25: 3195–3203.
210. Schmid, MG, Koidl, J, Wank, P, Kargl, G, Zoehrer, H, Guebitz, G. 2007. Enantioseparation by ligand-exchange using particle-loaded monoliths: Capillary-LC versus capillary electrochromatography. *J Biochem Biophys Methods* 70: 77–85.

211. Gatschelhofer, C, Schmid, MG, Schreiner, K, Pieber, TR, Sinner, FM, Guebitz, G. 2006. Enantioseparation of glycyl-dipeptides by CEC using particle-loaded monoliths prepared by ring-opening metathesis polymerization (ROMP). *J Biochem Biophys Methods* 69: 67–77.
212. Nakanishi, K, Soga, N. 1992. Phase-separation in silica sol-gel system containing polyacrylic acid. 2. Effects of molecular weight and temperature. *J Non-Cryst Solids* 139: 14–24.
213. Núñez, O, Nakanishi, K, Tanaka, N. 2008. Preparation of monolithic silica columns for high-performance liquid chromatography. *J Chromatogr A* 1191: 231–252.
214. Ishizuka, N, Kobayashi, H, Minakuchi, H, Nakanishi, K, Hirao, K, Hosoya, K, Ikegami, T, Tanaka, N. 2002. Monolithic silica columns for high-efficiency separations by high-performance liquid chromatography. *J Chromatogr A* 960: 85–96.
215. Chankvetadze, B, Yamamoto, C, Tanaka, N, Nakanishi, K, Okamoto, Y. 2004. High-performance liquid chromatographic enantioseparations on capillary columns containing monolithic silica modified with cellulose tris(3,5-dimethylphenylcarbamate). *J Sep Sci* 27: 905–911.
216. Chankvetadze, B, Ikai, T, Yamamoto, C, Okamoto, Y. 2004. High-performance liquid chromatographic enantioseparations on monolithic silica columns containing a covalently attached 3,5-dimethylphenylcarbamate derivative of cellulose. *J Chromatogr A* 1042: 55–60.
217. Qin, F, Xie, C, Yu, Z, Kong, L, Ye, M, Zou, H. 2006. Monolithic enantiomer-selective stationary phases for capillary electrochromatography. *J Sep Sci* 29: 1332–1343.
218. Mayr, B, Sinner, F, Buchmeiser, MR. 2001. Chiral beta-cyclodextrin-based polymer supports prepared via ring-opening metathesis graft-polymerization. *J Chromatogr A* 907: 47–56.
219. Ding, G, Tang, A. 2006. Preparation of norvancomycin-bonded chiral silica monolithic column for capillary electrochromatography and its applications. *Chin J Chromatogr* 24: 402–406.
220. Von Brocke, A, Wistuba, D, Gfrorer, P, Stahl, M, Schurig, V, Bayer, E. 2002. On-line coupling of packed capillary electrochromatography with coordination ion spray-mass spectrometry for the separation of enantiomers. *Electrophoresis* 23: 2963–2972.
221. Shamsi, SA, Miller, BE. 2004. Capillary electrophoresis-mass spectrometry: Recent advances to the analysis of small achiral and chiral solutes. *Electrophoresis* 25: 3927–3961.
222. Barceló-Barrachina, E, Moyano, E, Galceran, MT. 2004. State-of-the-art of the hyphenation of capillary electrochromatography with mass spectrometry. *Electrophoresis* 25: 1927–1948.

[illegible]

15 Recent Applications in Capillary Electrochromatography

Debby Mangelings and Yvan Vander Heyden

CONTENTS

15.1 Introduction ... 459
15.2 Enantioseparations in Regular Capillary Electrochromatographic Mode ... 461
15.2.1 Separations on Particle-Based Stationary Phases ... 461
15.2.2 Separations on Monolithic Stationary Phases ... 478
15.3 Enantioseparations in Pressurized Capillary Electrochromatographic Mode ... 481
15.3.1 Separations on Particle-Based Stationary Phases ... 481
15.3.2 Separations on Monolithic Stationary Phases ... 486
15.4 Enantioseparations in Open-Tubular Capillary Electrochromatography ... 486
15.5 Enantioseparations in Chip Capillary Electrochromatography ... 489
15.6 Concluding Remarks ... 490
Acknowledgment ... 492
List of Abbreviations ... 492
References ... 493

15.1 INTRODUCTION

Capillary electrochromatography (CEC) is a separation technique that theoretically combines the advantages of both liquid chromatography (LC) and capillary electrophoresis (CE). Due to the presence of a stationary phase (SP) inside a capillary and the use of an electro-osmotic flow (EOF) as a driving force for the mobile phase, the user can benefit from properties, such as a high sample loading capacity, high efficiencies, high peak capacities, and low sample and solvent consumptions [1]. Since its introduction as separation technique in the early 1970s [2], many applications have been developed in CEC. These applications find their origin from a variety of research areas, such as environmental analysis, food and natural products analysis, industrial analysis, forensic analysis, and pharmaceutical and biomedical analysis [3]. In the pharmaceutical domain, the separation of chiral compounds is most thoroughly studied.

Chiral separations are quite important in pharmaceutical analysis because a racemate of a drug compound mostly contains a therapeutically active enantiomer, while in organisms the other enantiomer mostly exhibits a different activity. Therefore, regulatory authorities [4,5] demand the development of single-enantiomer drugs when possible, and of analytical methods that are able to determine the enantiomeric purity of a pharmaceutically active compound. The separation of enantiomers can only be performed when the analyst is using a chiral environment; otherwise, no discrimination between both isomers will occur. Besides CEC, different separation techniques are used for chiral separations, e.g., LC [6–8], CE [9,10], gas chromatography [11], and supercritical fluid chromatography [12]. Both indirect and direct methods can be used to enable the separation of a chiral compound. In the indirect methods, a chemical reaction is used to convert the enantiomers into diastereoisomers, which can then be separated using conventional nonchiral separation techniques. The direct methods can be subdivided into two classes. In a first approach, a chiral selector is added to the mobile phase or running buffer, and is occasionally combined with a nonchiral SP (for chromatographic techniques). The second approach (in chromatography) uses chiral stationary phases (CSPs), where the chiral selector is immobilized or coated onto an inert carrier.

The chiral selectors used for the CSP are those applied in LC. The CSPs are classified into three groups based on the chemical structure of the chiral selectors involved and on the chiral recognition mechanism. The first group contains the low-molecular-weight selectors, such as the Pirkle-type CSPs, the ion-exchange and ligand-exchange CSPs. The second group involves the macrocyclic selectors, such as cyclodextrins (CDs), crown ethers, and macrocyclic antibiotics. The third group includes macromolecular selectors, such as polysaccharides, synthetic polymers, molecularly imprinted polymers (MIPs), and proteins.

In CEC, the three chiral separation options are possible and the technique is characterized by low selector, SP, or mobile-phase consumptions due to its miniaturized setup. Therefore, much research has already been devoted to develop CEC as an analytical separation technique for chiral compounds. However, on the level of SP development, CEC is still in its childhood, as it is still characterized by many practical drawbacks such as the need for frits for some column types, fragile columns, bubble formation during analysis, and the lack of SPs specifically developed for CEC. These drawbacks are most frequently seen when particle-based SPs, usually developed for HPLC columns, are applied for CEC column fabrication. The capillary is then filled with the SP that is fixed by means of two frits [13]. Monolithic SPs may provide a solution for some of these problems because they are synthesized inside the capillary and do not require frits to be retained [14]. However, column-to-column variation can be rather large with these SPs.

Monolithic columns can be roughly divided into polymeric- and silica-based columns. In the first group, a polymerization mixture containing monomers, cross-linkers, charge-providing agents, and porogenic solvents is used for stationary-phase synthesis; in the second group, sol–gel reactions from alkoxysilane precursors are used. Other SPs related to the monolithic group are the particle-glued phases.

Particles-based SPs are then used, but they are "glued" together and onto the capillary wall using a sol–gel process. Such columns do not need retaining frits. A third type of columns used in CEC are open-tubular (OT) columns, which contain a layer of SP coated or bonded onto the capillary wall. These columns also do no require retaining frits, but are often characterized by overloading problems and a relatively low stability as a function of time. The three types of columns used in CEC thus all have their advantages, but also drawbacks. Further information concerning these different types of columns can be found in the Chapter 14.

In this chapter, an overview is given of the applications of the last 5 years (2003–2008) in chiral separations by means of CEC, pressurized CEC (pCEC), OT-CEC, and chip electrochromatography. We focus on discussing the most interesting topics which demonstrate that CEC still develops as separation technique. They comprise approaches of column development, the use of CEC in the analysis of biological samples and screening approaches, fast separations in the context of high-throughput analysis, validation studies and the coupling to detection techniques other than the most frequently used UV/DAD detection. Minor importance is given to the discussion of classic method development, where influences of buffer concentration, pH, organic modifier concentration and type, and applied voltage are investigated. It must also be noted that the applications in electrokinetic chromatography (EKC) will not be discussed, because they usually are classified as CE approaches.

15.2 ENANTIOSEPARATIONS IN REGULAR CAPILLARY ELECTROCHROMATOGRAPHIC MODE

Using regular CEC mode means that analyses are performed either with both vials pressurized or without pressure at any vial. The mobile phase is then only driven by the EOF, and the resulting flow profile is flat.

An overview of all chiral compounds that were found separated using the regular CEC mode is given in Table 15.1. Below, the most interesting applications were selected for a more extensive discussion.

15.2.1 SEPARATIONS ON PARTICLE-BASED STATIONARY PHASES

Indirect chiral chromatographic separations are only scarcely described in the literature. They are inherent to several drawbacks, such as the need for an enantiopure reagent and the fact that the enantiomers are altered in a chemical reaction. The same is seen for direct methods developed on an achiral phase with the selector in the mobile phase because the chiral selector consumption is often rather high. The same trends are observed for CEC analyses. Nevertheless, some recent applications using these approaches are described in the literature. In [15], the indirect approach was used for the chiral separation of 5 β-blocking agents after conversion into diastereoisomers by means of (+)-1-(9-fluorenyl)ethyl chloroformate (FLEC). Three types of octadecyl silica (ODS) were tested as achiral phase and Lichrospher 100 RP18 was finally selected. All substances could be separated with good precision between runs, days, and columns; an LOD (limit of detection) of 1.025 μg/mL and a sample stability of five days were found. Relative standard deviations (RSDs)

TABLE 15.1
Separations in Regular CEC Mode

Substances	Type of Stationary Phase	Chiral Selector of CSP, in Mobile Phase or Imprinting Molecule	Mobile Phase/Applied Voltage	Detection	Refs.
Metoprolol, oxprenolol, pindolol, propranolol, atenolol, all derivatized with FLEC	Particle-based (commercial, Lichrospher RP18)	(Indirect method)	5 mM sodium borate pH 8 with 55%–80%ACN; 30 kV	DAD	[15]
Chlorthalidone	Particle-based (commercial, Hypersil)	Hydroxypropyl-β-CD (selector in mobile phase)	1 mM phosphate buffer pH 6.5 + 33 mM HP-β-CD/ACN (84/16); 25 kV	DAD	[16]
Levonorgestrel	Particle-based (in-house made)	Cellulose tris (3,5-dichlorophenyl)carbamate	5 mM ammonium acetate in methanol; 12 kV	DAD	[18]
Tröger's base, praziquantel, benzoin, ranolazine, alprenolol, metoprolol, propranolol, *trans*-stilbene oxide, α-dimethyl-dicarboxylbiphenyl derivative, 2 drug candidates	Particle-based (in-house made)	Cellulose 2,3-bisphenylcarbamate and cellulose 2,3-bis (3,5-dimethylphenylcarbamate)	100% EtOH, Hexane/EtOH/MeOH (35/25/8), or Hexane/EtOH/MeOH (35/15/10), all containing 14.2 mM acetic acid and 1.4 mM DEA; Voltages ranging between −7 and −25 kV	UV	[19]
Trans-stilbene oxide, warfarin, praziquantel, benzoin, Tröger's base, drug candidate	Particle-based (in-house made)	Cellulose 2,3-bisphenylcarbamate and cellulose 2,3-bis (3,5-dimethylphenylcarbamate)	ACN/2 mM phosphate pH 3.8 (60/40), ACN/2 mM phosphate pH 3.1 (50/50), or EtOH/water (95/5) containing 14.2 mM acetic acid and 1.4 mM DEA; Voltages ranging between −6 and −25 kV	UV	[19]
Acebutolol, alprenolol, atropine, betaxolol, bupranolol, ephedrine, fluoxetine, labetalol, metoprolol, mianserin, nadolol, oxprenolol, pindolol, promethazine, propiomazine, propranolol, sotalol, sulpiride, tertatolol, tetramisole, toliprolol, verapamil, lorazepam, oxazepam, temazepam	Particle-based (commercialized)	Cellulose tris(3,5-dimethylphenyl) carbamate = Chiralcel OD and amylose tris(3,5-dimethylphenyl) carbamate = Chiralpak AD	5 mM phosphate buffer pH 11.5/ ACN/HA (70/30/0.15): 15 or 5 kV	DAD	[20]

Acenocoumarol, warfarin, oxazepam, praziquantel, pindolol, tetramisole	Particle-based (commercialized)	Cellulose tris(3,5-dimethylphenyl) carbamate = Chiralcel OD, amylose tris(3,5-dimethylphenyl) carbamate = Chiralpak AD, cellulose tris-(4-methylbenzoate) = Chiralcel OJ or amylose tris [(*S*)-α- methylbenzylcarbamate] = Chiralpak AS	45 mM ammonium formate pH 3 (acidic, neutral, bifunctional compounds) or 5 mM phosphate pH 11.5 (basic, bifunctional, neutral compounds) mixed with 70% ACN. Voltage low pH: 20 kV, high pH: 10 kV	DAD	[21]
Warfarin, ibuprofen, hexobarbital, flurbiprofen, suprofen, methylphenobarbital, fenoprofen, indoprofen, coumachlor, 2-phenylbutyric acid, 2-phenylpropionic acid	Particle-based (commercialized)	Amylose tris(3,5-dimethylphenyl) carbamate = Chiralpak AD or cellulose tris-(4-methylbenzoate) = Chiralcel OJ	45 mM ammonium formate pH 2.9/ ACN (35/65); 15 kV	DAD	[22]
Acebutolol, alprenolol, ambucetamide, atropine, betaxolol, bisoprolol, bopindolol, bupranolol, carazolol, carteolol, celiprolol, chlorpheniramine, dimethindene, esmolol, ethopropazine, felodipine, fluoxetine, labetalol, lorazepam, mebeverine, meptazinol, metoprolol, mianserine, morfine, nadolol, nebivolol, nimodipine, oxazepam, oxprenolol, pindolol, praziquantel, promethazine, propiomazine, propranolol, sotalol, temazepam, tertatolol, tetramisole, toliprolol, *trans*-stilbene oxide, verapamil	Particle-based (commercial)	Cellulose tris(3,5-dimethylphenyl) carbamate = Chiralcel OD, amylose tris(3,5-dimethylphenyl) carbamate = Chiralpak AD, cellulose tris-(4-methylbenzoate) = Chiralcel OJ or amylose tris [(*S*)-α- methylbenzylcarbamate] = Chiralpak AS	5 mM phosphate buffer pH 11.5 mixed with 35%–70% ACN; Applied voltage between 5 and 25 kV	DAD	[23]
Levetiracetam	Particle-based	Amylose tris(3,5-dimethylphenyl) carbamate = Chiralpak AD	5 mM phosphate pH 11.5/ACN (30/70); 10 kV	DAD	[25]

(*continued*)

TABLE 15.1 (continued)
Separations in Regular CEC Mode

Substances	Type of Stationary Phase	Chiral Selector of CSP, in Mobile Phase or Imprinting Molecule	Mobile Phase/Applied Voltage	Detection	Refs.
Tröger's base, benzoin, praziquantel, warfarin, 3 α-dimethyl dicarboxyl biphenyl derivatives, 2 drug candidates, ranozaline, alprenolol, metoprolol, propranolol	Particle-based (in-house made)	Cellulose 3,5-dimethylphenylcarbamate	100% ethanol or mixtures of ethanol and hexane, containing 14.2 mM acetic acid and 1.4 mM DEA; Voltages between −15 and −30 kV (with and without applied pressure of 10–12 bar at inlet vial)	DAD	[26]
Tröger's base, drug candidate, benzoin, praziquantel, indapamide, pindolol, tetrahydropalmitine, warfarin	Particle-based (in-house made)	Cellulose tris(3,5-dimethylphenyl) carbamate	2 mM phosphate buffer pH 6.8, 3.2 or 10.8, mixed with 40% or 50% ACN, or 40% THF; 10 or 15 kV	DAD	[27]
DL-*m*-hydroxymandelic acid, 3-hydroxy-4-methoxymandelic acid, DL-mandelic acid, DL-*p*- hydroxymandelic acid, DL-2-phenyllactic acid, 4-chloro-DL-mandelic acid	Particle-based (in-house made)	Hepta-Tyr (glycopeptide antibiotic derivative)	50 mM ammonium acetate pH 6/water/MeOH/ACN (10/40/20/30); Voltages between 2.5 and 30 kV	DAD	[28]
Ketoprofen, indoprofen, suprofen, 2-[(4′-benzoyloxy-2′-hydroxy)phenyl] propionic acid, 2-[(5′-Benzoyl-2′hydroxy) phenyl]propionic acid	Particle-based (in-house made)	Vancomycin	5 mM ammonium formate/ACN (10/90); 25 kV	DAD	[29]
Carprofen, cicloprofen, etodolac, ibuprofen, ketoprofen, 2-[(4′-benzoyloxy-2′-hydroxy) phenyl]propionic acid, 2-[(5′-benzoyl-2′hydroxy)phenyl]propionic acid, 2-(4′-isobutylphenyl)-3-methylbutanoic acid, 2-(4′-isobutylphenyl)-butanoic acid, 2-(4′-isobutylphenyl)ciclopentylacetic acid	Particle-based (in-house made)	Hepta-Tyr	5 mM sodium phosphate pH 6/ACN (1/1); −25 kV	DAD	[29]
Loxiglumide	Particle-based (in-house made)	Hepta-Tyr	ACN/phosphate buffer pH 6 (1/1); 15 kV	DAD	[30]

Acebutolol, alprenolol, atenolol, fluoxetine, metoprolol, norfluoxetine, oxprenolol, pindolol, propranolol, salbutamol	Particle-based (in-house made, 3.5 or 5 µm)	Teicoplanin (mixed with achiral silica particles)	MeOH/ACN/ammonium acetate (60/40/0.05); 20 kV	DAD	[31]
Acebutolol, alprenolol, atenolol, fluoxetine, metoprolol, norfluoxetine, oxprenolol, pindolol, propranolol, salbutamol, tolperisone	Particle-based (in-house made, 3.5 or 5 µm)	Teicoplanin	MeOH/ACN/ammonium acetate (60/40/0.05); 20 kV	DAD	[31]
Mixture of mirtazapine, 8-OH-mirtazapine, and *N*-desmethyl mirtazapine	Particle-based (in-house made)	Vancomycin	100 mM ammonium acetate pH 6/ water/MeOH/ACN (5/15/30/50); 25 kV	DAD	[32]
Mianserin	Particle-based (commercial)	Vancomycin (Chirobiotic V)	9.4 mM TEAA buffer pH 4.8/ACN (37.5/62.5); 25 kV	DAD	[33]
Clenbuterol, ephedrine, flecainide, mefloquine, nifenalol, pronethalol, rimiterol, salbutamol, sotalol, terbutaline	Particle-based (in-house made)	(*R*)-2-(4-Allyloxy-3,5-dichlorobenzoylamino)-3-sulfo-propionic acid = CSP 1	ACN/MeOH (4/1) with 25 mM 2-amino-1-butanol and 50 mM formic acid; 15 kV	DAD	[34]
Acebutolol, alprenolol, atenolol, bunitrolol, bupranolol, celiprolol, clenbuterol, dipivefrine, ephedrine, flecainide, mefloquine, naphtylethylamine, nifenalol, orciprenaline, penbutolol, phenmetrazine, pronethalol, propranolol, rimiterol, salbutamol, sotalol, terbutaline	Particle-based (in-house made)	(*R*)-2-(4-Allyloxy-3,5-dichlorobenzoylamino)-3-methyl-3-sulfo-butyric acid = CSP 2	ACN/MeOH (4/1) with 25 mM 2-amino-1-butanol and 50 mM formic acid; 15 kV	DAD	[34]
Acebutolol, atenolol, benzetimide, bunitrolol, bupranolol, celiprolol, clenbuterol, dipivefrine, ephedrine, flecainide, mefloquine, naphtylethylamine, nifenalol, omeprazole, pantoprazole, penbutolol, phenmetrazine, pronethalol, rimiterol, salbutamol, talinolol, terbutaline	Particle-based (in-house made)	(*S*)-1-(4-Allyloxy-3,5-dichlorobenzoylamino)-1-*tert*-butylcarbamoyl-2-methyl-propane-2-sulfonic acid = CSP 3	ACN/MeOH (4/1) with 25 mM 2-amino-1-butanol and 50 mM formic acid; 15 kV	DAD	[34]

(*continued*)

TABLE 15.1 (continued)
Separations in Regular CEC Mode

Substances	Type of Stationary Phase	Chiral Selector of CSP, in Mobile Phase or Imprinting Molecule	Mobile Phase/Applied Voltage	Detection	Refs.
Bunitrolol, celiprolol, clenbuterol, dipivefrine, ephedrine, flecainide, mefloquine, nifenalol, penbutolol, pronethalol, rimiterol, sotalol, terbutaline	Particle-based (in-house made)	(2*S*)-2-[(2*R*)-2-(4-Allyloxy-3,5-dichloro-benzoylamino)-3-sulfopropionylamino]-4-methyl-pentanoic acid = CSP 4	ACN/MeOH (4/1) with 25 mM 2-amino-1-butanol and 50 mM formic acid; 15 kV	DAD	[34]
Ephedrine	Particle-based (in-house made)	(*R*)- or (*S*) 1-(4-Allyloxy-3,5-dichlorobenzoylamino)-1-*tert*-butylcarbamoyl-2-methylpropane-2-sulfonic acid	ACN/MeOH (80/20) containing 50 mM formic acid and 25 mM 2-amino-1-butanol; 15 kV	DAD	[35]
Alprenolol, atenolol, bamethane, benzetimide, bupranolol, carazolol, celiprolol, clenbuterol, dipivefrine, ephedrine, flecainide, mefloquine, nifenalol, norephedrine, omeprazole, orciprenaline, quinine + quinidine, penbutolol, pindolol, practolol, pronethalol, propafenone, propranolol, rimiterol, salbutamol, sotalol, talinolol, terbutaline	Particle-based (in-house made)	(*S,S*)- *N*-[*N*-(4-Allyloxy-3,5-dichlorobenzoyl)-leucyl]-2-amino-3,3-dimethylbutane sulfonic acid	ACN/MeOH (4:1), 50 mM formic acid, 25 mM 2-amino-1-butanol; 15 kV	DAD	[36]
Alprenolol, atenolol, bamethane, benzetimide, bupranolol, carazolol, celiprolol, clenbuterol, dipivefrine, ephedrine, flecainide, mefloquine, *O*-(*tert*-butylcarbamoyl) mefloquine, nifenalol, omeprazole, orciprenaline, quinine + quinidine, *O*-(*tert*-butylcarbamoyl)-quinine + *O*-(*tert*-butylcarbamoyl)-quinidine, penbutolol, pindolol, practolol, pronethalol, propafenone, propranolol, rimiterol, salbutamol, sotalol, talinolol, terbutaline, *E*/*Z*-flupentixol	Particle-based (in-house made)	(*R,S*) - *N*-[*N*-(4-Allyloxy-3,5-dichlorobenzoyl)-leucyl]-2-amino-3,3-dimethylbutane sulfonic acid	ACN/MeOH (4:1), 50 mM formic acid, 25 mM 2-amino-1-butanol; 10 kV	DAD	[36]

Alprenolol, atenolol, benzetimide, bupranolol, bunitrolol, carazolol, celiprolol, clenbuterol, mefloquine, *O*-(*tert*-butylcarbamoyl) mefloquine, nifenalol, quinine + quinidine, *O*-(*tert*-butylcarbamoyl)-quinine + *O*-(*tert*-butylcarbamoyl)-quinidine, penbutolol, pindolol, practolol, propafenone, propranolol, rimiterol, talinolol, terbutaline, *E*/*Z*-flupentixol	Particle-based (in-house made)	(*S*,*S*)- *N*-[*N*-(4-Allyloxy-3,5-dichlorobenzoyl)- leucyl]-2-pyrrolidinemethane sulfonic acid	ACN/MeOH (4:1), 50 mM formic acid, 25 mM 2-amino-1-butanol; 10 kV	DAD	[36]
Alprenolol, atenolol, bamethane, benzetimide, bunitrolol, clenbuterol, dipivefrine, ephedrine, mefloquine, *O*-(*tert*-butylcarbamoyl) mefloquine, nifenalol, norephedrine, orciprenaline, *O*-(*tert*-butylcarbamoyl)-quinine + *O*-(*tert*-butylcarbamoyl)-quinidine, penbutolol, practolol, pronethalol, sotalol, terbutaline, *E*/*Z*-flupentixol	Particle-based (in-house made)	(*R*,*S*)- *N*-[*N*-(4-Allyloxy-3,5-dichlorobenzoyl)- leucyl]-2-pyrrolidinemethane sulfonic acid	ACN/MeOH (4:1), 50 mM formic acid, 25 mM 2-amino-1-butanol; 10 kV	DAD	[36]
28 chiral compounds including β-blockers and nonsteroidal anti-inflammatory drugs	Particle-based (in-house made)	Mono-(8-benzenesulfonamidoquinoline-2-ylmethyl)-substituted cyclam-capped β-CD and 1,8-di-(2-hydroxymethylpyridine-6-ylmethyl)-substituted cyclam-capped β-CD	Varying percentages of ACN or MeOH with a 10 mM Tris-HCl buffer pH 8.6, with or without 2 mM $Ni(ClO_4)$; Voltages ranging between 10 and 20 kV	DAD	[37]
Flamprop, 2-phenoxypropionic acid, diclofop, haloxyfop, fluazipop	Particle-based (in-house made)	L-RNA aptamer	ACN/2 mM phosphate pH 6 with 10 mM NaCl and 5 mM $MgCl_2$ (15/85)	DAD	[38]
Ketoprofen	Particle-based, magnetic particles (in-house made)	Avidin	10 mM acetate pH 5; −10 kV	UV	[39]

(*continued*)

TABLE 15.1 (continued)
Separations in Regular CEC Mode

Substances	Type of Stationary Phase	Chiral Selector of CSP, in Mobile Phase or Imprinting Molecule	Mobile Phase/Applied Voltage	Detection	Refs.
Mandelic acid, 3-OH-mandelic acid, 4-OH-mandelic acid, 4-brommandelic acid, 4-methoxymandelic acid, 3-OH-4-methoxy mandelic acid, 3,4-di-OH-mandelic acid, atrolactic acid, 3-phenyllactic acid, citramalic acid, malic acid	Monolithic-polymeric (in-house made)	L-4-hydroxyproline derivative	5 mM $Cu(CH_3COO)_2$ and 50 mM NH_4CH_3COO, pH 4.4; −5 kV	DAD	[42]
Warfarin, bupivacain, thalidomide	Monolithic- polymeric (in-house made)	Vancomycin	20% ACN in 0.15% TEAA pH 4.6; 20 kV	DAD	[43]
DL Tryptophan	Monolithic-silica (in-house made)	BSA	5 mM phosphate buffer pH 7; 2 kV	UV	[44]
DL Tryptophan	Monolithic-silica (in-house made)	BSA	10 mM phosphate buffer pH 7; 2 kV	DAD	[46]
DNB-Leucine, DNZ-Leu, FMOC-Leu, FMOC-Ser, CC-Ala, CC-Ser, Dns-Ser, DBD-Leu	Monolithic-polymeric (in-house made)	*O*-9-(*tert*-Butylcarbamoyl)-11-[(2-methacryloyloxy) ethylthio]-10,11- dihydroquinine	ACN/MeOH (80/20) containing 0.4 M acetic acid and 4 mM TEA; −25 kV	DAD and Fluo-rescence	[47]
Mefloquine, mefloquine-tert-butylcarbamate	Monolithic- polymeric-MIP (in-house made)	(*S*)-*N*-(4-Allyloxy-3,5-dichlorobenzoyl)2-amino-3,3-dimethylbutanephosphonic acid	ACN/MeOH (80/20) with 25 mM formic acid and 12.5 mM 2-amino-1-butanol; 12 kV	DAD	[51]
Acebutolol, celiprolol, clenbuterol, ephedrine, isopenbutolol, isoxsuprine, mefloquine, mefloquine-*tert*-butylcarbamate, nifenalol, salbutamol, sotalol, talinolol	Monolith-silica (in-house made)	(*S*)-*N*-(4-Allyloxy-3,5-dichlorobenzoyl)2-amino-3,3-dimethylbutanephosphonic acid	ACN/MeOH (80/20) containing 25 mM formic acid and 12.5 mM 2-amino-1-butanol; 8 kV	DAD	[52]

Benzoin, 3-butyl-phtalide, indapamide, praziquantel, Tröger's base, drug candidate, *trans*-stilbene oxide, alprenolol, hydroxyzine, pindolol, ranolazine, tetrahydropalmitine	Monolith-silica (in-house made)	Cellulose tris(3,5-dimethylphenyl) carbamate	2 mM phosphate buffer pH 6.8 or 2 mM Na-dihydrogen phosphate + 10 mM DEA pH 9.6 mixed with 40% or 60% ACN; 10 kV	DAD	[53]
Tröger's base, ranolazine	Monolith-silica (in-house made)	Cellulose tris(3,5-dimethylphenyl) carbamate	MeOH containing 5 mM ammonium acetate; 20 kV	DAD	[53]
Benzoin, indapamide, praziquantel, Tröger's base, warfarin, 1,1′-bi-2-naftol, octahydro-1,1′-bi-2-naftol	Monolithic- polymeric (in-house made)	Cellulose tris(3,5-dimethylphenyl) carbamate	2 mM phosphate pH 2.7/ACN (65/35); −15 kV	DAD	[54]
Tetrahydropalmatine, pindolol	Monolithic- polymeric (in-house made)	Cellulose tris(3,5-dimethylphenyl) carbamate	2 mM phosphate pH 9.7/ACN (65/35); −15 kV	DAD	[54]
2-phenoxypropionic acid, diclofop, fluazipop, fenoxaprop, haloxyfop, mecoprop, dichlorprop	Monolith-silica (in-house made)	(+)-1-(4-Aminobutyl)-(5*R*,8*S*,10*R*)-terguride	5 mM TEA/acetic acid (ratio varying from 0.024 to 0.013) in ACN/MeOH (90/10); −15 kV	DAD	[55]
Dichlorprop + achiral clofibric acid as internal standard	Monolith-silica (in-house made)	(+)-1-(4-Aminobutyl)-(5*R*,8*S*,10*R*)-terguride	4 mM TEA-acetic acid in ACN/MeOH (9/1); −15 kV	DAD	[56]
N-benzyloxycarbonyl phosphinic pseudodipeptide methyl ester benzyloxycarbonyl- homophenylalanine and the corresponding *N*-2,4-dinitrophenyl-derivative with free C-terminal carboxylic group	Monolith-silica (in-house made)	*O*-9-(*tert*-Butylcarbamoyl) quinidine	ACN/MeOH (50/50) with 200 mM acetic acid, 200 mM formic acid, and 4 mM TEA; −15 or −20 kV	DAD	[57]
Propranolol, pindolol, metoprolol, atenolol, terbutaline, alprenolol	Monolith-silica (in-house made)	Vancomycin	MeOH/ACN/acetic acid/TEA (80/20/0.1/0.1); 10 kV	DAD	[58]
Thalidomide, benzoin	Monolith-silica (in-house made)	Vancomycin	10 mM TEA phosphate buffer pH 6.5/ACN (80/20); 10 kV	DAD	[58]
Cefadroxil	Monolith-polymer (in-house made)	β-CD (added to the mobile phase)	30 mM Na-acetate, 1% acetic acid, 20 mM β-CD in formamide pH 7/isopropanol (95/5)	UV	[59]

(*continued*)

TABLE 15.1 (continued)
Separations in Regular CEC Mode

Substances	Type of Stationary Phase	Chiral Selector of CSP, in Mobile Phase or Imprinting Molecule	Mobile Phase/Applied Voltage	Detection	Refs.
1-methyl-tryptophan, 4-chlor-phenylalanine, 4-fluor-phenylalanine, 5-brom-tryptophan, 5-fluor-tryptophan, 5-methyl-tryptophan 6-methyl-tryptophan, DOPA, ethionine, kynurenine, methyl-DOPA, *m*-tyrosine, *p*-tyrosine, phenylalanine, phenylserine, *p*-tyrosine, tryptophan, α-methyl-*m*-tyrosine, α-methyl-tryptophan	Monolith with embedded particles (in-house made)	L-4-Hydroxyproline	50 mM phosphate pH 4.5 with 1 mM Cu(II)sulfate; 5 kV	DAD	[60]
Mandelic acid, 3-OH-mandelic acid, *p*-OH-mandelic acid, 4-bromomandelic acid, *p*-methoxymandelic acid, 3-OH-4-methoxymandelic acid, 3,4-di-OH-mandelic acid, atrolactic acid, β-phenyllactic acid, 4-chloromandelic acid, malic acid, 4-OH-3-methoxymandelic acid, 3-(3-indolyl)-lactic acid	Monolith with embedded particles (in-house made)	L-4-Hydroxyproline	50 mM phosphate pH 4.5 with 1 mM Cu(II)sulfate; −2 kV	DAD	[60]
1,1′-bi-2,2′-naftol	Monolith-MIP (in-house made)	(*R*)-1,1′-bi-2,2′-Naftol (imprint molecule)	10 mM acetate pH 4/ACN (20/80); 5 kV	UV	[61]
Naproxen	Monolithic- polymeric-MIP (in-house made)	(*S*)-naproxen (imprint molecule)	50 mM acetate pH 3/ACN (20/80); 5 kV	UV	[63]
Ibuprofen	Monolithic- polymeric-MIP (in-house made)	(*S*)-Ibuprofen (imprint molecule)	ACN/phosphate pH 3.2 (50/50); 15 kV	UV	[64]

Ibuprofen	Monolithic- polymeric-MIP (in-house made)	(*S*)-Ibuprofen (imprint molecule)	MeOH/acetic acid (99/1) or 1% acetic acid in MeOH/ACN (70/30); 20 kV	UV	[64]
Tröger's base	Monolithic- polymeric-MIP (in-house made)	(5*S*,11*S*)Tröger's base (imprinting molecule)	ACN/5 mM acetate pH 6 (80/20); 10 kV	DAD	[65]
Tetrahydropalmitine	Monolithic- polymeric-MIP (in-house made)	L-Tetrahydropalmitine (imprinting molecule)	ACN/5 mM acetate pH 6 (85/15); 10 kV	DAD	[65]
Naproxen	Monolithic-silica-based hybrid MIP	(*S*)-Naproxen (imprinting molecule)	ACN/50 mM acetate buffer pH 3.7 (40/60); 15 kV	UV	[66]
Zolmitriptan	Monolithic-silica-based hybrid MIP	Zolmitriptan (imprinting molecule)	50 mM Tris HCl pH 5.4/ACN (30/70); 5 kV	UV	[67]
12 glycyl dipeptides	Particle-based (commercialized)	Teicoplanin aglycone = Chirobiotic TAG	0.2% TEAA in water, pH 4/EtOH/ACN (50/20/30); 15 kV	DAD	[91]
Carprofen, warfarin, coumachlor	Particle-based (commercialized)	(3*R*,4*S*)-4-(3,5-Dinitro benzamido) tetrahydrophenanthrene = 3*R*,4*S* Whelk O1	5 mM phosphate buffer pH 3/ACN (40/60); 30 kV	DAD	[92]
9 Dns-amino acids	Monolithic-silica (in-house made)	γ-CD	MES-tris buffer pH 8/MeOH (60/40); −15 or −18 kV	UV	[93]
Positional isomers *o*-, *p*-, and *m*-cresols, benzoin	Monolithic-silica (in-house made)	β-CD	50 mM phosphate buffer pH 7.5/MeOH (80/20); 12 kV	UV	[93]
Dns-DL-Glu, Dns-DL-Asp	Monolithic-silica (in-house made)	β-CD	50 mM phosphate buffer pH 7.5/MeOH (80/20); −12 kV	UV	[93]
9 diastereomeric dipeptides (4 stereoisomers), 2 chiral tripeptides, and 3 diastereomeric tripeptides (4 stereoisomers)	Particle-based (commercialized)	Teicoplanin aglycone = Chirobiotic TAG	Binary mixtures of TEAA pH 4.1 with different amounts of ACN, MeOH, or EtOH; ternary mixtures of the above; 15 kV	DAD	[94]

(continued)

TABLE 15.1 (continued)
Separations in Regular CEC Mode

Substances	Type of Stationary Phase	Chiral Selector of CSP, in Mobile Phase or Imprinting Molecule	Mobile Phase/Applied Voltage	Detection	Refs.
N-benzyl-DL-phenylalanine-β-naftylamide, *N*-benzyl-DL-leucine-β-naftylamide, trifluoroacetyl-DL-alanine-β-naftylamide, trifluoroacetyl-DL-methionine-β-naftylamide, *N*-benzoyl-DL-Thr-methylester, *N*-benzoyl-DL-phenylalaninol	Particle-based (in-house made)	3,5-Dinitrobenzoyl-(*R*)-naphtylglycine	1 mM borate pH 8.5 mixed with different amounts of ACN; 30 kV	DAD	[95]
Dns-DL-NorVal, Dns-DL-Met, Dns-DL-Val, Dns-DL-Ser, Dns-DL-Thr, Dns-DL-Phe, Dns-DL-Trp, Dns-DL-NorLeu, DL-*p*-hydroxyphenyl lactic acid, DL-indole-3-lactic acid, DL-β-phenyl lactic acid, DL-Leu-DL-Phe, DL-Leu-DL-Trp	Monolithic-silica (in-house made)	L-Hydroxyproline	ACN/0.5 mM Cu acetate, 50 mM ammonium acetate (7/3) adjusted to pH 6.5; −13.5 kV	UV	[96]

Note: Unless otherwise stated, the particles of the CSPs are silica based. Only compounds for which partial and baseline separations were reported, are included.

for EOF mobility, retention time of the first eluting diastereoisomer, and selectivity were below 1.0%, and for areas below 9.0%. In [16], the chiral separation of chlorthalidone enantiomers was developed using an achiral SP and hydroxypropyl-β-cyclodextrin (HP-β-CD) in the mobile phase. An earlier described separation [17] was optimized using a central composite design, varying the concentrations of CD and acetonitrile (ACN) in the mobile phase. At optimal conditions, a strong reduction in analysis time was seen compared to the initial separation, still maintaining an equally good separation between the enantiomers. Drawbacks encountered using this approach were that the system was highly selective, i.e., could not be applied on other substances, and that the selector consumption was quite high, due to the need of rinsing the columns by means of an HPLC pump.

Undoubtedly, most chiral separations, also in CEC, are obtained using the direct approach with CSPs. CSPs containing polysaccharide selectors are most frequently used because of their broad enantioselectivity. The CEC and capillary liquid chromatographic (CLC) separation of levonorgestrel, for instance, was studied in [18] using a cellulose tris(3,5-dichlorophenylcarbamate) CSP. Higher efficiencies and better resolutions were observed using the CEC mode. Moreover, the separation was complete at all selector concentrations tested (5, 15, and 25 (% w/w)), while the lowest concentration did not allow separation in CLC. The method was used to assay the enantiomeric impurity of levonorgestrel in a drug formulation down to the 0.1% level. For this application, CEC provided however less good results than CLC due to a higher baseline noise. In [19], cellulose 2,3-bisphenylcarbamate and cellulose 2,3-bis(3,5-dimethylphenylcarbamate) selectors were immobilized onto diethylenetriaminopropylated silica through a positively charged spacer, resulting in two CSPs. These were compared with another CSP that consisted of a cellulose 2,3-bisphenylcarbamate selector immobilized onto the more commonly used 3-aminopropyl silica. The CSP containing the first type of silica exhibited a reversed EOF for both aqueous and nonaqueous mobile phases. The silica support can thus be adapted to reverse the EOF on a SP. Comparison with the CSP made from the traditional silica using nonaqueous phases also pointed out that the EOF magnitude was larger using the new type of support, resulting in faster enantioresolution. Furthermore, the reversed EOF presented a possibility to resolve negatively charged species under aqueous conditions.

Polysaccharide-based CSPs are very useful to define separation strategies due to their broad enantiorecognition ability. A series of papers have been published by our group in the context of defining a generic CEC separation strategy [20–23]. In [20], two types of commercial polysaccharide-based CSPs, Chiralpak AD-RH and Chiralcel OD-H, were evaluated for their enantioselectivity in CEC. For basic and bifunctional compounds, a screening mobile phase could be defined, which was able to induce enantioselectivity for 86% of a test set consisting of 29 compounds. For acidic compounds, other conditions, i.e., where they appear uncharged, were needed. In a next paper [21], the differences between the reversed-phase (RP) and normal-phase (NP) materials of the above CSP were investigated. Simultaneously, different electrolytes were tested. At high pH, no significant differences between both RP and NP materials were observed, while at low pH a reversal of EOF was seen on the NP version of Chiralcel OD, indicating that the types of silica used are different. The best electrolytes were phosphate at high pH and ammonium formate at low pH. Further investigation focused on the separation of acidic molecules [22], investigating the effects of the ACN concentration, pH, buffer concentration, temperature, and applied voltage using an experimental design approach. This allowed defining a separation strategy consisting of a screening procedure and some optimization steps for the situations in which no, a partial, or a baseline resolution was observed after the screening. The strategy was applied on 15 compounds, and showed enantioselectivity for 11, of which 10 could be baseline separated after the optimization steps. In analogy with [22], a separation strategy was defined for nonacidic compounds [23] using the same CSP, and evaluated on a set of 48 substances. In the screening step, enantioselectivity was seen for 64% of the test set, and after the optimization steps, the success rate was elevated to 85%. A summary of the above strategies is presented in [24]. The proposed screening approach of [23] was used to develop an assay on a Chiralpak AD-RH column for the (*R*)-impurity in the presence of the levetiracetam enantiomer [25]. Calibration curves were constructed with and without taking an internal standard into account, and similar accuracies were obtained using both approaches. It was seen that the impurity was quantifiable down to a 0.08% concentration relative to the active enantiomer. Run-to-run repeatabilities gave acceptable results with RSD below 9.3% for retention time, resolution, ratio area impurity/area internal standard, and *R*-enantiomer peak area. The between-days variability was higher, indicating that it was better to make a calibration curve daily. The method was applied to a commercially available tablet containing levetiracetam. No impurity above the LOD was seen in the tablets. Spiked concentrations at the LOQ (limit of quantification) level could be quantified accurately in the presence of the major enantiomer.

An application that uses only the short-end of the capillary is described in [26]. Only the short section of the capillary situated behind the detection window is packed with polysaccharide-based CSP to enable the separation. This leads to a shorter separation path and may result in short analysis times. Cellulose 3,5-dimethylphenylcarbamate was chemically immobilized onto methacroyldiethylenetriaminopropylated silica, which generated a positively charged CSP. This CSP allowed generating a higher (anodic) EOF than those prepared from other types of silica. However, for basic compounds pCEC was necessary to make elution possible because their electrophoretic mobility is opposite to the EOF direction. Tetrahydrofuran (THF) and chloroform could be used as organic modifiers because of the immobilization of the selector and for some compounds a better separation was obtained using these solvents. The benefit of using THF/chloroform in the mobile phase was also observed in [27], where cellulose tris(3,5-dimethylphenylcarbamate) was chemically bonded onto vinylized silica gel, creating a CSP that also displayed a high EOF under acidic conditions. This was beneficial for the separation of acidic enantiomers, but also basic and neutral compounds could be separated in RP conditions. Both short- and long-end analyses, where the long section of the capillary is filled with CSP, were described.

Macrocyclic antibiotics are also frequently used as chiral selectors in CEC. Fanali et al. [28] used a CSP with a glycopeptide antibiotic selector, Hepta-Tyr, which belongs to the teicoplanin family. The CSP was prepared in-house and injections were performed at the short-end of the capillary. An effective separation length of 6.6 cm was used. The EOF was reversed with this type of CSP. Five hydroxy acids could be baseline separated, with the fastest separation being that of mandelic acid, in only 72 s. This indicates the potential advantage of filling the SP at the short-end of the capillary: It might increase the throughput considerably without a loss of separation capability. The Hepta-Tyr-based CSP of [28] was compared with a vancomycin-based CSP in [29] for the enantioseparation of acidic compounds. Both CSPs were mixed with ordinary or diol silica particles to generate a higher EOF. In this study, the long end of the capillary was used to prepare the column. The Hepta-Tyr CSP, which had a reversed EOF, was found more suitable for the separation of acidic compounds than the vancomycin CSP, which displayed a normal EOF. On the latter columns, less enantiorecognition was seen and separations were only possible at low pH. The separation of loxiglumide on the Hepta-Tyr-based CSP in both CEC and CLC modes using short-end analysis was studied in [30]. The separation was optimized in both separation modes by investigating the effects of several experimental parameters, and both modes provided baseline separation within an acceptable time, i.e., around 4 min. However, only the CLC method was validated because the required conditioning time for CEC appeared to be longer.

Two types of teicoplanin-based CSPs were tested in polar organic solvents mode CEC in [31]. In the first type, the chiral particles were mixed with regular silica particles in a 3/1 ratio. In the second type, only the chiral particles were used. Both 3.5 and 5 μm particles were used. Generally, good results were obtained with the two types of SPs tested. The presence of silica particles enabled the shortening of the analysis times, however, at the cost of resolution and efficiency. Overall, the best results were obtained using the 3.5 μm particles, as theoretically expected.

The separation of mirtazapine and two of its main metabolites (6 isomers in total) using a vancomycin-based CSP mixed with diol silica to increase the EOF was examined in [32]. After optimization of the analysis parameters such as buffer concentration and pH, organic modifier type and content, water content, temperature and applied voltage, the mixed mode CSP enabled a baseline separation of two peaks and a partial separation of the other four. A column with only vancomycin-based CSP enabled baseline separation of all the six isomers, with a slight increase in analysis time due to a lower EOF velocity. The method was validated and used for the analysis of spiked urine samples, which were first cleaned up by means of solid phase extraction (SPE). Recoveries ranging between 78% and 88% were obtained. To improve the quantification limits of mianserin enantiomers, peak compression effects were used in [33] on a Chirobiotic V (vancomycin) SP. Usually, continuous stacking effects are only applied with strong cation-exchange SPs. It was investigated whether this approach also could be used with chiral phases. An important criterion was that the analytes eluted close to the EOF marker, which was achieved by adjusting the mobile-phase composition. Stacking effects were obtained using a higher organic modifier content in the sample solution than in the mobile phase. With ACN, no stacking was observed, while with 2-propanol and THF, having low dielectric constants, peak compression effects were seen. Using peak compression, the LOQ improved with a factor 10, enabling to quantify one enantiomer in a 0.07% concentration relative to the second enantiomer.

A next category of chiral selectors applied in CEC are the low-molecular-weight selectors, which can have π-donor, π-acceptor, ion-exchange, or ligand-exchange properties. Four new types of low-molecular-weight selectors were synthesized and bonded onto thiol-modified silica particles in [34]. The selectors were sulfonic acid derivatives of cysteine which exhibited strong cation exchanging capacities (we refer to Table 15.1 for the exact names of the selectors). The synthesized CSPs were tested in nonaqueous mode with 28 basic compounds as test solutes. CSP 1 (see Table 15.1) exhibited the least enantiorecognition, with only 10 partially or baseline resolved compounds. With 13 partially or fully separated compounds, CSP 4 showed somewhat more enantioselectivity. Both CSPs 2 and 3 were enantioselective for 22 of the 28 compounds. The difference in enantioselectivity of the four selectors was attributed to the presence or absence of bulky groups next to strong interaction sites and their influence on electrostatic interactions.

A low-molecular-weight strong cation exchanger based on penicillamine sulfonic acid was immobilized onto silica particles and evaluated in [35]. This CSP was used to separate the enantiomers of ephedrine. After some method optimization, validation was carried out on a "well-packed" column. Acceptable results regarding run-to-run repeatability and linearity were obtained. Enantiomeric impurities below the 0.1% level could be accurately determined. However, when a "less good packed" column was used, method validation results were inadequate, indicating that the packing protocol is of tremendous importance. Notice that the difference between a well-packed and a less good packed column was not defined. Regarding the CSP itself, it can be mentioned that changing from the (*S*)-form of the selector to the (*R*)-enantiomer resulted in a reversal of elution order (Figure 15.1). This can

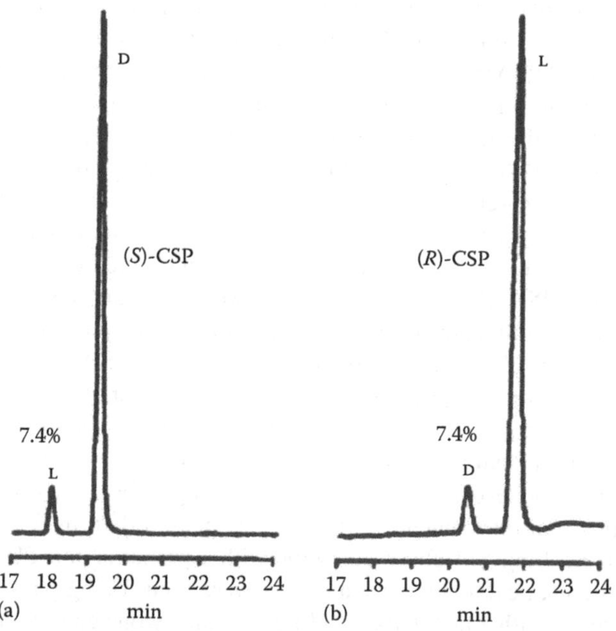

FIGURE 15.1 Chromatograms of the chiral separation of ephedrine, (a) on an (*S*)-CSP and (b) on an (*R*)-CSP. (Extracted from Bicker, W. et al., *Electrophoresis*, 24, 2532, 2003. With permission.)

be of practical importance when one wants to determine enantiomeric impurities because then it is best that the impurity elutes before the main peak and not in its tail. In [36], four new strong cation-exchange chiral selectors were synthesized, bonded to thiol-modified silica particles, and evaluated in nonaqueous CEC. All selectors were β-amino sulfonic-terminated dipeptide derivatives. Two displayed a high enantiorecognition ability, which was found to be highly dependent on the structure of the selector itself. These selectors (see Table 15.1) showed enantioselectivity toward 29 and 31 of 33 test compounds, respectively. The two other CSPs were found to be less enantioselective, with 22 and 20 partially or baseline separated compounds.

CD-based CSPs are also quite popular in separation science, but their application in particle-based CEC analysis seems to decrease, as only one application was found in the period 2003–2008. In [37], two substituted cyclam-capped-β-CD-bonded CSP were prepared and tested. MeOH, as an organic modifier, in the mobile phase provided the best enantioselectivity for the investigated substances. The recognition mechanism was assigned to inclusion in the CD cavity together with further host–guest-, hydrogen-, and dipolar interactions from the cyclam moiety. The sidearm of the cyclam also provided an extra ligand site for the solute. Broader enantioselectivity was seen in CEC mode compared to CE or LC. Addition of Ni^{2+} ions to the running buffer makes the selector positively charged, which can enhance electrostatic interactions with ionizable species and dipolar interactions with neutral solutes. A comparison with a crown-ether capped β-CD-based CSP pointed out that selectors with cyclam moiety had a superior separation performance.

Other applications that use new, but less known selectors are also described. The application of a biotinylated RNA aptamer, immobilized on streptavidin-modified particles as selector, is described in [38]. The columns were filled using a slurry-based packing procedure. The selector displayed stereoselective binding affinity for five herbicide analogs and the between-column reproducibility was excellent, as RSDs for retention times were below 0.2% for five columns. Moreover, the columns were stable, as a function of temperature and time, because after 700 injections no changes in retention factors were seen. The above was observed for ACN concentrations below 30%. Higher ACN contents resulted in deviating retention times and increased peak asymmetry after 400 injections. The addition of magnesium ions into the mobile phase stabilized the aptamer and could improve the enantioselectivity, but only when used in an optimal concentration range. The latter chapter indicates that unconventional selectors can also be of practical use in CEC and that it might be worthwhile to screen their separation capability. A special type of particle-based SPs was presented in [39]. Magnetic particles with a physically adsorbed avidin selector were brought inside a capillary and fixed by means of a magnet that was put at the outside of the capillary. Hence, the particles were fixed without the need for retaining frits, nor a difficult preparation procedure. This can improve the column-to-column variation. The usefulness of the resulting CSP was demonstrated by means of the separation of ketoprofen enantiomers. However, some drawbacks were also reported, such as a low-packing density, changes in EOF between packed and unpacked sections, and inhomogeneity in size and surface charge of the particles, reducing the efficiency.

15.2.2 Separations on Monolithic Stationary Phases

Various kinds of selectors are used in monolithic CSPs. They are often copolymerized in case of polymeric-based monoliths, or else chemically bonded/coated onto silica- or polymer-based support. Several overviews concerning the use of monolithic chiral SPs, and of which [40,41] are recent, have already been published. One concerns the approaches to prepare CD-based monoliths [40]. It is concluded that in general, gels give higher efficiencies but in terms of durability, rigid polymers and silica-based monoliths provide better results. Regarding polymeric monoliths, the ability of using a wide range of monomers is addressed, enabling the tailoring of the monoliths easily according to the needs. For both polymeric and silica monoliths, however, some skills are required to prepare reproducible SPs. This is also the main conclusion of another paper [41] in which monoliths with other selectors were discussed.

Polymeric-based monolithic phases are usually prepared by a one-pot reaction containing a polymerization mixture, which also includes a charge-providing monomer. The substitution of the charge-providing monomer by another with an opposite charge can be useful to analyze compounds with electrophoretic mobilities opposite to the EOF. The chiral selector can also be directly incorporated into the polymerization mixture, which presents an advantage of these types of SPs. In [42], negatively charged vinylsulfonic acid was replaced by positively charged diallydimethylammoniumchloride as charge-providing agent in the synthesis of a polymeric monolith. In this way, the EOF of the SP was superimposed on the electrophoretic mobility of some hydroxy acids (Table 15.1), which helped to reduce retention times. Analyses were conducted in ligand-exchange mode with Cu^{2+} as counterion in the mobile phase. The selector was an L-4-hydroxyproline derivative and its concentration was found to be crucial for enantioselectivity. In combination with the new charge-providing agent better peak shapes were seen. In [43], *N,N′*-diallyltartaramide was used as an alternative cross-linker to allyl glycidyl ether, in a preparation procedure for a chiral monolith using vancomycin as the selector. The new cross-linker could be converted easily into aldehyde groups containing reaction products through a treatment with periodate, resulting in cleavage of the cross-linker and the formation of two aldehyde groups. By this approach, an already existing synthesis procedure could be shortened and made more generally applicable, meaning that other selectors can also be incorporated. The enantioselectivity of the resulting CSP was demonstrated on warfarin, thalidomide, and bupivacaine. Repeatability experiments provided acceptable results (RSD between 0.2% and 0.4%) for retention times, selectivity, and retention factors, and somewhat higher values for resolution and theoretical plate numbers (RSD between 2.5% and 10.2%).

As already mentioned, varying types of selectors can be used to prepare monolithic SPs. Kato et al. [44] used a BSA encapsulated monolithic phase for the chiral separation of tryptophan enantiomers. The SP was constructed by a sol–gel method, which formed a silica matrix that trapped BSA as chiral selector during its formation. Generally, a poor sample loading capacity was seen, as is often observed with protein-based CSP [45]. A higher pH provided better separations due to pH dependent conformational changes of the selector. Injection repeatability and column-to-column reproducibility studies provided acceptable RSD values for retention times, ranging between 2.6% and 3.5%. In a next study [46], gelatin and chitosan were incorporated into the sol–gel procedure to evaluate whether better columns could be synthesized. Small amounts of gelatin and chitosan seem to create an environment around BSA that favors enantioseparation.

The groups of Lindner and Svec synthesized a quinine derivative as chiral selector and incorporated it into a polymeric monolithic matrix [47]. Its performance was compared with an earlier synthesized quinidine derivative, which was found useful in the enantioseparation of amino acid derivatives [48–50]. The new selector displayed better enantiorecognition capabilities, which was attributed to the presence of a more bulky residue at the carbamate group. Additionally, the new selector shows a reversal of the elution order. The different elution behavior is due to the fact that the selectors are derived from quinine and quinidine, which are diastereoisomers. The preparation of the monolithic CSP was also found highly reproducible, which is not evident when synthesizing polymeric monoliths.

Monoliths derivatized with a chiral strong cation exchanger as selector were studied in [51]. The partial substitution of lipophilic monomers, such as glycidyl methacrylate and methyl methacrylate, by 2-hydroxyethyl methacrylate appeared to have a positive effect on enantioselectivity, because nonspecific interactions of the analytes with the polymeric matrix were seen when using the two first monomers. The monolithic CSPs were compared with particle-based CSPs containing the same selector, and their performance was equally good. In extension of this study, a silica-based monolith with the same chiral selector was prepared [52] and compared with a particle-based column. Generally, the silica monolith gave a similar enantioselectivity compared to the corresponding particle-based CSP and the polymeric monolith from [51], although usually lower separation factors, efficiencies, and resolutions were reported. The type of counterion had the largest influence on separation, while 2-aminobutanol gave better results than diethylamine (DEA), triethylamine (TEA), and *N,N*-diisopropylethylamine. Overall, the column was robust, displayed an extended column lifetime, and gave stable currents during analysis, which are advantages over the particle-based column.

Silica-based monoliths coated with a cellulose tris(3,5-dimethylphenylcarbamate) selector were evaluated in [53]. The monoliths could be used with both aqueous and nonaqueous mobile phases, though the latter gave lower EOFs. Thirteen racemic mixtures were separated under aqueous conditions, while two racemates could be separated in nonaqueous CEC. Fast separations using short-end analysis were obtained for benzoin and a drug candidate in 160 and 90 s, respectively. Column-to-column reproducibility, which can be a problem in the preparation of monolithic SPs, was assessed and found to be satisfactory. Moreover, the monolith appeared stable using basic mobile phases, even though the structure was silica based. The same group synthesized a hydrophilic polyacrylamide monolith, which was afterwards coated with the above cellulose derivative as chiral selector [54]. The hydrophilic matrix theoretically should reduce the nonenantioselective interactions. The EOF velocity was stable; thanks to the incorporation of a charged monomer that was charged over the whole pH range. Nine acidic or basic analytes of twelve tested could be successfully resolved on the resulting CSP under both acidic and basic analysis conditions. The prepared columns provided acceptable precision for retention time and resolution (below 9.6%) regarding run-to-run repeatabilities and column-to-column reproducibility and could be used for 50 injections without any observable column deterioration.

In [55], the separation of seven 2-aryloxypropionic acids was developed on a polymeric monolith with a chemically bonded ergot alkaloid derivative as the chiral selector. This CSP was further used to separate the enantiomers of dichlorprop, an herbicide [56]. After some method optimization in terms of buffer concentration and analyzing voltage, dichlorprop racemate and the internal standard clofibric acid were separated within 5 min. The method was applied on incubated soil samples and it was demonstrated that the *R*-enantiomer of the herbicide was the more persistent.

Other applications on monolithic columns are described in [57], where a silica monolith derivatized with *O*-9-(*tert*-butylcarbamoyl)quinidine was used to separate phosphonic acid pseudopeptides, and in [58], where a silica-monolith modified with vancomycin was used for the separation of six basic analytes in nonaqueous CEC, and of thalidomide and benzoin in RP mode.

As in CEC with particle-based SPs, also with monolithic phases one can perform chiral separations with the selector added to the mobile phase. A capillary containing a nonchiral acrylate monolith combined with a mobile phase with β-CD as the chiral selector was used to separate cefadroxil enantiomers in [59]. Changes in pH were not found to influence the separation, while the addition of isopropanol increased Rs. Increasing the applied voltage resulted in a shorter analysis time, as expected. The RSDs for migration time and areas were below 3.2% and 5.1%, respectively.

An interesting approach to prepare CSPs that are particle based but which have the properties of a monolithic column in the sense of being prepared from a one-pot mixture and requiring no frits is presented in [60]. Silica particles containing chemically bonded L-4-hydroxyproline as selector are combined with a polymerization mixture of monomers, cross-linkers, and charge-providing agents. In this way, CSPs were created containing chiral particles, embedded in an achiral continuous bed. The choice of the charge-providing agent also presents a way to control the EOF direction: A cathodic EOF was needed for the separation of amino acids, while an anodic one enables the separation of hydroxyl acids, all in ligand exchange mode (Table 15.1). This approach for preparing fritless SPs should be, according to the authors, applicable on any silica-based CSP.

MIPs form another type of SPs which can be categorized as polymeric-based monoliths. MIPs are prepared by reaction of a polymerization mixture in the presence of an imprinting molecule, which is one enantiomer of the target molecule that needs to be separated. Although these SPs are frequently used, they exhibit some drawbacks in the sense that the resulting MIP is very selective, i.e., it can only separate one or a series of structurally analog compounds, thus eliminating their applicability for screening purposes. Furthermore, the separation efficiency of MIPs is frequently low. In [61], (*R*)-1,1′-bi-2,2′-naphtol was used as imprinting molecule enabling the chiral separation of the 1,1′-bi-2,2′-naphtol racemate. Several parameters were found to have an influence on the separation, i.e., the ratio imprinting molecule/methacrylic acid and the porogens in the polymerization mixture, and the pH and the ACN content of the mobile phase. The use of Tween 20 had a positive effect on resolution, an effect that was only seen when the chiral MIP was used, meaning that the enantiorecognition was not attributed to Tween 20. In [62], an attempt was made to separate two steroid isomers, 17-α-estradiol and 17-β-estradiol, which differed in the position of an OH group. For this purpose, a MIP was synthesized with β-estradiol as an imprinting molecule. However, only a slight difference in the retention times of both isomers was achieved. Other studies with MIPs describe the use of (*S*)-naproxen [63], (*S*)-ibuprofen [64], (5*S*, 11*S*) Tröger's base [65], and L-tetrahydropalmitine [65] as imprinting molecules for the analysis of their respective racemates.

Because of the shrinking or swelling that can occur with polymer-based MIPs, or the cracking and shrinking with silica-based MIPs, a novel approach was used in [66] for the preparation of silica-based hybrid MIPs. The methodology is based on a room temperature ionic liquid (RTIL) mediated, nonhydrolytic sol–gel process and a molecular imprinting technique, producing monoliths that do not require ageing and drying at high temperatures, and do not display cracking or shrinking. Moreover, an improved selectivity was seen, as naproxen enantiomers could be resolved with a resolution of 8.8. Compared with other MIP applications, peaks shapes were acceptable because the (*S*)-naproxen peak (i.e., the retained peak) still exhibited 12,000 plates/m. A MIP was prepared in [67] with zolmitriptan as imprinting molecule using the above preparation technology.

15.3 ENANTIOSEPARATIONS IN PRESSURIZED CAPILLARY ELECTROCHROMATOGRAPHIC MODE

pCEC is a mode of CEC, where only the inlet vial is pressurized during analysis. Consequently, a hydrodynamic flow is superimposed onto the EOF, which generally results in a faster elution. The flow profile is more parabolic than in CEC, and therefore the efficiencies in pCEC are frequently lower than in regular CEC mode. pCEC experiments are possibly executed on two types of instruments. Either a regular CE instrument or a modified CLC device can be used. In the modified CLC device, an HPLC pump pumps the mobile phase over the capillary column while an electrical field is applied simultaneously and this is commercially available. In the latter instrument, the pressure over the column is kept constant by means of a backpressure regulator to enable a constant linear velocity inside the CEC column.

An overview of the applications in the chiral separation field is given in Table 15.2. Although the number of papers dealing with chiral separations in pCEC is lower than in regular CEC, applications with both particle-based and monolithic columns were found.

15.3.1 Separations on Particle-Based Stationary Phases

The coupling of CEC to electrospray ionization (ESI) mass spectrometry (MS) and its application to enantioseparations was described in [68]. Actually, the mode the authors worked in was pCEC, as only the inlet of the capillary column can be pressurized when coupling to MS. The CEC–MS experiments were performed on a CE instrument using a Whelk-O1 CSP to separate warfarin and coumachlor enantiomers, the

TABLE 15.2
Separations in Pressurized CEC

Substances	Type of Stationary Phase	Chiral Selector of CSP, in Mobile Phase or Imprinting Molecule	Mobile Phase/Applied Voltage	Detection	References
Warfarin + coumachlor	Particle-based (commercial)	(3*R*,4*S*)-4-(3,5-Dinitro benzamido) tetrahydrophenanthrene = Whelk-O1	ACN/H_2O (70/30) containing 5 mM ammonium acetate; 20 kV	MS	[68]
Warfarin	Particle-based (commercialized)	(3*R*,4*S*)-4-(3,5-Dinitro benzamido) tetrahydrophenanthrene = Whelk-O1	ACN/H_2O (70/30) containing 5 mM ammonium acetate pH 4; 25 kV	MS	[69,71]
Oxprenolol, alprenolol, talinolol, atenolol, metoprolol, propranolol	Particle-based (commercialized)	Vancomycin = Chirobiotic V	MeOH/ACN/acetic acid/TEA (70/30/1.6/0.2); 25 kV	MS	[69,70]
Carteolol, pindolol	Particle-based (commercialized)	Vancomycin = Chirobiotic V	MeOH/ACN/acetic acid/TEA (70/30/1.6/0.2); 25 kV	MS	[70]
Promethazine, carteolol, celiprolol, albuterol	Particle-based (commercial)	Vancomycin = Chirobiotic V	MeOH/acetic acid/TEA (100/0.1/0.1) 250 V/cm	UV	[72]
Propranolol, isoprotenerol, epinephrine, synephrine, alprenolol, *N*-benzoyl-DL-phenylalanine-β-napthyl ester, DL-methionine-β-naphtylamide	Particle-based (in-house made)	Perphenylcarbamoylated-β-CD	5 mM phosphate buffer pH 6.3/MeOH (30/70); 15 kV	UV	[73]
6 α-Amido phosphonate derivatives	Particle-based (in-house made)	Mono (6-*N*-ethylenediamine-6-deoxy)perphenylcarbamoylated-β-CD	MeOH/water/acetic acid/TEA (75/25/0.3/0.2); 10 kV	UV	[74]

Mephobarbital, hexobarbital, pentobarbital, 5-ethyl-1-methyl-5-(*n*-propyl)-barbituric acid, 1-methyl-5-(2-propyl)-5-(*n*-propyl)-barbituric acid, thiopental, α-methyl-aphenylsuccinimide, MTH-proline, mecoprop methyl, fenoxaprop ethyl, diclofop methyl, carprofen, PCB 132, 2,2,2-trifluoro-(9-anthranyl)-ethanol	Monolith-particle-glued (in-house made)	Permethyl-β-CD	20 mM MES pH 6 or 10 mM Na-acetate mixed with 10%–70% MeOH; 15–25 kV	UV	[75]
DOPA, 4-chloro-phenylalanine, 2-fluoro-phenylalanine, 3-fluoro-phenylalanine, 4-fluoro- phenylalanine, 5-fluoro-phenylalanine, methyl-DOPA, phenylalanine, tryptophan, *m*-, *o*-, and *p*-tyrosine, alanine, ethionine, leucine, methionine, norleucine, norvaline, seleno-methionine, serine, valine, 12 glycyl dipeptides, 9 diastereomeric dipeptides	Monolith with embedded particles (in-house made)	Teicoplanin aglycone	15 mM TEAA pH 4.1/MeOH/ACN (40/40/20); 25 kV	UV	[76]
Mandelic acid, 3-OH-mandelic acid, *p*-OH-mandelic acid, 4-bromomandelic acid, *p*-methoxymandelic acid, 3-OH-4-methoxy mandelic acid, 3,4-di-OH-mandelic acid, atrolactic acid, 3-phenyllactic acid, 4-OH-3-methoxymandelic acid, 3-(4-OH-phenyl)-lactic acid, 3-(3-indolyl)lactic acid	Monolith with embedded particles (in-house made)	Ristocetin A	MeOH with 34.8 mM acetic acid and 7 mM TEA; −15 kV	UV	[76]
12 glycyl dipeptides	Monolith with embedded particles (in-house made)	Teicoplanin aglycone	0.2% TEAA pH 4.1/MeOH/ACN (40/40/20); 15 kV	DAD	[77]

Note: Unless otherwise stated, the particles of the CSPs are silica based. Only compounds for which partial and baseline separations were reported, are included.

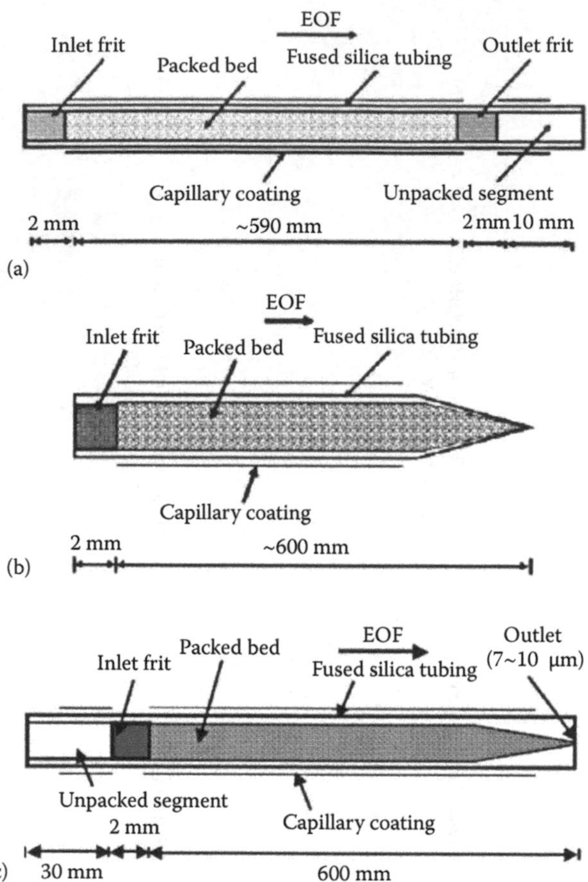

FIGURE 15.2 (a) Untapered and (b) externally tapered columns used in [68,69] for coupling CEC to MS detection. (From Zheng, J. and Shamsi, S.A., *Anal. Chem.*, 75, 6295, 2003. With permission.). (c) Internally tapered columns used in [68,69] for coupling CEC to MS detection. (From Zheng, J. et al., *Anal. Chem.*, 78, 1323, 2006. With permission.)

latter being the internal standard. Both tapered and untapered columns (Figure 15.2) were evaluated. With untapered columns, a short unpacked section of the capillary serves as electrospray tip, while tapered columns use either a reduced outer (externally tapered) or inner (internally tapered) capillary diameter, eliminating the need for an outlet frit. It was seen that the tapered columns provided reproducible retention times, acted as backpressure resistor, and hence suppressed bubble formation, and stabilized the electrospray into the detector. The CEC separation conditions and the MS parameters were optimized sequentially. The resulting method was successfully used to assay warfarin in SPE-extracted plasma samples. Injection had to be performed hydrodynamically, as electrokinetic injections gave a worse sensitivity. Overall, good linearity was obtained with an acceptable LOD. The technique was able to distinguish 1% of (*R*)-warfarin in the presence of 99% (*S*)-warfarin. The fabrication of capillaries to connect with an MS detector was further investigated, as it was seen that externally tapered columns [68] were difficult to prepare and to control, resulting in poor column-to-column reproducibility. A new, simple, and cheap fabrication procedure was presented in [69], producing internally tapered columns by simply heating the capillary end in a methane/oxygen flame during a short period while rotating it. The internal taper is prevented from breaking on the one hand, while debris accumulation inside the capillary is prevented on the other hand. The resulting columns were well compatible with MS detection because of lower noise levels and enhanced sensitivity compared with externally tapered columns. Moreover, experiments in polar organic solvents mode were possible with internally tapered columns, which was not the case with the other type. The developed chiral separation methods for β-blocking agents and warfarin were validated and allowed to detect as low as 0.1% enantiomeric impurity. Further investigation of the separation of β-blocking agents using CEC-ESI-MS was performed in [70]. Eight β-blockers were analyzed in one run on a vancomycin-based CSP using internally tapered columns, enabling the separation of 15 out of 16 enantiomers within 60 min on a 60 cm column, although the achiral separation of some β-blockers was also incomplete. After optimizing the mobile phase composition, only *R*-pindolol and *S*-propranolol remained unseparated. Alternative columns were used, i.e., containing 60 cm teicoplanin-based particles, 30 cm teicoplanin-30 cm vancomycin, 30 cm vancomycin-30 cm teicoplanin, and 60 cm mixed CSP, obtained by mixing teicoplanin and vancomycin particles in a 1/1 ratio. However, no better results were obtained than with the initially tested column. Calibration curves were constructed for atenolol, oxprenolol, metoprolol, and propranolol, and showed linearity in the 3–600 μM range. An LOD of 30 nM was achieved using MS detection, enabling the distinction of 0.1%–1% of the minor enantiomer in the presence of the major enantiomer. A further application of the warfarin separations from [68,69] was demonstrated in [71], where the developed method was used in the context of a clinical study.

Enantioseparations on a Chirobiotic V CSP in pCEC on a modified CLC system are described in [72]. Four compounds were separated in both CLC and pCEC mode, and in general higher resolutions, more theoretical plates, and shorter analysis times were achieved in the latter mode, indicating the advantage of applying a voltage over the capillary column. The influence of changing the backpressure regulator of the instrument (thus the linear velocity) on the efficiency was investigated, and it generally decreased as a function of linear velocity. This was unexpected, and it was attributed to the large dead volume in the instrument. Furthermore, retention factors changed as a function of the electrical field, while this is neither seen in regular CEC, nor in HPLC (as a function of flow rate). A mathematical model was therefore constructed, describing the retention factor in terms of experimental conditions in pCEC. This paper [72] indicates that performing experiments in pCEC on a modified CLC system is not as straightforward as initially expected. Several factors have to be taken into account, such as the instrumental dead volume, and the fact that retention factors change as a function of the electrical field.

Chiral separations of seven basic analytes on a perphenylcarbamoylated-β-CD using pCEC were described in [73]. A cathodic EOF was generated and its direction was the same as the pressure driven flow and the electrophoretic mobilities of the analytes, resulting in short analysis times. Using a triethylamine acetate (TEAA) electrolyte, retention times were decreasing as a function of pH due to a higher EOF velocity, followed by an increase due to a reduction in the electrophoretic mobilities as the analytes become less charged at higher pH. Using phosphate, no significant changes in elution time were seen because the electrophoretic mobilities were higher than the EOF. However, phosphate provided more and better enantioseparations. Another application on a modified CLC system describes the separation of α-amidophosphonates on perphenylcarbamoylated-β-CD based CSP [74].

15.3.2 Separations on Monolithic Stationary Phases

CD-modified silica particles (Chiradex) were packed inside a capillary and immobilized using a sol–gel solution as presented in [75]. The particles are glued together and are retained in the capillary without the need for frits. The resulting SP is often referred to as a particle-glued monolith in the literature. Separations were conducted in pCEC mode and compared with CLC separations. The results indicated that with shorter analysis times, higher theoretical plate numbers (and hence higher Rs) can be achieved using pCEC mode due to the superposition of the EOF onto the pressure-driven flow. The highest selectivity was obtained when the EOF and the pressure-driven flow directions were the same and when the analyte is uncharged, eliminating its electrophoretic mobility. Of course, analysis times become longer in this situation.

In [76], CSPs created from silica particles with either a teicoplanin aglycone selector or a ristocetin A selector, embedded in an achiral continuous bed were studied, similar to the approach presented in [60] where a ligand-exchange CSP was used. There was no need to pressurize both vials during analysis, thus it could be conducted on any CE instrument. The highest possible concentration of silica particles in the polymeric matrix was found to be 25%. These CSPs were tested using several analytes, such as aliphatic and aromatic acids, and dipeptides (Table 15.2). The phases showed acceptable enantioselectivity toward the selected analytes. Moreover, the EOF direction can be favored in the direction of the electrophoretic mobility of the analyzed substance by means of the used charge-providing agent in the polymerization mixture. In [77], particle-loaded monoliths were prepared with teicoplanin aglycone bonded particles. The achiral monolith was prepared using a ring opening metastasis reaction. This reaction is much faster than the regular polymerization of, for example, methacrylamide monoliths and provides SPs within 30 min. The selected analytes were 12 glycyl-dipeptides, which at least could be partially resolved. Compared with the particle-based monoliths from [76], these SPs showed a similar separation behavior, separation power, and robustness against high electrical fields and back pressures. It indicates that the achiral backbone does not have a significant influence on the chiral performance of the SP.

15.4 ENANTIOSEPARATIONS IN OPEN-TUBULAR CAPILLARY ELECTROCHROMATOGRAPHY

In OT-CEC, a layer of SP is coated or bonded to the capillary wall. Because the SP does not completely fill the capillary, shortcomings such as a low sample loading capacity and an instability of the SP over time are often observed. Nevertheless, this approach seems to remain quite popular in the chiral separation area, probably because of the easiness to prepare the columns.

Most applications were on CSPs created by means of a multilayer coating procedure. In [78], such procedure was used to create a chiral OT column with bilayers consisting of an anionic and a cationic polymer. Poly(diallyldimethylammonium chloride) was used as cationic part and poly(sodium *N*-undecanoyl-L-leucyl-valinate) as both anionic part and chiral selector. The effects of changes in the coating procedure were investigated. An increasing NaCl concentration resulted in slight increases in retention and a higher resolution, probably because the thickness of the polymer layer increases. The number of bilayers inside the capillary also had an influence on separation: analysis time, retention, and resolution increased at a higher number of bilayers. Regarding precision, good results (RSD below 0.2% for retention times) were observed during the first two days of use, afterward the RSD values increased, indicating that the column needs to be regenerated. Compounds such as 1,1′-binaphtyl-2,2′-dihydrogenphosphate, 1,1′-bi-2-naphtol, secobarbital, pentobarbital, and temazepam were partially or baseline separated using columns with two or three bilayers. A similar approach was used in [79]. Bilayers were constructed using either poly(sodium undecanoyl-L-leucylalinate) or poly(sodium undecanoyl-L-alanine leucinate) as both anionic polymer and chiral selector, and poly(L-lysine) as cationic compound. The separations of 1,1′-bi-2-naphthyl-2,2′-dihydrogen phosphate, 1,1′-bi-2-naphthol, 1,1′-binaphthyl-2,2′-diamine, labetalol, and sotalol were investigated on the resulting columns. Poly(sodium undecanoyl-L-alanine leucinate) appeared to be less enantioselective; thus, the study mainly focused on poly(sodium undecanoyl-L-leucylalinate). The NaCl concentration in the polymer deposition solution again had an effect on the resolution, and better results were obtained using NaCl only in the cationic polyelectrolyte solution. On the other hand, capillaries coated without NaCl in the deposition solution gave better reproducibility. Injection repeatability and capillary-to-capillary reproducibility were all below 1% for retention times of analytes and EOF, and the columns were stable for five days. Finally, these columns were successfully coupled to an ESI–MS detector. With the latter setup, the (partial) separation of sotalol enantiomers was demonstrated. A further application of these polyelectrolyte multilayer coatings concerns a stability study toward 0.1 and 1 M NaOH solutions using laser scanning confocal microscopy [80]. It was concluded that the coating was relatively stable toward these solutions. The 2- and 20-bilayer coatings could be completely removed after long exposure to a 1 M NaOH solution, i.e. for 3.5 and 9.5 h, respectively.

A capillary coated with multilayers of alternating lysozyme and γ-zirconium phosphate was used as CSP for the separation of D, L-tryptophan enantiomers in [81]. Different parameters, such as the content of isopropanol, applied voltage, temperature, and pH, were found to have an effect on the separation and were sequentially optimized.

Other coating procedures can also be used, such as a phospholipid coating, on which the chiral selector is immobilized afterwards. In [82], different coating procedures were evaluated for stability and repeatability on the chiral separation of tryptophan, phenylthiohydantoin (PTH)-serine, and PTH-threonine enantiomers. The SP was generated from a phospholipid coating with liposomes, followed by

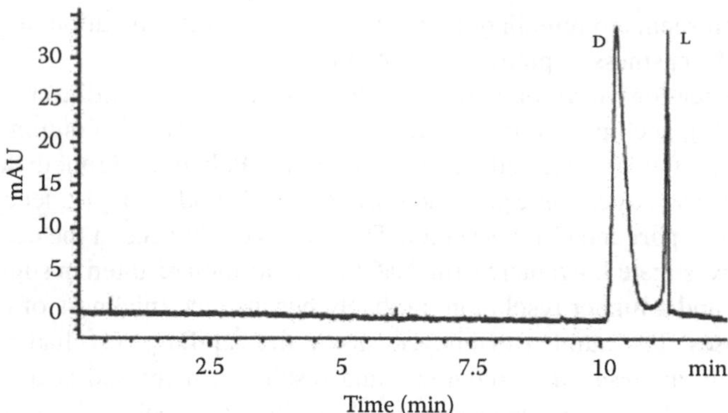

FIGURE 15.3 Separation of DL-tryptophan obtained in [83]. Analysis conditions: sample injection 10 s at 50 mbar, electrolyte: 20 mM Tris at pH 7.4, 20 kV, 25°C, UV detection at 214 nm. (Reproduced from Wiedmer, S. et al., *Anal. Biochem.*, 373, 26, 2008. With permission.)

immobilization of an avidin chiral selector onto the phospholipid layer. Liposomes prepared from 1,2-dipalmitoyl-sn-glycero-3-phosphocholine L-α-phosphatidylserine (PS) (90/10), 1,2-dipalmitoyl-sn-glycero-3-phosphoethanolamine-*N*-(Cap-biotinyl) PS (80/20), and 1,2-dipalmitoyl-sn-glycero-3-phosphoethanolamine-*N*-(biotinyl) PS (80/20) mixtures gave coatings that could be prepared in a repeatable manner displaying good resolution for the selected analytes.

In [83], the same approach was used to prepare phospholipid coatings starting from phosphatidylcholine-derived liposomes using bovine serum albumin (BSA), lysozyme, α-chymotrypsine, and avidin as chiral selectors. The selected analyte to evaluate the resulting columns was DL-tryptophan. Generally, of the four examined selectors, BSA was found superior, as faster and better separations were obtained. High efficiencies were only observed for the L-tryptophan peak (Figure 15.3).

Other applications found in chiral OT-CEC are summarized below. Mode-filtered light detection in OT-CEC was used in [84], according to a setup that was proposed in [85] (Figure 15.4). This is a fiber optic technique of refractive index detection where a detector is placed along the length of an optic fiber, situated in a capillary, to measure the intensity of the transmitted mode-filtered light and results, theoretically, in a lower background and signal-to-noise ratios. Using the sol–gel technique, the optic fiber and the capillary wall of a small annular column were modified with BSA as chiral selector, and different conditions of voltage/current were investigated to separate DL-tryptophan enantiomers. Enantioseparation was achieved in 320 μm ID capillaries with a 200 μm diameter optic fiber and 250 μm ID capillaries also with a 200 μm diameter optic fiber. Larger diameters of capillaries or optic fibers resulted in too high currents and no optical signals were detected.

In [86], a preparation procedure of an avidin-containing SP by means of a Schiff base formation reaction was presented. The quality of separation depended on the selector concentration used in the preparation, with better separations at higher avidin concentrations, but at the cost of efficiency. The column was successfully used

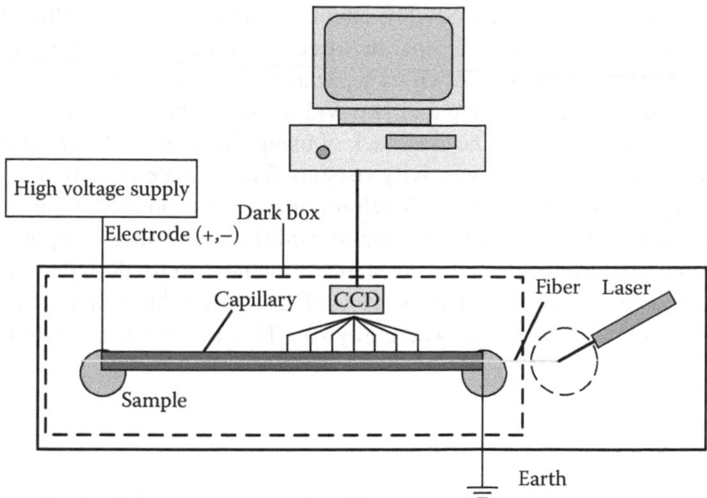

FIGURE 15.4 Setup of mode-filtered light detection [84,85]. Along the annular column, several detection channels are positioned that gather and transmit the mode-filtered light to a charge-coupled device (CCD). (From Zhou, Z. et al., *Analyst*, 126, 1838, 2001. With permission.)

for the separation of abscisic acid, ibuprofen, and ketoprofen enantiomers. Run-to-run repeatabilities for retention time, resolution, and peak area were satisfactory and day-to-day repeatability for resolution was acceptable (RSD 8.9%). The capillary was used during 50 days, which demonstrated that the used reaction provided stable columns. Finally, the columns were coupled to ESI–MS detection, but relatively low detection sensitivity and a bad precision were seen.

In [87], an OT-CEC column was prepared by coating the internal surface of the capillary with an in-house synthesized BSA-β-CD selector. The resulting column was used for the successful separation of ibuprofen, phenylalanine, and chlorpheniramine enantiomers by means of pCEC. A positive effect on the separation was seen at higher operating voltages. Comparison with columns containing only β-CD or BSA as selector to separate tryptophan indicated that the newly synthesized BSA-β-CD selector was responsible for enantiorecognition.

15.5 ENANTIOSEPARATIONS IN CHIP CAPILLARY ELECTROCHROMATOGRAPHY

The lab-on-chip technology has known an increasing number of applications in last years. Interest rises from its increased analysis speed and its potential for high-throughput screening. Although chip-based systems are most frequently used in electrophoresis mode for chiral separations, two applications were found in the literature, describing the use of a gel-based CSP.

In [88], the channel of a polydimethylsiloxane (PDMS) chip was filled with a polyacrylamide gel monolithic SP, which used allyl-γ-CD as both cross-linker of the gel and chiral selector of the SP. The enantioselectivity of the resulting CSP was demonstrated by its capability to separate three fluorescein isothiocyanate (FITC)-labeled dansyl (Dns)-amino acids with a separation potential of 3.6kV. A mixture of two of the analytes, i.e., Dns-Asp and Dns-Val, could be baseline enantioseparated in one run within only 100s. The drawback of using chip electrochromatography for chiral separations is that for sensitivity reasons fluorescence detection is required when no coupling to MS is made. Therefore, one is often obliged to label the analytes. The enantioselectivity of the system toward the derivatized compound is then evaluated and not toward the native substance. In a next study [89], the same group also presented the separation of FITC-labeled Dns-threonine using a PDMS chip containing a polyacrylamide gel where allyl-β-CD was used as cross-linker/chiral selector.

15.6 CONCLUDING REMARKS

When summarizing all applications considered in this chapter into a figure (Figure 15.5), one sees that in the last years there has been an important change in the column technology used to prepare CEC columns. Whereas in early 2000, most

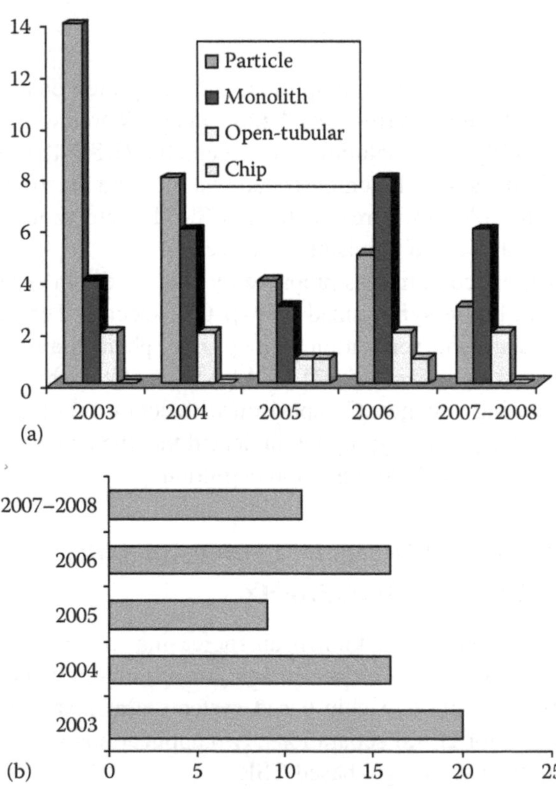

FIGURE 15.5 Classification of the applications according to (a) column technology, and (b) year published.

applications were using columns prepared from particle-based SPs, it is observed that from 2006 onward the monolithic column technology is mainly used. Several reasons for this can be given. First of all, some drawbacks inherent to particle-based columns, such as the need for frits and fragile columns, are eliminated using monolithic SPs. The next reason is that monoliths allow designing CSP that are especially developed for CEC, where the selector can be coated or bonded onto the monolithic backbone, or even copolymerized with the monolith in the case of polymeric monoliths. Particle-glued SPs and particle-containing monoliths also present interesting possibilities to overcome the shortcomings of particle-packed CEC columns. The number of applications that use OT columns remains relatively constant but represents only a minority. This will remain so in our opinion, since OT columns generally display limited sample loading capacities and SP stability. The same applies for chip electrochromatography where detection problems are a major issue, although we expect some increase in the number of applications in the next years. Regarding the number of CEC applications published the last years (Figure 15.5b), we see that it remains relatively constant, maybe with a slight decrease. However, this conclusion cannot be generalized to all CEC applications, as this chapter focusses on only a small part, i.e., chiral separations, is focused on in this chapter. Presumably, the number of applications in the chiral separation domain will again increase when users become convinced of the usefulness of monolithic column technology and when the technology itself becomes more robust and reproducible.

For the selectors used, we can state that most publications consider polysaccharide-, macrocyclic antibiotic-, CD-, and low-molecular-weight selectors for the separation of chiral molecules, although protein-based CSP and new types of selectors such as RNA aptamer and combined protein-CD selectors are also used.

Application oriented, various kinds were encountered, although most publications consider the preparation of a SP and a demonstration of its usefulness for chiral separations. Only a limited number of publications consider a further method development to separate a specific compound, followed by its validation and application to biological samples or pharmaceutical formulations. Acceptable precision results (for retention times, resolutions, number of theoretical plates, peak areas), recoveries, LOD, and LOQ are often reported, but the results are not satisfactory. Efforts to improve the LOD using peak compression effects are also described. In our opinion, the validation of developed methods and their application to real samples must continue in order to demonstrate the potential of CEC in analytical separation science. Many authors prove that it is possible to obtain good results, and this presents a criterion to introduce CEC as separation technique for routine analysis besides the currently used.

The use of pCEC as operating mode also increased in last years, not only because of the commercialization of an instrument specifically developed to operate in this mode, but because faster analyses often are obtained. However, a part of the efficiency obtained in regular CEC is lost due to the superposition of a pressurized flow onto the EOF. The presence of a loop injection system with the specialized pCEC instrument is an advantage for injection repeatability and a solution for injection problems with charged species that can occur in regular CEC. However, the external injector is, on the other hand, also responsible for a large dead volume, which can have a detrimental effect on the results. When performing pCEC on a CE instrument, the latter will, of course, not be the case.

The majority of applications use either UV or DAD detection. However, also the coupling to MS is possible. Practical aspects of coupling MS to CEC are described in a review on this topic [90]. However, as observed in this chapter, the number of applications in the chiral area is quite limited. The rather simple fabrication of CEC columns to be used with MS detection can present a future study area.

In conclusion, we can state that efforts still are made in separation science to make CEC a successful technique. Various kinds of applications convince us that the monolithic column technology presents a possible nice future for chiral CEC applications. However, better column-to-column reproducibility still is a goal that needs to be achieved. Regarding the developed separations, we feel that more effort should be invested in their validation and application to real-life biological or pharmaceutical samples. Finally, the coupling of CEC to MS and the generalization of its use also seems to be a challenging task.

ACKNOWLEDGMENT

D. Mangelings is a postdoctoral fellow of the Research Foundation Flanders (FWO).

LIST OF ABBREVIATIONS

ACN	acetonitrile
Ala	alanine
Asp	aspartic acid
BSA	bovine serum albumin
CC	carbazole-9-carbonyl
CD	cyclodextrin
CE	capillary electrophoresis
CLC	capillary liquid chromatography
CSP	chiral stationary phase
DAD	diode array detector
DBD	7-dimethylaminosulfonyl-1,3,2-benzoxadiazol-4-yl
DEA	diethylamine
DNB	3,5-dinitrobenzoyl
Dns	dansyl
DNZ	3,5-dinitrobenzyloxycarbonyl
DOPA	dihydroxyphenylalanine
EOF	electro-osmotic flow
ESI	electrospray ionization
EtOH	ethanol
FITC	fluorescein isothiocyanate
FLEC	(+)-1-(9-fluorenyl)ethyl chloroformate
FMOC	9-fluorenylmethoxycarbonyl
Glu	glutamic acid
HA	hexylamine
HP-β-CD	hydroxypropyl-β-CD

HPLC	high-performance liquid chromatography
Leu	leucine
LOD	limit of detection
LOQ	limit of quantification
MeOH	methanol
MES	2-(*N*-morpholino)ethanesulfonic acid
Met	methionine
MIP	molecularly imprinted polymer
MS	mass spectrometry
MTH	methylthiohydantoin
NorLeu	Norleucine
NorVal	Norvaline
NP	normal phase
ODS	octadecyl silica
PCB	poly-chloro-biphenyl
Phe	phenylalanine
PS	L-α-phosphatidylserine
PDMS	polydimethylsiloxane
PTH	phenylthiohydantoin
RNA	ribo nucleic acid
RP	reversed phase
RTIL	room temperature ionic liquid
Rs	resolution
RSD	relative standard deviation
Ser	serine
SP	stationary phase
SPE	solid phase extraction
TEA	triethylamine
TEAA	triethylamine acetate
THF	tetrahydrofuran
Thr	threonine
Trp	tryptophan
UV	ultraviolet
Val	valine

REFERENCES

1. Krull, IS, Stevenson, RL, Mistry, K, Swartz, ME. 2000. *Capillary Electrochromatography and Pressurized Flow Capillary Electrochromatography: An Introduction*. New York: HNB Publishing.
2. Pretorius, V, Hopkins, BJ, Schieke, JD. 1974. Electro-osmosis: A new concept for high-speed liquid chromatography. *J Chromatogr* 99: 23–30.
3. Huo, Y, Kok, WT. 2008. Recent applications in CEC. *Electrophoresis* 29: 80–93.
4. FDA's policy statement for the development of new stereoisomeric drugs, U.S. Food and Drug Administration, 1992 (online): http://www.fda.gov/cder/guidance/stereo.htm (accessed on 07/04/2008).
5. The International Conference on Harmonisation of Technical Requirements for Registration of Pharmaceuticals for Human Use Q6A, Current step 4 version 1999 (online): http://www.ich.org/ (accessed on 07/04/2008).

6. Vander Heyden, Y, Mangelings, D, Matthijs, N, Perrin, C. 2005. Chiral separations. In: *Handbook of Pharmaceutical Analysis by HPLC*, Ahuja, S, Dong, M (Eds.), pp. 447–498. Amsterdam, the Netherlands: Elsevier.
7. Ali, I, Aboul-Enein, HY. 2006. Impact of immobilized polysaccharide chiral stationary phases on enantiomeric separations. *J Sep Sci* 29: 762–769.
8. Thompson, R. 2005. A practical guide to HPLC enantioseparations for pharmaceutical compounds. *J Liq Chromatogr Rel Techn* 28: 1215–1231.
9. Ha, PTT, Hoogmartens, J, Van Schepdael, A. 2006. Recent advances in pharmaceutical applications of chiral capillary electrophoresis. *J Pharm Biomed Anal* 41: 1–11.
10. Gübitz, G, Schmid, MG. 2004. Recent advances in chiral separation principles in capillary electrophoresis and capillary electrochromatography. *Electrophoresis* 25: 3981–3996.
11. He, LF, Beesley, TE. 2005. Applications of enantiomeric gas chromatography: A review. *J Liq Chromatogr Rel Techn* 28: 1075–1114.
12. Phinney, KW. 2005. Enantioselective separations by packed column subcritical and supercritical fluid chromatography. *Anal Bioanal Chem* 382: 639–645.
13. Zhang, B, Bergstrom, ET, Goodall, DM, Myers, P. 2007. Single-particle fritting technology for capillary electrochromatography. *Anal Chem* 79: 9229–9233.
14. Klodzinska, E, Moravcova, D, Jandera, P, Buszewski, B. 2006. Monolithic continuous beds as a new generation of stationary phase for chromatographic and electro-driven separations. *J Chromatogr A* 1109: 51–59.
15. Aturki, Z, Vichi, F, Messina, A, Sinibaldi, M. 2004. Indirect resolution of β-blocker agents by reversed-phase capillary electrochromatography. *Electrophoresis* 25: 607–614.
16. Mangelings, D, Perrin, C, Massart, DL, Maftouh, M, Eeltink, S, Kok, WT, Schoenmakers, PJ, Vander Heyden, Y. 2004. Optimisation of the chlorthalidone chiral separation by capillary electrochromatography using an achiral stationary phase and cyclodextrin in the mobile phase. *Anal Chim Acta* 509: 11–19.
17. Lelievre, F, Yan, C, Zare, RN, Gareil, P. 1996. Capillary electrochromatography: Operating characteristics and enantiomeric separations. *J Chromatogr A* 723: 145–156.
18. Chankvetadze, B, Kartozia, I, Yamamoto, C, Okamoto, Y, Blaschke, G. 2003. Comparative study on the application of the capillary liquid chromatography and capillary electrochromatography for investigation of enantiomeric impurity of the contraceptive drug levonorgestrel. *J Pharm Biomed Anal* 30: 1897–1906.
19. Chen, X, Jin, W, Qin, F, Liu, Y, Zou, H, Guo, B. 2003. Capillary electrochromatographic separation of enantiomers on chemically bonded type of cellulose derivative chiral stationary phases with a positively charged spacer. *Electrophoresis* 24: 2559–2566.
20. Mangelings, D, Hardies, N, Maftouh, M, Suteu, C, Massart, DL, Vander Heyden, Y. 2003. Enantioseparations of basic and bifunctional compounds by capillary electrochromatography using polysaccharide stationary phases. *Electrophoresis* 24: 2567–2576.
21. Mangelings, D, Maftouh, M, Massart, DL, Vander Heyden, Y. 2004. Enantioseparations by capillary electrochromatography: Differences exhibited by normal and reversed-phase versions of polysaccharide stationary phases. *Electrophoresis* 25: 2808–2816.
22. Mangelings, D, Tanret, I, Matthijs, N, Maftouh, M, Massart, DL, Vander Heyden, Y. 2005. Separation strategy for acidic chiral pharmaceuticals with capillary electrochromatography on polysaccharide stationary phases. *Electrophoresis* 26: 818–832.
23. Mangelings, D, Discry, J, Maftouh, M, Massart, DL, Vander Heyden, Y. 2005. Strategy for the chiral separation of non-acidic pharmaceuticals using capillary electrochromatography. *Electrophoresis* 26: 3930–3941.
24. Mangelings, D, Maftouh, M, Massart, DL, Vander Heyden, Y. 2006. Generic capillary electrochromatographic screening and optimization strategies for chiral method development. *LC-GC Eur* 19: 40–47.
25. Mangelings, D, Saevels, J, Vander Heyden, Y. 2006. Enantiomeric impurity determination of levetiracetam using capillary electrochromatography. *J Sep Sci* 29: 2827–2836.

26. Chen, X, Qin, F, Liu, Y, Kong, L, Zou, H. 2004. Preparation of a positively charged cellulose derivative chiral stationary phase with copolymerization reaction for capillary electrochromatographic separation of enantiomers. *Electrophoresis* 25: 2817–2824.
27. Qin, F, Liu, Y, Chen, X, Kong, L, Zou, H. 2005. Capillary electrochromatographic separation of enantiomers under aqueous mobile phases on a covalently bonded cellulose derivative chiral stationary phase. *Electrophoresis* 26: 3921–3929.
28. Fanali, S, Catarcini, P, Presutti, C, Stancanelli, R, Quaglia, MG. 2003. Use of short-end injection capillary packed with a glycopeptide antibiotic stationary phase in electrochromatography and capillary liquid chromatography for the enantiomeric separation of hydroxyl acids. *J Chromatogr A* 990: 143–151.
29. Fanali, S, Catarcini, P, Presutti, C. 2003. Enantiomeric separation of acidic compounds of pharmaceutical interest by capillary electrochromatography employing glycopeptide antibiotic stationary phases. *J Chromatogr A* 994: 227–232.
30. Fanali, S, D'Orazio, G, Quaglia, MG, Rocco, A. 2004. Use of Hepta-Tyr antibiotic modified silica stationary phase for the enantiomeric resolution of D,L-loxiglumide by electrochromatography and nano-liquid chromatography. *J Chromatogr A* 1051: 247–252.
31. Catarcini, P, Fanali, S, Presutti, C, D'Acquarica, I, Gasparrini, F. 2003. Evaluation of teicoplanin chiral stationary phases of 3.5 μm and 5 μm inside diameter silica microparticles by polar-organic mode capillary electrochromatography. *Electrophoresis* 24: 3000–3005.
32. Aturki, Z, Scotti, V, D'Orazio, G, Rocco, A, Raggi, MA, Fanali, S. 2007. Enantioselective separation of the novel antidepressant mirtazapine and its main metabolites by CEC. *Electrophoresis* 28: 2717–2725.
33. Enlund, AM, Andersson, ME, Hagman, G. 2004. Improved quantification limits in chiral capillary electrochromatography by peak compression effects. *J Chromatogr A* 1028: 333–338.
34. Constantin, S, Bicker, W, Zarbl, E, Lämmerhofer, M, Lindner, W. 2003. Enantioselective strong cation-exchange molecular recognition materials: Design of novel chiral stationary phases and their application for enantioseparation of chiral bases by nonaqueous capillary electrochromatography. *Electrophoresis* 24: 1668–1679.
35. Bicker, W, Hebenstreit, D, Lämmerhofer, M, Lindner, W. 2003. Enantiomeric impurity profiling in ephedrine samples by enantioselective capillary electrochromatography. *Electrophoresis* 24: 2532–2542.
36. Hebenstreit, D, Bicker, W, Lämmerhofer, M, Lindner, W. 2004. Novel enantioselective strong cation exchangers based on sulfodipeptide selectors: Evaluation for enantiomer separation of chiral bases by nonaqueous capillary electrochromatography. *Electrophoresis* 25: 277–289.
37. Gong, Y, Lee, HK. 2003. Application of cyclam-capped-β-cyclodextrin-bonded silica particles as a chiral stationary phase in capillary electrochromatography for enantiomeric separations. *Anal Chem* 75: 1348–1354.
38. André, C, Berthelot, A, Thomassin, M, Guillaume, Y-C. 2006. Enantioselective aptameric molecular recognition material: Design of a novel stationary phase for enantioseparation of a series of chiral herbicides by capillary electrochromatography. *Electrophoresis* 27: 3254–3262.
39. Okamoto, Y, Ikawa, Y, Kitagawa, F, Otsuka, K. 2007. Preparation of fritless capillary using avidin immobilized magnetic particles for electrochromatographic chiral separation. *J Chromatogr A* 1143: 264–269.
40. Wistuba, D, Schurig, V. 2006. Comparison of monolithic approaches for enantioselective capillary electrochromatography involving cyclodextrins. *J Sep Sci* 29: 1344–1352.
41. Qin, F, Xie, C, Yu, Z, Kong, L, Ye, M, Zou, H. 2006. Monolithic enantiomer-selective stationary phases for capillary electrochromatography. *J Sep Sci* 29: 1332–1343.
42. Lecnik, O, Gübitz, G, Schmid, MG. 2003. Role of the charge in continuous beds in the chiral separation of hydroxyl acids by ligand-exchange capillary electrochromatography. *Electrophoresis* 24: 2983–2985.

43. Kornyšova, O, Jarmalavičiene, R, Maruška, A. 2004. A simplified synthesis of polymeric nonparticulate stationary phases with macrocyclic antibiotic as chiral selector for capillary electrochromatography. *Electrophoresis* 25: 2825–2829.
44. Kato, M, Matasumoto, N, Sakai-Kato, K, Toyo'oka, T. 2003. Investigation of the chromatographic performances and binding characteristics of BSA encapsulated capillary column prepared by the sol-gel method. *J Pharm Biomed Anal* 30: 1845–1850.
45. Millot, MC. 2003. Separation of drug enantiomers by liquid chromatography and capillary electrophoresis, using immobilized proteins as chiral selectors. *J Chromatogr B* 797: 131–159.
46. Kato, M, Saruwatari, H, Sakai-Kato, K, Toyo'oka, T. 2004. Silica sol-gel/organic hybrid material for protein encapsulated capillary column of capillary electrochromatography. *J Chromatogr A* 1044: 267–270.
47. Lämmerhofer, M, Tobler, E, Zarble, E, Lindner, W, Svec, F, Fréchet, JMJ. 2003. Macroporous monolithic chiral stationary phases for capillary electrochromatography: New chiral monomer derived from cinchona alkaloid with enhanced enantioselectivity. *Electrophoresis* 24: 2986–2999.
48. Lämmerhofer, M, Peters, E, Yu, C, Svec, F, Fréchet, JMJ. 2000. Chiral monolithic columns for enantioselective capillary electrochromatography prepared by copolymerization of a monomer with quinidine functionality. 1. Optimization of polymerization conditions, porous properties, and chemistry of the stationary phase. *Anal Chem* 72: 4614–4622.
49. Lämmerhofer, M, Svec, F, Fréchet, JMJ, Lindner, W. 2000. Chiral monolithic columns for enantioselective capillary electrochromatography prepared by copolymerization of a monomer with quinidine functionality. 2. Effect of chromatographic conditions on the chiral separations. *Anal Chem* 72: 4623–4628.
50. Lämmerhofer, M, Svec, F, Frechet, JMJ, Lindner, W. 2000. Monolithic stationary phases for enantioselective capillary electrochromatography. *J Microcol Sep* 12: 597–602.
51. Preinerstorfer, B, Lindner, W, Lämmerhofer, M. 2005. Polymethacrylate-type monoliths functionalized with chiral amino phosphonic acid-derived strong cation exchange moieties for enantioselective nonaqueous capillary electrochromatography and investigation of the chemical composition of the monolithic polymer. *Electrophoresis* 26: 2005–2018.
52. Preinerstorfer, B, Lubda, D, Lindner, W, Lämmerhofer, M. 2006. Monolithic silica-based capillary column with strong chiral cation-exchange type surface modification for enantioselective nonaqueous capillary electrochromatography. *J Chromatogr A* 1106: 94–105.
53. Qing, F, Xie, C, Feng, S, Qu, J, Kong, L, Ye, M, Zou, H. 2006. Monolithic silica capillary column with coated cellulose tris (3,5-dimethylphenylcarbamate) for capillary electrochromatographic separation of enantiomers. *Electrophoresis* 27: 1050–1059.
54. Dong, X, Wu, R, Dong, J, Wu, M, Zhu, Y, Zou, H. 2008. Polyacrylamide-based monolithic capillary column with coating of cellulose tris(3,5-dimethylphenylcarbamate) for enantiomer separation in capillary electrochromatography. *Electrophoresis* 29: 919–927.
55. Messina, A, Flieger, M, Bachechi, F, Sinibaldi, M. 2006. Enantioseparation of 2-aryloxypropionic acids on chiral porous monolithic columns by capillary electrochromatography. Evaluation of column performance and enantioselectivity. *J Chromatogr A* 1120: 69–74.
56. Messina, A, Sinibaldi, M. 2007. CEC enantioseparations on chiral monolithic columns: A study of the stereoselective degradation of (R/S)dichlorprop [2-(2,4-dichlorphenoxy) propionic acid] in soil. *Electrophoresis* 28: 2613–2618.
57. Preinerstorfer, B, Lubda, D, Mucha, A, Kafarski, P, Lindner, W, Lämmerhofer, M. 2006. Stereoselective separations of phosphinic acid pseudopeptides by CEC using silica monoliths modified with an anion-exchange-type chiral selector. *Electrophoresis* 27: 4312–4320.

58. Dong, X, Dong, J, Ou, J, Zhu, Y, Zou, H. 2007. Preparation and evaluation of a vancomycin-immobilized silica monolith as chiral stationary phase for CEC. *Electrophoresis* 28: 2606–2612.
59. Liu, H, Yu, A, Liu, F, Shi, Y, Han, L, Chen, Y. 2006. Chiral separation of cefadroxil by capillary electrochromatography. *J Pharm Biomed Anal* 41: 1376–1379.
60. Schmid, MG, Koidl, J, Wank, P, Kargl, G, Zöhrer, H, Gübitz, G. 2007. Enantioseparation by ligand-exchange using particle-loaded monoliths: Capillary-LC versus capillary electrochromatography. *J Biochem Biophys Methods* 70: 77–85.
61. Liu, ZS, Xu, YL, Wang, H, Yan, C, Gao, RY. 2004. Chiral separation of binaphtol enantiomers on molecularly imprinted polymer monolith by capillary electrochromatography. *Anal Sci* 20: 673–678.
62. Szumski, M, Buszewski, B. 2004. Molecularly imprinted polymers: A new tool for separation of steroid isomers. *J Sep Sci* 27: 837–842.
63. Xu, Y-L, Liu, ZS, Wang, HF, Yan, C, Gao, RY. Chiral recognition ability of an (S)-naproxen imprinted monolith by capillary electrochromatography. *Electrophoresis* 26: 804–811.
64. Deng, Q-L, Lun, Z-H, Gao, R-Y, Zhang, L-H, Zhang, W-B, Zhang, Y-K. 2006. (S)-ibuprofen-imprinted polymers incorporating γ-methacryloxypropyltrimethylsiloxane for CEC separation of ibuprofen enantiomers. *Electrophoresis* 27: 4351–4358.
65. Ou, J, Dong, J, Tian, T, Hu, J, Ye, M, Zou, H. 2007. Enantioseparation of tetrahydropalmatine and Trögers' base by molecularly imprinted monolith in capillary electrochromatography. *J Biochem Biophys Methods* 70: 71–76.
66. Wang, H-F, Zhu, Y-Z, Yan, X-P, Gao, R-Y, Zheng, J-Y. 2006. A room temperature ionic liquid (RTIL)-mediated, non-hydrolytic sol-gel methodology to prepare molecularly imprinted, silica-based hybrid monoliths for chiral separation. *Adv Mater* 18: 3266–3270.
67. Wang, H-F, Zhu, Y-Z, Lin, J-P, Yan, X-P. 2008. Fabrication of molecularly imprinted hybrid monoliths via a room temperature ionic liquid-mediated non-hydrolytic sol-gel route for chiral separation of zolmitriptan by capillary electrochromatography. *Electrophoresis* 29: 952–959.
68. Zheng, J, Shamsi, SA. 2003. Combination of chiral capillary electrochromatography with electrospray ionization mass spectrometry: Method development and assay of warfarin enantiomers in human plasma. *Anal Chem* 75: 6295–6305.
69. Zheng, J, Norton, D, Shamsi, SA. 2006. Fabrication of internally tapered capillaries for capillary electrochromatography electrospray ionization mass spectrometry. *Anal Chem* 78: 1323–1330.
70. Zheng, J, Shamsi, SA. 2006. Simultaneous enantioseparation and sensitive detection of eight β-blockers using capillary electrochromatography-electrospray ionization-mass spectrometry. *Electrophoresis* 27: 2139–2151.
71. Redman, AR, Zheng, J, Shamsi, SA, Huo, J, Kelly, EJ, Ho, RJY, Ritchie, D, Hon, YY. 2008. Variant CYP2C9 alleles and warfarin concentrations in patients receiving low-dose versus average-dose warfarin therapy. *Clin Appl Thrombosis/Hemostasis* 14: 29–37.
72. Yao, C, Tang, S, Gao, R, Jiang, C, Yan, C. 2004. Enantiomer separations on a vancomycin stationary phase and retention mechanism of pressurized capillary electrochromatography. *J Sep Sci* 27: 1109–1114.
73. Lin, B, Shi, Z-G, Zhang, H-J, Ng, S-C, Feng, Y-Q. 2006. Perphenylcarbamoylated β-cyclodextrin bonded-silica particles as chiral stationary phase for enantioseparation by pressure-assisted capillary electrochromatography. *Electrophoresis* 27: 3057–3065.
74. Zhou, A, Lv, X, Xie, Y, Yan, C, Gao, R. 2005. Chromatographic evaluation of perphenylcarbamoylated β-cyclodextrin bonded stationary phase for micro-high performance liquid chromatography and pressurized capillary electrochromatography. *Anal Chim Acta* 547: 158–164.

75. Wistuba, D, Banspach, L, Schurig, V. 2005. Enantiomeric separation by capillary electrochromatography using monolithic capillaries with sol-gel glued cyclodextrin-modified silica particles. *Electrophoresis* 26: 2019–2026.
76. Schmid, MG, Koidl, J, Freigassner, C, Tahedl, S, Wojcik, L, Beesley, T, Armstrong, DW, Gübitz, G. 2004. New particle-loaded monoliths for chiral capillary electrochromatographic separation. *Electrophoresis* 25: 3195–3203.
77. Gatschelhofer, C, Schmid, MG, Schreiner, K, Pieber, TR, Sinner, FM, Gübitz, G. 2006. Enantioseparation of glycyl dipeptides by CEC using particle-loaded monoliths prepared by ring opening metathesis polymerization. *J Biochem Biophys Methods* 69: 67–77.
78. Kapnissi, CP, Valle, BC, Warner, IM. 2003. Chiral separations using polymeric surfactants and polyelectrolyte multilayers in open-tubular capillary electrochromatography. *Anal Chem* 75: 6097–6104.
79. Kamande, MW, Zhu, X, Kapnissi-Christodoulou, C, Warner, IM. 2004. Chiral separations using a polypeptide and polymeric dipeptide surfactant polyelectrolyte multilayer coating in open-tubular capillary electrochromatography. *Anal Chem* 76: 6681–6692.
80. Kapnissi-Christodoulou, C, Lowry, M, Agbaria, RA, Geng, L, Warner, IM. 2004. Investigation of the stability of polyelectrolyte multilayer coatings in open-tubular capillary electrochromatography using laser scanning confocal microscopy. *Electrophoresis* 26: 783–789.
81. Geng, L, Bo, T, Liu, H, Li, N, Liu, F, Li, K, Gu, J, Fu, R. 2004. Capillary coated with layer-by-layer assembly of γ-zirconium phosphate/lysozyme nanocomposite film for open tubular capillary electrochromatography chiral separation. *Chromatographia* 29: 65–70.
82. Han, N-Y, Hautala, JT, Bo, T, Wiedmer, S, Riekkola, M-L. 2006. Immobilization of phopholipid-avidin on fused-silica capillaries for chiral separation on open-tubular capillary electrochromatography. *Electrophoresis* 27: 1502–1509.
83. Wiedmer, S, Bo, T, Riekolla, M-L. 2008. Phospholipid coatings for chiral capillary electrochromatography. *Anal Biochem* 373: 26–33.
84. Zuo, X, Wang, K, Zhou, L, Huang, S. 2003. Separation and determination synchronously by multichannel mode-filtered light capillary electrochromatography. *Electrophoresis* 24: 3202–3206.
85. Zhou, Z, Wang, K, Yang, X, Huang, S, Zhou, L, Qin, D, Du, L. 2001. Synchronization of separation and determination based on multichannel mode-filtered light detection with capillary electrophoresis. *Analyst* 126: 1838–1840.
86. Kitagawa, F, Inoue, K, Hasegawa, T, Kamiya, M, Okamoto, Y, Kawase, M, Otsuke, K. 2006. Chiral separation of acidic drug components by open tubular electrochromatography using avidin immobilized capillaries. *J Chromatogr A* 1130: 219–226.
87. Dai, R, Tang, L, Li, H, Deng, Y, Fu, R, Parveen, Z. 2007. Synthesis and characterization of β-CD derivatized bovine serum albumin protein as chiral selector in pressurized capillary electrochromatography. *J Appl Polym Sci* 106: 2041–2046.
88. Zeng, H-L, Li, H-F, Lin, J-M. 2005. Chiral separation of dansyl amino acids by PDMS microchip gel monolithic column electrochromatography with γ-cyclodextrin bonded in polyacrylamide. *Anal Chim Acta* 551: 1–8.
89. Zeng, H-L, Li, H-F, Wang, X, Lin, J-M. 2006. Development of a gel monolithic column polydimethylsiloxane microfluidic device for rapid electrophoresis separation. *Talanta* 69: 226–231.
90. Klampfl, C. 2004. Review of coupling capillary electrochromatography to mass spectrometry. *J Chromatogr A* 1044: 131–144.
91. Schmid, MG, Grobuschek, N, Pessenhofer, V, Klostius, A, Gübitz, G. 2003. Enantioseparation of dipeptides by capillary electrochromatography on a teicoplanin aglycone stationary phase. *J Chromatogr A* 990: 83–90.

92. Zheng, J, Shamsi, SA. 2003. Brush-type chiral stationary phase for enantioseparation of acidic compounds. Optimization of capillary electrochromatographic parameters. *J Chromatogr A* 1005: 177–187.
93. Chen, Z, Ozawa, H, Uchiyama, K, Hobo, T. 2003. Cyclodextrin-modified monolithic columns for resolving dansyl-amino acids enantiomers and positional isomers by capillary electrochromatography. *Electrophoresis* 24: 2550–2558.
94. Schmid, MG, Grobushek, N, Pessenhofer, V, Klostius, A, Gübitz, G. 2003. Chiral resolution of diastereoisomeric di-and tripeptides on a teicoplanin aglycone phase by capillary electrochromatography. *Electrophoresis* 24: 2543–2549.
95. Honzatko, A, Aturki, Z, Flieger, M, Messina, A, Sinibaldi, M. 2003. Chiral separations by capillary electrochromatography on multiple-interaction based stationary phases. *Chromatographia* 58: 271–275.
96. Chen, Z, Nishiyama, T, Uchiyama, K, Hobo, T. 2004. Electrochromatographic enantioseparation using chiral ligand-exchange monolithic sol-gel column. *Anal Chim Acta* 501: 17–23.

16 Chiral Separations with Microchip Technology

Markéta Vlčková, Franka Kalman, and Maria A. Schwarz

CONTENTS

16.1 Introduction 501
16.2 On-Chip Chiral Separation Principle 504
16.3 Microchip Device 504
16.3.1 Microchip Design 504
16.3.2 Injection 505
16.3.3 Detection 507
16.4 Resolution Enhancement on a Chip 508
16.5 Advances in On-Chip Chiral Separations 509
16.6 Conclusions and Future Perspectives 511
References 512

16.1 INTRODUCTION

Miniaturization is a modern trend for many technologies, including chemistry. The microchip technology for analytical separations is now well established and the number of on-chip applications has increased tremendously over the last decade. On a microchip, the consumption of sample and reagents is substantially reduced and the separation is accelerated compared to conventional separation techniques. The higher separation speed leads not only to a higher sample throughput but allows also a faster optimization of a separation method. The high sample throughput on a microchip can be further drastically increased using parallel chips, as the fabrication of the multiple separation units, rather than a single unit, does not significantly raise the cost of production. Furthermore, microchips have a great potential of integrating multiple analytical steps into one miniaturized device, leading eventually to the development of so-called micro-total analysis systems (μ-TAS) that would enable fast, fully automated analysis of complex samples.

The advantages of microchips over conventional separation systems can also be well exploited in the area of chiral separations. Since the first demonstration of feasibility of enantiomeric separations on the microchip in 1999 [1], various applications (summarized in Table 16.1) appeared in the literature, including two reviews, one

TABLE 16.1
Overview of On-Chip Chiral Separations

Analytes	pH/Chiral Selector	Detection Mode/ Derivatization Agent (If Used)	Comment	Year, References
Amino acids	9.2/γ-CD	Fluorescence/FITC	MEKC mode (SDS), folded separation channel	1999, [1]
Amino acids	9.4/γ-CD	Fluorescence/FITC	MEKC mode (SDS)	2000, [4]
Amphetamines	7.4–8.5/HS-γ-CD	Fluorescence/NBD-F	SDS below CMC as buffer additive, folded separation channel	2000, [5]
Tryptophan	9.1/α-CD	Conductivity	Isotachophoresis	2001, [6]
Neurotransmitters, ephedrine	6–12.9/CM-β-CD, HP-β-CD, M-β-CD	Amperometry	18-Crown-6 as buffer additive	2001, [7]
Tryptophan, thiopental	BSA on a membrane	UV detection	Chromatography	2001, [8]
Amines	9.0/HP-γ-CD	Fluorescence/FITC		2002, [9]
Gemifloxacin	4.0/18C_6H_4	Fluorescence native	Negative effect of Na^+ ions present in excess suppressed by EDTA	2002, [10]
Amines	9.2/HP-γ-CD	Fluorescence/FITC	Uncoated/PVA-coated channel	2003, [11]
Neurotransmitter metabolites, precursor	2.3–7.2/S-β-CD, CM-β-CD	Amperometry	18-Crown-6 or PAMAM dendrimer G1.5 as buffer additives	2003, [12]
Various basic and acidic drugs	2.5/HS-α, β, γ-CD	Imaged UV detection	Commercial microchip instrument (MCE-2010)	2003, [13]
Amino acids	≤9.0/HP-β-CD	Fluorescence/fluorescamine		2003, [14]
Dansyl-phenylalanine	7/HP-β-CD	Chemiluminescence/H_2O_2 + TDPO	Online coupling of derivatization reaction with chiral separation	2003, [15]
Dansyl-amino acids	2.5/HS-γ-CD	Fluorescence native	Subsecond separations	2004, [16]

Gemifloxacin	4.0/$18C_6H_4$	Fluorescence native	Urine, online cleaning of matrix	2004, [17]
Aminoindan	7.0/S-β-CD	Imaged UV detection	Commercial microchip instrument (MCE-2010)	2005, [18]
Amino acids	9.0/L-prolinamide, Cu (II)	Fluorescence/NBD-F	Ligand-exchange chiral separation, SDS below CMC as buffer additive	2005, [19]
Dansyl-amino acids, amino acids	9.0/allyl-γ-CD or allyl-β-CD immobilized	Fluorescence/FITC	Electrochromatography	2005, 2006, [20,21]
Tryptophan	7.4/BSA immobilized	Amperometry	Electrochromatography	2005, 2006, [22,23]
Basic drugs	9.3/α, β, γ-CD, HP-α, β, γ-CD, DM-β-CD	Fluorescence/FITC	Parallel processing	2005, [24]
Amino acids	10/not used	Fluorescence/OPA + TATG	On-chip integrated derivatization on forming diastereomers and their subsequent separation, indirect (without chiral selector), MEKC mode (SDS micelles)	2005, [25]
Glycidyl phenyl-ether	8.5/heptakis-6-S-β-CD	Fluorescence	On-chip integrated enzymatic reaction and chiral separation	2006, [26]
1-Cyclohexylethyl-amine	9.2/HP-γ-CD	Fluorescence/FITC	Comparison of uncoated/PVA-coated powder blasted or wet chemical etched chips	2006, [27]
trans-Cyclohexane-1, 2-diamine	2.5/DM-β-CD, $18C_6H_4$	Contactless conductivity	Two chiral selectors used concurrently	2006, [28]

Abbreviations: $18C_6H_4$, 18-crown-6 tetracarboxylic acid; BSA, bovine serum albumin; CD, cyclodextrin; CM-, carboxymethyl-; CMC, critical micellar concentration; DM-, dimethyl-; FITC, fluorescein isothiocyanate; HP-, hydroxypropyl-; HS-, highly sulfated-; M-, methyl-; MEKC, micellar electrokinetic chromatography; NBD-F, 4-fluoro-7-nitrobenzofurazane; OPA, *o*-phthaldialdehyde; PAMAM, polyamidoamine; PVA, poly(vinyl alcohol); S-, sulfated-; SDS, sodium dodecyl sulfate; TATG, 2,3,4,6-tetra-O-acetyl-1-thio-β-D-glucopyranose; TDPO, bis[2-(3,6,9-trioxadecanyloxycarbonyl)-4-nitrophenyl] oxalate.

very recent, devoted entirely to on-chip chiral separations [2,3]. Even though only model samples have been enantiomerically separated in most of the hitherto published applications, the short separation times achieved clearly indicate the great potential of the high-throughput on-chip measurements in the area of chiral analysis.

16.2 ON-CHIP CHIRAL SEPARATION PRINCIPLE

Most of the on-chip chiral separations have been performed as zone electrophoretic measurements. In analogy to capillary zone electrophoresis (CZE), the enantiomeric electrophoretic separation is based on different interaction of respective enantiomers with a chiral selector altering their migration velocity and thus leading to their mutual separation. In direct on-chip chiral separations (that are by far prevailing), the chiral agent is dissolved in the background electrolyte (a buffer solution) and the interaction of the chiral agent with the individual enantiomers takes place during the separation. For certain analytes, the enantiomeric separation can be achieved via ligand exchange with another enantiomeric species complexed to copper(II) ions acting as a chiral selector [19]. In indirect on-chip chiral separations, the enantiomers react with the chiral selector in forming stable diastereomeric compounds prior to their separation, which is performed subsequently in the absence of the chiral agent [25].

Chiral separations on microchips can also be realized in isotachophoresis (ITP) mode, as demonstrated by Olvecka et al. [6]. In the ITP mode, the chiral selector is dissolved in a leading electrolyte and its interaction with the racemic sample leads to the splitting of the sample zone into two consecutive zones of related enantiomers. The application of ITP mode for on-chip chiral separation can be advantageous, if the isolation of the individual enantiomers is desired.

On the other hand, immobilized chiral selectors are applied for on-chip chiral separations based on electrochromatography or chromatography modes. The separation principle of these two techniques is based on different distribution of analytes between two phases, the stationary one (sorbent) and the mobile one (eluent). In electrochromatography, an electric field is applied to induce the mobile phase flow governed by electroosmotic flow (EOF). In chromatography, an external pressure is used instead of electric field. Different interactions of the sample enantiomers with the selector present in the stationary phase cause a change in their retention inside the channel. The incorporation of the stationary phase to the microchip channel is easily realized by in situ synthesis. The selector can be physically adsorbed onto the stationary phase, entrapped within the phase or covalently attached. The advantage of performing on-chip chiral separations with an immobilized selector is its reapplicability and thus, enabling utilization of expensive chiral agents.

16.3 MICROCHIP DEVICE

16.3.1 Microchip Design

Microchip is a planar microfabricated separation device, made from glass or plastic, containing microchannels usually arranged into a cross shape with one elongated limb (called separation channel), where the separation process takes place

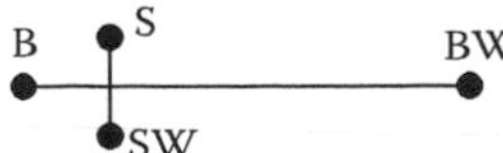

FIGURE 16.1 Schematic drawing of a typical microchip used for electrophoretic measurements. S, sample; SW, sample waste; B, buffer; BW, buffer waste.

(Figure 16.1). Typical channel depths are 15–50 μm, widths are 50–200 μm, and lengths of the separation channel are usually 0.5–10 cm. Fluid reservoirs are located at the ends of all channels; mostly there are two inlet reservoirs serving for introduction of sample and background electrolyte or eluent and two outlet reservoirs acting as a waste. In microchips for electrophoretic or electrochromatographic separations, electrodes are present in all reservoirs to enable a connection to a high voltage power supply for application of an electric field.

The design of a microchip for pressure-driven separations is more complicated since an additional unit is needed for inducing the fluid flow. The hydrodynamic flow on the microchip is typically achieved by connecting an external macrosyringe pump to the fluid reservoirs, yet various microfabricated pumps incorporated within the chip have also been developed [29]. However, the connection of the pump to a microfluidic device always presents a technical challenge and thus the number of pressure-driven separations performed on a microchip is in general very small. In fact, only one report on on-chip chiral separation based on chromatographic separation principle has been found in the literature [8]. Here, a homemade microfluidic chromatographic system was employed for separation of enantiomers using bovine serum albumin (BSA) as chiral selector adsorbed onto a membrane forming the stationary phase. Therefore, the following description of the realization of injection and detection on a microchip device refers only to common chips for electrophoretic separations.

16.3.2 Injection

The injection on the microchip with a cross design is accomplished through the intersection of the microchannels. The intersection is formed as a simple cross or, alternatively, as a so-called double tee, where the two arms of the sample channel are offset to make a larger injector region. The sample is injected mostly electrokinetically by applying an electric field across the sample channel. The portion of the sample present in the intersection represents the injection plug, which is subjected to separation in the subsequent separation phase by applying electric field across the separation channel.

However, without an additional voltage control of injection as well as separation phase, an uncontrollable sample leakage into the separation channel occurs due to diffusion through the cross [30]. Nevertheless, the on-chip measurements, including chiral separations, are still realizable in the simplest uncontrolled injection scheme, yet with lower efficiency and poorer reproducibility compared to the advanced voltage-controlled injection strategies. An overview of available schemes of electrokinetic injection is given in Figure 16.2.

The most often used voltage-controlled injection strategy for on-chip chiral separations is the so-called pinched injection [31]. In this injection scheme, pinching voltages are applied at the buffer inlet and outlet during injection. Thereby, a buffer flow toward the sample waste (SW) reservoir is induced in order to counteract the diffusion of analytes into the separation channel. In the following separation step, back

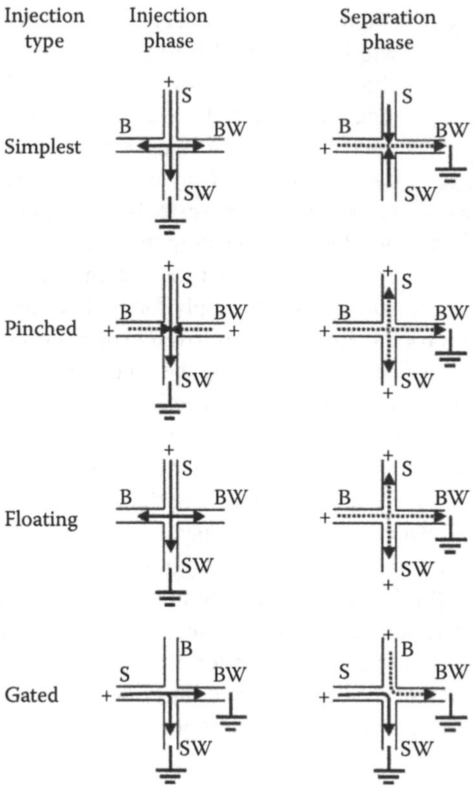

FIGURE 16.2 Overview of the voltage-controlled injection schemes developed for electrokinetic injection on the microchip. S, sample; SW, sample waste; B, buffer; BW, buffer waste. The arrows indicate the flow directions of the sample (solid line) and of the buffer (dotted line) during injection and separation phases.

voltages are applied at the sample inlet (SI) and SW reservoirs to draw the analytes back into the reservoirs, preventing sample leakage into the separation channel [32]. However, higher detection limits of the analytes are observed using pinched injection compared to the uncontrolled simple injection, since significantly lower amount of the sample is introduced by the pinched injection scheme [33].

As a compromise between injection reproducibility and detection sensitivity, the so-called floating injection was proposed [32] and also used for on-chip chiral separations [14]. In this approach, the injection step is carried out without the voltage control and only the subsequent separation step is controlled by applying the back voltage to the SI and SW reservoirs. The final concentration of the sample, injected using the floating injection, is increased by diffusion of the sample into the separation channel during the injection phase.

A different approach to the voltage control of the electrokinetic injection represents the so-called gated injection [34]. Unlike the pinched or floating injection, in the gated injection scheme the sample flows permanently, making a 90° turn at the injection cross toward the SW. Simultaneously, a continuous buffer stream is generated toward the SW and buffer waste (BW) reservoirs, preventing sample leakage into the separation channel. For introducing the sample, the buffer flow is interrupted allowing a plug of the sample to move into the separation channel. Such an approach allows a periodical sampling from a continuous flow of the analytes and is usually applied for on-chip integration of the chiral discrimination process with some preceding analytical step such as derivatization [15,25], online clean up [17], or enzymatic reaction [26].

However, all electrokinetic injection modes lead to discrimination among analytes due to their different electrophoretic mobilities. This can negatively affect the determination of enantiomeric ratios in case of injecting enantiomer-selector complexes having higher mutual mobility difference. The discrimination can be prevented using a pressure-driven sample injection, which has already been introduced for microchip electrophoresis [35] but not yet implemented for on-chip chiral separations.

16.3.3 Detection

On a microchip, the separated enantiomers are either detected at the end point of the separation channel or, alternatively, the whole separation channel is monitored by an optical imaging detector. Fluorescence is by far the most commonly used detection technique due to its high sensitivity. However, most of the analytes are not native fluorophores and, for this kind of detection, analytes have to be derivatized in a separate step prior to the separation. The analytes are mostly labeled outside the microanalytical system, although an on-chip coupling of derivatization reaction with the enantiomeric separation has already been reported [25]. The derivatization can also have a negative effect on the performance of a chiral separation method as it can alter the enantioselectivity of the particular analyte toward a given chiral selector.

Electrochemical detection represents the second most common type of detection for on-chip chiral separations. The advantages of this detection mode are a good sensitivity and a potential for miniaturization of the detector size enabling its integration into the chip. Among electrochemical detection modes, amperometric detection is the most popular for on-chip chiral measurements [7,12,22,23] due to the easy operation and minimal background-current contributions. Amperometry is based on detection of oxidation or reduction of analytes on the working electrode and is therefore restricted to electroactive species. In contrast, conductivity detection is universal, since it is based on measuring differences in conductivity between the background electrolyte and the analyte. The contactless variant of the conductivity detector, which does not require direct contact of the electrodes with the electrolyte solution, has been applied in on-chip detection of enantiomers [6,28].

UV/Vis absorption detection is not widely used on the chip because its detection sensitivity is low. However, due to the existence of commercial microchip station equipped with an imaging UV/Vis absorption detector (Shimadzu, MCE-2010), this kind of detection has also been applied for the detection in on-chip chiral separations [13,18]. The imaging detector enables direct determination of the optimal separation length as the progress of the chiral separation is monitored in the course of the measurement (Figure 16.3). The clear advantage of UV/Vis absorption as well as

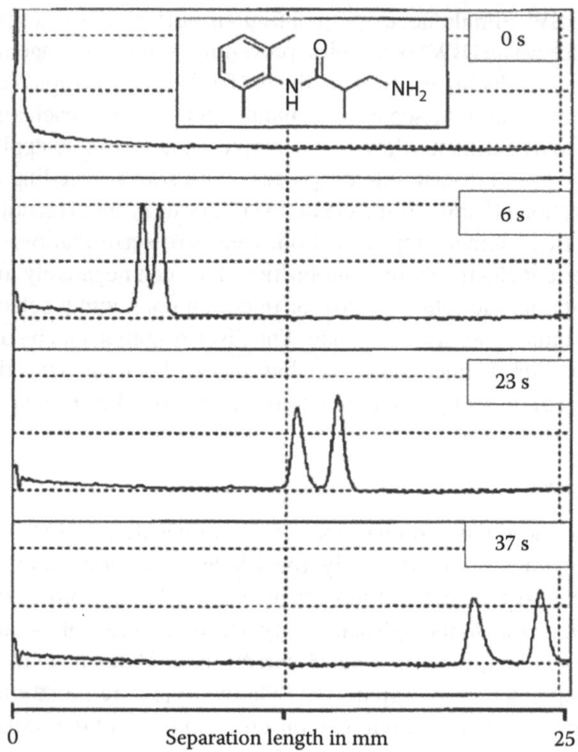

FIGURE 16.3 Imaging UV detection of the chiral separation of a tocainide derivative. Separation conditions: 5% HS-γ-CD, 25 mM triethylammonium phosphate buffer, pH 2.5; UV detection at 200 nm. (From Ludwig, M. et al., *Electrophoresis*, 24, 3233, 2003. With permission.)

electrochemical detection compared to fluorescence is the direct detection without tedious derivatization reactions.

In addition to the above mentioned detection methods, on-chip chemiluminescence detection has also been reported for the detection of enantiomers on the microchip [15]. Here, the derivatization reaction has been on-chip coupled with the subsequent chiral separation of dansyl-phenylalanine.

16.4 RESOLUTION ENHANCEMENT ON A CHIP

The separation lengths employed on microchips are usually restricted to several centimeters at most. In order to achieve the chiral separation on such a short separation length, the chiral selector used for the on-chip measurement must have a higher enantioselectivity in comparison with selectors used in conventional chiral CZE separations. The search for suitable chiral selectors to be employed for chiral separation on a microchip is therefore of an extreme importance. Several strategies have been developed to enhance the resolution if the enantioselectivity of any chiral agent available for chiral recognition is rather limited.

One strategy is to increase the effective separation length on the microchip while maintaining the electric field optimized to highest possible intensity, which still does not induce the detrimental Joule heating. Folded separation channels have been used to increase the separation length without enlarging the compact format of the chip [1,5]. Better separation can also be achieved by suppressing the EOF, usually by coating the inner wall of the separation channel. The permanent coating of the separation channel by poly(vinyl alcohol) (PVA) has been demonstrated to improve considerably the resolution of the amine enantiomers on the chip as well as the reproducibility of the injection [11,27]. Nevertheless, the most effective approach to enhance resolution is to increase the enantioselectivity of the chiral selector using various buffer additives (see Table 16.1). Sodium dodecyl sulfate (SDS) is the most popular buffer additive used to promote the separation on the microchip. SDS has been successfully applied for enhancing resolution in micellar electrokinetic chromatography (MEKC) mode [1,4] as well as in the concentration below its critical micellar concentration (CMC). Other buffer additives used to promote the separation on the microchip include crown ethers, achiral as well as chiral (acting as a second chiral selector), and dendrimers.

16.5 ADVANCES IN ON-CHIP CHIRAL SEPARATIONS

The main benefit of performing chiral separations on the microchip is the high separation speed. Even subsecond on-chip enantiomeric separation of several dansyl-amino acids has been reported [16]. The combination of low pH of the separation buffer leading to a negligible EOF with a highly sulfated γ-cyclodextrin (HS-γ-CD) as chiral selector, which is well-known for its extraordinary high enantioselectivity, made it possible to achieve the chiral separation in such a short time.

The potential of the chip for high-throughput screening in the area of chiral separations has clearly been demonstrated by Ludwig et al. [13]. In this report, 19 racemic compounds were enantiomerically separated on a chip, each in less than one minute, using various HS-CDs as chiral selectors. Imaged UV detection applied for these on-chip chiral separations enabled the determination of a suitable separation time in an easy way (see Figure 16.3). The high throughput of the chip for chiral separations can be further increased by increasing the number of separation channels on one chip as shown by Gao et al. [24]. Using parallel processing, various chiral selectors, samples, or separation buffers can be tested concurrently. In this way, the time required for the development of a given separation system might be reduced substantially.

Integration of more analytical steps into a single device is another promising feature of the microchip. The feasibility of online coupling of a chiral separation with some preseparation processes has already been shown by several research groups. Besides devices integrating a derivatization reaction with a subsequent on-chip chiral separation of a labeled analyte [15,25], two other types of a preceding analytical step (reaction and cleanup step, respectively) have been successfully coupled on-chip to a chiral separation. A microchip device connecting a microfluidic reactor with a subsequent enantiomeric separation has been applied for testing enantioselectivity of enzyme mutants created by directed evolution of the enzyme [26]. The product enantiomers from the on-chip enzymatic reaction have been separated from the educt enantiomers on the same microchip device. Furthermore, an online cleanup of the urine matrix prior to the on-chip chiral separation of gemifloxacin has been reported by Il Cho et al. [17]. Metal ions have been effectively removed in the first electrophoretic separation step followed by chiral separation of the drug in a second electrophoretic step (Figure 16.4).

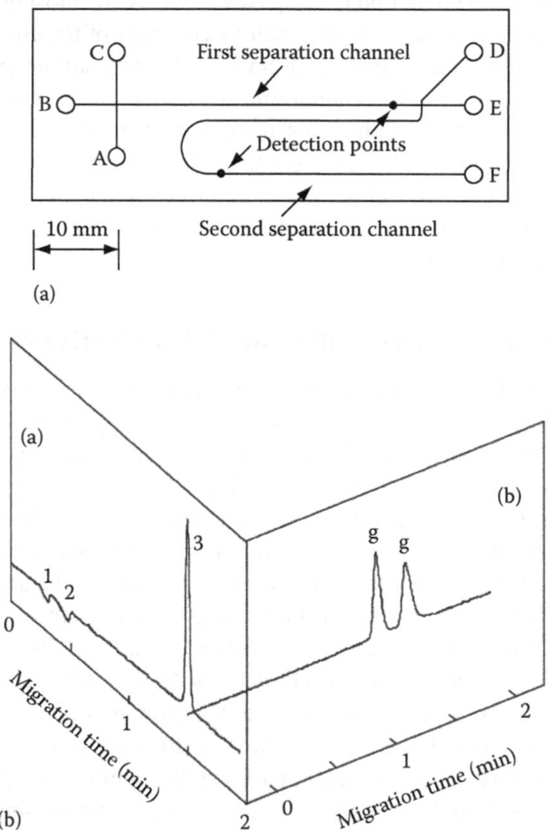

FIGURE 16.4 (a) Schematic drawing of the microchip used for online coupling cleanup and chiral separation of gemifloxacin. Reservoirs: B, running buffer; C, sample; A, sample waste; D, buffer waste; E, running buffer with chiral selector; F, buffer waste containing chiral selector. Channel depth was 35 μm, width at half-depth was 60 μm. Dots represent the detection points located at 32 and 38 mm from the first and second injection crosses, respectively. (b) Chiral separation of gemifloxacin dissolved in urinary solution. Peaks: (1) K^+, (2) Na^+, and (3) gemifloxacin racemate. Peaks for gemifloxacin enantiomers are denoted by "g." a, The removal of metal ions was performed in the first separation channel. b, The chiral separation was performed in the second separation channel using a running buffer of 50 mM bis-Tris/citric acid containing 50 μM $18C_6H_4$ (pH 4). Urinary solution was fivefold diluted using 50 mM bis-Tris/citric acid (pH 4). Gemifloxacin concentration was 100 μM in fivefold diluted urinary solution. (From Il Cho, S. et al., *J. Chromatogr. A.* 1055, 241, 2004. With permission.)

16.6 CONCLUSIONS AND FUTURE PERSPECTIVES

The miniaturization trend present in many technical domains can also be observed in the area of chiral separations as demonstrated by the current number of chiral separations performed in microfluidic separation devices (summarized in Table 16.1). On-chip chiral separations have been performed in electrophoretic, electrochromatographic as well as chromatographic mode. The electrophoretic mode is prevailing due to a simple arrangement of the measurement system. Although the chip for an on-chip electrophoretic chiral separation has often a modest cross design, the versatility of the chip makes more complex designs for advanced on-chip applications easily realizable.

The development of various voltage-controlled injection schemes improved the reproducibility of the injection on the microchip in a remarkable manner. The possibility of injection of a very narrow sample zone using a pinched injection scheme ensures that the separation efficiency is not sacrificed by too long injection lengths. Furthermore, the gated injection scheme enabling periodical sampling from continuous flow of the sample is well suited for online coupling of the chiral discrimination process with a preceding analytical step.

On the other hand, the detection still represents a weak point of the on-chip chiral separations. Most commonly used fluorescence detection has the drawback of necessary derivatization for most of the analytes. Furthermore, the labeling reaction may alter the enantioselectivity of the analytes toward the selector. In conventional separation systems, popular UV/Vis detection is not well suited for microchips due to its low sensitivity resulting from short optical path lengths. However, it is still applicable for on-chip chiral separations, if higher concentrations of analytes are available, as demonstrated by on-chip chiral separations performed on a commercial microchip instrument equipped with imaged UV/Vis detector [13,18]. Electrochemical detection is in general more suitable for microchip separations owing to the possible miniaturization of the detector and inherently a good sensitivity. Amperometric detection is the most commonly used mode of electrochemical detection on a chip due to its sensitivity and simplicity. However, it is restricted to electroactive species. In contrast, conductivity detection is universal and the applicability of the contactless conductivity detector for on-chip chiral separations has been demonstrated [28]. Even though this detection mode requires careful tuning of the composition of the background electrolyte to reach a compromise between separation and detection aspects, its wider utilization for on-chip detection can be expected in the future [36].

The main potential of chiral separations performed on the microchips is given by the extreme high sample throughput, which can even be multiplexed by employing microchip arrays. Moreover, the short time needed for on-chip chiral separation speeds up the development of a chiral separation method, including lower costs due to lower consumption of chemicals. All these features make the exploitation of chip technology for chiral separations attractive especially for commercial laboratories. Therefore, the application of on-chip chiral separations, for example, for routine tests on enantiomeric purity of pharmaceutical products, or for screening of large libraries of newly developed enantioselective catalysts can be imagined even in the near future. For a more distant future, the development of fully automated lab-on-a-chip analysis systems integrating also on-chip chiral separations can be foreseen. Such a μ-TAS would enable the complete analysis of samples of interest, including their chiral recognition, even outside specialized analytical laboratories.

REFERENCES

1. Hutt, LD, Glavin, DP, Bada, JL, Mathies, RA. 1999. Microfabricated capillary electrophoresis amino acid chirality analyzer for extraterrestrial exploration. *Anal Chem* 71: 4000–4006.
2. Belder, D, Ludwig, M. 2003. Microchip electrophoresis for chiral separations. *Electrophoresis* 24: 2422–2430.
3. Mangelings, D, Vander Heyden, Y. 2007. High-throughput screening and optimization approaches for chiral compounds by means of microfluidic devices. *Comb Chem High Throughput Screen.* 10: 317–325.
4. Rodriguez, I, Jin, LJ, Li, SFY. 2000. High-speed chiral separations on microchip electrophoresis devices. *Electrophoresis* 21: 211–219.
5. Wallenborg, SR, Lurie, IS, Arnold, DW, Bailey, CG. 2000. On-chip chiral and achiral separation of amphetamine and related compounds labeled with 4-fluoro-7-nitrobenzofurazane. *Electrophoresis* 21: 3257–3263.
6. Olvecka, E, Masar, M, Kaniansky, D, Johnck, M, Stanislawski, B. 2001. Isotachophoresis separations of enantiomers on a planar chip with coupled separation channels. *Electrophoresis* 22: 3347–3353.
7. Schwarz, MA, Hauser, PC. 2001. Rapid chiral on-chip separation with simplified amperometric detection. *J Chromatogr A* 928: 225–232.
8. Wang, PC, Gao, J, Lee, CS. 2002. High-resolution chiral separation using microfluidics-based membrane chromatography. *J Chromatogr A* 942: 115–122.
9. Belder, D, Deege, A, Maass, M, Ludwig, M. 2002. Design and performance of a microchip electrophoresis instrument with sensitive variable-wavelength fluorescence detection. *Electrophoresis* 23: 2355–2361.
10. Il Cho, S, Lee, KN, Kim, YK, Jang, JH, Chung, DS. 2002. Chiral separation of gemifloxacin in sodium-containing media using chiral crown ether as a chiral selector by capillary and microchip electrophoresis. *Electrophoresis* 23: 972–977.
11. Ludwig, M, Belder, D. 2003. Coated microfluidic devices for improved chiral separations in microchip electrophoresis. *Electrophoresis* 24: 2481–2486.
12. Schwarz, MA, Hauser, PC. 2003. Chiral on-chip separation of neurotransmitters. *Anal Chem* 75: 4691–4695.
13. Ludwig, M, Kohler, F, Belder, D. 2003. High-speed chiral separations on a microchip with UV detection. *Electrophoresis* 24: 3233–3238.
14. Skelley, AM, Mathies, RA. 2003. Chiral separation of fluorescamine-labeled amino acids using microfabricated capillary electrophoresis devices for extraterrestrial exploration. *J Chromatogr A* 1021: 191–199.
15. Liu, BF, Ozaki, M, Utsumi, Y, Hattori, T, Terabe, S. 2003. Chemiluminescence detection for a microchip capillary electrophoresis system fabricated in poly(dimethylsiloxane). *Anal Chem* 75: 36–41.
16. Piehl, N, Ludwig, M, Belder, D. 2004. Subsecond chiral separations on a microchip. *Electrophoresis* 25: 3848–3852.
17. Il Cho, S, Shim, J, Kim, MS, Kim, YK, Chung, DS. 2004. On-line sample cleanup and chiral separation of gemifloxacin in a urinary solution using chiral crown ether as a chiral selector in microchip electrophoresis. *J Chromatogr A* 1055: 241–245.
18. Kitagawa, F, Aizawa, S, Otsuka, K. 2005. Rapid enantioseparation of 1-aminoindan by microchip electrophoresis with linear-imaging UV detection. *Anal Sci* 21: 61–65.

19. Nakajima, H, Kawata, K, Shen, H, Nakagama, T, Uchiyama, K. 2005. Chiral separation of NBD-amino acids by ligand-exchange micro-channel electrophoresis. *Anal Sci* 21: 67–71.
20. Zeng, HL, Li, HF, Lin, JM. 2005. Chiral separation of dansyl amino acids by PDMS microchip gel monolithic column electrochromatography with gamma-cyclodextrin bonded in polyacrylamide. *Anal Chim Acta* 551: 1–8.
21. Zeng, HL, Li, HF, Wang, X, Lin, JM. 2006. Development of a gel monolithic column polydimethylsiloxane microfluidic device for rapid electrophoresis separation. *Talanta* 69: 226–231.
22. Bi, HY, Weng, XX, Qu, HY, Kong, JL, Yang, PY, Liu, BH. 2005. Strategy for allosteric analysis based on protein-patterned stationary phase in microfluidic chip. *J Proteome Res* 4: 2154–2160.
23. Weng, XX, Bi, HY, Liu, BH, Kong, JL. 2006. On-chip chiral separation based on bovine serum albumin-conjugated carbon nanotubes as stationary phase in a microchannel. *Electrophoresis* 27: 3129–3135.
24. Gao, Y, Shen, Z, Wang, H, Dai, ZP, Lin, BC. 2005. Chiral separations on multichannel microfluidic chips. *Electrophoresis* 26: 4774–4779.
25. Ro, KW, Hahn, JH. 2005. Precolumn diastereomerization and micellar electrokinetic chromatography on a plastic microchip: Rapid chiral analysis of amino acids. *Electrophoresis* 26: 4767–4773.
26. Belder, D, Ludwig, M, Wang, LW, Reetz, MT. 2006. Enantioselective catalysis and analysis on a chip. *Angew Chem Int Ed* 45: 2463–2466.
27. Belder, D, Kohler, F, Ludwig, M, Tolba, K, Piehl, N. 2006. Coating of powder-blasted channels for high-performance microchip electrophoresis. *Electrophoresis* 27: 3277–3283.
28. Gong, XY, Hauser, PC. 2006. Enantiomeric separation of underivatized small amines in conventional and on-chip capillary electrophoresis with contactless conductivity detection. *Electrophoresis* 27: 4375–4382.
29. Dutta, D, Ramachandran, A, Leighton, DT. 2006. Effect of channel geometry on solute dispersion in pressure-driven microfluidic systems. *Microfluid Nanofluid* 2: 275–290.
30. Crabtree, HJ, Cheong, ECS, Tilroe, DA, Backhouse, CJ. 2001. Microchip injection and separation anomalies due to pressure effects. *Anal Chem* 73: 4079–4086.
31. Zhang, C-X, Manz, A. 2001. Narrow sample channel injectors for capillary electrophoresis on microchip. *Anal Chem* 73: 2656–2662.
32. Jacobson, SC, Hergenroder, R, Koutny, LB, Warmack, RJ, Ramsey, JM. 1994. Effects of injection schemes and column geometry on the performance of microchip electrophoresis devices. *Anal Chem* 66: 1107–1113.
33. Roddy, ES, Xu, HW, Ewing, AG. 2004. Sample introduction techniques for microfabricated separation devices. *Electrophoresis* 25: 229–242.
34. Jacobson, SC, Koutny, LB, Hergenroder, R, Moore, AW, Ramsey, JM. 1994. Microchip capillary electrophoresis with an integrated postcolumn reactor. *Anal Chem* 66: 3472–3476.
35. Lee, NY, Yamada, M, Seki, M. 2004. Pressure-driven sample injection with quantitative liquid dispensing for on-chip electrophoresis. *Anal Sci* 20: 483–487.
36. Matysik, FM. 2008. Advances in amperometric and conductometric detection in capillary and chip-based electrophoresis. *Microchim Acta* 160: 1–14.

Index

A

Achiral and chiral surfactant mixtures
    enantiomeric separations, 210
    EOF velocity, 209
    MEKC separations, 210–211
    mixed micelle system, 209–210
Actaplanin A, 125
Active pharmaceutical ingredients (API), 314, 317
Affinity capillary electrochromatography (ACEC)
    definition, 143
    peptide selectors, 154
    protein selectors
        coating, 152–153
        fused-silica capillaries, 154
        silica particles, 151–152
        sol-gel method, 154
        tryptophan enantiomers, 151
Affinity capillary electrophoresis (ACE)
    capillary wall, 143–144
    peptide selectors
        deconvolution approach, 151
        DNP-D,L-amino acids, 150
        pseudo-stationary phases, 148
        simultaneous separation and enantioseparation, 149
    protein selectors
        cationic compounds, 148
        chiral selectors, 146
        OGCHI, 145–146
        partial filling technique, 145
        tolperisone enantiomers, 147
    UV detection interference, 144–145
Amino acids (AAs), 169
Anionic surfactants
    alkylglucoside, 204
    amino acid-derived, 204–206
    bile salts, 202–204
    miscellaneous, 207
    vesicle-forming, 206–207
Atmospheric pressure ionization (API) source
    APPI/APCI, 368–369
    CE–ESI–MS coupling, 368
    ESI electrospray, 367–368
Atmospheric pressure photoionization (APPI), 368
Avoparcin, 125
Aziridines
    enantioseparation electropherograms, 331
    residual plot, 332
    structural formulae, 330
    trans-diastereomers, 333–335

B

Background electrolyte (BGE)
    ACN-based, 296
    analytes adsorption, 272
    aqueous, 289
    buffer concentration, 280
    chiral selector
        chirally impure, 26
        and enantiomer, 28–29
        equilibrium concentration, 36–38
        pH, nature and ionic strength, 52
        side-interactions, 41
    composition changes, 36
    definition, 195
    electroosmotic transport velocities, 31
    HDMS-β-CD, 297
    with macrocyclic antibiotics, 42
    MeOH-based, 287
    molecules, 169
    non-volatile, 303
    organic modifiers, 125
    pH
        ionization state, 28
        separation system, 33
    reproducible mobility values, 167
    sample zone and, 301–302
    solvent
        buffers, 33
        properties, 278
    stability constants, 39
    thioridazine concentration, 273
Benzoin
    atropisomers/enantiomers, 206
    enantioseparations, 208
BGE, *see* Background electrolyte
Bovine serum albumin (BSA), 505
Buffer waste (BW) reservoirs, 507

C

Cahn–Ingold–Prelog method, 4
Camphorsulfonic acid (CSS), 282, 284
    quinine (QN), 281, 292
    selectivity, 296
Capillary electrochromatography (CEC)
    antibiotic uses, 126–129

chiral selectors, 122, 125–129
chromatographic and electrophoretic techniques, 185–186
definition, 393
drugs and herbicides, 130
enantioseparations
chip, 489–490
drugs, 50–51
flexibility, 17
open-tubular, 486–489
perphenylcarbam-oylated-β-CD-silica, 64
pressurized mode, 481–486
regular mode, 478–481
ligand exchange, 186
macrocyclic antibiotics, 124
monolithic phases
polymer-based, 65
silica-based, 63–65
open-tubular, 62–63
partial/complete chiral separation, 186–187
P-CEC, 63
plasma samples, 123
sol-gel process, 125
three chiral separation, 460
Capillary electrophoresis (CE), 271, 363
analytes separation modes, 116
CDs chiral recognition mechanisms, 49
vs. CEC mode, 477
chiral, 304, 370–371 (*see also* Quantitative analysis)
chiral separations examples, 336
in CSPs, 486
enantiomer
binding, 113
drug separation, 50–51
and ESI–MS, 221
flexibility, enantiomeric separations, 49
vs. HPLC, 316
lipid vesicles, 198
macrocyclic antibiotics, 130
methods and techniques, 196
in pCEC experiments, 481
protein and peptide selector modes
ACE, 143–151
ACEC, 151–154
use, 143
physical properties, 142
separations techniques, 460
vancomycin, 111
Capillary isotachophoresis (CITP), 109
Capillary liquid chromatographic (CLC)
devices, 481
modes, 475
and pCEC mode, 485–486
Capillary zone electrophoresis (CZE), 52, 91, 370
anti-inflammatory drugs, 125
chiral analysis, 116
chiral compounds, 116–121
chiral separations, 110
enantiomeric separation, 115
Carboxymethyl-β-CD (CM-β-CD), 375
Cationic surfactants, 204, 207–208
CD-mediated micellar electrokinetic chromatography (CD-MEKC), 60
CDs, *see* Cyclodextrins
CEC, *see* Capillary electrochromatography
CE, *see* Capillary electrophoresis
Charged resolving agent migration model (CHARM)
enantioselectivity and enantioresolution, 91–92
S-linezolid separation, 92
Chip CEC, enantioseparations
lab-on-chip technology, 489
PDMS, 489–490
Chiral CE
selectors, 371
techniques, 370–371
Chiral CE-MS
analyzers, 369–370
biological sample quantitation
and ESI analysis,
real blood samples, 381
optimal sprayer positioning, 379
parameters, 378–379
postcapillary infusion system, 379–380
SIM mode, 378
validation, 380–381
countercurrent technique, 374–377
coupling, 364–365
crown ethers, 377
and ESI analysis, ropivacaine, 372
2-hydoxypropyl-β-CD (HP-β-CD), 372
ionization sources
API, 367–369
ICP, 369
MALDI, 369
MEKC, 377–378
NACE, 378
PFT, 373–374
sheath–flow interface, 366
sheathless interface, 366–367
Chiral center, 3–4
Chiral coated stationary phase (CCP), 164
Chiral compounds, analytical classification, 23–25
Chiral cyclodextrin-modified MEEKC (CD-MEEKC), 326
Chiral drugs, body fluid analysis
classical techniques
LLE, 343
protein precipitation, 344

SPE, 343–344
detection system improvements
CE–MS coupling, 355
electrochemical, 354
ESI, 355
LIF, 354
mass spectrometry, 354–355
UV-Vis, 353–354
microextraction techniques
LPME, 347–349
SPME, 344–347
sensitivity improvements
ITP, 352–353
sample stacking, 350–352
sweeping, 352–353
validation
buffer composition, 356
capillary properties, 356–357
injection mode, 357
procedure, 355–356
temperature, 357
Chiral ligand exchange capillary electrophoresis (CLECE)
"all in solution"
complex formation, 168–169
definition, 168
metal and borate ion complexes, 171–172
tetrahedral coordination, 170–171
CEC, ligand exchange
electropherograms, 186, 188
partial/complete chiral separation, 186–187
cyclodextrins
characteristics, 172
chiral selectors, 175
diastereoisomeric complexes, 173
electropherograms, 173–174
LECE separations, 175–179
L-enantiomer, 174
separation mechanism, 175
tyrosine chiral separation, 175–176
micelles
copper(II) complexes, 185
electropherograms, 182
enantiomeric separations, 179
hydrophilic surface, 183
ligand exchange process, 180
sodium dodecyl sulfate (SDS), 183–184
stationary phases, 181–182
separation techniques, 163
theoretical considerations
chiral selecting system, 167–168
enantiomeric separation, 166
ligand properties, 165
mobility values, 167
normalizing factor, 168
Chiral MEEKC
analytes separation, 265–267
chiral components microemulsion, two or more
enantioselectivity, 241–242
independent processes, 243
stereochemical combinations, 242
synergistic effects, 244–245
cosurfactant-based chiral microemulsions
enantiomer separation, 257
enantioresolution, 256
EKC
chiral, 238–241
microemulsions, 238
oil in water microemulsion, 236–238
multiple-chiral-component microemulsions
cosurfactant stereochemical configuration, 259
dibutyl tartrate, 258–259
diethyl tartrate, 260–264
metoprolol resolution, 265
oil stereochemical configurations, 261
resolution improvements, 263
stereochemical composition, 264
surfactant DDCV, 258
synephrine enantiomers, separation, 262
N-undecenoyl-D-valinate (D-SUV)
binapthyl derivatives, 255
microemulsion polymer (MP), 256
polymeric chiral microemulsions, 254
oil-based chiral microemulsions, 256–258
single-chiral-component microemulsions, 241
surfactant-based chiral microemulsions
DDCV, 245–254
D-SUV, 254–256
Chiral mobile phase (CMP)
chromatography, 181
techniques and methods, 164
Chiral mobile phase additive (CMPA) approach, 281, 394–395
Chiral recognition mechanism
molecular modeling and ligand docking, 140
PGA and protein peptide, 141
Chiral selectors
chiral recognition mechanism, 140–141
description, 139–140
developments
diastereomers and enantiomers separation, 155
FTPFACE approach, 154–155
tryptophan enantiomers separation, 156
proteins and peptides uses and modes, CE
ACE, 143–151
ACEC, 151–154
Chiral stationary phases (CSPs), 65, 164, 186, 281, 285; *see also* Enantioseparations, CEC

ACEC, 151
approach, 395
in CEC (*see* Capillary electrochromatography (CEC))
classification, 460
Chiral switches
definition, 15
enantiomers *vs.* racemate, 15–16
Chromatographic and electrophoretic techniques, 185
CITP, *see* Capillary isotachophoresis
CLC, *see* Capillary liquid chromatographic
Coumachlor, 222
Critical aggregation concentration (CAC)
glucopyranoside-derived surfactants, 209
self-association, 215
surfactant values, 196–199
Critical micelle concentration (CMC), 180, 183, 196
Cyclodextrin modified MEKC (CD-MEKC)
applications, 214
enantiomer separation, 213
separation principles, 211
Cyclodextrins (CDs)
achiral surfactants and, 211
analyte mechanism, 49
applications
salbutamol, 298
timolol, 299
BGE pH and ionic strength
CE method, 93–94
effective selectivity, 95
EOF, 94–95
influence, 94
camphorsulfonate, 297
and carbohydrates, 58
CEC
monolithic phases, 63–65
open-tubular, 62–63
P-CEC, 63
charged derivatives, 292
amphoteric, 57
for analytes, 53–54
negative, 54
positive, 54–57
CLECE separation, 175–176
concentration influence
CHARM model, 91–92
enantiomers, 91
HP-β-CD concentration, 92–93
and crown ethers, 60–61
derivatization, 291–292
description, 47
dual systems
achiral additive, 97–98
analyte complexation, 98
applications, 100–103
CM-β-CD concentration, 58–59
enantioselectivity, 98–99
enantioseparation optimization, 58
interconversion, 100
separation system, 99
in electrophoresis, 41
enantioselective mechanisms, 291
hypothesis, structural, 177, 181
inclusion complexation and ligand-exchange, 61–62
injection mode, 96
and ion-pairing reagents, 60
LECE, 173
and linear oligosaccharides, 41–42
and macrocyclic antibiotics, 61
MEKC separations, 211–214
microfluid devices
FITC, 66
ITP and electrochromatography, 67
microchip CE system, 66–67
miscellaneous
EMO, 68–69
ITP system, 69–71
molecular structure, 172
native and neutral derivatives
hemispherodextrins and NSAIDs, 52–53
ITP and CZE, 52
negatively charged, 324, 327
nonaqueous medium, 67–68
organic solvents addition, 95–96
polymerized, 58
resolution power, 321
single isomer and randomly substituted, 315–316
starch enzymatic treatment, 47–49
structure, 290
and surfactants, 60
temperature, 96–97
ternary systems, 174
1,3,4-thiadiazine derivative, 292
trimipramine separation, 296
tryptophan enantiomers, 174–175
type
chiral recognition mechanism, 90–91
derivatives application, 88–89
enantioseparation, 87
variabilities, 319
CYP2C19 expression, 13–14
CYP2D6, 13
CZE, *see* Capillary zone electrophoresis

D

Dalgliesh concept, 26–27
Degree of substitution (DS), 321
Diisopropylidene ketogulonic acid (DIKGA), 275, 284, 287, 289, 302–303

Diode array detector (DAD), 315
Direct chiral discrimination models
feature, 26
selector and analyte three-point interaction, 26–27
on shape fit, 27–28
Dodecoxycarbonylvaline (DDCV)
based chiral microemulsions, 266
as chiral surfactant
ACES buffer, 249
achiral cosurfactant identity, 251–252
baseline stability, 252
cosurfactant identity, 253–254
cyclodextrins, 247
enantiomer separation, 246
enantioselectivity, 252–253
low-interfacial-tension oils, 247
micelles, 251
phosphate buffer system, 247–248
temperature, 248
van't Hoff plot, 249–250
zwitterionic buffer, 245
Dual selector systems
CDs
and carbohydrates, 58–59
and crown ethers, 60–61
and macrocyclic antibiotics, 61
surfactants, ion-pairing reagents and, 60
ligand-exchange
CD-borate-diol complexation, 62
host–guest interaction, 61–62

E

Electrical double layer (EDL), 276
Electrokinetic chromatography (EKC), 143
chiral
resolution, 239–241
separation mechanism, 239
microemulsion structure, 237
oil-in-water (o/w) microemulsion formulations
cosurfactants, 238
surfactant, 236
water-in-oil (w/o) systems, 237
separation principles, 199
Electromigration methods (EMs)
acidic compounds, 130
chiral selectors, 125–129
racemic compounds, 125
Electron pair acceptors, 163
Electroosmotic flow (EOF); *see also* Chiral stationary phases (CSPs)
BGE, 301
cathodic, 480
characteristics, 395
coated capillaries, 148
CSPs classification, 397
description, 395–396
flow profile, 122
HDB, 116
ion-pair selector system, 287
magnitude, 151
mobility, 473
NACE, 287, 301
neutral micelles migration, 200, 209
pH dependency, 397–398
phospholipid coating, 153
ξ-potential, 398–399
quaternary ammonium compounds, 56
reproducibility, 278
velocity, 396
Electrophoretic migration mechanisms, 199
Electrophoretic theory
BGE, 28–30
electrophoretic velocities, 31–32
enantiomers and ligands complexation, 28
separation selectivity, 30
Electrospray ionization (ESI), 367–368
Enantiomer migration order (EMO), 68, 176–177
Enantiomers; *see also* Background electrolyte (BGE); Cyclodextrins (CDs)
chiral discrimination, 33–34
shape fit with selector cavity, 27–28
three-point interaction, 26–27
drug separation, 50–51
EMO reversal, 182
eutomer and distomer, 5
and HSA separation, 152
impurity quantification, 102
indirect separations, 26
interaction, 15
isolation, 171
LECE separations, 175, 177
MEKC separations, 202
mobilities, 167
pharmacodynamics stereoselectivity
activities and properties, 6
impurity, 5–6
R-to S-ibuprofen inversion, 6–7
pharmacokinetic ratio, 14
separation, 166
simultaneous separation and, 149
single *vs.* racemate, 15–16
ternary complexes, 168
theory, electrophoretic (*see* Electrophoretic theory)
tolperisone, 147
tramadol, 376–377
tryptophan, 151, 174
Enantioseparations, CEC
approaches

CMPA, 394–395
CSP, 395
indirect, 394
EOF
characteristics, 395
CSPs classification, 397
description, 395–396
pH dependency, 397–398
velocity, 396
ξ–potential, 398–399
mephobarbital, 424–425
migration mechanism
and CLC separations, pronethalol, 400–401
linear migration velocity, 399–400
pressurized flow equation, 400
retention factor, 400, 402
separation factor, 403
mobile phase modes
EOF velocity, 407–408
hydrophobic interactions, 407–408
normal phase (NP) condition, 408–409
monolithic columns
enantioselective approaches, 426–428
hybrid particles, 440–441
organic polymer, 431–439
silica, 428–431
stationary phase types, 426
monolithic stationary phases
2-aryloxypropionic acids, 480
BSA encapsulated, 478–479
MIPs, 480–481
polymeric-based, 478
quinine derivative, 479
silica-based, 479–480
and MS
column designs, 442
instrumentations, 441–442
pCEC separation performance, 442–443
tip opening, 442
OT columns
advantage, 409
cellulose tris, 414–415
and EC capillary, 410–411
fused silica surface, 414
limitation, 409–410
MIP, 415
path length and sensitivity detection, 410
PEM-coated capillary scheme, 411–413
PLOT, 415
retention factors, 410
WCOT use, 414
packed capillary columns
benefit, 417
bubble formation, 416
cyclodextrin-based CSPs, 421, 423–424
EOF limitation approaches, 422–423
EOF velocity *vs.* field strength, 417–418
frits, 416
ion-exchange site, 425
macrocyclic antibiotics CSPs, 420, 423
multistep procedure, 415–416
particle diameters, 417
particulate CSP evaluation, 417–418
performance, 417
poly-*N*-acryloyl-l-phenylalanineethylester, 426
polysaccharide-based CSP, structure, 419, 422
polysaccharide selectors, 422
pore diameter effects, 420, 422
residual silanols, 424–425
risk, 416–417
thalidomide on vancomycin, comparison, 423–424
particle-based stationary phases
CD-based CSPs, 477
chiral compounds, 462–472
chlorthalidone enantiomers, 473
CSPs, 473–475
ephedrine chromatograms, 476
low-molecular-weight selectors, 475–476
ODS types, 461
RSDs, 461, 473
pressurized CEC (pCEC)
chromatographic performance, 404–405
commercial vancomycin phase, 404
zone dispersion
electro-driven and pressure-driven, 407
focusing effect and mass transfer, 406
kinetic plot methods, 405
Erythema nodosum leprosum (ENL), 7
European Medicines Evaluation Agency (EMEA), 15
European Pharmacopoeia (*Ph. Eur.*), 314, 316, 321, 323–324, 326, 332

F

Field-amplified sample stacking (FASS), 300, 350–351
Floating injection approach, 506
Flow-through partial-filling ACE (FTPFACE) approach, 154–156
Fluorescein isothiocyanate (FITC), 66, 374, 490
Fluorescence quenching method, 218
Food and drug administration (FDA), 355
Fourier transform infrared (FTIR) spectroscopy, 300
Fourier transform-ion cyclotron resonance (FT-ICR), 369
Free solution electrophoresis
chiral compounds
analytes and diastereoisomers, 24
chemical entity, 25

electrophoretic methods, 23–24
chiral selectors
CDs and oligosaccharides, 41–42
discrimination capability, 40–41
direct separations
chiral discrimination models, 26–28
electrophoretic theory, 29–32
stability constants, 34–40
theory application, 33–34
indirect separations
enantiomeric pairs, 26
selector and analytes, 25–26

G

Gas chromatography (GC), 17, 201, 316, 364
Gated injection approach, 506–507

H

Hepta-Tyr antibiotic, 130
Hexadimethrine bromide (HDB), 116
High-performance liquid chromatography (HPLC)
advantages, 196
chemical bonding methods, 153
vs. chiral CE, 315–316
chiral resolution, 113
CMPA, 281
DNB-Leu, 286–287
microbore, 124
silica gels, 151
Human serum albumin (HSA)
crystal structure, 141
enantioseparation, 152
Hybrid monoliths
description, 440
SEM images, 440–441
Hydroxyzine and cetirizine structure, 16

I

Inductively coupled plasma (ICP), 369
Interface, CE-MS
sheath-flow, 366
sheathless, 366–367
International Conference on Harmonization (ICH), 355
Ion-pair selectors
electrolytes and solvents
EOF, 289
negative and positive charged, 291
selector concentration profile, 289
enantiomer separation, 286
isoprenaline enantioseparation, 288
pharmaceutical applications, 289–290
selectivity, 287
Isotachophoresis (ITP); *see also* Cyclodextrins (CDs)
capillary and CZE coupling, 69
description, 352–353
principle, 67

L

Labetalol diastereoisomers relationship, 3
Large-volume sample stacking (LVSS), 352
Laser induced fluorescence (LIF), 354
Levobupivacaine, 16
Levodopa
D-dopa determination, 325–327
enantioseparation electropherograms, 324
L-and D-dopa, 323–324
peak area ratio and RMT, 325–326
Ligand exchange chromatography (LEC)
techniques, 164
tryptophan enantiomer separation, 173
Ligand exchange mechanism, 169, 179, 186, 273, 275, 299
Limit of detection (LOD), 319
Limit of quantitation (LOQ), 319–320
Liquid chromatography (LC), 17, 321, 364
Liquid–liquid extraction (LLE), 343
Liquid-phase microextraction (LPME) technique
device, 348
modes, 348–349
vs. SDME, 347–348
stereoselective methods, 349

M

Macrocyclic antibiotics
capillary electromigration methods
anti-inflammatory drugs, 125
semisynthetic glycopeptide, 130
electromigration separation modes
CEC (*see* Capillary electrochromatography)
CZE (*see* Capillary zone electrophoresis)
enantiomers detection
MS signal, 115
partial-filling countercurrent method, 114
UV nonabsorbing compounds, 116
enantioresolution mechanism (EM)
anionic compounds, 112
chemical structure, 113
enantiorecognition capability, 111–112
uncharged compounds, 113–114
properties, 110–111
Mass spectrometry (MS), 115, 364
Matrix-assisted laser desorption/ionization (MALDI), 369
Micellar electrokinetic chromatography (MEKC), 109, 180, 288, 377–378

CD-modified
applications, 214
enantioseparations, 213
SDS *vs.* chiral analyte molecules, 211
separations, chiral, 212–213
three-component system, 211–212
chiral separation mechanism, 201–202
drugs, enantiomer separation, 50–51
enantioseparations, polymerized chiral surfactants
chemical structures, 215, 217
chiral micelle–analyte interactions, 220
copolymers, 220–221
hydrophobic compounds, 214
light scattering measurements, 218
micellar solutions, 219–220
molecular micelles, 215–216, 218
polymer micelles, 215–216
PSP, 215
surfactant molecules, 219
mass spectrometric detection, 221–222
modes, chiral separation, 196
scheme, 60
separation principles
micelles, 199–200
migration window, 200
retention factor, 201
surfactants, 199
silica capillaries, 196
surfactant-based chiral selectors
achiral and, 209–211
anionic, 202–207
cationic, 207–208
neutral, 209
zwitterionic, 208
surfactants, self-assembly
geometric parameters, 196
micelle shapes, 197–198
self-assembly structures, 198–199
ULVs and MLVs, 198
Microchip design
hydrodynamic flow, 505
structure, 504–505
Microchip electrophoresis (ME), 185
Microemulsion electrokinetic chromatography (MEEKC), 315, 326–330
achiral, 236, 238
chiral (*see* Chiral MEEKC)
microemulsions, 236
resolution, 240
Microemulsion polymer (MP), 256
Micro-total analysis systems (μ–TAS), 501
Molecularly imprinted polymers (MIPs), 343; *see also* Enantioseparations, CEC
chiral recognition mechanism, 460
preparation, 480
MS, *see* Mass spectrometry
Multilamellar vesicles (MLVs), 197–198

N

N-2-Acetamido-2-aminoethanesulfonic acid (ACES), 246–247, 249
Neutral surfactants, 204, 209
Nonaqueous capillary electrophoresis (NACE)
advantages, 271–272, 280
charged derivatives, 292
chiral
achiral separations, 278
detection limit, 299–304
EOF, 287
organic solvents and electrolytes, 276–277
chiral selectors, 274–275, 299–300
EOF, 289
ion-pair selectors, 281, 292
native CDs, 290–291
Nonaqueous media, chiral separations
BGEs
acidity scale, 279
aqueous, 274
buffer preparation, 280
flammable and toxic solvents, 278
organic solvent composition, 278–279
pH*, organic solvents, 279–280
CD-based selectors, 293–295
enantiomers, mobility difference, 272
ion-pair selectors
alkaloid derivatives, 286
chiral separations, 282–284
CSS, 281
negatively and positively charged, 281
miscellaneous chiral selectors, 300
NACE
advantages, 271–272, 280
chiral selectors, 274–275
detection limit, 299–304
disadvantages, 280
thioridazine concentration, 273
Nuclear magnetic resonance (NMR), 49

O

On-chip chiral separations
advances
benefit, 509
online coupling cleanup feasibility, 509–510
applications, 501–503
injection, microchip
floating approach, 506
gated approach, 506–507
intersection, 505
pinched approach, 505–506
microchip design, 504–505
microchip detection
advantages, 507
imaging UV/V, 507–508

perspectives
features, 511–512
separation modes, 511
principle, 504
resolution enhancement
SDS, 509
separation lengths, 508–509
Open-tubular capillary electrochromatography (OTCEC), 153
enantioseparations
column preparation, 489
mode-filtered light detection, 488
multilayer coating procedure, 487
phospholipid coating, 487–488
Schiff base formation reaction, 488–489
Open-tubular (OT) columns
advantage, 409
cellulose tris, 414–415
and EC capillary, 410–411
fused silica surface, covalent binding, 414
limitation, 409–410
MIP, 415
path length and sensitivity detection, 410
PEM-coated capillary, scheme, 411–413
PLOT, 415
retention factors, 410
WCOT use, 414
Organic polymer monoliths
in situ approach
bonded β-cyclodextrin, 435–436
cross-linking effects, 433–434
EDMA and HEMA, 432
imprinting process, 434–435
individual polymer microglobules, 432
polyacrylamide matrices, 435
polymethacrylate-type organic polymer monolith, 432–433
porogen and monomer compositions, effects, 434
vinylsulfonic acid, role, 437
postmodification strategy
O-(*tert*-butylcarbamoyl) mefloquine separation, 438–439
surface structure, 439
terguride-modified column, 437–438
Ovoglycoprotein from chicken egg whites (OGCHI), 146
chiral selector, 145
tolperisone enantiomers, 147

P

Packed CEC (P-CEC), 63
Partial filling countercurrent technique (PFT-CC)
CE–ESI–MS analysis, 376
CM-β-CD, HS-CD and SBE-β-CD, 375
MS detection, 374
scheme, 114
tramadol enantiomers, 376–377
vancomycin, CE–UV, 377
Partial filling technique (PFT), 368, 373–374
Particle-loaded technique, 186
Penicillin G-acylase (PGA)
chiral recognition mechanism, 141
chiral selector, 148
Pharmacological importance, chiral separations
analytical tools, 17
capillary electrophoresis
advantages and disadvantages, 17–18
definition, 17
drug-related factors
enantiomers interaction, 15
input rate, 14–15
interspecies differences, 12
isomerism
constitutional, 2
stereoisomers, 2–3
patient-specific factors
age, 12–13
disease states, 14
gender, 13
genetic factors, 13–14
pharmacodynamics stereoselectivity
description, 5
enantiomer, 5–6 (*see also* Enantiomers)
R- and *S*-ibuprofen, 6–7
pharmacokinetics stereoselectivity, 7
absorption, 8
distribution, 8–9
product, 10–11
renal excretion, 12
substrate, 9–10
single enantiomer *vs.* racemate
chiral switches definition, 15
levobupivacaine, 16
Phenothiazines
electropherogram, 101
electrophoretic mobility variations, 100
Pinched injection approach, 505–506
Polydimethylsiloxane (PDMS), 489
Porous-layer open tubular (PLOT) column, 415
Positively charged derivatives, CDs
in acidic and neutral compounds, 54
cationic β-CD, 55–56
6-HPTMA-β-CD synthesis, 56–57
quaternary ammonium, 56
selectors and analytes dierction, 56–58
Pressurized CEC (pCEC), enantioseparations
on monolithic stationary phases, 486
particle-based stationary phases
applications, 482–483
chirobiotic V CSP, 485
perphenylcarbamoylated-β-CD, 485–486
warfarin and coumachlor, 481, 484–485
Pseudostationary phases (PSPs)
analytes, 260–262

chiral micelles, 201
electrophoretic mobility, 214, 240
microemulsion, 238, 246
migration time, 239
retention time, 247
surfactant aggregates, 196
surfactants, 236

Q

Quantitative analysis
chiral CD-modified CE methods
aziridine derivatives, 330–335
levodopa, 323–326
tropa alkaloids, 326–330
chiral CE, advantages and disadvantages
CDs derivatives, 315–316
vs. HPLC, 316
Ph. Eur., 316–317
sensitivity, 315
enantiomeric purity testing, 322–323
validation methods
accuracy and, 318
LOD and LOQ, 319–320
precision, 318–319
range and linearity, 320
robustness, 320–322
specificity, 317–318
quantitative Nuclear magnetic resonance (qNMR) spectroscopy, 316

R

R-citalopram, 6
Relative migration times (RMTs), 325
Relative standard deviations (RSDs), 461
R-enantiomer, 6–16
Restrict access media (RAM), 343
Ring-opening metathesis polymerization (ROMP), 125
Room temperature ionic liquid (RTIL), 481
R/S convention, 4

S

S-3-amino-1,2-propanediol (SAP), 172
Sample stacking
acetonitrile, 351
FASS model, 350–351
LVSS, 352
Sample waste (SW) reservoirs, 505–506
Selected ion monitoring (SIM) mode, 370, 375, 378
S-enantiomer, 6
Separation selectivity, chiral, 23–24
Sheath–flow interfaces, coaxial sheath-liquid and liquid junction, 366
Sheathless interface, 366–367
Silica monoliths
chiral selectors, 429
electron microscopy image, 428–429
immobilization procedures, 430
in situ protein-encapsulation, 431
limitation and silanization, 429
mesopore formation, 428
Single-drop microextraction (SDME) technique, 344
Single-walled carbon nanotubes (SWNTs), 156
Sodium dodecyl sulfate (SDS), 114, 236–238, 247, 256–257, 509
micellar system, 206, 214
selector concentration, 184
separation, 183–184
surfactants, 211
Solid-phase extraction (SPE) technique, 343–344
Solid-phase microextraction (SPME) technique
advantages and drawbacks, 346
CE analysis and interface, 346–347
direct and headspace extraction, 344–345
fibers employed, 345
Stability constant, electrophoretic chiral separations
credible, 39
determination, 34
raw data and use
chiral selector equilibrium concentration, 36–38
electrophoretic transport, 35
reproducibility, 35–36
stoichiometric, 36
reversible complexation, 39–40
and separation selectivity values, 38
Stationary phase (SP); *see also* Capillary electrochromatography (CEC)
achiral, 473
chirobiotic V, 475
EOF, 473, 478
nonchiral, 460
sol-gel method, 478–479
Stir bar sorptive extraction (SBSE) technique, 344
Sulfobutylether-β-CD (SBE-β-CD), 375
Supercritical fuid chromatography (SFC), 63
Surfactant N-undecenoyl-D-valinate (D-SUV), 254–255, 267

T

Thin layer chromatography (TLC), 17
Time-of-flight (TOF), 369
transient Isotachophoresis (t-ITP), 300–301
Trichoderma reesei, 146
Triethylamine acetate (TEAA), 485
Tropa alkaloids, 326–330

U

Unilamellar vesicles (ULVs), 197–198

V

Vancomycin
chemical structure, 111
chiral stationary phase, 122
ionic compound resolution, 112
Vancomycin silica stationary phase (Van-SP), 122–123
Van Deemter equation, 405
Verapamil, 210
Voltage-controlled injection schemes, 505–506

W

Warfarin, 210, 219, 222

Z

Zwitterionic surfactants, 204, 208

Milton Keynes UK
Ingram Content Group UK Ltd.
UKHW021925071024
449327UK00022B/1706